Smart Innovation, Systems and Technologies

Volume 194

The Smart Innovation, Systems and Technologies book series encompasses the topics of knowledge, intelligence, innovation and sustainability. The aim of the series is to make available a platform for the publication of books on all aspects of single and multi-disciplinary research on these themes in order to make the latest results available in a readily-accessible form. Volumes on interdisciplinary research combining two or more of these areas is particularly sought.

The series covers systems and paradigms that employ knowledge and intelligence in a broad sense. Its scope is systems having embedded knowledge and intelligence, which may be applied to the solution of world problems in industry, the environment and the community. It also focusses on the knowledge-transfer methodologies and innovation strategies employed to make this happen effectively. The combination of intelligent systems tools and a broad range of applications introduces a need for a synergy of disciplines from science, technology, business and the humanities. The series will include conference proceedings, edited collections, monographs, handbooks, reference books, and other relevant types of book in areas of science and technology where smart systems and technologies can offer innovative solutions.

High quality content is an essential feature for all book proposals accepted for the series. It is expected that editors of all accepted volumes will ensure that contributions are subjected to an appropriate level of reviewing process and adhere to KES quality principles.

**** Indexing: The books of this series are submitted to ISI Proceedings, EI-Compendex, SCOPUS, Google Scholar and Springerlink ****

More information about this series at http://www.springer.com/series/8767

Debahuti Mishra · Rajkumar Buyya ·
Prasant Mohapatra · Srikanta Patnaik
Editors

Intelligent and Cloud Computing

Proceedings of ICICC 2019, Volume 1

Springer

Editors
Debahuti Mishra
Department of Computer Science
and Engineering
Siksha 'O' Anusandhan Deemed
to be University
Bhubaneswar, Odisha, India

Prasant Mohapatra
Department of Computer Science
University of California
Davis, CA, USA

Rajkumar Buyya
University of Melbourne
Melbourne, VIC, Australia

Srikanta Patnaik
Department of Computer Science
and Engineering
Siksha 'O' Anusandhan Deemed
to be University
Bhubaneswar, Odisha, India

ISSN 2190-3018 ISSN 2190-3026 (electronic)
Smart Innovation, Systems and Technologies
ISBN 978-981-15-5973-0 ISBN 978-981-15-5971-6 (eBook)
https://doi.org/10.1007/978-981-15-5971-6

This Springer imprint is published by the registered company Springer Nature Singapore Pte Ltd.
The registered company address is: 152 Beach Road, #21-01/04 Gateway East, Singapore 189721, Singapore

Preface

The 1st International Conference on Intelligent and Cloud Computing (ICICC-2019) was organized by the Department of Computer Science & Engineering, ITER, Siksha 'O' Anusandhan (Deemed to be) University on 16–17 December 2019. ICICC-2019 has provided a suitable platform for academicians, practitioners, researchers and experts to showcase their work and findings as well as exchange ideas and future directions while sharing their experiences with each other. ICICC-2019 has focused on recent research trends and advancements in cloud computing and its importance in diversified application fields. Intelligent and cloud computing integrated with other relevant techniques not only adds values to crucial decisions but also provides new direction to advancement in key technical areas such as adoption of Internet of things, distributed computing, cloud computing, blockchain, big data, etc. It provides significant insights that support managers in various sectors such as human resource developments, resource management, biological data processing, marketing strategies, supply chain management, logistics, business operations, financial markets, etc., thus making it a potential research area. The conference has received a good response with a large number of submissions. The total number of relevant papers accepted for publication has been broadly divided into five major parts: (i) intelligent and cloud computing, (ii) software engineering, (iii) wireless sensor network, (iv) IoT and its application, and (v) AI and machine learning.

First part, i.e. intelligent and cloud computing, attracted significant attention in both research and industry in the current era. Cloud computing enlarges the field of distributed computing by providing elastic services, large data storage and processing, applications, and high-performance computation in addition to the business cost reduction.

Second part, service-oriented software engineering, incorporates the best features of both the services and cloud computing paradigms, offering many advantages for software development and applications, but also exacerbating old concerns. Services and cloud computing have garnered much attention from both industry and academia because they enable the rapid development of large-scale distributed applications in areas such as collaborative research and development, e-business, healthcare, grid-enabled applications, enterprise computing infrastructures, military applications, and homeland security. These computing paradigms have made it easier and more economical to create everything from simple commercial software to complex mission-critical applications.

Third part deals with the integration of Wireless Sensor Networks (WSN), cloud computing and big data technologies which offer vast opportunities to solve many real-world problems. The combination of WSN and cloud computing environment-based application can collect and process large amount of data from the beginning until the end of the process loop.

There are another two tracks, namely, Internet of Things (IoT) and Artificial Intelligence (AI), which have been included in Volume 2.

ICICC-2019 has not only encouraged submission of unpublished original articles in the field of intelligent and cloud computing issues but also considered the several cutting edge applications across organizations and firms while scrutinizing the relevant papers.

Bhubaneswar, India — Debahuti Mishra
Melbourne, Australia — Rajkumar Buyya
Davis, USA — Prasant Mohapatra
Bhubaneswar, India — Srikanta Patnaik

Contents

About the Editors

Prof. Dr. Debahuti Mishra received her B.E. degree in Computer Science and Engineering from Utkal University, Bhubaneswar, India, in 1994; her M.Tech. degree in Computer Science and Engineering from KIIT Deemed to be University, Bhubaneswar, India, in 2006; and her Ph.D. degree in Computer Science and Engineering from Siksha 'O' Anusandhan Deemed to be University, Bhubaneswar, India, in 2011. She is currently working as a Professor and Head of the Department of Computer Science and Engineering, at the same university.

Dr. Rajkumar Buyya is a Redmond Barry Distinguished Professor and Director of the Cloud Computing and Distributed Systems (CLOUDS) Laboratory at the University of Melbourne, Australia. He is also the founding CEO of Manjrasoft, a spin-off company that commercializes university innovations in cloud computing. He served as a Future Fellow of the Australian Research Council from 2012 to 2016. He has authored over 625 publications and seven textbooks, including "Mastering Cloud Computing", published by McGraw Hill, China Machine Press, and Morgan Kaufmann for Indian, Chinese and international markets, respectively.

Dr. Prasant Mohapatra is the Vice Chancellor for Research at the University of California, Davis. He is also a Professor at the Department of Computer Science and served as the Dean and Vice-Provost of Graduate Studies at the University of California, Davis, from 2016 to 18. He was also an Associate Chancellor in 2014–16, and the Interim Vice-Provost and CIO of UC Davis in 2013–14. Further, he was the Department Chair of Computer Science from 2007 to 13 and held the Tim Bucher Family Endowed Chair Professorship during that period. He has also been a member of the faculty at Iowa State University and Michigan State University.

Dr. Srikanta Patnaik is a Professor at the Department of Computer Science and Engineering, Faculty of Engineering and Technology, SOA University, Bhubaneswar, India. Dr. Patnaik has published 100 research papers in international journals and conference proceedings. Dr. Patnaik is the Editor-in-Chief of the International Journal of Information and Communication Technology and

International Journal of Computational Vision and Robotics, published by Inderscience Publishing House, England, and also Editor-in-Chief of a book series on "Modeling and Optimization in Science and Technology", published by Springer, Germany.

Intelligent and Cloud Computing

M-Throttled: Dynamic Load Balancing Algorithm for Cloud Computing

Amrutanshu Panigrahi, Bibhuprasad Sahu, Saroj Kumar Rout, and Amiya Kumar Rath

Abstract Cloud computing environment can also be called as Internet-based computing process in which there is no limitation of work. There are multiple numbers of data center (DC) available for solving multiple user requests coming from a different user base (UB). The data center is capable of negotiating multiple instructions simultaneously. But the instructions are submitted to the DC randomly. Thus, there is chance of overload for a particular DC. Hence, the load balancing plays a vital role in cloud computing to maintain the performance of the computing environment. In this research article, we have implemented throttled, round-robin, and the shortest job first loadbalancing algorithm. Also, we have proposed one more algorithm called M-throttled which has the high performance compared to others. We have taken different parameters such as overall response time and DC processing time for comparison. These are simulated by taking closest data center policy in CloudSim environment.

Keywords Data center · User base · Throttled · Round-robin · M-throttled · CloudSim

1 Introduction

Cloud computing is a dynamic system in which shared resources, data, and programs are given by the customer's prerequisite at the explicit time. It is a term which is commonly utilized if there should arise an occurrence of web. Load adjusting in distributed computing structures is very a test now. Persistently, a passed on procedure is required, since it isn't in every case basically plausible to keep up at least one inactive system which is similar to the active one to satisfy the required requests. In cloud

A. Panigrahi (✉) · B. Sahu · S. K. Rout
Department of CSE, Gandhi Institute for Technology, Bhubaneswar, Odisha, India
e-mail: amrutansup89@gmail.com

A. K. Rath
Department of CSE, VSSUT, Deputed to NAAC, Burla, Bangalore, Odisha, India

D. Mishra et al. (eds.), *Intelligent and Cloud Computing*,
Smart Innovation, Systems and Technologies 194,
https://doi.org/10.1007/978-981-15-5971-6_1

computing, the cloud can be observed as the cloud which provides the availability of resources on the Internet for different users. The end users use these resources as per the requirement. Simultaneously multiple users utilize these resources simultaneously. Cloud computing provides various services such as IaaS, SaaS, and PaaS to handle different kinds of requests coming from the end users [1]. The request coming from the end users are not homogeneous and are submitted to different data centers for execution randomly. Hence, it is very difficult to handle the heterogeneous requests by an overloaded data center. Hence, load balancing is becoming the most challenging factor in the field of cloud computing. So it is required to disseminate the stack evenly at each data center to realize an efficient client fulfillment and proper resource usage proportion. The main objectives of this mechanism are as follows [2, 3]:

- To make all the DCs equally loaded.
- To keep the performance constant.
- To maintain the execution speed in heavy traffic.

It is the method for redistributing the hard and fast burden to every processor or DCs or to make resource use fruitful and furthermore to improve the reaction time of the assignment, at the same time ousting a condition in which a bit of the DCs is over stacked, while some others are under stacked. The basic intriguing focuses while becoming such calculation are estimation of load, assessment of load, security of different systems, execution of structure, and relationship between different DCs [4]. This heap considered can be regarding CPU load, measure of memory utilized, postponement, or system load. A site or a web application can be acquired by a lot of clients at any time. It ends up troublesome for a web application to deal with all these client asks for at one time. It might even outcome in framework breakdowns. For a site proprietor, whose whole work is reliant on his entryway, the sinking feeling of site being down or not available additionally brings lost potential clients. Here, the heap balancer assumes a vital job. Cloud load balancing is the way toward disseminating remaining tasks at hand and figuring resources crosswise over at least one server. This sort of dissemination guarantees most extreme throughput in least response time [5, 6]. In this research paper, we have implemented throttled, RR, and shortest job first algorithm for calculating the efficiency of the cloud computing environment while performing the load balancing. Along with these algorithms, we have proposed one more algorithm M-throttled with an improvised performance of the network with the presence of the heavy traffic at different instances of time. The influencing parameters such as response time and process time for UB and DC are considered for the evaluation of different algorithms.

2 Literature Survey

Load balancing ensures the principal work in providing a QoS in the cloud environment. It also has been generating liberal energy for the exploration arrange. There are tons of philosophies that have adjusted to the store modifying issue in conveyed processing. The choice involves varying customary procedures without including any swarm knowledge computations. Different load adjusting methodologies were proposed starting late and each based on various edges of figuring and techniques, e.g., using a central burden modifying approach for virtual machines [13], the arranging system on weight altering of virtual machine (VM) resources subject to inherited computations [14], a mapping course of action reliant on multi-resource burden changing for virtual machines [15], different dispersed calculations for VMs [16], weighted least-association methodology [17], and two-stage booking calculations [18]. Also, a few techniques for burden adjusting were displayed for different cloud applications, for instance, an administration-based work set for extensive scale stockpiling [19], information focus the executive design [15], and a heterogeneous cloud.

The mediocre contains approaches like [8] swarm learning figuring, counterfeit honey bee state [9, 10], and PSO [11, 12], which has the fine granularity result for the dynamic behavior of appropriated processing. In [9], the author represented an estimation for weight movement of an extraordinary weight with a changed technique of ACO. In [7], a load advertisement adjusting framework was proposed in light of underground creepy crawly state and complex framework speculation in an open appropriated registering class. This is the first occasion when that ACO and complex frameworks were united into weight modifying in conveyed processing what's increasingly procured extraordinary execution. In [8], the author offered a response for weight altering in the cloud by ACO, to grow or confine different execution parameters, for instance, CPU weight and memory limit. In [9, 10], the author displayed a novel technique for burden altering reliant on fake bumble bee settlement. PSO was in like manner grasped for weight altering in dispersed registering, for instance [11, 12].

3 Existing Algorithms

3.1 Throttled Load Balancing

In this method [16], heap balancer keeps up a database of VM with their present states (Accessible/Occupied). At any instant when a demand to allocate another VM from the DC controller reaches, it processes the record table from best till the most readily accessible VM is detected. As soon as the VM is discovered, the Heap Balancer restores the corresponding VM's ID to the DC controller. The DC controller sends the demand to VM with the identified by the corresponding ID. The DC controller informs the heap balancer of the new allotment. The heap balancer refreshes the allotment table by augmenting as needs be. When the VM wraps up the demand and the data center controller gets the reaction cloudlet, it advises the heap balancer of the VM de-assignment. The heap balancer de-assign the equivalent VM whose Id is as of now imparted. The objective of this method is to determine the response time of every VM as VMs are having different capacities corresponding to processing efficiency.

$$RT = F_t - A_t + T_d$$

where RT = response time, F_t= finish time, A_t= arrival time, and T_d = transmission delay.

3.2 *Round-Robin Load Balancing*

It is the least complex calculation that utilizes the idea of time slice [8]. In this method, time is separated into numerous slices also, every datacenter is provided with a specific time slice and inside that defined amount of time the hub will play out its activities. The data center controller doles out the demand to a rundown of VMs on a pivoting premise. The first demand is allotted to an available VM which will be determined haphazardly from the gathering and afterward DC controller doles out the resulting demands in a roundabout request. If the VM is appointed the demand, then the VM is shifted to the end of the rundown. RR calculation chooses the heap on arbitrary premise and also, in this manner, prompts a circumstance where a few hubs are intensely stacked and some are daintily stacked.

3.3 *SJF Load Balancing*

Shortest Job First (SJF) [17] planning is a need and non-preemptive booking. In Non-preemptive methods, when the procedures are allocated for a time to a processor then the processor can't be taken by the other, until the procedure is finished in the execution section. This calculation appropriates the heap haphazardly by first checking the extent of the procedure and after that exchanging the heap to a virtual machine, which is gently stacked. All things considered that procedure measure is least; this procedure will first need to execute whether we guess most reduced estimated process executed in least time. The heap balancer spreads the heap on to various hubs known as spread range technique. Shortest Job First (SJF) calculation can be said to be ideal with a normal holding up time negligible, which improves the framework execution.

4 Proposed Algorithm: M-throttled

In this calculation, the heap balancer keeps up a rundown table of VMs and the amount of requesting and by assigned to VM. At first, all VMs have zero portions.

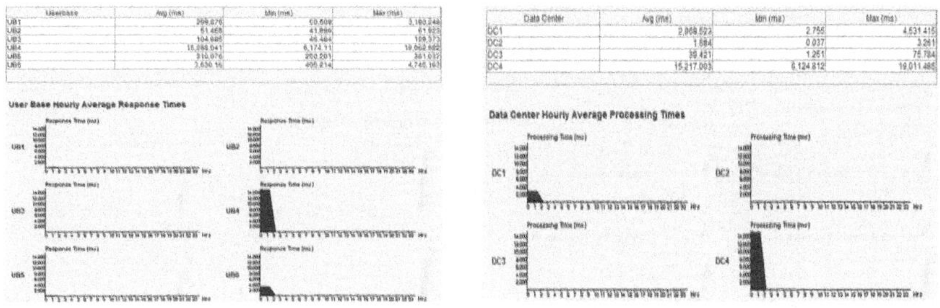

Fig. 1 UB RT and DC PT using RR algorithm

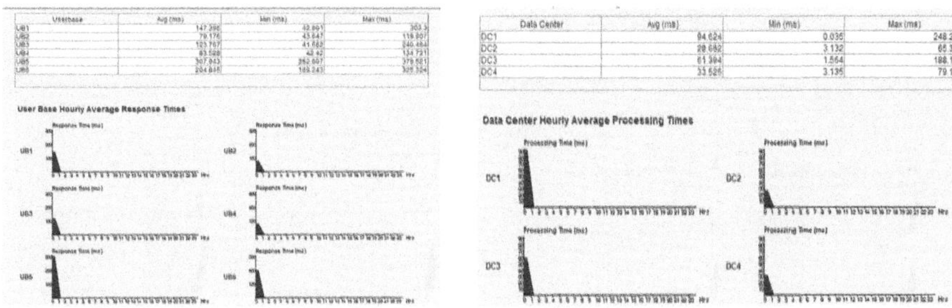

Fig. 2 UB RT and DC PT using throttled algorithm

Right when an interest to assign another VM from the DC controller comes to, it forms the rundown table and perceives the base stacked VM. If there are multiple, the chief recognized is picked. The heap balancer reestablishes the machine IDs to the DC controller. The DC controller provides the interest to the VM recognized by its identified ID. The DC controller illuminates heap balancer about the newly generated task. The heap balancer invigorates the task table augmenting the distribution with that VM. Right when the VM wraps up the interest and DC controller gets the interested cloudlet, it educates the heap balancer regarding the corresponding VM's appropriation. The heap balancer revives the dissemination table by diminishing the assignment of the VM by one. In M-throttled algorithm, a correspondence exists between the stack balancer and the data center controller for reviving the rundown table provoking an overhead that makes the delay in giving response to the showed up sales (Figs. 1, 2, 3, and 4).

5 Implementation

In this simulation, six datacenters DC1, DC2, DC3, DC4, DC5, and DC6 with closed data set policy have been adapted during the simulation. In this, different arrangement parameters can be set like number of user bases, number of tasks produced by

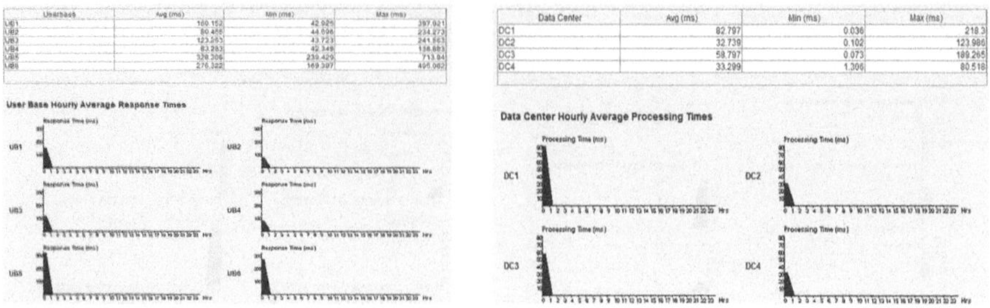

Fig. 3 UB RT and DC PT using SJF algorithm

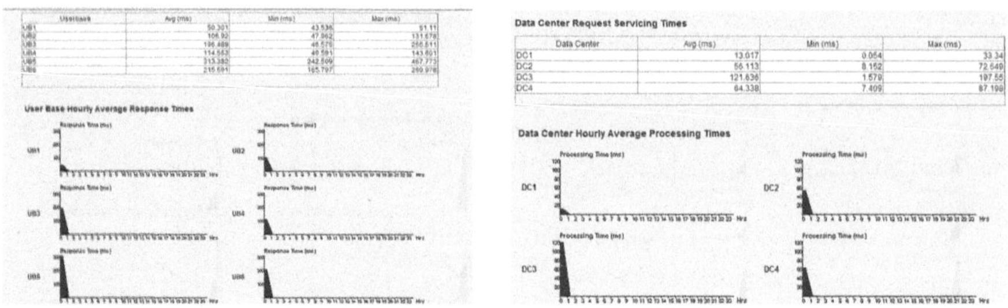

Fig. 4 UB RT and DC PT using M-throttled algorithm

each user base per hour, number of VMs, number of processors, processing speed, available bandwidth, etc. Based on the parameters, the result incorporates reaction time, handling time, fetched, etc. UB response time (RT) and DC processing time (PT) are considered for the evaluation purpose (Tables 1 and 2) [16].

6 Conclusion

We have studied the concepts of load balancing and its vital effects on cloud computing environment. Different algorithms provide the solution for the existing problem of load balancing among different data centers to maintain the efficiency of the network. The performance of the strategies such as throttled, round-robin, shortest job first, and M-throttled time has been studied by taking some influencing parameters such as response time and processing time. A comparison has been done on the basis of some predefined parameter such as UB response time and DC processing time. With the presence of heavy traffic in each region the UB response time and the DC processing time, the M-throttled algorithm performs much better than that of other existing algorithms.

Table 1 DC configuration

Parameter	Value used
VM image size	10000
VM memory	1024 Mb
VM band	1000
DC-Arch	X86
DC-OS	Linux
DC-Machine	20
DC-Memory/machine	2048 Mb
DC-Storage	100000 Mb
DC-Band	10000
DC-Processors/Machine	4
DC-Speed	100MIPS
DC-Policy	Time Shared/Space Shared
DC grouping UB based	1000
DC grouping request based	100
Instruction length	250

Table 2 Region configuration

Cloud analyst region id	Users (M)
0	4.4
1	1.1
2	2.6
3	1.3
4	0.3
5	0.8

References

1. Velte, A.T., Velte, T.J., Elsenpeter, R.C., Elsenpeter, R.C.: Cloud Computing: A Practical Approach, p. 44. McGraw-Hill, New York (2010)
2. Randles, M., Lamb, D., Taleb-Bendiab, A.: A comparative study into distributed load balancing algorithms for cloud computing. In: 2010 IEEE 24th International Conference on Advanced Information Networking and Applications Workshops, pp. 551–556. IEEE (2010)
3. A Vouk, M.: Cloud computing–issues, research and implementations. J. Comput. Inf. Technol. **16**(4), 235–246 (2008)
4. Alakeel, A.M.: A guide to dynamic load balancing in distributed computer systems. Int. J. Comput. Sci. Inf. Secur. **10**(6), 153–160 (2010)
5. http://www.ibm.com/press/us/en/pressrelease/22613.wss
6. http://www.amazon.com/gp/browse.html?node=20159001
7. Randles, M., Odat, E., Lamb, D., Abu-Rahmeh, O., Taleb-Bendiab, A.: A comparative experiment in distributed load balancing. In: 2009 Second International Conference on Developments

in eSystems Engineering, pp. 258–265). IEEE (2009)
8. Shah, M.M.D., Kariyani, M.A.A., Agrawal, M.D.L.: Allocation of virtual machines in cloud computing using load balancing algorithm. Int. J. Comput. Sci. Inf. Technol. Secur. (IJCSITS) **3**(1), 2249–9555 (2013)
9. Moges, M., Robertazzi, T.G.: Wireless sensor networks: scheduling for measurement and data reporting. IEEE Trans. Aerosp. Electron. Syst. **42**(1), 327–340 (2006)
10. Pallis, G.: Cloud computing: the new frontier of internet computing. IEEE Internet Comput. **14**(5), 70–73 (2010)
11. Armbrust, M., Fox, A., Griffith, R., Joseph, A.D., Katz, R., Konwinski, A., Lee, G., Patterson, D., Rabkin, A., Stocia, I., Zaharia, M.: Above the Clouds: A Berkeley View of Cloud Computing, pp. 1–23. EECS Department, University of California (2009)
12. Bhadani, A., Chaudhary, S.: Performance evaluation of web servers using central load balancing policy over virtual machines on cloud. In: Proceedings of the Third Annual ACM Bangalore Conference, p. 16. ACM (2010)
13. Rimal, B.P., Choi, E., Lumb, I: A taxonomy and survey of cloud computing systems. In: 2009 Fifth International Joint Conference on INC, IMS and IDC, pp. 44–51 (2009)
14. Zhang, Z., Zhang, X.: A load balancing mechanism based on ant colony and complex network theory in open cloud computing federation. In: 2010 The 2nd International Conference on Industrial Mechatronics and Automation, vol. 2, pp. 240–243. IEEE (2010)
15. Hiranwal, S., Roy, K.C.: Adaptive round robin scheduling using shortest burst approach based on smart time slice. Int. J. Comput. Sci. Commun. **2**(2), 319–323 (2011)
16. Wickremasinghe, B., Calheiros, R.N., Buyya, R.: Cloudanalyst: a cloudsim-based visual modeller for analysing cloud computing environments and applications. In: 2010 24th IEEE International Conference on Advanced Information Networking and Applications, pp. 446–452. IEEE (2010)
17. Waheed, M., Javaid, N., Fatima, A., Nazar, T., Tehreem, K., Ansar, K.: Shortest job first load balancing algorithm for efficient resource management in cloud. In: International Conference on Broadband and Wireless Computing, Communication and Applications, pp. 49–62. Springer, Cham (2018)
18. Buyya, R., Ranjan, R., Calheiros, R.N.: Modeling and simulation of scalable cloud computing environments and the cloudsim toolkit: challenges and opportunities. In: 2009 International Conference on High Performance Computing and Simulation, pp. 1–11. IEEE (2009)
19. Bo, Z., Ji, G., Jieqing, A.: Cloud loading balance algorithm. In: Proceedings of the 2010 2nd International Conference on Information Science and Engineering (ICISE), pp. 5001–5004. Hangzhou, China (2010)

iGateLink: A Gateway Library for Linking IoT, Edge, Fog, and Cloud Computing Environments

Riccardo Mancini, Shreshth Tuli, Tommaso Cucinotta, and Rajkumar Buyya

Abstract In recent years, the Internet of Things (IoT) has been growing in popularity, along with the increasingly important role played by IoT gateways, mediating the interactions among a plethora of heterogeneous IoT devices and cloud services. In this paper, we present *iGateLink*, an open-source Android library easing the development of Android applications acting as a gateway between IoT devices and edge/fog/cloud computing environments. Thanks to its pluggable design, modules providing connectivity with a number of devices acting as data sources or fog/cloud frameworks can be easily reused for different applications. Using *iGateLink* in two case studies replicating previous works in the healthcare and image processing domains, the library proved to be effective in adapting to different scenarios and speeding up development of gateway applications, as compared to the use of conventional methods.

Keywords Internet of Things · Gateway applications · Edge computing · Fog computing · Cloud computing

1 Introduction

Recently, the Internet of Things (IoT) has gained significant popularity among both industry and academia, constituting a fundamental technology for creating novel computing environments like smart cities and smart healthcare applications, which pose higher requirements on the capabilities of modern computing infrastructures [1, 2]. Cloud computing allowed offloading complex and heavyweight computations,

R. Mancini · S. Tuli (✉) · R. Buyya
Cloud Computing and Distributed Systems (CLOUDS) Lab, School of Computing and Information System, The University of Melbourne, Melbourne, Australia
e-mail: shreshthtuli@gmail.com

R. Mancini · T. Cucinotta
Scuola Superiore Sant'Anna, Pisa, Italy

S. Tuli
Department of Computer Science and Engineering, Indian Institute of Technology, Delhi, India

D. Mishra et al. (eds.), *Intelligent and Cloud Computing*,
Smart Innovation, Systems and Technologies 194,
https://doi.org/10.1007/978-981-15-5971-6_2

Fig. 1 Example scenario in which one or more IoT devices communicate to the cloud/fog through a gateway device

including big-data processing pipelines, to remote data centers [1]. However, the exponential growth of connected IoT devices and the forecast in the produced data volumes for the upcoming years [3] pushed industry and academia to look into optimized solutions, where virtual machines are dropped in favor of more lightweight containers [4] and, more importantly, *decentralized solutions* are employed, where computing happens at the *edge* of the network, giving rise to the fog computing paradigm. This leads to reduced latency, deployment costs, and improved robustness [5], so in recent years many fog/cloud frameworks have been proposed leveraging computing resources both at the edge of the network and in cloud data centers [6, 7].

In nowadays IoT and fog computing frameworks, a crucial component is the *gateway* device which enables communication of users, sensors, and actuators with edge devices and cloud resources [8]. Gateway devices could be small embedded computers, smart routers, or even smartphones. In IoT, the number of sensors and actuators has increased tremendously in the last few years [9, 10], including advanced gateway devices that emerged in recent fog environments [11]. Even though fog computing frameworks greatly simplify engineering gateway functionality, they do not focus on making generic gateway interfaces for seamless integration with diverse applications, user needs, and computation models.

Contributions. This paper presents the design and implementation of *iGateLink*, an open-source modular fog–cloud gateway library easing the development of applications running on Android-based IoT gateway devices. It provides common core functionalities, so as to let developers focus on the application-specific code, for example, implementing communications with specific sensors or protocols to exchange data with specific fog or cloud systems. *iGateLink* is specific to the mentioned IoT-fog use case but generic enough in order to allow simple extensions to be used in different IoT-fog integrated environments as shown in Fig. 1. It is also easy to use through a simple API and supports integration of different frameworks within the same application, including the possibility to run the required execution locally.

iGateLink has been applied to two use cases, one dealing with an oximeter-based healthcare application for patients with heart diseases, and the other one for low response time object detection in camera images. The presented framework proved to be effective for reducing development complexity and time, when comparing with existing practices in coding IoT gateway applications.

2 Related Work

Due to the vast heterogeneity in the Internet of Things, the importance of the IoT gateway in enabling the IoT has been acknowledged by several works [12–14]. Some authors propose IoT gateways to act only as routers, e.g., encapsulating raw data coming from *Bluetooth* devices in *IPv6* packets to be sent to the cloud [15]. However, in order to enable more complex scenarios, offload computations, and/or save network bandwidth, IoT gateways need to become "smart" by pre-processing incoming data from the sensors [16]. In this context, the use of smartphones as IoT gateways has been proposed [17, 18]; however, those works do not take into consideration the most recent fog and edge computing paradigms.

Regarding efforts to integrate the IoT with fog and edge computing, Aazam et al. [8] proposed a smart gateway-based communication that utilizes data trimming and pre-processing, along with fog computing in order to help lessen the burden on the cloud. Furthermore, Gia et al. [19] developed a Smart IoT gateway for a smart healthcare use case. Finally, Tuli et al. [6] developed *FogBus* an integration framework for IoT and fog/cloud. However, their work does not provide a generic application able to integrate IoT and fog/cloud frameworks, building their application from the ground up, tailored to their specific use case.

Finally, it is worthwhile to note that none of the previously mentioned works focuses on the design and implementation of the software that is required to run in the IoT gateway, especially in the case of an Android device, in order to make it generic and adaptable to many different scenarios, as this paper does.

3 System Model and Architecture

The principles discussed in Sect. 1 have been addressed by using a modular design, with a generic core on which different modules can be loaded.

A variation of the publish/subscribe paradigm [20] has been used, where the components collecting data from sensors and the ones sending it to the fog/cloud are the publishers and are called *Providers*, while the subscribers are the auxiliary components that manage the execution of the publishers or UI components that show incoming data to the user. The publish/subscribe paradigm is realized through an intermediate component that stores the incoming data from the *Providers*, called *Store*, and notifies the subscribers, called *Triggers*, using the *observer* design pattern. A *Store* can also be thought of as the *topic* to which the *Provider publishes* data, while the *Triggers* are its *subscribers*.

In the considered scenario, a *Provider* may be started either as a result of a user interaction (e.g., a button click) or as a consequence of external event (e.g., incoming Bluetooth data). It can also be started with some input data or with no data at all. The proposed model does not assume any of the aforementioned cases, employing a generic design, able to adapt many different scenarios. While the input of a *Provider* can vary, the result is always data that needs to be stored in a *Store*. Whenever a *Provider* stores new data to a *Store*, all *Triggers* associated with the *Store* are executed. The most common use of a *Trigger* is to start another *Provider* but it could also be used, for example, to update the UI.

These design choices enabled (1) *modularity*, since *Providers* and *Triggers* can be independently and easily plugged and unplugged; (2) *flexibility*, since this model enables even more complicated use cases than the one mentioned above; and (3) *ease of use*, thanks to code reusability.

Furthermore, several *Providers* providing the same data (i.e., publishing to the same *Store*) can be active simultaneously, for example, *Providers* for different fog/cloud frameworks and/or local execution. In this case, it is useful to define a new component, the *Chooser*, whose function is to select a specific *Provider* among a list of equivalent ones in order to produce data. By doing so, it is possible to, for example, use another *Provider* when one is busy or to fallback to another *Provider* if one fails.

3.1 Implementation Details

Based on the model described above, we have implemented the *iGateLink* library for Android devices. The library is open-source and available at https://github.com/Cloudslab/iGateLink. From a high-level point of view (Fig. 2), the library is composed of a platform-independent core written in Java, an Android-specific module which extends the core to be efficiently used in Android and several extension modules that provide complimentary functionalities.

Core. The core of the library is composed of the following classes: *ExecutionManager*, *Data*, *Store*, *Provider*, *Chooser*, and *Trigger*. Refer to Fig. 3 for a conceptual overview of their interactions. The *ExecutionManager* class coordinates the other components and provides an API for managing them. The *Data* class is not properly a component but is the base class that every data inside this library must extend. It is characterized by an `id` and a `request_id`: the former must be unique between data of the same *Store*; the latter is useful for tracking data belonging to the same request. The *Store* class stores *Data* elements and provides two operations: `retrieve` for

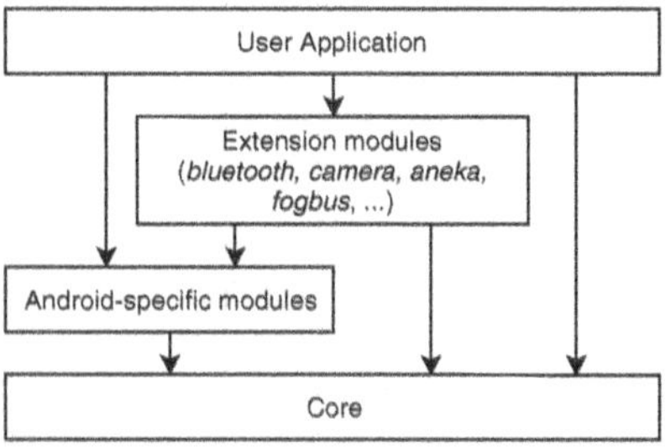

Fig. 2 High-level overview of the library software components

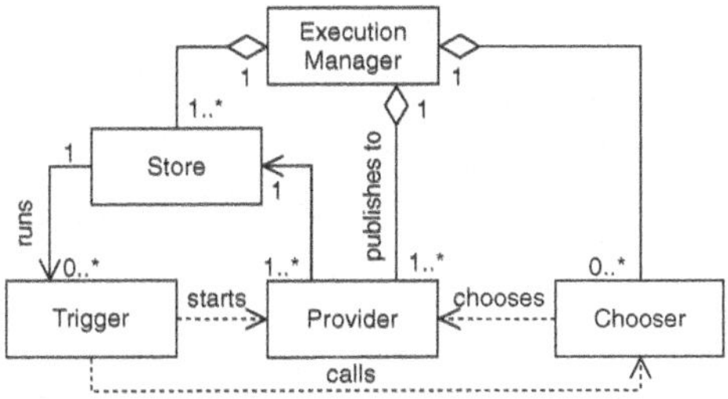

Fig. 3 Simplified UML class diagram of the core components

retrieving previously stored *Data* and `store` for storing new data. Every *Store* is uniquely identified by a `key`. There can be multiple *Stores* for the same *Data* type. Furthermore, a *Store* can have one or more *Triggers* associated with it that are called whenever new data is stored. For example, a common use for a *Trigger* is starting a *Provider* with the recently stored *Data* from another *Provider*. The *Provider* class takes some *Data* in input and produces some other *Data* in output. Every *Provider* is uniquely identified by a `key`. There can be multiple *Providers* for the same *Store* but a *Provider* can only use one *Store* at a time. A *Provider* can be started by calling its `execute` method either through *ExecutionManager*'s `runProvider` or `produceData`. In the latter case, the user specifies only which *Data* it wants to be produced (by means of a *Store* key) and a suitable *Provider* will be executed. In case there are two or more *Providers* for the same *Store*, a *Chooser* will be used.

Android-specific modules. When developing such an application on Android, a common problem is that using *worker threads* for time-consuming tasks that need to execute in the background still lets the Android runtime kill the app as needed, if not properly managed. This is addressed by providing the *AsyncProvider* class, which makes use of the *AsyncTask* class.[1] Furthermore, the *ExecutionManager* is hosted within an Android *Service*, which, if configured correctly as a foreground service, prevents it from being killed by the Android runtime.

4 Case Studies

In order to test the library and demonstrate its applicability to real applications, two case studies have been developed. They both reproduce existing works, namely, FogBus [6] and EdgeLens [7]. The first application connects to a *Bluetooth* oximeter to collect and analyze its data. The second application takes a photo and sends it to the fog or cloud for object detection.

In the original works, both applications have been developed using *MIT App Inventor*, which eases and speeds up the development but provides only a very simple API that cannot be adapted to more complicated use cases. Furthermore, every

[1] It provides a simple API for scheduling a new task and executing a custom function on the main thread before and after the execution.

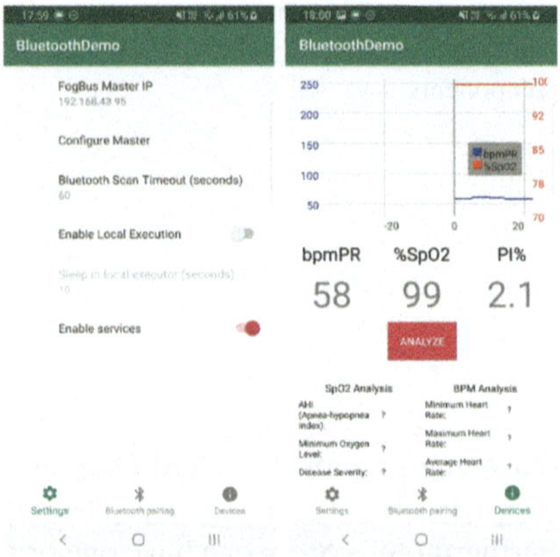

Fig. 4 Oximeter demo app: on the left, configuration screen; on the right, live data and analysis results screen

application needs to be built from the ground up even though many components may be in common, especially when using the same input method or framework. By developing the applications using *iGateLink*, both modularity and fast development can be achieved.

Oximeter-based healthcare application The *bluetoothdemo* application can be used to detect hypopnea in a patient by collecting data from an oximeter. The application consists of four screens: (1) the configuration screen (left picture in Fig. 4); (2) the Bluetooth device pairing screen; (3) the Bluetooth device selection screen through which the user chooses the device whose data to show; and (4) the data and analysis result screen (right picture in Fig. 4).

Data is collected in real time from the oximeter using *Bluetooth LE*. In order to do that, the *Bluetooth* module has been developed which provides an implementation of the *Provider* which registers to the *GATT* characteristics in order to receive push notifications from the *Bluetooth* device. Raw data received from the oximeter is then converted to a *Data* object and stored.

The user can then upload the data to FogBus [6] by tapping the "Analyze" button in order to get the analysis results. Data is sent using an HTTP request to the *FogBus* master node which forwards the request to a worker node (or the cloud) and then returns the result to the application. This simplifies the development of the application, since only one HTTP request is required.

Object detection application The *camerademo* application can be used to take a photo and run an object detection algorithm on it, namely, Yolo [21]. The user interface is very simple, with just a screen showing the camera preview and a button (Fig. 5). When the button is clicked, the photo is taken and sent to the fog/cloud for object detection. When the execution is terminated, the resulting image is downloaded and shown to the user.

In order to integrate Android camera APIs with *iGateLink*, the *camera* module has been developed, which provides an implementation of a *Provider*, namely, *CameraProvider*, that takes a photo when its *execute* method is called. When the photo is

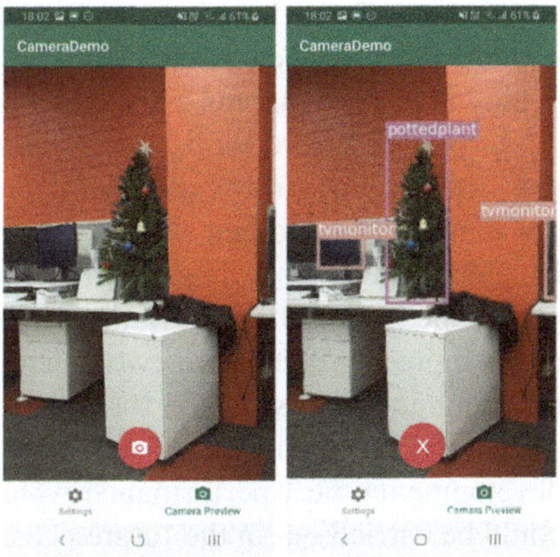

Fig. 5 Object detection demo app: on the left, preview screen; on the right, result screen

stored, two providers are executed: *BitmapProvider* that converts the photo from an array of bytes to a *Bitmap* object so that it can be plug in an *ImageView* and shown to the user, and one of *EdgeLensProvider* or *AnekaProvider* that executes the object detection on either EdgeLens [7] or Aneka [22]. The result of the object detection is, again, an array of bytes so it goes through another *BitmapProvider* before being shown to the user.

The *EdgeLensProvider* is responsible for uploading the image to the *EdgeLens* framework [7] and downloading the result once the execution is completed. *EdgeLens* has a similar architecture to *FogBus* but, differently from it, communication between client and worker is direct, instead of being proxied by the master. The usual *EdgeLens* workflow is (1) query the master to get the designated worker, (2) upload the image to the worker, (3) start execution, and (4) download the result when the execution is terminated.

The *AnekaProvider* is responsible for execution in Aneka [22] through its REST APIs for task submission. The image is first uploaded to an FTP server (which could be hosted by the master node itself), then the object detection task is submitted to *Aneka*, whose master node chooses a worker node to submit the request to. The worker downloads the image from the FTP server, runs the object detection on it, and uploads the resulting image back to the FTP server. In the meanwhile, the client repeatedly polls the master node in a loop, waiting for the submitted task to complete. When it does, the client finally downloads the result which is displayed to the user.

5 Conclusions and Future Work

In this paper, we presented a new Android library, *iGateLink*, which enables developers to easily write new Android applications linking IoT, edge, fog, and cloud computing environments, as it comes with a set of highly reusable modules for common IoT scenarios.

We demonstrated, by means of two use cases, that the library can adapt to different applications and is easy to use, increasing the engineering simplicity of application deployments and making such systems robust and easy to maintain.

As part of future work, more modules could be added to the library to cover the most common use cases, for example, the current version of the library does not include any module providing support for Bluetooth devices (currently only Bluetooth low energy is supported), built-in sensors (for example, accelerometer, gyroscope, magnetometer, luminosity), touch events, and audio recording. Furthermore, more fog/cloud frameworks could be integrated within the library, making it like plug-and-play software for end users.

Finally, while this paper focuses on the advantages, the library brings in terms of development time, a performance evaluation, and a comparison with existing systems could be carried out in the future.

References

1. Yi, S., Li, C., Li, Q.: A survey of fog computing: concepts, applications and issues. In: Proceedings of the 2015 Workshop on Mobile Big Data, pp. 37–42. ACM (2015)
2. Gill, S.S., Tuli, S., Xu, M., Singh, I., Singh, K.V., Lindsay, D., Tuli, S., Smirnova, D., Singh, M., Jain, U., et al.: Transformative effects of IoT, blockchain and artificial intelligence on cloud computing: evolution, vision, trends and open challenges. Internet Things 100118 (2019)
3. Ericsson Mobility Report (2019). https://www.ericsson.com/en/mobility-report
4. Cucinotta, T., Abeni, L., Marinoni, M., Balsini, A., Vitucci, C.: Reducing temporal interference in private clouds through real-time containers. In: 2019 IEEE International Conference on Edge Computing (EDGE), pp. 124–131 (2019)
5. Bonomi, F., Milito, R., Zhu, J., Addepalli, S.: Fog computing and its role in the internet of things. In: Proceedings of the First Edition of the MCC Workshop on Mobile Cloud Computing, pp. 13–16. ACM (2012)
6. Tuli, S., Mahmud, R., Tuli, S., Buyya, R.: FogBus: A blockchain-based lightweight framework for edge and fog computing. J. Syst. Softw. **154**, 22–36 (2019)
7. Tuli, S., Basumatary, N., Buyya, R.: EdgeLens: Deep learning based object detection in integrated IoT , fog and cloud computing environments. In: 4th IEEE International Conference on Information Systems and Computer Networks (2019)
8. Aazam, M., Huh, E.N.: Fog computing and smart gateway based communication for cloud of things. In: 2014 International Conference on Future Internet of Things and Cloud, pp. 464–470. IEEE (2014)
9. Singh, D., Tripathi, G., Jara, A.J.: A survey of internet-of-things: Future vision, architecture, challenges and services. In: 2014 IEEE World Forum on Internet of Things (WF-IoT), pp. 287–292. IEEE (2014)
10. Lee, I., Lee, K.: The internet of things (iot): Applications, investments, and challenges for enterprises. Bus. Horizons **58**(4), 431–440 (2015)
11. Whiteaker, J., Schneider, F., Teixeira, R., Diot, C., Soule, A., Picconi, F., May, M.: Expanding home services with advanced gateways. ACM SIGCOMM Comput. Commun. Rev. **42**(5), 37–43 (2012)
12. Chen, H., Jia, X., Li, H.: A brief introduction to iot gateway. In: IET International Conference on Communication Technology and Application, pp. 610–613 (2011)
13. Datta, S.K., Bonnet, C., Nikaein, N.: An iot gateway centric architecture to provide novel m2m services. In: 2014 IEEE World Forum on Internet of Things (WF-IoT), pp. 514–519. IEEE (2014)

14. Kang, B., Kim, D., Choo, H.: Internet of everything: a large-scale autonomic iot gateway. IEEE Trans. Multi-Scale Comput. Syst. **3**(3), 206–214 (2017)
15. Zachariah, T., Klugman, N., Campbell, B., Adkins, J., Jackson, N., Dutta, P.: The internet of things has a gateway problem. In: Proceedings of the 16th International Workshop on Mobile Computing Systems and Applications, pp. 27–32. ACM (2015)
16. Saxena, N., Roy, A., Sahu, B.J., Kim, H.: Efficient iot gateway over 5g wireless: a new design with prototype and implementation results. IEEE Commun. Mag. **55**(2), 97–105 (2017)
17. Kamilaris, A., Pitsillides, A.: Mobile phone computing and the internet of things: a survey. IEEE Internet Things J. **3**(6), 885–898 (2016)
18. Aloi, G., Caliciuri, G., Fortino, G., Gravina, R., Pace, P., Russo, W., Savaglio, C.: A mobile multi-technology gateway to enable iot interoperability. In: 2016 IEEE First International Conference on Internet-of-Things Design and Implementation (IoTDI), pp. 259–264. IEEE (2016)
19. Gia, T.N., Jiang, M., Rahmani, A.M., Westerlund, T., Liljeberg, P., Tenhunen, H.: Fog computing in healthcare internet of things: a case study on ecg feature extraction. In: CIT/IUCC/DASC/PICom 2015, pp. 356–363. IEEE (2015)
20. Eugster, P.T., Felber, P.A., Guerraoui, R., Kermarrec, A.M.: The many faces of publish/subscribe. ACM Comput. Surv. (CSUR) **35**(2), 114–131 (2003)
21. Redmon, J., Divvala, S., Girshick, R., Farhadi, A.: You only look once: unified, real-time object detection. In: Proceedings of the IEEE Conference on Computer Vision and Pattern Recognition, pp. 779–788 (2016)
22. Vecchiola, C., Chu, X., Buyya, R.: Aneka: a software platform for .net-based cloud computing. High Speed Large Scale Sci Comput **18**, 267–295 (2009)

Hetero Leach: Leach Protocol with Heterogeneous Energy Constraint

Amrutanshu Panigrahi, Bibhuprasad Sahu, and Sushree Bibhuprada B. Priyadarshini

Abstract Wireless Sensor Networks (WSNs) are used for checking and information gathering from the physical world in various applications, for instance, condition watching, developing the board, following animals or items, social protection, transportation, and general home frameworks. Nowadays, WSNs are pulling in unfathomable thought in research. The objective of this research is to focus on clustering based routing approach in WSN. We have explained different existing clustering based routing algorithms for WSN. Initially, we have implemented one well-known algorithm LEACH with 100 no of node in a 500 × 500 flat grid. Also, we have introduced one new algorithm called HETERO LEACH which removes the disadvantages of LEACH protocol. The leach protocol works in homogeneous energy constraint, while the HETERO LEACH can work with heterogeneous energy constraints. On the other hand, all these routing protocols have been implemented and simulated using MATLAB. Subsequently, these protocols have been simulated with different parameters such as Number of packets to CH, No of Alive Nodes, and Dead Node to prove their functionality and to find out their behavior in different sorts of sensor networks. The result shows the comparison of these two protocols and the best protocol by taking the energy constraint into consideration. And finally, we have shown that the hetero leach works much better than as compared to the leach protocol.

Keywords Leach · Apteen · WSN · CH · Hetero Leach

A. Panigrahi (✉) · B. Sahu
Department of CSE, Gandhi Institute for Technology, Bhubaneswar, Odisha, India
e-mail: amrutansup@gmail.com

S. B. B. Priyadarshini
Department of Computer Science & Information Technology, Siksha 'O' Anusandhan, Deemed to Be University, Bhubaneswar, Odisha, India

D. Mishra et al. (eds.), *Intelligent and Cloud Computing*,
Smart Innovation, Systems and Technologies 194,
https://doi.org/10.1007/978-981-15-5971-6_3

1 Introduction

Wireless Sensor Network or WSN [1], are generally used for checking and for also information gathering from the physical world in various applications, for instance, condition watching, developing the board, following animals or items, social protection, transportation, and general home frameworks. WSN [2], involves an ordinarily enormous number of sensor center points, in like manner called bits. These bits are accumulated to form the data which will be transmitted towards different network elements called nodes. Sink focus focuses gather and procedure the got information so as to make it open to the client. Notwithstanding the path that in a little WSN one-jump correspondence to the sink can be executed, when all is said and done, a multi-ricochet diagram must be considered [3]. For this situation, conventional bits are responsible for executing a planning custom so as to move the data towards the sink. Since bits ought to control in doubt work unattended for quite a while, they have uncommon essentialness obstacles [4, 5]. This significantly impacts the structure of a WSN and unequivocally on the planning custom. Since correspondence is an expensive asset to the degree essentialness use, an animal control message sending section (i.e., flooding) is the time when all is said and done senseless. Or then again perhaps, the course of action of the planning custom [6], is a central edge that ought to consider tradeoffs between transmission power and sending techniques so as to give relentless quality and nature of association. Moreover, since a piece should crash in perspective on battery weariness or unmistakable reasons, a fit controlling convention to be enough adaptable to respond to a misstep by reconfiguring the edge work [7, 8].

As progression makes, WSN is snatching open passages for application in flexible affiliations [9]. There are two unmistakable ways to deal with show flexibility in a Mobile WSN environment. Subsequent techniques are there to keep up static sinks, while sensor focuses are versatile, for instance, when added to creatures in the following applications [10, 11]. For this condition, a static sink can be utilized to amass following data set away in the sensor focus focuses when the creatures are in its range. At last, the two systems can be joined, giving every single focus point get to the WSN be adaptable [12]. For instance, in a private situation for created individuals or for individuals with insufficiencies, sensors added to them can offer data to the cell phones of the accomplice work compel [13]. Despite the extra intricacy of planning customs for MWSN, pass on capacity brings the chance of diminishing the measure of jumps to the sink focus point. As exhibited by [14–16], the likelihood of having something near a sensor focus point in the degree of sink increments with the correspondence extend, the speed of the middle point, and the measure of sinks, accomplishing a reduction of the dormancy. Regardless, high portability conditions could avoid different transmissions to sufficiently pass on messages [17, 18].

2 Routing Algorithm

2.1 LEACH

In [19], the author had presented Low Energy Adaptive Clustering Hierarchy, a progressive routing calculation for sensor systems, called Low Energy Adaptive Clustering Hierarchy (LEACH). LEACH organizes the hubs in the system into little bunches and picks one of them for cluster head. At that point, the group head totals and packs the data got from every one of the hubs and sends it to the corresponding base station. The hubs picked as the group head channel out more vitality when contrasted with different hubs as it is required to send information to the base station which might be far found. Henceforth, filter utilizes irregular turn of the hubs required to be the bunch heads to equitably convey vitality utilization in the system. After various reenactments, it was discovered that just 5% of the absolute number of hubs needs to go about as the bunch heads [10]. The TDMA/CDMA is being utilized to diminish between the group and intra-bunch impacts. This convention is utilized where a consistent observation by the sensor hubs are there as the primary objective, because the information accumulation is concentrated at the BS and is being performed occasionally.

- **Operations**

LEACH operations are conducted in two phases:

1. Setup phase
2. Steady phase

In the initial stage, the clusters are formed with the corresponding cluster heads. Mean while in the second phase the data is sensed and sent back to the different base station. The steady phase takes much longer time as compared to the setup phase, as in this phase all information is gathered and distributed to the different base station.

(a) **Setup phase**: In the setup stage, a foreordained portion of hubs, p, pick themselves as group heads. It is finished by a limit denoted as T(n). The limit esteem relies on the ideal rate to turn into a bunch head-p, the current round r, and the arrangement of hubs that have not turned into the group head in the last 1/p rounds. The formulae are as follows:

$$T(n) = \frac{p}{1 - p \times \left(r \times \bmod \frac{1}{p}\right)} \forall n \in G \quad (1)$$

Here p no of nodes sends the advertisement message to different nodes to participate in their group. The nodes receiving the message has to send an acknowledgment message to raise their flags. The node which sends the advertisement message will be elected as the CH. Similarly, different node has to participate in different clusters. In the wake of getting the affirmation message, contingent on the quantity of hubs under their group and the kind of data required by the framework (in which the WSN is setup), the bunch heads make a TDMA plan and allots every hub a schedule vacancy, in which it can transmit the detected information. The TDMA plan is communicated to all the bunch individuals. In the event that the extent of any bunch turns out to be excessively vast, the group head may pick another bunch head for its bunch. The bunch head picked for the current round can't again turn into the group head until the various hubs in the system haven't turned into the bunch head.

(b) **Steady stage**: In this stage, the available information will be sensed from different channels and sends them to individual cluster heads by using the TDMA technique. The data will be received by the cluster head in packet format. After receiving all packets, the cluster head combines it and sends it back to the intendant recipients. The nodes present in the cluster will receive the message by implementing the CDMA technique.

2.2 Hetero Leach

In HETERO LEACH, once the cluster heads are taken then in each defined time period the CH first communicates by considering the following parameters:

Thresholds: It includes two parameters called a hard farthest point (HL) and a fragile cutoff (DL). HL is a particular estimation of a quality past in which a center point can be enacted to transmit data. DL is a little change in the estimation of a quality which can trigger a center point to transmit data again.

Schedule: It utilizes the TDMA technique for scheduling the message to different cluster-based nodes.

TC: It is commonly known as counter time which is the extreme time span between two consecutive communications. In a sensor compose, close to centers fall in a comparable gathering, sense tantamount information, and try to send their data at the same time, causing possible accidents. We present a TDMA schedule with the ultimate objective that each center point in the pack is doled out a transmission opening, as showed up in Fig. 1.

Important Features

The primary highlights of these algorithms are:

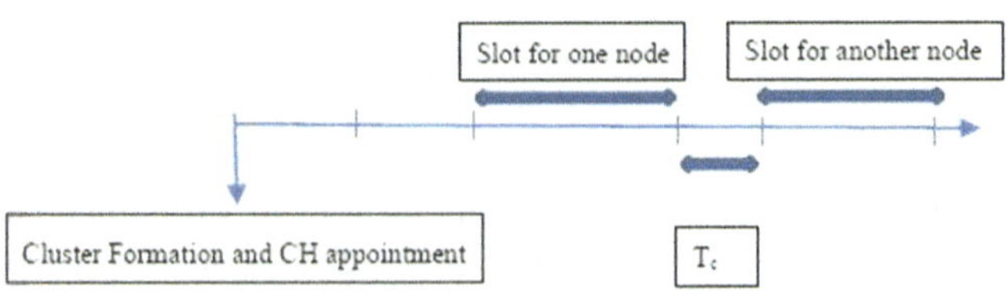

Fig. 1 Timeline for HETERO LEACH

1. By sending intermittent information, it gives the client a total image of the system. It additionally reacts right away group change time bunch development edge schedule opening for a hub I TDMA schedule and parameters to extraordinary changes, accordingly making it receptive to time basic circumstances. Hence, it consolidates both proactive and receptive approaches.
2. It offers adaptability of enabling the client to set the time interim (TC) and the limit esteems for the properties.
3. Vitality utilization can be constrained by the tally time and the edge esteems. The half breed system can imitate a proactive system or on the other hand a responsive system, by appropriately setting the check time also, the limit esteems.

3 Simulation and Result

For LEACH and HETERO LEACH WSN routing protocol, we have used Matlab Simulator A simulation environment having 50nodes,100 nodes in 500 × 300 flat grid has been created with random position. The performance of each protocol has been analyzed by considering some influencing parameters like Cluster Head, Alive Nodes, Dead Nodes, Number of Base Stations. The performance details are as follows:

CH

In a hierarchical protocol in which the maximum number of nodes transmit to cluster heads, and then it aggregates and compresses the data and forwards it to the related base station (sink). As heavy amounts of data need to be transmitted during the communication and Cluster Head is wholly responsible for delivering the packets to the intended destination. Random distribution of cluster heads in Leach makes the Network overloaded.

Alive Nodes

This value indicates the number of active nodes participating in the communication which defines the lifetime of the network. LEACH is a proactive routing protocol, hence the number of alive nodes is higher as compared to the HETERO LEACH protocol.

Dead Node

Dead node is nothing but the routing holes present in the communication path. Routing hole means the node which takes part in the communication path goes dead during the communication of the packet (Figs. 2, 3, 4, and 5).

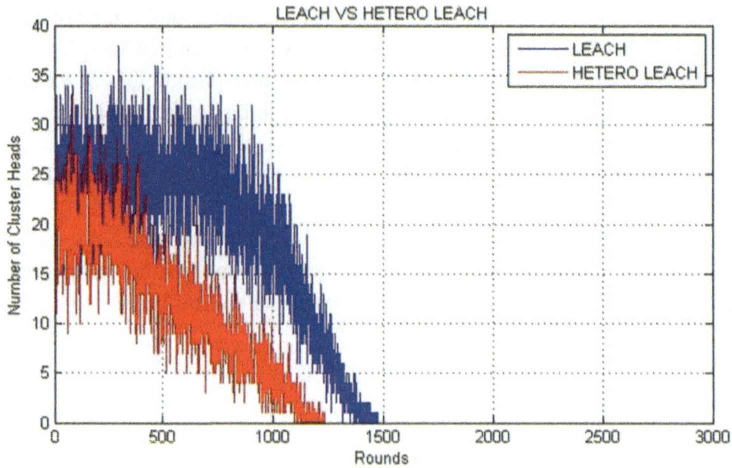

Fig. 2 Number of CH LEACH versus HETERO LEACH

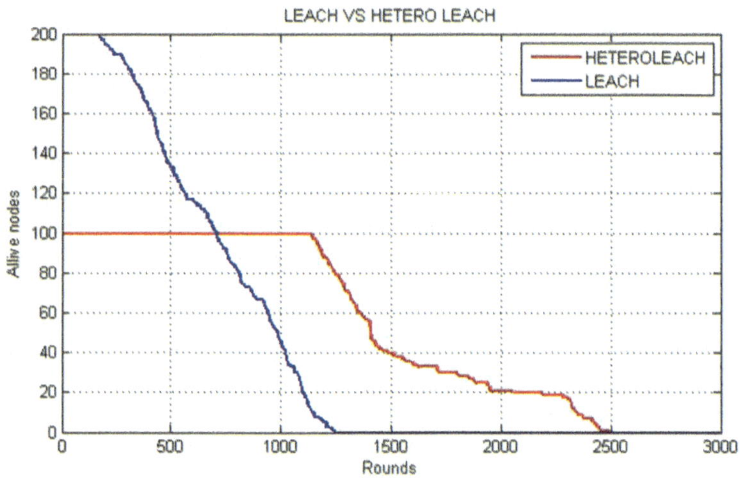

Fig. 3 Number of alive node LEACH versus HETERO LEACH

4 Conclusion

Based on the parameters, the HETERO LEACH performs better because it deals with the hybrid network with heterogeneous energy level. LEACH protocol is a proactive routing protocol, therefore, the number of nodes associated with the communication will be higher. And due to these characteristics, the number of dead nodes for LEACH protocol is also high because the probability of the node for becoming dead increases as well with the increase in the number of rounds (Table 1).

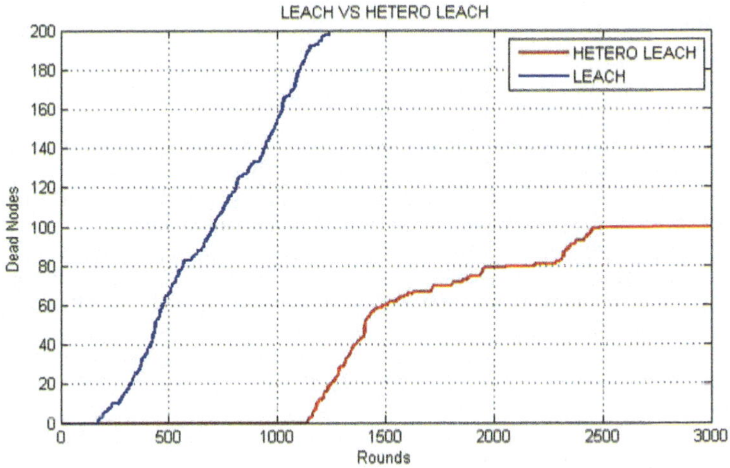

Fig. 4 Number of dead node LEACH versus HETERO LEACH

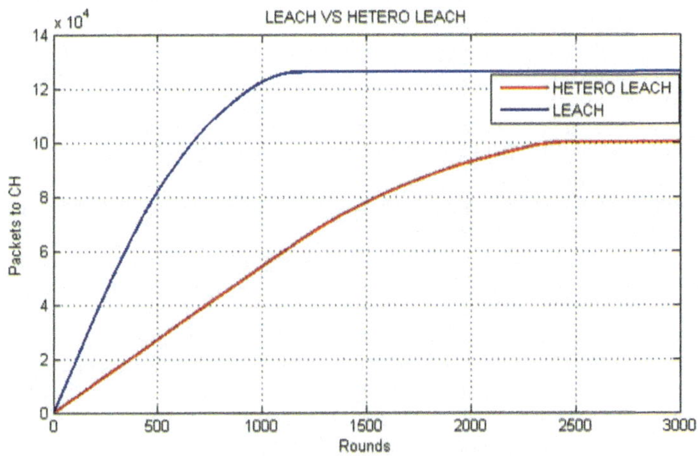

Fig. 5 Number of packets to CH LEACH versus HETERO LEACH

Table 1 Comparison of LEACH versus HETERO LEACH

Characteristic	HETERO LEACH	LEACH
Network type	Hybrid network	Proactive network
No of dead node	High	Moderate
No of alive node	High	Moderate
No of packets to BS	High	Moderate
No of packets to CH	Moderate	High
No of CH	High	Moderate
Energy efficiency	High	High

References

1. Akyildiz, I.F., Su, W., Sankarasubramaniam, Y., Cayirci, E.: A survey on sensor networks. IEEE Commun. Mag. **40**(8), 102–114 (2002)
2. Ogundile, O., Alfa, A.: A survey on an energy-efficient and energy-balanced routing protocol for wireless sensor networks. Sensors **17**(5), 1084 (2017)
3. Calzado, C.G.: Contributions on agreement in dynamic distributed systems (Doctoral dissertation, Universidad del País Vasco-Euskal Herriko Unibertsitatea) (2015)
4. Munir, S.A., Ren, B., Jiao, W., Wang, B., Xie, D., Ma, J.: Mobile wireless sensor network: architecture and enabling technologies for ubiquitous computing. In: 21st International Conference on Advanced Information Networking and Applications Workshops (AINAW'07), vol. 2, pp. 113–120. IEEE (2007)
5. Burgos, U., Gómez-Calzado, C., Lafuente, A.: Leader-based routing in mobile wireless sensor networks. In: Ubiquitous Computing and Ambient Intelligence, pp. 218–229. Springer, Cham (2016)
6. Al-Karaki, J.N., Kamal, A.E.: Routing techniques in wireless sensor networks: a survey. IEEE Wirel. Commun. **11**(6), 6–28 (2004)
7. Choi, J.Y., Yim, S.J., Huh, Y.J., Choi, Y.H.: A distributed adaptive scheme for detecting faults in wireless sensor networks. WSEAS Trans. Commun **8**(2), 269–278 (2009)
8. Behera, A., Panigrahi, A.: Determining the network throughput and flow rate using GSR and AAL2R (2015). arXiv:1508.01621
9. Burgos, U., Soraluze, I., Lafuente, A.: Evaluation of a fault-tolerant wsn routing algorithm based on link quality. In: Proceedings of the 4th International Conference on Sensor Networks, pp. 97–102 (2015)
10. Crowcroft, J., Segal, M., Levin, L.: Improved structures for data collection in wireless sensor networks. In: IEEE INFOCOM 2014-IEEE Conference on Computer Communications, pp. 1375–1383. IEEE (2014)
11. Akkari, W., Bouhdid, B., Belghith, A.: Leach: low energy adaptive tier clustering hierarchy. Procedia Comput Sci **52**, 365–372 (2015)
12. Gómez-Calzado, C., Casteigts, A., Lafuente, A., Larrea, M.: A connectivity model for agreement in dynamic systems. In: European Conference on Parallel Processing, pp. 333–345. Springer, Berlin (2015)
13. Liang, Y., Yu, H.: Energy adaptive cluster-head selection for wireless sensor networks. In: Sixth International Conference on Parallel and Distributed Computting Applications and Technologies (PDCAT'05), pp. 634–638. IEEE (2005)
14. Manjeshwar, A., Agrawal, D.P.: APTEEN: a routing protocol for enhanced efficiency in wireless sensor networks. In: Null, p. 30189a. IEEE (2001)
15. Kafi, M.A., Challal, Y., Djenouri, D., Doudou, M., Bouabdallah, A., Badache, N.: A study of wireless sensor networks for urban traffic monitoring: applications and architectures. Procedia Comput. Sci. **19**, 617–626 (2013)
16. Ko, J., Lu, C., Srivastava, M.B., Stankovic, J.A., Terzis, A., Welsh, M.: Wireless sensor networks for healthcare. Proc. IEEE **98**(11), 1947–1960 (2010)
17. Wieselthier, J.E., Nguyen, G.D., Ephremides, A.: Algorithms for energy-efficient multicasting in static Ad Hoc wireless Networks, pp. 251–263 (2001)
18. Broch, J., Maltz, D.A., Johnson, D.B., Hu, Y.C., Jetcheva, J.G.: A performance comparison of multi-hop wireless ad hoc network routing protocols. In: MobiCom, vol. 98, pp. 85–97 (1998)
19. Heinzelman, W.R., Chandrakasan, A., Balakrishnan, H.: Energy-efficient communication protocol for wireless microsensor networks. In: Proceedings of the 33rd Annual Hawaii International Conference on System Sciences, p. 10. IEEE (2000)
20. Gómez-Calzado, C., Lafuente, A., Larrea, M., Raynal, M.: Fault-tolerant leader election in mobile dynamic distributed systems. In: 2013 IEEE 19th Pacific Rim International Symposium on Dependable Computing, pp. 78–87. IEEE (2013)

21. Anitha, R.U., Kamalakkannan, P.: Enhanced cluster based routing protocol for mobile nodes in wireless sensor network. In: 2013 International Conference on Pattern Recognition, Informatics and Mobile Engineering, pp. 187–193. IEEE (2013)

A Semantic Matchmaking Technique for Cloud Service Discovery and Selection Using Ontology Based on Service-Oriented Architecture

Manoranjan Parhi, Sanjaya Kumar Jena, Niranjan Panda, and Binod Kumar Pattanayak

Abstract As we are moving ahead with cloud computing, preference for cloud services is getting increased day by day. These services mostly seem to be significantly identical in their functionality excepting their key attributes like storage, computational power, price, etc. As of now, there is no uniform specification for defining a service in the cloud domain. In order for the specification of the identical operations and publication of the services on the websites, different cloud service providers tend to use completely different vocabulary. The process of requesting for a cloud service becomes merely a challenging task as a result of increasing number of selection parameters and QoS constraints. Hence, a reasoning mechanism is very much required for service discovery that could resolve the resemblance appearing across different services with reference to the respective cloud ontology. In this paper, a semantic matchmaking technique is proposed for most relevant cloud service discovery and selection procedure based on Service-Oriented Architecture (SOA).

Keywords Cloud computing · Cloud service discovery and selection · Quality of Service (QoS) · Cloud ontology · Service-Oriented Architecture (SOA)

M. Parhi (✉) · S. K. Jena · N. Panda · B. K. Pattanayak
Department of Computer Science and Engineering, Siksha 'O' Anusandhan (Deemed to be University), Bhubaneswar, Odisha, India
e-mail: manoranjanparhi@soa.ac.in

S. K. Jena
e-mail: sanjayjena@soa.ac.in

N. Panda
e-mail: niranjanpanda@soa.ac.in

B. K. Pattanayak
e-mail: binodpattanayak@soa.ac.in

D. Mishra et al. (eds.), *Intelligent and Cloud Computing*,
Smart Innovation, Systems and Technologies 194,
https://doi.org/10.1007/978-981-15-5971-6_4

1 Introduction

As an innovative modern technology, Cloud Computing tends to leave a significant impact on the Information Technology (IT) industry as a whole in the recent past, where large and popular enterprises like IBM, Google, Amazon Web Service, and Microsoft continue to strive in providing a relatively more robust cloud computing services that are cost-effective. However, the properties such as no up-front investment, lowered operating cost, enhanced scalability, reduced risks in business, and less maintenance expenses make cloud computing extensively attractive for its clients and the business owners as well [1, 2]. It is worth mentioning that the cloud services published on the web by various service providers can be conveniently accessed by intended customers with the help of web-portals. As mentioned earlier that several cloud services possess similar functionality, and hence it becomes important to identify the appropriate service which can necessarily comply with the desired service requirements as requested by the customers. Therefore, customers, most of the time find it difficult in making the choice of the most suitable cloud service provider to fulfill their objectives due to the following reasons [3]:

1. There are no standardized naming conventions implemented by different cloud service providers. For example, Amazon Web Service named its computing services as "EC2 Compute Unit", whereas the same services provided by GoGrid are commonly known as "Cloud Servers".
2. Description of the services that are not formatted, pricing strategies along with the rules pertaining to SLA are used by different cloud service providers as displayed in their respective websites.
3. The modifications incorporated in the above mentioned information are not reported to the users that make it difficult to achieve manually, and compare the configuration of the services as reported in the websites of different cloud service providers along with the documentation that represents the only source of information that is available.
4. Traditional search engines on the internet such as Google, MSN, Yahoo, and others are not necessarily capable of performing reasoning, as well as comparison of various relations, that persist among several categories of cloud services along with their configuration.

There are many service discovery techniques proposed in the literature in the context of cloud computing based on QoS under Service-Oriented Architecture (SOA) [4]. But, it lacks in the uniformity of concepts in these works. Similarly, logical solutions are provided by different studies without considering the current status of cloud service providers. Thus, there arises a necessity of an intelligent strategy for cloud service discovery in order for searching an accurate service that possesses higher accuracy and more importantly, searching swiftly with respect to the criteria as imposed by the user request.

In this paper, an efficient semantic service matchmaking technique is proposed that necessarily takes into consideration the preferences of the customer in order for calculation of similarity level in the relevance of descriptions of services provided by two service providers. Here, the matching technique focuses on selecting an appropriate cloud service provider with a higher success rate, thereby consuming lesser response time.

The remaining part of this paper is organized as follows. In Sect. 2, the background of this research work, i.e., Service-Oriented Computing is explained briefly. The workflow of the proposed model for service discovery and selection followed by the description of corresponding cloud ontology is elaborated in Sect. 3. In Sect. 4, the proposed semantic matchmaking technique is explained along with algorithms followed by a case study. At the end, Sect. 5 concludes the paper with the contributions made in this work and possible extensions in the future.

2 Service-Oriented Computing

Service-Oriented Computing (SOC) represents a computing standard that incorporates a set of concepts, methods, as well as principles, which facilitate Service-Oriented Architecture (SOA) that contributes to the design of software applications those mostly rely upon independent component services along with desired standard interfaces [5, 6]. Here, SOC is referred to as an innovative approach to software development carried out by virtue of using a loosely coupled system and independent of tightly coupled monolithic systems.

2.1 Service-Oriented Architecture

The Service-Oriented Architecture (SOA) depicted in Fig. 1, focuses on determining the interactions taking place among three distinct components as follows:

- A "Service Provider" represents a node in the network that provides an interface for the web services and simultaneously, resolves the service requests generated by consumers.
- A "Service Requester" represents a node in the network that is responsible for discovering and invoking the desired web service that is intended for the realization of a business solution.
- A "Service Registry", also regarded as the broker of the service, represents a node in the network that forms a repository of services intended for description of the interfaces to the available services which are published on the network by the service provider.

The interactions among these distinct components induce three specific operations such as publishing, finding, and binding of a web service.

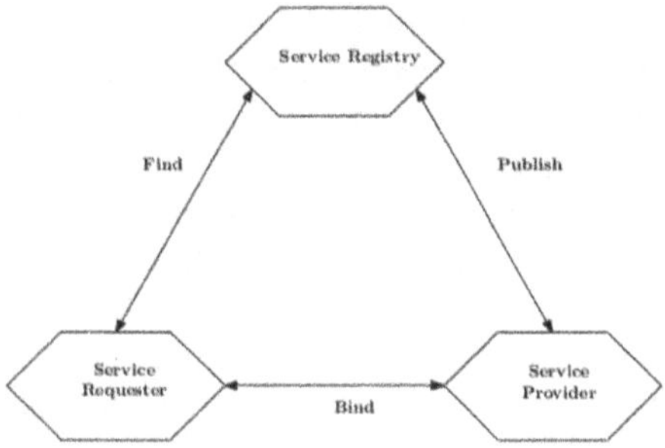

Fig. 1 Service-oriented architecture

3 Proposed Model

The proposed model is based on Service-Oriented Architecture (SOA) as shown in Fig. 2, comprises of the following phases as described below.

3.1 Client Request

First of all, the client generates a request for discovering the desired cloud service making use of a graphical interface. The client needs to make a choice of values corresponding to the functional, as well as nonfunctional attributes, like Central

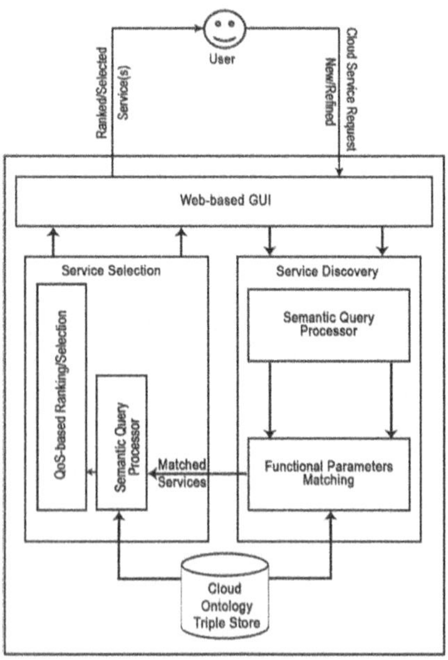

Fig. 2 Proposed model for cloud service discovery and selection

Processing Unit, EC2 Compute Unit, Random Access Memory, Hard Disk space, service availability, network bandwidth, cost, service rating, and so on.

3.2 Cloud Service Discovery

In the next step, the request from the client is executed by a software component named as Semantic Query Processor that principally relies on Simple Protocol and RDF Query Language (SPARQL) [7]. Then a matching process is carried out between the values of user specified technical parameters with those values stored in the proposed cloud ontology triple store by applying the proposed semantic matchmaking technique.

3.3 Cloud Service Selection

Then the values matched functionality in the previous step are further processed by another Semantic Query Processor making use of SPARQL and semantic rules in order for matching of predefined values of QoS with those values that are available in the proposed cloud ontology triple store. Once the matching process is accomplished, the requested ranked cloud service is presented to the client via GUI, and with this, the entire process is terminated.

3.4 Cloud Ontology Triple Store

In this model, the Cloud ontology Triple Store is developed on the basis of the cloud IaaS domain model. This ontology describes cloud infrastructure services and their functional and QoS attributes. It is represented as semantic registry under Service-Oriented Architecture for service registration, discovery, and selection. The cloud service domain knowledge is obtained from various resources, cloud ontology [8], cloud taxonomy [9], and the industry based standards [10]. This ontology is represented in terms of RDF (Resource Description Framework) which is based on Subject-Predicate-Object expression known as Triple Store. All the classes along with the subclasses of the proposed cloud ontology Triple Store are depicted in Fig. 3, which is created using Protege Ontology editor [11].

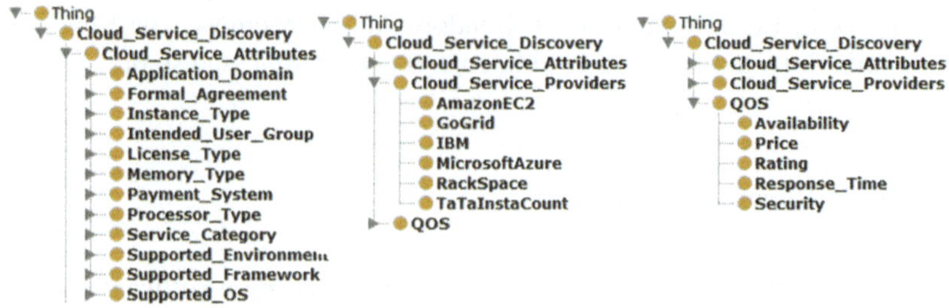

Fig. 3 The proposed cloud ontology comprises of classes and subclasses

4 Proposed Matchmaking Technique

The proposed semantic matchmaking technique is explained as follows:

In the proposed framework, the Semantic Query Processor acts as a major component, which takes client request (functional parameters) from a graphical interface during the service discovery phase and generates a query in terms of SPARQL to find semantically equivalent cloud services. The process of matchmaking is illustrated in Algorithm 1, which begins with a set of iterations which are applied to the over all services associated with the resources referring to client request. The *getServicesBySemanticAnnotation()* method is used for retrieving the services and their semantic score is calculated with the help of *semanticSimilarity()* method. The *semanticSimilarity()* method is used to compute the semantic relevance which lies among the resources of the service published (from $CR\_service$) and request of the resources (from CR_query). If the *semanticSimilarity()* method returns a positive value, then a logarithm function is applied to score which is finally obtained in order to reduce the score value.

Referring to the Tversky similarity model [12], the semantic relevance between two different resources is calculated by *semanticSimilarity()* method (Algorithm 2). While comparing two resources, the method getConcepts() returns their cloud concepts and the method getProperties() returns the list of their object and data properties from the proposed cloud ontology Triple store. Then the method TverskySimilarity Measure() (Algorithm 3) is invoked for calculating the degree of relevance on the basis of common and different features with respect to the compared sets. The method semanticSimilarity() returns the final score value by finding the average among the matching score of cloud concepts, object, and data properties.

In this work, the Tversky's model is used for matching the cloud services based on semantics. This model is treated as one of the best feature-based similarity model which takes into account the features those are shared by two concepts, thereby distinguishing the features specific to each. Further, it can be explained with a case study as follows:

Let us consider the cloud resource rs representing Infrastructure as a Service (IaaS), which is a part of the cloud request made by the client and the cloud resource rs' representing Network as a Service (NaaS), which is a part of the cloud service

Algorithm 1 Semantic Cloud service Matchmaker

Input: Receives a collection of cloud resources CR_query explaining the intended service functionalities
Output: Generates a list of cloud services $CS' \subseteq CS$ which are semantically matched

```
function cloudServiceMatchmaker(CR_query)
CS=getServicesBySemanticAnnotation(CR_query)
for each cs ∈ CS do
   score − value = 0
   CR_service = getAnnotationCloudResources(cs)
   for each rs' ∈ CR_service do
      for each rs ∈ CR_query do
         score − value = score − value + semanticSimilarity(rs, rs')
      end for
   end for
   if score − value > 0 then
      cs.semanticscore − value=log(1+score − value)
      CS' = CS'U{cs}
   end if
end for
return CS'
end function
```

Algorithm 2 semanticSimilarity()

Input: Receives rs and rs' which represents two cloud resources
Output: Generates a similarity score which refers semantic relevance between two resources rs and rs'

```
function semanticSimilarity(rs, rs')
C_rs=getConcepts(rs)
C_rs'=getConcepts(rs')
P_rs=getPropeties(rs)
P_rs'=getPropeties(rs')
S_cloud_concepts=TverskySimilarityMeasure(C_rs, C_rs')
S_cloud_properties=TverskySimilarityMeasure(P_rs, P_rs')
sim − score = AVERAGE (S_cloud_concepts, S_cloud_properties)
return sim − score
end function
```

Algorithm 3 TverskySimilarityMeasure()

Input: Receives F_1 and F_2 as two list of cloud features
Output: Compute a similarity cloud_sim between 0 and 1

```
function TverskySimilarityiMeasure(rs, rs')
C=CloudCommonFeatures(F1 and F2)
UF1=CloudUniqueFeatures(F1 and F2)
UF2=CloudUniqueFeatures(F2 and F1)
cloud_sim = |C| / (|C|+|UF1|+|UF2|)
return cloud_sim
end function
```

Table 1 Client request versus result obtained after discovery

Parameters	Client request	Result obtained after discovery
Cloud service category	Infrastructure as a service	Network as a service
Operating system	Windows 10	Windows vista
Cost ($)	0.02	0.03
RAM (GB)	8	6
Bandwidth (GB/s)	16	14
EC2 compute unit	2	4
Processor type	Core i5 (6th Generation)	Core i3 (6th Generation)
Number of cores	8	4
CPU clock (GHz)	2.4	3.6
Hard disk drive (GB)	200	500

advertisement published by the provider. Both the resources are concepts and annotated within the same ontology. In this case, both IaaS and NaaS are similar with respect to their common features such as Network, Security, and Firewall and dissimilar with respect to the features like servers, OS, virtualization, Memory, Data Center, etc.

Let F_1 be the set of features of Infrastructure as a Service (IaaS) = {Servers, Memory, OS, Virtualization, Network, Data Center, Security, Firewall}

Let F_2 be the set of features of Network as a Service (NaaS) = {VPN, Mobile Network Virtualization, Network, Firewall, Security}

Let CF be the Features which are shared or common in between them, i.e., {Network, Firewall, Security}

UF_1={OS, Virtualization, Servers, Memory, Data Center}

UF_2={VPN, Mobile Network Virtualization}

Now, the cloud similarity measure between IaaS and NaaS is computed as

$$cloud\_sim = \frac{|CF|}{|CF| + |UF_1| + |UF_2|} = \frac{3}{3+5+2} = 0.3$$

The result as shown in Table 1, represents the comparison between the client request and the responses generated after discovery.

After the completion of the discovery phase, the next phase is initiated by another Semantic Query Processor, called the selection phase. During this phase, a list of most suitable cloud services are obtained based on nonfunctional QoS requirements of a user. However, some complex QoS requirements may not be satisfied by some of these services. Hence, a ranking system is proposed for the services which are matched in the previous phase. This ranking system comprises of some semantic rules which are written using SWRL (Semantic Web Rule Language) [13]. These rules describe the alternative choice of services on the basis of QoS constraints. These rules use an open-source OWL-DL reasoner named as *pallet* [14] which provides the features

Table 2 Semantic rules

Semantic_Rule1:
CloudProviders(?s), hasServiceAvailability(?s, ?availability), hasServiceCost(?s, ?cost), hasServiceRating(?s, ?score), higherThan(?cost, 0.2), lowerThan(?availability, 90.0), lessThan(?score, 6.0) → hasServiceRemoved(?s, true)
Semantic_Rule2:
CloudProviders(?s), double[≥ 0.1, ≤ 0.2](?cost), double[≥ 6.0, ≤ 8.0](?score), double[≥ 90.0, ≤ 95.0](?availability), hasServiceAvailability(?s, ?availability), hasServiceRemoved(?s, false), hasServiceCost(?s, ?cost), hasServiceRating(?s, ?score)→CloudProviderMatchedList(?s), hasServiceRank(?s, "2"^^*int*)
Semantic_Rule3:
CloudProviders(?s), hasServiceAvailability(?s, ?availability), hasServiceRemoved(?s, false), hasServiceCost(?s, ?cost), hasServiceRating(?s, ?score), higherThan(?availability, 95.0), greaterThan(?score, 8.0), lowerThan(?cost, 0.1)→CloudProviderMatchedList(?s), hasServiceRank(?s, "1"^^*int*)

to check the consistency and integrity of ontology, determines and calculates the classification hierarchy, describes the inference rules, and responds to the SPARQL queries.

In this work, some semantic rules as shown in Table 2, are written using SWRL that describes the ranking of the desired service on the basis of QoS attributes in the cloud IaaS domain. Let us consider the three rules (Semantic_Rule1, Semantic_Rule2, Semantic_Rule3) that have been defined in such a way that Semantic_Rule1 > Semantic_Rule2, Semantic_Rule1 > Semantic_Rule3. It indicates that the first semantic rule has high precedence over the second and third semantic rules. In other words, it can be said that Semantic_Rule1 must be executed before Semantic_Rule2 and Semantic_Rule3. Semantic_Rule1 removes any such service having cost higher than 0.2$/hour, availability lower than 90%, and service rating lower than 6.0 in a ten-point rating scale. Hence, Semantic_Rule2 and Semantic_Rule3 are to be applied only on cloud services which satisfy Semantic_Rule1. Semantic_Rule2 assigns a rank of two points to any cloud service whose cost lies between 0.1$/hour and 0.2$/hour, availability between 90% and 95%, and service rating between 6.0 to 8.0, where Semantic_Rule3 generates only one point rank to the cloud services those have a cost lower than 0.1$/hour, availability higher than 95%, and service rating higher than 8.0.

5 Conclusion and Future Work

In this paper, a model has been proposed that integrates a semantic matchmaking technique with the proposed cloud ontology Triple Store. This model primarily contributes towards the description of the cloud service along with their attributes in a standardized and consistent way using ontology. This ontology assists the users to discover the desired suitable service as per the requirements as specified in users request. With the help of a graphical user interface, a user can select the request for discovery of the desired cloud service. Consequently, the request can be handled on the basis of proposed cloud ontology and the corresponding reasoning rules, and finally, the most relevant service is obtained based on QoS parameters. In the future, it is planned to implement the entire system considering the security and scalability issues along with semantic matchmaking for PaaS and SaaS types of cloud services.

References

1. Zhang, Q., Cheng, L., Boutaba, R.: Cloud computing: state-of-the-art and research challenges. J. Internet Serv. Appl. **1**(1), 7–18 (2010)
2. Dillon, T., Chen, W., Chang, E.: Cloud Computing: Issues and Challenges. In: 24th IEEE International Conference on Advanced Information Networking and Applications, pp. 27–33. IEEE Press, New York (2010)
3. Parhi, M., Pattanayak, B.K., Patra, M.R.: A multi-agent-based framework for cloud service discovery and selection using ontology. J. Serv. Oriented Comput. Appl. **12**(2), 137–154 (2018). https://doi.org/10.1007/s11761-017-0224-y
4. Hayyolalam, V., Kazem, A.A.: A systematic literature review on QoS-aware service composition and selection in cloud environment. J. Netw. Comput. Appl. **110**, 52–74 (2018)
5. Huhns, M.N., Singh, M.P.: Service-oriented computing: key concepts and principles. IEEE Internet Comput. **9**(1), 75–81 (2005)
6. Petrenko, A.I.: Service-oriented computing in a cloud computing environment. Comput. Sci. Appl. **1**(6), 349–358 (2014)
7. SPARQL 1.1 Query Language. https://www.w3.org/TR/sparql11-query/
8. Al-Sayed, M.M., Hassan, H.A., Omara, F.A.: Towards evaluation of cloud ontologies. J. Parallel Distrib. Comput. **126**, 82–106 (2019)
9. Hoefer, C.N., Karagiannis G.: Taxonomy of Cloud Computing Services. IEEE GLOBECOM Workshop on Enabling the Future Service Oriented Internet, pp. 1345–1350 (2010)
10. NIST Cloud Computing Standards Roadmap. https://www.nist.gov/sites/default/
11. Protege Ontology Editor. http://protege.stanford.edu/
12. Tversky, A.: Features of similarity. CPsych. Rev. 327–352 (1977)
13. SWRL: A Semantic Web Rule Language Combining OWL and RuleML. www.w3.org/Submission/SWRL/
14. Sirin, E., Parsia, B., Grau, B.C., Kalyanpur, A., Katz, Y.: Pellet: a practical OWL-Lreasoner. J. Web Semant. **5**(2), 51–53 (2007)

An Intelligent Approach to Detect Cracks on a Surface in an Image

Saumendra Kumar Mohapatra, Pritisman Kar, and Mihir Narayan Mohanty

Abstract Crack detection is one of the vital tasks in monitoring structural health and ensuring structural safety. This paper provides a smart way to recognize cracks in the structure and analyze its property. In this model, Min-Max Grey Level Discrimination (M2GLD) method is applied for intensity adjustment. It uses Otsu model for the preprocessing task of the image. The aim of this grey intensity tuning system is to increase the accuracy of the detection of cracks. From the result, it can be noticed that the proposed approach is providing a satisfactory result for detecting the cracks in a digital image.

Keywords Cracks · Enhancement · Otsu method · Image thinning · Crack width

1 Introduction

One of the major concerns for making sure the protection, strength, and serviceability of structures are cracks. These cracks occur in the foundation of a structure. These occur over time and should not be considered as the fault of workers or structural shape.

The reason is that when cracks happen, they tend to create a decrease in the successful loading area which brings about the increase of stress and subsequent failure of the concrete or other structures. Mostly for concrete components, cracks make harmful and sarcastic chemicals to go through into the structure, and it damages their integrity.

Image Binarization is mostly used for recognition of text and medical Image Processing, also useful for crack detection. This is because they both have similar properties and have distinguishable lines and curves. The Otsu method which uses

S. K. Mohapatra · P. Kar
Deaprtment of Computer Science and Engineering, Chennai, India

M. N. Mohanty (✉)
Department of Electronics and Communication Engineering, ITER, Siksha 'O' Anusandhan (Deemed to be University), Bhubaneswar, India
e-mail: mihirmohanty@soa.ac.in

D. Mishra et al. (eds.), *Intelligent and Cloud Computing*,
Smart Innovation, Systems and Technologies 194,
https://doi.org/10.1007/978-981-15-5971-6_5

image binarization technique for rambling; the cause is that binarization of image depends on the feature of the image, characteristics of the background, and associated parameters.

In the proposed work, a digital image handling model that consequently recognizes and breaks down cracks on the building surface is presented. The projected model additionally performs different estimations of split attributes including the territory, edge, width, length, and direction. At the focal point of the designed model, a picture upgrade calculation called Min-Max Grey Level Discrimination (M2GLD) is advanced as a pre-preparing venture to improve the Otsu binarization approach, trailed by shape investigations for correcting the break location execution. The break distinguished by the proposed methodology is contrasted and procured by the ordinary strategy.

2 Literature Review

Crack is a significant sign of the damage of any structure. Identification of cracks is frequently required in the phase of building upkeep. Likewise, assessments of the auxiliary respectability dependent on break examinations become considerable for the administration life expectation of structures. In the last few years a lot of work have been done by the researchers for the crack detection and some of them are discussed here (Fig. 1).

Adhikari et al. [1], have designed a replica that numerically speaks to the cracks deformities; their projected methodology was additionally fit for break evaluation and detection. In [2], authors have designed a picture based mechanized crack detection model for post-fiasco construction appraisal; in light of the numerical test, the researchers have shown that their proposed technique can achieve extraordinary advantages in a post-calamity investigation of building components. Authors in [3], have planned an automatic crack detection technique that analyzes and detects cracks on the concrete surface; crack identification is perceived from a picture of a solid surface, and crack examination ascertains the qualities of the recognized splits, for example, split width, length, and unfulfilled obligations.

All things considered, the exhibition of the above crack recognition models is regularly decayed by the mind-boggling surface of black-top asphalt and concealing states of the digital images [4].

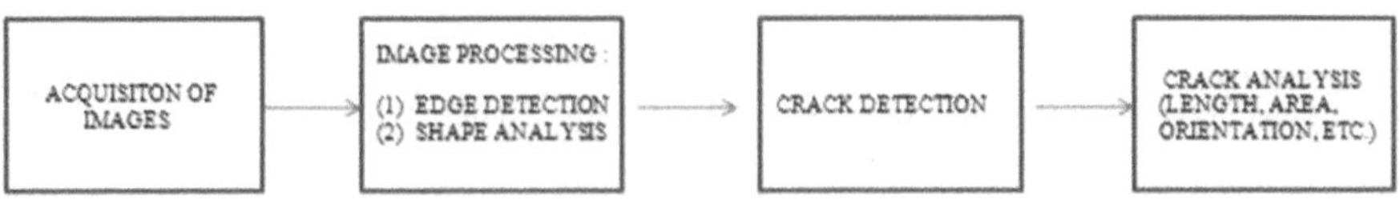

Fig. 1 General model of crack detection in image processing

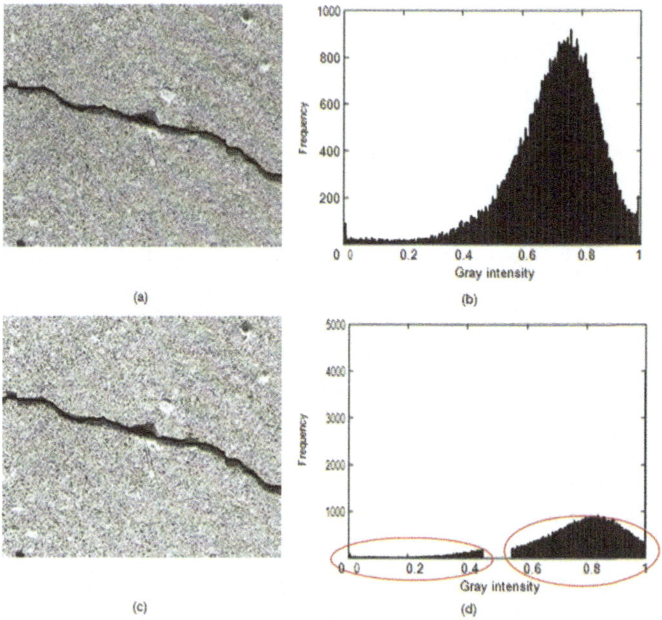

Fig. 2 Enhancement using the projected technique **a** Raw image **b** Histogram of raw image **c** Enhanced image **d** Enhanced image Histogram

Because of its straightforwardness, low computational expense, and thresholding ability, Otsu strategy is broadly utilized in the latest works of crack identification [5–9]. The computational effectiveness of the Otsu technique is because of the way that the interclass distinguishableness is utilized to figure the ideal estimation of the limit of dim power. Adding to it, the base blunder strategy for the Otsu technique depends on the Bayes grouping mistake [10]. On account of such confinements of the surveyed strategies and the benefits of the Otsu strategy, Otsu's picture thresholding calculation has been chosen to be utilized in this examination.

3 Min-Max Gray Level Discrimination Method

It very well may be checked on that the image obtained by a digital camera, the quantity of light in various areas of the surface organization may fluctuate radically. Along these lines, the foundation brilliance of a picture isn't uniform. Additionally, building surfaces frequently highlight low complexity, uneven enlightenment, and serious commotion unsettling influence. To address this wonder, it is important to look at the improve discovery execution of cracks (Fig. 2).

The distinguished lines and curves contained in the crack is a special feature of the crack due to which the grey-scale rate of it is often a local least amount inside an

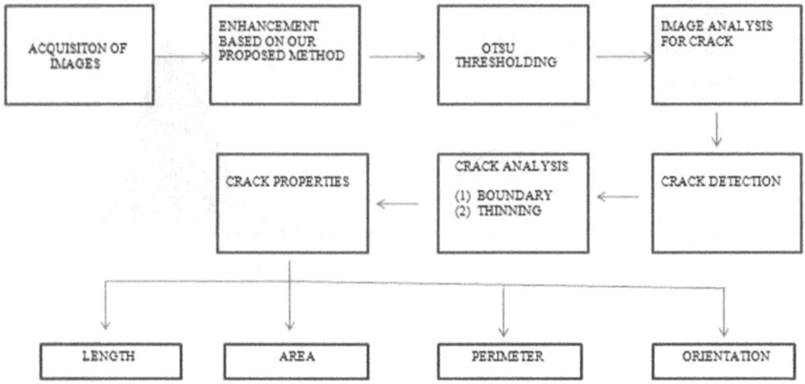

Fig. 3 Proposed flow diagram to detect crack and it's characteristic

image. For splitting the pixels into the crack and non-crack groups, it is helpful for working out an method for enhanced discriminating the two-pixel groups of curiosity.

$$G_Q(x, y) = \min(G_O\_\max, G_O(x, y)C_A) \text{if } G_O(x, y) > G_O\_(x, y) + \lambda(G_O\_\max - G_O\_\min) \quad (1)$$

$$G_Q(x, y) = \max(G_O\_\min, G_O(x, y)/C_A) \text{ if } G_O(x, y) \leq G_O\_(x, y) + \lambda(G_O\_\max - G_O\_\min) \quad (2)$$

where $G_Q(x, y)$ indicates the attuned grey intensity of the pixel at position (x, y), C_A is the adjusting ratio, $G_O\_\max$ and $G_O\_\min$ are the maximum and minimum values of the grey intensity of the original image, and λ represents a margin parameter.

The fundamental idea of the M2GLD is that, in this technique, the non-crack pixels value becomes higher and the pixel value of crack pixels decreases. This method followed by Otsu's approach can help to differentiate the crack area from non-crack regions. In Fig. 3, it is compared with the simple Otsu's approach. Herein, the parameters C_A and λ are set to be 1.05 and 0.52, respectively. This model also helps for converting an apparently unimodal image.

4 Proposed Model for Crack Detection

In the following section of the Image Processing Model, a complete description is given about the proposed model. In the wake of being built, the model can be applied to detect and investigate cracks on the surface of different components in the building, for instance, concrete beam, slab, door, wall, and a brick wall covered by mortar. The model design is shown in Fig. 4. The proposed model is designed by

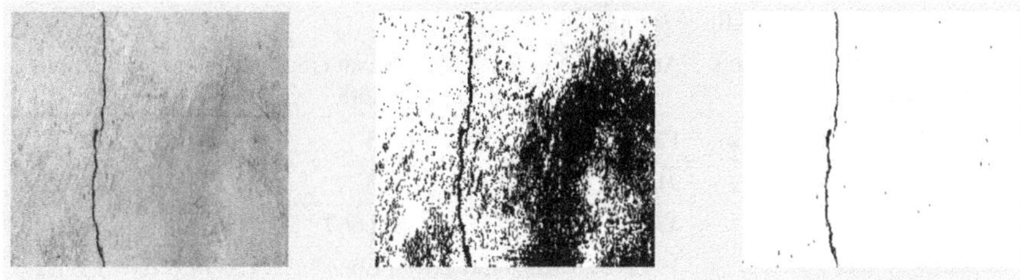

(a) (b) (c)

Fig. 4 1st test image **a** Raw image **b** After applying Otsu technique **c** After applying M2GLD technique

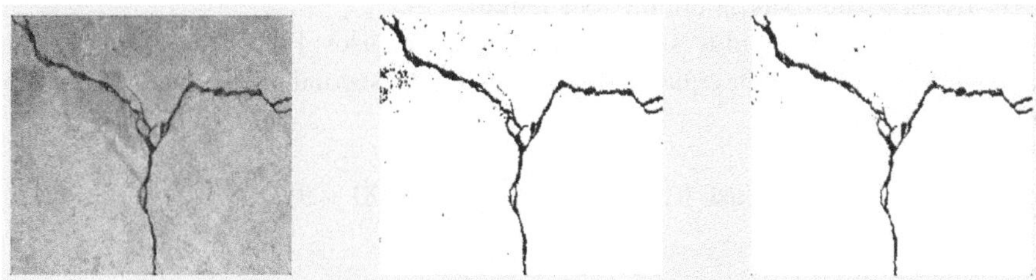

(a) (b) (c)

Fig. 5 2nd test image **a** Raw image **b** After applying Otsu technique **c** After applying M2GLD technique

(a) (b) (c)

Fig. 6 3rd test image **a** Raw image **b** After applying Otsu technique **c** After applying M2GLD technique

using the MATLAB environment. The first picture gained from the digital camera is considered as the mode input (Figs. 5 and 6).

The preprocessing of the image is categorized into two stages: initially, objects without a specific amount of pixels (M_O) are discarded. It is built by applying the below formula

Table 1 Crack properties of three sample digital images

Testing image	Crack objects	Area	Orientation	Mean crack width	Perimeter	Length
1	2 (1st Crack)	1237.5	90.008	5.5	461	225
	(2nd Crack)	910.0017	90.107	3.5	527.0010	260.0005
2	1	4705	–	6.6667	1424.8	705.7465
3	1	19977	69.9480	12.6667	3179.6	1577.1

$$ARI = \frac{L_{\text{MAJOR}}}{L_{\text{MINOR}}} \tag{3}$$

where L_{MAJOR} and L_{MINOR} are the axis length.

The crack estimate width is categorized into two cases: First is the orientation of crack which is less than or equal to 45 degree and the second one is crack orientation larger than 45 degree

$$\text{Case 1}: \ W(r) = N_V(r).\sin(90 - \theta)$$

$$\text{Case 2}: \ W(r) = N_H(r).\sin(\theta)$$

where $N_V(r)$ and $N_H(r)$ are amount of crack pixel present in the vertical and horizontal direction in the region r with orientation equals to. Using these values, we can calculate the length of the crack using the given Eq. (4)

$$L_C = \frac{(\text{Perimeter} - 2X\, W_{\text{AVG}})}{2} \tag{4}$$

where W_{AVG} is the average crack width. A model that utilizes Otsu technique erroneously distinguishes a split item in Fig. 1; it additionally neglects to recognize split pixels in picture Fig. 3. Actually, the model outfitted with the M2GLD has effectively perceived Fig. 2 as no split; it likewise effectively recognizes existing break pixels in images. Besides, the break objects found by the proposed methodology unmistakably look like the genuine split examples in the first pictures is considered by the advance digital camera. Those realities affirm that robust model is, in fact, helpful to common sense the uses of split identification in building structures. Likewise, the consequences of break examinations are given in Table 1.

5 Conclusion

The proposed work is a digital image handling model that distinguishes cracks absconds on the outside of the structure. The new model utilizes a picture upgrade calculation called Min-Max Grey Level Discrimination (M2GLD) for getting better Otsu technique. From the result, it can be noticed that the proposed technique is performing better for crack detection.

References

1. Adhikari, R.S., Moselhi, O., Bagchi, A.: Image-based retrieval of concrete crack properties for bridge inspection. Autom. Constr. **39**,180–194 (2014)
2. Thatoi, D., Guru, P., Jena, P.K., Choudhury, S., Das, H.C.: Comparison of CFBP, FFBP, and RBF networks in the field of crack detection. Model. Simul. Eng. **13** (2014)
3. Zakeri, H., Nejad, F.M., Fahimifar, A.: Image based techniques for crack detection, classification and quantification in asphalt pavement: a review. Archiv. Comput. Methods Eng. **24**, 935–977 (2016)
4. Ying, L., Salari, E.: Beamlet transform-based technique for pavement crack detection and classification. Comput.-Aided Civil Infrastruct. Eng. **25**, 572–580 (2010)
5. Rimkus, A., Podviezko, P., Gribniak, V.: Processing digital images for crack localization in reinforced concrete members. Procedia Eng. **122**, 239–243 (2015)
6. Alam, S.Y., Loukili, A., Grondin, F., Roziere, E.: Use of the digital image correlation and acoustic emission technique to study the effect of structural size on cracking of reinforced concrete. Eng. Fract. Mech. **143**, 17–31 (2015)
7. Ebrahimkhanlou, A., Farhidzadeh, A., Salamone, S.: Multifractal analysis of crack patterns in reinforced concrete shear walls. Struct. Health Monit. **15**, 81–92 (2016)
8. Yu, T., Zhu, A., Chen, Y.: Efficient crack detection method for tunnel lining surface cracks based on infrared images. J. Comput. Civil Eng. **31**(3) (2017)
9. Bose, K., Bandyopadhyay, S.K.: Crack detection and classification in concrete structure. J. Res. **2**, 29–38 (2016)
10. Kamaliardakani, M., Sun, L., Ardakani, M.K.: Sealed-crack detection algorithm using heuristic thresholding approach. J. Comput. Civil Eng. **30**(1) (2016)

A Study of Machine Learning Techniques in Short Term Load Forecasting Using ANN

Saroj Kumar Panda, Papia Ray, and Debani Prasad Mishra

Abstract The electrical transient burden estimating has been increased as one of the important essential fields of investigations used for effectual, as well as solid activity of intensity structure. This presumes an extremely important job within the area of booking, opportunity investigation, stack stream examination, assemble, as well as upkeep of intensity structure. This paper tends to audit on as of late distributed research take a shot at various variations of the counterfeit neural system in the field of momentary burden guaging. Specifically, the mixture systems which are a mix of neural system with stochastic learning methods, i.e., Back propagation (BP), Fuzzy logic (FL), Hereditary calculation (HC), Particle Swarm Optimization (PSO) which have been effectively useful for short term load forecasting (STLF) is talked about completely.

Keywords Short Term Load Forecasting (STLF) · Artificial Neural Network (ANN) · Back Propagation (BP) · Fuzzy Logic (fl) · Hereditary Calculation (HC) · Particle Swarm Optimization (PSO)

1 Introduction

To provide electricity vitality toward the client within the protected and monetary way, the electricity organization meets numerous practical, as well as specialized difficulties inactivity. Among these difficulties booking, load stream examination, arranging, and control of electric vitality framework are generally noticeable. Burden

S. K. Panda · P. Ray
VSSUT, Burla 768018, India
e-mail: saroj.panda89@gmail.com

P. Ray
e-mail: papiavssut@gmail.com

D. P. Mishra (✉)
IIIT, Bhubaneswar 751003, India
e-mail: debani11@gmail.com

D. Mishra et al. (eds.), *Intelligent and Cloud Computing*,
Smart Innovation, Systems and Technologies 194,
https://doi.org/10.1007/978-981-15-5971-6_6

anticipating be important rising fields of the investigation used for significant, as well as the testing field, in the most recent couple of years. Burden determining might be characterized as the proportion of precision of the distinction between the genuine and anticipated estimation of future burden requests.

Estimating power requests will help in improving the beginning up expense of producing units, and can likewise ready to spare the interest in the development of the required number of intensity offices. It can likewise check the unsafe activity, fluctuating interest, request of turning store, and defenselessness to disappointments. Burden gauging gives the most significant data to control conveyance and arranging. In [1], it likewise assumes a significant job in the vitality of the executive's framework. Power framework booking, load stream investigation, everyday activity, and effectiveness are some of the intriguing fields that can be investigated by burden determining. The estimation of the burden request is noteworthy as it will support the generation and dissemination of electric power. In [2], by under-evaluating the heap request, it has a negative result on interest reaction and consequently on power establishment. Likewise, this under-estimation of burden results in trouble to deal with the over-burden situation. Essentially, the extra calculation influences the establishment, as well as subsequently the effectiveness of the framework. With the end goal of burden gauging, a few procedures have just been applied during the most recent couple of decades [3, 4]. Comprehensively, the heap estimating systems can be isolated into two classes, for example, parametric or nonparametric methods. The straight relapse, auto backward moving normal (ARMA), general exponential method, and stochastic time arrangement procedures are a few instances of parametric strategy. The primary disadvantage of this system is its ability in the unexpected difference in any kind of condition or else common alters. Nonetheless, its deficiency be overwhelmed via effect non-constraint support method given this possibility to worldwide inquiry [5, 6]. Among these computerized reasoning based strategies, the fake neural system has risen as one of the most noticeable procedures that get considerably more consideration of specialists. In [7, 8], the capacity to comprehend the mind-boggling connections, versatile control, picture denoising, basic leadership under vulnerability, and forecast examples makes ANN a groundbreaking entertainer than recently actualized strategies. Here ANN is examined with BP, FL, HC, and PSO.

In Sect. 2, a dialog on ANN in STLF has been finished. In segment 3, distinct variations of ANN, that incorporates the customary and cross breed neural system procedures that have effectively applied to STLF is depicted. At long last, the finish of the present audit work has been introduced in segment 4.

2 ANN in STLF

In [9, 10], ANN can be characterized since the profoundly associated exhibit of rudimentary processors identifies neurons, as well as a total portrayal of it. This takes after its source as of the soul mind that has an enormous amount of neurons interrelated within profoundly unpredictable, nonstraight, and shaping exceptionally huge parallel systems. A fake neural system (ANN) with an info layer, at least one shrouded layer and one yield layer is known as a multilayer perceptron (MLP) in [11, 12].

2.1 Multilayer Perceptron Network (MLP)

Multilayer perceptron arranges mainly due to the mainstream neural system sort, as well as the large portion of the announced neural system. The fundamental component (neuron) of the system is a perceptron. This is a calculation component, which delivers its yield through taking a straight blend of the info signals, as well as by changing this through a capacity known as movement work. The yield of the perceptron as an element of the information signs would thus be able to be composed

$$U_j = \varphi\left(\pi \sum_{i=1}^{N} W_{ij} X_i + B_j\right) \tag{1}$$

where

U_j	output of the network
$X_1, X_2, \ldots, X_i \ldots X_N$	input signal
W_{ij}	neuron weight
B_j	another neuron weight
φ	activity function

Potential types of movement work are direct capacity, step work, calculated capacity, and hyperbolic digression function. The MLP organize comprises of a few layers of neurons. Each neuron within a given layer is associated with each neuron of the following film. There are no input associations. A three-layer MLP organize is represented in Fig. 1, which allows the signal from input to hidden layer and hidden layer to output with adjustment of weight and activity function and produce output.

The Fig. 1, allows the signal from input to hidden layer and hidden layer to output with adjustment of weight and activity function. The mathematical equation helps to

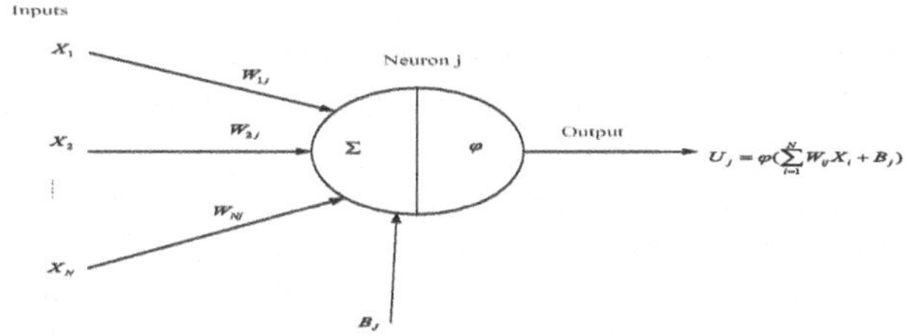

Fig. 1 Feedforward neural network with an invisible hidden layer

get the best simulation result in STLF. Wight adjustment means it decides how much influence the input will have on the output. The activity function of a node defines the output of that node given an input or set of inputs.

While an N-dimensional information vector is nourished toward the system, an M-amplitude yield vector be created. The system can exist as comprehended because single capacity beginning the N-amplitude contribution area toward the M-amplitude yields area. This capacity is able to write within the structure

$$Y = f(X, W) = S(W_N S(W_{N-1} S(\ldots(S(W_1 X)\ldots)))) \tag{2}$$

where

Y output vector
X input vector
W_i neuron weight
i hidden layer

2.2 Learning

The system loads are fair through preparing the system. It is believed in the structure study by models. The consideration is to provide the structure input flags, as well as required yields. The structure generates a single yield flag, as well as it goes restrictive for the addition of two times of the distinction among the actual yields. Forthcoming, we think this ability the whole of squared blunders. The study is completed by over and over encouraging the information yield examples to the system. A single complete introduction of the whole preparing position is known as age. The study method is generally executed on an age-by-age premise until the stack balance out, as well as the total of squared mistakes, joins to some base esteem. The regularly used study estimation used for the MLP structure is the backpropagation calculation. This is an exacting strategy used for carrying out angle drop technique in the burden area, wherever the incline of the whole of squared blunders amid separate to the stacks be approximated through engendering the mistake flags in back within the structure. The induction of the estimation is given, for instance. Likewise, some particular strategies to quicken the combination are clarified there. An all the more dominant estimation be gotten through using an estimation of Newton's strategy known as Levenberg-Marquardt. The subsidiaries of every sum of squared blunder (every preparation case) toward every structure mass be approximated, as well as gathered within a single network. This grid speaks to the Jacobian of the limited capacity. The Levenberg-Marquardt estimation is used within this effort to arranging the MLP structure. The learning of the system is only assessing the model parameters. On account of the MLP replica, the reliance of the yield on the replica constraint is anyway exceptionally confounded rather than the most normally utilized scientific models (for instance relapse replica). This is the motivation behind why the iterative study is necessary for the preparation put to discover appropriate parameter values.

There is no real technique to making certain of detecting the worldwide least of the aggregate of squared mistake. Then again, the entangled nonlinear nature of the information yield reliance makes it workable for a solitary system to adjust to a lot bigger size of unexpected relations in comparison to for instance relapse models. That is the reason the term learning is utilized regarding neural system models of this sort.

2.3 Generalization

The preparation goes for limiting the mistakes of the system yields concerning the information yield examples of the preparation set. The accomplishment in this does not be that as it may demonstrate anything about the execution of the system after the preparation. Progressively significant is the achievement in speculation. An ordinary issue amid system replica be overfitting, likewise called retention in the system writing. This implies the system learns the information yield examples of the preparation set, and yet unintended relations are put away in the synaptic loads. Generalization is impacted through three issues: the magnitude, as well as the proficiency of the preparation set, the model structure (engineering of the system), and the physical unpredictability of the current issue. The last of these can not be controlled, so the way to counteract overfitting is constrained to influencing the initial two factors. The bigger the preparation set, the more outlandish the overfitting is. In any case, the preparation set should just incorporate information yield designs that effectively mirror the genuine procedure being displayed. In this manner, all invalid and unimportant information ought to be barred. The impact of the replica system inside the speculation is able to found within two different traditions. To begin with, the choice of the info factors is basic. The info space ought to be diminished to a sensible size contrasted with the extent of the preparation set. On the off chance that the component of the information space is expansive, at that point, the arrangement of perceptions can be unreasonably meager for legitimate speculation. Thusly, no pointless info factors ought to be incorporated, because the system can learn conditions on them that don't generally exist in the real process. Then again, all variables affecting the yield ought to be incorporated. The bigger the quantity of gratis constraint inside the replica, the almost certain the overfitting be. At that point, we talk about high-parameterization. Each concealed layer neuron brings a specific quantity of gratis constraints within the replica, to maintain a strategic distance from over-parameterization, the amount of shrouded film neurons ought not to be excessively huge. There is an unpleasant standard guideline in favor of a 3-film MLP. Assume H = amount of shrouded film neurons, N = dimension of the info film, M = dimension of the yield film, T = dimension of the preparation position. The amount of gratis constraint is generally $W = H(N + M)$. It ought to exist littler exclusive the span of the preparation position, ideally as regards T/5. Along these lines, the extent of the shrouded layer ought to be around

$$H \simeq \frac{T}{5(N+M)} \tag{3}$$

2.4 MLP Organizes in Burden Estimating

The thought behind the utilization of MLP models in burden estimating is straightforward: it is accepted that future burden is subject to past burden and outside elements (for example temperature), and the MLP organize is utilized to surmised this reliance. The contributions to the system comprise of those temperature esteems and past burden esteems, and the yield is the objective burden esteems (for instance a heap estimation of a specific hour, load estimations of numerous future hours, the pinnacle heap of multi-day, the all-out heap of multi-day, etc.). Therefore, the structure of an MLP model for burden gauging can be viewed as a nonlinear framework distinguishing the proof issue. The deciding of the model structure comprises choosing the info factors and choosing the system structure. The parameter estimation is completed via preparing the system on burden information of the history.

3 Proposed Work

Hybridization of various procedures with ANN that has been effectively applied to transient burden estimating is depicted in this area.

3.1 Back Propagation (BP) by ANN

For the algorithm as of the past investigate job distributed, a rear spread calculation has consistently be measured because the customary preparing of a neural system used for burden determining issues. In [13], it has utilized the closeness degree parameter to recognize the proper authentic burden information as preparing a set of neural systems. The neural system amid rear engendering force preparing calculation was likewise proposed in the previously mentioned paper for burden estimating to lessen preparing instant, as well as toward progress combination rapidity. In [14], It has been displayed an Artificial Neural Network (ANN) prepared by the Artificial Immune System (AIS) knowledge calculation in favor of momentary burden estimating model. This calculation has explicit advantages, for example, precision, speed of union, monetary, and recorded information necessary for preparing and so forth. The real advantage of this calculation over back engendering calculation is as far as progress in mean normal rate blunder.

3.2 Fuzzy Logic by ANN

ANN is commonly used to arrange a huge information informational collection. There be the technique to disentangle framework structure and upgrade gauging accuracy in [15]. It is used to stipulate the enrollment capacities for momentary burden, as well as the heap determining outcome can be upgraded up to a limited degree [16]. It displayed a gauging technique depending on a comparative day approach in [17]. Within the previously mentioned manuscript, the impact of warmth, as well as mugginess amid the burden issue, used for the choice of comparable years be considered. In [18], the use of the intermission sort-2 furry reason scheme is used for STLF. Sort-2 furry reason scheme, amid an additional level of opportunity, is demonstrated to be a phenomenal instrument used for taking care of various fake vulnerabilities used for civilizing the expectation precision.

3.3 Hereditary Calculation (HC) by ANN

The hereditary calculation is a heuristic hunt method, to be broadly utilized to locate the ideal arrangement. At the same time as the HA be hybridized amid ANN, This is utilized toward all-inclusive streamline quantity of info neurons, as well as the amount of neurons, in the shrouded film of the neural system engineering in [19]. Hereditary calculations (HCs), that take care of streamlining issues utilizing the technique for development and natural selection, additionally enhances the loads among neurons of ANN. It has additionally exhibited how to prepare the counterfeit neural system utilizing improved Genetic calculation in [20]. The hereditary calculation amid the mix of a few additional, for example, PSO, furry rationale, and so forth to lessen the blunder in the expectation of burden request in [21, 22].

3.4 Particle Swarm Optimization (PSO) by ANN

It is a populace based subsidiary free calculation created. In [23], The searching pursuit support calculation is effectively useful toward a few constant enhancement issues in the various rising meadow. The built-up another preparation technique for the spiral premise work (RBF) neural system, because of quantum carried on PSO. The PSO based RBF neural system model for burden gauging in [24]. The planned another PSO calculation amid a versatile dormancy burden issue, as well as joined Chaos with PSO [25].

4 Conclusion

Inside this manuscript, we have displayed the ongoing distributed job taking place in various crossbreed neural systems that have been effectively useful toward transient burden determining. From the work announced by various specialists, it may very well be presumed that the computerized reasoning based anticipating calculations are demonstrated to be the potential methods for this difficult activity of nonstraight time arrangement expectation. Diverse arbitrary inquiry procedures, for example, GA, PSO, BFO, AIS which are fit for worldwide learning abilities had likewise been featured in mix with ANN for this difficult and intriguing issue. The proposed methods that demonstrate their capacity in anticipating electrical burden which at last diminished the operational expense of intensity framework and builds the productivity of activity.

References

1. Saini, L.M., Soni, M.K.: Artificial neural network-based peak load forecasting using conjugate gradient methods. IEEE Trans. Power Syst. **17**, 907–912 (2002)
2. Walczak, S.: An empirical analysis of data requirement for financial forecasting with neural networks. J. Manag. Inf. Syst. **17**, 203–222 (2001)
3. Mandal, P., Senjyu, T., Urasaki, N., Funabashi, T., Srivastava, K.A.: A novel approach to forecast electricity price for PJM using neural network and similar days method. IEEE Trans. Power Syst. **22**(4) (2007)
4. Ron, K.: A study of cross-validation and bootstrap for accuracy estimation and model selection. Proc. Fourteenth Int. Joint Conf. Artif. Intell. **2**(12), 1137–1143 (1995)
5. Chang, J., Luo, Y., Su, K.: GPSM: a generalized probabilistic semantic model for ambiguity resolution. In: Annual Meeting of the ACL. Association for Computational Linguistics Morristown NJ, pp. 177–184 (1992)
6. Devijver, P.A., Kittler, J.: Pattern Recognition: A Statistical Approach. Prentice-Hall, London (1982)
7. Ray, P., Mishra, D.: Signal processing technique based fault location of a distribution line. In: 2nd IEEE International Conference on Recent Trends in Information Systems (ReTIS), pp. 440–445 (2015)
8. Ray, P., Mishra, D.: Artificial intelligence based fault location in a distribution system. In: 13th International Conference on Information Technology (ICIT), pp. 18–23 (2014)
9. Charytoniuk, W., Box, E.D., Lee, W.J., Chen, M.S., Kotas, P., Olinda, P.V.: Neural-network-based demand forecasting in a deregulated environment. IEEE Trans. Ind. Appl. **36**, 893–898 (2000)
10. Kim, K.H., Youn, H.S., Kang, Y.C.: Short-term load forecasting for special days in anomalous load conditions using neural networks and fuzzy inference method. IEEE Trans. Power Syst. **15**, 559–565 (2000)
11. Irie, B., Miyake, M.: Capability of three-layered perceptrons. In: Proceedings of IEEE International 21. Conference on Neural Networks, pp. 641–648. San Diego, USA (1988)
12. Meng, J.L., Sun, Z.Y.: Application of combined neural networks in nonlinear function approximation. In: Proceedings of the Third World Congress on Intelligent Control and Automation, pp. 839–841. Hefei, China (2000)
13. Sun, W., Zou, Y.: Short term load forecasting based on bp neural network trained by PSO. In: Proceedings of the Sixth International Conference on Machine Learning and Cybernetics, pp. 2863–2868. Hong Kong (2007)

14. El-Desouky,A.A., El-Kateb, M.M.: Hybrid adaptive techniques for electric-load forecast using ANN and ARIMA. IEEE Proc. Gener. Transm. Distrib. **147**, 213–217 (2000)
15. Srinivasan, S.S.D., Tan, C.S., Cheng, C.S., Chan, E.K.: Parallel neural network-fuzzy system strategy for short-term load forecasting: System implementation and performance evaluation. IEEE Trans. Power Syst. **14**, 1100–1106 (1999)
16. Bakirtzis, A.G., Theocharis, J.B., Kiartzis, S.J., Satsios, K.J.: Short-term load forecasting using fuzzy neural networks. IEEE Trans. Power Syst. **10**, 1518–1524 (1995)
17. Daneshdoost, M., Lotfalian, M., Baumroonggit, G., Ngoy, J.P.: Neural network with fuzzy set-based classification for short-term load forecasting. IEEE Trans. Power Syst. **13** (1998)
18. Homaifar, A., McCormick, E.: Simultaneous design of membership functions and rule sets for fuzzy controller using genetic algorithms. IEEE Trans. Fuzzy Syst. **3**, 129–139 (1995)
19. Ray, P., Panda, S.K., Mishra, D.: Short-term load forecasting using genetic algorithm. In: 4th Springer International Conference on Computational Intelligence in Data Mining (ICCIDM), pp. 863–872 (2017)
20. Pham, D.T., Karaboga, K.: Intelligent Optimization Techniques, Genetic Algorithms, Tabu Search, Simulated Annealing and Neural Networks. Springer, New York (2000)
21. Wang, X., Elbuluk, M.: Neural network control of induction machines using genetic algorithm training. IAS Ann. Meet. **3**, 1733–1740 (1996)
22. Lu, W.Z., Fan, H.Y., Lo, S.M.: Application of evolutionary neural network method in predicting pollutant levels in downtown area of Hong Kong. Neurocomput. **51**, 387–400 (2003)
23. Panda, S.K., Ray, P., Mishra, D.: Effectiveness of PSO on short-term load forecasting. In: 1st Springer International Conference on Application of Robotics in Industry Using Advanced Mechanisms (ARIAM), pp 122–129 (2019)
24. Kennedy, J.: The particle swarm: social adaptation of knowledge. In: Proceedings of the 1997 International Conference on Evolutionary Computation, pp. 303–308. Indianapolis, Indiana, USA (1997)
25. Kennedy, J., Eberhart, R.C.: Particle swarm optimization. In: Proceedings of the 1995 IEEE International Conference on Neural Networks, pp. 1942–1948. Perth, Australia (1995)

A Review for the Development of ANN Based Solar Radiation Estimation Models

Amar Choudhary, Deependra Pandey, and Saurabh Bhardwaj

Abstract Solar radiation estimation is the most vital part of solar system design. The solar system may be optimized if the radiation is estimated well in advance. Solar radiation is measured by devices such as Pyranometer, Pyrheliometer, Solarimeter, Radiometer, etc., installed at meteorological stations. Due to the unavailability of meteorological station at the locations of interest, solar radiation is estimated by means of estimation models. These models may be broadly classified as Mathematical/Statistical/Empirical models and the Soft Computing based models. These models accept meteorological variables such as wind speed, ambient temperature, relative humidity, cloud cover, etc., and geographical variables such as latitude, longitude, and altitude as input and provides Global Solar Radiation(GSR) at the output. Radiation estimation models are statistically tested and compared. The main aim of this paper is to briefly study and compare the Artificial Neural Network (ANN) based models. The paper deals with the basics of ANN along with its scope in solar radiation estimation. The study indicates that Artificial Neural Network (ANN) based models have significantly better accuracy than others. After the study of models, research gaps have also been pointed out in this paper to draw the attention of researchers.

Keywords Multilayer perceptron · Radial bias function · Solar radiation · Artificial neural network · Levenberg-marquardt algoritm · Renewable and sustainable energy

---

A. Choudhary (✉) · D. Pandey
Department of Electronics and Communication Engineering, Amity School of Engineering and Technology, Amity University, Lucknow 226010, Uttar Pradesh, India
e-mail: amar.giet.ece@gmail.com

D. Pandey
e-mail: dpandey@lko.amity.edu

S. Bhardwaj
Department of Electronics and Instrumentation Engineering, Thapar Institute of Engineering and Technology, Thapar University, Patiala 147001, Punjab, India
e-mail: saurabh.bhardwaj@thapar.edu

D. Mishra et al. (eds.), *Intelligent and Cloud Computing*,
Smart Innovation, Systems and Technologies 194,
https://doi.org/10.1007/978-981-15-5971-6_7

1 Introduction

Renewable energy is the demand of the twenty-first century due to the limited availability of traditional sources of energy and due to the advantages of renewable energy over them. Among all other types of renewable energy, solar energy is more desirable due to its abundance. Quantity of solar radiation needs to be estimated well before to design the solar systems and its optimum utilization. Solar radiation is measured by various devices like pyranometer, pyrheliometer, etc., installed at various meteorological stations. Data obtained from these devices are the accurate one. These devices are the costlier one and are installed at selected cities or areas only. However, better prediction data from various locations needs to be collected for estimation purposes. Due to this limitation, solar radiation estimation is carried out with various meteorological variables like day, sunshine duration, wind speed, cloud cover, ambient temperature, etc., and with geographical location parameters like latitude, longitude, altitude, etc. Two broad categorization of solar radiation estimation models may be made. The first is the mathematical modeling like linear model, quadratic model, cubic model, Angstrom Prescott Model, Rietveld Model, etc., while the second is the soft computing based modeling like fuzzy logic models, ANN models, etc. After a wide comparative study of both the types of modeling ANN modeling is found to have better accuracy than others in solar radiation estimation. However, due to the presence of hidden layers computational requirements increases. This paper is put in the order as: basics of ANN are briefed in Sects. 2, 3 deals with the importance of ANN in solar radiation estimation. In Sect. 4, twelve ANN-based solar radiation estimation models have been briefly reviewed. In Sect. 5, a comparative study of all twelve models have been carried out. Section 6, deals with gaps found in solar radiation modeling after the review of models. Finally, the conclusions of the review are briefed in Sect. 7.

2 Basics of Artificial Neural Network (ANN)

Neurons are one of the fundamental part of the human brain. It provides the facility to apply foregoing experiences to the current action. ANN is the part of Artificial Intelligence (AI) suitable best for simulation, pattern recognition, nonlinear function estimation. ANN is the better technology for providing methods to reslove composite and vague problems. They gain experiences from past examples, able to work with noisy and incomplete data too. ANN, once trained, becomes able to perform estimation at a high rate. They are especially useful in system modeling [1].

Activation function, learning algorithm (training), and architecture are the parameters by which ANN can be characterized. The interconnection between neurons is described by architecture. ANN architecture comprises of the input layer, one or more hidden layers, and the output layer. Figure 1 represents the basic structure of ANN.

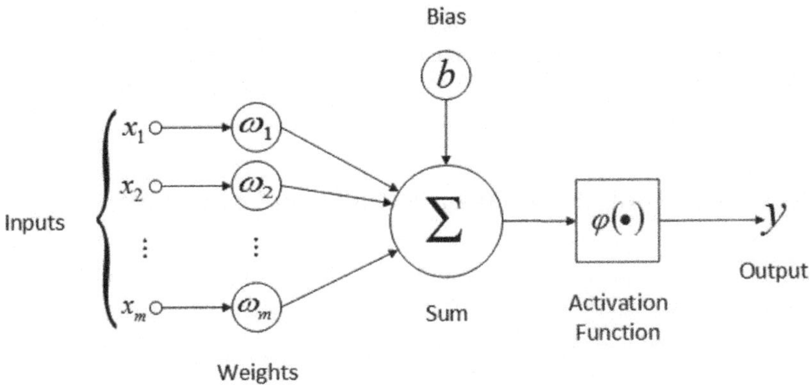

Fig. 1 Basic structure of ANN

Table 1 Steps of ANN prediction methodology

Steps	Action
1	Selection of variables for input and output
2	Separation of data in two sets: training and testing
3	Development of artificial neural networks models
4	Selection of training variables
5	Training of artificial neural networks models
6	Computation of errors
7	Selection of model with minimum MAPE/RMSE/MBE

The layers of the network are connected by communication link termed as weight which places their effect at the time of passing the information. These weights are determined by virtue of the learning algorithm. On the basis of the learning algorithm, ANN may be categorized as fixed weight ANN, Unsupervised ANN , and Supervised ANN [2]. Based on the input activity level, the activation function establishes the relation between the output of a neuron to its input. Table 1, lists various steps involved in the prediction methodology using Artificial Neural Network (ANN).

3 ANN for Solar Radiation Estimation

ANN is a superb tool for solar radiation estimation. It has a higher prediction perfection than Linear, Nonlinear, and Fuzzy models [3]. In solar radiation estimation using ANN, output, and input variables are selected at first. The output variable is generally daily, monthly, yearly average solar radiation, whereas input data are latitude, longitude, altitude, sunshine duration, air temperature, relative humidity, day of the year, wind speed, wind direction, atmospheric pressure, global solar radiation. ANN models accurately estimate solar radiation than linear and nonlinear model, Angstrom model, fuzzy logic models, and other empirical regression models [4–6].

4 Review of Global Radiation Estimation Using ANN

Mohandes et al. [7] taken latitude, longitude, altitude, and sunshine duration at the input parameter with a result of 16.4% accuracy. They used the data of 31 stations of Saudi Arabia to train the model, whereas data of 10 stations were used for testing purposes.

Kemmoku et al. [8] used the data (atmospheric pressure and temperature) in Omaezaki Japan from 1998–1993. The mean error was about 20% for multi stage and 30% for single stage neural network. The range of the hidden layer was 8 to 18.

Azeez [9] used the parameters sunshine duration, relative humidity, and maximum ambient temperature for input and got solar irradiation at the output. R = 99.96, MAPE = 0.8512, RMSE = 0.0028 was obtained which shows better results. The data were taken from Gasau, Nigeria and monthly average solar radiation was estimated. They considered feed-forward back propagation artificial neural network.

Mishra et al. [10] used longitude, latitude, relative humidity ratio, mean sunshine, rainfall ratio, and month of a year as the input parameter. The RMSE value of 7 to 29% was obtained for radial bias function (RBF) and 0.8 to 5.4% was obtained for multilayer perceptron (MLP). Direct solar radiation for 8 stations in India was estimated in this.

Elminir et al. [11] received the estimation accuracy of 94.5% using one hidden layer and sigmoid transfer function in the back propagation algorithm with the cloud cover, ambient temperature, relative humidity, airflow velocity and direction as input to ANN model. Solar radiation in different spectrum bands was determined with data (2001–2002) of Helwan, Egypt.

Chatterjee and Keyhani [12] used the Levenberg-Marquardt (LM) algorithm for testing with a random selection of hidden layers and neurons. At epoch 7, the best value of RMSE obtained was 3.2033. Fourteen inputs were given to the model and the solar radiation on the tilted surface was estimated.

Rehman and Mohandes [13] predict direct normal solar radiation and diffuse solar radiation with MAPE of 0.016 and 0.41, respectively, using the Radial Bias Function (RBF). The input data were day, ambient temperature, relative humidity, and global solar radiation (GSR).

Reddy [14] predicts monthly average daily and hourly global solar radiation by ANN. They used solar radiation data of 11 locations of India (South and North) for training the network and data 02 locations for testing the network. The MAPE of 4.1% was obtained for estimated hourly global radiation. The obtained result indicates that ANN is a fairly good choice for solar radiation estimation comparing to other regression models.

Benghanem and Mellit [15] predict daily global solar radiation of AI-Madinah (Saudi Arabia) by means of Radial Basis Function (RBF) with a combination of days, relative humidity, sunshine duration, and air temperature. The result with a correlation coefficient of 98.80% is obtained for the input of sunshine duration and air temperature to the model.

Moustris et al. [16] used the data for Greece locations to get missing mean, global, and diffuse solar irradiance range. Hourly data of sunshine duration, air temperature, relative humidity, and latitude were used. The correlation coefficient found to be statistically significant at 99%. Thus, the obtained result by ANN has good agreement with the actual data.

Rao et al. [17], developed six solar radiation estimation models using ANN by varying meteorological variables from one to six. They analyzed all the models using statistical tools. They obtained Relative Root Mean Square Error (RRMSE) of 3.96% with max and min radiation and extraterrestrial radiation. They concluded that Artificial Neural Network (ANN) models outperform the considered empirical/mathematical models. Superior performance was obtained by them with less number of easily available inputs for any location.

Jahani et al. [18] provided a contrast between the performance and suitability of different models for the prediction of daily global solar radiation in Iran. From obtained results (median of R2, MBE, RMSE for MLP-GA(n) 0.92, 38.4, and 185.5 $J/cm^2/day$) they concluded that Artificial Neural Network (ANN) models coupled with the genetic algorithm are most suitable for the stations of interest.

Malik and Garg [19], used Artificial Neural Network to forecast long term solar irradiance for 67 cities in India, with the help of feedforward network using back propagation as the learning algorithm. They collected data from Atmospheric Science Data Centre (ASDC) at NASA Langley Research Centre and Centre for Wind Energy Technology (CWET), Chennai, India. They considered nineteen different input parameters to predict solar radiation. They received the regression value of 94.7% for 18 hidden neurons and 78.86% for 30 hidden neurons.

5 Comparative Study of All the Models

See Table 2.

6 Identified Research Gap in Solar Radiation Estimation

Research gaps have been obtained by the literature review of various models in this paper. Gaps are briefed below

- Authors normally undertake 3–4 input meteorological parameters and draw the conclusion after training and testing. Some parameters have better correlation with

Table 2 Comparison of all the reviewed models

S. No.	Model	Input	Output
1	Mohandes	Latitude, longitude, altitude, and sunshine duration	Accuracy 16.4%
2	Kemmoku	Atmospheric pressure and temperature	Mean error 20% for multi stage and 30% for single stage
3	Azeez	Sunshine duration, relative humidity, and maximum ambient temperature	R = 99.96, MAPE = 0.8512, RMSE = 0.0028
4	Mishra	Longitude, latitude, relative humidity ratio, mean sunshine, rainfall ratio, and month	RMSE = 7–29% for Radial Bias Function (RBF) and 0.8–5.4% for (MLP)
5	Elminir	Cloud cover, ambient temperature, relative humidity, wind velocity, and direction	94.5%
6	Chatterjee	Irradiation of each month, latitude, and the ground reflectivity of the location	RMSE = 3.2033
7	Rehman	Day, ambient temperature, relative humidity, and global solar radiation.	MAPE: 0.016 and 0.41
8	Reddy	Hourly solar radiation	MAPE: 4.1%
9	Benghanem	Air temperature, sunshine duration, and relative humidity	Correlation coefficient: 98.80%
10	Moustris	Hourly data of sunshine duration, air temperature, relative humidity, latitude	Correlation coefficient: 99%
11	Rao K., D. V. Siva	Maximum and minimum radiation and extraterrestrial radiation	Relative root mean square error of 3.96%
12	Babak Jahani	Duration of sunshine hours and diurnal air temperature	Median of R_2, MBE, RMSE for MLP-GA(n) 0.92, 38.4, and 185.5 J/cm^2/day
13	Hasmat Malik, Siddharth Garg	Used nineteen input parameters	Maximum Regression value of 94.65% for 18 neurons and minimum of 78.86% for 30 neurons

the output. So, they are widely chosen by the authors. However, the conclusion may be the best one where the maximum of parameters are incorporated.

- Authors generally use 3–4 year of meteorological data for training and testing the model. However, more data may be used to get firm conclusions.
- Effect of degradation on solar panels needs to be considered at the time of testing the network by ANN.
- Optimization of training of ANN models may be performed using techniques like genetic algorithm, annealing techniques, etc.
- Estimation of solar radiation at the tilted surface deserves more exploration.

7 Conclusion

A brief study of ten solar radiation estimation models are made here. Solar radiation is required to be estimated well in advance for optimizing the solar energy utilizing systems. The study indicates that ANN is the better choice for solar radiation modeling due to its accuracy comparing to other regression models like linear, nonlinear, and other soft computing based modeling like fuzzy logic modeling, etc. Among all meteorological parameters, sunshine duration and air temperature have a correlation coefficient of 97.5% in solar radiation estimation. So, meteorological parameters need to be selected appropriately to get the results with better accuracy.

References

1. Kalogirou, S.A.: Artificial neural networks in renewable energy system applications: a review. Renew. Sustain. Energy Rev. PERGAMON 373–401 (2001). https://doi.org/10.1016/s1364-0321(01)00006-5
2. Dorvlo, Atsu S.S., Jervase, Hoseph A., Lawati, Ali A.I.-: Solar radiation estimation using artificial neural networks. Appl. Energy **71**, 307–319 (2002). https://doi.org/10.1016/s0306-2619(02)00016-8
3. Yadav, A.K., Chandel, S.S.: Solar radiation prediction using artificial neural network techniques: a review. Renew. Sustain. Energy Rev. (2013). https://doi.org/10.1016/j.rser.2013.08.055i
4. Senkal, O., Kuleli, T.: Estimation of solar radiation over Turkey using artificial neural network and satellite data. Appl. Energy **86**(7–8), 1222–1228 (2009). https://doi.org/10.1016/j.apenergy.2008.06.003
5. Senkal, O.: Modeling of solar radiation using remote sensing and artificial neural network in Turkey. Energy **35**(12), 4795–4801 (2010). https://doi.org/10.1016/j.energy.2010.09.009
6. Jiang, Y.: Computation of monthly mean daily global solar radiation in China using artificial neural networks and comparison with other empirical models. Energy **34**(9), 1276–1283 (2009). https://doi.org/10.1016/j.energy.2009.05.009
7. Mohandes, M., Rehman, S., Halawani, T.O.: Estimation of global solar radiation using artificial neural networks. Renew. Energy **14**(1–4), 179–184 (1998). https://doi.org/10.1016/s0960-1481(98)00065-2

8. Kemmoku, Y., Orita, S., Nakagawa, S., Sakakibara, T.: Daily insolation forecasting using a multi-stage neural network. Sol. Energy **66**(3), 193–199 (1999). https://doi.org/10.1016/s0038-092x(99)00017-1
9. Azeez, M.A.A.: Artificial neural network estimation of global solar radiation using meteorological parameters in Gusau, Nigeria. Archiv. Appl. Sci. Res. **3**(2), 586–595 (2011)
10. Mishra, A., Kaushika, N.D., Zhang, G., Zhou, J.: Artificial neural network model for the estimation of direct solar radiation in the Indian zone. Int. J. Sustain. Energ. **27**(3), 95–103 (2008). https://doi.org/10.1080/14786450802256659
11. Eminir, H.K., Areed, F.F., Elsayed, T.S.: Elimination of solar radiation components incident on Helwan site using neural networks. Sol. Energy **79**(3), 270–279 (2005). https://doi.org/10.1016/j.solener.2004.11.006
12. Chatterjee, A., Keyhani, A.: Neural network estimation of micro grid maximum solar power. IEEE Trans. Smart Grid **3**(4), 1860–1866 (2012). https://doi.org/10.1109/tsg.2012.2198674
13. Rehman, S., Mohandes, M.: Splitting global solar radiation into diffuse and direct normal fractions using artificial neural networks. Energy Source Part A: Recovery, Utilization Environ. Effects **34**(14), 1326–1336 (2012). https://doi.org/10.1080/15567031003792403
14. Reddy, K.S., Ranjan, M.: Solar resources estimation using artificial neural networks and comparison with other correlation models. Energy Convers. Manag. **44**(15), 2519–2530 (2003). https://doi.org/10.1016/s0196-8904(03)00009-8
15. Benghanem, M., Mellit, A.: Radial basis function network-based prediction of global solar radiation data: application for sizing of a stand-alone photovoltaic system at Al-Madinah, Saudi Arabia. Energy **35**(9), 3751–3762 (2010). https://doi.org/10.1016/j.energy.2010.05.024
16. Moustris, K., Paliatsos, A.G., Bloutsos, A., Nikolaidis, K., KoronakiI, Kavadias K.: Use of neural networks for the creation of hourly global and diffuse solar irradiance data at representative locations in Greece. Renew. Energy **33**(5), 928–932 (2008). https://doi.org/10.1016/j.renene.2007.09.028
17. Rao, K., Siva Krishna, D.V., Premalatha, M., Naveen, C.: Analysis of different combinations of meteorological parameters in predicting the horizontal global solar radiation with ANN approach: a case study. Renew. Sustain. Energy Rev. **91**, 248–258. https://doi.org/10.1016/j.rser.2018.03.096
18. Jahani, B., Mohammadi, B.: A comparison between the application of empirical and ANN methods for estimation of daily global solar radiation in Iran. Theor. Appl. Climatol. https://doi.org/10.1007/s00704-018-2666-3
19. Malik, H., Garg, S.: Long-term solar irradiance forecast using artificial neural network: application for performance prediction of Indian cities. In: Malik, H. et al. (ed.) Springer Nature Singapore Pte Ltd., Applications of Artificial Intelligence Techniques in Engineering, Advances in Intelligent Systems and Computing, vol. 697 (2018). https://doi.org/10.1007/978-981-13-1822-1_26

Analysis and Study of Solar Radiation Using Artificial Neural Networks

Deepa Rani Yadav, Deependra Pandey, Amar Choudhary, and Mihir Narayan Mohanty

Abstract Solar energy has greater advantages as it is both renewable and sustainable sources. Artificial Intelligence (AI) is an emerging trend, which makes machines learn and adhere like humans. ANN is the excellent and one of the main tools used in machine learning. The neural part in artificial neural network suggests brain-inspired systems, similar to the way we humans learn. ANN is basically used for finding patterns which are very complex for the programmer to extract and teach the machines to recognize. ANN-based models have been successfully upskilled to have different solar radiation variables, so as to improve the existing empirical and statistical approaches that are being used in solar radiation estimation. ANN has various applications in almost all areas like in aerospace, automotive, defense, mathematics, engineering, medicine, economics, meteorology, psychology, neurology, etc. Along with so many applications, ANN can also be used for the prediction of solar radiation. Radiations are the rays received by earth, analyzing the amount of radiation received will be helpful in efficient utilization of solar energy.

Keywords ANN (artificial neural network) · Solar radiations · AI

D. R. Yadav · D. Pandey · A. Choudhary (✉)
Department of Electronics and Communication Engineering, Amity School of Engineering and Technology, Amity University, Lucknow 226010, Uttar Pradesh, India
e-mail: amar.giet.ece@gmail.com

D. R. Yadav
e-mail: deepayadav201092@gmail.com

D. Pandey
e-mail: dpandey@lko.amity.edu

M. N. Mohanty
Department of Electronics and Communication Engineering, Institute of Technical Education and Research, S'O'A University, Bhubaneswar 751030, Odisha, India
e-mail: mihir.n.mohanty@soa.ac.in

D. Mishra et al. (eds.), *Intelligent and Cloud Computing*,
Smart Innovation, Systems and Technologies 194,
https://doi.org/10.1007/978-981-15-5971-6_8

1 Introduction

With increased energy demands, renewable energy sources have become very popular. Among renewable energy source alternatives, solar energy is the most significant and advantageous. It is a huge energy source meeting the rapid growing energy demands. Solar energy potential is beyond imagination [1]. It is advantageous as it is renewable and sustainable source; in short, we can say that one can never run out of this source [2].

Because of growing and strong increase in solar power generation, the prediction of solar radiation is acquiring utmost importance nowadays. Prediction is basically a rough calculation which is important so as to have accurate idea of solar radiation distribution over a particular geographic coverage area [3].

ANNs are the new tools in the computation world that have the solution to almost all the complex real-world problems and have extensive utilization and applications. Artificial neural networks are more efficient as compared to other computational models and methods. It is more accurate also as it is brain-inspired system which is based on examples and experiences [4].

ANN has numerous applications in almost all areas. Solar radiation prediction based on ANN modeling comprises ANN models having input–output mapping of different meteorological parameters. ANN-based models brought improvements in existing empirical and statistical approaches and have the capability to model both linear and non-linear networks. ANN-based modeling is done for solar radiation prediction so as to have accurate idea of radiation over particular geographic location for efficient utilization.

2 ANN Fundamentals

2.1 What is ANN?

ANN stands for artificial neural networks; neural term tells that it is a brain-inspired technique. The meaning of ANN is there in the name itself; it performs and operates similar to the way the human brain operates and performs. ANNs are amazing tools that can even find the patterns that are very complex that human programmer cannot find easily [5].

2.2 Nodal Configuration

An artificial neural network is a group of nodes that are interconnected in a configuration that represents the neurons of animal brain. The nodal configuration represents circular nodes including input nodes that act as source and output node that act as

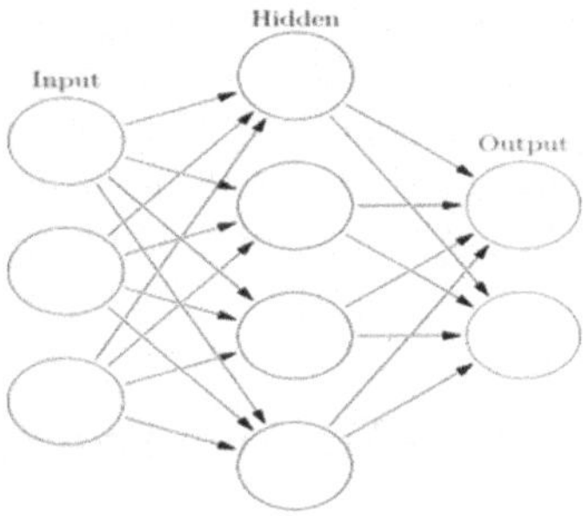

Fig. 1 Nodal configuration of ANN with three layers (input, output, hidden)

destination, with the arrows depicting the information flow. Each circular node is basically an artificial neuron. As they are brain-inspired system, so they basically learn from examples (from data given to them), they are not specifically programmed to do a specific task (Fig. 1).

2.3 ANN Types

There are various types of neural networks; these multiple neural networks have different specifications, different levels of complexity, and different operating criteria. The various types are as follows:

- Feedforward neural network: It is the most basic and simplest form of neural network. As its name suggests, feedforward means the data/information will flow in one direction only from source to destination.
- Recurrent neural network: The other basic type is recurrent neural networks; it is the most widely used network. Here, the information will flow in multiple directions not just in one direction as in feedforward case. They are most widely used because of their capability that they can perform complex tasks, for example, they can be used for language recognition.
- There are various other networks also like convolution neural networks, Boltzmann machine network, Hopfield networks, and many others. Each type is designed to perform specific task with different complexity levels. Choosing appropriate type depends upon the type of task to be performed, upon specific application. This also may happen that in certain application, multiple networks are required, for example, voice recognition [6].

2.4 Learning Stuff

ANN is the most flexible and user-friendly system. They are far beyond advanced to Artificial intelligence. As its name suggests that it is as brain-inspired system, it performs and learns the way our human brain learns and performs. As we human learn from our mistakes, we learn and we become more accurate with more experiences; in this similar way, artificial neural network learns and performs with the help of experiences, that is, the more data they will get, the more they will be able to perform well and hence will be more accurate.

3 ANN Working Principle and Architecture

3.1 Introduction

ANN works in a way similar to working of human brain that learns and functions through examples, experiences, and data received. So the basis for ANN is the brain. Since the basis for ANN is the brain-inspired system, let us first understand how brain learns and processes:

Step 1: Dendrites receive the external signals.

Step 2: This received signal is processed in neuron cell body.

Step 3: Processed signal is converted into output signal and transferred to axon.

Step 4: Output signal is received by the dendrites of the next neuron through the synapse (Fig. 2).

ANN is the simple structural configuration of how brain neuron works; here in the structure w1, w2, and w3 give the strength to incoming signal, i.e., the input signal.

Now let us see how ANN functions (Fig. 3).

Let us take an example to illustrate the working of neurons in ANN to have better clarity of the concept.

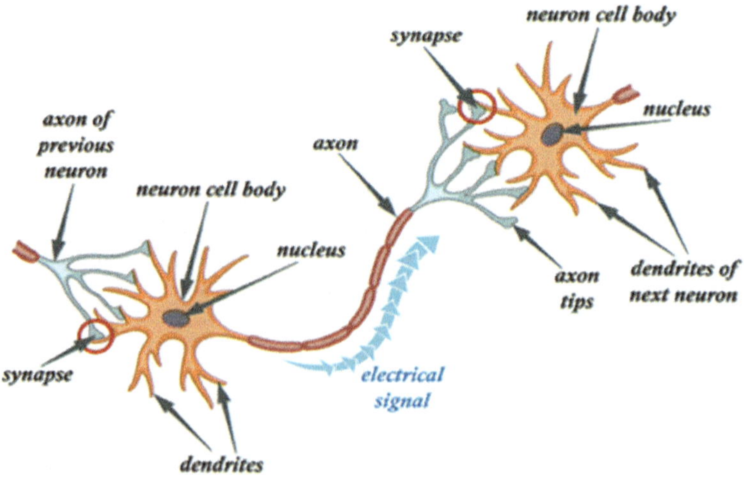

Fig. 2 Human brain neuron structure

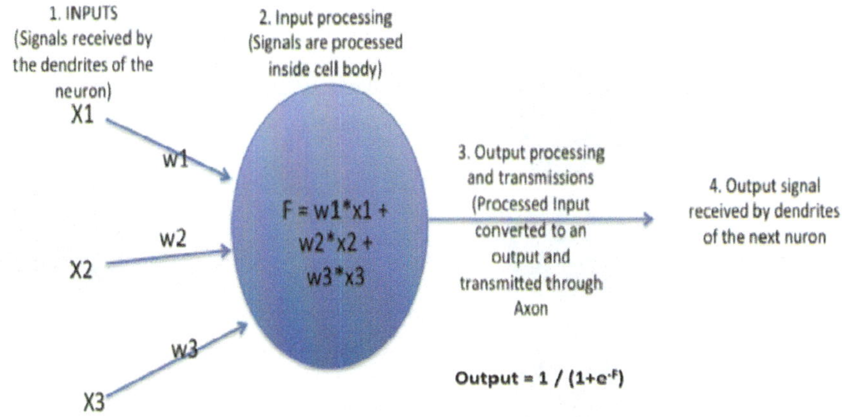

Fig. 3 Structural configuration of working of neuron

For example, a bank assessing to provide a loan to the customer, by predicting the defaulters, the related data is listed in Table 1.

The table is arranged in columns mentioning the customer id, age, debt ratio, monthly income, loan defaulter (1 means default, 0 means not a defaulter), and a default prediction column. Default prediction is done through column X in the table, and entry close to 1 predicts a defaulter (more chances of).

Now illustrating this table into simple structure of ANN so as to have connectivity and better understanding of the working principle involved (Fig. 4).

Illustrating the Architecture and Explaining the Key Points Involved:

The architecture comprises multiple layers: input layer, hidden layer, and output layer. Hidden layer is the intermediate layer that plays an important role; being an intermediate one, it passes on the useful information and ignores the redundant useless information, thus making the system more efficient. The 03 layer is the final

Table 1 Data

Customer ID	Customer age	Debt ratio (% of income)	Monthly income ($)	Loan Defaulter YES: 1 NO: 2 (column W)	Default Prediction (column X)
1	45	0.80	9120	1	0.76
2	40	0.12	2000	1	0.66
3	38	0.08	3042	0	0.34
4	25	0.03	3300	0	0.55
5	49	0.02	63,588	0	0.15
6	74	0.37	3500	0	0.72

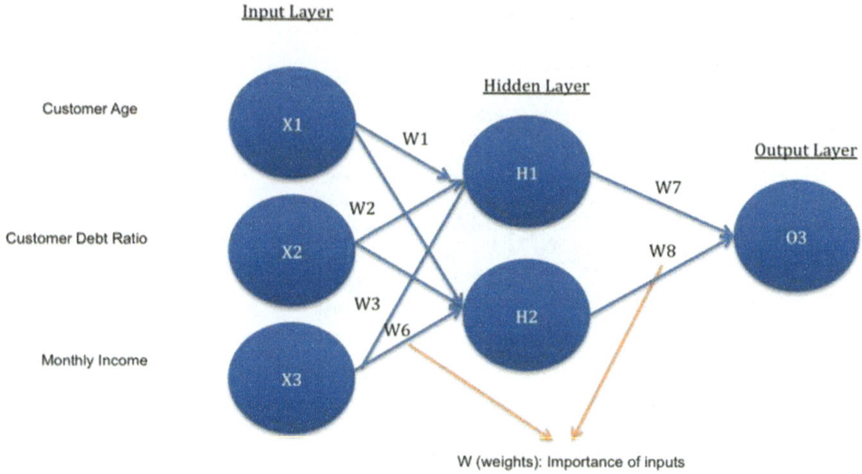

Fig. 4 Structural 3 layer configuration for bank assessing the customer to provide loan

layer whose value is either 0 or 1; value exactly 1 or closer to 1 (say 0.85) indicates a default. This is an example of feedforward neural network, since the information is flowing in one direction W, i.e., the weights are the deciding factor of this architecture, with the optimal value of W; model will be good and accurate, and W minimizes the error prediction, for example, if W1 is 0.65 and W2 is 0.95, then higher importance is associated with W2, i.e., attached to x2 (debt ratio) than x1 (age) in predicting h1.

4 Applications, Advantages, and Disadvantages

4.1 Applications

ANNs are amazing new tool with wonderful properties, so it has numerous applications in banking, finance, agriculture, defense, engineering, automotive, economics, meteorology, and so many others. List of some of its applications among the wide range is as follows [5]:

- **Character recognition and image processing**: ANN is playing a major role in security world. It has the ability to receive multiple inputs and then process them (in the hidden part of architectural configuration); this is helpful in fraud detection, for example, in handwriting recognition as in the area of banking. It is also helpful in face recognition, for example, defense area, social media, and cancer detention.
- **Forecasting**: Artificial neural network is capable of extracting unseen features and information. ANN application in forecasting is useful in many areas such as economics, monitory policy, business decisions, etc.

- **Medical diagnosis**: ANN has its various tremendous advances and applications in medical world also, for example, cancer diagnose—lung cancer, prostate cancer, etc. It can also identify the level of severance, if it is a highly invasive cancer or less invasive.
- **Disaster management**: ANN is used for reliability analysis of zones that are near to have disaster and for disaster settlements also.

4.2 Advantages

ANN is basically human brain modeling of simple neuron-type structural configuration. As we can say that a human brain consists of thousands of parallel processing, similarly this brain-inspired system is multi-layered neuron structure of so many processes taking place in parallel.

ANN-based models used for solar radiation prediction are better models with more accurate results, and it also reduces complex calculations involved in estimation and modeling used for prediction [9]. List of some of its advantages is as follows:

- **Information storage**: It has a new concept of information storage; it stores the information on the entire network, whereas other systems store the information on a database, so it ultimately prevents the loss of data and provides more flexibility to the user.
- **Fault tolerance**: It has greater fault tolerance capability, i.e., if, for example, few cells of the neuron structure get corrupted, then this failure will not affect the entire network, and hence output will be generated normally without getting affected by those corrupted cells.
- **Distributed memory**: It has a distributed memory in the sense that since it is a brain-inspired system, whose processing is based on experiences and collected set of data, so this distributed memory of ANN makes it successful tool in the complex world.
- **Gradual corruption**: Every process has certain errors and corruptions but the advantage with ANN is that it is not affected if few cells get corrupted as discussed earlier, instead it experiences relative degradation with time due to corruption and does not get immediately affected by such faults.
- **Machine learning**: As ANN is a part of machine mimicking of human brains, it can make machines to adhere and learn as we human learn, thus making machine to make decisions and learn through experiences.
- **Parallel Processing**: ANN has tremendous strength, as human brain thinks of multiple processes at a time; similarly, ANN is capable of handling and performing multiple tasks at a time.

4.3 Disadvantages

As nothing is completely risk-free, ANN also has both pros. and cons. List of some of its disadvantages is as follows:

- **Hardware-dependent**: ANN requires structured processor for processing the task needed; hence, it has hardware dependency.
- **Unexplained behavior**: ANN is an advanced tool with application almost in every area and solutions to the complex problems but sometimes it gives the solution without even telling how and why? This unexplained behavior reduces network trust.
- **No proper network structure**: It has no predetermined network structure. The structure is on the basis of trials, errors, and on the basis of data set and experiences.
- **Problem showing difficulty**: ANN has a limitation that it only deals with numerical values and understands the numeric only. Hence, every problem before applying to ANN network has to be converted to ANN-friendly form, i.e., in a numerical form which is the responsibility of a programmer.

5 Methodology

5.1 ANN for Predicting Solar Radiation

ANNs are the intelligent tools that can be used to model any linear and non-linear systems using its structural configuration that consists of input layer, hidden layer, and output layer. Within these layers, ANN includes large set of inputs at both input and hidden layers so as to have minimum error, and hence more accuracy and better results [7].

There are various ways and models for solar radiation estimation because of growing energy demands, but none of the ways exactly has accurate radiation estimation because solar radiation data is inconsiderate and is continuously changing in seasons, months, and even in a day. Therefore, many of the techniques have done estimation on hourly basis, and this data is recorded over a month, since the estimation is done hourly so provides more accuracy.

For the ANN modeling, used for solar radiation estimation, the inputs used for such computations involve geographical parameters and meteorological parameters, and the output of such computations is based on solar radiation on different time spans.

5.2 *Data Considered*

For predicting the global solar radiation, the data considered for the study consists of five input parameters, NASA worked for collecting these data over 5 years, and thus data collection is done for a particular geographical location. List of various input parameters is as follows [3]:

Minimum temperature—(degrees centigrade).

Maximum temperature—(degrees centigrade).

Average temperature—(degrees centigrade).

Relative humidity—(percentage).

Wind velocity—(m/s).

The daily estimation of record of min temperature, max temperature, average temperature, humidity, and wind velocity is being used for solar radiation prediction and the estimation is done on hourly basis and the output parameters consist of extraterrestrial radiation ($MJ/m^2/day$).

ANN is a numerical-based methodology; this technique is also referred to as "black box" because on the basis of collected set of input data, it will give solution to any complex problem but it will not give the explanation of the solution.

5.3 *Operating Stages*

The stages of operation consist of two stages:

- **Training stage**: In this stage of processing, ANN experiences on the basis of collected input data set and learns through it (since it is a brain-inspired system with the similar neuron functioning as in human brain) and then stores the required necessary data pattern of the information provided.
- **Testing stage**: As its name specifies, it is a testing which means it will test the data to provide the desired output on the basis of input data applied.

5.4 *Performance*

ANNs are efficient tools with numerous applications that have solutions in almost all the areas and complex world. Moreover, it has better performance than any other estimation technique. The performance is affected and is not good in the case if input data provided is not enough or less (since it requires collected set of input data to have less error and more accurate results) and the other reason may be the mismatch in the training and testing stages/stations [8].

6 Comparison Between ANN-Based and Existing Modeling

- ANN-based modeling works specifically with numerical data only, whereas there is no such compulsion with other modeling techniques. This makes ANN-based modeling little less flexible.
- ANN has distributed memory and exhibits parallel processing since it is brain-inspired system which makes them to handle multiple tasks at a time with greater accuracy, whereas other modeling does not exhibit so.
- Machine mimicking and fault tolerance capabilities of ANN-based modeling make them learn and adhere like human and enables gradual corruption and stable results, whereas other modeling techniques do not possess machine mimicking capabilities [9].

7 Conclusion

Artificial neural networks are the excellent tools; the neural part depicts that it is brain-inspired systems. Neural part tells that the functioning of ANN and its operating will be similar to the operating of human brain. ANN performs on the basis of collected set of data and experiences so as to have less errors and better accuracy. It takes numerical values only.

ANN has its applications in almost all the areas including defense, meteorology, aerospace, etc. and finds solution to the complex problems that are very complex for the programmers also. Solar technology is an attractive topic, and it is advantageous because it is sustainable and renewable source so one can never run out of this source. Estimation and prediction are important so as to have recorded idea of solar radiation at particular geographical location. Estimation based on ANN modeling has less errors, and it is more accurate since processing is done on numeric values, and various sets of input data are taken for processing. With the advantageous feature of fault tolerance, parallel processing, and machine learning, this brain-inspired technique provides best solution to every complex problem than any other methodology.

References

1. Deepa, Y.R., Deependra, P.: Analytical study of ANN modeling based estimation of solar radiation. In: IOSR Journal of Engineering (IOSR JEN). National seminar cum workshop on "Data Science and Information Security 2019" (2019)
2. Tsoutsos, T., Frantzeskaki, N., Gekas, V.: Environmental impacts from the solar energy technologies. Energy Policy **33**(3), 289–296 (2005). https://doi.org/10.1016/s0301-4215(03)00241-6
3. Ampratwum, D.B., Dorvlo, A.S.S.: Estimation of solar radiation from the number of sunshine hours. Appl. Energy **63**(3), 161–167 (1999). https://doi.org/10.1016/s0306-2619(99)00025-2
4. Kumar, N., Sharma, S.P., Sinha, U.K., Nayak, Y.: Prediction of solar energy based on intelligent ANN modeling. Int. J. Renew. Energy Res. (S. P. Sharma et al.) 6(1) (2016)

5. Yadav, A.K., Chandel, S.S.: Solar radiation prediction using ANN techniques: a review. Renew. Sustain. Energy Rev. 5(4) (2001). Elsevier science Ltd.
6. Basheer, I., Hajmeer, M.: Artificial neural networks: fundamentals, computing, design, and application. J. Microbiol. Methods **43**(1), 3–31 (2000). https://doi.org/10.1016/s0167-7012(00)00201-3
7. Zhang, J., Zhao, L., Deng, S., Xu, W., Zhang, Y.: A critical review of the models used to estimate solar radiation. Renew. Sustain. Energy Rev. **70**, 314–329 (2017). https://doi.org/10.1016/j.rser.2016.11.124
8. Kumari Namrata, S.P., Sharma, S.B.L.S.: Comparison of different models for estimation of Global solar radiation in Jharkhand region. Smart Grid Renew. Energy **4**, 348–352 (2013)
9. Ibeh G.F., Agbo G.A., Agbo, P.E., Ali, P.A.: Application of artificial neural networks for global solar radiation forecasting with temperature. Pelagia Res. Lib. Adv. Appl. Sci. Res. **3**(1), 130–134 (2012)

A Scheme to Enhance the Security and Efficiency of MQTT Protocol

S. M. Kannan Mani, M. Balaji Bharatwaj, and N. Harini

Abstract Internet of Things (IoT) ecosystem is built to collect information from the environment via Internet and sensors that are active for a particular IoT. These are implemented in various areas like monitoring, quadcopters, etc. Regarding protocols that are related to IoT, MQTT is one of the popular message queuing protocols. In this paper, digital signature cryptography and end-to-end encryption have been compared. This paper will bring about the efficiency of digital signature over end-to-end encryption in the field of IoT. A secure communication scheme capable of operating at a reduced overhead rate compared to individual packet encryption scheme has been proposed. The scheme can benefit in resource constrained environment like IoT.

Keywords IoT (internet of things) · AES (advanced encryption standard) · RSA—(rivest-shamir-adleman) · ECC (elliptical curve cryptography) · DoS (denial of service) · QoS (quality of service) · MQTT (MQ telemetry transport)

1 Introduction

The revelation in the Internet has extended its boundary to comprise a diverse computer devices. Internet of Things (IoT) ecosystem is built to collect pieces of information from an Environment via Internet and sensors like motion, temperature, proximity and pressure sensors. Internet of Things is getting implemented in

S. M. Kannan Mani · M. Balaji Bharatwaj · N. Harini (✉)
Department of Computer Science and Engineering, Amrita School of Engineering, Coimbatore, India
e-mail: n_harini@cb.amrita.edu

S. M. Kannan Mani
e-mail: CB.EN.U4CSE16617@cb.students.amrita.edu

M. Balaji Bharatwaj
e-mail: CB.EN.U4CSE16607@cb.students.amrita.edu

Amrita Vishwa Vidyapeetham, Coimbatore, India

D. Mishra et al. (eds.), *Intelligent and Cloud Computing*,
Smart Innovation, Systems and Technologies 194,
https://doi.org/10.1007/978-981-15-5971-6_9

a variety of verticals like wearable, street-cameras, quadcopters and quality monitoring systems. Many challenges are introduced when these stand alone devices are interconnected to the network. With security being in the front as failure to provide this will need to exposure of information to an unknown audience. Unfortunately, the capability limit of these low-end IoT devices makes many traditional security [1, 2] methods in applicable to secure the IoT [3] systems. This fact opens the door for attacks and exploits which target IoT services and Internet. Currently IoT [4] network relies on security protocols [5] using a trusted suit of cryptography algorithms like AES, RSA, ECC, etc. Due to the restriction in resource [6] and consumption environment of the IoT devices, they do not run full-fledged security mechanisms. Increasing risk in vulnerabilities, compromising sensitive information motivated the devices manufacturers and communication services providers along with the researchers to achieve a high degree of security and privacy by designing systems to control the flow of information between devices [7]. Presents a scheme that uses cryptographic-based technique on individual packets to thwart spoofing attacks [8]. The authors claim that the presented scheme is efficient because of adoption of ECC. In this paper, an attempt has been made to thwart the spoofing attacks using an end-to-end encryption mechanism that uses encryption of the channel instead of encrypting individual packets. The experimentation procedure clearly brought out the efficiency of the proposed scheme compared to the individual packet encryption mechanism. The observations on experimentation clearly conveyed the ability of the proposed scheme in terms of providing enhanced security capability while being still in the resource constraint environment. The rest of the paper is organised as follows: Literature review is explained in Sects. 2, 3 presents the architecture and working of the proposed scheme, Sect. 4 explains the experiment results and Conclusion and scope of future task is explained in Sect. 5.

2 Literature Review

The nature of the IoT has made it a hacking target for all the widespread Cybercriminals. Attacks on secrecy and authentication, service attacks on network availability have been frequently reported as major security [9] issues in IoT network. The object layer generally represents the devices that are connected in the network to collect and process information. The security issues in this layer include physical security of sensing devices. IoT cannot provide a security system due to the diversity of the security devices connected, limited energy and weak protective capability of sensing nodes. At the network layer, IoT is characterised by a large number of devices with heterogeneous characteristics and hence a wide variety of formats in the data collected, a huge amount of data information stored and high rate of voluminous data transfer that generally lead to network congestion resulting in Denial of Service [10] (DoS) attack. Sybil, wormhole acknowledgement, etc. are examples of other attacks in this layer. In the application layer, communication between devices in the network are generally supported by routing protocols like Low Power Wireless Personal Area

Network (LOWPAN), Constrained Application Protocol (CoAP) [11], IPv6. One of the major issues in this layer is DoS attack.

2.1 Security Requirements of IoT

IoT devices have now began to pervade everyone's life. There will be a day very soon when all of us will become more dependent on this network for our day to day activities. This is evident from the fact that industries are now moving towards smart vehicular management, smart home, smart classroom management, etc. Any system is secure if it satisfies integrity, confidentiality and non-repudiation requirements. Integrity refers to non-modification of the data during the transit, confidentiality refers to data being delivered only to right persons and non-repudiation refers to assurance that someone cannot deny the action performed. The challenges [12] in enhancing the security of IoT arises from the factors like heterogeneity, resource constraints, large scale, privacy and trust management policies, etc. It is important to understand that all IoT layers are vulnerable to attacks. It is reported that most of the IoT attacks happen due to lack of appropriate authentication and authorization procedures.

2.2 Privacy Concerns in IoT

The present Internet is not basically defined for ensuring security between IoT devices. Cyber-attackers are talented in such a way that they can attack not only public network but also private networks like cars and smart homes. Access control is an important factor to be taken care to fight against malicious penetration. Hence improper monitoring could lead to data being exposed to malicious parties rather than being revealed to a smaller group. Information privacy is one of the main concerns in IoT Network security. Conventional securing methods are not essentially suitable for the IoT environment due to the unique characteristic of these devices. This demands the solution to be comprehensive and original to secure the data transfer between IoT devices.

2.3 MQ Telemetry Transport (MQTT)

MQTT protocol is designed to connect the network and devices along with middleware and applications. Possible connections include client to server, server to server and client communication patterns using a one-to-one, one-to-many or many-to-many model. The protocol designed to work on its TCP/IP (Transmission Control Protocol/Internet Protocol) port number 1883. It uses a publish–subscribe pattern to facilitate using brokers like Mosquitto, HiveMQ and Eclipse Paho. The protocol itself is well designed for resource constraint devices and the protocol is message oriented which allows communication under a tag called topic. This lightweight protocol uses three entities namely subscriber, publisher and broker to facilitate communication. The broker entity facilitates communication between subscribers and publishers under a common topic of interest. The broker is also responsible for managing and assigning the permissions assigned to the publishers and subscribers, managing the topic list under which communication can take place. In a three-tier system, one can model the publisher or subscriber as a client which is to establish communication with its peer entities through a broker entity only. MQTT client set a topic into the messages. It has the overall responsibility routing of information to the MQTT broker such as publish message to concerned users, subscribe to topics of interest under which message should be received, unsubscribe from subscribed topics and disconnect from the broker. Broker takes responsibility for the sharing/distribution of information and more importantly capable for receiving all messages from publisher, and filter messages from concerned clients performing actions like accept client requests, receives published messages from client, process request from client like subscribe and unsubscribe and send messages to interested users.

2.4 Architecture of MQTT

The following are the participants in the experimental setup (Fig. 1).

Publisher sends messages to subscribers to one or many subscribers. **Subscriber** is the end user who is interested in receiving communication from a specific publisher under a specific topic. **Broker** is a middleware that facilitates communication between publishers and subscribers.

The MQTT protocol was introduced to collect data from multiple devices and transport to an infrastructure capable of analysing the data. The inherent lightweight nature of the protocol makes it ideal for remote monitoring in scenarios where the

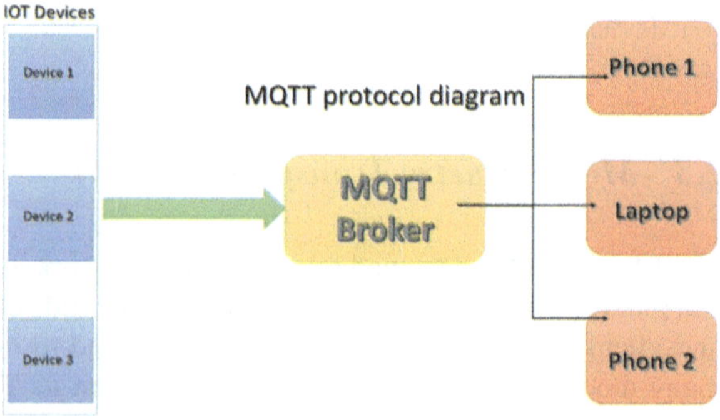

Fig. 1 Architecture of MQTT

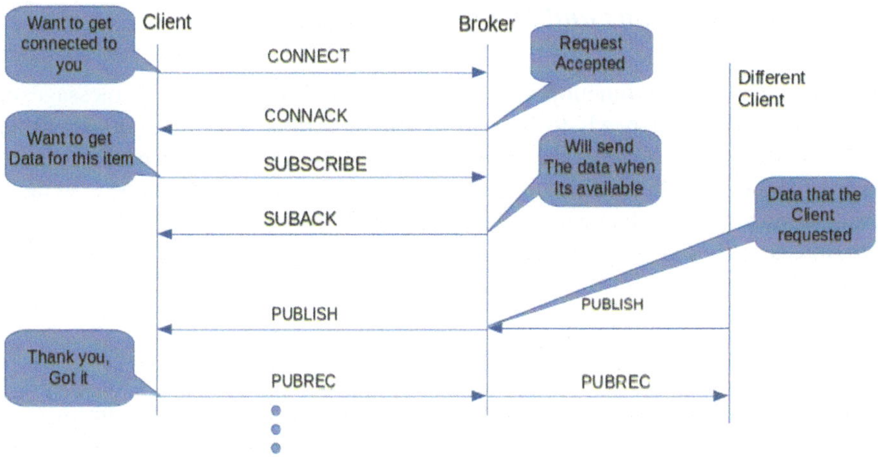

Fig. 2 Working of MQTT

network bandwidth is limited. The steps used in the communication process are depicted in Fig. 2.

2.5 *MQTT Quality of Service*

The MQTT allows efficient distribution of information, increased scalability and collection of voluminous data with less bandwidth. This protocol draws the Quality of Service (QoS) levels for both the communicating entities subscribers and publishers. It supports quality of services in three levels which are described are **At most once** which is the message is sent at most once and broker, client doesn't provide delivery of acknowledgement, **At least once** is the data can be sent more than once until it receives an acknowledgement; **Exactly once** is the message is sent exactly once by using four-way handshaking between the sender and the receiver. These doesn't affect the TCP data transmission. It is only used between MQTT sender and receivers.

2.6 *Security Issues with MQTT*

Message Expiry. MQTT doesn't contain message expiry, so in that case, a cyber-criminal can get the messages of the actual client but avoiding client from receiving the message. Broker is overloaded with messages in case nobody picks up the message and it degrades the overall performance.

2.7 *Hash Functions and Digital Signatures*

A cryptographic hash function is a hash function that converts the message of haphazard length that needs to be sent to the subscriber, into a fixed-size string of bytes. This string is called the hash value. The hash value must have these three basic properties are the hash value must be easily calculated, It should be extremely computationally difficult to calculate the message from the obtained hash value, and two messages should not have the same hash value.

2.7.1 Properties of Hash Function

Pre-image resistance. If a hash function produces a hash digest k to a message m, then it must be computationally hard to calculate a value x that produces the hash value k. This property is to protect from the attackers who only has the hash value and wants to find the message.

Second Pre-image resistance. If a hash function f produces $f(x)$ for a message x, then it should be computationally hard to find any other message y such that f(y) = f(x). This property is there to prevent attackers who has both input value and hash function and wanted to replace the message with any other message that has the same hash value.

Collision Resistance. It should be really hard to find two messages that have either the same or different length that has the same hash value. Hash function compresses a given message to a fixed hash length, it is impossible to have a collision. This property makes the attackers to find two messages that have the same hash.

2.7.2 Digital Signatures

In real life, the message signatory can be fixed by a handwritten sign to a document. Similarly, a technique that fixes an entity to a digital data is digital signature. The authenticity of the document can be verified by the receiver or any third-party people. Digital signature is a calculable cryptographic value from the provided data in the document and a secret key is known only to the person who signs the document. The assurance that the message came from the concerned sender to the receiver can get higher by the use of digital signature and non-repudiation can be avoided. This technique uses the modulo arithmetic to sign a message digitally. RSA is widely used for signing documents and encryption because of its ability to mitigate many forms of attacks that exist in the cyber world. To implement the required cryptographic functions this algorithm uses private and public key pair. Key generation, signing and verification of authentication are the three stages of signing algorithm for the involved entity. The algorithm guarantees confidentiality by the use of three phases, namely, registration, encryption and decryption (Table 1).

Table 1 Legend—Rivest, Shamir, and Adelman

S. no.	Label	Explanation
1	A	An enormous prime number
2	B	An enormous prime number
3	k	Value obtained by multiplying *A* and *B*
4	X	Public key, GCD(X, ø(k)) = 1, where $1 \leq X \leq ø(n) - 1$
5	D	Private key
6	M	Simple text
7	C	Cipher text

Algorithm as proposed by Rivest, Shamir and Adelman
Stage 1.

1. A and B are produced arbitrarily
2. Compute K=A.B
3. Obtain X (Select a number that does not have any division factor common with (A-1) or (B-1)
4. Compute D
   $D \equiv X^{-1} \bmod ø(k)$

The generated key is possessed by both the entities at the end of this stage, where {X,k} is a public key pair and {D} is a private key.
Stage 2. Compute the hash of the message as h= H(msg) using any Hashing algorithm. Sign the message using the equation

$$S \equiv (\textit{Hashtag of the message})^{E} \bmod k$$

Stage 3. Formula to decrypt using the private key D,

$$V \equiv (\textit{Reversed hash})^{D} \bmod k$$

2.7.3 Properties of Digital Signature

Message authentication. When the concerned receiver uses the public key provided by the sender to verify the signature, the receiver is assured that this has been sent by the concerned sender who has the secret key and wasn't sent by any other person.

Data-Integrity. The digital signature is created from the data provided by the sender. If an attacker tries to tamper with the data, the receiver's validation of the digital signature will fail. So the message can be denied by the receiver safely stating that the integrity of the data has been compromised.

Non-repudiation. The signature has been created by the signer and the public and private key has been created by the signer. If any disputes arise anytime, the receiver can use this digital signature and the data as an evidence to the third party.

2.8 Channel Encryption

Setting up a confidentiality for data prevents a message from being overheard but not necessarily from being tampered. Setting up a secure channel makes data resistant to tampering but not from being overheard. Research has proposed different ways like using quantum cryptography as a means to create a secure channel for data exchange which are theoretically free from eavesdropping, interception and tampering. In practice an authenticated channel built using digital signatures and encryption [12] is used for enhancing the security.

2.9 Summary of Findings

The present model of the WWW does not meet the requirements of the modern environment of IoT. The Swift development in the IoT domain has attracted people to this fictional space that actually lacks security services for the simple reason that they are based on traditional architecture of WWW. Undeniably, these dynamic environments are now appraised as an encouraging technology to ease the cooperative link between devices and people for their day to day activities. Albeit a diverse solution is presented by the literature to guarantee privacy and security concerns. The survey of the literature clearly brings out the nature of the MQTT protocol [13] to facilitate efficient transmission of data in a limited bandwidth, but this transmission is prone to many forms of attacks. This set off a need for identification and examination of the actual communication protocols for vulnerabilities to put forward alleviation to strengthen the protocol performance.

2.10 Problem Statement

The existing literature on IoT device communications using MQTT protocol are nearly based on cryptographic algorithms capable of operating with reduced keyspace. However, these methods performed encryption and decryption individually on each packet thereby increasing the overhead. To address this issue, a channel based encryption mechanism has been adopted and its benefits in comparison with existing schemes have been evaluated. The objective of the work is to reduce the communication overhead associated with mqtt protocol without compromise in security.

	Existing schemes [7]	Expected outcome from proposed scheme
Packet encryption and decryption	✓	✓
Entity of encryption and decryption	Packet	Channel
Disadvantage	More overhead	Reduced overhead

3 Proposed Scheme

The messages sent via MQTT protocol can be tapped or the messages can be read by a man in the middle attack because there is absolutely no way the messages are encrypted. Open source software like MQTT lens can be used to read the messages sent via a channel. This scheme operates in two phases, namely, registration and authenticated communication.

3.1 Registration Process

The registration of subscribers and publishers with a common MQTT broker is the prerequisite for the commencement of communication process. During the registration process, the IP address of the publisher and subscriber are registered with the broker. The first and most important step is to make sure that broker, subscriber, and publisher are on the same network. Then, the next step is to start the Mosquitto program (Fig. 3). The system that has Mosquitto running will act as a broker. The IP address of that particular machine must be noted. With the IP address, the subscriber and publisher must connect to the broker. Although different types of publish–subscribe schemes can be used like topic based or hierarchy based, etc., the experimentation was carried out based on topic-based scheme. The data published in the system is unstructured. But each event is labelled with an identifier (Topic). Subscribers issue subscription containing issues they are interested in. The topic can thus be represented as a channel connecting the subscriber and publisher.

The steps to be followed for commencing the communication process until expiration of subscription are Generate key pair, Generate Certified Authority file with the available key pair, Create a certificate that takes the Certified Authority file, Generate the key pair for the server side, Create a server certificate file with the server key and mention the domain name as given in certified authority certificate, Verify the server certificate and certificate authority certificate and create a common signature, and Use the certificate authority certificate file in the publisher side for server authentication and for further communication of messages. In the (Figs. 4, 5) proposed scheme, as the decryption of the message happens on the side of the server, the subscriber can be thin client, thus the scheme suitable for IoT devices which has limited resources.

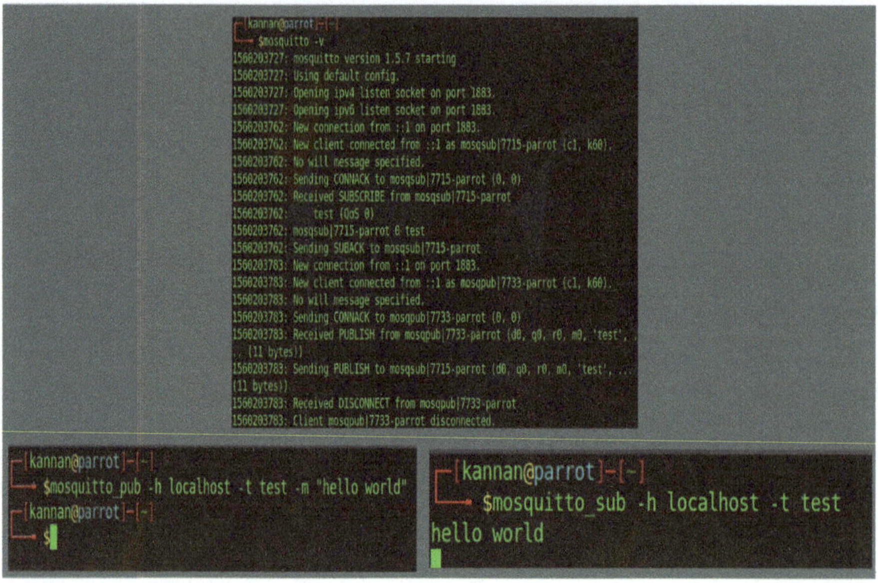

Fig. 3 Broker, publisher and subscriber communication

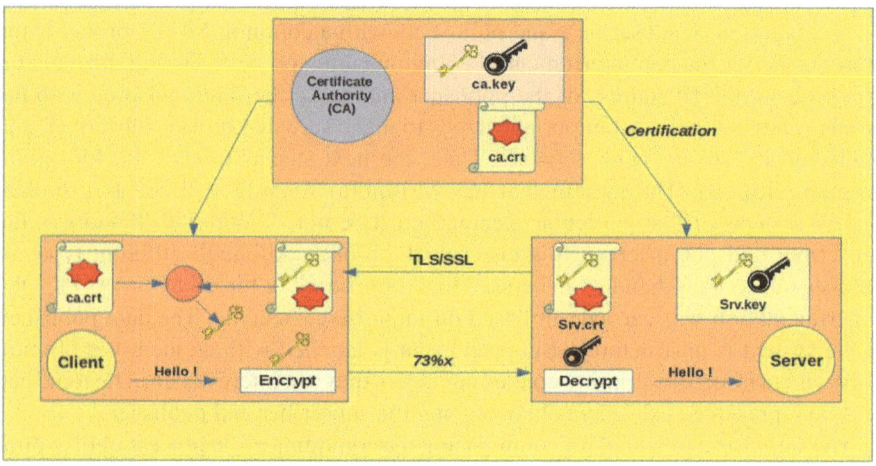

Fig. 4 TLS handshake with keys and certificate

The experimentation was also done to verify the scenario where the broker turns malicious. In this case, the communication between the entities would fail because of the frequent update of key pairs.

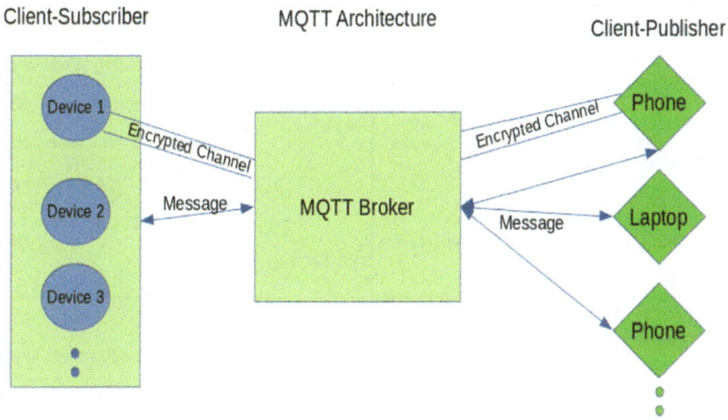

Fig. 5 Proposed MQTT architecture

4 Result and Discussions

The controlled environmental setup involving publisher, subscriber and Mosquitto broker was created. Figure 6 demonstrates the procedure to configure the MQTT broker. On mere configuration, message exchanges are facilitated only in an insecure fashion as communication through port 1883 are by default not encrypted shown in Fig. 6. To enhance the security of the communication the entities need to be authenticated [14] frequently using certificates. Figure 7 states the generation of the Certified Authorities key pairs. The snippet of the code used for automating the key generation process is shown in Fig. 8. It is mandatory that the communicating entities

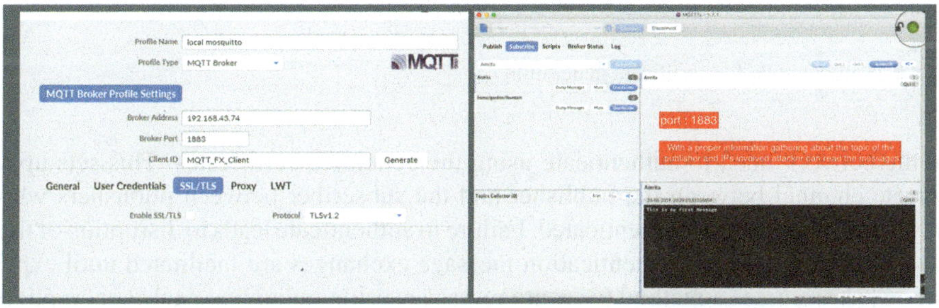

Fig. 6 The broker configuration and default communication is not encrypted

```
cafile /usr/local/Cellar/mosquitto/1.5.3/etc/mosquitto/ca_certificates/ca.crt
certfile /usr/local/Cellar/mosquitto/1.5.3/etc/mosquitto/ca_certificates/server.crt
keyfile /usr/local/Cellar/mosquitto/1.5.3/etc/mosquitto/ca_certificates/server.key
```

Fig. 7 Generated keys and certificates for mosquitto broker

```
if [ ! -f $CACERT.crt ]; then
        $openssl req -newkey rsa:${keybits} -x509
-nodes $defaultmd -days $days -extensions v3_ca
-keyout $CACERT.key -out $CACERT.crt -subj
"${CA_DN}"
        echo "Created CA certificate in $CACERT.crt"
        $openssl x509 -in $CACERT.crt -nameopt
multiline -subject -noout
        chmod 400 $CACERT.key
        chmod 444 $CACERT.crt
        chown $MOSQUITTOUSER $CACERT.*
        echo "Warning: the CA key is not encrypted;
store it safely!"
fi
```

```
if [ $kind == 'server' ]; then
    if [ ! -f $SERVER.key ]; then
            echo "--- Creating server key and signing
request"
            $openssl genrsa -out $SERVER.key
$keybits
            $openssl req -new $defaultmd \
                    -out $SERVER.csr \
                    -key $SERVER.key \
                    -subj "${SERVER_DN}"
            chmod 400 $SERVER.key
            chown $MOSQUITTOUSER
$SERVER.key
fi
```

```
echo "--- Creating and signing server certificate"
$openssl x509 -req $defaultmd \
                -in $SERVER.csr \
                -CA $CACERT.crt \
                -CAkey $CACERT.key \
                -CAcreateserial \
                -CAserial "${DIR}/ca.srl" \
                -out $SERVER.crt \
                -days $days \
                -extfile ${CNF} \
                -extensions JPMextensions
        rm -f $CNF
        chmod 444 $SERVER.crt
        chown $MOSQUITTOUSER
$SERVER.crt
```

```
if [ ! -f $CLIENT.key ]; then

            echo "--- Creating client key and signing
request"
    CNF=`mktemp /tmp/cacnf-req.XXXXXXXX` || { echo
"$0: can't create temp file" >&2; exit 1; }
            $openssl genrsa -out $CLIENT.key
$keybits
    $openssl req -new $defaultmd \
                    -out $CLIENT.csr \
                    -key $CLIENT.key \
                    -config $CNF
            chmod 400 $CLIENT.key
fi
```

```
if [ -f $CLIENT.csr -a ! -f $CLIENT.crt ]; then
        CNF=`mktemp /tmp/cacnf-cli.XXXXXXXX` || {
echo "$0: can't create temp file" >&2; exit 1; }
    echo "--- Creating and signing client certificate"
    $openssl x509 -req $defaultmd \
                -in $CLIENT.csr \
                -CA $CACERT.crt \
                -CAkey $CACERT.key \
                -CAcreateserial \
                -CAserial "${DIR}/ca.srl" \
                -out $CLIENT.crt \
                -days $days \
                -extfile ${CNF} \
                -extensions JPMclientextensions
        rm -f $CNF
        chmod 444 $CLIENT.crt
fi
```

Fig. 8 Snippet code for certificate generation

in the network need to authenticate using the generated certificates. This sets up a private channel between the publisher and the subscriber between publishers who have been successfully authenticated. Failure to authenticate leads to disruption of the service. On successful authentication message exchanges are facilitated until expiration [7] discusses a method to secure messages using individual packet encryption scheme. Although this scheme offers higher level of security the overhead involved in the process is high. The proposed scheme achieves the same level of security with reduced overhead thus making it better for practical applications. The properties of non-repudiation, integrity are ensured by the usage of RSA digital signature scheme. The publisher who grasp the right key alone is capable of signing the message. The time stamp attribute is associated with every message to ensure that replay of message by an attacker is prevented. The confidentiality is ensured by the use of keys that are available with subscribers to decode the message. In case of suspicion on an entity, the key revocation can be used by CA. It can be observed that channel encryption technique exhibits an improvised performance compared to the scheme encrypting every data packet. The experimentation revealed that the proposed scheme was able to achieve 33.3% improvement in terms of speed and reliability.

5 Conclusions

In this paper, a secure communication scheme capable of operating at a reduced overhead rate compared to individual packet encryption scheme has been proposed. The proposed scheme can be useful in applications designed to handle mobility of devices in network, data processing in cloud, etc. This scheme finds its application in resource constraint domains that involves devices communicating using MQTT protocol. Experimentation clearly revealed that this scheme is highly beneficial at times of peak loads on server as it removes individual packet encryption. The above fact is evident from the results in Figs. 9, 10. The results of experimentation associated with communicating a message "on 33" encrypted using Fernet symmetric key algorithm showed a 33.3% reduction in the delivery time. It can be seen that a high performance can be achieved in environments demanding continuous message communication between publisher and subscriber. The presented scheme makes use of public key cryptography schemes efficiently in order to be accepted by the MQTT communication model without attaching additional messages (overhead). When compared to similar work in the field, our solution in the security message overhead is minimal. As a future extension, it is planned to test the scheme to evaluate its capability in terms of mitigating different forms of attacks on encrypted channel.

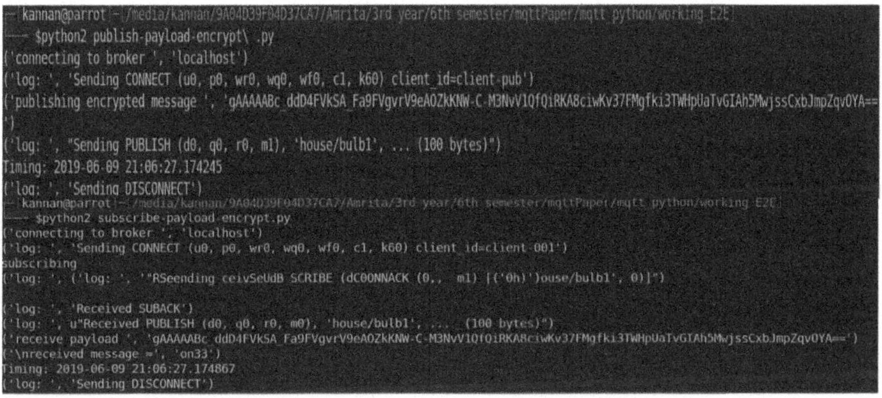

Fig. 9 End-to-end encryption timing

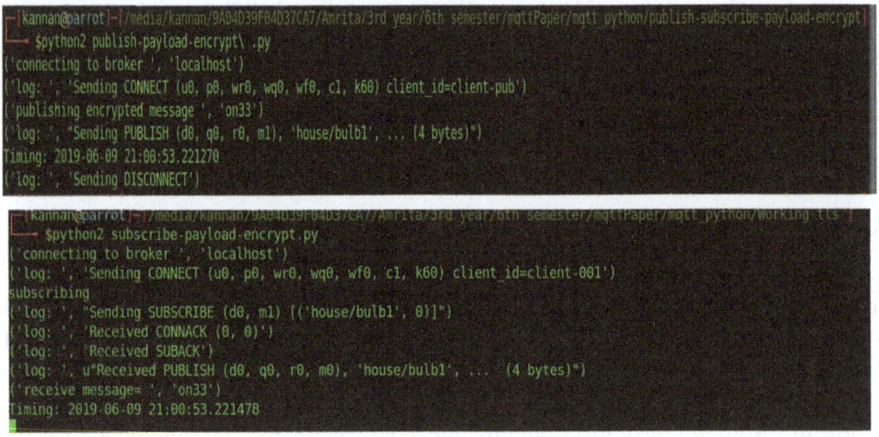

Fig. 10 SSL encryption timing

References

1. Padmanabhan, T.R., Harini, N., Shyamala, C.K.: Cryptography and Security, 1st edn. Wiley, India (2011)
2. Qadeer, M.A., Iqbal, A., Zahid, M., Siddiqui, M.R.: Network traffic analysis and intrusion detection using packet sniffer. In: Second International Conference on Communication Software and Networks, pp. 313–317 (2010)
3. Zhang, Z.K., Cho, M.C.Y., Wang, C.W., Hsu, C.W., Chen, C.K., Shieh, S.: IoT security: ongoing challenges and research opportunities. In: IEEE 7th İnternational Conference on Service-Oriented Computing and Applications, pp. 230–234. IEEE (2014)
4. Yokotani, T., Sasaki, Y.: Comparison with HTTP and MQTT on required network resources for IoT. In: International Conference on Control, Electronics, Renewable Energy and Communications (ICCEREC), pp. 1–6. IEEE (2016)
5. Mahmoud, R., Yousuf, T., Aloul, F., Zualkernan, I.: Internet of Things (IoT) security: current status, challenges and prospective measures. In: 2015 10th International Conference for Internet Technology and Secured Transactions (ICITST), pp. 336–341. IEEE (2015)
6. Kraijak, S., Tuwanut, P.: A survey on internet of things architecture, protocols, possible applications, security, privacy, real-world implementation and future trends. In: 16th International Conference on Communication Technology (ICCT), pp. 26–31. IEEE (2015)
7. Jayan, A.P., Harini, N.: A scheme to enhance the security of MQTT protocol. Int. J. Pure Appl. Math. **119**(12), 13975–13982 (2018)
8. Yassein, M.B., Shatnawi, M.Q., Aljwarneh, S., Al-Hatmi, R.: Internet of Things: survey and open issues of MQTT protocol. In: International Conference on Engineering & MIS (ICEMIS), pp. 1–6. IEEE (2017)
9. Zamfir, S., Balan, T., Iliescu, I., Sandu, F.: A security analysis on standard IoT protocols. In: 2016 International Conference on Applied and Theoretical Electricity (ICATE), pp. 1–6 IEEE (2016)
10. Elhoseny, M., Shankar, K., Lakshmanaprabu, S.K., Maseleno, A., Arunkumar, N.: Hybrid optimization with cryptography encryption for medical image security in Internet of Things. Neural Comput. Appl. 1–15 (2018)
11. Singh, M., Rajan, M.A., Shivraj, V.L., Balamuralidhar, P.: Secure MQTT for İnternet of Things (İoT). In: Fifth International Conference on Communication Systems and Network Technologies, pp. 746–751. IEEE (2015)

12. Babar, S., Stango, A., Prasad, N., Sen, J., Prasad, R.: Proposed embedded security framework for İnternet of Things (iot). In: 2nd International Conference on Wireless Communication, Vehicular Technology, Information Theory and Aerospace & Electronic Systems Technology (Wireless VITAE) , pp. 1–5. IEEE (2011)
13. Niruntasukrat, A., Issariyapat, C., Pongpaibool, P., Meesublak, K., Aiumsupucgul, P., Panya, A.: Authorization mechanism for MQTT-based internet of things. In: International Conference on Communications Workshops (ICC), pp. 290–295. IEEE (2016)
14. Templeton, S.J., Levitt, K.E.: Detecting spoofed packets. In: Proceedings DARPA Information Survivability Conference and Exposition, vol. 1, pp. 164–175. IEEE (2003)

Software Fault Prediction Using Random Forests

Kulamala Vinod Kumar, Priyanka Kumari, Avishikta Chatterjee, and Durga Prasad Mohapatra

Abstract In this paper, we present a software fault prediction model using random forests. Software fault prediction identifies the faulty regions in a software product early in its lifecycle and hence improves the quality attributes such as reliability of the software. Random forest is an ensemble learning method for classification. Random forests contain many of decision trees and the result is the function of these decision trees. The proposed approach is applied on software defect prediction datasets collected from PROMISE software engineering repository and evaluated the performance using precision, recall, accuracy, and F-measure. The results of random forest model are compared with other models such as support vector machines, backpropagation neural networks, decision trees. Based on comparison, it is observed that random forest model is superior to the other models.

Keywords Fault prediction · Defect · Random forests · Reliability

1 Introduction

Software Development Life Cycle (SDLC) has many phases such as requirements analysis, design, testing, deployment, and maintenance. Hence, lifecycle of a software product is exposed to faults; the faults could be due to syntax error, incorrect

K. V. Kumar (✉) · P. Kumari · A. Chatterjee · D. P. Mohapatra
National Institute of Technology, Rourkela 769008, India
e-mail: 517cs1007@nitrkl.ac.in

P. Kumari
e-mail: 218cs3438@nitrkl.ac.in

A. Chatterjee
e-mail: 218cs3318@nitrkl.ac.in

D. P. Mohapatra
e-mail: durga@nitrkl.ac.in

D. Mishra et al. (eds.), *Intelligent and Cloud Computing*,
Smart Innovation, Systems and Technologies 194,
https://doi.org/10.1007/978-981-15-5971-6_10

requirement, or the design errors. If the faults are detected in later stages of software, then more effort and time is required for the rework process. Thus, the prediction of defective modules in software is a crucial task. Identifying faulty modules in early stages of SDLC will help project managers of software, to decide the modules which need more rigorous testing. In the software engineering domain, Software Fault Prediction (SFP) in early stages improves the reliability of software.

Many machine learning techniques have been used to predict the faulty modules in software. Still, software fault prediction is a challenging task in software engineering domain. Software fault prediction can be considered as classification problem in machine learning, which will predict a module is faulty or not. These models take fault prediction datasets as input and get trained. After training the models, they are tested using testing data to check model performance. Then, these models can be used to predict the faulty modules in software. Machine learning techniques such as support vector machines [1], K-nearest neighbors (KNN) [2], Naive Bayes [3–5], etc. are some of the methods applied to predict faulty modules in software. In this paper, Random Forest (RF) algorithm is used to predict the fault prone software modules.

1.1 Random Forests (RF)

Random forest algorithm is a supervised classification algorithm. Random forests generate many classification trees randomly. The outcome of the random forest will be the function of all these classification trees. In a random forest algorithm, each tree grows by considering random number of cases and random number of features (variables). If there are F features, at each node a number $f \ll F$ is specified, the best split on these "f" is used to split the node [6]. Generally in a random forest, the more strength of the individual trees the less error rate occurs [7], but increasing the correlation between trees increases the error rate.

The working of random forest is explained as follows:

Step 1. "Select random samples from a given dataset".
Step 2. "Construct a decision tree for each sample and get a prediction result from each decision tree".
Step 3. "Perform a vote for each predicted result".
Step 4. "Select the prediction result with the most votes as the final prediction".

Rest of the paper is organized as follows: Sect. 2 presents literature review. Section 3 describes the methodology of this work. Section 4 shows the implementation details with results. Section 5 concludes the paper with thoughts on future work.

2 Literature Review

Soe et al. [8] developed software defect prediction models using random forest classifier. The developed model is evaluated by applying it to PROMISE public datasets, and an average accuracy of 92% was achieved. The random forest algorithm generated hundred trees while developing a model. The results showed that the random forest with hundred trees is more stable for software defect prediction.

Ibrahim et al. [9] employed two algorithms to software fault prediction that are a Bat-based search Algorithm (BA) is used for selecting the best features and the Random Forest algorithm (RF) to predict the fault proneness in a module. From the results, it was observed that random forest classifier has proved its efficiency with the other classifiers. But the BA has the superiority on the tested datasets.

Jacob [10] described a model based on the random forest with feature selection to give better accuracy. They applied feature subset selection with correlation to select the best subset of features. The resulted features are given as input of random forest classification model to give better accuracy in software fault prediction. The algorithm gave six features as best features. The experiments were carried on public datasets of PROMISE repository.

Catal [11] investigated 90 software fault prediction papers published between the year 1990 and 2009. He reviewed the literature in the field of software fault prediction using both machine learning-based and statistical-based approaches. This paper investigated the previous work based on software metrics, methods, datasets, performance evaluation metrics, and experimental results perspectives.

Dhankhar et al. [12] described software fault prediction models using different machine learning techniques such as bayesian network, multi-layered perceptron, and naive bayes classifier. As a result, a software fault prediction framework is proposed. This framework is applied on 10 public domain datasets from PROMISE Repository. In this study, it is observed that metric-based classification is useful. Furthermore, the results obtained are compared using selected performance criteria such as True Positive Rate (TPR) and False Positive Rate (FPR). Output showed that the neural network classification models are more superior to the other network models.

3 Proposed Methodology

The goal of this work is to build a random forest model to predict faulty modules in a software product and evaluate its performance. The input to the model is the software fault prediction datasets and the output is whether a module is faulty or not. The input datasets contain software metrics extracted at module levels for different softwares. For example, ar1 dataset is having 121 modules. For each module 29 software metrics (features) are extracted as shown in Table 2.

Figure 1 shows the flow of proposed approach. We have taken the fault prediction datasets from PROMISE Repository [13]. The datasets considered are shown in

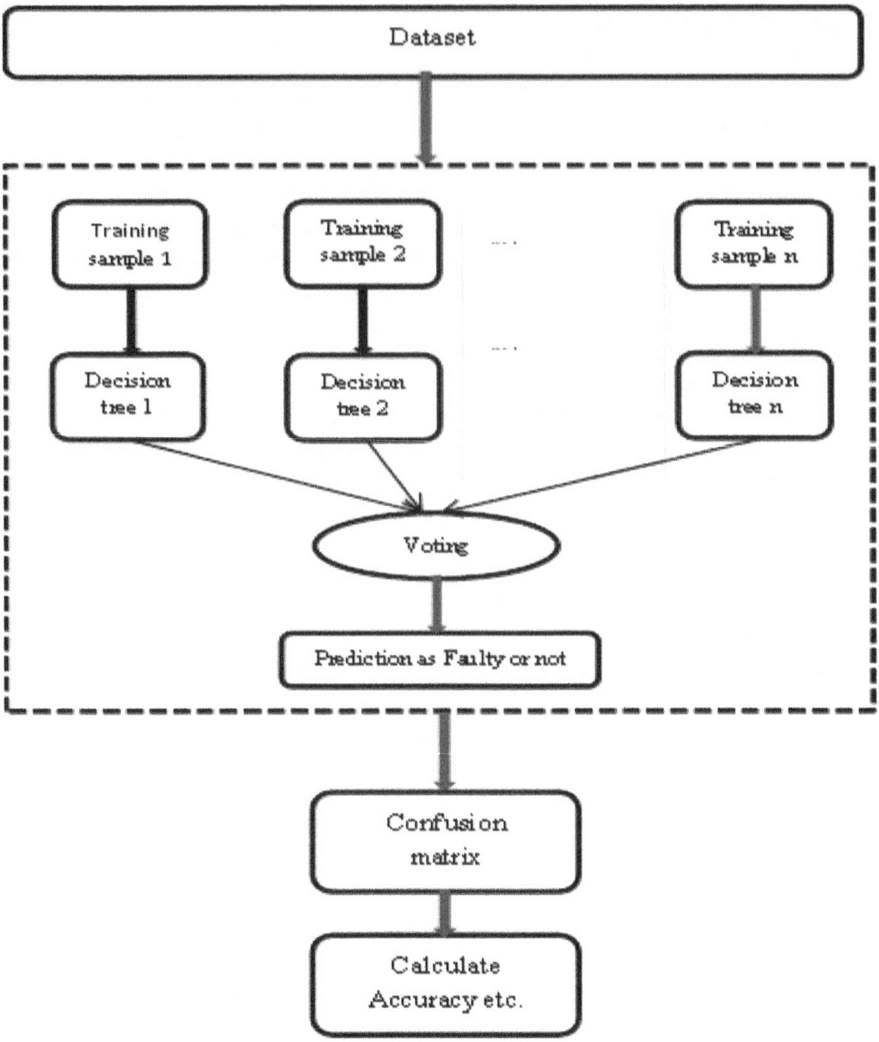

Fig. 1 Approach to software fault prediction using random forests

Table 1. These datasets are normalized using min-max normalization within the range [0, 1] as shown in Eq. (1).

$$y = (x - \min)/(\max - \min) \tag{1}$$

Here, min and max are the minimum and maximum values in X, where X is the set of observed values of x, and y is resulted value.

These normalized datasets are input to the random forest model. Random forest algorithm samples the data randomly and constructs decision trees as shown in Fig. 1.

Table 1 Characteristics of datasets considered

Dataset	Total No. of modules	No. of faulty modules	No. of non-faulty modules
ar1	121	9	112
ar3	63	8	55
ar5	36	8	28

Then the output of the Random forest is decided by voting process in consideration with the decision of all decision trees. We have applied a tenfold cross validation while constructing the model. In cross validation the data is split into 10 parts, then we train the model with 9 out of 10 parts of the data and we test the model with the remaining 1 part of the data. This process is repeated 10 times. The random forest algorithm is applied to understand the trend in the data and after this the confusion matrix is generated from which the performance metrics such as the model accuracy, precision, recall, and F-measure are obtained.

3.1 Datasets

See Tables 1, 2 and Fig. 1.

3.2 Evaluation Criteria

We consider the performance measures such as Accuracy, Precision, Recall, and F-measure to evaluate the proposed random forest model to predict a module as faulty or non-faulty. These performance measures are calculated from "confusion matrix". Precision is the ratio of correctly predicted faulty modules to the total predicted faulty modules. The formula for precision is shown in Eq. (2).

$$\text{Precision} = \text{TP}/(\text{TP} + \text{FP}) \tag{2}$$

where TP is "True Positive" and FP is "False Positive". Recall is the ratio of correctly predicted faulty modules to the actual number of existing faulty models. Mathematically, Recall can be expressed as shown in Eq. (3).

$$\text{Recall} = \text{TP}/(\text{TP} + \text{FN}) \tag{3}$$

where TP is "True Positive", FN is "False Negative". F-measure is weighted average of Precision and Recall. It takes both false positives and false negatives into account,

Table 2 Metrics considered in the datasets

Sl. No.	Feature
1	"total_loc numeric"
2	"blank_loc numeric"
3	"comment_loc numeric"
4	"code_and_comment_loc numeric"
5	"executable_loc numeric"
6	"unique_operands numeric"
7	"unique_operators numeric"
8	"total_operands numeric"
9	"total_operators numeric"
10	"halstead_vocabulary numeric"
11	"halstead_length numeric"
12	"halstead_volume numeric"
13	"halstead_level numeric"
14	"halstead_difficulty numeric"
15	"halstead_effort numeric"
16	"halstead_error numeric"
17	"halstead_time numeric"
18	"branch_count numeric"
19	"decision_count numeric"
20	"call_pairs numeric"
21	"condition_count numeric"
22	"multiple_condition_count numeric"
23	"cyclomatic_complexity numeric"
24	"cyclomatic_density numeric"
25	"decision_density numeric"
26	"design_complexity numeric"
27	"design_density numeric"
28	"normalized_cyclomatic_complexity numeric"
29	"formal_parameters numeric"

and it is defined as shown in Eq. (4)

$$\text{F} - \text{measure} = \frac{(2 * \text{Recall} * \text{Precision})}{(\text{Recall} + \text{Precision})} \tag{4}$$

Accuracy is the most intuitive metric to measure the performance of a fault prediction model. It is defined as ratio of correctly predicted samples to the total samples [14]. The formula for accuracy is shown in Eq. (5).

$$\text{Accuracy} = (\text{TP} + \text{TN})/(\text{TP} + \text{FN} + \text{TN} + \text{FP}) \quad (5)$$

where TP is "True Positive", FP is "False Positive", TN is "True Negative" and FN is "False Negative".

4 Implementation and Results

We have implemented our proposed approach for predicting software reliability (using RF) in MATLAB software. The proposed method is applied on ar1, ar2, and ar3 datasets collected from PROMISE repository. We have also implemented some other existing algorithms such as SVM with three kernels (linear, polynomial, RBF), Decision trees, and BPNN for predicting software reliability on ar1, ar2, and ar3 datasets. We have compared the performance of the proposed approach with the existing approaches, by computing the performance measures such as accuracy, F-measure, precision, recall. The accuracy measure of various algorithms on datasets ar1, ar3 and ar5 is given in Table 3.

F-measure for various algorithms on the datasets ar1, ar3, and ar5 is given in Table 4.

Precision measure for various algorithms on the datasets ar1, ar3, and ar5 is given in Table 5.

Recall measure for various algorithms on the datasets ar1, ar3, and ar5 is given in Table 6.

Table 3 Accuracy of BPNN, decision tree, random forest, and SVM

Algorithm	ar1	ar3	ar5
SVM (Linear)	81.96	62.5	83.3
SVM (Degree 2)	83.6	56.25	83.3
SVM (RBF)	90.7	90.5	72.2
Decision Tree	90.16	90.625	77.8
BPNN	91.8	90.625	70.2
Random Forest	94.6	94.73	81.8

Table 4 F-measure of BPNN, decision tree, random forest, and SVM

Algorithm	ar1	ar3	ar5
SVM (Linear)	0.846	0.702	0.838
SVM (Degree 2)	0.856	0.653	0.838
SVM (RBF)	0.871	0.862	0.605
Decision Tree	0.8705	0.912	0.787
BPNN	0.8788	0.862	0.61
Random Forest	0.919	0.952	0.837

Table 5 Precision of BPNN, decision tree, random forest, and SVM

Algorithm	ar1	ar3	ar5
SVM (Linear)	0.842	0.821	0.522
SVM (Degree 2)	0.883	0.776	0.847
SVM (RBF)	0.878	0.828	0.846
Decision Tree	0.84	0.941	0.825
BPNN	0.843	0.821	0.52
Random Forest	0.893	0.894	0.782

Table 6 Recall of BPNN, decision tree, random forest, and SVM

Algorithm	ar1	ar3	ar5
SVM (Linear)	0.82	0.625	0.833
SVM (Degree 2)	0.836	0.563	0.813
SVM (RBF)	0.901	0.906	0.713
Decision Tree	0.902	0.687	0.778
BPNN	0.918	0.906	0.722
Random Forest	0.919	0.895	0.636

Results from Tables 3, 4, 5, and 6 show that according to accuracy and F-measure criteria; random forest has the highest score among the others. For precision random forest and decision tree classifiers have nearly same score and that is greater than the remaining scores among the mentioned algorithms. For recall measure, backpropagation neural network has the highest score among the remaining, and random forest has the next best score. Considering all these measures, random forest shows better results among the other mentioned algorithms (SVM with linear, polynomial with degree 2, RBF kernel, Decision tree, BPNN) for predicting software reliability.

5 Conclusion

In this paper, software fault prediction model is developed using random forest algorithm. We have applied the developed model on three datasets available publicly from PROMISE repository. The performance evaluation is carried out using measures such as accuracy and F-measure metrics. The developed random forest model is compared with other models such as support vector machines (Linear, polynomial, RBF kernels), BPNN, and decision trees. From the results, it is inferred that the developed random forest model predicts the faulty modules in a better way than other related models. In future we will consider more datasets and we will apply some feature selection methods for more generalized results.

References

1. Elish, K.O., Elish, M.O.: Predicting defect-prone software modules using support vector machines. J. Syst. Softw. **81**(5), 649–660 (2008)
2. Perreault, L., Berardinelli, S., Izurieta, C., Sheppard, J.: Using classifiers for software defect detection. In: 26th International Conference on Software Engineering and Data Engineering, SEDE, pp. 2–4 (2017)
3. Alsaeedi, A., Khan, M.Z.: Software defect prediction using supervised machine learning and ensemble techniques: a comparative study. J. Softw. Eng. Appl. **12**, 85–100 (2019)
4. Wang, T., Li, W.-H.: Naive bayes software defect prediction model. In: 2010 International Conference on Computational Intelligence and Software Engineering, pp. 1–4. IEEE (2010)
5. Jiang, Y., Cukic, B., Menzies, T.: Fault prediction using early lifecycle data. In: The 18th IEEE International Symposium on Software Reliability (ISSRE'07), pp. 237–246. IEEE (2007)
6. https://www.datacamp.com/community/tutorials/random-forests-classifier-python
7. Ho, T.K.: Random decision forests. In: Proceedings of 3rd International Conference on Document Analysis and Recognition, vol. 1, pp. 278–282. IEEE (1995)
8. Soe, Y.N., Santosa, P.I., Hartanto, R.: Software defect prediction using random forest algorithm. In: 2018 12th South East Asian Technical University Consortium (SEATUC), vol. 1, pp. 1–5. IEEE (2018)
9. Ibrahim, D.R., Ghnemat, R., Hudaib.: Software defect prediction using feature selection and random forest algorithm. In: 2017 International Conference on New Trends in Computing Sciences (ICTCS), pp. 252–257. IEEE (2017)
10. Jacob, S.G.: Improved random forest algorithm for software defect prediction through data mining techniques. Int. J. Comput. Appl. **117**(23) (2015)
11. Catal, C.: Software fault prediction: a literature review and current trends. Expert Syst. Appl. **38**(4), 4626–4636 (2011)
12. Dhankhar, S., Rastogi, H., Kakkar, M.: Software fault prediction performance in software engineering. In: 2015 2nd International Conference on Computing for Sustainable Global Development (INDIACom), pp. 228–232. IEEE (2015)
13. https://tunedit.org/repo/PROMISE/DefectPrediction
14. Kulamala, V.K., Maru, A., Singla, Y., Mohapatra, D.P.: Predicting software reliability using computational intelligence techniques: a review. In: 2018 International Conference on Information Technology (ICIT), pp. 114–119. IEEE (2018)

A Power Optimization Technique for WSN with the Help of Hybrid Meta-heuristic Algorithm Targeting Fog Networks

Avishek Banerjee, Victor Das, Arnab Mitra, Samiran Chattopadhyay, and Utpal Biswas

Abstract A novel approach toward the minimization of energy consumption along with maximization of the coverage area in any Fog Network consisting of wireless sensor area networks (WSNs) is presented in this research, which further results in energy efficiency in the concern network. An ant colony optimization (ACO)-based technique at random deployment has been considered for our proposed research in simulation. Results obtained in simulation and related analyses have ensured the efficiency of our proposed approach.

Keywords Fog computing · Wireless sensor area network (WSN) · Minimized energy consumption of WSN · Energy efficiency · Green computing

A. Banerjee · V. Das
Department of Information Technology, Asansol Engineering College, Asansol, West Bengal, India
e-mail: avishekbanerji@gmail.com

V. Das
e-mail: victordas.2005@gmail.com

A. Mitra (✉)
Department of Computer Science & Engineering, Siksha 'O' Anusandhan (Deemed to be University), Bhubaneswar, Odisha, India
e-mail: arnabmitra@soa.ac.in

S. Chattopadhyay
Department of Information Technology, Jadavpur University, Kolkata, West Bengal, India
e-mail: samirancju@gmail.com

U. Biswas
Department of Computer Science & Engineering, Kalyani University, Kalyani, West Bengal, India
e-mail: utpalbiswas@klyuniv.ac.in

D. Mishra et al. (eds.), *Intelligent and Cloud Computing*,
Smart Innovation, Systems and Technologies 194,
https://doi.org/10.1007/978-981-15-5971-6_11

1 Introduction

The advances in information technology have resulted in the rise of low-cost and co-operative computing over geographically separated locations. For this reason, the simple networked computing architecture of the past has transformed into today's Cloud Computing, and further, it has evolved into Fog (also known as Edge) Computing, where some computing are allowed to be performed at the edge devices of the network along with the server. As described in [1], besides the computation at edge, Fog Computing is also responsible for heterogeneity, location consciousness, low-latency, real-time processing. For this reason, several Internet of Things (IoTs)-based applications such as Smart Grids, Smart City, Wireless Sensor and Actuator Network (WSAN), cutting edge defense technology, e.g., the surveillance at war fields etc. may be modeled today involving the Fog Networks. A typical Fog Computing architecture involving wireless sensor area network (WSN) and wireless-actuator area network (WAN) is presented in the following Fig. 1. Figure 1 is inspired by the architecture of the Fog Network of [2].

From Fig. 1, WSN is found to be an integral component of Fog Network. It is already described that resource-constrained (i.e., in view of limited computing facility and limited power availability) Fog Networks (e.g., WSN) are potentially capable to be used toward the sensors-based surveillances in the war-affected area(s) [3].On the other hand, Green Computing is one of the primary focuses among researchers to provide sustainable and environment-friendly technology-based solutions. For this reason, energy-efficient uses of WSN are a primitive concern among researches [4, 5]. For this reason, in this work, we are going to describe how the energy efficiency of WSN can be enhanced. As the Fog Network includes sensor network(s), so we can claim that the energy efficiency in the WSN module should directly enhance the total energy efficiency of the concerned Fog Network. A brief discussion on WSN is following next.

Fig. 1 A typical Fog Network involving WSN (inspired from [2])

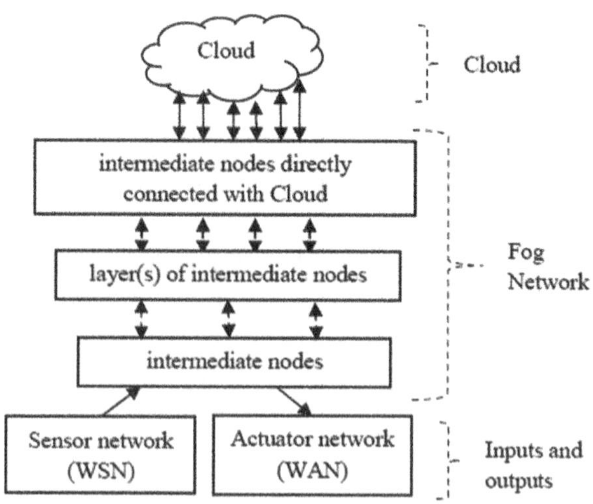

The nodes found in any WSN usually consist of three different types of units. The first unit is referred to as the Processing Unit which consists of a very tiny type of processor (usually a micro-controller). The second unit is the power backup unit, and the last but not the least unit is the communication unit. The power backup in WSN referred to as the tiny battery backup unit, where most frequently the alkaline batteries are used. In the case of alkaline batteries, the Zinc metal (Zn) reacts with Manganese di-Oxide (MnO_2), and this type of batteries have a better performance with respect to high shelf life as well as higher energy density red cell voltage. The cell voltage required in case of a alkaline battery is 1.5 V for getting energy density in the range from 0.4 to 0.59 MJ/kg. This is highly suitable for the design of tiny WSN nodes.

To enhance efficiency, several optimizations approaches have been presented by researchers since the past couple of years. Among several others, nature-inspired ant colony optimization (ACO) technique has acquired attention due to its precision toward the optimal results. In Ant Colony Optimization, several artificial ants build solutions which are considered toward the optimization problem and exchange data about the quality of these results via a communication media, "pheromone trail, which is reminiscent of the one adopted by real ants" [6]. The original ACO algorithm, acknowledged as the Ant System, was presented in [7]. A brief discussion on ACO is followed next.

An ant produces a solution by repeatedly applying a state transition rule and further the solution is enhanced by a local search algorithm. Then the ant adapts "… the amount of pheromone on the visited edges by applying a local pheromone updating rule" [6]. Once all ants have finished their operations, "the amount of pheromone is modified by applying a global updating rule" [6]. ACO activity may be realized with the following two equations Eqs. 1 and 2.

$$\tau_{ij} = \{ \begin{matrix} (1-\rho)\cdot\tau_{ij} + \rho\cdot\Delta\tau_{ij}, & \text{if (i, j)} \in \text{best solution} \\ \tau_{ij} & \text{otherwise} \end{matrix} \tag{1}$$

The local pheromone updating rule is shown in Eq. 2.

$$\tau_{ij} = \{\tau_{ij}.(1-\varphi) + \varphi.\tau_0 \tag{2}$$

The major contributions of this research as presented in this paper are as followed.

(a) A novel approach has been presented toward the minimization of energy consumption along with maximization of the coverage area WSN(s);
(b) an enhanced ACO algorithm has been presented to support our proposed enhancement;
(c) a theoretical approach has been presented to ensure Green Computing in WSN.

Rest of the paper organization is as follows. Related works have been briefly described in Sect. 2, the proposed solution methodology has been described in Sect. 3, a numerical result analysis has been presented in Sect. 4, and concluding remarks and future research direction are described in Sect. 5.

2 Related Works

It has been mentioned in Sect. 1 that energy efficiency is a primary concern in present days Fog Network based solutions. Several researchers and practitioners have presented different solutions to ensure minimized energy consumption. Among several existing state of the art literature, the efforts found in [3–5] have concisely presented a framework for energy efficiency in WSN/IoT (Internet of Things). A detailed discussion related to the probable applications and studies on Sensor Cloud (SC), WSNs, and related issues regarding green SC were presented in [4]. In another literature, we have found the detailed study on several evolutionary algorithms toward its possible uses in Green Computing [5]. Besides, an approach toward energy harvesting in autonomous sensors and WSNs has been described in [3].

In another research, Gajalakshmi and Srikanth [4] in 2016 discussed the different shortest path techniques by using ACO. Gao et al. [5] adopted a hybrid method called HM-ACOPSO which combines ant colony optimization (ACO) and particle swarm optimization (PSO) to schedule an efficient moving path for the mobile agent. Chu et al. [7] improved the traditional ant colony algorithm and proposed a modified ACO algorithm for lower energy consumption in a sensor network. Akbas et al. [8] developed a realistic WSN link layer model which is built on the top of the empirically verified energy dissipation characteristics of Mica2 motes and WSN channel models.

3 Proposed Solution Methodology

Our proposed research is based on an enhanced version of ACO algorithm. The proposed ACO based design has been presented in the following Fig. 2. Due to page limitation, Fig. 2 has not been explained in texts.

Our main aim is to minimize the energy consumption of the WSN nodes. There are so many factors that we have to concentrate on. To design an efficient WSN network, we have categorized the network design into three subsections and those are as followed.

3.1 Design Efficient WSN Nodes

While designing efficient WSN nodes it should keep in mind that the node should not consume higher energy food processing as well as communication purpose. To process data, the WSN node requires some energy and also data from one node to another WSN node which requires some amount of energy. In this scenario, we have to design the WSN in such a way so that we can minimize the WSN energy consumption.

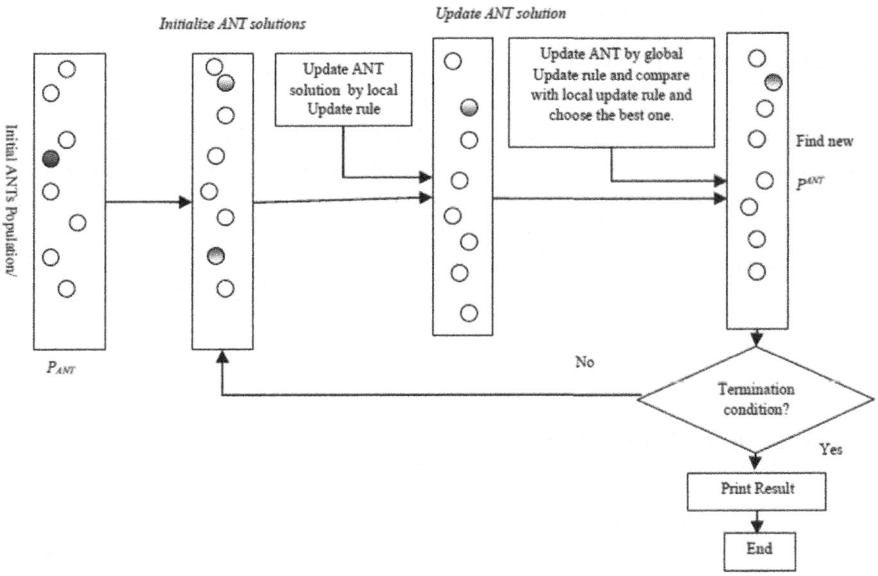

Fig. 2 A typical block diagram of the modified ACO algorithm (inspired from [6])

3.2 *Clusterization of the Target Area*

Choosing the target area is an essential task while designing a WSN. After choosing that target area, the area has been divided into some clusters for better monitoring purposes. In our experiment, we have used two types of clusterization policies and those are as follows:

3.2.1 Using Square Pattern Clusterization Strategy

In the case of the square pattern clusterization strategy, the target area is divided into equal square blocks. In this type of strategy, we have used the approximation method to calculate the number of the block along the length or breadth of the target area, because the target area might not be a uniform one.

3.2.2 Using Hexagonal Pattern Clusterization Strategy

In this type of hexagonal pattern clusterization strategy, the entire target area is divided into uniform hexagonal wax cell of honeycomb-like blocks.

3.3 Deployment of WSN Nodes Using Spiral Pattern WSN Node Deployment Policy

Deployment of WSN nodes is the second step toward the design of WSN network building. The deployment is done in such a manner so that we can get a very efficient network. Though there exist so many strategies, in our experiment, we have used Spiral pattern WSN node deployment policy.

Spiral pattern WSN node deployment policy is such a kind of policy where deployment is done starting from a corner of the target area and ultimately it goes to the core section of the network.

4 Numerical Result Analysis

Using the ACO algorithm, choosing the optimized path for the minimization of energy consumption for transmitting the data as well as receiving data: ACO algorithm is used to choose the optimized path for the minimization of energy consumption for transmitting the data. The linear problem is as described in [9]. The energy consumption during successful data transmission between the cluster head to the sink node has been calculated and minimized using the below-mentioned equations:

$$f(C, D) = Minimization\left(Energy_{transmission}^{Total}(C, D)\right) \tag{3}$$

Subject to

$D \leq D_0$, for free-space propagation model (in the case of Fog Network we need not consider ground propagation model).

Where,

$$Energy_{transmission}^{Total}(C, D) = Energy_{transmission}^{Total}(C, D) + Energy_{transmission}^{Total}(C)$$

$$\begin{aligned} Energy_{transmission}^{Total}(C, D) = {} & Energy_{transmission}^{SinkNode-ClusterHead}(C, D) \\ & + Energy_{transmission}^{ClusterHead-ClusterHead}(C, D) \end{aligned}$$

$$\begin{aligned} Energy_{Receive}^{Total}(C) = {} & Energy_{Receive}^{SinkNode-ClusterHead}(C) \\ & + Energy_{Receive}^{ClusterHead-ClusterHead}(C) \end{aligned}$$

For Free-space propagation model

$$\begin{aligned} Energy_{transmission}^{SinkNode-ClusterHead}(C, D) = {} & Energy_{Electronic-energy}^{SinkNode-ClusterHead}(C) * C \\ & + Energy_{Amplifier}^{SinkNode-ClusterHead}(C, D) * C \end{aligned}$$

$$Energy_{Receive}^{SinkNode-ClusterHead}(C) = Energy_{Electronic-energy}^{SinkNode-ClusterHead}(C) * CEnergy_{Amplifier}^{SinkNode-ClusterHead}(C, D) = Energy_{fs}^{SinkNode-ClusterHead} * D^2$$

Therefore, the problem can be defined as the objective function as follows.

$f(C, D) = Minimization\left(Energy_{transmission}^{Total}(C, D)\right)$ subject to $D <= D_o$, where D_o is the threshold transmission distance.

Where

$$Energy_{transmission}^{Total}(C, D) = Energy_{transmission}^{Total}(C, D) + Energy_{Receive}^{Total}(C)$$

$$Energy_{transmission}^{Total}(C, D) = Energy_{transmission}^{SinkNode-ClusterHead}(C, D) + Energy_{transmission}^{ClusterHead-ClusterHead}(C, D)$$

$$Energy_{Receive}^{Total}(C) = Energy_{Receive}^{SinkNode-ClusterHead}(C)$$

$$Energy_{transmission}^{SinkNode-ClusterHead}(C, D) = Energy_{Electronic-energy}^{SinkNode-ClusterHead}(C) * C + Energy_{Amplifier}^{SinkNode-ClusterHead}(C, D) * C$$

$$Energy_{Receive}^{Total}(C) = Energy_{Receive}^{SinkNode-ClusterHead}(C)$$

$$Energy_{Receive}^{SinkNode-ClusterHead}(C) = Energy_{Electronic-energy}^{SinkNode-ClusterHead}(C) * CEnergy_{Amplifier}^{SinkNode-ClusterHead}(C, D) = Energy_{fs}^{SinkNode-ClusterHead} * D^2$$

$Energy_{Amplifier}^{SinkNode-ClusterHead}$ = energy required for the transmitting data packets between sink node and cluster head for the amplifier to maintain an acceptable signal-to-noise ratio in order to transfer data messages reliably.

$Energy_{Electronic-energy}^{SinkNode-ClusterHead}$ = Electronic energy degenerated during the transmission between sink node and adjacent cluster head.

Energy = amount of energy used by each node at the time of transmitting data packets.

$Energy_{Receive}$ = energy used for receiving data packets.

Measurement of distance between two cluster heads is done using the following formula

$$D_{ab} = \sqrt{(a1 - a2)^2 + (b1 - b2)^2}$$

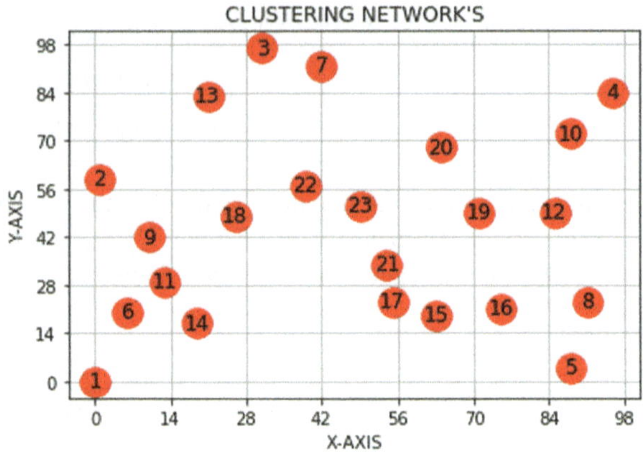

Fig. 3 Deployment of WSN nodes using spiral pattern deployment policy

where $(a1, b1)$ and $(a2, b2)$ are coordinates of reference nodes and D_{ab} is the distance measured between two adjacent cluster heads and the notation D_{ab} and D are the same.

For the sake of our experiment, we have considered the nature of all nodes as static Terrestrial WSNs [10], where we can do the deployment of the nodes as per our requirement which may be called organized deployment. In the time of deployment, we also carefully calculated the feasible range of transmitting the power of nodes and optimized the numbers of nodes to form the cell of the WSN network. In our experiment the structure of cell is hexagonal and the justification this type of shape has been defined previously. In this way, we have calculated that the boundary cell should have more nodes than that of internal nodes of the network. In the below-mentioned figure we have depicted the structure of the network which is called Grid representation (see Fig. 3). The following parameter values are used in the experiment for simulating the system [9].

In Table 1, the scale of energy is different for the initial energy, $E_{electronic-energy}$ and E_{fs} and those scales are Joule, Nano Joule, and Peta-Joule, respectively. So for maintaining equivalency all calculations have been done in Peta-Joule in Tables 2 and 3. In this section, we have plotted the best path of shortest distances (see Fig. 4) obtained from the ACO algorithm by solving the equations in Sect. 5 on the basis of the data supplied in Table 1. As the energy consumption is directly proportional to

Table 1 Parameters for simulation

Parameter	Value	Parameter	Value
Size of target area	100×100 m^2	Data packet size (k)	512 bytes
Total number of clusters	36	Max no. of sensor nodes (in the network)	23
Initial energy	1 J	$E_{electronic-energy}$	50 nJ/bit
E_{fs}	10 pJ bit^{-1} m^{-2}	Deployment time interval in millisecond	10

Table 2 Data for the communication between sink node and first cluster head

Path number_(sink node)- > (row number, Column number)	Calculated threshold distance (d0) in meter	Obtained distance (dxy) between sink node and cluster head in the feasible range (d0 to d_{ab}) after applying ACO algorithm	Sink node coordinate (x1,y1) for d0	Cluster node coordinate (x2, y2) for d0	Sink node co-ordinate (x1, y1) for d_{ab}	Cluster node coordinate (x2, y2) for d_{ab}	Minimized E_{tx}^{SN-CH} (in pJ)	Calculated E_{tx}^{SN-CH} (in pJ)	Calculated E_{rx}^{SN-CH} (in pJ)
p1_(0,0)->(r1,c1)	2.134374778	2.09065174	0, 0	1,668	0, 0	14.2, 11.2	2292349156	2291668164	2275555556
p2_(0,0)->(r1,c2)	3.081205472	3.04417071	0, 0	2,779	0, 0	24.5, 11.5	2309717333	2310553600	
p3_(0,0)->(r1,c3)	2.134374778	2.09065174	0, 0	1,668	0, 0	34.2, 11.2	2292349156	2291668164	
p4_(0,0)->(r1,c4)	3.081205472	3.04417071	0, 0	2,779	0, 0	44.4, 11.4	2309717333	2310553600	
p5_(0,0)->(r1,c5)	2.134374778	2.09065174	0, 0	1,668	0, 0	54.6, 11.6	2292349156	2291668164	
p6_(0,0)->(r1,c6)	3.081205472	3.04417071	0, 0	2,779	0, 0	64.6, 11.6	2309717333	2310553600	
p7_(0,0)->(r1,c7)	2.134374778	2.09065174	0, 0	1,668	0, 0	74.5, 11.5	2292349156	2291668164	
Path number_(sink node)->(row number, Column number)	Total energy consumption before optimization for specific node (in pJ)			Total energy consumption after optimization for specific node (in pJ)			Difference between total energy consumption before optimization and total energy consumption after optimization for one node		
p1_(0,0)->(r1,c1)	4567904711			4567223719			680991.8		
p2_(0,0)->(r1,c2)	4586109156			4585272889			8362666.67		
p3_(0,0)->(r1,c3)	4567904711			4567223719			680991.8		
p4_(0,0)->(r1,c4)	4586109156			4585272889			8362666.67		

(continued)

Table 2 (continued)

Path number_(sink node)- > (row number, Column number)	Calculated threshold distance (d0) in meter	Obtained distance (dxy) between sink node and cluster head in the feasible range (d0 to d_{ab}) after applying ACO algorithm	Sink node coordinate (x1,y1) for d0	Cluster node coordinate (x2, y2) for d0	Sink node co-ordinate (x1, y1) for d_{ab}	Cluster node coordinate (x2, y2) for d_{ab}	Minimized E_{tx}^{SN-CH} (in pJ)	Calculated E_{tx}^{SN-CH} (in pJ)	Calculated E_{rx}^{SN-CH} (in pJ)
p5_(0,0)->(r1,c5)	4567904711			4567223719			680991.8		
p6_(0,0)->(r1,c6)	4586109156			4585272889			8362666.67		
p7_(0,0)->(r1,c7)	4567904711			4567223719			680991.8		
Energy saving using ACO algorithm after optimizing the consumed energy for the communication between sink node and cluster head of every path							786279629.8		

Table 3 Data for the communication between intermediate cluster heads and sink no

Path number_(sink node)- > (row number, Column number)	Calculated threshold distance (d0) in meter	Distance (d_{xy}) between sink node and cluster head in the feasible range (d0 to d_{xy}) after applying ACO algorithm	Minimized E_{tx}^{CH-CH}(in pJ)	Calculated E_{tx}^{CH-CH}(in pJ)	Calculated E_{rx}^{CH-CH}(in pJ)
p1_(r1,c1)->(r2,c2)	0.942	0.86164	22427803238	22433152×10^3	224×10^8
p1_(r2,c2)->(r3,c1)		0.93155	22432498278		
p1_(r3,c1)->(r4,c2)		0.87381	22428594586		
p1_(r4,c2)->(r5,c1)		0.93270	22432578560		
p1_(r5,c1)->(r6,c2)		0.88599	22429397402		
p1_(r6,c2)->(r7,c1)		0.93401	22427803238		
p2_(r1,c3)->(r2,c4)	0.956	0.88213	22800582453	22805619237	22771916800
p2_(r2,c4)->(r3,c3)		0.93255	22432507686		
p2_(r3,c3)->(r4,c4)		0.88295	22428994560		
p2_(r4,c4)->(r5,c3)		0.93240	22432600422		
p2_(r5,c3)->(r6,c4)		0.87990	22429904986		
p2_(r6,c4)->(r7,c3)		0.93211	22432646298		
p3_(r1,c5)->(r2,c6)	0.488	0.45030	11639081041	1164165218	11641652183
p3_(r2,c6)->(r3,c5)		0.48394	11641349064		
p3_(r3,c5)->(r4,c6)		0.45820	11639599027		
p3_(r4,c6)->(r5,c5)		0.48386	11641343624		
p3_(r5,c5)->(r6,c6)		0.45662	11639494685		

(continued)

Table 3 (continued)

Path number_(sink node)- > (row number, Column number)	Calculated threshold distance (d0) in meter	Distance (d_{xy}) between sink node and cluster head in the feasible range (d0 to d_{xy}) after applying ACO algorithm	Minimized E_{tx}^{CH-CH}(in pJ)	Calculated E_{tx}^{CH-CH}(in pJ)	Calculated E_{rx}^{CH-CH}(in pJ)
p3_(r6,c6)->(r7,c5)		0.48371	11641333022		
p4_(r1,c7)->(r2,c8)	0.388	0.361	9260615635	11641652183	9248768000
p4_(r2,c8)->(r3,c7)		0.385	9262247376		
p4_(r3,c7)->(r4,c8)		0.360	9260533358		
p4_(r4,c8)->(r5,c7)		0.384	9262206423		
p4_(r5,c7)->(r6,c8)		0.363	9260739605		
p4_(r6,c8)->(r7,c7)		0.384	9262210677		
p5_(r1,c9)->(r2,c10)	0.47128	0.43475	11237167742	1123950099	11223040000
p5_(r2,c10)->(r3,c9)		0.46739	11239368221		
p5_(r3,c9)->(r4,c10)		0.43551	11237217348		
p5_(r4,c10)->(r5,c9)		0.46731	11239362789		
p5_(r5,c9)->(r6,c10)		0.44238	11237667841		
p5_(r6,c10)->(r7,c9)		0.46708	11239347032		
p6_(r1,c11)->(r2,c12)	0.26556	0.24455	6332157132	6333583852	6324224000
p6_(r2,c12)->(r3,c11)		0.26312	6333407406		
p6_(r3,c11)->(r4,c12)		0.24799	6332381768		
p6_(r4,c12)->(r5,c11)		0.26320	6333413072		

(continued)

Table 3 (continued)

Path number_(sink node)- > (row number, Column number)	Calculated threshold distance (d0) in meter	Distance (d_{xy}) between sink node and cluster head in the feasible range (d0 to d_{xy}) after applying ACO algorithm	Minimized E_{tx}^{CH-CH}(in pJ)	Calculated E_{tx}^{CH-CH}(in pJ)	Calculated E_{rx}^{CH-CH}(in pJ)
p6_(r5,c11)->(r6,c12)		0.24541	6332212987		
p6_(r6,c12)->(r7,c11)		0.26360	6333441076		
p7_(r1,c13)->(r2,c14)	0.34847	0.32428	8309126340	8310777774	8298496000
p7_(r2,c14)->(r3,c13)		0.34559	8310569349		
p7_(r3,c13)->(r4,c14)		0.32259	8309015705		
p7_(r4,c14)->(r5,c13)		0.34521	8310542628		
p7_(r5,c13)->(r6,c14)		0.32485	8309163351		
p7_(r6,c14)->(r7,c13)		0.34526	8310546246		
Path number_(sink node)->(row number, Column number)	Total energy consumption before optimization for specific node (in pJ)		Total energy consumption after optimization for specific node (in pJ)		Difference between total energy consumption before optimization and total energy consumption after optimization for one node
p1_(r1,c1)->(r2,c2)	44833152×10^3		44827803238		5348761.6
p1_(r2,c2)->(r3,c1)			44832498278		653721.6
p1_(r3,c1)->(r4,c2)			44828594586		4557414.4
p1_(r4,c2)->(r5,c1)			44832578560		573440
p1_(r5,c1)->(r6,c2)			44829397402		3754598.4

(continued)

Table 3 (continued)

Path number_(sink node)- > (row number, Column number)	Calculated threshold distance (d0) in meter	Distance (d_{xy}) between sink node and cluster head in the feasible range (d0 to d_{xy}) after applying ACO algorithm	Minimized E_{tx}^{CH-CH}(in pJ)	Calculated E_{tx}^{CH-CH}(in pJ)	Calculated E_{rx}^{CH-CH}(in pJ)
p1_(r6,c2)->(r7,c1)			44832670310		481689.6
p2_(r1,c3)->(r2,c4)	45577536037		45572499253		5036783.645
p2_(r2,c4)->(r3,c3)			44832507686		644313.6
p2_(r3,c3)->(r4,c4)			44828994560		4157440
p2_(r4,c4)->(r5,c3)			44832600422		551577.6
p2_(r5,c3)->(r6,c4)			44829904986		3247014.4
p2_(r6,c4)->(r7,c3)			44832646298		505702.4
p3_(r1,c5)->(r2,c6)	23266100183		23263529041		2571141.906
p3_(r2,c6)->(r3,c5)			23265797064		303119.106
p3_(r3,c5)->(r4,c6)			23264047027		2053156.504
p3_(r4,c6)->(r5,c5)			23265791624		308559.3477
p3_(r5,c5)->(r6,c6)			23263942685		2157497.549
p3_(r6,c6)->(r7,c5)			23265781022		319160.8443
p4_(r1,c7)->(r2,c8)	18511224177		18509383635		1840541.827
p4_(r2,c8)->(r3,c7)			18511015376		208800.1864
p4_(r3,c7)->(r4,c8)			18509301358		1922818.867
p4_(r4,c8)->(r5,c7)			18510974423		249753.7311

(continued)

Table 3 (continued)

Path number_(sink node)- > (row number, Column number)	Calculated threshold distance (d0) in meter	Distance (d_{xy}) between sink node and cluster head in the feasible range (d0 to d_{xy}) after applying ACO algorithm	Minimized E_{tx}^{CH-CH}(in pJ)	Calculated E_{tx}^{CH-CH}(in pJ)	Calculated E_{rx}^{CH-CH}(in pJ)
p4_(r5,c7)->(r6,c8)			18509507605		1716571.341
p4_(r6,c8)->(r7,c7)			18510978677		245499.2978
p5_(r1,c9)->(r2,c10)	22462690099		22460207742		2086485.066
p5_(r2,c10)->(r3,c9)			22462408221		258040.6272
p5_(r3,c9)->(r4,c10)			22460257348		1757968.826
p5_(r4,c10)->(r5,c9)			22462402789		249753.7311
p5_(r5,c9)->(r6,c10)			22460707841		2004799.947
p5_(r6,c10)->(r7,c9)			22462387032		208800.1864
p6_(r1,c11)->(r2,c12)	12657807852		12656381132		1426719.638
p6_(r2,c12)->(r3,c11)			12657631406		176445.8496
p6_(r3,c11)->(r4,c12)			12656605768		1202083.201
p6_(r4,c12)->(r5,c11)			12657637072		170779.3449
p6_(r5,c11)->(r6,c12)			12656436987		1370864.091
p6_(r6,c12)->(r7,c11)			12657665076		142775.681
p7_(r1,c13)->(r2,c14)	16609273774		16607622340		1651433.898
p7_(r2,c14)->(r3,c13)			16609065349		208425.0255
p7_(r3,c13)->(r4,c14)			16607511705		1762069.447

(continued)

Table 3 (continued)

Path number_(sink node)- > (row number, Column number)	Calculated threshold distance (d0) in meter	Distance (d_{xy}) between sink node and cluster head in the feasible range (d0 to d_{xy}) after applying ACO algorithm	Minimized E_{tx}^{CH-CH}(in pJ)	Calculated E_{tx}^{CH-CH}(in pJ)	Calculated E_{rx}^{CH-CH}(in pJ)
p7_(r4,c14)->(r5,c13)			16609038628		235146.1827
p7_(r5,c13)->(r6,c14)			16607659351		1614422.606
p7_(r6,c14)->(r7,c13)			16609042246		231528.0384

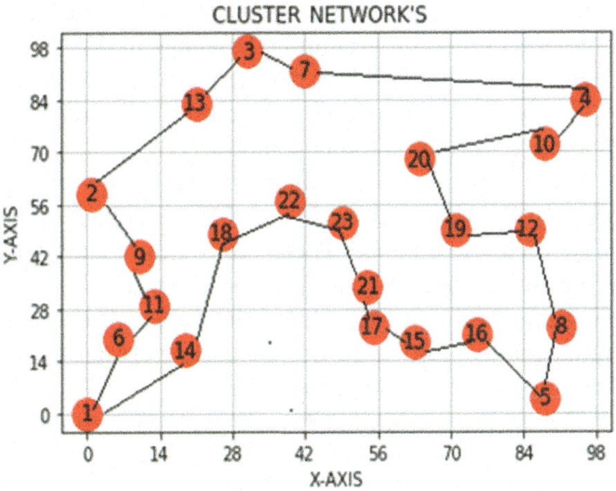

Fig. 4 Path Covered in the cluster representation of the network

distance between nodes that is why we have calculated maximum coverage area. We obtained the coordinates of sink node and cluster head as depicted in Tables 2 and 3. Table 1 shows the communication between sink node and cluster head, whereas Table 2 shows the communication between adjustment cluster heads. In the below diagram, we are going to show the first phase of path selection until we get the minimum consumed energy for the shortest distance between nodes. Here the first phase denotes the consumption of full initial energy (here 1 J) of a node so that we can declare the node as dead node and in the second phase we will not consider that node as participating nodes.

Among two types of propagation models, i.e., two ray propagation model and free-space propagation models, the second one is related to distance between communicating nodes; therefore in our experiment we have chosen only free-space propagation model to show the result. The representation of a node path has been denoted using Path number, source node coordinator, and cell index of destination node. Here cell index is denoted as row and column number of the designated cell. The first cell has been considered as the nearest cell to the sink node and it has been denoted as (r1, c1) which means row number one and column number one. Please note that 'r' represents root node and 'c' represents cluster head.

For example, p1_(0,0)- > (r1, c1) denotes the communication path number one and it is the communication between the source node (that is the sink node) and the destination node (cell of the row number one and column number one).

As the path number is associated with a specific communication path therefore we did not give each and individual node number in short, so when the new node of the cell (r_1, c_1) is selected the path number obviously will be changed. The numerical result related to the supplied data from Table 1 data has been calculated and represented in Table 2 (for Free-space Propagation model ($d_{ab} \leq d_O$)).

The proposed ACO algorithm has been run for 50 times to get the best result among those run. The problem has been solved considering a different set of random numbers in the feasible set of constraints. The proposed algorithm has been solved using the Python programming language in the Anaconda environment of Windows operating system. In this experiment/simulation, a run is considered a successful run if we obtained the solution of the problem, either the same valued result or better than the known best-found solution.

From Table 2, we can calculate total energy saving for the first phase of communication between sink node and different cluster head to cover the whole area that is 786,279,628.8 pj/sec. Therefore, we have calculated that 2–3 days of the lifetime of the whole network can be saved with respect to un-optimized Wireless Sensor Network. In the second phase, the path will be changed after consumption of the initial energy store of the corresponding cluster head and so on up to the full energy draining of the selected cluster head.

The research work has been compared with the work of Lande and Kawale [9]. There is a remarkable improvement in energy saving using the proposed algorithm as well as network in comparison to the work of Lande and Kawale [9].From the paper it has been calculated that a maximum 1–2 days of lifetime can be achieved, whereas using the suggested algorithm and network 2–3 days of the lifetime of the whole network can be saved with respect to un-optimized Wireless Sensor Network [9].

5 Conclusion and Future Work

The main aim of this paper is to minimize the energy consumption as well as maximize the target coverage area in Wireless Sensor Networks. To minimize the power consumption of a WSN at first, we believe that we have received success to minimize the traversal path to cover each cell of the particular path as well as the traversal path between sink node and cells. The minimizing technique that we have used is ant colony optimizations. In this network configuration, we have used only random deployment of WSN nodes. For this reason, the minimization of energy consumption is achieved as a result of minimized traversal path, which further enhances energy efficiency of WSN (i.e., Fog Network). Thus, we believe that our proposed approach facilitates Green Computing in Fog Network as it ensures energy efficiency in WSN.

As already described, we have presented our work based on random deployment only. In future, we have a plan to extend this presented work and compare its performance with reference to the other types of deployment processes (e.g., S—pattern deployment and spiral deployment).

Acknowledgements Authors sincerely thank the anonymous reviewers for their suggestions which have further enhanced the quality of this manuscript.

References

1. Bonomi F, Milito R, Zhu J, Addepalli S (2012) Fog computing and its role in the internet of things. In: Proceedings of the first edition of the MCC workshop on mobile cloud computing, August 2012. ACM, pp 13–16
2. Oma, R., Nakamura, S., Duolikun, D., Enokido, T., Takizawa, M.: An energy-efficient model for fog computing in the internet of things (IoT). Internet of Things **1**, 14–26 (2018)
3. Qian, H., Sun, P., Rong, Y.: Design proposal of self-powered WSN node for battle field surveillance. Energy Proc **16**, 753–757 (2012)
4. Gajalakshmi G, Srikanth GU (2016) A survey on the utilization of Ant Colony Optimization (ACO) algorithm in WSN. In: Proceedings of the international conference on information communication and embedded systems (ICICES), February 2016. IEEE, pp 1–4
5. Gao, Y., Wang, J., Wu, W., Sangaiah, A.K., Lim, S.J.: A hybrid method for mobile agent moving trajectory scheduling using ACO and PSO in WSNs. Sensors **19**(3), 575 (2019)
6. Banerjee A, Chattopadhyay S, Mukhopadhyay AK, Gheorghe G (2016) A fuzzy-ACO algorithm to enhance reliability optimization through energy harvesting in WSN. In: Proceeding of the international conference on electrical, electronics, and optimization techniques (ICEEOT), March 2016. IEEE, pp 584–589
7. Chu, K.C., Horng, D.J., Chang, K.C.: Numerical optimization of the energy consumption for wireless sensor networks based on an improved ant colony algorithm. IEEE Access **7**, 105562–105571 (2019)
8. Akbas, A., Yildiz, H.U., Tavli, B., Uludag, S.: Joint optimization of transmission power level and packet size for WSN lifetime maximization. IEEE Sens J **16**(12), 5084–5094 (2016)
9. Lande SB, Kawale SZ (2016) Energy efficient routing protocol for wireless sensor networks. In: Proceedings of the 8th international conference on computational intelligence and communication networks (CICN), December 2016. IEEE, pp 77–81
10. Srivastava N (2010) Challenges of next-generation wireless sensor networks and its impact on society. https://arxiv.org/abs/1002.4680

A Model for Probabilistic Prediction of Paddy Crop Disease Using Convolutional Neural Network

Sujay Das, Ritesh Sharma, Mahendra Kumar Gourisaria, Siddharth Swarup Rautaray, and Manjusha Pandey

Abstract Agriculture is the backbone of the human society as it is an essential need of every creature present on this planet. Rice cultivation plays a vital role in the life of humans, especially in the Indian Subcontinent. So, it is necessary for humans to protect the importance and productivity of agriculture. The IT industry is very significant for the agriculture and especially for the healthcare of agricultural industry. Machine Learning and Artificial Intelligence is a buzzword in the IT industry and now these buzzwords have helped a lot in the field of agriculture. These technologies give foundation to various aspects of decision making and can be productive for every industry. This paper presents a model that uses the CNN algorithm of Deep Learning for the prediction of paddy crop disease. This model also finds the probability of the occurrence of disease which can be helpful to take some vital decisions related to plant's health.

Keywords Convolutional neural network · Image processing · Rice plant · Deep learning · Disease detection

S. Das (✉)
School of Electronics Engineering, KIIT Deemed to be University, Bhubaneswar, India
e-mail: dasujoy1999@gmail.com

R. Sharma · M. K. Gourisaria · S. S. Rautaray · M. Pandey
School of Computer Engineering, KIIT Deemed to be University, Bhubaneswar, India
e-mail: sharmaritesh3312@gmail.com

M. K. Gourisaria
e-mail: mkgourisaria2010@gmail.com

S. S. Rautaray
e-mail: siddharthfcs@kiit.ac.in

M. Pandey
e-mail: manjushafcs@kiit.ac.in

D. Mishra et al. (eds.), *Intelligent and Cloud Computing*,
Smart Innovation, Systems and Technologies 194,
https://doi.org/10.1007/978-981-15-5971-6_12

1 Introduction

Rice is also known as one of the chief grains of India. India has the largest area under rice cultivation, as it is one of the important food crops. Facts say that India is one of the leading producers of rice crops. But nowadays it has been a serious challenge to grow healthy rice crop because of various factors that leads to severe diseases on the plant. Those various factors are climatic conditions, pollution, human faults, etc. This paper focuses on the detection of the severe diseases that often occur in rice plants. It is important to identify the disease from the early stage and gets rid of it as it will be almost impossible for the plant to survive once the disease becomes more severe and the possibilities of the disease to spread will also increase.

The proposed model detects and classifies among the common disease that the plant might be suffering from. For the proposed model Convolutional Neural Networks is chosen because it works effectively on images. The dataset which is chosen contains the images of various infected leaves of paddy crop. The proposed model first augments the image and then trains on it. Then with the help of the image, it finds the probability of what disease the plant might be suffering from [1].

The proposed model is coded in Python programming language as it is one of the best choices for Data Science and Artificial Intelligence. The proposed model runs on the IPython Terminal as it provides the best performance for mathematical computations. The dataset that we have used is from UCI Machine Learning repository. It contains images of 3 types of diseases in paddy crop.

This paper is organized as follows:—Sect. 1 Discusses about the introduction, Sect. 2 discuss about the related work done in the field of crop prediction, Sect. 3 throws light on the different diseases that the model classifies, Sect. 4 explains how the model is implemented, Sect. 5 discusses about the algorithm, Sect. 6 discusses the results and Sect. 7 discusses the future work.

2 Literature Survey

Jayanthi et al. [2] proposed a model on analysis of automatic rice disease classification using Image processing techniques. Their paper has presented a detailed study of the different image classification algorithms.

Nithya et al. [3] proposes a model on symptoms based paddy crop disease prediction and recommendation system using big data. The diseases that are related are gathered from different websites and blogs. They are analyzed through Hadoop, hive tools and HiveQl. The collected documents are represented in the form vector using Vector Space model and calculate weight of the vector based on the TF-IDF ranking.

Suraksha et al. [4] proposes a technique for prediction paddy crop disease using data mining and image processing. In this paper, they have proposed a model for image processing followed by feature extraction and data mining techniques.

Rajmohan et al. [5] proposes a technique for smart paddy crop disease prediction using deep CNN and SVM classifier. In this paper, they have proposed a model for image processing followed by feature extraction and SVM classifier for classification. In which they have selected 250 models from which 50 are used for training and rest for testing In this paper they have developed a mobile app which clicks the image, zooms and crops it then uploads the image and the person receives a notification.

Barik [6] proposes a technique for region identification of Rice Diseases using image processing. In this paper, they have presented a model for identification of disease and region of infection using image processing and classification techniques like SVM classifier and Naive Bayes classifier. Prediction is done and severity various disease is found and then it is classified into different categories.

Jagan Mohan et al. [7] presents a technique for disease detection of plants using canny edge detection. In this paper they have proposed a model that uses canny edge detection algorithm to track the edge and then get the histogram value to detect the disease. It is concluded that this model periodically monitors the cultivated field.

Arumugam [8] proposes a model a predictive modeling approach for improving paddy crop productivity using data mining techniques. Their works aims in providing a predictive modeling which will help farmers to get high yield of paddy crops. They have used K-Means clustering and various decision tree classifiers and have applied them to meteorological data.

3 Paddy Crop Disease Classification

This research is done to find out what type of disease the rice plant might be suffering from. The diseases that the rice plant is suffering from summed up into three families that are Bacterial, Viral and Fungal. The major three types of diseases are targeted which are as follows:

3.1 Brown Spot

This disease can be recognized by the brown spots that form on the leaves of the rice plant. This disease can be identified by its symptoms like the death of seedlings, death of large areas of the leaf, brown spots or black spots on the grains. It comes under the fungal class. The damage that the disease causes are both quality and quantity [9].

3.2 Bacterial Leaf Blight

This disease can easily be recognized by looking at the yellow and white strips on the leaves. The yield loss can range up to 70% in the case of the rice plants. It can easily be known that the rice plant is suffering from the disease by looking at the youngest leaf which will be yellow if the plant will be suffering from bacterial leaf blight [10].

3.3 Leaf Smut

This is a minor disease and can be recognized by the appearance of minute, sooty, dull and angular patches on the leaves. The fungus almost covers the entire leaf surface of the older leaves. Leaf smut is caused by Entyloma orzyae. The spores reaching the leaves near the soil level causes infection. High nitrogen rates enhance the disease. The disease can be controlled by doing clean cultivation and by growing resistant varieties [11].

4 Implementation of the Model

Neural Networks are a set of complex algorithms, modeled after the human brain, that is designed to recognize patterns. They interpret sensory data through a kind of machine perception, labeling or clustering raw input. The patterns that are recognized by the neural networks are numerical which are contained in vectors, into which all the real-world data be it images, sounds, text or time series must be translated. The Convolutional Neural Networks are the best fit for working with images [12]. This model has been implemented in CNN using multiple layers. The detailed explanation of the steps is given below:

4.1 Image Generation and Augmentation

Image augmentation is a great way by which can be used to expand the data set artificially and expand the size of the training data set by generating different modified versions of the image present in the dataset. There are a lot of techniques for image augmentation out of which horizontal flip, vertical flip, shearing and brightening of images are performed.

4.2 Convolution Steps

In terms of mathematics, convolution is a function derived from two given functions by integration which expressed how the shape of one is modified by the other. The model gets an image as its input in the form of matrices. In the convolution step, a filter or kernel matrix is taken and convolution is performed with the input image matrix by sliding the filter over the input image. In this model, two convolution layers are used along with activation function ReLU [13]. Let us consider,

x: Input.
a^k: After convoluted image.
k: Index of kernel (weight filter).
W: Kernel (weight filter).
b: Bias.
E: Cost function.
For Forward Propagation:

$$a_{ij}^{(k)} = \sum_{s=0}^{m-1}\sum_{t=0}^{n-1} W_{st}^{(k)} x_{(i+s)(j+t)} + b^{(k)} \tag{1}$$

For Backward Propagation:

$$\frac{\partial E}{\partial W_{st}^{(k)}} = \sum_{i=0}^{M-m}\sum_{j=0}^{N-n} \frac{\partial E}{\partial a_{ij}^{k}} \frac{\partial a_{ij}^{k}}{\partial W_{st}^{(k)}} = \sum_{i=0}^{M-m}\sum_{j=0}^{N-n} \frac{\partial E}{\partial a_{ij}^{k}} x_{(i+s)(j+t)} \tag{2}$$

$$\frac{\partial E}{\partial b^{(k)}} = \sum_{i=0}^{M-m}\sum_{j=0}^{N-n} \frac{\partial E}{\partial a_{ij}^{k}} \frac{\partial a_{ij}^{k}}{\partial b^{(k)}} = \sum_{i=0}^{M-m}\sum_{j=0}^{N-n} \frac{\partial E}{\partial a_{ij}^{k}} \tag{3}$$

4.3 Pooling Step

One of the most important building blocks in CNN is the Pooling Step. One of the main objectives behind applying pooling is to reduce the spatial size of the image. The depth of the image remains intact as pooling is performed on each dimension. In this model, the technique max-pooling has been implemented [14].

Forward Propagation:

$$a_{ij} = max\left(0, x_{(i+s)(j+t)}\right) \tag{4}$$

Backward Propagation:

$$\frac{\partial E}{\partial x_{(i+s)(j+t)}} = \frac{\partial E}{\partial a_{ij}^{(k)}} \frac{da_{ij}^{(k)}}{dx_{(i+s)(j+t)}} = \begin{cases} \frac{\partial E}{\partial a_{ij}^{(k)}} & (a_{ij}^{(k)} = x_{(i+s)(j+t)}) \\ 0 & (Otherwise) \end{cases} \tag{5}$$

4.4 Fully Connected Layers

At the end of a CNN, the output of the last Pooling Layer acts as input to the so called Fully Connected Layer. A fully connected layer is used to map a matrix to a one-dimensional vector. A matrix gets converted into a long vector using a flatten() function and linear operations are performed in Dense() layer.

Relu() and Softmax() are used as Activation function.

Forward Propagation:

$$a_{ij} = ReLU(x) = max(0, x_{ij}) \tag{6}$$

Backward Propagation:

$$\frac{\partial E}{\partial x_{ij}} = \frac{\partial E}{\partial a_{ij}} \frac{\partial a_{ij}}{\partial x_{ij}} = \begin{cases} \frac{\partial E}{\partial a_{ij}} & (a_{ij} \geq 0) \\ 0 & (Otherwise) \end{cases} \tag{7}$$

4.5 Disease Prediction

This model classifies the image into three major diseases. So, at the last layer of the model, so there are three layers one for each. Each node will give the probability of the respective disease present in the image which is given as input. This output can be taken into consideration and some important decisions can be made based on these outputs. As the output of the proposed model is probability, hence it can be used for taking decisions related to health of the plant.

4.6 Process Flow Model

See Fig. 1.

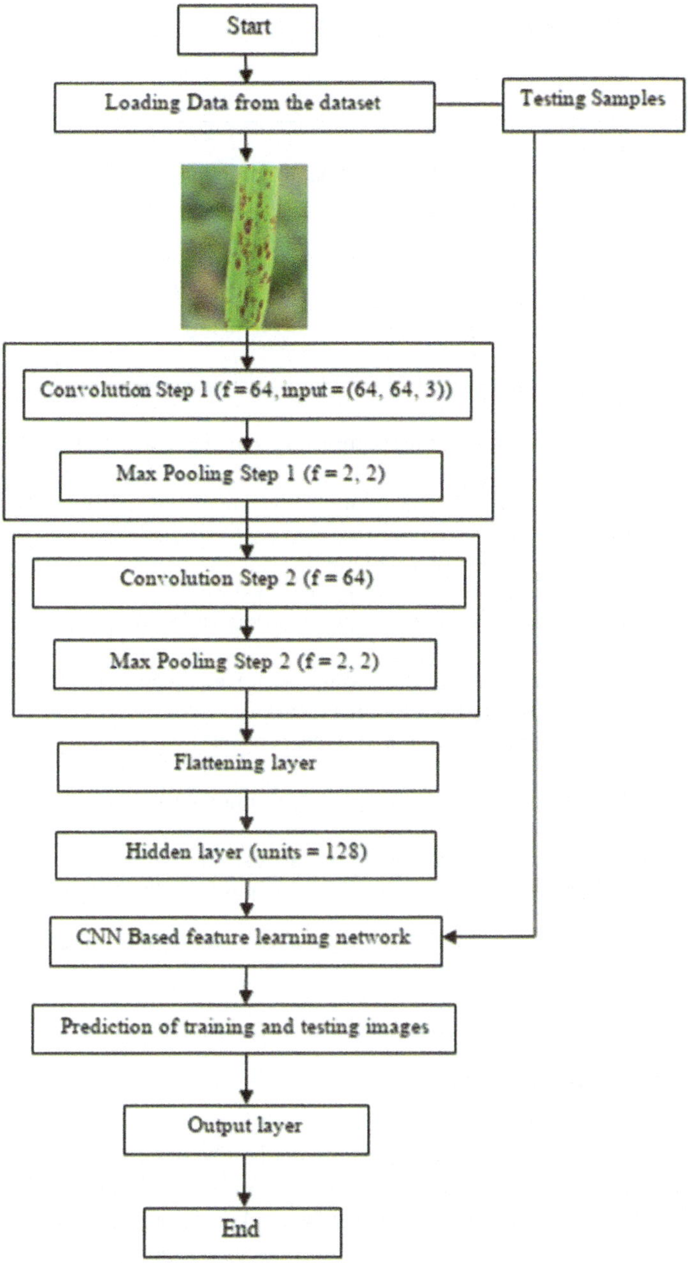

Fig. 1 Process flow model

5 Algorithm of the Model

PCDP Algorithm:
Input: A dataset with images of infected paddy crop leaves.
Output: A predicted value of disease.

Start
Step 1. Import necessary libraries
Step 2. Add a 2-dimensional convolution layer using conv2D() in the neural network
Step 3. Add a 2-dimensional max pooling layer using maxpooling2D() in the neural network
Step 4. Add a 2-dimensional convolution layer using conv2D() in the neural network.
Step 5. Add a 2-dimensional max pooling layer using maxpooling2D() in the neural network.
Step 6. Add a flattening layer using Flatten()
Step 7. Add a fully connected layer with an activation function of ReLu
Step 8. Add a dense layer using dense() with an activation function of Softmax
Step 9. Perform image augmentation using ImageDataGenerator() by Horizontal flip, vertical flip, brightening, shearing
Step 10. Load the images as training and test sets and load them in different Variables
Step 11. Fit the data and train the model on training dataset using fit_generator()
Step 12. Test the accuracy of the model on the test set.
Step 13. Predict the disease on a random image and print the output.
Stop

6 Experimental Results

After all the training and testing done, the results will be obtained from the last layer of the neural network. The work has focused on the major 3 kinds of paddy crop diseases. So, the output layer of the neural network consists of 3 nodes which will give the output of respective disease to occur. By using these probabilistic results, some vital decision and strategies can be taken which can help to cure the plant (Figs. 2, 3 and 4).

Fig. 2 Bacterial leaf blight

Fig. 3 Brown spot

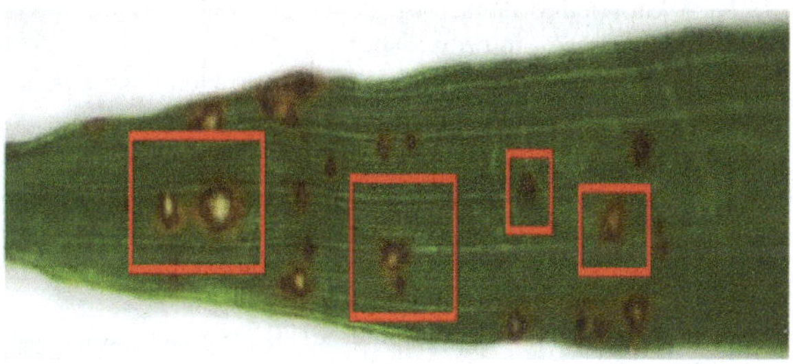

Fig. 4 Leaf smut

7 Conclusions and Future Work

Cultivation plays a very important role in the lives of human beings. It employs humans and it also provides food to humans as well as animals. As it important not only for humans to survive and also for animals so it must be protected this model uses Convolutional Neural Network to extract the feature from the leaves and predicts what type of disease the plant is suffering from. This model follows the principle of classification and prediction. There are a lot of scopes for this model to deploy on apps and websites. The person will have to upload a well-clicked image of the plant leaf in the website and immediately the predicted result can come along with the precaution measures as well as cure for this disease. Similarly an app can also be developed to function in the same way just as the website.

References

1. Dhaygude SB, Kumbhar NP (2013) Agricultural plant leaf disease detection using image processing. Int J Adv Res Electr Electron Instrum Eng 2(1)
2. Jayanthi G, Archana KS, Saritha A Analysis of automatic rice disease classification using image processing techniques. Int J Eng Adv Technol (IJEAT) 8(3S):15–20
3. Nithya S, Savithri S, Thenmozhi G, Shanmugham K (2017) Symptoms based paddy crop disease prediction and recommendation system using big data analytics. Int J Comput Trends Technol (IJCTT)
4. Suraksha IS, Sushma B, Sushma RG, Susmitha K, Uday Shankar SV (2016) Disease prediction of paddy crops using data mining and image processing techniques. Int J Adv Res Electr Electron Instrum Eng 5(6)
5. Rajmodhan R, Pajany M, Rajesh R, Raghuraman D, Prabu U (2018) Smart paddy crop disease identification using deep convolutional neural network and SVM classifier. Int J Pure Appl Math 118(15)
6. Barik L (2018) A survey on region identification of rice disease using image processing. Int J Res Sci Innov 5(1)
7. Jagan Mohan K, Balasubramanian M, Palanivel S (2016) Detection and recognition of diseases from paddy plat leaf. Int J Comput Appl 144(12)
8. Arumugam A (2017) A predictive modeling approach for improving paddy crop productivity using data mining techniques. Turk J Electr Eng Comput Sci
9. Rice knowledge bank. https://www.knowledgebank.irri.org/training/fact-sheets/pest-management/diseases/item/brown-spot
10. Rice knowledge bank. https://www.knowledgebank.irri.org/training/fact-sheets/pest-management/diseases/item/bacterial-blight?category_id=326
11. Rice knowledge management portal. https://www.rkmp.co.in/content/leaf-smut-2
12. Badage A (2018) Crop disease detection using machine learning: indian agriculture. Int Res J Eng Technol 5(9)
13. Jaswal D, Sowmya V, Soman KP (2014) Image classification using convolutional neural networks. Int J Adv Res Technol 3(6)
14. Scherer D, Muller A, Behnke S (2010) Evaluation of pooling operations in convolutional architectures for object recognition. In: International conference on artificial neural networks (ICANN)

A Hybridized ELM-Elitism-Based Self-Adaptive Multi-Population Jaya Model for Currency Forecasting

Smruti Rekha Das, Bidush Kumar Sahoo, and Debahuti Mishra

Abstract Forecasting currency market plays an important role basically in all aspects of international financial management. Based on the claimed poor performance and less efficiency of the popular forecasting methods, an Extreme Learning Machine (ELM)—elitism-based self-adaptive multi-population Jaya model—is designed with the possibilities of getting maximum prediction accuracy. The model has evaluated by using the exchange rate data of USD to INR and USD to EURO as in most of the countries USD is used as the base currency. The experimented model has explored its prediction ability over six-time horizon, which includes the 6 days between 1 day and 1 month in advance. For a standardized comparison, two broadly accepted prediction models such as standard ELM and Back-Propagation Neural Network (BPNN) has been recommended for this research work. Along with these two standard models, a Jaya optimized technique applying on both ELM and BPNN is also considered for experimentation. The simulated result depicted the outstanding performance of the proposed ELM-elitism-based self-adaptive multi-population Jaya model over ELM, BPNN, ELM-Jaya, and BPNN-Jaya.

Keywords Extreme learning machine (ELM) · Jaya · Currency forecasting · Back-propagation neural network (BPNN) · Prediction

---

S. R. Das (✉) · B. K. Sahoo
Gandhi Institute for Education and Technology, Bhubaneswar, India
e-mail: smrutirekhadas@gietbbsr.com

B. K. Sahoo
e-mail: bidushksahoo@gietbbsr.com

D. Mishra
Siksha 'O' Anusandhan (Deemed to be) University, Bhubaneswar, India
e-mail: debahutimishra@soa.ac.in

D. Mishra et al. (eds.), *Intelligent and Cloud Computing*,
Smart Innovation, Systems and Technologies 194,
https://doi.org/10.1007/978-981-15-5971-6_13

1 Introduction

Nowadays, exchange rate forecasting plays a vital role as it provides plenty of opportunities to generate money, but to be successful in this market better prediction model is to be designed. In the last decades, various forecasting models have designed in order to get better prediction accuracy, but the findings show its inability to predict due to the nonlinearity that exists in the financial data. Hence, a good [1] amount of nonlinear prediction model has been proposed in the last few years, with the intention of making better performance than simple random walk models. Among all theses prediction models, neural network is found to be a better prediction model with the encouraging results. The neural network model is having the inherent capabilities to handle the complex nonlinearity that exists in financial data. But the basic drawback in [2] back-propagation method is that the network is overtrained, and the learning speed is also very slow. It also suffers from iterative tuning using such [3] learning algorithm. These drawbacks of feed-forward neural network method can be overcome by Extreme Learning Machines (ELM), where at each iteration the hidden layer is not necessarily to be tuned. Hence, it is comparatively faster than the implementation of conventional popular learning methods.

This study is motivated to explore the predictive ability of ELM, as this algorithm provides a good generalization process of performing a task to a very great degree of [3] faster learning speed. Though ELM is having the potential to give better prediction in a state of being correct, the input weights and biases, which basically selected randomly, can create some nonoptimal solution. Hence, elitism-based self-adaptive multi-population Jaya optimization technique is used in this study, to optimize the input weight matrix and biases. Jaya is a metaheuristic-based optimization algorithm, which solves the constraint; additionally, the unconstrained optimization problem and the beauty of this algorithm is that any type of algorithm-specific parameter cannot control the algorithm; rather, it is controlled by some common controlling parameters. Though it is an efficient optimization technique, its search mechanism can be developed by using the subpopulation search scheme with elitism [4]. Here, for dividing the population into subpopulation, adaptive scheme is used. Through the division of the total population into a finite number of subpopulations, the diversity of search space can be improved [5]. But the changes in dividing the total population into a finite number of subpopulations, the self adapts that is used, depends upon the [4] strength of the solution.

This study has given emphasis over prediction of exchange rate of USD to INR and USD to EURO over six-time horizon such as 1 day, 3 days, 5 days, 7 days, 15 days, and 1 month in advance. For future prediction of the above exchange rate data, the proposed ELM-elitism-based self-adaptive multi-population Jaya is considered in this experimental work. For a significant proof of better accuracy, a comparison is established among the proposed model and standard BPNN and ELM along with optimized BPNN and ELM. BPNN and ELM models are used for comparison as they have proven their efficiency over prediction of time series data and also have accepted widely in various application domains. For a standard comparison, Jaya algorithm is used for optimizing the weights and biases of both BPNN and ELM.

The paper is organized in the following manner; Sect. 2 deals with the details of exchange rate data and, Sect. 3 describes the technologies that have adopted for the experimental purpose. Section 4 expresses the details of data used in this experimental work and the experimental analysis along with the resulting findings, and finally Sects. 5 and 6 cover the discussions and conclusions of the work with the future scope, respectively.

2 Methodology Adopted

The experimental work has carried out based on the ELM and Jaya, where the efficiency of ELM is merged with the multi-population based on the searching scheme of Jaya. Here, basically the subpopulation concept is adopted with the intention of improving the diversity of the search space [5], but for division of a number of subpopulations [4], the self-adaptive concept, which is totally based on the strength of the solution is used. In this section, some details about ELM and Jaya are described.

2.1 ELM

ELM follows the generalized principle of feed-forward neural network, but the basic difference of ELM from feed-forward neural network is that in ELM the hidden layer nodes are not necessarily be updated in the next iteration. The basic steps to implement ELM are as follows[6, 7]:

Step 1: The input and output data are given in training input, where S = $\{(X_k, Y_k)|X_k \in R^n,\ Y_k \in R^n, k = 1, 2, 3, 4, \ldots, N\}$, $g(x)$ is the activation function, and L is the hidden layer nodes.
Step 2: w_j is the input weight and b_j is the bias, where j = 1, 2,......, L which is randomly assigned to the experimentation.
Step 3: The output matrix H is calculated by using Eq. (1)

$$H = \begin{bmatrix} g(w_1.x_1 + b_1) & \ldots & g(w_L.x_1 + b_L) \\ \vdots & \ddots & \vdots \\ g(w_1.x_N + b_1) & \ldots & g(w_L.x_N + b_L) \end{bmatrix}_{N \times L} \tag{1}$$

Step 4: Output β is calculated by using the Eq. (2)

$$\beta = H^{\dagger} Y \tag{2}$$

Here, $Y = [Y_1, Y_2, \ldots\ldots\ldots Y_N]^T$ and $H^{\dagger}$ is the Moore–Penrose generalized inverse of H.

2.2 *Jaya*

Jaya is a metaheuristic learning algorithm, developed by Venkata Rao [8], where very minimum numbers of parameters are required by the algorithm. It does not require any type of parameter, which is specified by some algorithm; rather, it focuses on the parameter, which is to be controlled commonly. The basic concept is that the solution always moves toward the best by selecting the best one and keeping away the failure by back away from the worst solution. The complexity of Jaya is very less as it moves toward the best with a very simple step of movement. After finding the best and worst value using the objective function, Eq. (3) is used to update the position toward the best.

$$x'_{j,k,i} = x_{j,k,i} + r_{1,j,i}\left[\left(x_{j,\text{best},i}\right) - \left|\left(x_{j,k,i}\right)\right|\right] - r_{2,j,i}\left[\left(x_{j,\text{worst},i}\right) - \left|\left(x_{j,k,i}\right)\right|\right] \quad (3)$$

3 Dataset Description

Two different exchange rate datasets such as USD to INR and USD to EURO have been well thoughtout for this experimental work. The range of data for USD/INR has considered for the period of May 4, 2000–June 9, 2018 and for USD/EURO, September 14, 2000–June 9, 2018. The detailed elaboration of the datasets is shown in Table 1.

4 Experimental Analysis

This section analyzes the basic work of predicting the exchange rate data in the following manner. At first, it has described the entire work in a pictorial manner in the form of schematic layout, which is presented in Fig. 1, and then the algorithm of the proposed model is delineated in detail. The result analysis part is followed after

Table 1 Description of data samples and data range

Datasets	Samples	Range of data	Training sample	Test sample
USD to INR	4531	04/05/2000–09/06/2018	3171	1360
USD to EURO	4072	14/09/2000–09/06/2018	2850	1222

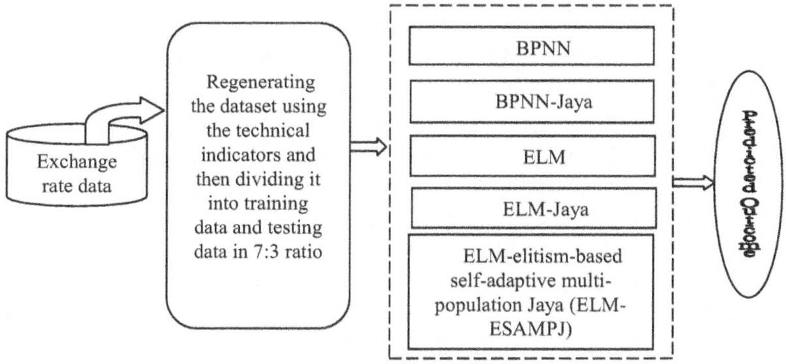

Fig. 1 Diagrammatic layout of the proposed model

this with the actual graph versus the predicted outcome graph and the Mean Square Error (MSE) result during the period of training.

The working flow of the entire work is depicted in Fig. 1, where it can be observed that the five experimented models used for prediction such as ELM, BPNN, ELM-Jaya, BPNN-Jaya, and ELM-elitism-based self-adaptive multi-population Jaya algorithm have considered for this study. The currency exchange rate data USD to INR and USD to EURO are well thoughtout for future price prediction. The regeneration of the dataset is carried out by using some technical indicators. The details of the technical indicator with its formula have referred from [9, 10]. After regenerating the dataset, all the experimented prediction model has trained with 70% of the data, which is chosen for experimentation, and the 30% data have been used for testing.

4.1 *Abstract View of the Proposed Model*

See Fig. 1.

4.2 *Proposed ELM-ESAMPJ Algorithm for Exchange Rate Analysis*

Algorithm: Exchange rate analysis model

Input: Dataset containing the available input data along with the calculated technical indicators, population number (*P*), elite size (*ES*), and stopping criteria.
Output: Predicted_opening_Price.
nP_i is the new population, O_i is the objective value of the new population, and V is the weight vector.

Step 1: Divide the dataset into 70% for training and 30% for testing.
Step 2: Set K random weight population, each of size $1 \times N_c$
$P_i = \{V^i_1, V^i_2, V^i_3, \ldots, V^i_{Nc}\}$ for $i = 1, 2, 3, \ldots, K$
Step 3: Initial candidate solution is created by using the ELM as the fitness function for the entire population P to generate global-best and global-worst solution using the steps 7–9, where the minimum objective value gives the global best and maximum objective value gives the global worst.
Step 4: While $i <$ Maximum Iteration
Step 5: The population K is divided into M subpopulation $P_1, P_2, P_3, \ldots, P_M$, Initially, the value of M is decided as 2 and replace the global-worst solutions (equal to ES) of the inferior group with solutions of the global best (elite solutions).
Step 6: For each subpopulation P_j, $J = 1$ to M.
Find global_best P_j where $\arg_{\min}(O_j)$
Find global_worst P_j where $\arg_{\max}(O_j)$
by calculating the error value in ELM.
Step 7: Using Eqs. (1) and (2), we can calculate the β value.
Step 8: $obt\_output = (test\_input \times P_i) \times \beta_i$
Step 9: $O_i = MSE(obt\_output, test\_output)$
Step 10: For each subpopulation P_j.
Step 11: $nP_j = P_j + rand(1, N_c) \times (global\_best - |P_j|) - rand(1, N_c) \times (global\_worst - |Pj|)$
Step 12: Find nO_j *by* step 8–12.
Step 13: If $O_j > nO_j$.
Step 14: Replace P_j with nP_j.
Step 15: Merge the entire subpopulation $P_1, P_2, P_3, \ldots, P_M$ into P
(nO_i is the objective value of the current best solution of the entire population and O_i is the objective value of the previous best solution of the entire population).
If $nO_i < O_i\ O_i$
$M = M+1;$
Else if $M > 1$
$\mathrm{M} = \mathrm{M} - 1$
End if
Step 16: Repeat steps 5–18 until stopping criteria satisfy; Otherwise, follow the steps of:

(a) Replace the duplicate solutions with randomly generated new solutions,
(b) For re-dividing the populations, go to Step 3.

Step.17: global_best is considered as final weight W_{global_best} and store the corresponding β_{global_best}
Step 18: For unknown Input_data (I') find the Predicted_opening_Price by the following equation:

$$\text{Predicted\_opening\_Price} = \left(I' \times w_{global\_best}\right) \times \beta_{global\_best}$$

4.3 Result Analysis

The actual graph versus the predicted outcome graph for exchange data of USD to INR mentioning 1 day, 3 days, 5 days, 7 days, 15 days, and 1 month in advance is depicted in Fig. 2. From the figure itself, it can be observed that the actual graph is very close to the predicted graph, and also it has been marked that the number of increasing days is inversely proportional to the performance. The simulated graph of the actual versus the predicted outcome graph for all the models experimented in this study is observed, but due to lack of space only the proposed model results are given. The predicted result of the rest of the experimented model can be analyzed through the MSE result during training, which is given in Tables 2 and 3.

The controlling parameters considered for experimentation are totally application-oriented. It differs from application to application; no fixed value can be assigned and

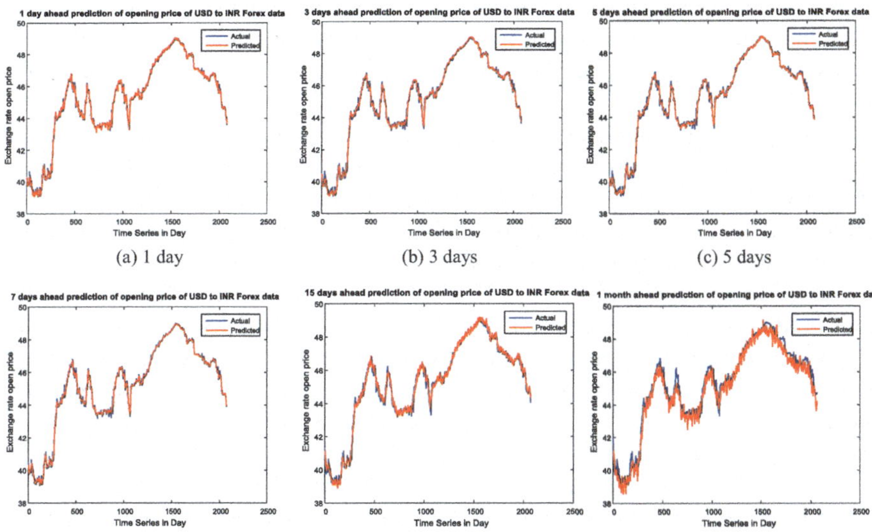

(a) 1 day (b) 3 days (c) 5 days

(d) 7 days (e) 15 days (f) 30 days

Fig. 2 Prediction of opening price of USD to INR forex data using ELM-elitism-based self-adaptive multi-population Jaya Algorithm for 1 day, 3 days, 5 days, 7 days, 15 days, and 1 month in advance

Table 2 MSE during training for USD/INR Forex data

Methods	1 day	3 days	5 days	7 days	15 days	30 days
ELM-ESAMPJ	**0.007567**	**0.011043**	**0.010825**	**0.008392**	0.037364	**2.784651**
ELM-Jaya	0.009453	0.019183	0.019923	0.012912	**0.020234**	3.576454
ELM	0.034523	0.052436	0.026453	0.127632	0.173542	3.345462
BPNN-Jaya	1.352611	2.657453	5.834634	4.473974	7.760956	11.84683
BPNN	5.876059	11.76586	12.69865	10.76876	31.50946	23.94547

Table 3 MSE during training for USD/EURO Forex data

Methods	1 day	3 days	5 days	7 days	15 days	30 days
ELM-ESAMPJ	**0.00634521**	**0.012875**	0.017856	**0.0093665**	**0.067342**	3.674853
ELM-Jaya	0.00843521	0.027864	**0.016565**	0.032451	0.081209	**3.647532**
ELM	0.0435233	0.056735	0.078364	0.234354	0.146352	3.654657
BPNN-Jaya	2.8576411	2.784653	6.758644	4.8937465	8.8926361	10.68678
BPNN	8.3542673	12.45642	14.22918	9.9686754	27.578654	25.74974

whenever a comparison is carried out among all the experimented model, it should be standardized; hence, 100 numbers of iteration and 100 numbers of population are considered, to make the comparison homogenized. For ELM, the number of hidden layer nodes is 12 and simultaneously in BPNN, 3 numbers of hidden layers with 4 number of nodes in each layer are well thoughtout. The elite size is considered as 5, and the value of subpopulation number is initially well thoughtout as 2, which increases or decreases basically to explore and to exploit the feature of the search space, respectively.

From Table 2, it can be observed that ELM-elitism-based self-adaptive multi-population Jaya is giving significantly better results with minimum MSE for all the experimented days except 15 days, where ELM-Jaya is showing better. The above analysis was for exchange rate data of USD to INR; similarly, it can be observed for experimented exchange rate data of USD to EURO from Table 3, where it has been observed that ELM-elitism-based self-adaptive multi-population Jaya is showing better prediction performance for 1 day, 3 days, 7 days, and 15 days except 5-day and 30-day ahead prediction. From the MSE result during training and actual graph versus the predicted outcome graph obtained from all the experimented prediction model, it can be clearly noticed that the prediction performance of ELM-elitism-based self-adaptive multi-population Jaya is showing significantly improved result compared to the rest of the prediction model, considered here in this study.

The MSE during training cannot be the only parameter to evaluate the performance. Hence, some performance evaluation measures such as Root Mean Square Error (RMSE), Mean Absolute Percentage Error (MAPE), and MSE during testing are considered in evaluating the prediction performance faultlessly. The performance measures are referred from [9, 10], where the mathematical formulation of MSE, RMSE, and MAPE is given in detail. From the result of these performance measures given in Table 4, it can be clearly understood that ELM-elitism-based self-adaptive multi-population Jaya is showing improved performance compared to the rest experimented prediction model.

Table 4 MSE during testing of USD/INR Forex data for measuring the performance

Methods	MSE	RMSE	MAPE
ELM- ESAMPJ	**0.006743**	**0.34424**	**0.11591**
ELM-Jaya	0.009623	0.45432	0.13981
ELM	0.037643	0.49563	0.13999
BPNN-Jaya	1.372341	3.87654	2.45632
BPNN	5.762982	7.98765	5.56433

5 Discussions

This experimental work has focused on future price prediction on USD/INR and USD/EURO forex data as EURO and USD; both are found to be very important international reserve currency and in many of the countries these two currencies are considered as the base currency. For prediction purpose, neural network-based methods are considered, as it is having the inherent capabilities to handle nonlinear data. Among so many neural network-based models, ELM is found to be faster and less complex, as it is free from iterative tuning. Though ELM is having the potential to handle nonlinear data with better prediction performance, the input weights and biases which are selected randomly can create some nonoptimal solution. Hence, some global optimization techniques such as Jaya and its modification have been used for optimizing the weights and biases of the prediction models. ELM-elitism-based self-adaptive multi-population Jaya [4] is the modified Jaya, where the entire population is divided into subpopulation, as put into optimization to the entire population may generate complex; hence, to divide the total population into a number of subpopulations is a good idea with the intention of getting a better outcome. To take a decision about the number of subpopulations in an adaptive technique, elitism-based self-adaptive method based on changing the solution depending on strength is used. The motivation behind this adapts with the changing strength of the solution to explore and exploit the feature of the search space. This study has looked at the future price prediction of six-time horizon in advance.

6 Conclusion and Future Work

ELM-elitism-based self-adaptive multi-population Jaya has been projected for the prediction of exchange rate data of USD to INR and USD to EURO considering six-time horizon, 1 day to 1 month in advance. The proposed model has compared with some widely accepted models such as ELM, BPNN, ELM-Jaya, and BPNN-Jaya for efficiency measurement. The potential ability of ELM is efficiently merged with multi-population search scheme of Jaya focusing on the changing strength of the solution. It can be clearly noticed from the simulated result that the proposed model outperforms over the rest experimented prediction model. This work can be extended in the future by applying for a different application domain.

References

1. Lisi, F., Schiavo, R.A.: A comparison between neural networks and chaotic models for exchange rate prediction. Comput. Statis. Data Anal. **30**(1), 87–102 (1999)
2. Huang, G.-B., Zhu, Q.-Y., Siew, C.-K.: Extreme learning machine: a new learning scheme of feedforward neural networks. Neural Netw. **2**, 85–990 (2004)
3. Huang, G.-B., Zhu, Q.-Y., Siew, C.-K.: Extreme learning machine: theory and applications. Neurocomputing **70**(1–3), 489–501 (2006)
4. Rao, R.V., Saroj, A.: An elitism-based self-adaptive multi-population Jaya algorithm and its applications. Soft Comput. **23**(12), 4383–4406 (2019)
5. Das, S.R., Mishra, D., Rout, M.: A hybridized ELM using self-adaptive multi-population-based Jaya algorithm for currency exchange prediction: an empirical assessment. Neural Comput. Appl. 1–24 (2018)
6. Tang, J., Deng, C., Huang, G.-B.: Extreme learning machine for multilayer perceptron. IEEE Trans. Neural Netw. Learn. Syst. **27**(4), 809–821 (2015)
7. Huang, G-B., Zhu, Q.-Y., Siew, C.-K.: Extreme learning machine: a new learning scheme of feedforward neural networks. Neural Netw. **2**, 985–990 (2004)
8. Rao, R.V., Rai, D.P., Ramkumar, J., Balic, J.: A new multi-objective Jaya algorithm for optimization of modern machining processes. Adv. Prod. Eng. Manag. **11**(4), 271 (2016)
9. Das, S.R., Mishra, D., Rout, M.: A hybridized ELM-Jaya forecasting model for currency exchange prediction. J. King Saud Univ.-Comput. Inf. Sci. (2017). (Article in Press)
10. Nayak, R.K., Mishra, D., Rath, A.K.: A Naïve SVM-KNN based stock market trend reversal analysis for Indian benchmark indices. Appl. Soft Comput. **35**, 670–680 (2015)

A Fruit Fly Optimization-Based Extreme Learning Machine for Biomedical Data Classification

Pournamasi Parhi, Jyotirmayee Naik, and Ranjeeta Bisoi

Abstract In high-dimensional biomedical datasets, the main objective is to map an input feature space to a predetermined class labels with less execution time and with high classification accuracy. Recently, many researchers have proposed numbers of machine learning methods in the area of biomedical classification. However, accurate classifier modeling is still an unresolved problem for the researchers. Accurate diagnosis of different diseases is very much important in the biomedical area so it is required to design a precise classifier. Therefore, in this work a new hybrid classifier named as fruit fly optimization (FFO)-based extreme learning machine (ELM) is proposed to classify the biomedical data. The presented classifier performance is also compared with various classifiers such as SVM and ELM. These classifiers are validated using various performance indices like sensitivity, G-mean, accuracy, F-Score, precision and specificity. The results prove that ELM-FFO outperforms other models.

Keywords Support vector machine (SVM) · Extreme learning machine (ELM) · Fruit fly optimization (FFO) · Biomedical data · Classification

P. Parhi (✉) · R. Bisoi
Department of Computer Science and Engineering, Siksha'O'Anusandhan Deemed to be University, Bhubaneswar, Odisha, India
e-mail: pournamsi.parhi@gmail.com

R. Bisoi
e-mail: ranjeeta.bisoi@gmail.com

J. Naik
Department of Electrical Engineering, Siksha'O'Anusandhan Deemed to be University, Bhubaneswar, Odisha, India
e-mail: jyotirmayeenaik@soa.ac.in

D. Mishra et al. (eds.), *Intelligent and Cloud Computing*,
Smart Innovation, Systems and Technologies 194,
https://doi.org/10.1007/978-981-15-5971-6_15

1 Introduction

Machine learning is one of the most popular research areas among researchers. Nowadays, biomedical engineering is one of the most demanding research field because it's related to health science. Biomedical engineering is useful and beneficial for the department of health science; therefore, it is a key area of research for many researchers. Biomedical data classification is one of the most burgeoning research areas among them due to high-dimensional data complexity of medical data. So, till now a number of machine learning techniques have been proposed to deal with data complexity to diagnose and detect various types of diseases correctly. Recently, various classification algorithms such as Artificial Neural Network (ANN), Support Vector Machine (SVM) and Fuzzy set theory are applied by the researchers for biomedical data classification. Some existing literatures are given in this paper. In [1], SVM is enforced with various kernel functions to classify biomedical data accurately. Computational complexity and high storage requirement are the main disadvantages of this algorithm. Recently, many researchers used a hybrid model (which is a combination of two or more models) for biomedical classification. SVM integrates with Genetic Algorithm (GA) to classify multiclass tumor data in [2]. Result shows that this hybrid model outperforms the conventional model performance. Least Square Support Vector Machine (LSVM) with Particle Swarm Optimization is presented for classification of fault in [3]. However, computational complexity and large memory requirement are the main drawbacks of SVM. So, to reduce the computational time and to enhance the classification performance, a new classifier named as Extreme Learning Machine (ELM) is presented in this work for biomedical data classification. ELM is a uni-layer feedforward network in which the input weights are chosen randomly and output weights are calculated by using the Moore–Penrose pseudo inverse least squares. In ELM, weight selection influences the model performance. Therefore, to obtain the optimal weight, recently, a number of optimization algorithms such as Particle Swarm Optimization (PSO) [4], Genetic Algorithm (GA) [5], Moth-Flame Optimization (MFO) [6], and Cuckoo-search [7] have been proposed by many researchers. The disadvantages of these methods are they may trap in local optimum point and their lower convergence speed. To overcome these drawbacks, in this paper a new optimization algorithm named as Fruit Fly Optimization (FFO) is presented to enhance the optimization performance. As discussed in the above literature, the hybrid model's performance is better than that of conventional models. Therefore, in this paper a new hybrid model named as fruit fly optimization-based extreme learning machine (ELM-FFO) is presented for biomedical classification, where the random weights of ELM are optimized using FFO to enhance the proposed model accuracy. In this study, the proposed model performance is tested with three types of biomedical datasets (Parkinson datasets, Indian liver patient datasets and Indian diabetes datasets). Again, to evaluate the superiority of the proposed model, it is also compared with other models like SVM and ELM. The detail framework is illustrated in Fig. 1.

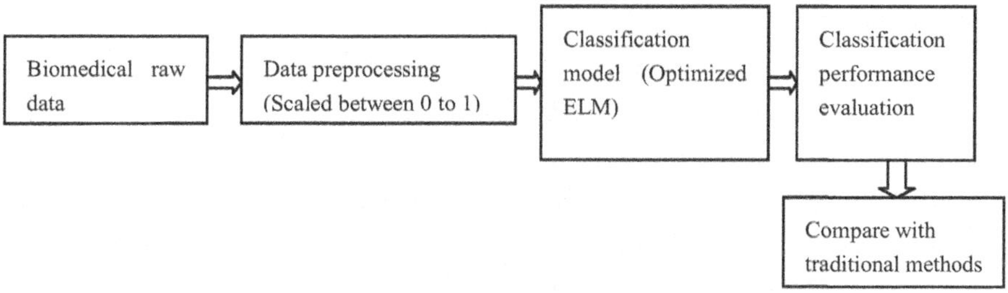

Fig. 1 Detailed framework of the proposed work

After a brief introduction to the considered work in Sects. 1 and 2, the paper describes different machine learning approaches for biomedical data classification. Section 3 explains details about the considered datasets and different performance measures. Further, the validation of various classification techniques is included in Sect. 3. Section 4 concludes the proposed study.

2 Machine Learning Techniques for Biomedical Data Classification

2.1 Extreme Learning Machine (ELM)

ELM is developed by Huang [8, 9]. This is the most well-known learning technique to train single hidden-layer feed-forward network (SLFN). Better generalization performance and faster learning, speed is the main advantage of ELM. The computation of ELM is free from iterations which makes this algorithm faster with lesser computational time. The ELM architecture is pictorially demonstrated in Fig. 2.

The output of SLFN having Z hidden nodes can be expressed as

$$f_Z(k) = \sum_{i-1}^{Z} \beta_i a_i k = \sum_{i=1}^{Z} \beta_i A\left(w_i, b_i, k\right), k \in R^d,\ \beta_i \in R \tag{1}$$

where a is the activation function and can be calculated as

$$a_i = A\left(w_i, b_i, k\right) = a(w_i k + b_i), w_i \in R^d,\ b_i \in R \tag{2}$$

$$\sum_{i=1}^{Z} \beta_i A\left(w_i, b_i, k\right) = p_j, j = 1,\ 2,\ 3 \ldots M \tag{3}$$

Equation (3) can be reformulated as

Fig. 2 ELM architecture

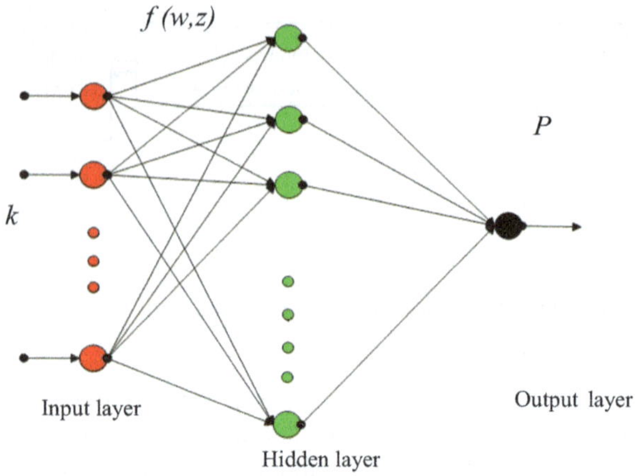

$$D\beta = P \tag{4}$$

where

$$D(w_1 \ldots w_z, b_1 \ldots b_z, k_1 \ldots k_z) = \begin{pmatrix} a(w_1 k_1 + b_1) \cdots a(w_Z k_1 + b_Z) \\ \vdots \quad \ddots \quad \vdots \\ a(w_1 k_M + b_1) \ldots a(w_Z k_M + b_Z) \end{pmatrix} \tag{5}$$

$$\beta = \begin{bmatrix} \beta_1^P \\ \vdots \\ \beta_M^P \end{bmatrix} \tag{6}$$

$$P = \begin{bmatrix} p_1^P \\ \vdots \\ p_M^P \end{bmatrix} \tag{7}$$

D denotes the hidden layer. In Ref. [8], it has been proved that for an arbitrary value ($\in > 0$), if the activation function present in the hidden layer is infinitely differentiable and $Z \leq M$, then the weights present in the input layer and the biases present in the hidden layer can be calculated and the single layer feedforward network can solve the linear system optimization problem by applying the below equation:

$$\left\| D\hat{\beta} - P \right\| = \text{minimize}_{\beta} \| D\beta - P \| \tag{8}$$

where $\hat{\beta}$ can be calculated as

$$\hat{\beta} = (D^T P) = (D^T D)^{-1} D^T P \tag{9}$$

The Moore–Penrose generalized inverse of D is denoted as D^T.

The ELM algorithm steps are given as follows:

(1) Randomly select the weights w_i and biases b_i present in input and hidden layers, respectively.
(2) Calculate the output matrix of hidden layer D.
(3) Compute weight present in the output layer $\hat{\beta}$ by applying Eq. (9).

2.2 Fruit Fly Optimization (FFO)

FFO is a metaheuristic optimization algorithm which is based on the food searching behavior of a fruit fly [10]. The strong vision and osphresis of fruit fly can help it to get the source of food from a long distance even if it is 40 km away from its position. The smell organ of fruit fly is so strong that it can get the smell of all fragrances floating in the air. By taking the advantage of it, fruit fly can reach close to the food source from where the smell is coming and then by using its strong vision fruit fly can get the exact location of the food and move toward that direction. Fruit fly optimization technique for solving the optimization problem is developed based on the above characteristics of fruit fly which are described in the following steps:

Step 1: Initialize population size of fruit fly swarm and also initialize maximum number of iterations.

Step 2: Initialize randomly location of fruit fly swarm as

$$initialize\_X\_axis$$

$$initialize\_Y\_axis$$

Step 3: By using osphresis, each fruit fly 'l' updates their position in the direction of food as

$$X_l = X\_axis + Random\_Value \tag{10}$$

$$Y_l = Y\_axis + Random\_Value \tag{11}$$

Step 4: Since the distance of the food location from the origin is unknown, the distance of each fruit fly from the origin is calculated as

$$D_l = Sqrt\left(x_l^2 + y_l^2\right) \tag{12}$$

Step 5: Then the judgment value of smell concentration (U_l) is calculated as

$$U_l = \frac{1}{D_l} \tag{13}$$

Step 6: Replace judgment value of smell concentration (U_l) with the judgment function of smell concentration and then computed smell concentration value of each fruit fly as

$$Smell_l = fun(U_l) \tag{14}$$

Step 7: The fruit fly having maximum smell concentration value among the fruit fly swarm and its corresponding indices is detected as

$$[BestSmell \;\; BestIndex] = Maximum(Smell) \tag{15}$$

Step 8: Find out the best fruit fly and its corresponding position by considering the smell concentration value, then the entire swarm is updated with their position by using their vision and moves in the direction of that best fly.

$$SmellBest = BestSmell \tag{16}$$

$$X\_axis = X(Best\_Index) \tag{17}$$

$$Y\_axis = Y(Best\_Index) \tag{18}$$

Step 9: Repeat the steps from 2 to 7, and then compare the smell concentration with the smell concentration obtained from previous iteration. If the current smell concentration is superior to the previous one, then implement step 8.

2.3 The Proposed FFO-ELM Models

The above-discussed FFO is applied to ELM to optimize the random input weights of ELM and to provide the best results. These steps are repeated till it reaches the stopping criteria as given below:

Step 1: Initialization of variables like population, maximum iteration, upper and lower bound, etc. and initialization of random position for the fruit flies.
Step 3: The fitness of each fruit fly is evaluated by the help of accuracy obtained from ELM. Further, fruit fly with the best fitness value is calculated and considered as the global optimum.
Step 4: If the fitness of the best individual in the swarm is better than the global optimum, then the global optimum is updated with this new value.
Step 5: The iteration t is updated as t = t + 1 up to reach the maximum number of iterations, else go to step 3.

3 Result and Discussion

3.1 Data Description

In this proposed model, three different types of healthcare datasets (Parkinson, Indian diabetes and Indian liver patients) are used for classification. The Indian diabetes dataset has 768 samples and 8 features. For this dataset, 500 samples (i.e. 65% of the total data samples) are taken for training purpose and rest 268 samples (i.e. 25% of the total data samples) are taken for testing purpose. But in case of Parkinson dataset, 140 samples (i.e. 70% of the total data samples) are considered for the training purpose and the rest 57 samples (i.e. 30% of the total data samples) are considered for the testing purpose. Similarly, the Parkinson dataset has 23 features. Likewise, Indian liver patient dataset having 10 features is divided for training and testing purposes as 400 samples (i.e. 70% of the total data samples) and 183 samples (i.e. 30% of the total data samples), respectively.

The biomedical data are scaled on a scale of 0 to 1 by using the following formula:

$$k_i = \frac{x_i - \min(x)}{\max(x) - \min(x)} \tag{19}$$

3.2 Performance Evaluation

The performance for the discussed classification algorithms is tested using 5 types of performance measurement such as G-mean, accuracy, sensitivity, F-Score, precision and specificity.

$$\text{Sensitivity} = \text{True Positive}/(\text{True Positive} + \text{False Negative}) \tag{20}$$

$$\text{Specificity} = \text{True Negative}/(\text{True Negative} + \text{False Positive}) \tag{21}$$

$$\text{G - mean} = \text{sqrt}(\text{Sensitivity} \times \text{Specificity}) \tag{22}$$

$$\text{F - Score} = \frac{(2 \times \text{Sensitivity} \times \text{Specificity})}{(\text{Sensitivty} + \text{Specificity})} \tag{23}$$

$$\text{Precision} = \text{True Positive}/(\text{True Positive} + \text{False Positive}) \tag{24}$$

Table 1 Performance measures of classifiers

Data	Methods	Accuracy	Sensitivity
Diabetes	*SVM*	0.9370	0.9557
	ELM	0.9510	0.9754
	ELM-FFO	0.9590	0.9803
Parkinson	*SVM*	0.8770	0.9459
	ELM	0.8950	0.9730
	ELM-FFO	0.9120	1
Liver	*SVM*	0.9180	0.9704
	ELM	0.9290	0.9778
	ELMFFO	0.9400	0.9852

3.3 Result Analysis

Here in this study, ELM-FFO is applied for the biomedical data classification. Three datasets (i.e. Parkinson, diabetes and Indian liver datasets) are used to test the efficiency of the proposed model. The proposed model is also compared with other models like SVM and ELM. The obtained results of all the classifiers are given in Tables 1 and 2. From this table result, it can be concluded that the proposed ELM-FFO outperforms the others models.

The convergence graphs for the proposed ELM-FFO are shown in Fig. 3a–c for diabetes, Parkinson and liver datasets, respectively. From the graphs, it proves that the proposed ELM-FFO converges very fast and accurately for the considered biomedical datasets.

Table 2 Performance measures of classifiers

Data	G-Mean	Specificity	F-Score	Precision
Diabetes	0.9154	0.8769	0.8769	0.9604
	0.9248	0.8769	0.8769	0.9612
	0.9353	0.8923	0.8923	0.9660
Parkinson	0.8423	0.7500	0.7500	0.8750
	0.8542	0.7500	0.7500	0.8780
	0.8660	0.7500	0.7500	0.8810
Liver	0.8649	0.7708	0.7708	0.9225
	0.8798	0.7917	0.7917	0.9296
	0.8974	0.8125	0.8125	0.9366

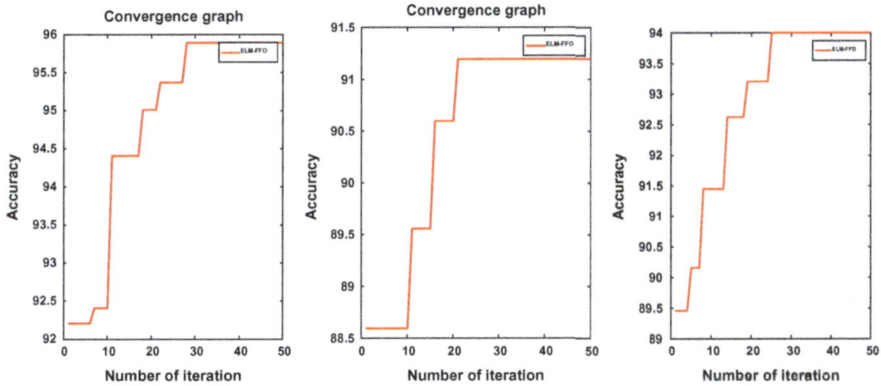

Fig. 3 Convergence graph comparison for different datasets: **a** Indian diabetes, **b** Parkinson, **c** Liver

4 Conclusion

In this work, ELM-FFO is presented for the biomedical data classification. The performance of the proposed model is compared with other classifiers such as SVM and ELM. Various performance indices are applied to validate the effectiveness of the models. From the results, it can be concluded that the performance of the proposed ELM-FFO is superior among all the discussed models for biomedical classification.

References

1. Aydïlek, İ.B.: Examining effects of the support vector machines kernel types on biomedical data classification. In: 2018 International Conference on Artificial Intelligence and Data Processing (IDAP), pp. 1–4. IEEE (Sep 2018)
2. Peng, S., Xu, Q., Ling, X.B., Peng, X., Du, W., Chen, L.: Molecular classification of cancer types from microarray data using the combination of genetic algorithms and support vector machines. FEBS Lett. **555**(2), 358–362 (2003)
3. Deng, W., Yao, R., Zhao, H., Yang, X., Li, G.: A novel intelligent diagnosis method using optimal LS-SVM with improved PSO algorithm. Soft Comput. **23**(7), 2445–2462 (2019)
4. Ling Chen, H., Yang, B., Jing Wang, S., Wang, G., Zhong Li, H., Bin Liu, W.: Towards an optimal support vector machine classifier using a parallel particle swarm optimization strategy. Appl. Math. Comput. 239, 180–197 (2014)
5. Huang, C.L., Wang, C.J.: A GA-based feature selection and parameters optimizationfor support vector machines. Expert Syst. Appl. **31**(2), 231–240 (2006)
6. Li, C., Hou, L., Sharma, B.Y., Li, H., Chen, C., Li, Y., Zhao, X., Huang, H., Cai, Z., Chen, H.: Developing a new intelligent system for the diagnosis of tuberculous pleural effusion. Comput. Methods Programs Biomed. **153**, 211–225 (2018)
7. Mohapatra, P., Chakravarty, S., Dash, P.K.: An improved cuckoo search based extreme learning machine for medical data classification. Swarm Evol. Comput. **24**, 25–49 (2015)
8. Wang, D., Huang, G.B.: Protein sequence classification using extreme learning machine. In: Proceedings 2005 IEEE International Joint Conference on Neural Networks, vol. 3, pp. 1406–1411. IEEE (July 2005)

9. Huang, G.B., Zhu, Q.Y., Siew, C.K.: Extreme learning machine: a new learning scheme of feedforward neural networks. Neural Netw. **2**, 985–990 (2004)
10. Pan, W.T.: A new fruit fly optimization algorithm: taking the financial distress model as an example. Knowl.-Based Syst. **26**, 69–74 (2012)

Partial Reconfiguration Feature in Crypto Algorithm for Cloud Application

Rourab Paul and Nimisha Ghosh

Abstract The implementation of crypto algorithm in general-purpose computing platforms involves security protocols, secured storage, digital certificates, secure execution, digital right management, etc. The present researchers undertook systematic design explorations to implement various crypto issues in hardware in concurrence with the generation changes taking place in FPGA technology including the latest dynamic reconfigurable embedded platform. The partial reconfigurable hardware platform is an advanced and latest version in the growth path of FPGA technology and implementation of various crypto issues in such a system is an important area of today's research interest. The performance of a software crypto implementation in processors usually takes extensive execution time, while the same implemented in hardware takes much lesser time as well as performs in much cost-effective manner in respect of resource usage and power. Hence, a new approach has to be adopted to design such systems based on metrics involving resource usage, throughput, and power along with adequate security. This new solution to security leads to an issue of optimization of crypto processors which can balance both the hardware and security issues of modern crypto systems. This paper explores three generations of crypto solutions. The upgradation issue of the generation of crypto solutions comes from the changing trend of FPGA technology as well as the crypto systems. The first-generation crypto system operated in processors has less efficiency and is more vulnerable to security, whereas the second generation operated in embedded hardware comes with more security, poor flexibility, and higher design complexity. The third generation using PR platform provides an optimum balance between first- and second-generation crypto solutions leading to high flexibility time and high efficiency. In this paper, we implement third-generation crypto hardware in a fog system which is placed between IoT device and cloud. The adoption of fog in

R. Paul (✉) · N. Ghosh
Computer Science and Engineering Department, Siksha 'O' Anusandhan (Deemed to be University), Bhubaneswar, Odisha, India
e-mail: rourabpaul@soa.ac.in

N. Ghosh
e-mail: nimishaghosh@soa.ac.in

D. Mishra et al. (eds.), *Intelligent and Cloud Computing*, Smart Innovation, Systems and Technologies 194,
https://doi.org/10.1007/978-981-15-5971-6_16

existing architecture reduces the processing cost of the cloud. ISE14.4 suite with ZYNQ7z020-clg484 FPGA platform is used to implement third-generation crypto core.

Keywords IoT · Cloud · Fog · Cryptography · FPGA

1 Introduction

Partial reconfiguration is the latest feature of Field Programmable Gate Array (FPGA) platform where any portion of the circuit can be changed dynamically, while the rest of the circuit is functional. This feature serves an excellent scope to reduce system power and resource with a little penalty of reconfiguration time. Designers store all components in secondary memory and according to the task requirement required component will be downloaded into the FPGA configuration memory. For high-speed applications, crypto applications like SSL, TLS, IPSec, Networking and Cryptography library (NaCl), and other security protocols find their implementation in hardware platform. Software-based security protocols are compromised with flexibility and execution time. To achieve more flexibility is hardware developer places all algorithms implemented in FPGA/ASIC platform. The combination of crypto algorithms, chosen for a particular session, gets enabled by an enable port deactivating others. Though the redundant crypto cores are not used during a particular session, still it consumes static power and resources for the entire run time. This implies a research challenge in achieving a balance of the power-resource parameters and at the same time preserving the algorithm flexibility. Few papers described in this survey use partial reconfiguration feature in a very efficient approach to save resource and power consumption preserving the hardware flexibility. The main contribution of this proposed article is as follows:

- The paper proposes an architecture where IoT devices can communicate with cloud through fog.
- The proposed fog system can perform heavyweight pre-processing to reduce the cloud expenses.
- The fog platform consists of general-purpose processor, embedded processor, and partial reconfiguration area. Binary files of various crypto algorithms are stored in a secondary memory. The binary file of chosen crypto algorithm will be implemented in PR area. This technique implies excellent flexibility in FPGA platform.

The organization of this paper is stated below. Section 2 discussed different generations of crypto implementation. In Sects. 3 and 4, we have stated about literature survey and proposed architecture, respectively. The research is concluded in Sect. 5.

2 Generations of Crypto Core

Figure 1 defines three generations [1] of security processing architectures based on comparison of performance and power consumption efficiencies, flexibility, and design turnaround time. It has also been observed that the three generations of crypto systems defined above can also be suitably connected with the three generations of embedded system technology defined earlier. The first-generation crypto system performs its processing activities exclusively in the processors embedded in the system using high-level language only. Such activities can be undertaken in the first- and second-generation embedded systems and one can achieve fast design time and flexibility with the software platform, but not that efficient in terms of their performance and power consumption. In both the first and second generations of embedded system, the embedded processor can also off-load cryptographic computations to custom-specific cryptographic hardware, which can be architected to achieve high performance and consume lower energy. This architecture is stated as the second-generation crypto solution which includes embedded processors, interfacing with crypto coprocessors on a processor or high-speed bus with special-purpose crypto hardware core in the pipeline. The second-generation crypto solutions compromise the flexibility and design times of the first-generation crypto solution. In the third-generation crypto solutions implemented in the third-generation embedded system technology proposed by [2], the benefits of the first- and second-generation crypto system can be balanced. It has high flexibility and efficiency. These approaches have a common characteristic that the embedded processor can off-load large portions of the security protocol to the security protocol processor. Since they are based on repro-

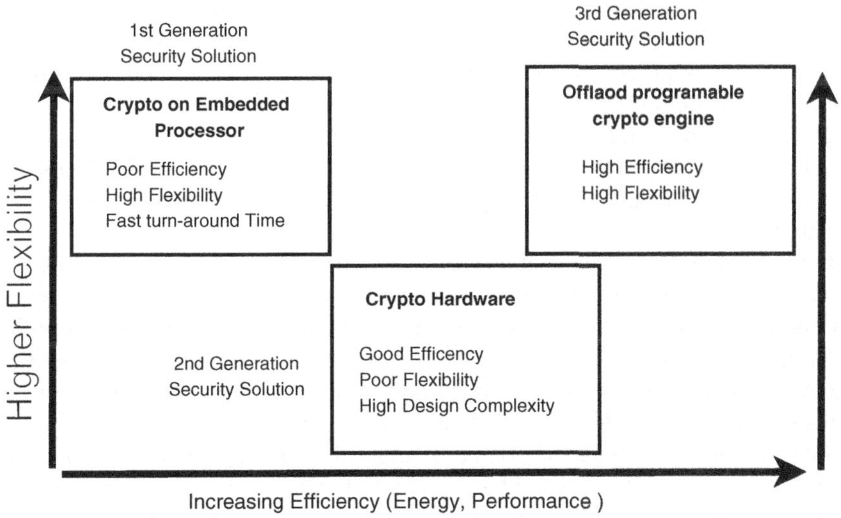

Fig. 1 Taxonomy of security processing architectures

grammable hardware, third-generation crypto solutions are capable of performing the bookkeeping requirements to support multiple concurrent security processing. Today, demands are increasing in respect of designing low-power applications in resource-constrained embedded systems with high throughput. For such kind of budgeted embedded applications, it is necessary to optimize design architecture maintaining optimal balance among throughput, resources, and power without compromising security issues of crypto algorithms. The dynamically and partially reconfigurable FPGA provides unique scopes to design such embedded devices combining the benefits of the flexibility of general-purpose processors along with the performance efficiency of dedicated hardware. FPGA gives unique solutions where various hardware flexibilities and hardware features like softcore and hardcore processors, CLBs, DSP blocks, distributed block RAM, and programmable IOs are present that make this platform more suitable for crypto applications.

3 State of the Art

Paper [3] proposes an efficient software/hardware implementation for the Advanced Encryption Standard (AES) crypto algorithm which states the design and performance testing algorithm for embedded platforms. This article enables an FPGA to dynamically reconfigure itself under the control of an embedded processor. Hardware acceleration increases the performance of embedded systems designed on FPGA significantly. Allowing an FPGA-based MicroBlaze processor to self-select the coprocessor cores used can help reduce the area requirements and increase the versatility of a system. In this work, three AES designs are implemented for different key sizes, such that AES-128, AES-192, and AES-256. Here, three AES binary files are placed on secondary memory offline, and according to the requirement of the user, the Microblaze softcore processor downloads chosen AES bit file from memory using a standard *Internal Configuration Access Port* IP provided by Xilinx.

The main advantage of the article [3] appears in the capacity of the proposed architecture to modify or change the key size without stopping the current operation of the system. As a result, this system is capable to raise the security of the AES crypto algorithm.

Article [4] proposed a security threat relevant to partial reconfiguration of FPGAs together with PUFs systems. The bitstreams of FPGA always susceptible to issue of piracy, tampering, reverse engineering, and hardware trojans. The mentioned problem can be sorted out by authenticated encryption (AE) which guarantees both the confidentiality and the authenticity of the ciphertext. PUF might be a solution to this attack which assures device-specific unique identity. They introduced a family of boards named as Side-channel Attack Standard Evaluation Board which placed a compact and secure delay-based PUF to achieve better throughput. Other existing solutions like PUFs and PL-PUF generate an N-bit response from an N number of challenges, which instantly generates attack resistant key.

Article [5] states that partial reconfiguration in FPGAs can improve the area and power consumption of cryptographic algorithms. In most of the applications, the keys are fixed during a cipher session, so that several blocks, like multipliers or adders, can be altered. This approach is faster and also consumes lesser resource and power. In this solution, the changes in the key are performed through a partial reconfiguration. For a case study, they used International Data Encryption Algorithm (IDEA).

In article [6], it observed that the dynamic reconfiguration reduces the resource power consumption which makes the FPGA platform suitable for crypto application. This work implemented an AES cryptographic algorithm using dynamic partial reconfiguration. The main merit of this work is to modify the size of the AES key without hampering the normal operation flow of the system which leads to more secure AES implementation. Additionally, the proposed work also optimized the reconfiguration time of the AES core. When the result of this paper is compared with other AES implementations available in literature, it obtained the good efficiency (throughput/area).

This paper [7] presented the implementation of a self-reconfigurable platform which is capable of implementing secure partial reconfiguration of Xilinx FPGAs using ICAP IP and embedded processors. Zeineddini and Gaj [7] show that FPGA could be reconfigured by encrypted partial bitstream stored in secondary memory. The said process can use embedded OS also. The design generates partial bitstream using the difference-based flow.

This work, [8], presented an implementation of the AES cryptographic algorithm using partial and dynamic reconfiguration. Two languages, Handel-C and VHDL, are used to implement AES. The implementation is JBits-based pipelined and parallel reconfigure. In summary, this paper has proposed combining architecture which used Handel-C, VHDL, JBits, partial, and dynamic reconfiguration, and a pipelined and parallel implementation. This solution is rare in the literature. The same proposed design methodology also has scope in other crypto algorithms. The results in [8] achieved less throughput, less resource consumption, and less latency. The proposed architecture achieved best throughput/area among 30 FPGA-based AES implementations.

Article [9] proposed a CryptoBooster which has tried to achieve maximum data throughput. CryptoBooster is a modular coprocessor which is dedicated to crypto application. It is designed to be implemented in FPGA with partial reconfiguration features. The CryptoBooster works as a coprocessor with a general-purpose computer to increase the speed of cryptographic operations. A session-based memory is responsible for storing session information. The said session is generated by few parameters such that the algorithm used, the cryptographic packets, the key(s), the initial vector(s) for block chaining, and other information. The CryptoCore module is divided into three parts, such that Cypher Core which executes encryption algorithm, Session Adapter to manage session parameter, and Session Control to control central controller for session management. The Cypher Core and Session Adapter modules are intelligent parts and can be queried by the Session Control. Hence, it is possible to exchange these modules without altering the control mechanism. All these modules communicate together and with the parts outside the Crypto Core using

Core Link which are unidirectional point-to-point connections. The CryptoBooster is implemented using the VHDL. Thus, the design can be synthesized without major problems for FPGAs as well as for VLSI technology.

In FPGA, there are full reconfiguration and partial reconfiguration. Full reconfiguration is the common method currently used. The configuration of FPGA is replaced by a new configuration file each time the algorithm is changed. Partial reconfiguration allows to reconfigure part where the algorithm is implemented on the FPGA. This resulted in a much shorter interrupt of service compared to full reconfiguration. The Crypto Booster takes advantage of partial reconfiguration.

4 Architecture and Implementation

The proposed architecture has three main parts such that Low Processing (LP) device, fog, and cloud. The LPs assemble raw data from sensors. The sensor data is communicated to local fog processor. The fog processor computes application-specific pre-processing. Finally, the fog processed data is passed to cloud for further actions. The architecture of proposed platform is shown in Fig. 2.

4.1 Low Processing Platform (LP)

The LP collects raw data from different sensors which is passed to a fog platform through Wi-Fi router. The sensor data are encrypted and hashed by lightweight cryptographic algorithm which protects data from attackers. The details of this communication are discussed in [10].

4.2 Fog

The fog is used for two purposes **I**. Fog acts as a pre-processing platform to reduce expensive cloud transactions and computations. The LPs are continuously collecting a huge amount of data from a different type of sensors but all sensors data might not be required for some specific application. The fog is a general-purpose processor-based platform which can compute application-specific processing to reduce the number of transactions to the cloud. Let us take an example where heart-related health care data are being sent to fog processor. Here, all the heart data are not important, because the application wants to work only on those data which are prone to cardiovascular diseases. To remove the irrelevant data, we introduce few computations on fog which detects only those data which are prone to cardiovascular diseases, and then those data will be sent to cloud for storage or the other processing. The adoption of fog reduces the number of transactions toward the cloud which makes the solution less

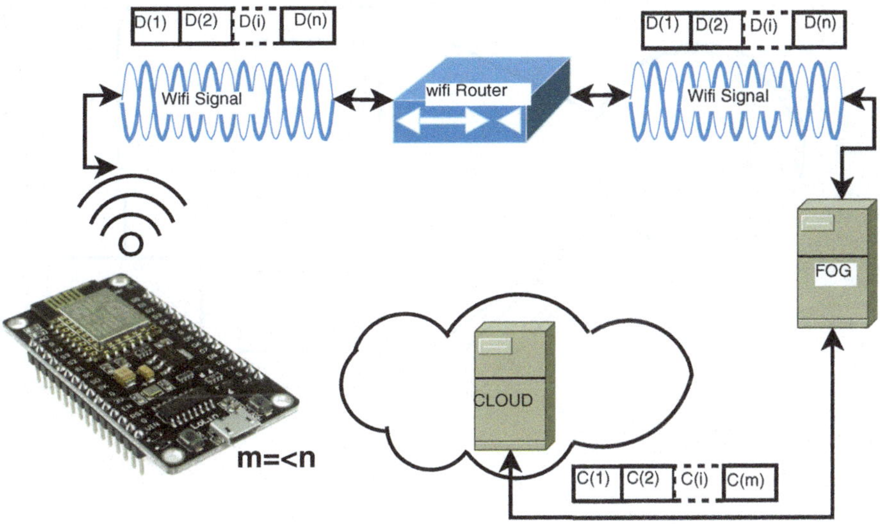

Fig. 2 Architecture of proposed platform

expensive. **II**. The fog acts as manager between LP and cloud which processes security applications on data. It is to be noted that the fog is a high processing platform that has a powerful processor and storage. The fog has six components such that general-purpose processor (GPP), dedicated processor, secondary memory, internal configuration access port (ICAP), PR area, and PR controller.

4.2.1 General-Purpose Processor (GPP)

The fog consists of general-purpose processor with a standard operating system to execute all heavy tools. Additionally, an FPGA is connected with the GPP through the PCB bus. The GPP can communicate with an embedded processor placed inside the FPGA.

4.2.2 Dedicated Processor (GPP)

The fog has a softcore dedicated processor or embedded processor in FPGA area. Xilinx MicroBlaze softcore processor is used as a dedicated processor to control the internal configuration access port (ICAP), PR controller, crypto core, and PR area.

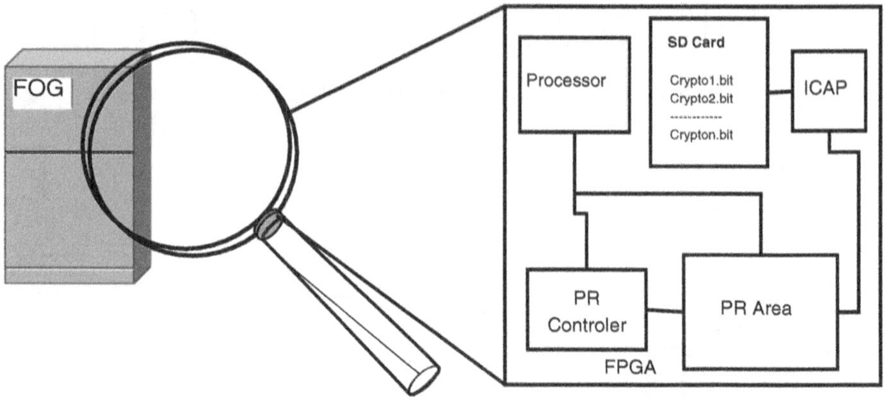

Fig. 3 System architecture

4.2.3 Secondary Memory

Secure digital (SD) card is used as a secondary memory to store the binary files of different crypto algorithms.

4.2.4 Internal Configuration Access Port (ICAP)

Internal configuration access port (ICAP) is an IP provided by Xilinx to download the binary files of different crypto algorithms from SD card to PR area is shown in Fig. 3.

4.2.5 PR Area

The whole FPGA is partitioned into two areas such that static area and PR area. The hardware inside the static area cannot be changed during the run time, whereas hardware inside PR area can be altered in run time. In our design, crypto algorithm is placed in PR area and an embedded processor, PR controller, and ICAP are in the static area of FPGA. In this paper, we have implemented AES, RC4, Grain, RSA, and DSA crypto algorithms in the said PR area.

4.2.6 PR Controller

As per Xilinx suggestion during the partial reconfiguration time, the clock signal of the hardware placed inside PR area should be stopped and the PR hardware should be in the reset state. To control the clock and reset signal of the PR crypto hardware, we designed a PR controller IP. The PR controller IP is connected with the standard processor which can be controlled by the dedicated processor.

4.3 *Cloud*

The proposed cloud in this architecture can store or can process sensor data depending on the demand from a particular application. This cloud architecture is conventional, and the communication between cloud and fog is encrypted.

5 Conclusion

In this paper, we have explored partial reconfiguration issues which is an obvious requirement in crypto implementation. Before the PR era, the security processor serves flexibility with compromised power and resource but while PR came into the FPGA platform, the heavy crypto algorithm with excellent flexibility is possible to implement in such a budgeted embedded platform. We proposed a secure communication between IoT and cloud through a fog platform. The adoption of fog reduces overheads on the cloud which makes this platform cheaper compared to existing literature.

References

1. Ravi, S., Raghunathan, A., Kocher, P., Hattangady, S.: Security in embedded systems: design challenges. ACM Trans. Embed. Comput. Syst. **3**(3), 461–491 (2004)
2. Kocher, P., Lee, R., McGraw, G., Raghunathan, A.: Security as a new dimension in embedded system design. In: Proceedings of the 41st Annual Design Automation Conference, DAC'04 (Moderator: Ravi, S.), pp. 753–760. ACM, New York, NY, USA (2004)
3. El Abidine Alaoui Ismaili, Z., Moussa, A.: Self-partial and dynamic reconfiguration implementation for AES using FPGA. CoRR, abs/0909.2369 (2009)
4. Hori, T.K.Y., Satoh, A.: Tackling the security issues of fpga partial reconfiguration with physical unclonable functions. In: International Conference on Reconfigurable Systems and Algorithms (2012)
5. Gonzalez, I., Lopez-Buedo, S., Gomez, F.J., Martinez, J.: Using Partial Reconfiguration in Cryptographic Applications: An Implementation of the IDEA Algorithm, pp. 194–203. Springer, Berlin (2003)
6. Wankhade, S., Mahajan, R.: Dynamic partial reconfiguration implementation of AES algorithm. Int. J. Comput. Appl. **97**(3), 15–18 (2014). Full text available
7. Zeineddini, A.S., Gaj, K.: Secure partial reconfiguration of FPGAs. In: Proceedings. 2005 IEEE International Conference on Field-Programmable Technology, December 2005., pp. 155–162 (2005)

8. Granado-Criado, J.M., Vega-Rodríguez, M.A., Sánchez-Pérez, J.M., Gömez-Pulido, J.A.: A new methodology to implement the AES algorithm using partial and dynamic reconfiguration. Integr. VLSI J. **43**(1), 72–80 (2010)
9. Paul, R., Saha, S., Pal, C., Sau, S.: Novel architecture of modular exponent on reconfigurable system. In: 2012 Students Conference on Engineering and Systems (SCES), March 2012, pp. 1–6 (2012)
10. Paul, R., Ghosh, N., Sau, S., Chakrabarti, A., Mahapatra, P.: Iot based smart access controlled secure smart city architecture using blockchain (2019)

Proactive Content Caching for Streaming over Information-Centric Network

Shatarupa Dash, Suvendu Kumar Dash, and Bharat J. R. Sahu

Abstract Video streaming applications have emerged as the most dominant traffic sources over the Internet. In addition, the edge cloud is accelerating the video sharing process with the help of social networks. Information-Centric Networking (ICN) is a promising paradigm for future fast Internet. ICN is location independent (free from IP address), and serve "One data for one Interest". Video streaming applications over ICN need to send multiple Interest packets for video segments and enhancement layers due to changing network conditions. Moreover, media presentation file can be inaccurate to use with ICN. In this paper, we propose a proactive caching scheme for video streaming over ICN. The proposed video streaming approach for ICN not only supports connection-less streaming with limited Interest packets but also reduces the video content access delay up to 50% by proactively caching video contents close to the clients.

Keywords Information-Centric Networking (ICN) · Video streaming · Layer encoding · Interest packets

1 Introduction

With the advancement of high-speed network access technology and the exponential growth of mobile devices, video streaming service has become the dominant traffic source in the current Internet [1]. Moreover, the emergence of the social network and edge cloud technology has accelerated the use of video streaming by social sharing leading to redundant traffic flow over the same network [2, 3].

S. Kumar Dash
Oretes Consulting Pvt. Ltd., Bhubaneswar, Odisha, India

S. Dash · B. J. R. Sahu (✉)
Department of Computer Science and Engineering, Siksha 'O' Anusandhan Deemed to be University, Bhubaneswar, Odisha, India
e-mail: bharatjyotisahu@soa.ac.in

D. Mishra et al. (eds.), *Intelligent and Cloud Computing*,
Smart Innovation, Systems and Technologies 194,
https://doi.org/10.1007/978-981-15-5971-6_17

Reducing bandwidth consumption and redundancy elimination have become the vital goals for performance improvement of future cloud based Internet. Information-Centric Networking (ICN) is emerging as a future Internet paradigm with a natural design for the goals above. By focusing on the content-centric design rather than host- centric design, ICN is changing the connectivity landscape.

The ICN architecture includes an in-network caching and a content-based routing as main features where every content has its name as an identifier. The content-based routing mechanism delivers content request messages to routers holding the contents. The content request message, called the Interest packet in ICN, visits intermediate routers along the path toward the designated server. The content message, called the data packet, is returned along the reverse path of the Interest. The content may be provided by the content server or by the cache storage in an intermediate router and can be transparently replicated in an intermediate router.

Recently, Scalable Video Coding (SVC) has emerged as a promising video encoding technique. Segmented videos are encoded into different layers in SVC. In SVC the video segments are formatted into multiple layers: a Base Layer (BL) that provide the lowest bit rate for the particular stream, and multiple Enhancement Layers (EL) transmitted separately depending on the network conditions [4]. The higher EL offer higher resolutions and enhancement of the video quality. The quality of video on user side depends on the network state. The layered approach allows for adaptation to the network conditions.

Multiple enhancements layer along with the base layer are cached at a different location in the network [4]. Video and audio come from different layers in case of multi-layer coding requiring better synchronization algorithm. Authors in [5] suggest that the caching priority should go to the base layer to avoid freezing. In case of HTTP streaming, a series of segments might be returned with a particular request. This is the contradiction of "one Interest, one data packet" principle of ICN. When channel quality is high, all layers must be sent to the client for better quality of experience. However, the client needs to specify each segment of streaming while using ICN. Consequently, the network will get excessive Interest packets. Moreover, the rate adaption mechanism will keep changing the requirement of enhancement layers. As the cache location of each layer might differ, the user will experience high jitter value. Moreover, like HTTP streaming, it's hard to maintain a Media Presentation Description (MPD) that keeps tracks of all the media representation due to location independence in ICN.

The existing literature mostly focuses on shaping the caching and forwarding to suit the adaptive streaming procedure. In this work, we tried to develop a proactive caching for streaming applications. Our work is inspired by the use of ant colony algorithm in [6] and cache partition in [7]. We propose an efficient proactive caching algorithm inspired by the ant colony process to improve the quality of streaming over ICN. The proposed procedure proactively caches video segments close to the users improve delay and jitters.

2 System Model

Let, $\mathbf{V} = \{v_1, v_2, \ldots, v_i, \ldots v_m\}$ represents the set of videos used in overall network and $P_i, i = 1, 2, \ldots m$ represent popularity of v_i. Each video v_i is a combination of multiple base layers $BL^s_{v_i}, s = 1, 2, \ldots b$, and each $BL^s_{v_i}$ has L enhancement layers $EL^s_{v_i}(l), l = 1, 2, \ldots L$. The BL is the basic video segment with lowest quality and addition of each EL to the BL, enhance the quality of the video. Higher ELs are fetched when channel condition is good.

The incoming Interests may query a data that has been requested earlier by others or a new data. Basically, nodes try to cache most popular data in ICN network. The popularity distribution of contents are described by a generalized Zipf distribution function as follows:

$$P_i = \frac{\sum_{i=1}^{m} \frac{1}{i^\gamma}}{m^\gamma}, \quad 0 < \gamma < 1.2 \tag{1}$$

The BL segments follow in-network caching. The ELs of each video are initially cached randomly. Depending upon subsequent usage they are cached at different nodes.

The number of Interests $|I_{v_i}|$ needed to stream video v_i, with b segments considering best channel condition is given by

$$|I_{v_i}| = b(1 + L) \tag{2}$$

Subsequently, since the BLs and ELs are stored at different nodes, total hops traveled by the Interest is given by

$$\sum_{s=0}^{b} d_{BL^s_{v_i}} + \sum_{l=0}^{L} d_{EL^s_{v_i}(l)} \tag{3}$$

where $d_{BL^s_{v_i}}$ and $d_{EL^s_{v_i}(l)}$ represents the distance of BL and EL from the client respectively. Let $t_{BL^s_{v_i}}$ and $t_{EL^s_{v_i}(l)}$ denote time requires to fetch the BL and EL from its cached location to client. The average access time for retrieving all contents is given by

$$\sum_{i=0}^{b} t_{BL^s_{v_i}} + \sum_{i=0}^{b} t_{EL^s_{v_i}(l)} \tag{4}$$

For improved buffering and rate adaption, $t_{BL^s_{v_i}}$ and $t_{EL^s_{v_i}(l)}$ should be as minimum as possible. Due to limited storage, not all video elements can be cache close to the user. Moreover, accessing a less popular video will consume a longer duration than a popular video. Thus, pre-fetching some BL and EL is the only option to reduce content access delay.

The default ICN procedure sends Interest packets to the network searching all possible nodes for a single data until it finds it or until the Interest lifetime expires.

Thus, the Interest from the client node traverses deep into the network. Sending Interest for each segment is not only time-consuming but also cause uncertain access time. Thus, the content fetching for streaming needs a better solution.

3 Proposed Solution

The proactive caching is applicable for all ongoing streaming; multiple video segments need to be cached for a short time irrespective of the content store caching policy. Thus, we propose a dynamic cache to implement our proactive caching. In this section, we describe the proactive caching algorithm with an example of an ant colony process.

3.1 Content Store (CS)

By default ICN support in-network caching. The nodes cache the most popular data in their content stores. In our proactive caching, contents must be cached on the path to the client. Consider a scenario where the ongoing streaming video is not popular hence not cached near the user. In such case, the caching policy may not consider caching the video content. As the goal of caching is to achieve bandwidth saving and redundancy elimination, the existing policy may not be altered to support pre-fetching. Moreover, the pre-fetched video only needs to be cached for a short duration. Thus, we propose to divide the content store into two parts: (1) Static (normal) cache to cache content based on caching policy, (2) Streaming Cache to support continuous video streaming. Streaming Cache may hold streaming data for a specific period of time [8].

3.2 Ant Colony Process

In the natural world, ants are species that March randomly, and upon finding food, return to their colony laying down pheromone trails. When other ants find such pheromone trails, they are likely not to keep traveling at random, but to follow the path, to return and to reinforce it, if they eventually find any food. The pheromone trail evaporates over time. Thus, its attractive strength reduces. In a longer path, pheromone evaporates quickly. A short path gets walked over more often, and thus the pheromone density becomes higher on shorter trails than longer ones. Pheromone dissipation also has the benefit of avoiding the convergence to a locally optimal solution.

In the existing streaming architectures, the MPD file tells about the location of content segments. Since ICN is location independent and the BLs and ELs are cached at unknown nodes, it is difficult to maintain MPD file. Thus, the client has to generate Interests for each BL and EL. The Interest generation must follow the available BL and EL. We assume that each node in the network can generate Interest for next Content in the sequence. BL segment 1, denoted as BL^1 is followed by BL^2 and $\{EL^1(1), EL^1(2), \ldots EL^1(l)\}$. While the $\{EL^1(1), EL^1(2), \ldots EL^1(3)\}$ should follow the BL^1 without delay, BL^2 can relatively tolerate delay based on buffer condition. Subsequently, BL^2 will be followed by the corresponding Els and BL^3, and so on.

To explain the proactive caching procedure, we describe its similarity to an ant colony process and called the proposed work as AntStream. BL and EL of a requested streaming represent the ants. The Interest for next content represents the pheromone. The client requesting the video is assumed to be the food source. The mapping is summarized as

1. Ant → BL/EL
2. Pheromone → Interest
3. Food Source → Client watching video.

(1) In any streaming, the first content to be queried is the BL of first segment, i.e., BL^1. Based on popularity the BL^1 should be available close to the client as compared to other BLs and ELs. When BL^1 is being queried, the very first data forwarding node (i.e., node caching BL^1) would generate Interest (pheromone) for next content, i.e., BL^2 and $\{EL^1(1), EL^1(2), \ldots EL^1(l)\}$, if they are not in its CS.
(2) Any intermediate nodes, through which BL^1 passes would generate similar Interest (pheromone) and forward to other nodes including one who just sent BL^1. Since the next data is requested before the client needs them, we say this as the pre-fetching step.
(3) Any node receiving the EL would forward them to the downstream without delay. But, the next BL can be delayed by t time based on the time gap between two consecutive BLs forwarding.
(4) The node close to the client, may not generate an immediate pheromone if there is no space for pre-fetching. Instead, it will wait for the client to send next Interest.
(5) As the pheromones evaporate the path becomes obsolete. Similarly, a node caching BL^2 in stream cache may consider dropping it, if the time difference since BL^1's dispatch is more than the play duration. This will help to avoid the situation when the client leaves the streaming session or skips the video.

Before the client requests next video content, it is gradually pulled by the nearby nodes. Hence, the content is available close to the user with less time as compared to any caching strategy. If the new content access time are $t(new)_{BL^s_{v_i}}$ and $t(new)_{EL^s_{v_i}(l)}$. We can say

$$
\begin{aligned}
t(new)_{BL^s_{v_i}} &\le t_{BL^s_{v_i}}, \\
t(new)_{EL^s_{v_i}(l)} &\le t_{EL^s_{v_i}(l)}, \\
\sum_{i=0}^{b} t(new)_{BL^s_{v_i}} + \sum_{i=0}^{b} t(new)_{EL^s_{v_i}(l)} &\le \sum_{i=0}^{b} t_{BL^s_{v_i}} + \sum_{i=0}^{b} t_{EL^s_{v_i}(l)}
\end{aligned}
\tag{5}
$$

4 Experimental Results

We have implemented our proposal with system simulator and evaluated its performance on a butterfly-like network with 100 nodes. The Interest size and data size considered in our simulation are 20 bytes and 1400 B, respectively. The link capacity for each link is 10 MB/s. Link delay of each link is considered as unit time (1 ms), to keep it consistent with hop-count. We distribute 1000 video data contents among all the network nodes and initialize the forwarding Interest base (FIB). During the simulation, we added 1% new data in 1 sec interval. The total simulation time is 200 s for each simulation. The simulation results are the average value of 10 simulations. Each video segment has a maximum of three ELs. Each video needs 10 BLs to complete. BLs have play duration of 10 s. We compared our proposed procedure with on-path and off-path caching. The on-path caching scheme cache video segments in the streaming path according to their usage sequence. The off-path caching may use popularity-based caching decision for individual nodes, thus the placement can be random sometimes.

Figure 1a presents the average turnaround time of the data packets of the video being requested. It is the time between when the client generates an Interest packet and the time when the client gets that data packet. Since data are pre-fetched to a nearby node, proposed AntStream shows up to 6ms turnaround time. The on-path caching and the off-path caching shows up to 12 ms delay. Initially, on-path caching performs better than off-path caching as data are cached on the path to the consumer. But when videos are played longer, off-path caching shows better performance than off path caching as some videos cached randomly are available at less hop count.

Figure 1b shows the turnaround time for each Interest of video access. In our proposed AntStream, the data are always available very close to the client. Thus, the client generates Interest that is satisfied at a nearby node ($\le$6 ms). Initially, on-path caching shows better result as frequently used initial videos are cached close to the user. Later parts of the video are encounter near the origin server. The off-path caching is random, hence the turnaround time varies greatly.

The proposed AntStream greatly reduce the jitter value during streaming as contents are available close to the consumer proactively. The maximum jitter is 3 ms in the proposed algorithm, whereas on-path caching has a maximum jitter value of 5 ms. The on-path caching shows smooth jitter but the delay experience is higher. Off-path caching also shows worse jitter value (up to 5 ms), as it has more uncertainty in data discovery. The results are depicted in Fig. 1c.

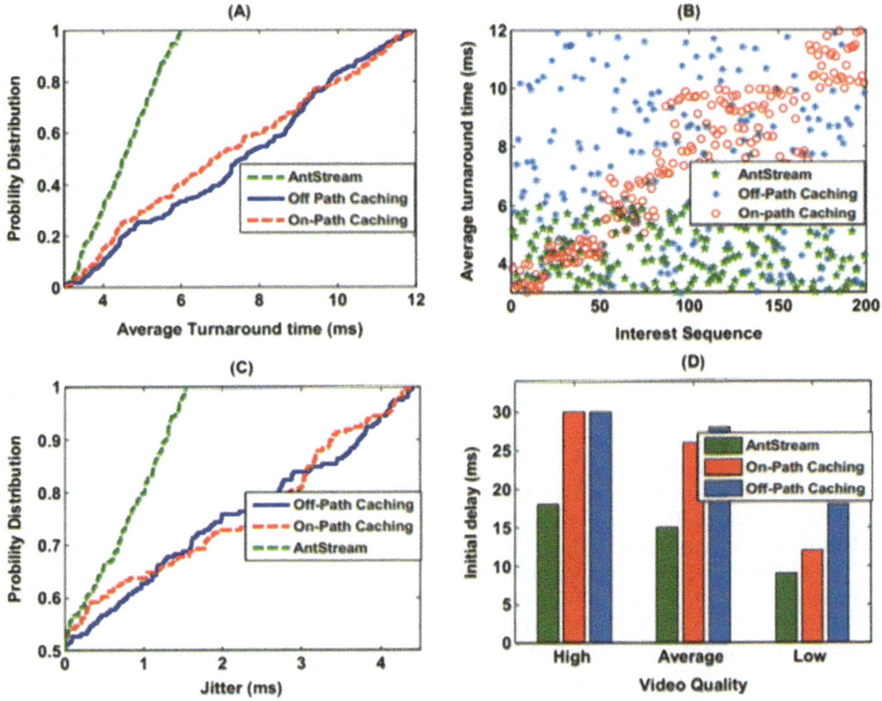

Fig. 1 Simulation results: **a** Average turnaround time for different Interests, **b** average turnaround time distribution for 200 Interests, **c** Jitter **d** Initial delay for video playback

We consider three types of videos and study the initial delay for video playback: high quality with three ELs, average quality with two ELs, and low quality with one EL. The result of our simulation is shown in Fig. 1d. The proposed AntStream shows lowest initial delay (18 ms) for high quality videos, whereas on-Path and off-Path caching results 28 ms and 30 ms, respectively. The reduction is due to the proactive fetching process happening before the clients send an Interest for the same data.

Overhead: The number of Interests sent in the uplink by AntStreaming is less than that of the traditional ICN procedure. The AntStreaming only forwards Interest proactively. As the default ICN procedure cancels duplicate Interests, there is no additional Interest overhead. Additional requirements in the proposed system are the procedures to generate subsequent Interests. Moreover, some pre-fetched data might not be utilized if the user has skipped the video before completion, given by $r\Delta t$, where r represents the rate of pre-fetching, and Δt is the waiting time until next Interest generation.

5 Conclusion

In this paper, we highlighted the problems in ICN to support streaming. We proposed an ant colony inspired process to pre-fetch the future video data before the client requests them. Simulation results show up to 50% improvement in delay and 66% improvements in jitter value as compared to both on-path and off-path caching. The intermediate node sending an Interest for pre-fetching may contradict the ICN principle which needs more investigation for practical implementation.

References

1. Cisco Visual Networking Index: Forecast and Trends, 2017–2022. White Paper (2019)
2. Han, B., Wang, X., et al.: pp. 375–380, Turin (2013)
3. Dash, S., Sahu, B.J.R., Saxena, N.: Social network aware caching for 5G radio access network. IETE Tech. Rev. **34**(sup1), 52–61 (2017)
4. Lederer, S., et. al.: Adaptive Video Streaming over Information-Centric Networking (ICN), RFC 7933 (2016). https://tools.ietf.org/html/rfc7933
5. Petrangeli, S., et al.: Towards SVC-based adaptive streaming in information centric networks. In: IEEE International Conference on Multimedia & Expo Workshops (ICMEW), pp. 1–6 (2015)
6. Jianhui, L.: Ant colony optimization-inspired ICN routing with content concentration and similarity relation. IEEE Commun. Lett. **21**(6), 1313–1316 (2017)
7. Li, W., et al.: Bitrate adaptation-aware cache partitioning for video streaming over information-centric networks. In: IEEE 43rd Conference on Local Computer Networks (LCN), Chicago, IL, USA, pp. 401–408 (2018)
8. Hsu, J., et al.: A partial cache for multimedia content in named data networking. In: International Conference on Platform Technology and Service, Jeju, pp. 37–38 (2015)

Performance Analysis of ERWCA-Based FLANN Model for Exchange Rate Forecasting

Arup Kumar Mohanty, Sarbeswara Hota, Ajit Kumar Mahapatra, and Debahuti Mishra

Abstract Currency exchange rate forecasting has been the focus of research in the field of engineering and mathematics since the last decade. As the exchange rate data are nonlinear and dynamic in nature, variants of artificial neural network (ANN) models are the common choice for developing forecasting models. To overcome the drawbacks of neural network models, different nature-inspired optimization algorithms have been proposed. In this paper, the FLANN model is used for the exchange rate prediction and ERWCA algorithm is used for the optimal weight determination of FLANN model. The SGD to INR exchange rate data has been collected for a specific period for the simulation purpose, and the dataset is divided in the ratio 8:2 for training and testing purposes. The performance of the proposed FLANN/ERWCA model is compared with FLANN-PSO and FLANN-WCA models. The experimental results demonstrated that the proposed model outperformed the other two models for 1-day and 7-day ahead in SGD to INR exchange rate forecasting.

Keywords Exchange rate · Trigonometric expansion · Water cycle algorithm · Mean absolute error

A. K. Mohanty · A. K. Mahapatra
Department of CSIT, Siksha 'O' Anusandhan Deemed to be University, Bhubaneswar, India
e-mail: arupmohanty@soa.ac.in

A. K. Mahapatra
e-mail: ajitmahapatra@soa.ac.in

S. Hota (✉)
Department of CA, Siksha 'O' Anusandhan Deemed to be University, Bhubaneswar, India
e-mail: sarbeswarahota@soa.ac.in

D. Mishra
Department of CSE, Siksha O Anusandhan Deemed to be University, Bhubaneswar, India
e-mail: debahutimishra@soa.ac.in

D. Mishra et al. (eds.), *Intelligent and Cloud Computing*,
Smart Innovation, Systems and Technologies 194,
https://doi.org/10.1007/978-981-15-5971-6_19

1 Introduction

The economic condition of a country is influenced by a lot of factors. The international trading plays a major role in deciding the economy and it is dependent on currency exchange rate [1]. Exchange rate is highly dynamic and nonlinear in nature. So forecasting of exchange rate has become an important research item in the field of engineering and financial mathematics [2]. Various statistical methods have been used in financial forecasting domain. But these methods couldn't map the nonlinear behavior of the financial time series data. Models using artificial neural networks and its variants are more useful in predicting the financial time series data [3, 4].

The literature study indicates that different neural network models, that is, multilayer perceptron (MLP) [5, 6], radial basis function neural network (RBFNN) [7], functional link artificial neural network (FLANN) [8], and extreme learning machine (ELM) [9] have been used in exchange rate forecasting. For the currency exchange rate prediction between Indian rupees to Japanese Yen, US dollar to British pound, the authors in [10] developed two novel ANN models involving nonlinear inputs and simple ANN structure with one or two neurons, that is, a FLANN model and cascaded FLANN model for forecasting of exchange rate. The exchange rate forecasting performance of the cascaded FLANN model was compared with the FLANN and least mean square (LMS) based prediction model through simulation. It was shown that the cascaded FLANN model was relatively better as compared to FLANN and LMS-based forecasting model. The authors in [9] used an ELM model hybridized with Jaya optimization technique for the currency exchange rate prediction and compared the prediction performance with MLP and FLANN-based models. In [11], the authors used ANN model for the exchange rate prediction.

In order to cope with the limitations associated with the traditional training technique process in FLANN model, different evolutionary and swarm intelligence-based algorithms have been used for training the FLANN model in financial time series forecasting task. The objective of this paper is to use the evaporation rate-based water cycle algorithm (ERWCA) with the FLANN, that is, FLANN/ERWCA model for the 1-day and 7-day ahead forecasting of Singapore Dollar to Indian Rupees (SGD to INR) exchange rate. The performance of FLANN/ERWCA model is compared with the FLANN-PSO and FLANN-WCA models in exchange rate forecasting. The paper is organized as follows. Section 2 describes the methods used in this work. Section 3 deals with the simulation study and Sect. 4 concludes the paper.

2 Methodology

This section describes the FLANN model and ERWCA method. The different steps of FLANN/ERWCA model are depicted in this section.

2.1 FLANN Model

FLANN model is a single-layer flat neural network model consisting of an input layer and one output layer without any hidden layer. This model is suitable for nonlinear financial time series data prediction task as the functional expansion property provides nonlinearity [12]. In this work, the input pattern is expanded through trigonometric expansion to provide nonlinear elements. Various researchers have used trigonometric expansion in FLANN model as expansion function scheme. The output of a node is calculated using the weighted linear combination of the expanded inputs and the activation function. The error is calculated as the difference between actual and predicted output. Using this error value, the weights are updated using LMS algorithm.

2.2 Evaporation Rate-Based Water Cycle Algorithm (ERWCA)

The ERWCA approach is variation of WCA, which is based on the natural water cycle mechanism. The flow of water from streams to rivers and then to sea is simulated in this approach [13]. WCA has been applied for solving multiobjective problems [14]. The authors in [15] used WCA in solving water reservoir system. The ERWCA approach is based on the evaporation rate associated with WCA [16].

The ERWCA consists of different phases, that is, initialization of population, flow of streams into river and sea, evaporation condition, and raining process. Table 1 describes the different notations used in this approach.

- **Initialization of Population**

Equation (1) describes each stream. One stream is randomly chosen as sea and some of the streams are randomly chosen as river.

Table 1 Notations used in ERWCA approach

Notation	Description
N_{sr}	Total numbers of sea and river
$x_1, x_2, x_3 \dots x_n$	Weights of the FLANN model
N_{pop}	Size of the population
N_{stream}	Number of streams
NS_n	Assignment of streams to rivers and sea
s_{max}	Random number
LB	Lower bound
UB	Upper bound

$$\text{Stream} = [x_1, x_2, x_3 \ldots x_n] \tag{1}$$

Equation (2) defines the number of sea and rivers.

$$N_{\text{sr}} = \text{Number of Rivers} + 1 \tag{2}$$

Equation (3) calculates the cost using the fitness function.

$$C_i = \text{Cost}_i = f\left(x^i_{1,}x^i_{2,\ldots}x^i_{\text{Nvar},}\right), i = 1, 2, 3, \ldots, N_{\text{pop}} \tag{3}$$

The number of streams is defined using Eq. (4).

$$N_{\text{stream}} = N_{\text{pop}} - N_{\text{sr}} \tag{4}$$

Equation (5) describes the process of allocating streams to rivers and sea.

$$NS_n = \text{round}\left\{\frac{\text{Cost}_n t}{\sum_{i=1}^{Nsr} \text{Cost}_i} \times N_{\text{Streams}}, n = 1, 2, \ldots N_{\text{sr}}\right. \tag{5}$$

- **Flow of streams into rivers and sea**

Equation (6) describes the updating of positions of streams and Eq. (7) describes the position of rivers.

$$X^{i+1}_{\text{Stream}} = X^i_{\text{Stream}} + \text{rand} \times C \times \left(X^i_{\text{River}} - X^{i+1}_{\text{Stream}}\right) \tag{6}$$

$$X^{i+1}_{\text{River}} = X^i_{\text{River}} + \text{rand} \times C \times \left(X^i_{\text{Sea}} - X^i_{\text{River}}\right) \tag{7}$$

In this case rand gets value within [0, 1].

- **Evaporation condition and Raining Process**

The following code decides whether evaporation process starts or not.

If$X_{\text{sea}} - X_{\text{river}} < s_{\max} i = 1, 2, \ldots N_{\text{sr}-1}$ Then evaporation and raining process stops.

Equation (8) specifies the decreasing value of $s_{\max}$.

$$s^{i+1}_{\max} = s^i_{\max} - \frac{s^i_{\max}}{\text{maxiteration}} \tag{8}$$

New streams are generated due to raining process. It is described using Eq. (9).

$$X_{\text{stream}}^{\text{new}} = \text{LB} + \text{rand} \times (\text{UB} - \text{LB}) \tag{9}$$

- **Evaporation rate for ERWCA.**

Some of the streams are evaporated before joining into sea. Equation (10) defines the evaporation rate (ER) that enables the evaporation process in ERWCA approach.

$$\text{ER} = \frac{\text{Sum}(NSn) \times \text{rand}()}{N_{\text{sr}} - 1}, \; n = 2, 3, \ldots, N_{\text{sr}} \tag{10}$$

2.3 *FLANN/ERWCA Model*

The ERWCA processes are used to optimize the weights of FLANN model. The mean square error (MSE) is used as the objective function using Eq. (11).

$$\text{MSE}(n) = \frac{1}{N} \sum_{n=1}^{N} \text{err}^2(n) \tag{11}$$

where the error is calculated using Eq. (12).

$$\text{err}(n) = x(n) - y(n) \tag{12}$$

$x(n)$ and $y(n)$ represent the actual currency and the predicted currency value, respectively. The FLANN/ERWCA model is described in Fig. 1.

3 Simulation Study

This section discusses the dataset preprocessing, training, and testing the model. The performance of the FLANN/ERWCA model is compared with FLANN-WCA and FLANN-PSO models. The exchange rate data of Singapore Dollar to Indian Rupee (SGD to INR) from November 4, 2003 to December 3, 2018 are collected in this paper. The data are normalized using min–max normalization. The different technical indicators used for this work are simple moving average (SMA), momentum, exponential moving average (EMA), average true range (ATR), relative strength index (RSI), Willium %R, moving average convergence divergence (MACD), %k (stochastic oscillator). A sliding window of size 12 is considered for generating the input pattern. For training purpose, 80% data and for testing purpose, 20% data are taken in this paper. The weights between the input layer and output layer of the

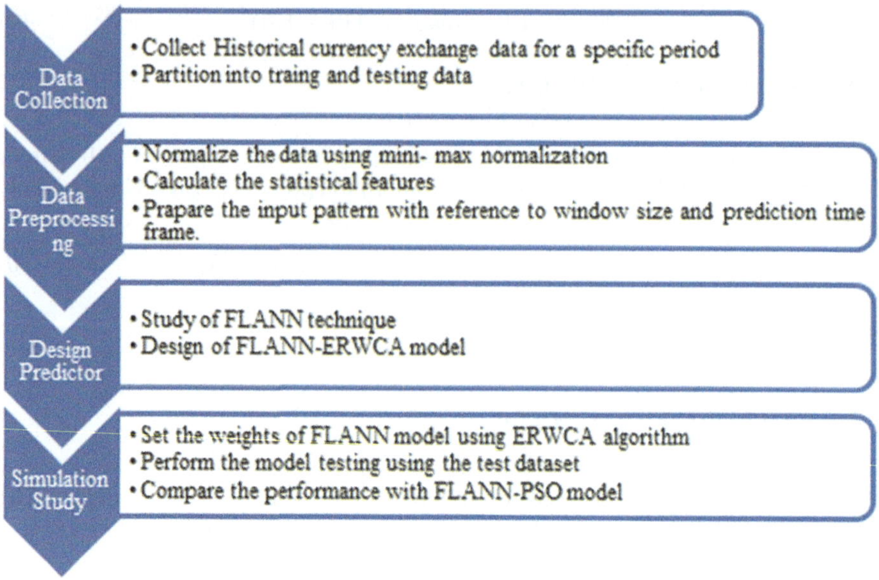

Fig. 1 Workflow of FLANN/ERWCA model

FLANN model are optimized using different evolutionary and swarm-based algorithms, that is, PSO, WCA, and ERWCA. The population size is 100 and number of iterations is 100. The number of streams, rivers, and sea are 80, 19, and 1, respectively. Each input pattern is applied sequentially in the FLANN model to determine the predicted price. The error values in each input pattern are used to calculate the RMSE after one iteration. The RMSE values are plotted against each iteration in Fig. 2 for 1-day and 7-day ahead predictions.

The weights are set after the training process. In this work to measure the performance of the models, mean average error (MAE), root mean square error (RMSE) and mean absolute percentage error (MAPE) are taken as the performance measures during the testing process. Tables 2 and 3 describe these performance measures for the 1-day and 7-day ahead price prediction for different models.

During training the model, it is observed that RMSE of the training patterns achieve minimum value. From the Tables 2 and 3, it is found that the FLANN-ERWCA model produced better result as compared to the other two models in forecasting the SGD to INR exchange rate for 1-day and 7-day ahead forecasting.

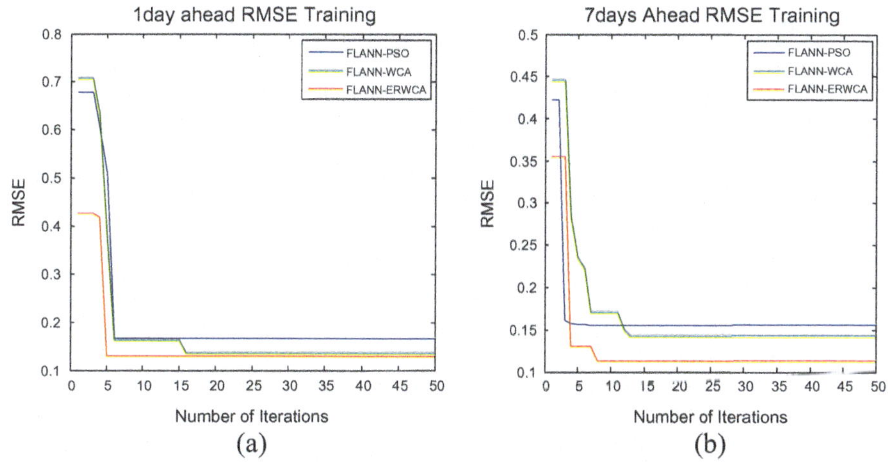

Fig. 2 Training time convergence graph for FLANN with PSO, WCA, and ERWCA (**a**) 1 day (**b**) 7 days ahead prediction

Table 2 1 day ahead prediction

Models	MAE	RMSE	MAPE
FLANN-PSO	0.153467	0.156768	0.319065
FLANN-WCA	0.147167	0.150607	0.305937
FLANN/ERWCA	0.14216	0.145725	0.295519

Table 3 7 days ahead prediction

Models	MAE	RMSE	MAPE
FLANN-PSO	0.159967	0.163137	0.332609
FLANN-WCA	0.157367	0.160588	0.327191
FLANN/ERWCA	0.157067	0.160294	0.326566

4 Conclusion

This paper describes the FLANN/ERWCA model for the currency exchange rate forecasting. The SGD to INR exchange rate data has been collected for a specific period and preprocessed. The different technical indicators are associated with the exchange rate data. The proposed model is trained with training dataset and the RMSE values are plotted for the comparison of convergence characteristics. During testing, the MAE, RMSE, and MAPE values are compared for measuring the performance of different models. It is observed that the proposed FLANN/ERWCA model outperformed the FLANN-PSO and FLANN-WCA model in 1-day and 7-day ahead exchange rate forecasting.

References

1. Anastasakis, L., Mort, N.: Exchange rate forecasting using a combined parametric and nonparametric self-organizing modelling approach. Expert Syst. Appl. **36**(10), 12001–12011 (2009)
2. Henrique, B.M., Sobreiro, V.A., Kimura, H.: Literature review: machine learning techniques applied to financial market prediction. Expert Syst. Appl. **124**, 226–251 (2019)
3. Gardner, E.: Exponential smoothing: the state of the art. J. Forecast. **4**(1), 1–28 (1985)
4. Rojas, I., Valenzuela, O., Rojas, F., Guillén, A., Herrera, L.J., Pomares, H., Pasadas, M.: Soft-computing techniques and ARMA model for time series prediction. Neurocomputing **71**, 519–537 (2008)
5. Tealab, A.: Time series forecasting using artificial neural networks methodologies: A systematic review. Future Comput. Inf. J. **3**, 334–340 (2018)
6. Galeshchuk, S.: Neural networks performance in exchange rate prediction. Neurocomputing **172**, 446–452 (2016)
7. Yu, L., Lai, K.K., Wang, S.: Multistage RBF neural network ensemble learning for exchange rates forecasting. Neurocomputing **71**(16), 3295–3302 (2008)
8. Majhi, R., Panda, G., Sahoo, G.: Efficient prediction of exchange rates with low complexity artificial neural network models. Expert Syst. Appl. **36**(1), 181–189 (2009)
9. Das, S.R., Mishra, D., Rout, M.: A hybridized ELM-Jaya forecasting model for currency exchange prediction. J. King Saud Univ. Comput. Inf. Sci. (2017)
10. Majhi, R., Panda, G., Sahoo, G.: Development and performance evaluation of FLANN based model for forecasting of stock markets. Expert Syst. Appl. **36**, 6800–6808 (2009)
11. Jena, P.R., Majhi, R., Majhi, B.: Development and performance evaluation of a novel knowledge guided artificial neural network (KGANN) model for exchange rate prediction. J. King Saud Univ.-Comput. Inf. Sci. **27**(4), 450–457 (2015)
12. Dehuri, S., Roy, R., Cho, S.B., Ghosh, A.: An improved swarm optimized functional link artificial neural network (ISO-FLANN) for classification. J. Syst. Softw. **85**(6), 1333–1345 (2012)
13. Eskandar, H., Sadollah, A., Bahreininejad, A., Hamdi, M.: Water cycle algorithm–A novel metaheuristic optimization method for solving constrained engineering optimization problems. Comput. Struct. **110**, 151–166 (2012)
14. Sadollah, A., Eskandar, H., Kim, J.H.: Water cycle algorithm for solving constrained multi-objective optimization problems. Appl. Soft Comput. **27**, 279–298 (2015)
15. Haddad, O.B., Moravej, M., Loáiciga, H.A.: Application of the water cycle algorithm to the optimal operation of reservoir systems. J. Irrig Drain. Eng. **141**, 5 (2014)
16. Sadollah, A., Eskandar, H., Bahreininejad, A., Kim, J.H.: Water cycle algorithm with evaporation rate for solving constrained and unconstrained optimization problems. Appl. Soft Comput. **30**, 58–71 (2015)

Multi-document Summarization Using Deep Learning

Abhishek Upadhyay, Khan Ghazala Javed, Rakesh Chandra Balabantaray, and Rasmita Rautray

Abstract Due to the advancement of the Internet nowadays, a lot of people are mostly dependent on the Web to get the required information. As data is increasing exponentially, there is a high chance of duplication of data; it is difficult and tedious for the manual reading of all the documents as well as the rejection of the duplicates and extraction of useful information. One of the solutions to this issue is "Text Summarization," through which a huge volume of data can be read quickly; but it is very hard to summarize documents manually, thus necessitating the use of an automatic tool to perform this task. Abstractive Text Summarization is one such automated technique of producing a short and accurate summary of a document while preserving essential information and comprehensive meaning. In this paper, Abstractive Text Summarization using Deep Learning with Attention Mechanism has been proposed. The designed framework removes duplicate data and generates new sentences by rephrasing them or adding words originally absent in the source text. Experimental results on the dataset, Amazon Fine Food Review, are evaluated by utilizing performance metrics such as Rouge scores.

Keywords Summarization · LSTM · Abstract summary · Deep learning

A. Upadhyay · K. G. Javed · R. C. Balabantaray
Department of CSE, International Institute of Information Technology, Bhubaneswar, India
e-mail: A118001@iiit-bh.ac.in

K. G. Javed
e-mail: A118004@iiit-bh.ac.in

R. C. Balabantaray
e-mail: rakeshbray@gmail.com

R. Rautray (✉)
Department of CSE, ITER, Siksha 'O' Anusandhan Deemed to be University, Bhubaneswar, Odisha, India
e-mail: rashmitaroutray@soa.ac.in

D. Mishra et al. (eds.), *Intelligent and Cloud Computing*,
Smart Innovation, Systems and Technologies 194,
https://doi.org/10.1007/978-981-15-5971-6_20

1 Introduction

Automatic summarization Text refers to the procedure of shortening a large volume of text to create a summary point of source document while retaining its major key features [1, 2]. Text Summarization is parted into two categories, namely, extractive and abstractive Text Summarization. The Extractive Text Summarization is a technique of Text Summarization that extract few sentences directly from the source document to summarize the original document [3], whereas Abstractive Text Summarization is a method in which novel sentence is generated with the help of natural language generation techniques [4].

Extractive Text Summarization models were foremost planned to extract the elemental linguistics data present in the text and to summarize the original text. With the advancement in the performance of the system and new development and increasing effectiveness and capabilities of deep learning theory, Abstractive Text Summarization models that map the set of input into another set of output have been proposed known as sequence-to-sequence models. In past, the problem which involves sequential information can be solved using sequence-to-sequence models [5]; it portrays, input of fixed length to output of fixed length where input and output length may differ. It has been used widely in many problems such as machine translation, voice recognition, and video captioning [2]. Although in many recognized datasets, the current Text Summarization models have obtained very good results, but all the problems are not solved by these models. Semantic and Syntactic structures are two very important elements for assessing Text Summarization models, but only one aspect is the emphasis of these two model types. The old models of the Extractive Text Summarization are coarse-grained and have some isolated sentences in the summaries. The benefit of Extractive Text Summarization models is that with the syntactic structure criteria, the sentences present in the summaries are consistent. The disadvantage of Extractive Text Summarization models, however, is that the sentences in the summaries are not semantically coherent. The current Abstractive Text Summarization models are incompressible and summarizes of the model with semantic elements phrases (phrases, word, etc.). The advantage of Abstractive Text Summarization models is comprehensive semantics, because models given trains the bond between words and after training it will produce keywords sequence based upon the bond between the words.

In Abstractive Summarization the summary which is generated is very compact and precise as well as it does not rely on the source document length in summarization. Furthermore, an important aspect in Text Summarization is to efficiently compress the source document to text in such a lossy manner such that the important features are preserved for source documents. Recently, Abstractive summarization is implemented with the help of Deep Learning where input given is a long sequence of words and output is a short and precise summary. Due to the involvement of sequential information, Sequence-to-Sequence modeling is used. The Seq2Se2 model can be built on a given problem which involves sequential information. Machine Translation, Voice recognition, Sentiment Classification can be applications of sequential information. In Machine Translation, the input text is given in one language and the output produced is text in another language [2].

I love food → J'aime la nourriture.

Seq2Seq model intent to map a fixed length input with a fixed length output, where the input and output length may be different [6]. The dataset which is used for evaluating the accuracy is Amazon Fine Food Review which is an open access dataset that includes 568,454 reviews of 256,059 users on 74,258 products up to October 2012. The review includes ratings, user, and products information and review as plain text [7]. The rest of the section is organized as follows: in Sect. 2, related work on the topic of Text Summarization is discussed; Sect. 3 contextualizes the model proposed with the previous related work; Sect. 4 depicts the results, compared with the other existing systems, and the last section concludes the paper with the future scope of the devised model.

2 Related Work

Summarization is a task to take a single document as input or a collection of several documents and generate a compact and smooth summary preserving the crucial information. Based on methodology, summarization can be classified into Extractive and Abstractive Summarization. Extractive Summarization is an easy method for shortening the text, important and key words from the input text are copied to generate the desired summary.

In summarization a lot of past work has been done on extractive, and it consists of finding important sentences in the input, and summary is recreated. Some of the appreciable works are done before [1, 3, 5]. Conversely, Humans try to rephrase the original document in their own words. As such, the essence of human summaries is mostly abstractive and hardly contains a copy of original sentences from the source document. Summarization is called Abstractive when it generates a compact and fluent summary of source document while preserving its important feature [4]. It is much more difficult which involved advanced technique for meaning representation, content organization, and paraphrasing. Abstractive summarization has been specifically used in machine translation based model [8], but as deep learning has been popularized in natural language processing (NLP) tasks, it has influenced more work in this area.

Researchers have found this framework as a tempting, as well as a substitute to abstractive summarization. To get more accurate summary convolutional models were used by the author for encoding the input text, and a context-sensitive attentional feed-forward neural network producing appreciable results on Giga-word corpus and standard DUC datasets [9]. In addition to this work, another author used an identical convolutional model for the encoding, but decoder is replaced by Recurrent Neural Network generating more enhancement in accuracy on both datasets [4]. Some more remarkable works are also done in [1, 4]. The first and easy Abstractive

Text Summarization model was sequence-to-sequence model which gives promising results on standard dataset as provided in [6].

3 Proposed Model

Text Summarization is classified broadly into three steps: Data Preprocessing, Word embedding, and train-test to evaluate model as shown in Fig. 1.

In the Data Preprocessing, Standard text processing technique has been used to drop all the unwanted symbols, characters, etc. Certain steps are to be applied to raw data: Converting everything to lower case, remove Html tags, eliminate punctuations and special character, and remove stop words. In the Word embedding, the dataset is split into training and validation set, the tokenizer is prepared for text as well as for summary; a tokenizer converts a word sequence into an integer sequence. Then to train as well as test the data, the input text is fed into the proposed model, i.e., a 3-stacked Long Short-Term Memory (LSTM) encoder and decoder, with attention mechanism, due to their capability of acquisition of long-term dependencies by mitigating vanishing gradient problem. The Encoder–Decoder framework is designed to work in two phases: Training and Inference phase is shown in Fig. 2.

In the Training phase, the entire input sequence is provided to the encoder at a rate of one word per time step. At every time step, the present contextual information in the given input sequence is captured and processed; subsequently, the decoder gets trained in next-word prediction using the previous word as a basis.

The goal of the Inference phase is to evaluate new source sequences having no known target sequence. As shown in Fig. 3, firstly, the entire sequence as input is encoded and the decoder is initialized with the encoder's internal states then the

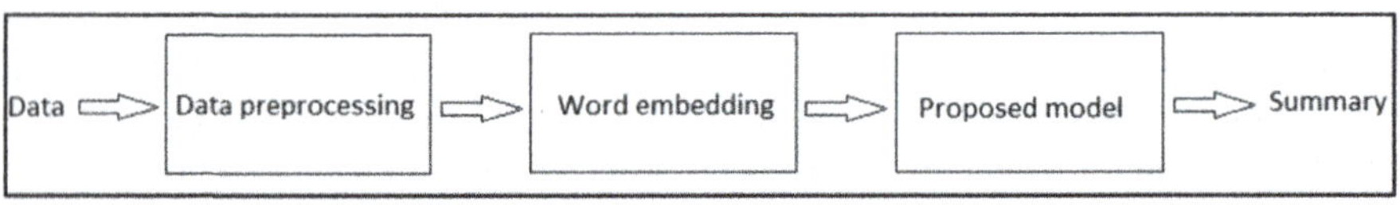

Fig. 1 Flow diagram of summarization

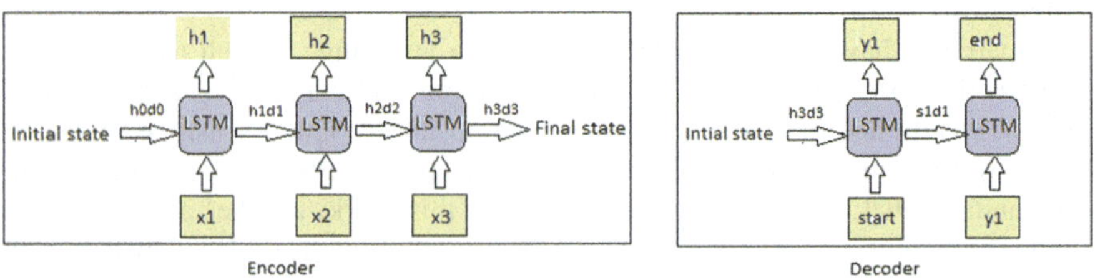

Fig. 2 Training phase

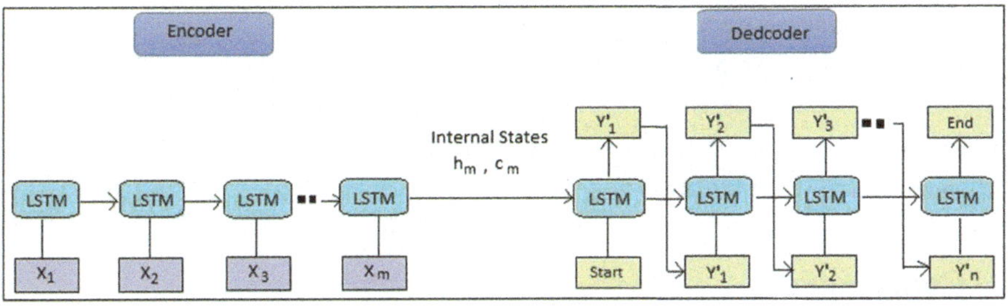

Fig. 3 Inference phase

initial token is passed as decoder input. After that, for one time step, the decoder with internal states is run and the outcome with the possibility for the subsequent word. The word which gives maximal possibility will be chosen. Then in the next time step the sampled word is passed as a decoder input and the internal states are updated with the current time, these steps are carried out until the final token is generated or target sequence maximum length is reached.

4 Experiment and Results Analysis

The evaluation is performed on the Amazon Fine Food review dataset, which has open access and includes 568,454 reviews of 256,059 users on 74,258 products up to October 2012. The review contains the information of the user as well product, ratings, and a plain text review [7]. The ROUGE is known as Recall-Oriented Understudy for Gistng Evaluation. It is a set of metrics used for calculating summarization accuracy of texts as well as calculating machine translation. It compares an automatically generated summary against a collection of original summaries (mostly human-generated). There are different categories of rouge as ROUGE-N known as recall-based measure and it compares n-grams. A streak of n-grams, ROUGE-L compares the two sequences of text on the basis of longest common subsequence (LCS) algorithm. The chance of similarity of summarization is more as long as the LCS between two summary sentences are bigger, and compared to the previous one this metric is more adaptable, but the flaw in this measure is that all n-grams must be consecutive. ROUGE-S metric examines both bi-grams and uni-grams; it is known as skip-bigram. In bi-grams words can be inserted between first and last word, so sequences of words need not be consecutive [10]. Table 1 concluded that our approach gives better rouge 2 and rouge SU4 score than ID24 summarization system.

Table 1 Comparing our model with system ID24

No. of summaries	Proposed model		System ID24	
	Rouge 2	Rouge SU4	Rouge 2	Rouge SU4
50 summaries	0.19519	0.2060	0.05244	0.1117
100 summaries	0.18628	0.1989	0.13502	0.1901

5 Conclusion

In this paper, our model uses deep learning with attention mechanism to summarize the document. Preprocessing of the data has been carried out, then it is converted into tokenizer and fed to the proposed model for training phase and inference phase. Finally, extensive experiments on the dataset, Amazon Fine Food Review is evaluated and the result has been compared to the existing system in terms of ROUGE score which shows that our model shows promising result compared to other system ID24. The improvement can be made using bi-directional LSTM which can capture context from both ways which gives proper context vector. While performing training and testing with more dataset, better efficiency can be achieved.

References

1. Jiwei Li, F., Minh-Thang Luong, S., Dan Jurafsky, T.: A hierarchical neural autoencoder for paragraphs and documents. CoRR (2015). https://arxiv.org/abs/1506.01057
2. Sankaran, B.F., Mi, H.S., AlOnaizan, T.Y., Ittycheriah, A.F.: Temporal attention model for neural machine translation (2016). ArXiv e-prints
3. Kam-Fai Wong, F., Mingli Wu, S., Wenjie Li, T.: Extractive summarization using supervised and semi-supervised learning. In: Proceedings of the 22nd Annual Meeting of the Association for Computational Linguistics, pp. 985–992 (2008)
4. Sumit Chopra, F., Michael Auli, S., Rush, A.M.T.: Abstractive sentence summarization with attentive recurrent neural networks. HLT-NAACL (2016)
5. Caglar Gulcehre, F., Sungjin Ahn, S., Ramesh Nallapati, T., Bowen Zhou, F., Yoshua Bengio, F.: Pointing the unknown words. In: Proceedings of the 54th Annual Meeting of the Association for Computational Linguistics (2016)
6. Ramesh Nallapati, F, Bing Xiang, S., Bowen Zhou, T.: Sequence-to-sequence rnns for text summarization. In: ICLR workshop (2016). https://arxiv.org/abs/1602.06023
7. McAuley, J.F., Leskovec, S.J.: From amateurs to connoisseurs: modeling the evolution of user expertise through online reviews (2013)
8. Dzmitry Bahdanau, F., Jan Chorowski, S., Dmitriy Serdyuk, T., Philemon Brakel, F., Yoshua Bengio, F.: End-to-end attention based large vocabulary speech recognition. CoRR (2015). https://arxiv.org/abs/1508.04395
9. Rush, A.M.F., Sumit Chopra, S., Jason Weston, T.: A neural attention model for abstractive sentence summarization. CoRR (2019). https://arxiv.org/abs/1509.00685
10. Lin, C.Y., Och, F.J.: Looking for a few good metrics: ROUGE and its evaluation. NTCIR Workshop (2004)

A Convolutional Neural Network Framework for Brain Tumor Classification

Sanjaya Kumar Jena and Debahuti Mishra

Abstract Machine learning has proven to be of enormous importance over the last couple of years. The recent boom known as convolutional neural networks (CNNs) began outperforming other established models across a variety of areas like data analysis, health care, and computer vision. CNN is also widely deployed in industry and academia. CNN-based models have a huge potential for medical image analysis, medical data analysis, medical diagnosis, and healthcare. In this paper, a CNN framework is presented for *Glioma* tumor classification of brain MRI image. The classification accuracy is compared to multilayered neural network and KNN ($k = 2$) models. The proposed CNN framework outperforms the other models irrespective of its complexity.

Keywords Machine learning · Convolutional neural network · KNN · Multilayered neural network · MRI

1 Introduction

In recent times, machine learning plays a vital role in classification, regression type of problem around us. A subclass of machine learning, deep neural net, makes learning in multiple levels of layers for feature representation that provides more attention in various domains such as data analysis, health care, industry, and academic, respectively. The deep neural net structure basically contains more hidden layers in extending the single layer feedforward artificial neural net. CNN is a kind of deep neural net based framework for better feature representation in different application domains. In recent years, CNN has proven better performance in different machine learning

S. Kumar Jena (✉) · D. Mishra
Department of Computer Science and Engineering, Siksha 'O' Anusandhan Deemed to be University, Bhubaneswar, Odisha 751030, India
e-mail: sanjayjena@soa.ac.in

D. Mishra
e-mail: debahutimishra@soa.ac.in

D. Mishra et al. (eds.), *Intelligent and Cloud Computing*, Smart Innovation, Systems and Technologies 194,
https://doi.org/10.1007/978-981-15-5971-6_21

applications like medical image analysis, natural language processing, image dataset classification, large scale documents analysis, and computer vision [1].

The abnormal cell growth in brain is termed as brain tumor that affects the person very badly. The tumor disrupts the proper working of the brain and that leads to life-threat. A recurrent spreading and more dangerous is the malignant tumor in contrast to benign tumor. In medical domain, various techniques such as CT scan, Ultrasound, PET, and MRI are employed to detect the case of tumor and its treatment. Brain MRI image is found to be more suitable for the diagnosis and treatment of tumors. In recent times, MRI has shown as an effective measure for clinical research and brain tumor detection. Specifically, machine learning approaches like artificial neural net, multiplayer perceptron, CNN are getting more effective in detection and classification of tumors and provides better diagnosis results [2].

CNN is a deep learning neural net architecture that performs complex feature mapping operations through different sizes of filters with additional stride and pooling operations [3, 4]. A hybrid method for tumor classifier using feature extraction through 2D DWT, feature reduction through normalization-PCA and NN classifier has been discussed in [5].

This paper is described in the given ways: related work in Sects. 2 and 3 focused on the proposed implementation model. Section 4 described the experimental setup and simulation results, and Sect. 5 discussed the conclusion of the work.

2 Related Work

Artificial neural network, deep neural network, and a subclass of deep learning networks play an important role in medical health care to identify brain tumor and provide efficient measures for diagnosis. The most complex and vital organ of the human body is the brain, which works with billions of cells. An abnormal group of cells created inside the brain due to uncontrolled division of cells inside or around the brain causes brain tumors. Basically, brain tumors are classified as *benign* and *malignant*, and malignants are found to be more dangerous.

A deep neural net approach has been made to classify brain tumor using image segmentation, feature extraction, feature reduction, and classification [6]. They reported a classification accuracy of 96.97% on the dataset based on Harvard medical School [7]. Image classification framework employing local patterns in binary to extract image features locally are discussed in [8]. They employ local patterns in binary to extract image features locally with feature and decision fusion levels to follow the decision of the classifier. An ensembled-based classifier has been proposed for better accuracy in [9].

Support vector machine based tumor detection with inputted image preprocessing, noise removal using diffusion filter has been discussed in [10]. Different manifold phases like preprocessing and segmentation features extraction using first-order and second-order Gray level Co-occurrence Matrix from segmented brain MR images and classification using feed forwarded artificial neural network has been discussed in [11]. The proposed classifier found better accuracy on AANLIB database. An automatic brain MRI segmentation method based on convolutional neural networks using small filter of size 3×3 has been proposed in [12]. The proposed model was validated on BRATS-2013 dataset and got the first position for the complete, core, and enhancing regions in dice similarity coefficient metric (0.88, 0.83, 0.77) for the challenge dataset.

A visualization and clear understanding of the convolutional neural network have been given in [13]. The authors have discussed the impressive classification performance of convolution networks on Imagenet benchmark and also gave the realization of each output of the convolution layer. The intermediate feature map layers and their role in the classifier have been widely discussed in their work. The two paths based convolutional neural network and FreeSurfer, to extract image features for MRI image analysis in predicting Alzheimer's disease with ELM has been discussed with accuracy 86.1% in [14]. They made a comparison with svm and random forest to have accuracies of 79.87% and 75.0%, respectively. In recent years different kinds of convolutional network have been proposed and their performances are studied in image classification to improve the classification accuracy and to reduce the error rate [15–18].

3 Proposed Model

The description of the proposed model shown in Fig. 1 is as follows.

3.1 Input Image

The original images of size $256 \times 256 \times 1$ are resized to $36 \times 36 \times 1$. So input image of size $36 \times 36 \times 1$ is fed to the model for simulation purpose.

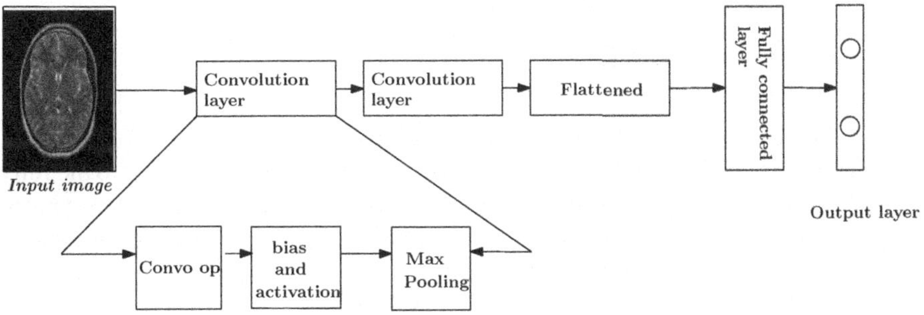

Fig. 1 Convolutional neural network based classification model

3.2 Convolution Layer

The convolution layer in the proposed model consists of (i) convo operation, (ii) bias and activation, (iii) maximum pooling. This layer extracted the feature maps from the image. *Convo operation*- images are represented in multidimensional array. The multidimensional array is convolved with a kernel or filter to generate output feature map. For simplicity image size $6 \times 6 \times 1$ and kernel size 3×3 generate output 4×4 matrix. The convo operation is shown diagrammatically as

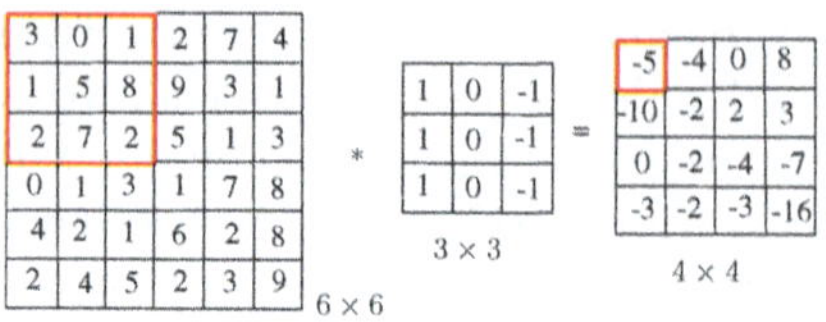

The red bordered matrix is convolved with the kernel as $3 \times 1 + 0 \times 0 + 11 \times -1 + 1 \times 1 + 5 \times 0 + 8 \times -1 + 2 \times 1 + 7 \times 0 + 2 \times -1 = 5$. The remaining cells of the 4×4 output matrix will be computed by sliding the kernel one cell over the 6×6 matrix to get all the cell values in the output.

To each of the output bias and activation are (nonlinearity) added and all of them are stacked together to form one convolution layer.

3.3 Pooling Layer

Pooling is another building block in convolution networks that is used to reduce the size of the feature representation. Pooling can be either *max pool* or *average pool* which is used to summarize the subregions of the features. The working of max pool and average poll are shown as

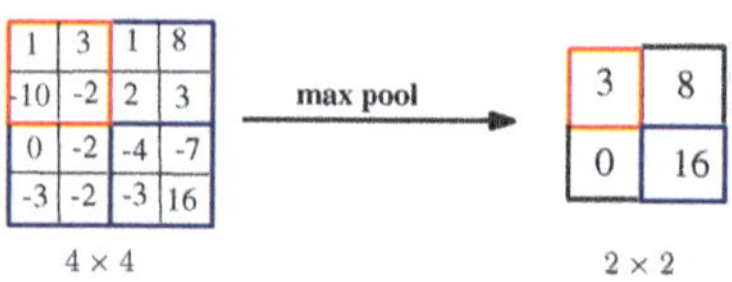

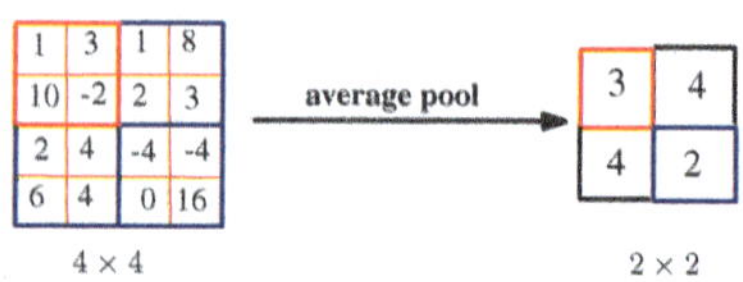

3.4 Fully Connected Layer

It is the feed forward neural network. These are the last few layers in the network. The output from the last pooling or convo layer is flattened and connected to the fully connected layer. This layer forms a vectorized feature map also called as dense layer. Many numbers of fully connected layers can be added depending on the requirement. In the proposed model a fully connected layer is used over the dataset.

3.5 Output Layer

This is the output layer in the network, from which the classification can be made. This layer outputs into different number of classes as required by the network.

4 Experimental Setup and Results

4.1 Dataset Description

The proposed model is demonstrated on brain tumor dataset collected from [7]. These dataset hold T_2 weighted brain tumor MR images in the format *gif* with size 256×256, In the dataset, the images are in *transaxial plane*. The brain disease *Glioma* dataset that contains 122 images are experimented in the proposed model for classification. The images are resized to size $36 \times 36 \times 1$. There are 72 abnormal and 50 normal cases of tumor present in the dataset.

4.2 Initial Kernel Weights

The kernel or filter weights are initialized randomly in the model for experiment. Various sizes of the kernel have been tested on the samples dataset and the kernel size of 3×3 was found suitable in this work.

4.3 Activation Function

Here, two types of activations are considered in the proposed model. The activation *Relu*, $f(x) = max(0, x)$ is used in the convolution layers and the *sigmoid*, $f(x) = \frac{1}{1+e^{-x}}$ in the output layer as the sigmoid activation that squashes output numbers to 0 and 1.

4.4 Loss Function Computation

The loss function has been selected as *binary-cross entropy loss* or *log loss* to enhance the accuracy of the convolutional model. As the proposed model is binary level classification, so log loss is treated as a better candidate and is also being used widely. The log loss for a training sample x_i is expressed as $Logloss =$

$-\frac{1}{n}\sum_{i=1}^{n} l_i \log(p(l_i)) + (1 - l_i) \times \log(1 - p(l_i))$, where l_i is the label for training sample x_i, and $p(l_i)$ is the predicted probability of sample x_i for all n training samples.

4.5 Training

The proposed model is trained using gradient descent algorithm. The training loss is minimized through gradient descent computation in each iteration. The model is employed with stochastic gradient optimizer for both cases (i) using multilayer network and (ii) convolution neural network.

4.6 Model Parameters

The different parameters and parameter sizes of the convolution neural network model are listed below as

(i) Original image size resizes to $36 \times 36 \times 1$ that acts as input to the model.
(ii) *Covolution layer-1*: 50 filters of size 3×3 output the feature map of $34 \times 34 \times 50$ size. Here, total number of trainable parameters are $(3 \times 3 + 1(biasunit)) \times 50(filters) = 500$.
(iii) *Max pooling Layer-1*: Size of each 2×2 output the feature map of size $17 \times 17 \times 50$ with total number of parameters 0. Here, no parameters are trained.
(iii) *Covolution layer-2*: 150 filters of size 3×3 output feature map of $15 \times 15 \times 150$. Here total number of trainable parameters are 67650.
(iv) *Max pooling Layer-2*: Size of each 2×2, output the feature map of size $7 \times 7 \times 150$ with total number of parameters 0. Here, no parameters are trained.
(v) *Fully connected layer*: Flattened to 7350 size of no trainable parameters, here total 512 nodes are considered along with 3763712 parameters. The last fully connected layer contains 513 parameters. So, the total number of trainable parameters in the model is 3832375. The output dense or fully connected layer contains 513 parameters. In the model, any padding or stride is considered to feature map.

The proposed model summary is shown in the Fig. 2.

The details of the multilayer network over the above dataset isare expressed as

(i) Total number of training sample:122 of 1296 attributes, and total number of hidden layers are 5. Size of hidden neurons are 120, 120, 120, 120, and 8, respectively. The activation functions used in the hidden units are *ReLu*, and the *sigmoid* activation is used in the output layer. The multilayer network is trained using gradient descent of total 10 epoch of batch size 5.

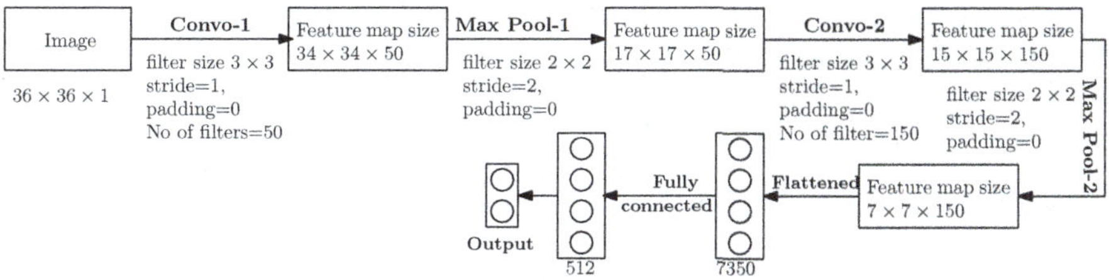

Fig. 2 The implemented summary of the proposed model and their activation sizes

Table 1 Performance metrics of CNN, Multilayer network, KNN $k = 2$ classifiers

Methodology	Classification accuracy in %
Proposed model	98.76
Multilayer network	97.84
KNN, $k = 2$	97.72

4.7 Simulation Results

In the convolution neural network, it has been taken 10 epoch each containing 122 images of size $36 \times 36 \times 1$. The batch size considered as 5 and the stochastic gradient descent optimizer is used. Python 3.6 is being used for the simulation purpose. The accuracy is given in Table 1.

The multilayer network of 5 hidden layers and KNN ($k = 2$) models are used for comparison purpose. The multilayer network and KNN model accuracy are also given in Table 1. It is found from the experimental analysis that the proposed CNN model performs better classification rate over the other two classifiers. The proposed CNN model has also been compared with the work of [19, 20] with a reported accuracy of 95.7% and 98.52%, respectively.

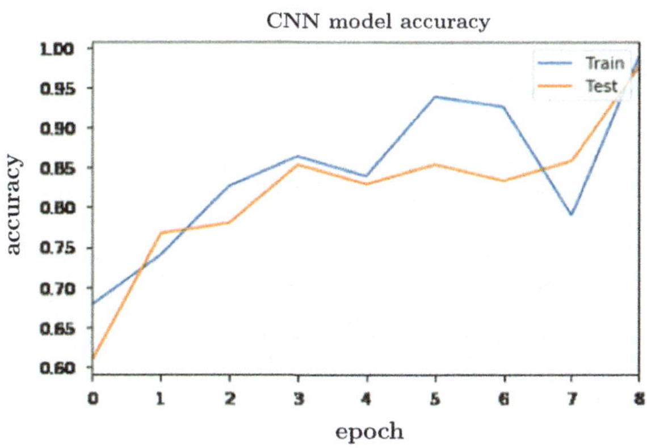

Fig. 3 The training and testing accuracy for the proposed model

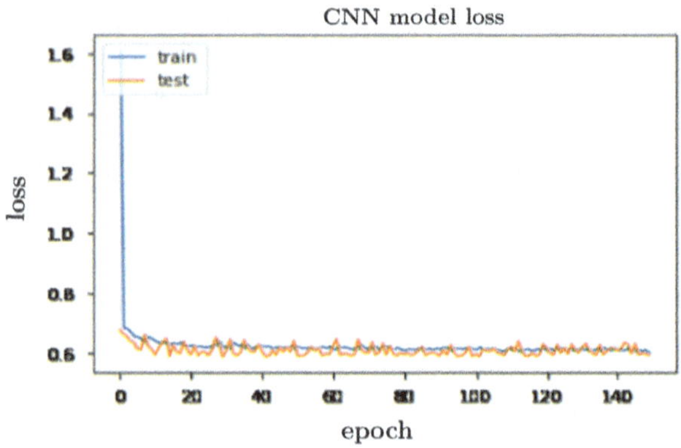

Fig. 4 The training and testing loss for the proposed model

The convergence graph for training accuracy and testing accuracy as well as the corresponding loss are shown in the following graphs: Fig. 3, and in Fig. 4 respectively.

5 Conclusion

In doing the above experimental evaluation, it is concluded that convolution neural network outperforms over the multilayer network of 5 hidden layers as well as the KNN ($k = 2$). Again the multilayer network requires the image data to be converted into pixels values. The number of trainable parameters is found to be more in convolutional neural network than multilayer network. The KNN ($k = 2$) model takes more execution time over the dataset. It is observed that the computational complexity of convolution is more as compared to multilayer network. Hence the same work can be deployed using GPU to reduce the workload of the underlined CPU. In future work, more delicate convolutional neural network approaches will be explored to enhance the performance of the classification.

References

1. Albawi, S., Mohammed, T. A., Al-Zawi, S.: Understanding of a convolutional neural network. In 2017 International Conference on Engineering and Technology (ICET) (pp. 1-6). IEEE (2007)
2. Rathi, V.G.P., Palani, S.: Brain tumor detection and classification using deep learning classifier on MRI images. Research Journal of Applied Sciences, Engineering and Technology **10**(2), 177–187 (2015)
3. Pan, Y., Huang, W., Lin, Z., Zhu, W., Zhou, J., Wong, J., Ding, Z.: Brain tumor grading based on neural networks and convolutional neural networks. In 2015 37th Annual International

Conference of the IEEE Engineering in Medicine and Biology Society (EMBC) (pp. 699-702). IEEE (2015)

4. Litjens, G., Kooi, T., Bejnordi, B. E., Setio, A. A. A., Ciompi, F., Ghafoorian, M., ... Sánchez, C. I.: A survey on deep learning in medical image analysis. Medical image analysis, 42, 60-88 (2017)
5. Zhang, Y., Dong, Z., Wu, L., Wang, S.: A hybrid method for MRI brain image classification. Expert Systems with Applications **38**(8), 10049–10053 (2011)
6. Mohsen, H., El-Dahshan, E.S.A., El-Horbaty, E.S.M., Salem, A.B.M.: Classification using deep learning neural networks for brain tumors. Future Computing and Informatics Journal **3**(1), 68–71 (2018)
7. http://med.harvard.edu/AANLIB
8. Li, W., Chen, C., Su, H., Du, Q.: Local binary patterns and extreme learning machine for hyperspectral imagery classification. IEEE Transactions on Geoscience and Remote Sensing **53**(7), 3681–3693 (2015)
9. Rokach, L.: Ensemble-based classifiers. Artificial Intelligence Review **33**(1–2), 1–39 (2010)
10. Saii, M., Kraitem, Z.: Automatic brain tumor detection in MRI using image processing techniques. Biomedical statistics and informatics **2**(2), 73–6 (2017)
11. Kumar, P., Vijayakumar, B.: An Efficient Brain Tumor MRI Segmentation and Classification Using GLCM Texture Features and Feed Forward Neural Networks. World Journal of Medical Sciences **13**(2), 85–92 (2016)
12. Pereira, S., Pinto, A., Alves, V., Silva, C.A.: Brain tumor segmentation using convolutional neural networks in MRI images. IEEE transactions on medical imaging **35**(5), 1240–1251 (2016)
13. Zeiler, M. D., Fergus, R.: Visualizing and understanding convolutional networks. In European conference on computer vision (pp. 818-833). Springer, Cham (2014)
14. Lin, W., Tong, T., Gao, Q., Guo, D., Du, X., Yang, Y., et. al.: Alzheimer's Disease Neuroimaging Initiative. Convolutional neural networks-based MRI image analysis for the alzheimer's disease prediction from mild cognitive impairment. Frontiers in neuroscience, 12 (2018)
15. LeCun, Y., Bottou, L., Bengio, Y., Haffner, P.: Gradient-based learning applied to document recognition. Proceedings of the IEEE **86**(11), 2278–2324 (1998)
16. Krizhevsky, A., Sutskever, I., Hinton, G. E.: Imagenet classification with deep convolutional neural networks. In Advances in neural information processing systems (pp. 1097-1105) (2012)
17. Szegedy, C., Liu, W., Jia, Y., Sermanet, P., Reed, S., Anguelov, D., Rabinovich, A.: Going deeper with convolutions. In Proceedings of the IEEE conference on computer vision and pattern recognition (pp. 1-9) (2015)
18. He, K., Zhang, X., Ren, S., Sun, J.: Deep residual learning for image recognition. In Proceedings of the IEEE conference on computer vision and pattern recognition (pp. 770-778) (2016)
19. Seetha, J., Raja, S.S.: Brain tumor classification using convolutional neural networks. Biomedical & Pharmacology Journal **11**(3), 1457 (2018)
20. Abiwinanda, N., Hanif, M., Hesaputra, S. T., Handayani, A., Mengko, T. R.: Brain tumor classification using convolutional neural network. In World Congress on Medical Physics and Biomedical Engineering 2018 (pp. 183-189). Springer, Singapore (2019)

Load Balancing in Cloud Computing Environment Using CloudSim

Saumendra Pattnaik, Jyoti Prakash Mishra, Bidush Kumar Sahoo, and Binod Kumar Pattanayak

Abstract Cloud computing has been built on the advancement of distributed computing and virtualization concept for offering effortless accessibility of the resources from various sources. It offers efficient access to many organizations to obtain services from the cloud service providers. The concept of virtualization is its advantage which leads to its extensive application. Through virtualization concept, it offers services to the users on demand basis. So, the cloud service providers consider the resource allocation to various servers, which needs to be done efficiently. So, we need cloud simulation and modeling tool for creating cloud as per necessity in cloud computing. Load balancing using these simulators is the task which is considered in this paper. CloudSim tool is used for verification of the load balancing modifying the number of brokers or number of virtual machines.

Keywords CloudSim · Virtual machine · Cloud computing · Broker · Cloudlet

S. Pattnaik (✉) · B. Kumar Pattanayak
Department of Computer Science and Engineering, ITER, Siksha 'O' Anusandhan (Deemed to be University), Bhubaneswar, India
e-mail: saumendrapattnaik@gmail.com

B. Kumar Pattanayak
e-mail: binodpattanayak@soa.ac.in

J. Prakash Mishra
Computer Science and Engineering Department, Gandhi Institute for Education and Technology, Baniatangi, Bhubaneswar, India
e-mail: jpmishra@gietbbsr.com

B. Kumar Sahoo
Department of Computer Science and Engineering, GIET, (Affiliated to BPUT), Baniatangi, India
e-mail: bidush.sahoo@gmail.com

D. Mishra et al. (eds.), *Intelligent and Cloud Computing*, Smart Innovation, Systems and Technologies 194,
https://doi.org/10.1007/978-981-15-5971-6_22

1 Introduction

Cloud computing is a mechanism where the connection of servers is hosted for providing access to store, manage, and process the data centrally instead of accessing the data through local servers. The cloud resources are used for multiple allocations by the cloud providers as per the client requirements with reduced cost and flexible resource utilization. So, the organizations use the cloud concept for reducing the infrastructure cost as well as the software resource costs. Virtual Machine, CPU Resource, Memory, etc. are various resources present in cloud environment situated in the data center. Cloud computing concept is used for describing a mixture of computing concepts which involves huge number of hosts linked by the communiqué media like Internet. Cloud computing is valuable in distributed computing over the network with a capacity for executing a program or application in numerous linked computers concurrently. The simulated hardware concerns for load balancing.

Load balancing is the process of distributing workloads and computing resources to improve the performance of the system. It allows the users to manage the demands of application or workload by allocating resources among multiple computers, networks, or servers. Load balancing refers to the distribution of various resources among different users with respect to their requests in a uniform manner in such a way that no node is supposed to be overloaded and neither be idle. Load balancing represents a vital task in cloud computing environment like in any other distributed computing online environment. The main goal of load balancing in cloud computing is to achieve maximum resource utilization, maximize throughput, maximize fault tolerance, minimize response time and avoid overload.

Load balancing in conventional computing environments such as parallel, grid, or distributed computing, can be implemented using scheduling algorithms classified as static, dynamic, and a mixture of both with respect to their characteristics [5]. Static Load Balancing Algorithm can be used for small distributed environments that operate with significantly higher Internet speed wherein the communication delays are negligibly small. However, Dynamic Load Balancing Algorithms are used for relatively larger distributed computing environments wherein the focus is made on reduction of communication delays as well as the execution time. Mixed Load Balancing Algorithms attempt to achieve a symmetrical distribution of the assigned computing tasks thereby reducing the communication cost of all the nodes in the distributed computing environment. Keeping in view the above classification, cloud computing can be considered to belong to the second category meaning that the aspect of load balancing in cloud computing environment tends to rely on dynamic scheduling algorithms. Hence, it can be assumed that a scheduling algorithm that is to be used for cloud computing needs to comply with the dynamic demands for various services thereby achieving an optimized load balancing at the scheduling time. The principal advantage of using dynamic load balancing mechanism is that the failure of a node in the computing environment may affect the system performance but the system will not be disrupted as a result of it.

2 Related Work

For modeling the Grid-based environments, various simulators were proposed earlier. But, for the multilayer service generalizations, the existing toolkits are incapable of segregating the (SaaS, PaaS, and IaaS) which is required by the cloud computing environments. For modeling the virtualized environment, no support is actually provided. So, CloudSim toolkit is a suitable platform for modeling the virtualized cloud environment, which contains the elements used to generate data centers, hosts, brokers, virtual machines along with the service requests [5].

For modeling and simulation, CloudSim is generally utilized in cloud computing systems. Various models are estimated by CloudSim with the values obtained corresponding to response time along with cost saving by utilizing a variety of configurations [7].CloudSim even permits multiple Data Centers simulation for facilitating a summary on federation with related policies regarding self-utilization of applications and immigration of virtual machines (VMs) used for improving reliability [8]. It maintains system as well as behavior modeling of cloud system components like data centers, and VMs with resource stipulating policies. It employs basic application stipulating methods which may be expanded with simplicity and restricted efforts. This toolkit holds modeling along with simulation of cloud computing settings comprising with sole as well as networked cloud environments [9].

CloudSim toolkit is created for simulating large-scale cloud environment and verifying cloud environment with different deployment platforms with configurations. For simulating distributed computing environments, Cloud Reports is a graphic tool based on the cloud computing environment. Cloud Reports can also be described as a graphical form of CloudSim with accessible environment, repeatability, and customizability [10].

3 Objective of Load Balancing in Cloud Computing

Proper load balancing with resource allocation takes a vital role in cloud computing environment. If the resources are improperly allocated, then some servers may have high loads and other servers may have low loads which may guide for extra energy utilization. So, resource allocation is a major issue as the cloud can be scaled starting from only two servers to thousands of servers which are impossible for deploying and testing the cloud. Thus, the essentiality of simulators occurs for testing the cloud as per the requirement which can consume less time, cost, and energy.

4 Necessity of Load Balancing

One of the major issues that cloud computing is facing nowadays is proper load balancing among the nodes or systems. There are many load balancing algorithms which are followed by different researchers in order to maintain fairness in the load distribution among various systems. The workload in cloud computing can be identified by some basic characteristics like

1. The processing model used by the cloud provider to identify the workload categories.
2. The cloud architecture by which the load can be distributed among the needy clients.
3. The resource requirements by the clients on regular basis and the list of requirements of the different clients.

The cloud computing environment always keeps track of its processing speed for meeting the unpredictable client needs, also which need to be considered carefully. So, for verifying the allocation of resources that is scalable up to large number of servers, cloud simulation and modeling tool is needed for taking care of the cloud with requirement basis. For providing dependable, safe, fault-tolerant, sustainable, and expandable computational services, cloud computing appeared as an important equipment that is available with Software, Infrastructure, or Platform as services. In an environment where various tests will be performed, the use of simulation tools provides an opportunity for evaluating the proposition before the software development. Exclusively in cloud computing environments, the simulation-based approaches provide noteworthy advantages where utilizing the infrastructure acquires pay-as-you use process. It also permits the cloud customers for testing the services in recurring and controlling environment with no charges, for improving the performance blocks prior to deploying it in actual Clouds [7].

The cloud computing simulation environment permits assessment of various kinds leasing of resources, situations with various loads along with pricing allocations in the side of cloud provider. It guides the providers for reducing the resource utilization pricing keeping in mind increasing the returns. By utilizing CloudSim, the developers may concentrate on precise system design concerns which need to be tested instead of worrying with the ground level data associated to Cloud-based infrastructures along with the services [7].

5 Verification of Load Balancing Using Cloud Simulator

For experimenting in the cloud computing environment, CloudSim is a simulation software for enabling the developers. It is an application that provides flawless modeling and simulation along with testing of application and cloud computing services. As a result of the issues that the accessible distributed system simulators are invalid

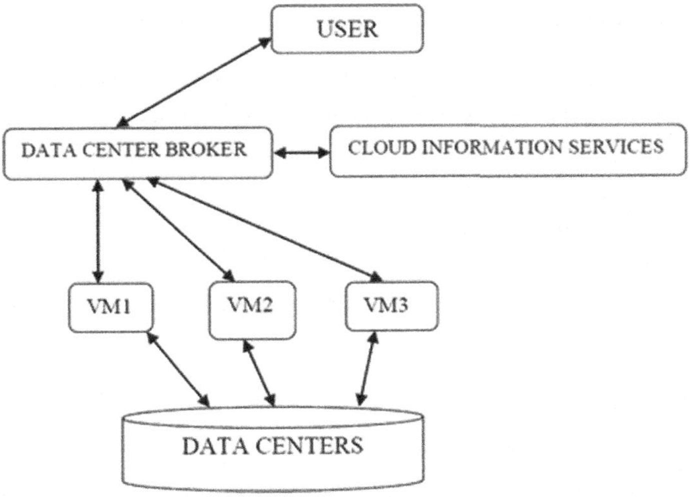

Fig. 1 Various Parts of CloudSim Connections

to the cloud computing environment, the evaluation of the result of cloud stipulating regulatory, application workload, services, and representations with resources results models under different systems, and user configurations with the needs are hard for attaining. So, CloudSim can be utilized in order to crossover these issues [9]. CloudSim also provides support system with behavior modeling of cloud system components like data centers, virtual machines (VMs), and resource allotment regulatory [9].

CloudSim employs basic application stipulating methods which may enlarge without any difficulty in partial attempts. In the CloudSim architecture that is described in Fig. 1, the data center is the major part of the cloud that contains set of host(s), virtual machine(s).

5.1 Various Features of CloudSim Tool

The subsequent are some of the characteristics of CloudSim tool.

1. It supports modeling and simulation in large-scale cloud computing data centers.
2. It supports modeling and simulation of energy-aware computational resources.
3. It supports modeling and simulation of data center network topologies and message-passing applications.
4. It supports modeling and simulation of united clouds.

5. It supports dynamic insertion of simulation elements, stop and resume of simulation.
6. It supports user-defined planning for allocation of hosts to virtual machines and planning to give host resources to virtual machines results.

5.2 Primary Elements of CloudSim Tool

The 4 primary elements that exist in CloudSim tool are

1. Data center
2. Virtual Machine
3. Broker
4. Cloudlet.

The data centers in CloudSim hold various resources for providing it to the clients or users. The job of virtual machines is to virtually allot the resources to the clients or users. The job of broker is to facilitate the available resources to the clients. The cloudlets are to customize the tasks of the cloud providers.

5.3 Algorithm Utilized in CloudSim Tool for Load Balancing

The algorithm used in the CloudSim simulator provides the idea for creating data center, host, and virtual machines. The beneficiary of the cloud should be able to access the proper amount of resources as per its requirement in usage basis. The algorithm that is utilized in CloudSim is

1. *step 1*
   Data center creation is done by using create Datacenter () with its various numberings or naming.
2. *step 2*
   Broker creation is done by using create Broker() with its ID.
3. *step 3*
   Create virtual machines using an array of virtual machine listings. Virtual machine ID, broker ID are required as arguments in virtual memory object creation.
4. *step 4*
   The virtual machines are submitted to the virtual memory list.
5. *step 5*
   The Cloudlet is created by the virtual machines and the data centers.
6. *step 6*
   The load or efficiency is calculated according to the request assigned to the virtual machines.

6 Experimental Results

The algorithm is utilized in the simulator for proper load balancing for resource allocations. The efficiency of the load balancing can be measured here. The snapshot of the source code of the virtual machine creation, load allocation for the nodes that consists of various resources is provided in Fig. 2.

The user is able to generate data center, number of hosts along with capability (MIPS), and RAM through the code which is illustrated in Fig. 2. In addition, this code shows that the data centers, broker, etc. can be created by the CloudSim simulator. HTML reports of every simulation and raw data files which can be effortlessly introduced by third-party applications like Octave [9]. The users are able to generate simulation environment that consists of data center, number of hosts, and RAM of preferred arrangement in the output as illustrated in Fig. 3.

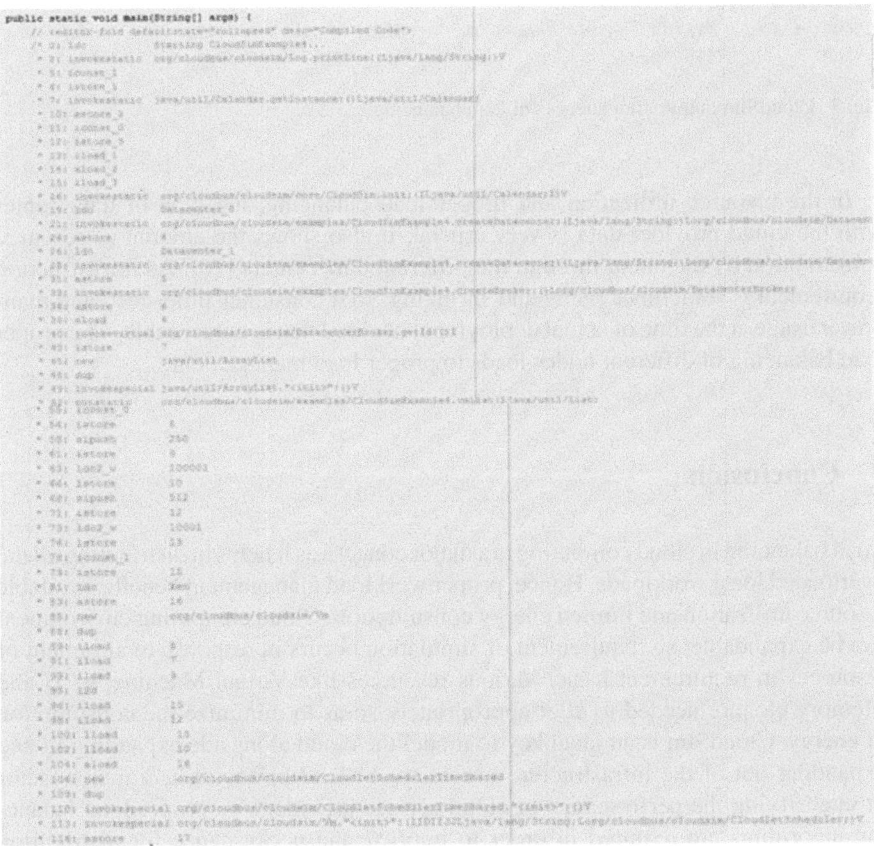

Fig. 2 Snapshot of source code of load balancing in CloudSim

```
Initialising...
Starting CloudSim version 3.0
Datacenter_0 is starting...
Broker is starting...
Entities started.
0.0: Broker: Cloud Resource List received with 1 resource(s)
0.0: Broker: Trying to Create VM #0 in Datacenter_0
0.1: Broker: VM #0 has been created in Datacenter #2, Host #0
0.1: Broker: Sending cloudlet 0 to VM #0
400.1: Broker: Cloudlet 0 received
400.1: Broker: All Cloudlets executed. Finishing...
400.1: Broker: Destroying VM #0
Broker is shutting down...
Simulation: No more future events
CloudInformationService: Notify all CloudSim entities for shutting down.
Datacenter_0 is shutting down...
Broker is shutting down...
Simulation completed
Simulation completed.

========== OUTPUT ==========
Cloudlet ID    STATUS    Data center ID    VM ID    Time    Start Time    Finish T
    0          SUCCESS        2              0       400       0.1         400.1
```

Fig. 3 CloudSim output after successful load balancing

In the resource utilization and proper load balancing, the type of data center with the cloud provider data is very crucial. It also shows the amount of resource consumption by the client through the different data centers [10]. As infrastructure requirement is sometimes rising and shrinking, so it consumes time, cost, and manpower usage at the time of actual deployment of cloud that is unfavorable. The proper load balancing of different nodes leads to proper load management.

7 Conclusion

Load balancing in cloud computing is a major concern as it helps in distributing all the overloaded local workloads. Hence, proper workload management benefits available resource utilization and limited energy consumption. Cloud computing environment can be expandable, so requirement of simulation occurs in response to allotment of resources in requirement basis. Various resources like Virtual Machine, CPU and Memory etc. are needed to allot appropriately so as to minimize the consumption of energy. CloudSim is an ideal key to model the cloud alongside expanding in and expanding out of the infrastructure necessity. As load balancing is a major factor in intensifying the performance of cloud service providers, so proper load balancing algorithms are required in order to improve the performance for distribution dynamically.

References

1. Calheiros, R. N., Ranjan, R., Beloglazov, A., De Rose, C. A. and Buyya, R.,: CloudSim- A Toolkit for Modeling and Simulation of Cloud Computing Environments and Evaluation of Resource Provisioning Algorithms, Software. Practice and Experience. 41(1): 2350. Wiley. January (2011)
2. Buyya, R., Ranjan, R. and Calheiros, R. N., Modeling and Simulation of Scalable Cloud Computing Environments and the CloudSim Toolkit: Challenges and Opportunities, 7th High Performance Computing and Simulation Conference (HPCS) IEEE Computer Society, 2009, 1-11, (2009)
3. Pakize, S. R., Khademi, S. M., Gandomi , A.: Comparison Of Cloud-Sim, Cloud Analyst and Cloud Reports Simulator in Cloud Computing. International journal of Computer Science & Network Solutions. 2(5). (2014)
4. Calheiros, R. N., Ranjan, R., Anton Beloglazov, César, A. F., Buyya, R.CloudSim: A Toolkit for the Modeling and Simulation of Cloud Resource Management and Application Provisioning Techniques.,(1995)
5. Vincent, A. K., Senthil, M. N. and Jaiganesh, M. Optimization of Cloud Resource Service Adaptability using Genetic Algorithms. International Conference on IT and Intelligent Systems (ICITIS'2013) August 28-29, (2013)
6. Devi, K., R., Sujan, S.: A Survey on Application of Cloudsim Toolkit in Cloud Computing. International Journal of Innovative Research in Science, Engineering and Technology. 3(6). (2014)
7. Colheiro R. N., Cloud Report: A CloudSim - based Tool for Modelling and Analysis of Large Scale Cloud Computing Environments, MEDC Project by Rajkumar Buyya project supervisor Report, (2013)
8. Liu, Y., Chen, L., Xu, K., Zhang, Y.: Application Modes. Architecture and Challenges for Cloud Educational System, Computer Research and Development (2010)
9. Zhou Y, Liu X, Wang X, Xue L, Liang X, Liang S.: Business process centric platform-as-a-service model and technologies for cloud enabled industry solutions. In: Proceedings of the 3rdInternationalConference on Cloud Computing; Miami, Florida (USA), 534?537, (2010)
10. Gangwar, H., Date, H., Ramaswamy, R.: Understanding determinants of cloud computing adoption using an integrated TAM-TOE model. Journal of Entrepreneurship Information Management **28**(1), (2015)

Assessment of Load in Cloud Computing Environment Using C-means Clustering Algorithm

Aradhana Behura and Sushree Bibhuprada B. Priyadarshini

Abstract Load balancing is an important concept which helps to raise the throughput. It helps to increase the performance of the system by assigning virtual machine for the execution of user request in less time. It assists in raising user satisfaction level and reducing user response time. As we know, the demand for cloud services raises day by day, which lead to load balancing as a major problem. The technique helps to distribute the workload among all available servers so that user can get their resource in less time. Cloud load balancing introduced the distribution of workload traffic and demands that reside over the Internet. It takes advantage of the cloud extensibility and possesses alertness to meet redirected workload demands and to improve long-term opportunity. The purpose of load balancing is to continue the system constancy and develop the performance considerably. We propose a load balancing technique by using c-means clustering to handle tasks efficiently which conserve less amount of energy.

Keywords Load balancing · C-means clustering · Task consolidation · Optimization technique · Cloud computing

1 Introduction

Power utilization in cloud technology is a multifaceted challenge, which can directly reduce the utilization costs. Task consolidation is a technique to raise resource usage. However, the greatest use of sources does not imply that there is convenient utilization of power and some resources which are unused resources that also consume significant quantity of energy. Current lessons show that the power consumption of unused resources is 1–19%. Therefore, the idle resources are allotted with some task to make

A. Behura
Veer SurendraSai University of Technology, Sambalpur, Burla, India
e-mail: aradhanabehura@gmail.com

S. B. B. Priyadarshini (✉)
Siksha 'O' Anusandhan Deemed to be University, Bhubaneswar, India
e-mail: bimalabibhuprada@gmail.com

D. Mishra et al. (eds.), *Intelligent and Cloud Computing*,
Smart Innovation, Systems and Technologies 194,
https://doi.org/10.1007/978-981-15-5971-6_23

completely effective use of the idle period so that there will be overall minimum energy consumption. The advancement of technologies like software applications and system gadget has built cloud technology a desired computing standard.

Cloud can be designed and handle the application accessible at any time. Cloud is greatly profitable because it runs at greater efficiency with optimal usage and provides load balancing that makes it more predictable and cloud just requires an Internet connection. It doesn't need to fit a software to handle application. Some of the issues one must consider in cloud computing are data privacy, security requirement, and training requirement. Cloud computing value proposition key factors are business model innovation, IT management simplification, and high service quality. All these factors are liable for upcoming progress. In cloud planning, phase key factors are IT planning development, business architecture development, and transfiguration plan development. In IT architecture development, we find the approach that promotes the business procedure and the technical knowledge required to support business applications and data systems. A set of quality assurance refers to the unproductive requirement such as disaster recovery, safety, accuracy, etc. The benefits of relevant cloud computing mode depend on these unproductive features. In business planning growth, we recognize the prospect that might be leading to cloud computing application from business point of view. In transformation phase, we develop all kinds of plans that are needed to transform current business to cloud computing modes. Cloud provider gives maintenance and technological service. They need to ensure the quality of service and for selecting cloud technology providers on the basis of SLA which defines the level of service the provider will meet.

2 Literature Review

Cloud technology is broadly used for its variety of services, namely, IaaS, PaaS, and SaaS according to Buyya [1]. It is very much important for proper utilization of the resources. However, effective resource utilization does not imply energy efficiency according to Hsu et al. [2]. The main purpose is reducing the total number of resource utilization that results in increasing resource usage of active resources. For example, information and communication technology resource consumption is about 9% of whole power consumption in US every year, and it is presumed to reach about 51% in the upcoming year [3–16]. Using coal and natural gas, 66% of electricity is generated. As a result, some practitioner's main purpose is to reduce power consumption [5, 6], and another objective is to increase resource utilization [7, 8, 16]. Maximum of the current experimental works [2, 9, 10] pay particular attention on objective that allots the client tasks to VM that results in reducing energy consumption. Moreover, it turned off the unutilized virtual machine for preserving energy. Crow search and honeybee algorithm can be used to optimize load balancing. By observing smart behavior of crow, researchers can propose new efficient load balancing algorithm. An application software-based proposal that includes two techniques, namely, load dispatching and dynamic provisioning, has been suggested by Chen et al. [12]. The first one aims to share load among active resources and the second one objective to switch on the least number of system resources. To deduce the optimum configuration, an automatic controller is used by Tesfatsion et al. [14]. Two energy-effective task consolidation algorithms are used to reduce the energy consumption, and it has been suggested by Lee et al. [8]. The algorithm is purely based on cost function that chooses a VM for a ready task. This algorithm in this paper is different from [2, 8]. In the following manner, energy is different from [9] by allocating the task in ascending order of fitness value. Power consumption is defined as *E*, and *E* is directly proportional to $C_{eff}V^2f_{clk}$ where V is the supply voltage, C_{eff} is the effective switching capacitance, f_{clk} is relevant to supply voltage as clock frequency f_{clk} is directly proportional to $(V-V_k)/V$, and there is a relationship among frequency, energy, and voltage which lead to dynamically adjust frequency and voltage according to the current workload to conserve energy. The power ingesting, which varies in CPU utilization, can be mathematically represented as

$$E(u) = (P_{\max} - P_{\min}) + U/100 + P_{\min} \tag{1}$$

Here, $P_{\max}$ and $P_{\min}$ are consumptions of power at peak load and inactive load. The $E(u)$ is the energy consumed by CPU at the deployment $U\%$ and U is the present value of process operation. Figure 1 throws light toward complete architecture of cloud computing and by viewing this, user can able to communicate with cloud service provider (CSP). Figure 2 consists of set of users, and they send the request for demanding the resource to the service provider. The virtual machine executes the request and sends the result to the service provider. Then finally, the service provider will send the output to the corresponding user accordingly. Figure 3 throws light toward a novel scheduling algorithm used in cloud computing platform.

3 Proposed Approach

3.1 Resource Allocation

Centralized resource manager can handle or manage the cloud computing resources efficiently according to the proposed algorithm. The task is allotted to require VMs by centralized resource manager. VMs give the resources of cloud datacentre to the user/application. Virtual machines are used to meet the execution time support application and resource requirement. In a specific running time, a presentation for essential resources can be made which can be obtained through two steps: resource provisioning, i.e., to schedule the request to the physical resources, and VM provisioning.

The VM mechanism has become prominent in upcoming years in environment of cloud technology and in datacentre the reason is due to security isolation and server consolidation. One or more than one VM can be assigned or setup or placed on a

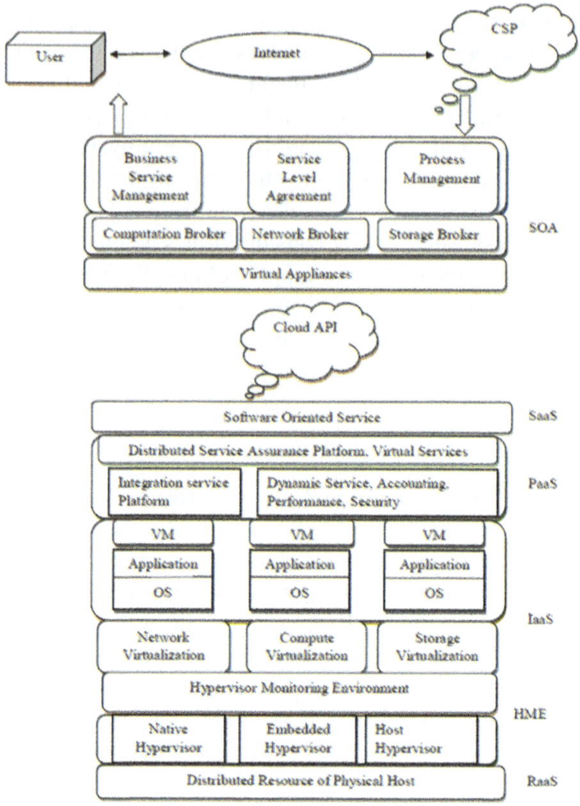

Fig. 1 Architecture of cloud computing environment

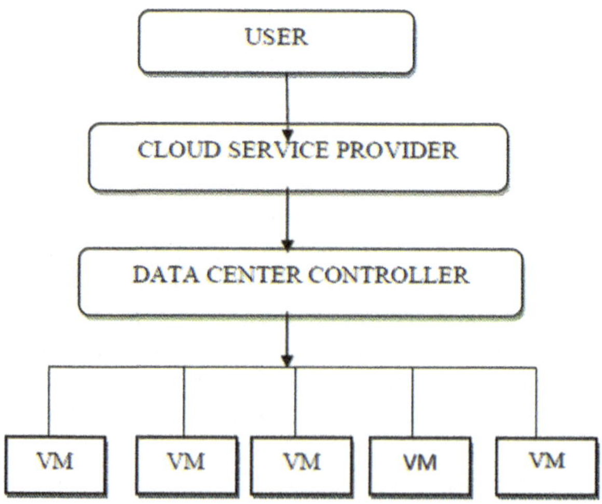

Fig. 2 Load balancing in cloud technology

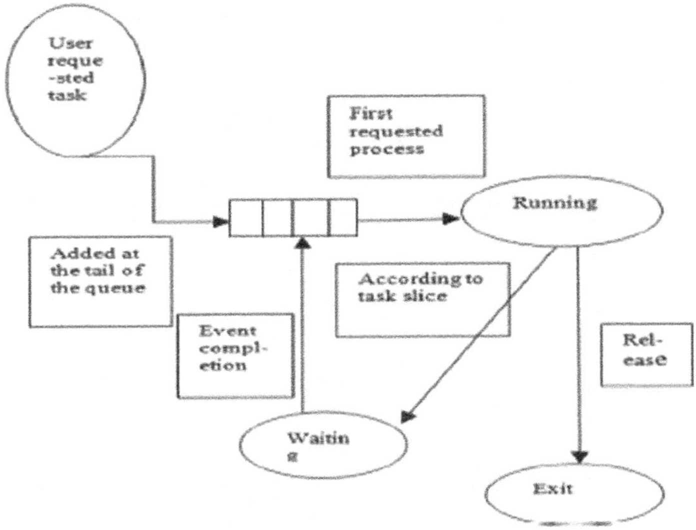

Fig. 3 Scheduling algorithm in cloud technology

physical appliance that comes across the condition of VM. To maximize resource utilization results in the improvement of IT service customization, reliable services, and rationalization of QoS, i.e., quality of services, maintenance, etc. The graphical representation of the proposed algorithm is shown in Figs. 4 and 5.

3.2 Algorithm

Step 1: Verify the capacity of the virtual machine and check whether it is balanced or not which is shown in Fig. 3.
Step 2: Apply C-means clustering algorithm to group the task as per their priority.
Step 3: If the task is greater than capacity, then load is not balanced, so we have to check another virtual machine which has capacity to handle the load which is described in Fig. 4.
Step 4: Check the virtual machine capacity whether it is low ordered, high ordered, or balanced according to task priority.
Step 5: Sort them using quick sort algorithm.
Step 6: Update the new weight of the virtual machine.

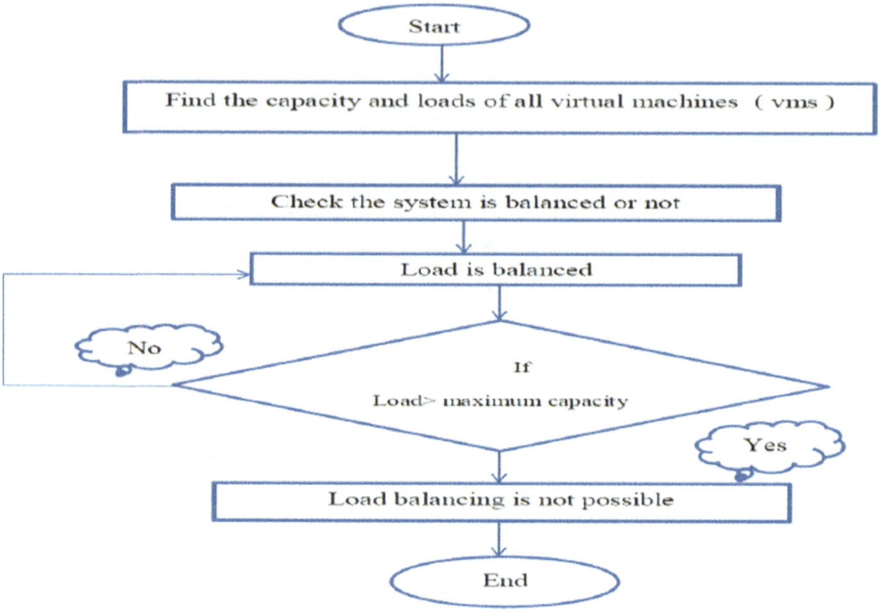

Fig. 4 Load balancing policy (Phase-1)

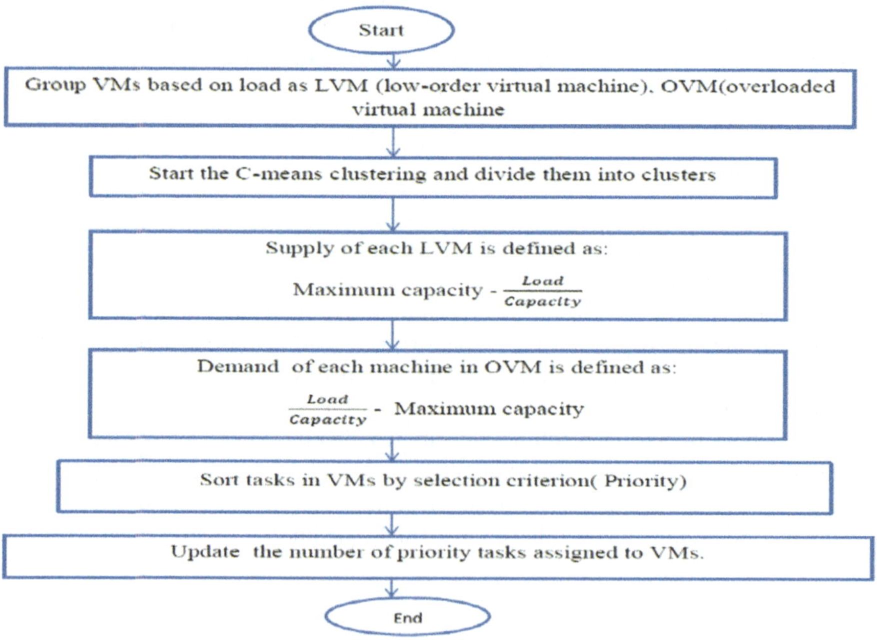

Fig. 5 Load balancing policy (Phase-2)

4 Result Analysis

The proposed algorithm has been compared with the existing algorithm with the help of extensive experiments by using CloudSim platform. From the results, the proposed algorithm found to have better accuracy and less energy consumption. The energy consumptions in the proposed algorithm and the existing one are found to be 2388.95 and 2533.058, respectively. The efficiency of the proposed algorithm is increased by approximately 6%.

5 Conclusion

The proposed work has been observed from the experimental result, and there will be less energy consumption in the proposed algorithm as associated to existing approach. By relating the two experimental results, it is initiated that the proposed algorithm outperforms better than the existing algorithm in total number of task completion and energy consumption. Here, we group the task according to their priority and estimate the energy consumption on arrival of task to make the scheduling taxation efficient. The impact of increasing the number of tasks on execution time and imbalancing degree are shown in Fig. 6 and Fig. 7 respectively.

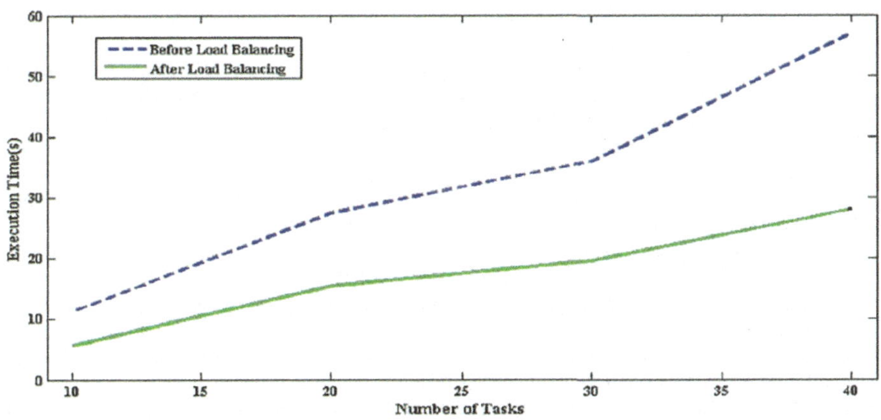

Fig. 6 Number of tasks versus execution time

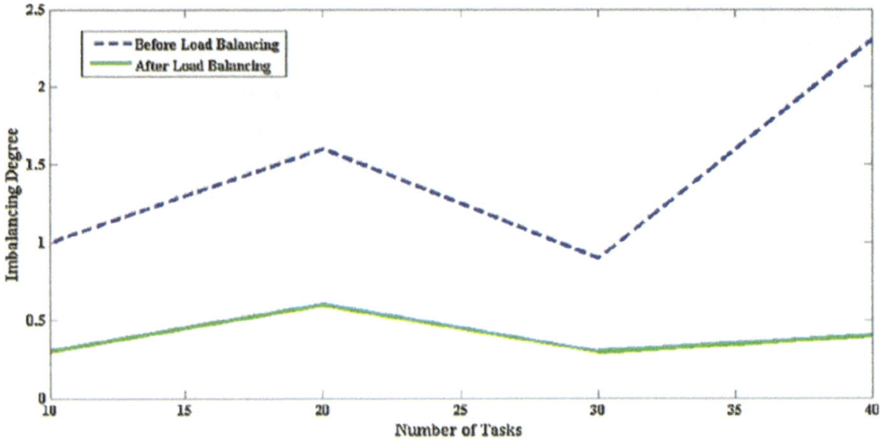

Fig. 7 Number of tasks versus imbalance degree

References

1. Buyya, R., Yeo, C.S., Venugopal, S., Broberg, J., Brandic, I.: Cloud computing and emerging it platforms: vision, hype and reality for delivering computing as the 5th utility. Future Gener. Comput. Syst. **25**, 599–616 (2009)
2. Hsu, C., Slagter, K.D., Chen, S., Chung, Y.: Optimizing energy consumption with task consolidation in clouds. Inf. Sci. **258**, 452–462 (2014)
3. Mills, M.P.: The cloud begins with coal: big data, big networks, big infrastructure and big power. Technical report, National Mining Association, American Coalition for Clean Coal Electricity (2013)
4. Hohnerlein, J., Duan, L.: Characterizing cloud datacenters in energy efficiency, performance and quality of service. In: ASEE Gulf-Southwest Annual Conference, The University of Texas, San Antonio, American Society for Engineering Education (2015)
5. Panda, S.K., Jana, P.K.: Efficient task scheduling algorithms for heterogeneous multi-cloud environment. J. Supercomput. **71**, 1505–1533 (2015)
6. Li, J., Qiu, M., Ming, Z., Quan, G., Qin, X., Gu, Z.: Online optimization for scheduling preemptable tasks on iaas cloud system. J. Parallel Distrib. Comput. **72**, 666–677 (2012)
7. Sanjaya, K.P., Gupta, I., Jana, P.K.: Task scheduling algorithms for multi-cloud systems: allocation-aware approach. Inf. Syst. Front. 1–19 (2017). Springer, SCIE
8. Khemka, B., Friese, R., Pasricha, S., Maciejewski, A.A., Siegel, H.J., Koenig, G.A., Powers, S., Hilton, M., Rambharos, R., Poole, S.: Utility driven dynamic resource management in an over subscribed energy-constrained heterogeneous system. In: 28th IEEE International Parallel and Distributed Processing Symposium Workshops, pp. 58–67 (2014)
9. Lee, Y.C., Zomaya, A.Y.: Energy efficient utilization of resources in cloud computing systems.J. Supercomput. **60**, 268–280 (2012)
10. Panda, S.K., Jana, P.K.: An efficient energy saving task consolidation algorithm for cloud computing. In: Third IEEE International Conference on Parallel, Distributed and Grid Computing, pp. 262–267 (2014)
11. Fan, X., Weber, W., Barroso, L.A.: Power provisioning for a warehouse-sized computer. In: The 34th Annual International Symposium on Computer Architecture, pp. 13–23. ACM (2007)
12. Chen, G., He, W., Liu, J., Nath, S., Rigas, L., Xiao, L., Zhao, F.: Energy-aware server provisioning and load dispatching for connection-intensive internet services. In: 5th USENIX Symposium on Networked Systems Design and Implementation, pp. 337–350 (2008)

13. Srikantaiah, S., Kansal, A., Zhao, F.: Energy aware consolidation for cloud computing. In: International Conference on Power Aware Computing and Systems, pp. 1–5 (2008)
14. Tesfatsion, S.K., Wadbro, E., Tordsson, J.: A combined frequency scaling and application elasticity approach for energy-efficient cloud computing. Sustain. Comput. Inf. Syst. **4**, 205–214 (2014)
15. Chen, H., Zhu, X., Guo, H., Zhu, J., Qin, X., Wu, J.: Towards energy-efficient scheduling real-time tasks under uncertain cloud environment. J. Syst. Softw. **99**, 20–35 (2015)
16. Panda, S.K., Jana, P.K.: SLA-based task scheduling algorithms for heterogeneous multi-cloud environment. J. Supercomput. Springer, SCI **73**(6), 2730–2762 (2017)

Parallel Implementation of Algebraic Curve for Data Security in Cloud Computing Environment

Sanjaya Kumar Jena, Manoranjan Parhi, and Sarbeswara Hota

Abstract In recent years data security plays an important role in cloud environment. The running time of secure algorithms has found to be time- consuming and lengthy in key size. In this paper, we present a parallel implementation to reduce the running time to achieve the same level of security using hyperelliptic curve. Due to small operand size as compared to other cryptographic methods like RSA and elliptic curve, hyperelliptic curve was found to be more suitable for security applications. Our work comprised of two parts: (i) sequential implementation of group addition of elliptic curves and hyperelliptic curves of genus-2 over prime fields, (ii) parallel implementation of the sequential version to speed-up the computational time. Here, thread-level parallelism has been carried out using GNU MP, openMP libraries, and a speed-up of $\approx 32\%$ achieved on dual core processor over the sequential implementation.

Keywords Hyperelliptic curve · Genus · GNU MP · OpenMP · Sequential · Parallel

1 Introduction

Security in cloud computing is quite essential to secure data from other malicious access as cloud services allow person and organizations to use resources being managed by third parties at remote locations [1]. Information security techniques are

---

S. Kumar Jena (✉) · M. Parhi
Department of Computer Science and Engineering, Siksha 'O' Anusandhan Deemed to be University, Bhubaneswar, Odisha, India
e-mail: sanjayjena@soa.ac.in

M. Parhi
e-mail: manoranjanparhi@soa.ac.in

S. Hota
Department of Computer Application, Siksha 'O' Anusandhan Deemed to be University, Bhubaneswar, Odisha, India
e-mail: sarbeswarahota@soa.ac.in

D. Mishra et al. (eds.), *Intelligent and Cloud Computing*,
Smart Innovation, Systems and Technologies 194,
https://doi.org/10.1007/978-981-15-5971-6_24

more important to make data security within the cloud environment covering various cloud dimensions. Algebraic curve cryptography is an essential basic technique among many cryptographic tools. The best public-key cryptography currently available based on the key size is the RSA and elliptic curve cryptography. However, a small subset of algebraic curves are elliptic curves used in cryptographic application. It is being motivated by the research study that a large group of algebraic curves, called hyperelliptic curves are required to create more effective cryptosystem. However, the hyperelliptic curve group operation is quite complex. So, a fast group operation is required to construct a secure system based on hyperelliptic curves.

The objective is to study the arithmetics of elliptic and hyperelliptic curves. The explicit formula for hyperelliptic curves with respect to the prime fields has been studied [2]. The explicit formula is derived from the concept of Cantor's algorithm and Harley's algorithm [3, 4].

Finally, we have parallelized the group addition of hyperelliptic curves taking into the account of the multi-core processors. Here, we have tried to optimize the timing of addition of hyperelliptic curves.

The various reasons for which we are influenced are as follows: (i) Algebraic curve cryptography results in small operand size as compared to other public-key cryptography like RSA. (ii) The hyperelliptic curve cryptosystems of genus ≥ 2 also have small operand size as compared to elliptic curves. (iii) Idea of parallelization enhances the efficiency of group operations under the finite field.

This paper is narrated in given ways: literature review in Sects. 2 and 3 focused on the proposed implementation model, implementation process of the proposed model in Sects. 4 and 5 described the experimental results and analysis, Sect. 6 elaborate the performance comparison, and Sect. 7 discussed the conclusion of our work.

2 Related Work

A faster cryptographic algorithm is highly essential for better data security in the cloud environment to save the computational time. In this section, we have studied various algebraic curve technique for security applications. One of the algebraic curve, hyperelliptic curve, is used in cryptographic applications. The group operation of this curve depends mainly on the explicit formula in doing addition and doubling. These explicit formulae are themselves composed of addition, multiplications(M), squaring(S), and inversions(I) over the given finite field. Moreover, explicit formulae are specific to a kind of genus. So, different explicit formulae are derived for different genus K.Nagao proposed ideas to speed-up the group operations of the hyperelliptic curves in their Jacobian over Cantor's algorithm [5].

A fast algorithm called Harley algorithm on finite fields of odd characteristics, but limited to genus-2 curves over odd characteristics fields and the input and output divisor classes are represented in Mumford's representation.[3]. The multiplication methods such as Karatsuba multiplication, Chinese Reminder Theorem, and the Newton iteration were used in that algorithm. A generalized Harley algorithm over finite fields of any characteristics and any genus took I + 26M for addition, I +31M for doubling, where I- an inversion cost and M-a multiplication cost [6].

The Explicit formula performs the algebraic curve group operation to achieve the fewest number of operations under any characteristics [2, 7] . They had considered three different coordinates namely, Affine, Projective, and a new coordinate system, suitable for different environment. Again Tanja Lange proposed a generalized Montgomery arithmetic for genus two curves by using the divisor class representation in using only their first polynomial. That feature helps for practical implementation to obtain faster group operations.

The Montgomery's trick of simultaneous inversions had made a speed-up in the hyperelliptic curve operation [8]. Hyperelliptic curve arithmetic further improved over odd characteristics with a small change in the order of a specific operation [9].

Xinxin fan and Gauang Gong proposed explicit formula on the hyperelliptic curve of genus two, which saves multiplications and each time a field squaring during computation of arithmetic in most frequent cases [10].

3 Proposed Model

Considering the mathematical foundations behind the group operations of elliptic and hyperelliptic curves, the proposed model is given as in Fig. 1.

3.1 Explicit Addition Formulation

The formulation of the explicit formula is tabulated as (Table 1);

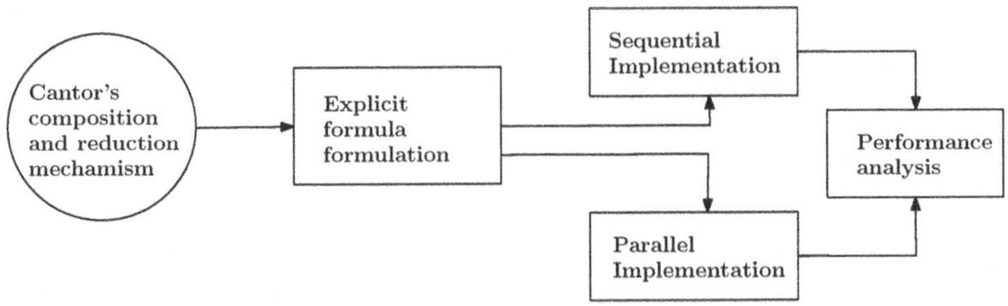

Fig. 1 Proposed model for sequential and parallel implementation of algebraic curve

Table 1 Explcit addition of HEC genu-2, deg u_1 = deg $u_2 = 2$

Input	Consider $d_1 = (p_1, q_1)$ and $d_2 = (p_2, q_2)$ are two reduced divisors on hyperelliptic curve $y^2 = f(x)$, genus-2 $p_1 = x^2 + p_{11}x + p_{10}$; $p_2 = x^2 + p_{21}x + p_{20}$; $q_1 = q_{11}x + q_{10}$; $q_2 = q_{21} + q_{20}$; and $f = x^5 + f_4x^4 + f_3x^3 + f_2x^2 + f_1x + f_0$;	
Output	A reduced divisor $d' = d_1 + d_2 = (p', q')$ of weight 2, where $p' = x^2 + p'_1x + p'_0$; $q' = q'_1x + q'_0$;	
Step	Computation	Cost
1	Computing resultantc of p_1, p_2: $m_1 = p_{11} - p_{21}, m_2 = p_{20} - p_{10}, m_3 = p_{11}m_1 + m_2$; $c = m_2m_3 + m_1^2p_{10}$;	$3M + 1S$
2	Computing almost inverse of $p_2\%p_1(iv = c/p_2\%p_1)$: $iv_1 = m_1, iv_0 = m_3$;	
3	Computing $t' = ct \equiv (q_1 - q_2)iv\%p_1$: $o_0 = q_{10} - q_{20}, o_1 = q_{11} - q_{21}, o_2 = iv_0o_0, o_3 = iv_1o_1$; $t'_1 = (iv_0 + iv_1)(o_0 + o_1) - o_2 - o_3(1 + p_{11})$; $t'_0 = o_2 - p_{10}o_3$; if $t'_1 = 0$ follow Cantor's algorithm	5M
4	Computing $t'' = x + t_0/t_1 = x + s'_0/t'_1$ and t_1: $o_1 = (ct_1)^{-1}(= 1/c^2t_1), o_2 = co_1(= 1/t'_1), o_3 = t_1^2o_1(= t_1)$; $o_4 = co_2(= 1/t_1), o_5 = o_4^2; t''_0 = t'_0o_2$;	5M+2S+I
5	Computing $n' = t''p_2 = x^3 + n'_2x^2 + n'_1x + n_0$: $n'_2 = p_{21} + t''_0, n'_1 = p_{21}t''_0 + p_{20}, n'_0 = p_{20}t''_0$;	2M
6	Computing $p' = (t(n + 2q_2) - k)/p_1 = x^2 + p'_1x + p'_0$: $p'_0 = (s''_0 - p_{11})(s''_0 - m_1) - p_{10} + n'_1 + (2q_{21})o_4+$ $(2p_{21} + m_1 - f_4)o_5$; $p'_1 = 2t''_0 - m_1 - o_5$;	3M
7	Computing $v' \equiv -(n + q_2)\%p' = q'_1x + q'_0$: $o_1 = n'_2 - p'_1, o_2 = p'_1o_1 + p'_0 - n'_1; q'_1 = o_2o_3 - q_{21}$; $o_2 = p'_0o_1 - n'_0; q'_0 = o_2o_3 - q_{20}$;	4M
Total	Computational complexity (Here M=Multiplication, S=squaring, I=Inversion)	I+3S+22M

4 Implementation

This section includes the sequential and parallel implementation of hyperelliptic curves of genus-1 and genus-2 under the prime field of odd characteristics [11, 12].

4.1 Sequential Implementation Methodology

The group operation for sequential implementation is described as (i) Input divisors are determined using the reduced divisor. (ii) Input polynomials are determined using Mumford's representation from the reduced divisor. (iii) Explicit formulae are divised depending on the degree of the input polynomials. The exclusive steps of the derivation are given below in the Algorithm 24.1.

Algorithm 24.1 Genus 2 group addition in frequent case

Input: $divisor_1 = div(u_1, v_1), divisor_2 = div(u_2, v_2)$
Output: $divisor_3 = d_1 + d_2 = div(u_3, v_3)$

1. (exact division) $k = \frac{f-v_1^2}{u_1}$, $s \equiv \frac{v_2-v_1}{u_1}$ mod u_2, $l = su_1$
2. (exact division) $u' = \frac{k-s(l+2v_1)}{u_2}$
3. $u_3 = u'$ made monic, and $v_3 \equiv -(l + v_1)$ % u_3

4.2 Parallel Implementation Methodology

This algorithm is devised on the explicit formula derived form the cantor's composition and reduction algorithm in special cases. Let P be the explicit formula consisting of multiplications, additions, squarings, one inversion operations, and several intermediate variables are involved. The following procedures are performed on P: (i) Complex multiplication is represented in two-operand format. (ii) Repeated use of variable names is eliminated. (iii) Reduction in the depth of multiplication (iv) Proper labeling of the algorithm. We assigned labeling to addition as addition step, multiplication as multiplication step, and unique names to the intermediate variables used in the algorithm. The Algorithm 24.2 describes the steps of parallel algorithm [13];

Algorithm 24.2 Genus 2 hyperelliptic curve parallel addition

Input: $divisor_1 = div(u_1, v_1)$, $divisor_2 = div(u_2, v_2)$. where $u_1 = u_{10} + u_{11}z + z^2$, $v_1 = v_{11}z + v_{10}$, $u_2 = u_{20} + u_{21}z + z^2$ and $v_2 = v_{20} + v_{21}z$. genus-2 curve $C : y^2 = f(z)$
Output: $d_3 = div(u_3, v_3) = d_1 + d_2$, $u_3 = z^2 + u'_1z + u'_0$, and $v_3 = v'_1z + v'_0$.
Buffered intermediates: u_{11}, u_{10}, v_{11}, v_{10}, u_{20}, u_{21}, v_{21}, and v_{20}.

addition-1 $inv_1 = u_{11} - u_{21}; q_1 = u_{20} - u_{10}; q_2 = v_{10} - v_{20}; q_3 = v_{11} - v_{21};$
multiplication-1 $q_4 = u_{11}inv_1; q_5 = inv_1^2; q_6 = inv_1q_3;$
addition-2 $q_7 = q_1 + q_4; q_8 = q_2 + q_3; q_9 = inv_1 + q_7; q_{10} = 1 + u_{11};$
multiplication-2 $q_{11} = u_{10}q_5; q_{12} = q_1q_7; q_{13} = q_8q_9; q_{14} = q_{10}q_6; q_{15} = u_{10}q_6; q_{16} = q_7q_9;$
addition-3 $r = q_{11} + q_{12}; s'_1 = q_{13} - q_{16} - q_{14}; s'_0 = q_{16} - q_{15};$
multiplication-3 $q_{17} = rs'_1; q_{18} = s_1^{\prime 2}; q_{19} = r^2; q_{20} = rs'_0;$
Inversion $q_{21} = q_{17}^{-1};$
multiplication-4 $q_{22} = rq_{21}; q_{23} = q_{21}q_{18}; q_{24} = q_{21}q_{19}; s''_0 = q_{21}q_{20};$
addition-4 $l'_2 = u_{21} + s''_0; q_{25} = s''_0 - u_{11}; q_{26} = s''_0 - inv_1; q_{27} = 2v_{21};$
multiplication-5 $q_{28} = q_{24}^2; q_{29} = u_{21}s''_0; l'_0 = u_{20}s''_0; q_{30} = q_{25}q_{26}; q_{31} = q_{27}q_{24};$
addition-5 $l'_1 = q_{29} + u_{20}; q_{32} = 2u_{21} + inv_1 - f_4; u'_1 = 2s''_0 - inv_1 - q_{28}; q_{33} = l'_2 - u'_1;$
multiplication-6 $q_{34} = q_{28}q_{32}; q_{35} = u'_1q_{33}; q_{36} = q_{33}q_{23}; q_{37} = l'_0q_{23}$
addition-6 $u'_0 = q_{30} - u_{10} + l'_1 + q_{31} + q_{34}; q_{38} = q_{35} + u'_0 - l'_1;$
multiplication-7 $q_{39} = q_{23}q_{38}; q_{40} = u'_0q_{36};$
addition-7 $q_{41} = q_{40} - q_{37}; v'_1 = q_{39} - v_{21}; v'_0 = q_{41} - v_{20};$

5 Experimental Results and Analysis

The basic field operations (multiplications and inversions) timing using the library GNU MP [14] on Intel Core 2 Duo 2.20 GHz processor, 2GB RAM under Linux carrying 1,000,000 times execution of each operation is shown in the Table 2.

5.1 *Sequential Implementation of Elliptic Curve Addition*

Point addition on Elliptic curve type $y^2 = f(x)$, $f(x) = x^3 - ax + b$, where a and b are curve parameters taken from the standards specification [15]. Different prime field sizes and curves are collected from the source [15, 16]. The timing results are shown in the Table 3.

5.2 *Sequential Implementation of Hyperelliptic Curve Addition*

The below Table 4 shows the timing of hyperelliptic curve group addition under the same specification as in Sects. 5.1 and 6, respectively.

Table 2 Basic field operations timing in μs

Bits /op	32	48	64	96	123	129	163	192	224	256	384	521
Mul(M)	0.06	0.129	0.139	0.146	0.164	0.203	0.212	0.245	0.32	0.338	0.508	0.875
Inv(I)	0.219	0.444	0.473	0.516	0.585	0.764	0.91	0.929	1.148	1.246	1.772	3.343

Table 3 Sequential timings of elliptic curve addition

Prime field size in bits	Group order in bits	Group addition in μs
163	**163**	**16.30**
167	167	18.60
179	179	16.74
184	184	16.79
191	**191**	**17.2**
192	192	17.96
195	195	19.60

Table 4 Sequential timings of hyperelliptic curve addition, genus-2

Prime field size in bits	Group order in bits	Group addition in μs
64	**127**	**13.839**
67	133	16.706
76	152	18.417
81	**162**	**18.327**
84	168	19.358
92	184	19.471
94	188	20.681
97	194	22.379

5.3 *Parallel Implementation of Hyperelliptic Curve Addition*

The timings of hyperelliptic curve group addition are shown in Table 5.

6 Performance Comparison of Sequential and Parallel Implementation

Here sequential and parallel computing results are compared. For example, a particular group order size of 162 bits, sequential implementation took 18.327 µs and parallel implementation took 12.34 µs in Table 6. So, a speed-up of (18.327–12.44)/18.327 $\approx$ 32% is achieved. A comparison of sequential vs parallel implementation is shown Fig. 2. The different types of curves are of the following types: (i) Elliptic curve of the form $y^2 = x^3 - ax + b$, where a, b are curve parameters.

Table 5 Timings of parallel addition of hyperelliptic curve, genus-2

Prime field size in bits	Group order in bits	Group addition in μs
64	127	10.81
67	133	13.10
76	152	13.67
81	162	12.44
84	168	14.47
92	184	13.62
94	188	14.88
97	194	16.68

Table 6 Sequential and parallel comparison of addition

Prime field Size in bits	Group order in bits	Addition in sequential μs	Addition in parallel μs	Ratio of sequential versus parallel
64	127	13.839	10.81	1.28
67	133	16.706	13.10	1.27
76	152	18.417	13.67	1.34
81	**162**	**18.327**	**12.44**	**1.47**
84	168	19.358	14.47	1.33
92	184	19.471	13.62	1.42
94	188	20.681	14.88	1.38
97	194	22.379	16.68	1.34

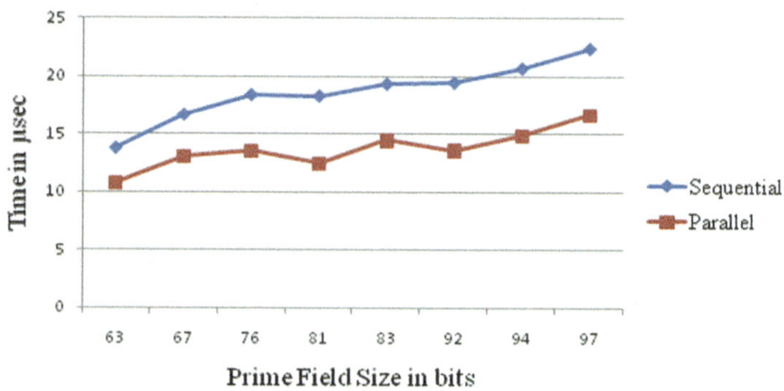

Fig. 2 Sequential versus parallel implementation

(ii) Hyperelliptic curve of the form $y^2 = x^5 + f_4x^4 + f_3x^3 + f_2x^2 + f_1x + f_0$, and $y^2 = x^5 + f_3x^3 + f_2x^2 + f_1x + f_0$, where f_4, f_3, f_2, f_1 and f_0 are curve parameters.

7 Conclusion

We have carried out the parallel implementation of the addition of genus-2 curves and achieved a maximum of $\approx 32\%$ speed-up. In our implementation we have used the GNUMP 4.3.2 library prefixed with function *mpz* and openMP. The GNUMP functions prefixed with *mpz* creates overhead due to preallocation of output operands. It can be further optimized by using the GNUMP prefixed function *mpn* for low level parallel implementation. Due to it's less operand size, it can be effectively used in cloud where memory and timings are constraints.

References

1. Zhang, Q., Cheng, L., Boutaba, R.: Cloud computing: state-of-the-art and research challenges. J. Internet Serv. Appl. **1**(1), 7–18 (2010)
2. Lange, T.: Formulae for arithmetic on genus 2 hyperelliptic curves. Appl. Algebra Eng. Commun. Comput. **15**(5), 295–328 (2005)
3. Harley, R.: Fast Arithmetic on Genus 2 Curves (2000). http://cristal.inria.fr/harley
4. Cantor, D.G.: Computing in the Jacobian of a hyperelliptic curve. Math. Comput. **48**(177), 95–101 (1987)
5. Nagao, K. I.: Improving group law algorithms for Jacobians of hyperelliptic curves. In International Algorithmic Number Theory Symposium, pp. 439–447. Springer, Heidelberg (2000)
6. Sugizaki, H., Matsuo, K., Chao, J., Tsujii, S.: 3-3 a generalized harley algorithm for genus two hyperelliptic curves. J. Natl. Instit. Inf. Commun. Technol. **52**(1/2) (2005)
7. Lange, T.: Efficient arithmetic on genus 2 hyperelliptic curves over finite fields via explicit formulae. IACR Cryptol. ePrint Archive **121** (2002)
8. Matsuo, K., Chao, J., Tsujii, S.: Fast genus two hyperelliptic curve cryptosystems. In: Symposium on Cryptography and Information Security-SCIS, IEICE Japan, pp. 497–502 (2001)
9. Takahashi, M.: Improving harley algorithms for Jacobians of genus 2 hyperelliptic curves. In Proceedings of SCIS, ICICE, Japan (2002)
10. Fan, X., Gong, G.: Efficient explicit formulae for genus 2 hyperelliptic curves over prime fields and their implementations. In International Workshop on Selected Areas in Cryptography, pp. 155–172. Springer, Berlin (2007)
11. Avanzi, R.M.: Aspects of hyperelliptic curves over large prime fields in software implementations. In: International Workshop on Cryptographic Hardware and Embedded Systems, pp. 148–162. Springer, Berlin (2004)
12. Menezes, A., Zuccherato, R., Wu, Y.H.: An Elementary Introduction to Hyperelliptic Curves, p. 155. Faculty of Mathematics, University of Waterloo (1996)
13. Mishra, P.K., Sarkar, P.: Parallelizing explicit formula for arithmetic in the Jacobian of hyperelliptic curves. In: International Conference on the Theory and Application of Cryptology and Information Security, pp. 93–110. Springer, Berlin (2003)
14. GNU MP: The GNU Multiple Precision Arithmetic Library. GNU Developers, 4.3.2 (2010). http://gmplib.org/
15. Digital Signature Stantdard, FIPS PUBS, National Institute of Standards and Technology (NIST) (2009)
16. IEEE Standard Specifications for Public-Key Cryptography: IEEE Std **1363–2000**, 1–228 (2000). https://doi.org/10.1109/IEEESTD.2000.92292

A Design Towards an Energy-Efficient and Lightweight Data Security Model in Fog Networks

Arnab Mitra and Sayantan Saha

Abstract A proposal targeting an efficient and lightweight data security model in Fog Networks is presented in this research paper. Cellular Automata generated patterns are emphasized for uses in proposed data security involving authentication and cryptography processes. Proposed security approach ensures simplicity as both authentication and cryptography are primarily dependent on the session which is controlled by the internal clock of the sender-receiver module of the Fog Network. Additional investigations and studies on Cellular Automata based said models ensure low-energy consumption and inexpensive physical implementation capability, and hence concluded as an efficient choice towards modeling of Green Computing in Fog Networks.

Keywords Fog network · Data security · Lightweight authentication · Lightweight cryptography · Cellular automata based model · Energy efficiency · Green computing

1 Introduction

The rise of Information Technology and its transformations have significantly played a crucial role towards the transformation of human society into a digital society. Several jobs/works of the society are today executed over networked based architecture, facilitating connections and co-operation among geographically isolated locations. Thus, the earlier networked computing has evolved into cutting edge Fog Computing. By properties, the computing facilities at edge devices as well as server are found over Fog Networks. A typical Fog Network is presented in Fig. 1. Further

A. Mitra · S. Saha (✉)
Department of Computer Science and Engineering, Siksha 'O' Anusandhan (Deemed to be University), Bhubaneswar, Odisha, India
e-mail: sayantansaha@soa.ac.in

A. Mitra
e-mail: arnabmitra@soa.ac.in

D. Mishra et al. (eds.), *Intelligent and Cloud Computing*, Smart Innovation, Systems and Technologies 194,
https://doi.org/10.1007/978-981-15-5971-6_25

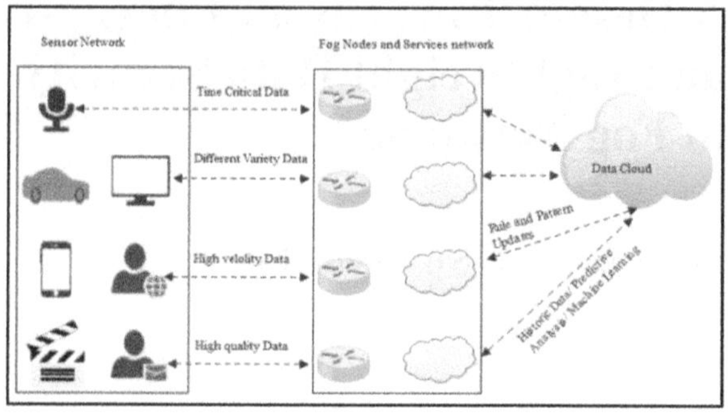

Fig. 1 A typical fog network

detailed literature related to the Fog Computing architecture and characteristics may be found over [1, 2].

From Fig. 1 it is observed that the data is required to be communicated between several nodes of a Fog Network, which potentially invites data risks [3, 4]. To deal with such scenario and to enhance the data security over networks, researchers and practitioners have suggested the uses of authentication [5] and cryptography [6]. Unfortunately, conventional authentication and cryptographic approaches towards enhanced data security are not efficient enough in view of modern days' IoT (Internet of Things)/Fog Computing based applications. Conventional data security approaches (authentication and cryptography) require a reasonable amount of power, power of processing and memory capabilities, which are often rarely possible to satisfy in IoT/Fog Network. To deal with this problem, lightweight data security approaches have been introduced over conventional approaches [7, 8]. A typical lightweight authentication and lightweight cryptography approaches are presented in following Figs. 2 and 3. Reader(s) is(are) requested to study [5, 6] to have more insights about authentication and cryptography.

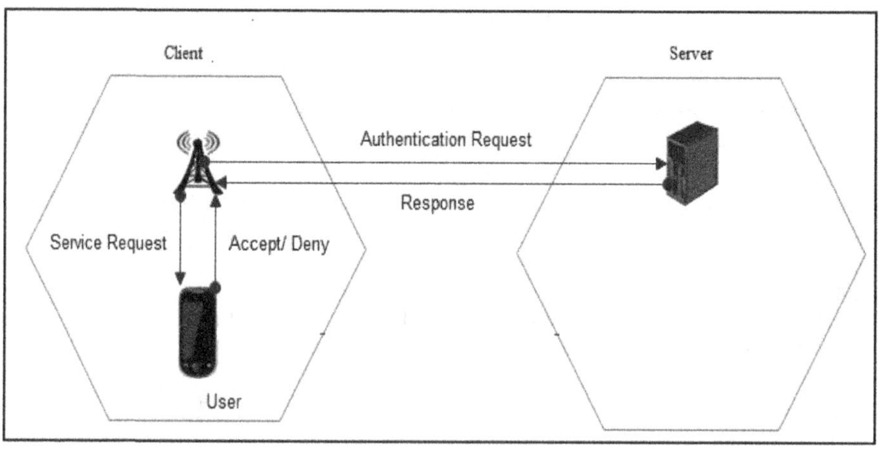

Fig. 2 A typical lightweight authentication approach

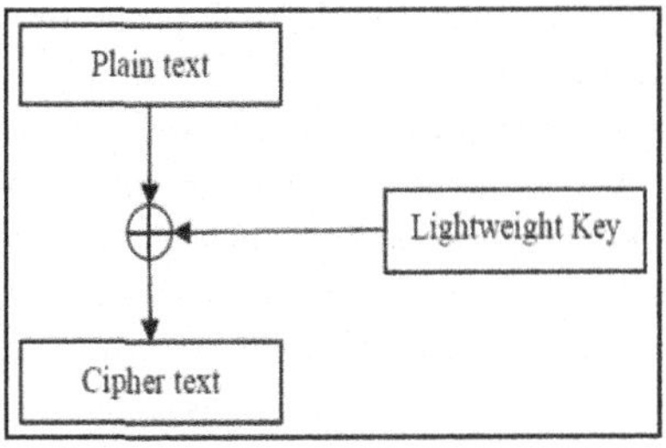

Fig. 3 A typical lightweight encryption in proposed cryptographic approach

On the other hand, green (low-energy/low-power consuming) and environmentally sustainable computing is one of the primary focus among researchers [9, 10]. For this reason, several modelling have been presented by researchers, which further facilitate low-cost physical modelling using Cellular Automata (CA) [6, 9–11].

Though, we found several lightweight data security approaches in [7, 8, 12–15], unfortunately we have not found any literature towards lightweight data security involving energy efficient low-cost physical modelling. We believe an energy efficient low-cost physical modelling towards lightweight data security should have an advantage towards modelling Green Computing in Fog Networks. For this reason, the scope for CA based modelling towards lightweight data security in Fog Networks is investigated and presented in this research. We request the reader(s) of this paper to go through [16] to study CA basics and its dynamics.

Major contributions of this research are as followed.

(a) CA based lightweight authentication in Fog Networks is presented;
(b) CA based lightweight cryptography in Fog Networks is presented;
(c) low-energy requirement for CA based modelling in Fog Networks is ensured.

Rest of the paper is organized as followed. State of the art literature survey is presented in Sect. 2; background research works are briefly described in Sect. 3; our proposed approach along with related discussions is briefly presented in Sect. 4; concluding remarks and future work are presented in Sect. 5.

2 Related Works

In [1], the Fog Computing was described as an evolution of Cloud Computing to facilitate location awareness, low latency, mobility, heterogeneity, real-time applications etc. with the help of a very large number of nodes (components). For this reason, several IoT based services (e.g., connected/joined vehicle, Smart City, Smart Grids, WSANs (wireless sensor and actuators networks), etc.) in [1] were explored as an application area of Fog Computing. Further, a detailed study related to the Fog Computing architecture was presented in [2].

It is found in [1, 2], that the computing in any Fog Network involves data mobility between several nodes of that network. For this reason, the data mobility in Fog

Computing networks potentially invites the data security and privacy issues. Potential security issues in view of IoT were investigated and reported by several researchers over time. Among several others, a detailed study related to the security issues in IoT was presented by Suo et al. in [3], and thereafter by Ni et al., in [4].To overcome the said shortfall and enhance data security and privacy, researchers suggested the uses several data security and privacy techniques such as uses of cryptography [6] and authentication [5]. In our further studies we found in [7] that the conventional data security and privacy methods do not fit well in view of IoT/Fog Computing networks due to a high constraint on limited availability of power (from energy source) and processing power. To overcome the said limitation authors introduced lightweight data security and privacy methods over the existing conventional data security and privacy methods in [7]. Besides, brief surveys on lightweight cryptography were presented in different literatures [8, 12]. Beyond the lightweight cryptography [14, 15], past researchers did also focus on lightweight authentication. A novel lightweight authentication scheme was presented in [13].

On the other hand, CA are quite popular among researchers as a dynamic modelling tool towards complex problems [16]. CA based data security and privacy techniques e.g., cryptography was presented in [6, 17]; CA based authentication was also presented in [5]. Uses of CA towards Prime numbers in view of enhanced data security and privacy were presented in [11, 18]. Thereafter, a detailed analysis in [19] on the CA based Prime generating sequence explored that the said sequence may be used as a source for "… more or less uniform noise…" [19] "… with a tendency of "weak periodicity"…" [19]. Besides the CA based modelling of data security and privacy techniques, scopes for CA were also investigated towards consideration in Green Computing. CA based several Green Computing model in Cloud were presented in [9, 10]. Based on the reference works of [20–22], it was explored in [9, 10] that the basic component for each CA cell is a D flip-flop [16], which requires a very low-power consumption (minimum $1.17E-07$ w to maximum $1.20E-05$ w at $1.0V_{dd}$ for several D flip-flops at $CMOS$ technology; and again a minimum of $1.12E-06$ w to a maximum of $1.20E-05$ w at $2.0V_{dd}$ for several D flip-flops at $CMOS$ technology) [20, 21].

From our studies, we concluded that a low-power consuming physical modelling of lightweight data security and privacy technique may have advantages in Fog Network. In spite of our hard efforts we were unable to find out literature(s) focused on physical modelling of data security and privacy technique in Fog Network with an advantage of Green Computing technique. For this reason, in present research CA based modelling of lightweight data security and privacy is investigated and presented. Most important past researches those are considered for our proposed work, are briefly presented in following Sect. 3.

3 Background

CA-based authentication in Cloud environment using OTPs (one-time passwords) was presented in [5]. Each set of OTP contains a controllable and finite number of member (e.g., 1, 2, 4, 8, 16 etc.) in it. An investigation confirmed that the characteristics polynomial involved towards the generation of OTPs was $(1 + x)$ or,$(1 + x)^n$ which was further found as primitive/recursive-primitive in nature (refer controllable generation of OTP set of [5]). Further p-value analysis in statistical RUNs test confirmed the presence of randomness in CA generated set of OTPs. In another research, a randomly populated set of Primes was obtained with CA based scheme [11]. In [19], a competitive time complexity $O(n) + O(k * \log^2 n)$ was observed for the involved primality testing by Fermat's primality testing method for the CA generated Prime numbers [5]. Based on the uses of existing CA-based authentication [5] and primes generation [11], we presented our proposal for a lightweight data security model, which is discussed in following Sect. 4.

4 Proposed Approach and Related Discussions

Presented research in this section is a work in progress by nature. Hence a limited study and analysis have been completed so far.

Proposed lightweight data security is modularized into two sub-modules: a lightweight authentication part and thereafter a lightweight cryptography part. In our proposed design, we emphasize on the inclusion of CA generated patterns as described in [5] for authentication part and CA patterns as described in [11] for cryptography purpose. CA patterns were considered for our approach as CA exhibits suitability towards VLSI (Very Large-Scale Integration) technology and supports parallel computing [16]. Further, CA based model results for a very little power consumption [9, 10].

4.1 Lightweight Authentication Using CA Based Scheme

Proposed authentication scheme emphasizes the uses of equally populated set of random OTPs (one-time passwords) as presented in [5]. Set of OTPs are made available at both sender and receiver end. One OTP at each time strap (based on the internal system clock) for each communication session is to be used towards authentication. Sender(s) is(are) allowed further to continue data encryption upon each successful authentication by the receiver. The authentication of sender is solely dependent on the internal clock of the sender-receiver module of Fog Network. Hence it is not easy to predict the OTP at each session. Proposed system flowchart for lightweight authentication scheme is presented in Fig. 4.

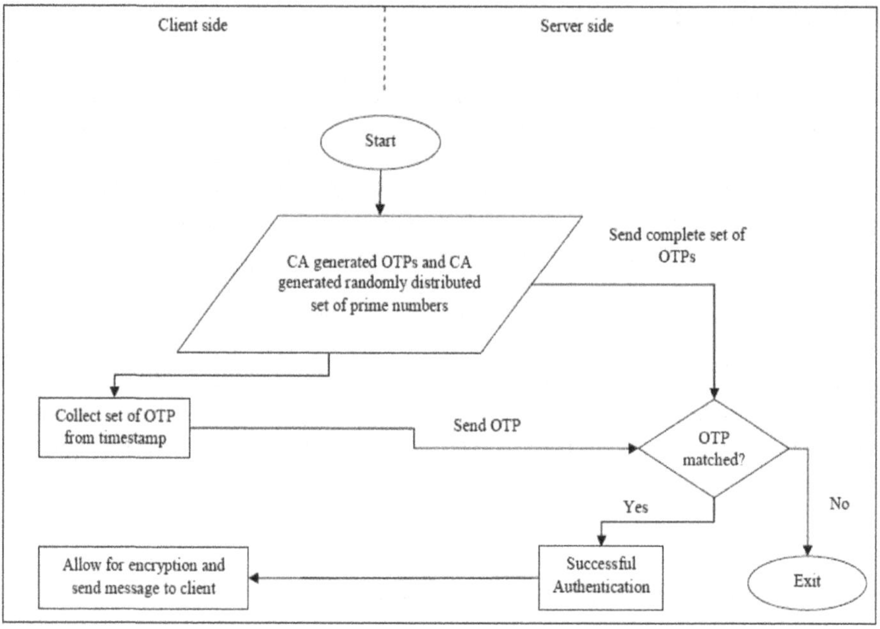

Fig. 4 Proposed system flowchart for lightweight OTP based authentication

4.2 Lightweight Cryptography Using CA Based Scheme

Proposed cryptography scheme emphasizes the uses of prime number as obtained with [11]. In our proposal, we emphasize on the uses of simple encryption. Set of primes as achieved from [11] is made available at both sender and receiver end. Sender(s) is(are) only allowed to continue data encryption upon successful authentication by the receiver at each session. One unique prime number from the CA generated randomly populated set of Primes is responsible for the encryption/decryption process, which is once again to be selected based on the internal clock. The cryptography phase is once again solely dependent on the internal clock of the sender-receiver module of the Fog Network. The sender will only communicate the value of time strap as public key along with the encrypted message (cipher text). Upon receipt, receiver need to find the decryption key first from the Set of Primes (which is only internally available to sender and receiver only) before message decryption. Hence it is again not easy to predict the encryption/decryption key for each session. Proposed system flowchart for lightweight cryptography scheme is presented in Fig. 5.

An illustration based on our presented proposed approach is followed next.

Illustration 1.

A complete CA-based data security is shown in following two figures Fig. 6 (targeting authentication) and Fig. 7 (targeting cryptography). For OTP-based authentication,

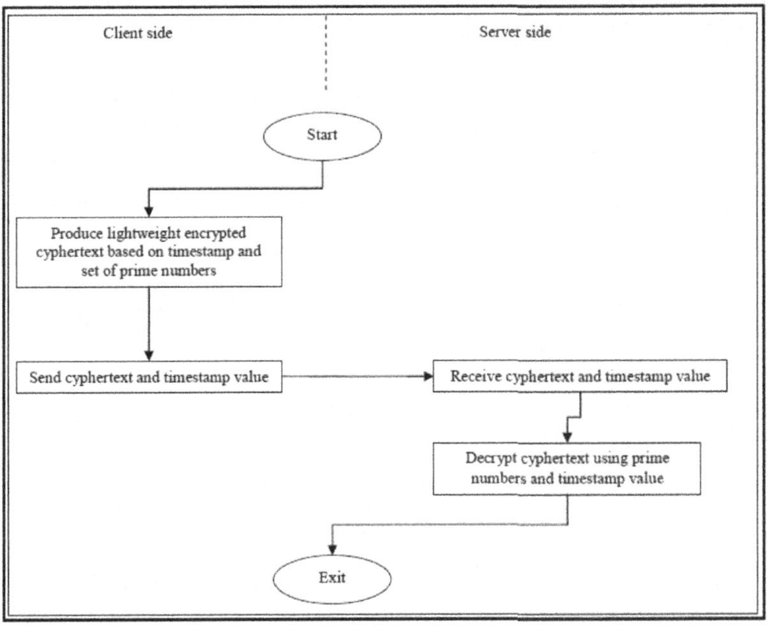

Fig. 5 Proposed system flowchart for lightweight cryptography

Fig. 6 An illustration for OTP-based authentication using CA generated pattern

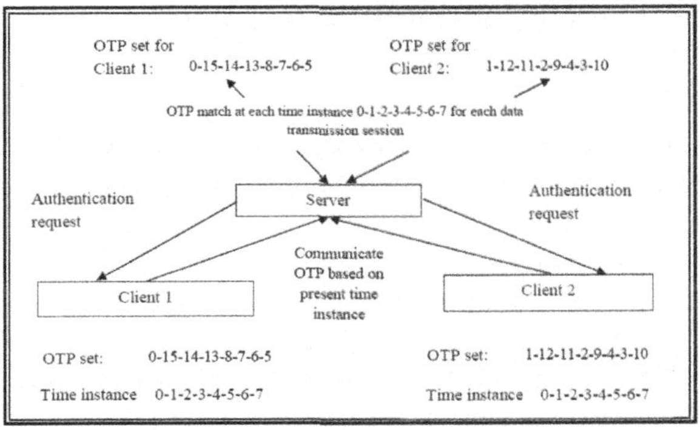

we described an uniform CA scenario with <153, 153, 153, 153> at null boundary scenario to generate two equally populated set of OTPs {0, 15, 14, 13, 8, 7, 6, 5}, and {1, 12, 11, 2, 9, 4, 3, 10} as shown in [5]. For simplicity, we presented simple client–server architecture with only one server connected with two clients.

For complete cryptography, we considered a CA scenario with <110, 110, 110, 204> at null boundary scenario to generate a randomly populated set of primes {3, 7, 13} as described in [11]. For simplicity, we presented simple client–server architecture. Presented cryptography is allowed to be followed if and only if it is authenticated as described in Fig. 6.

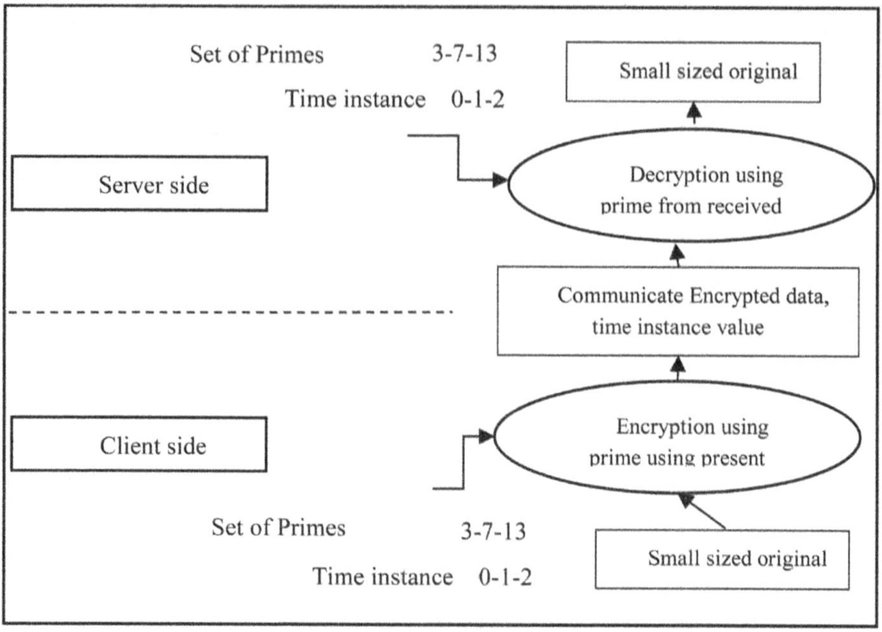

Fig. 7 An illustration for complete cryptosystem using CA generated pattern

5 Conclusions and Future Work

A work in progress towards the lightweight data security in Fog Networks has been described in this paper. Proposed approach is modularized into two sub-modules to present in depth lightweight design of (i) authentication, and (ii) cryptography; both based on CA generated patterns. Hence, proposed design ensures low-power consumption, suitability towards VLSI technology and support for parallel computing. Additionally, selection of the random OTP and, or random encryption/decryption key in proposed system is solely dependent on the uses of session which is further dependent on internal system clock of the sender-receiver pair of the Fog Network. Thus, it is capable to ensure privacy and results for an enhanced data safety. Hence, proposed lightweight data security is potentially applicable towards physical modelling in Green Computing in Fog Networks.

It is already mentioned that presented design is in work in progress. Hence, in future we plan to extend our work and test the efficiency of our proposed approach with reference to the existing lightweight data security approaches at different security attacks scenario.

Acknowledgments Authors sincerely acknowledge the reviews received from the anonymous reviewers, which have further helped to enhance the quality of this paper.

References

1. Bonomi, F., Milito, R., Zhu, J., Addepalli, S.: Fog computing and its role in the internet of things. In: Proceedings of the first edition of the MCC workshop on Mobile cloud computing, August 2012, pp. 13–16. ACM (2012)
2. Khakimov, A., Muthanna, A., Muthanna, M.S.A.: Study of fog computing structure. In: Proceedings of the 2018 IEEE Conference of Russian Young Researchers in Electrical and Electronic Engineering (EIConRus), January 2018, pp. 51–54. IEEE (2018)
3. Suo, H., Wan, J., Zou, C., Liu, J.: Security in the internet of things: a review. In: Proceedings of the2012 International Conference on Computer Science and Electronics Engineering, March 2012, vol. **3**, pp. 648–651. IEEE (2012)
4. Ni, J., Zhang, K., Lin, X., Shen, X.S.: Securing fog computing for internet of things applications: challenges and solutions. IEEE Commun. Surv. Tutor. **20**(1), 601–628 (2017)
5. Mitra, A., Kundu, A., Chattopadhyay, M., Chattopadhyay, S.: A cost-efficient one time password-based authentication in cloud environment using equal length cellular automata. J. Ind. Inf. Integr. **5**, 17–25 (2017)
6. Nandi, S., Kar, B.K., Chaudhuri, P.P.: Theory and applications of cellular automata in cryptography. IEEE Trans. Comput. **43**(12), 1346–1357 (1994)
7. Buchanan, W.J., Li, S., Asif, R.: Lightweight cryptography methods. J. Cyber Secur. Technol. **1**(3–4), 187–201 (2017)
8. Eisenbarth, T., Kumar, S., Paar, C., Poschmann, A., Uhsadel, L.: A survey of lightweight-cryptography implementations. IEEE Des. Test Comput. **24**(6), 522–533 (2007)
9. Mitra, A., Kundu, A.: Energy efficient CA based page rank validation model: a green approach in cloud. Int. J. Green Comput. (IJGC) **8**(2), 59–76 (2017)
10. Mitra, A.: On investigating energy stability for cellular automata based pagerank validation model in green cloud. Int. J. Cloud Appl. Comput. (IJCAC) **9**(4), 66–85 (2019)
11. Mitra, A., Kundu, A.: Cost optimized set of primes generation with cellular automata for stress testing in distributed computing. Proc. Technol. **10**, 365–372 (2013)
12. Katagi, M., Moriai, S.: Lightweight cryptography for the internet of things. Sony Corp. **7**, 10 (2008)
13. Lee, J.Y., Lin, W.C., Huang, Y.H.: A lightweight authentication protocol for internet of things. In: 2014 International Symposium on Next-Generation Electronics (ISNE), May 2014, pp. 1–2. IEEE (2014)
14. Diro, A.A., Chilamkurti, N., Kumar, N.: Lightweight cybersecurity schemes using elliptic curve cryptography in publish-subscribe fog computing. Mob. Netw. Appl. **22**(5), 848–858 (2017)
15. Lu, R., Heung, K., Lashkari, A.H., Ghorbani, A.A.: A lightweight privacy-preserving data aggregation scheme for fog computing-enhanced IoT. IEEE Access **5**, 3302–3312 (2017)
16. Chaudhuri, P.P., Chowdhury, D.R., Nandi, S., Chattopadhyay, S.: Additive Cellular Automata: Theory and Applications, vol. 1. Wiley (1997)
17. Blackburn, S.R., Murphy, S., Paterson, K.G., Nandi, S., Chaudhuri, P.P.: Comments on "Theory and applications of cellular automata in cryptography" [with reply]. IEEE Trans. Comput. **46**(5), 637–639 (1997)
18. Mitra, A.: Selection of cost-effective prime source towards possible uses in Fog Computing. In: Proceedings of the2018 2nd International Conference on Data Science and Business Analytics (ICDSBA), September 2018, pp. 19–24. IEEE (2018)
19. Mitra, A., Kundu, A.: Analysis of sequences generated by ELCA-type cellular automata targeting noise generation. In: Proceedings of the19th International Conference on System Theory, Control and Computing (ICSTCC), October 2015, pp. 883–888. IEEE (2015)

20. Mehta, K., Arora, N., Singh, B.P.: Low power efficient D flip flop circuit. In: Proceedings of theInternational Symposium on Devices MEMS, Intelligent Systems & Communication (ISDMISC), pp. 16–19. IJCA (2011)
21. Oskuii, S.T.: Comparative study on low-power high-performance flip-flops. Unpublished Master of Technology Thesis, Linköping University (2004)
22. https://www.divaportal.org/smash/get/diva2:19406/FULLTEXT01.pdf. Accessed 05 Sept 2019

Fuzzification-Based Routing Protocol for Wireless Ad Hoc Network

Santosh Kumar Das, Nikhil Patra, Biswa Ranjan Das, Shom Prasad Das, and Soumen Nayak

Abstract The applications of Wireless Ad hoc Network (WANET) increases rapidly due to its effectiveness. Each node of this network behaves as router to send/receive data packets. In WANET, nodes contain limited capacity of energy, i.e., fail during data transmission, so, due to this issue, network lifetime decreases. In this paper, an intelligent fuzzification-based routing protocol is proposed to solve this issue. Fuzzification is the method of fuzzy logic to reduce the uncertainties of the WANET. At last, the work is evaluated in Java programming language in terms of effectiveness.

Keywords Wireless ad hoc network · Network lifetime · Routing · Fuzzy logic · Floyd–Warshall algorithm

1 Introduction

In the last few years, usages of wireless network rapidly increase for its efficiency and robustness [1, 2], so it contains some variations like Wireless Ad hoc Network (WANET), which is one of the variations of the wireless network. WANET is a multi-hop system network. Its nature is randomness, so its topology is dynamic. Figure 1

S. K. Das (✉) · N. Patra · B. R. Das · S. P. Das
School of Computer Science and Engineering, National Institute of Science and Technology (Autonomous), Institute Park, Pallur Hills, Berhampur 761008, India
e-mail: sunsantosh2007@rediffmail.com

N. Patra
e-mail: nick004007@gmail.com

B. R. Das
e-mail: biswaraju12@gmail.com

S. P. Das
e-mail: shom@nist.edu

S. Nayak
Department of Computer Science and Engineering, SOA University, Bhubaneswar 751030, India
e-mail: soumennayak@soa.ac.in

D. Mishra et al. (eds.), *Intelligent and Cloud Computing*,
Smart Innovation, Systems and Technologies 194,
https://doi.org/10.1007/978-981-15-5971-6_26

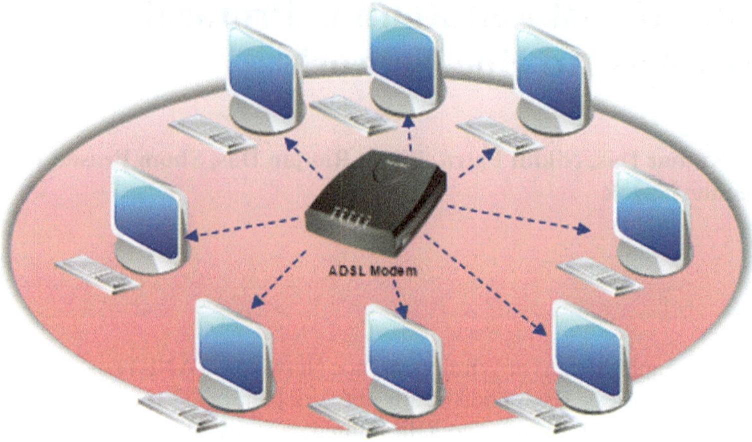

Fig. 1 Wireless ad hoc network

shows an example of WANET. In this figure, multiple computers are connected with an Asymmetric Digital Subscriber Line (ADSL) modem. This model is connected with WISP that indicates a wireless identification and sensing platform for works as a router. Basically, it is used to create a wireless environment for accessing and sharing data and information. It connects all the devices within a radio range.

There are several advantages of WANET such as more flexibility, better mobility, easily turn up/down, and more economical. Due to abovementioned advantages, there are several applications of WANET as follows:

(a) **Military Arena**: It allows in the military battleground to maintain the information of the WANET by using headquarters, vehicles, and soldiers.
(b) **Personal Area Network**: It is used in a bit range with some nodes for personal purposes.
(c) **Industry Sector**: It is also used in business purpose as commercial applications, emergency situation, and rapid development of the society.
(d) **Bluetooth**: It helps data send/receive within a limited range by using mobile phones and laptops.

Although there are several applications of WANET, it has some limitations like nodes of the network consist of low capacity battery which is not sufficient for any operation. During data transmission, route fails and transmission is restarted, so, in the current paper, an intelligent fuzzy-based routing protocol is designed for enhancing network lifetime.

The rest of the paper is divided as follows. Section 2 highlights some existing work related to some routing techniques. Section 3 shows detailed analysis parts of the proposed method. Section 4 indicates simulation analysis part. Finally, Sect. 5 concludes the paper.

2 Related Works

In the past few years, some works have been designed related to routing protocol of WANET. Some of them are as follows. Das and Tripathi [3] proposed a non-linear based routing technique for hybridization network. This is a combination of static and dynamic nodes. In this network, static and dynamic nodes' determination is decided based on the requirement of the users. In this work, the author uses geometric programming that is non-linear programming. This non-linear programming helps to manage different conflicting parameters of the network and derive the optimal routing of the network. Abdallah [4] proposed smart partial flooding routing algorithms for ad hoc network. In this paper, two 3D geographical routing algorithms are used for maintaining two things like overhead and flooding in the network. Das and Tripathi [5] proposed an intelligent routing protocol for transparent heterogeneous ad hoc network. Transparent ad hoc network indicates a network that consists of two phases, i.e., user and administrator. These two networks are connected by transparency system. In this proposal, the author uses a game theory technique to optimize the network parameters. In this system, the utilities of the game are optimized with the help of linear mathematical model and enhance the network lifetime. Hayes and Ali [6] proposed a routing protocol named RASeR for mobile sensor networks. In this paper, topology changing issue is solved with the help of global time-division multiple accesses using fixed nodes. Singh et al. [7] designed distributed routing algorithms for WANET. The proposed work is based on two existing routing protocols such as DSDV and AODV where performance of these protocols is analyzed based on pause time of the nodes' mobility. In [8], the authors designed an energy-based routing system for WANET where each node is mobile. In this protocol, the author uses an extended version of the fuzzy logic, i.e., known as intuitionistic fuzzy logic. The proposed method uses some network parameters for optimizing and reducing its uncertainty. It also uses two decision-making techniques such as entropy and preference ranking organization method for enrichment of evaluations. These two methods are ranking-based technique which helps to intelligent route of the network. In [9], the authors designed an artificial intelligence system for routing in WANET. In this work, the ad hoc network, i.e., used hybrid which is a fusion of static and dynamic nodes to resolve the issue of the user. Here, the author uses some of the artificial techniques such as geometric programming, fuzzy logic, and optimset. The author uses in this proposal some of the conflicting objective functions to optimize the network metrics. Finally, all the metrics help to enhance the network lifetime of the hybrid ad hoc network. In [10], the authors designed a fuzzy-based routing technique for ad hoc network. This routing technique is based on several constraints of the network where nature of the each constraint is conflicting one to another, so all the parameters are combined with the help of fuzzy inference system. The proposed fuzzification and defuzzification methods help to estimate the network parameters and enhance the network lifetime of the network. In [11], the authors designed an opportunistic-based routing protocol for WANET. The proposed work reduces congestion of the network by reducing energy of the nodes. It also helps to maintain backlog traffic of the network and helps to increase bandwidth of the network. Finally, it helps to reduce average end-to-end delay and routing overhead and enhance throughput of the network. In [12], the authors designed a multicast routing system for WANET. In this proposal, the selected parameters are energy and distance. Both the parameters are very crucial in terms of network lifetime. These two parameters are fuzzified in terms of generating linguistic variables. These linguistic variables help to design fuzzy rule-based system of the network. It helps to design fuzzy output parameter which is based on two input parameters, i.e., energy and distance. The name of the output parameter is a reward. It helps to removed ambiguity of the route selection and estimates uncertainties of the network. In [13], the authors designed a fuzzy-based multicast routing system WANET using four parameters hop count, packet, energy, and route lifetime. In these parameters, energy, hop count, and packet are input parameters, and route lifetime is the output parameter. Among all input parameters, energy is the crucial parameter, so, by using this energy parameter, fuzzy decision-maker enhances the network lifetime of the ad hoc network. Saini and Sharma [14] proposed a system for routing in WANET. It contains detailed analysis of existing works that categorized on the basis of some features. Fu et al. [15] designed a multiple path system for managing the environment of the sensor network. In this paper, messages are controlled and manage in harsh situation with multiple path systems. Hence, it increases the performance of sensor networks.

3 Proposed Work

The proposed method is based on fuzzy logic in which three input constraints are used such as residual energy, control packet, and distance and one output parameter, i.e., stability. Each input parameter is divided into three linguistic variables for fuzzification. Figure 2 shows the proposed network model where each node is categorized into three linguistic variables based on energy level of the node such as low, medium, and high. Tables 1, 2, 3, and 4 show membership functions of input and output variables.

In the proposed method, first and second input parameters have the same linguistic variables as low, medium, and high, but for the third input parameter are short, medium, and long. The output parameter is stability which defines the nature of the route from source to destination nodes. It has five linguistic variables such as very bad, bad, medium, good, and very good. All the input parameters are processed by the fuzzy logic. The architecture of the fuzzy logic is shown in Fig. 3 that consists of three steps (i) fuzzification, (ii) inference engine, and (iii) defuzzification. The complete process of the fuzzy logic system is shown in Algorithm 1. The proposed method is partially completed work which is simulated in Java programming language which is described in the next section.

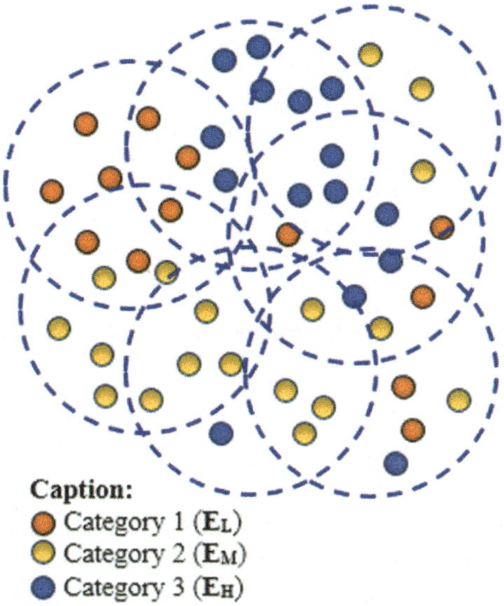

Fig. 2 Proposed network model

Table 1 Membership functions of input parameter energy

Linguistic values	Notation	Range
Low	E_L	$[E_{L-}, E_{L+}]$
Medium	E_M	$[E_{M-}, E_{M+}]$
High	E_H	$[E_{H-}, E_{H+}]$

Table 2 Membership functions of input control packet

Linguistic values	Notation	Range
Low	C_L	$[C_{L-}, C_{L+}]$
Medium	C_M	$[C_{M-}, C_{M+}]$
High	C_H	$[C_{H-}, C_{H+}]$

Table 3 Membership functions of input parameter distance

Linguistic values	Notation	Range
Short	D_S	$[D_{S-}, D_{S+}]$
Medium	D_M	$[D_{M-}, D_{M+}]$
Long	D_L	$[D_{L-}, D_{L+}]$

Algorithm 1. Fuzzy Logic Process

Step 1: Analyze the problem

Table 4 Membership functions of output parameter stability

Linguistic values	Notation	Range
Very bad	S_{VB}	$[S_{VB-}, S_{VB+}]$
Bad	S_B	$[S_{B-}, S_{B+}]$
Medium	S_M	$[S_{M-}, S_{M+}]$
Good	S_G	$[S_{G-}, S_{G+}]$
Very Good	S_{VG}	$[S_{VG-}, S_{VG+}]$

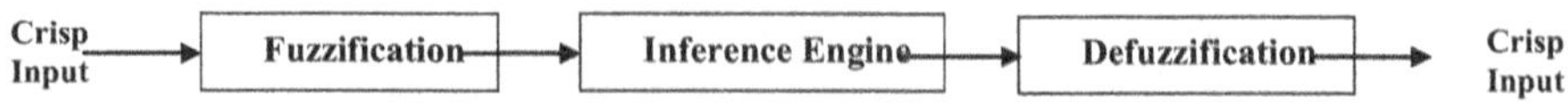

Fig. 3 Architecture of the fuzzy logic

Step 2: Analyze the input and output parameters
Step 3: Decide the membership functions of input and output parameters
Step 4: Analyze the linguistic variables of input and output parameters
Step 5: Apply fuzzification method for input parameters
Step 6: Decide antecedent and consequent of the rule base system
Step 7: Decide operators of the antecedent
Step 8: Design rule base system with input and output parameters
Step 9: Apply Mamdani controller on the rule base system
Step 10: Evaluate fuzzy value that is produced by the controller
Step 11: Apply defuzzification on the produced fuzzy value
Step 12: Deploy converted crisp value in the desired problem

4 Simulation Analysis

The performance evaluation of the proposed method is based on some software requirement specifications and some hardware system configurations that are shown in Tables 5 and 6. Figure 4 shows tabular information of the node details including node id, residual energy of the node, node position (i.e., *x*- and *y*-axes), control packet, and radio energy where residual energy, control packet, and node positions are random numbers. Figure 5 shows node distance between one node to another node. Here, self-loop is not considered which is denoted by zero. Figure 6 shows shortest distance of the nodes from one node to another node based on Floyd–Warshall algorithm. Here, also self-loop is not considered, and it is denoted by zero.

Table 5 Software requirement specification

Software	Specification
Windows OS	10
Java	12.0.2
MS Office	2013

Table 6 Hardware system configuration

Hardware	Specification
Monitor	HP
Processor	Intel(R) Core(TM) i5-7200U
Speed	2.50 GHz 2.71 GHz
Hard disk	1 TB
Keyboard	HCL
RAM	8.00 GB
Mouse	HCL

```
C:\Users\Asus\Desktop\Btech project>javac Experiment.java

C:\Users\Asus\Desktop\Btech project>java Experiment
Enter the number of nodes:
10
The Details of each nodes:
Node Residual Energy    X-Coordinate    Y-Coordinate  Control Packet           Radio Signal
n1       178                 211              58            82                    200
n2       441                   1             170           100                    200
n3       364                 113             100             8                    200
n4       241                 133             131            39                    200
n5       172                  57             220            60                    200
n6       194                  17              21            44                    200
n7       176                 287              31            14                    200
n8       367                 231              80             9                    200
n9       120                  21             211            87                    200
n10       164                  72              35            69                    200
```

Fig. 4 Node details in tabular format

```
The initial adjacency matrix is:
       0     238     106     106     223     197      80      29     243     140
     238       0     132     137      75     149     317     246      45     152
     106     132       0      36     132     124     187     119     144      76
     106     137      36       0     117     159     183     110     137     113
     223      75     132     117       0     202     297     223      37     185
     197     149     124     159     202       0     270     221     190      56
      80     317     187     183     297     270       0      74     321     215
      29     246     119     110     223     221      74       0     247     165
     243      45     144     137      37     190     321     247       0     183
     140     152      76     113     185      56     215     165     183       0
```

Fig. 5 Node distance details

The final adjacency matrix is:

0	238	106	106	223	196	80	29	243	140
238	0	132	137	75	149	317	246	45	152
106	132	0	36	132	124	186	119	144	76
106	137	36	0	117	159	183	110	137	112
223	75	132	117	0	202	297	223	37	185
196	149	124	159	202	0	270	221	190	56
80	317	186	183	297	270	0	74	320	215
29	246	119	110	223	221	74	0	247	165
243	45	144	137	37	190	320	247	0	183
140	152	76	112	185	56	215	165	183	0

Fig. 6 Node shortest distance

5 Conclusion and Future Work

The proposed paper consists of an intelligent routing technique using a fuzzy logic system. This method contains three input parameters such as residual energy, control packet, distance, and output parameter stability. The fuzzy logic is used in this protocol to control the conflicting behavior of the WANET and reduce the uncertainties of the network. The current proposal is work in progress, so future work is to complete this project and compare it with some existing routing techniques. At last implement, routing protocol will validate in terms of network metrics through simulation as well as mathematical analysis.

References

1. Lin, Z., Wang, P.: A review of data sets of short-range wireless networks. Comput. Commun. **147**, 138–158 (2019)
2. Das, S.K., Samanta, S., Dey, N., Kumar, R.: Design Frame-works for Wireless Networks, pp. 1–439. Springer, Lecture Notes in Net-works and Systems. ISBN: 978-981-13-9573-4
3. Das, S.K., Tripathi, S.: Energy efficient routing formation algorithm for hybrid ad-hoc network: A geometric programming approach. Peer-to-Peer Netw. Appl. **12**(1), 102–128 (2019)
4. Abdallah, A.E.: Smart partial flooding routing algorithms for 3D ad hoc networks. Procedia Comput. Sci. **94**, 264–271 (2016)
5. Das, S.K., Tripathi, S.: Adaptive and intelligent energy efficient routing for transparent heterogeneous ad-hoc network by fusion of game theory and linear programming. Appl. Intell. **48**(7), 1825–1845 (2018a)
6. Hayes, T., Ali, F.H.: Robust Ad-hoc Sensor Routing (RASeR) protocol for mobile wireless sensor networks. Ad Hoc Netw. **50**, 128–144 (2016)
7. Singh, J., Bhardwaj, M., Sharma, A.: Experimental analysis of distributed routing algorithms in ad hoc mobile networks. Procedia Comput. Sci. **57**, 1411–1416 (2015)
8. Das, S.K., Tripathi, S.: Intelligent energy-aware efficient routing for MANET. Wirel. Netw. **24**(4), 1139–1159 (2018b)
9. Das, S.K., Tripathi, S.: Energy efficient routing formation technique for hybrid ad hoc network using fusion of artificial intelligence techniques. Int. J. Commun. Syst. **30**(16), e3340, 1–16 (2017)
10. Yadav, A.K., Das, S.K., Tripathi, S.: EFMMRP: design of efficient fuzzy based multi-constraint multicast routing protocol for wireless ad-hoc network. Comput. Netw. **118**, 15–23 (2017)
11. Qin, Y., Li, L., Zhong, X., Yang, Y., Ye, Y.: Opportunistic routing with admission control in wireless ad hoc networks. Comput. Commun. **55**, 32–40 (2015)

12. Das, S.K., Yadav, A.K., Tripathi, S.: IE2M: Design of intellectual energy efficient multicast routing protocol for ad-hoc network. Peer-to-Peer Netw. Appl. **10**(3), 670–687 (2017)
13. Das, S.K., Tripathi, S., Burnwal, A.P.: Fuzzy based energy efficient multicast routing for ad-hoc network. In: Proceedings of the 2015 Third International Conference on Computer, Communication, Control and Information Technology (C3IT), pp. 1–5. IEEE (2015, February)
14. Saini, T.K., Sharma, S.C.: Prominent unicast routing protocols for mobile ad hoc networks: criterion, classification, and key attributes. Ad Hoc Netw. **89**, 58–77 (2019)
15. Fu, X., Fortino, G., Pace, P., Aloi, G., Li, W.: Environment-fusion multipath routing protocol for wireless sensor networks. Inf. Fusion **53**, 4–19 (2020)

Fuzzy-Based Strategy Management in Wireless Ad Hoc Network

Santosh Kumar Das, Aman Kumar Tiwari, Somnath Rath, Manisha Rani, Shom Prasad Das, and Soumen Nayak

Abstract Today's applications of wireless communication increase rapidly due to its flexibility. Wireless Ad hoc Network (WANET) is one of the variations of the wireless network. The topology of the network is dynamic, so, in this network, parameter variations raise frequently. In this paper, a game theory based on strategy management technique is proposed for WANET. Game theory is a mathematical modeling which is used to manage conflicting strategy of the situation between players. The proposed method is simulated in Python for managing strategy of the nodes.

Keywords Wireless ad hoc network · Strategy management · Game theory · Cooperative · Non-cooperative

S. K. Das (✉) · A. K. Tiwari · S. Rath · M. Rani · S. P. Das
School of Computer Science and Engineering, National Institute of Science and Technology (Autonomous), Institute Park, Pallur Hills, Berhampur 761008, India
e-mail: sunsantosh2007@rediffmail.com

A. K. Tiwari
e-mail: amantiwaari2016@gmail.com

S. Rath
e-mail: srath822@gmail.com

M. Rani
e-mail: manisharani2745@gmail.com

S. P. Das
e-mail: shom@nist.edu

S. Nayak
Department of Computer Science and Engineering, SOA University, Bhubaneswar 751030, India
e-mail: soumennayak@soa.ac.in

D. Mishra et al. (eds.), *Intelligent and Cloud Computing*,
Smart Innovation, Systems and Technologies 194,
https://doi.org/10.1007/978-981-15-5971-6_27

1 Introduction

Wireless network has several variations in terms of their applications and usage [1, 2]. Wireless Ad hoc Network (WANET) is one of the variations of the wireless network. It is used in several areas such as medical, school and college, hospital, market, business, etc. Although it has several advantages and applications, it has several drawbacks like each node contains few energy capacities of battery, so this battery is not suitable for any application. Hence, several types of issues are raised such as route breakage, bandwidth problems, fault and intrusion problems, resource allocation, anomaly enhancement, outlier, etc. Sometimes, some security and privacy issues are also occurred due to mentioned issues [3]. Some works exist for outlier detection or anomaly detection, fault diagnosis, intrusion detection, and mobility prediction [4, 5]. Most of the outlier detection and prediction algorithms are outperformed with the help of machine learning algorithms [6, 7]. The reason is robustness of the algorithm in order to optimize [8–10]. The current paper is proposed for the network strategy management technique for WANET, so that network lifetime increases simultaneously in terms of several network metrics.

The remaining of the paper is illustrated as follows: Sect. 2 discusses some reviews that highlight some of the existing techniques. Section 3 describes the details of the proposed method. Section 4 describes the simulation analysis parts. The last section, i.e., Sect. 5, concludes the proposal.

2 Related Works

In the past few years, some strategy-based works have been proposed in the context of WANET based on several strategies of management as Saini and Sharma [11] designed a prominent unicast technique for WANET. This article consists of several types of routing protocol works that are arranged based on particular features. Fu et al. [12] proposed a technique for more than one path. This paper contains discussion about multiple paths in harsh situation. Bai et al. [13] proposed a method for detecting wormhole attacks. It detects almost 100% wormhole attacks even in a wireless network with very poor connectivity. Dhief et al. [14] proposed a detailed analysis of surveillance systems for WANET. It basically consists of several analyses and their solutions for disaster based on forest. Li et al. [15] proposed a clustering-based technique for wireless mesh network which is based on IoT system. In this paper, existing errors of the network are solved by optimizing the network traffic and data generation. Lin et al. [16] used an ad hoc network based on software system. It is an energy-based technique which includes green energy system, security system, and privacy system. Chai et al. [17] proposed hybrid wireless mesh networks using load-aware cooperative hybrid routing protocol. In this paper, a load-aware cooperative hybrid routing protocol is adapted to cover the peculiarities of routers and clients and also considers load in routing. Das and Tripathi [18] proposed a routing method for static and dynamic ad hoc networks using non-linear programming. In this system, static and dynamic nodes' determination is decided based on the requirement of the users. In this work, non-linear programming helps to manage different conflicting parameters of the network and derive the optimal routing of the network. Das and Tripathi [19] proposed a routing protocol for transparent ad hoc network. In this proposal, the author uses a game theory technique to optimize the network parameters. In this system, fusion of linear programming and game theory is used to enhance the network's lifetime. Tripathi and Das [20] designed a routing protocol for reducing energy consumption where each node is mobile. In this protocol, the author uses an extended version of the fuzzy logic, i.e., known as intuitionistic fuzzy logic. The proposed method uses some network parameters for optimizing and reducing its uncertainty. It uses two decision-making techniques such as entropy and preference ranking organization method for enrichment of evaluations. In [21], the authors designed an artificial intelligent-based routing technique for ad hoc network. In this work, the ad hoc network used hybrid which is a fusion of static and dynamic nodes to resolve the issue of the user. In this paper, some of the conflicting objective functions are optimized. In [22], a fuzzy-based routing technique is designed for ad hoc network. This routing technique is based on several constraints of the network where nature of the each constraint is conflicting one to another. The proposed fuzzification and defuzzification methods help to estimate the network parameters and enhance the network lifetime of the network. In [23], a multicast routing method is designed for ad hoc network. In this proposal, the route is selected by using two parameters, i.e., energy and distance. It helps to design fuzzy output parameter for removing ambiguity of the route selection and estimates uncertainties of the network. Das et al. [24] designed a fuzzy-based multicast routing protocol using parameters hop count, packet, energy, and route lifetime. By using these parameters, fuzzy decision-maker enhance the network lifetime of the ad hoc network.

3 Proposed Work

In this paper, an intelligent strategy management technique is proposed for WANET. Figure 1 shows structure of ad hoc network. In this figure, there are different nodes available. The topology of the network is dynamic, and it is infrastructureless, so parameters of the WANET vary frequently. Due to this variation, capability of the nodes also changes frequently. Hence, there are different colored nodes are illustrated in this figure.

The proposed method is based on game theory-based optimization technique. In this mathematical modeling, two types of games are involved: (i) cooperative game theory and (ii) non-cooperative game theory. In cooperative game theory, one player cooperates with other player in terms of payoff. But in non-cooperative game theory, one player does not cooperative with other player in terms of payoff. In this proposal, there are two players considered as Player A and Player B in terms of crucial parameter of the network, i.e., residual energy. Payoff matrix of both players

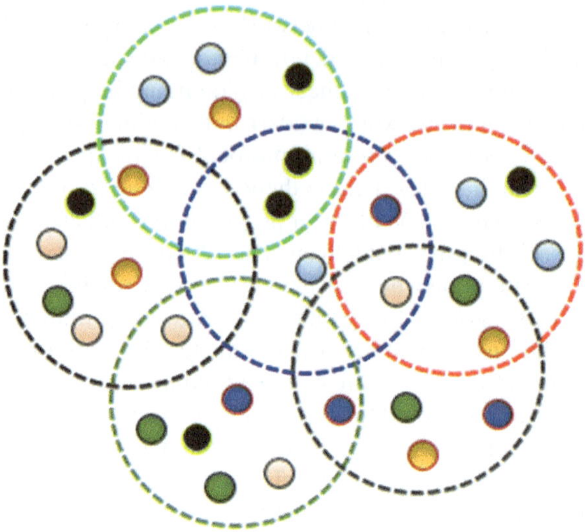

Fig. 1 Structure of the WANET

is shown in Fig. 2. Player A contains all the nodes those having energy level greater than or equal to 50% and Player B contains all the nodes those having energy level less than 50%. Both the players are competitive for transmitting the control packet within the network for creating the routes.

Player A's strategies are Strategy *i*, where *i* lies between 1 and *m* and Player B's strategies are Strategy *j*, where *j* lies between 1 and *n*, and (S_{A1}, S_{B1}), (S_{A1}, S_{Bj}),....,

		Player B		
		Strategy 1	**Strategy j..**	**..Strategy n**
Player A	**Strategy 1**	S_{A1}, S_{B1}	S_{A1}, S_{Bj}	S_{A1}, S_{Bn}
	Strategy i . .	S_{Ai}, S_{B1}	S_{Ai}, S_{Bj}	S_{Ai}, S_{Bn}
	Strategy m	S_{Am}, S_{B1}	S_{Am}, S_{Bj}	S_{Am}, S_{Bn}

Fig. 2 Payoff matrix of the proposed method

(S_{Am}, S_{Bj}), (S_{Am}, S_{Bn}) are different playoff values of the matrix. Strategy will design based on two network lifetimes of the WANET that consists of several conflicting metrics such as packet delivery ratio, throughput, packet loss, end-to-end delay, jitter, routing overhead, etc. The nature of these metrics is categorized into two types: first is increasing nature and second is decreasing nature. The parameters considered in this paper are energy, mobility, control packet, and bandwidth. Initially, each parameter is considered as crisp value and after that convert it into fuzzy value with the help of fuzzification method. In this proposal, triangular membership function is used as fuzzification using (x, $\mu_A(x)$) where x is the element and $\mu_A(x)$ is the membership value of the element x within the set A. The current work of this paper is not completed; it will complete as future work for designing feasible and optimal strategies of the network.

4 Simulation Analysis

The proposed scheme is evaluated in Python programming language in context of several network parameters. Figure 3 shows node deployment scenario where column 1 indicates node id, column 2 indicates energy level of the nodes, column 3 indicates mobility of the nodes, and column 4 indicates control packet of the nodes. Each node is divided into two parts, i.e., Player A and Player B based on threshold value of the energy level, i.e., shown in Figs. 4 and 5.

Details of ALL Nodes

Node Id	Energy	Mobility	Control Packet
0	49	73	11
1	60	19	26
2	69	49	48
3	67	91	23
4	70	68	51
5	13	50	44
6	11	65	62
7	7	12	29
8	79	94	48
9	16	75	83
10	73	57	61
11	32	28	41
12	67	40	77
13	61	43	31
14	54	53	14
15	54	96	35
16	61	92	2
17	66	74	33
18	52	28	55
19	47	9	33
20	3	50	43
21	52	23	10
22	17	5	83
23	2	44	4
24	32	70	86
25	76	98	91
26	54	91	13
27	96	60	95
28	46	21	51
29	5	90	59
30	51	61	22
31	35	40	3
32	10	29	21
33	21	4	71
34	8	61	65
35	58	98	49
36	19	51	17
37	93	83	4
38	50	84	76
39	98	3	95
40	6	76	53
41	10	47	52
42	39	52	94
43	91	34	17
44	61	18	26
45	43	62	44
46	23	77	11
47	72	84	15
48	2	42	36
49	83	70	68

Fig. 3 Node deployment scenario

PLAYER A

Node Id	Energy	Mobility	Control Packet
0	49	73	11
5	13	50	44
6	11	65	62
7	7	12	29
9	16	75	83
11	32	28	41
19	47	9	33
20	3	50	43
22	17	5	83
23	2	44	4
24	32	70	86
28	46	21	51
29	5	90	59
31	35	40	3
32	10	29	21
33	21	4	71
34	8	61	65
36	19	51	17
40	6	76	53
41	10	47	52
42	39	52	94
45	43	62	44
46	23	77	11
48	2	42	36

Fig. 4 Nodes of Player A

5 Conclusion and Future Work

The proposed work is based on "work in progress" category. The main key element of this paper is game theory in the non-cooperative environment. The topology and nature of the WANET are dynamic, so in this paper, non-cooperative game theory technique is applied. The paper contains an intelligent strategy management technique for WANET. This strategy will manage based on non-cooperative situation. In non-cooperative situation, parameters of the network vary frequently, so it will incorporate with the help of linear programming technique for managing the objective functions and constraints for both players, Player A and Player B. And finally, we will find the value of the game that will decide the profit and loss of the game.

PLAYER B

Node Id	Energy	Mobility	Control Packet
1	60	19	26
2	69	49	48
3	67	91	23
4	70	68	51
8	79	94	48
10	73	57	61
12	67	40	77
13	61	43	31
14	54	53	14
15	54	96	35
16	61	92	2
17	66	74	33
18	52	28	55
21	52	23	10
25	76	98	91
26	54	91	13
27	96	60	95
30	51	61	22
35	58	98	49
37	93	83	4
38	50	84	76
39	98	3	95
43	91	34	17
44	61	18	26
47	72	84	15
49	83	70	68

Fig. 5 Nodes of Player B

References

1. Lin, Z., Wang, P.: A review of data sets of short-range wireless networks. Comput. Commun. **147**, 138–158 (2019)
2. Das, S.K., Samanta, S., Dey, N., Kumar, R.: Design Frameworks for Wireless Networks, pp. 1–439. Springer, Lecture Notes in Networks and Systems. ISBN: 978-981-13-9573-4
3. Karati, A., Islam, S.H., Biswas, G.P.: A pairing-free and provably secure certificateless signature scheme. Inf. Sci. **450**, 378–391 (2018)
4. Jain, P.K., Pamula, R.: Two-step anomaly detection approach using clustering algorithm. In: International Conference on Advanced Computing Networking and Informatics, pp. 513–520. Springer, Singapore (2019)
5. Mishra, G., Agarwal, S., Jain, P.K., Pamula, R.: Outlier detection using subset formation of clustering based method. In: International Conference on Advanced Computing Networking and Informatics, pp. 521–528. Springer, Singapore (2019)
6. Kumari, P., Jain, P.K., Pamula, R.: An efficient use of ensemble methods to predict students academic performance. In: 2018 4th International Conference on Recent Advances in Information Technology (RAIT), pp. 1–6. IEEE (March 2018)
7. Punam, K., Pamula, R., Jain, P.K.: A two-level statistical model for big mart sales prediction. In: 2018 International Conference on Computing, Power and Communication Technologies (GUCON), pp. 617–620. IEEE (Sept 2018)

8. Das, S.P., Padhy, S.: A novel hybrid model using teaching–learning-based optimization and a support vector machine for commodity futures index forecasting. Int. J. Mach. Learn. Cybern. **9**(1), 97–111 (2018)
9. Das, S.P., Padhy, S.: Unsupervised extreme learning machine and support vector regression hybrid model for predicting energy commodity futures index. Memetic Comput. **9**(4), 333–346 (2017)
10. Das, S.P., Padhy, S.: A new hybrid parametric and machine learning model with homogeneity hint for European-style index option pricing. Neural (2017)
11. Saini, T.K., Sharma, S.C.: Prominent unicast routing protocols for mobile ad hoc networks: criterion, classification, and key attributes. Ad Hoc Netw. **89**, 58–77 (2019)
12. Fu, X., Fortino, G., Pace, P., Aloi, G., Li, W.: Environment-fusion multipath routing protocol for wireless sensor networks. Inf. Fusion **53**, 4–19 (2020)
13. Bai, S., Liu, Y., Li, Z., Bai, X.: Detecting wormhole attacks in 3D wireless ad hoc networks via 3D forbidden substructures. Comput. Netw. **150**, 190–200 (2019)
14. AL-Dhief, F.T., Sabri, N., Fouad, S., Latiff, N.A., Albader, M.A.A.: A review of forest fire surveillance technologies: mobile ad-hoc network routing protocols perspective. J. King Saud Univ. Comput. Inf. Sci. **31**(2), 135–146 (2017)
15. Li, J., Silva, B.N., Diyan, M., Cao, Z., Han, K.: A clustering based routing algorithm in IoT aware wireless mesh networks. Sustain. Cities Soc. **40**, 657–666 (2018)
16. Lin, H., Hu, J., Xu, L., Tian, Y., Liu, L., Blakeway, S.: A trustworthy and energy-aware routing protocol in software-defined wireless mesh networks. Comput. Electr. Eng. **64**, 407–419 (2017)
17. Chai, Y., Shi, W., Shi, T.: Load-aware cooperative hybrid routing protocol in hybrid wireless mesh networks. AEU-Int. J. Electr. Commun. **74**, 135–144 (2017)
18. Das, S.K., Tripathi, S.: Energy efficient routing formation algorithm for hybrid ad-hoc network: a geometric programming approach. Peer-to-Peer Netw. Appl. **12**(1), 102–128 (2019)
19. Das, S.K., Tripathi, S.: Adaptive and intelligent energy efficient routing for transparent heterogeneous ad-hoc network by fusion of game theory and linear programming. Appl. Intell. **48**(7), 1825–1845 (2018a)
20. Das, S.K., Tripathi, S.: Intelligent energy-aware efficient routing for MANET. Wirel. Netw. **24**(4), 1139–1159 (2018b)
21. Das, S.K., Tripathi, S.: Energy efficient routing formation technique for hybrid ad hoc network using fusion of artificial intelligence techniques. Int. J. Commun. Syst. **30**(16), e3340, 1–16 (2017)
22. Yadav, A.K., Das, S.K., Tripathi, S.: EFMMRP: design of efficient fuzzy based multi-constraint multicast routing protocol for wireless ad-hoc network. Comput. Netw. **118**, 15–23 (2017)
23. Das, S.K., Yadav, A.K., Tripathi, S.: IE2M: design of intellectual energy efficient multicast routing protocol for ad-hoc network. Peer-to-Peer Netw. Appl. **10**(3), 670–687 (2017)
24. Das, S.K., Tripathi, S., Burnwal, A.P.: Fuzzy based energy efficient multicast routing for ad-hoc network. In: Proceedings of the 2015 Third International Conference on Computer, Communication, Control and Information Technology (C3IT), pp. 1–5. IEEE (Feb 2015)

A TWV Classifier Ensemble Framework

Sidharth Samal and Rajashree Dash

Abstract Instead of relying on a single classifier's prediction, combining many classifiers using ensemble method to produce the final class label is a major topic of debate among researchers in diverse areas. But choosing the individual classifiers and their precedence in the ensemble is a compelling issue. Hence, in this investigation, a TOPSIS-based multi-criteria decision-making approach is advocated for determining a series of individual base classifiers on the basis of multiple assessment criteria for developing a weighted voting classifier ensemble. Again the weights are assigned to the classifiers for voting based on their TOPSIS ranking. The suggested ensemble framework is justified for predicting the stock index price movements as increasing or decreasing movement over BSE SENSEX and S&P 500 stock indices datasets. The results reveal that the TOPSIS-based weighted voting ensemble approach is producing better classification results compared to the majority voting ensemble.

Keywords Stock index movement prediction · TOPSIS · Majority voting · Weighted voting

1 Introduction

Predicting upcoming movements of stock index prices is attracting many investors as well as researchers who wish to reveal the underlying patterns in the stock market data and utilize it for obtaining profit. However, due to the nonlinearity and volatility of stock market data, it is very challenging to compose a robust, efficient, and superior prediction model that will be able to outperform pre-existing models. In literature, an extensive amount of machine learning approaches such as Support Vector Machines

S. Samal (✉) · R. Dash
Computer Science and Engineering Department, Siksha 'O' Anusandhan (Deemed to be University), Bhubaneswar, India
e-mail: sidharthsamal95@gmail.com

R. Dash
e-mail: rajashree_dash@yahoo.co.in

D. Mishra et al. (eds.), *Intelligent and Cloud Computing*,
Smart Innovation, Systems and Technologies 194,
https://doi.org/10.1007/978-981-15-5971-6_28

(SVM) [1], K-Nearest Neighbor (KNN) [2], Neural Networks [3–6], Linear Regression (LR) [7], Decision Tree (DT) [8], and so on are suggested for the prediction of future stock index movements. As these models focus on anticipating the increasing and decreasing price movements, we can consider this procedure as a binary classification problem.

Recently, in several classification problems, ensemble has gained wide acceptance due to its improved performance compared to single classifiers [9–12]. Through ensemble, the predictions of multiple classifier models are combined to generate a more accurate, robust, and reliable output. In most of the approaches, the assemblage of base classifiers is done either using Majority Voting (MV) or weighted voting. In MV, out of distinct classifiers' prediction, the most voted class label is nominated as the final class label. After the training of base classifiers, MV requires no further parameter calibration. In weighted voting, different weights are assigned to different classifiers, and the product of weights and prediction of individual classifiers form the final class label. The prediction of individual classifiers affects the weights of votes and also the overall output. However, in real world, none of the classifiers is appeared as the best one for all application domains. So determining the base classifiers for ensemble and allocating relevant weight to each classifier for voting has always presented itself as a critical affair. In the majority of the investigation, base classifiers are selected and the weights are assigned based on accuracy of classifier. But the selection of base classifiers and weight assignment based on other classification performance criteria has substantial impact on the final output of the ensemble framework. In this regard, screening classifiers for ensemble present itself as a Multi-Criteria Decision-Making (MCDM) problem [13–16], where the selection of base classifiers and assignment of weights to individual classifiers should be established on diverse performance evaluation criteria.

In this investigation, a prevalent MCDM approach designated as Technique for Order of Preference by Similarity to Ideal Solution (TOPSIS) is proposed for determining a batch of base classifiers evaluated on the basis of multiple evaluation criteria for developing a weighted voting classifier ensemble. Initially, prediction of seven popular classification schemes models such as KNN, MLP, SVM, Radial Basis Function network (RBF), NB, DT, and LR are evaluated and ranked on the basis of four classification performance measures such as accuracy, precision, recall, and F-measures using TOPSIS approach. Then, the top five of them are combined using a weighted voting ensemble scheme. Again, the weights are assigned to the classifiers for voting based on their TOPSIS ranking. The proposed weighted voting classifier ensemble based on TOPSIS is also differentiated with the conventional MV ensemble scheme. The MV approach is also applied to the classification output of the top 5 classifiers selected by TOPSIS ranking. The proposed ensemble framework is authenticated on top of the stock index price data of S&P 500 and BSE SENSEX stock indices collected between a time interval of January 1, 2015–November 13, 2017. From the result, it can be observed that TOPSIS-based Weighted Voting (TWV) classifier ensemble framework is producing better classification results compared to the MV ensemble.

The rest of this paper is arranged in the following manner. Section 2 indicates the working principle of the proposed TWV classifier ensemble approach for stock index price movement forecasting. An analysis of the empirical results collected from the proposed model is presented in Sect. 3. In the end, Sect. 4 holds the concluding observation.

2 Proposed TWV Classifier Ensemble

This section of paper provides a comprehensive description of stock index price movement prediction employing the proposed TWV classifier ensemble framework. The proposed ensemble framework is represented in Fig. 1. Initially, the collected stock index prices are preprocessed. In this step, six popular technical indicators, in particular, Moving Average Convergence and Divergence (MACD), Larry William's R%, Simple Moving Average (SMA), Stochastic K%, D%, and Relative Strength Index (RSI), are derived from historical stock prices and normalized to generate suitable input samples to be used by classifiers. Then through comparison of two adjacent closing prices, the up (class 1) and down (class 0) movements are calculated as follows:

$$\begin{aligned} &\text{If}\,\text{Cprice(d)} > \text{Cprice}\,(\text{d}-1)\,\text{then}\,\text{IM(d)} = 1\,(\text{increase}) \\ &\qquad \text{else}\,\text{IM(d)} = 0\,(\text{decrease}) \end{aligned} \tag{1}$$

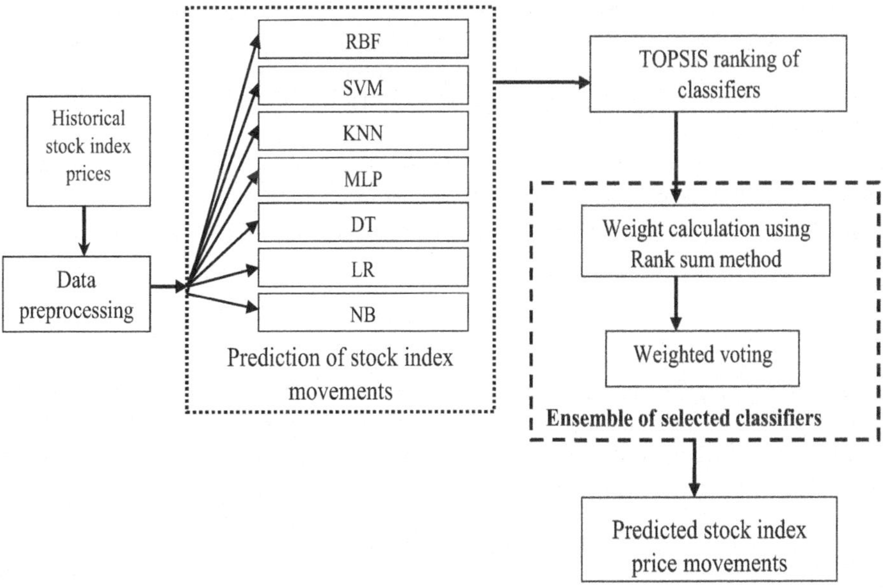

Fig. 1 Proposed TWV ensemble framework for stock index price movement prediction

where IM(d) = index movement on day d, and Cprice(d) = closing price on day d.

After accumulating all input–output vectors through data preprocessing, these are partitioned into two sets, one for training and validation and another for testing purpose. Then, all the seven classifier models are trained individually over the same dataset.

Individual classification models are assessed on the basis of four performance evaluation criteria using the following formulas:

$$\text{Accuracy} = \frac{\text{TP} + \text{TN}}{\text{TN} + \text{TP} + \text{FP} + \text{FN}} \tag{2}$$

$$\text{F} - \text{measures} = \frac{2 \times \text{Precision} \times \text{Recall}}{\text{Precision} + \text{Recall}} \tag{3}$$

$$\text{Precision} = \frac{\text{TP}}{\text{FP} + \text{TP}} \tag{4}$$

$$\text{Recall} = \frac{\text{TP}}{\text{FN} + \text{TP}} \tag{5}$$

where TP, FP, TN, and FN represent true positive, false positive, true negative, and false negative, respectively.

After listing the four criteria of seven classifiers, these are rearranged as an MCDM problem, so that the classifiers can be ranked by using TOPSIS approach and after that the top-ranked classifiers can be selected for ensemble. In TOPSIS approach, the best performance values corresponding to each criterion represent an ideal solution, and worst values of each criterion are considered as the negative ideal solution. Ranking of models is specified by comparing its performance measures with the ideal and negative ideal solution [13]. The steps of TOPSIS-based ranking of classifiers are adopted from [15, 16].

From the TOPSIS ranking, the top 5 classifiers are nominated as base classifiers for the ensemble framework. Then, the weights of each base classifier are allocated on the basis of their respective TOPSIS ranking using the following formula:

$$\text{WC}_i = \frac{n - \text{rank}(C_i) + 1}{\text{sum}} \tag{6}$$

where WC_i = weightage of classifier C_i, n = base classifiers count in the ensemble, rank (C_i) = TOPSIS rank of ith classifier, and sum = aggregation of all ranks.

The TWV classifier ensemble framework predicts the final class labels as follows:

$$\text{WV}(x) = \arg\max_j \sum_{i=1}^{n} \text{WC}_i \times O_{ji} \tag{7}$$

where $WV(x)$ = Final class labels after weighted voting, WC_i = weight of ith classifier in the ensemble, n = classifiers count, j = number of categories, and O_{ji} = class label.

3 Experimental Setup and Result Analysis

This research aims to determine and designate ranks to a group of base classifiers with respect to diverse performance measures along with assigning weights to them based on those ranking to further combine the predicted class labels using those weights by a weighted voting approach. Simulation has been done over stock indices of BSE SENSEX and S&P 500 collected from 1/1/2015 to 13/11/2017. Going through the data preprocessing step, the original historical stock index prices provide a total of 685 samples, which are then divided into two sets. First set contains 485 vectors, each one containing 6 normalized technical indicator values and 1 class value, which is used for training of individual classifiers. The second set including 200 vectors is kept for testing purpose. The study includes seven classifiers such as KNN, MLP, RBF, SVM, DT, NB, and LR. Further, the SVM model is tested with five diverse kernel functions and KNN with three distinct K values for both the datasets. The outputs corresponding to four evaluation measures such as Accuracy (E1), Precision (E2), Recall (E3), and F-measures (E4) for the seven classifiers are calculated using Eqs. (2–5) and are represented in Table 1.

Corresponding to individual criterion, the classifier producing the best result is underlined in the table. From analysis, it is found that ranking of classifiers will vary corresponding to individual criterion, because not a single classifier is producing the best value for all criteria. So it represents a realistic environment of decision-making form multiple criteria. Hence, in the next step, a TOPSIS approach is applied for evaluating the classifiers on multiple evaluation measures. With this approach, two decision matrices of size 7×4 are derived from Table 1 for both datasets. The WSDM, ideal and negative ideal solution, separation measures, relative closeness, and ranking

Table 1 Outcome of evaluation measures for different classifiers

Classifiers	BSE				S&P			
	E1	E2	E3	E4	E1	E2	E3	E4
DT	0.6500	0.5750	0.5610	0.5679	**0.8100**	0.7763	0.7375	0.7564
KNN	0.7350	0.6835	0.6585	0.6708	0.8250	0.7711	**0.8000**	**0.7853**
SVM	**0.8200**	0.7347	**0.8780**	**0.8000**	**0.8100**	0.7917	0.7125	0.7500
RBF	**0.8200**	**0.8194**	0.7195	0.7662	**0.8100**	0.7692	0.7500	0.7595
NB	0.6350	0.6154	0.2927	0.3967	0.6450	0.6452	0.2500	0.3604
LR	0.7550	0.7463	0.6098	0.6711	0.8050	**0.8475**	0.6250	0.7194
MLP	0.7330	0.6804	0.7293	0.6892	0.7490	0.8145	0.5800	0.6232

of classifiers obtained by applying the steps of TOPSIS on decision matrices are also represented in Tables 2 and 3 for BSE and S&P datasets, respectively. Rank 1 represents the alternative having a highest preference and rank 7 represents the alternative having lowest preference.

After ranking the classifiers, instead of using all classifiers in ensemble, the top-ranked five classifiers obtained from TOPSIS such as SVM, RBF, MLP, KNN, LR for BSE SENSEX dataset and KNN, RBF, DT, SVM, LR for S&P 500 dataset are used in ensemble eliminating the other weak learners. The weights assigned to the selected classifiers using Eq. (6) are represented in Table 4. Then, the predicted class labels are generated by weighted voting using Eq. (7) for both the datasets. The proposed approach is also compared with the MV-based ensemble approach. The class labels post-MV ensemble is computed using the following formula:

Table 2 TOPSIS-based ranking of classifiers for BSE SENSEX data

Classifiers	WSDM				S+	S−	C+	Rank
	E1	E2	E3	E4				
DT	0.1331	0.0311	0.0645	0.0970	0.0655	0.0426	0.3943	6
KNN	0.1505	0.0370	0.0757	0.1146	0.0385	0.0665	0.6333	4
SVM	0.1679	0.0398	0.1009	0.1367	0.0046	0.1039	0.9577	1
RBF	0.1679	0.0444	0.0827	0.1309	0.0191	0.0895	0.8239	2
NB	0.1300	0.0333	0.0336	0.0678	0.1041	0.0022	0.0206	7
LR	0.1546	0.0404	0.0701	0.1147	0.0404	0.0649	0.6167	5
MLP	0.1501	0.0369	0.0838	0.1178	0.0320	0.0738	0.6976	3
A+	0.1679	0.0444	0.1009	0.1367				
A−	0.1300	0.0311	0.0336	0.0678				

Table 3 TOPSIS-based ranking of classifiers for S&P 500 data

Classifiers	WSDM				S+	S−	C+	Rank
	E1	E2	E3	E4				
DT	0.1567	0.0378	0.0845	0.1237	0.0097	0.0915	0.9042	3
KNN	0.1596	0.0376	0.0917	0.1284	0.0037	0.1003	0.9642	1
SVM	0.1567	0.0386	0.0817	0.1227	0.0122	0.0891	0.8793	4
RBF	0.1567	0.0375	0.0860	0.1242	0.0086	0.0928	0.9153	2
NB	0.1248	0.0314	0.0287	0.0589	0.1006	0.0000	0.0000	7
LR	0.1558	0.0413	0.0717	0.1177	0.0231	0.0797	0.7753	5
MLP	0.1449	0.0397	0.0665	0.1019	0.0395	0.0613	0.6082	6
A+	0.1596	0.0413	0.0917	0.1284				
A−	0.1248	0.0314	0.0287	0.0589				

Table 4 Weight assignment to classifiers based on TOPSIS ranking for weighted voting

BSE		S&P	
Classifier	Weight	Classifier	Weight
SVM	0.3333	**KNN**	0.333
RBF	0.2667	**RBF**	0.2667
MLP	0.2000	**DT**	0.2000
KNN	0.1333	**SVM**	0.1333
LR	0.0667	**LR**	0.0667

Table 5 Comparison of MV and TWV ensemble

Dataset	Model	Accuracy	F-measure
BSE	MV	0.7850	0.7226
	TWV	**0.8250**	**0.8017**
S&P	MV	0.7500	0.6711
	TWV	**0.8150**	**0.7939**

$$\mathrm{MV}(x) = \arg\max_{j} \sum_{i=1}^{n} O_{ji} \tag{8}$$

where $\mathrm{MV}(x)$ = final class label of MV, n = number of base classifiers, j = number categories (0 and 1), and O_{ji} = class label of base classifiers (0 or 1).

Finally, Table 5 shows the accuracy and F-measure obtained by the proposed TWV classifier ensemble and MV classifier ensemble for both the datasets. Analyzing the result, we can observe that TWV ensemble model exceeds MV ensemble model in terms of prediction performance for both datasets. It also points out that the outcome of classifiers is reliant on data. It changes with changing dataset. So, it is inadequate to pick a sole classifier for diversified dataset.

4 Conclusion

This study focuses on the application of the TOPSIS-based MCDM framework for the assessment of seven commonly used classifiers such as SVM, RBF, KNN, LR, DT, NB, and MLP to forecast the upcoming movements of stock index prices. The classifiers are ranked by looking at their outcome on four performance measures such as accuracy, precision, recall, and F-measure. Then, a set of base classifiers are selected based on their TOPSIS ranking for developing a weighted voting classifier ensemble. Instead of assigning weights randomly or only with respect to a single performance criterion such as accuracy or F-measure, the weights are assigned to the classifiers for voting based on their TOPSIS ranking. A comparative study of TWV ensemble with respect to popular MV-based classifier ensemble is also provided with detailed results over two stock index datasets such as BSE SENSEX and S&P 500. The outcomes disclose that TWV ensemble framework performs superior contrary to MV.

References

1. Huang, W., Nakamori, Y., Wang, S.Y.: Forecasting stock market movement direction with support vector machine. Comput. Oper. Res. **32**(10), 2513–2522 (2005)
2. Imandoust, S.B., Bolandraftar, M.: Application of k-nearest neighbor (knn) approach for predicting economic events: Theoretical background. Int. J. Eng. Res. Appl. **3**(5), 605–610 (2013)
3. Guresen, E., Kayakutlu, G., Daim, T.U.: Using artificial neural network models in stock market index prediction. Expert Syst. Appl. **38**(8), 10389–10397 (2011)
4. Mostafa, M.M.: Forecasting stock exchange movements using neural networks: empirical evidence from Kuwait. Expert Syst. Appl. **37**(9), 6302–6309 (2010)
5. Dash, R., Dash, P. K.: A comparative study of radial basis function network with different basis functions for stock trend prediction. In: IEEE Power, Communication and Information Technology Conference (PCITC), pp. 430–435 (2015)
6. Dash, R., Dash, P. K.: Stock price index movement classification using a CEFLANN with extreme learning machine. In: IEEE Power, Communication and Information Technology Conference (PCITC), pp. 22–28 (2015)
7. Abdulsalam, S.O., Adewole, K.S., Jimoh, R.G.: Stock trend prediction using regression analysis–a data mining approach. ARPN J. Syst. Softw. **1**(4), 154–157 (2011)
8. Tiwari, S., Pandit, R., Richhariya, V.: Predicting future trends in stock market by decision tree rough-set based hybrid system with HHMM. Int. J. Electr. Comput. Sci. Eng. **1**(3) (2010)
9. Yang, L.: Classifiers selection for ensemble learning based on accuracy and diversity. Procedia Eng. **15**, 4266–4270 (2011)
10. Kim, H., Kim, H., Moon, H., Ahn, H.: A weight-adjusted voting algorithm for ensembles of classifiers. J. Korean Stat. Soc. **40**(4), 437–449 (2011)
11. Zhang, Y., Zhang, H., Cai, J., & Yang, B.: A weighted voting classifier based on differential evolution. In: Abstract and Applied Analysis, vol. 2014. Hindawi (2014)
12. Kuncheva, L.I., Rodríguez, J.J.: A weighted voting framework for classifiers ensembles. Knowl. Inf. Syst. **38**(2), 259–275 (2014)
13. Kou, G., Lu, Y., Peng, Y., Shi, Y.: Evaluation of classification algorithms using MCDM and rank correlation. Int. J. Inf. Technol. Decis. Mak. **11**(01), 197–225 (2012)
14. Mehdiyev, N., Enke, D., Fettke, P., Loos, P.: Evaluating forecasting methods by considering different accuracy measures. Procedia Comput. Sci. **95**, 264–271 (2016)
15. Dash, R., Samal, S., Rautray, R., Dash, R.: A TOPSIS approach of ranking classifiers for stock index price movement prediction, pp. 665–674. Soft Computing in Data Analytics, Springer, Singapore (2019)
16. Dash, R., Samal, S., Dash, R., Rautray, R.: An integrated TOPSIS crow search based classifier ensemble: in application to stock index price movement prediction. Appl. Soft Comput. 105784 (2019)

Recent Advancements in Continuous Authentication Techniques for Mobile-Touchscreen-Based Devices

Rupanka Bhuyan, S. Pradeep Kumar Kenny, Samarjeet Borah, Debahuti Mishra, and Kaberi Das

Abstract Mobile devices have proliferated into every area of activity in our lives today. Other than its use as a tool for communication and entertainment, it has also exhibited its importance as a device for conducting financial transactions, maintaining social profiles, storing and sharing confidential data. Owing to this criticality, there is an imminent need to provide security to this device at all levels. One type of security is Onetime User Authentication (OUA) for logging in to the device; the other being Continuous Authentication (CA) which is used for continuously and unobtrusively authenticating a user while the device is being used. If any unauthorized user bypasses the first type of security barrier, the second level of security, i.e., CA can act as the last line of defense for protecting the device. Although a good number of works have been carried out in this domain during the past decade, in this paper, we primarily elaborate on the direction of progress and developments that have taken place in the recent past.

Keywords Mobile devices · Touchstrokes · User authentication · Continuous authentication · Features

---

R. Bhuyan · S. P. K. Kenny
Department of Computer Science, St. Joseph University, Dimapur, Nagaland, India
e-mail: rupanka@yahoo.com

S. P. K. Kenny
e-mail: kenny_isles@yahoo.co.in

S. Borah
Department of Computer Applications, Sikkim Manipal Institute of Technology, Majhitar, Rangpo, Sikkim, India
e-mail: samarjeetborah@gmail.com

D. Mishra (✉) · K. Das
Siksha 'O' Anusandhan (Deemed to be) University, Bhubaneswar, Odisha, India
e-mail: debahutimishra@soa.ac.in

K. Das
e-mail: kaberidas@soa.ac.in

D. Mishra et al. (eds.), *Intelligent and Cloud Computing*,
Smart Innovation, Systems and Technologies 194,
https://doi.org/10.1007/978-981-15-5971-6_29

1 Introduction to Continuous Authentication (CA)

1.1 The Fundamental Idea

Mobile touchscreen devices have become dominant in the digital space [1]. The recent times have seen rapid growth in the ubiquity and pervasiveness of these devices in all areas of our lives. Business-related communication, social transactions, entertainment, and information access and sharing in a ubiquitous manner are just some of the uses of these smart portable devices [2]. Given the importance attached to these little devices, it is beyond doubt that the security of these devices is a prime concern. In line with this, we define mobile user authentication as a process of verifying a user's legitimate right to access a mobile device [3]. It can be a nontrivial and herculean task to authenticate users for the use of a mobile device. Because of the portability and pervasiveness of cell phones, the common approach of password-based verification may not be relevant any more. For instance, a user's password may be easily observed by others in a public place through various means. Additionally, if the device is lost or stolen after the genuine user has completed the (one time) login procedure, then all its data would be easily accessible to the person who found or stole it.

Owing to all these factors, there has been a growing research interest in Continuous Authentication (CA) of users to protect private or important data on mobile devices against unauthorized access.

1.2 Touchstrokes

Touchstroke analysis is a technique for capturing and assessing the touch behavior of a user on mobile devices fitted with touchscreens, such as smartphones, digital tablets, etc. When a user interacts with these devices a specific type of digital signature can be generated based on certain features of the user's touch behavior. These features are generally very subtle and mostly undetectable by human users.

1.3 Continuous Authentication (CA) Using Touchstrokes

As touchscreen-based mobile devices are quite common nowadays, a lot of user touch stroke behavior information can be collected for security and authentication purpose. Besides, this can be done unobtrusively, i.e., without requiring any user intervention surreptitiously. But, most importantly, there is the need to continuously monitor whether or not the device is being used by the genuine user. This is where Continuous Authentication (CA) or Post-Login Authentication, as mentioned in some literature, is relevant. CA can provide continuous security by constant and discreet monitoring of touch strokes to distinguish between usage by the genuine and illegitimate users.

2 Basic Concepts in Touchstroke-Based CA

2.1 *Advantages*

There are some inherent advantages specific to touchstrokes which needs a mention. Since every user's touchstroke behavior is unique, touchstrokes have the property of distinctiveness. They provide a second layer of security over other common security methods. There can be continuous monitoring. Unlike biometric profiles, touchstroke profiles can be revoked (in the unlikely situation where it is compromised) and recreated. Touchstrokes are not dependent on environmental factors, such as illumination, noise, etc. Touchstroke data is guaranteed to be available since touchscreen devices cannot be operated without touchstrokes. Transparency is another important factor. Lastly, there is no requirement for any additional hardware to capture touchstrokes.

2.2 *Features*

Touchstrokes are captured in the form of features. Although the number and type of features may vary widely depending on the nature of the Touchstroke CA under question, some commonly used features may be (1) keystroke features which include time (downtime, uptime), size (area covered), coordinates (x, y), pressure, etc. (2) motion features which may comprise accelerometer data, gyroscope data, and game rotation data.

In some works, the feature values are used in raw form while in others they are preprocessed before being fed to the algorithm. Yet, in some other cases, a new set of features are derived from the originally available features resulting in over tens of features.

3 Recent Works

In the subsections below under Sect. 3, we discuss the most recent works carried out in the field of Continuous Authentication techniques pertinent to mobile devices. We have segregated the different types of methods according to their mode of operation and scope. Also, a classification diagram is provided in Fig. 1.

3.1 *Methods in General*

Among the most general methods of Continuous Authentication for mobile devices, Alariki [4] carried out a study of touch behavior on the Android OS using—touch

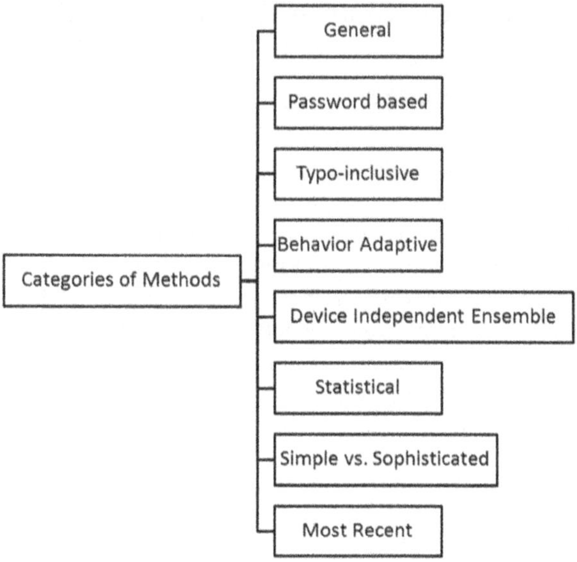

Fig. 1 Figure caption and image above the caption

direction, pressure, size, and acceleration as features where a Random Forest (RF) algorithm was used resulting in 98.14% classification accuracy. In a similar direction, Al-Obaidi [5] applied a Statistical Distance-to-Median Anomaly Detector system on a combined feature set of 'keystroke timing' and 'touch screen' features for authentication. Corpus [6] uses a Neural Network (NN) on features using accelerometer biometrics keystroke dynamics for user authentication. In a method involving two steps, Prabhakar [7] used PSO (Particle Swarm Optimization) extract features from a 16-parameter dataset and finally applied Support Vector Machine (SVM) for training and classification. To simulate a real-world scenario, Temper [8] created a banking app for the Android platform to capture touch stroke features (including di-graphs, speed, pressure, and size) to perform authentication by a Vaguely Quantified Nearest Neighbour (VQNN), which is a type of Fuzzy Roughset Nearest Neighbour (FRNN). In an interesting case, Meng [9] observed the performance instability of Machine Learning (ML) techniques and proposed a technique which selects the most well-performing ML technique from among a set of techniques trained on aggregated features. Of course, in another work of Meng [10], a Particle Swarm Optimization—Radial Basis Function Network (PSO-RBFN) is observed to be used for extracting 21 features from web browsing data.

A Euclidean Distance based classification involving 103 features captured from touch strokes generated by sitting, walking and in-hand activities were carried out by Roh [11]. In another study, Putri [12] implemented a technique including a Bayes Net along with an RF (Random Forest) to typing, swiping, pinching, and tapping touch stroke gestures through a transparently running app.

3.2 Password-Based Methods

Some of the works involved passwords playing a major role. Antal [13] who correlated difficult passwords with higher accuracy in his RF-based classifier. Kambourakis [14] trained an RF and a KNN algorithm on a combination of 10-character password and paraphrases, respectively. Salem [15] combined a complex 11-character password along with non-timing features which include (i) pressure, (ii) size, and (iii) position, to classify authentic users with the help of a Multilayer Perceptron (MLP) Backpropagation algorithm. Takahashi [16] captured 10-character password in touch devices in moving scenarios (which involved walking users and users in moving cars) to classify authentic users using a number of ML techniques. In another attempt involving passwords, Zhang [17] proposed a method using two separate Radial Basis Function based Neural Networks to model time-dependent characteristics and pressure-dependent characteristics independently from a user inputted 10-character password in touch screen devices; genuine users were identified based on the combined results of both the RBF NNs.

An enhancement to the body of works involving passwords was contributed by Zhou [18], who introduced the HATS (Harmonized user Authentication based on ThumbStroke dynamics) scheme. The method combines together password, gesture, keystroke, and touch dynamics based authentication methods and captures data with a specially designed virtual keyboard, namely, the ThumbStroke, which is operated using the thumb. This keyboard had all the alphabets separated by angles.

3.3 Typo-Inclusive Methods

In normal mode of operations, a user may make typing mistakes (usually known as typos). Most of the works in the domain of CA do not consider this. In a singular work, Alshanketi [19] contributed toward a Random Forest based authentication approach which took into account typing errors which crept into users touch strokes.

3.4 Behavior-Adaptive Methods

It is a fact that touch stroke behavior of a user may change in due course of time due to various factors. There are contributors who have addressed this issue. Palaskar [20] introduced the concept of training and retraining of changing user touch stroke behavior samples over time. This was done with the help of a Vote-Based Stroke Classification technique using a Random Forest. In a similar direction, Lu [21] developed a transparent app named Safeguard, which collects slide movements and pressure parameters; it incorporated a Support Vector Machine (SVM) and could adapt to changing behavioral touch stroke patterns of users.

3.5 Device-Independent Methods

Wang [22] captured hand-crafted features which reflect subtle user behavior patterns; these features are app independent and are preserved across devices.

3.6 Ensemble Methods

Ensemble methods generally comprise of an assortment of multiple methods used in succession or simultaneously to obtain the solution to the problem at hand. Kumar [23] incorporated ensemble techniques involving multiple methods including elliptic envelope, isolation forest, local outliers factor, and 1C-SVM (and their fusion) on genuine user samples only.

3.7 Statistical Methods

Notwithstanding the popularity of Machine Learning methods applied in this domain, a few works have been carried out using Statistical techniques. For instance, Gautam [24] applied Median Vector Proximity on keystroke features and obtained better results over 14 standard ML methods. In another direction, Roy [25] trained a Continuous left–right Hidden Markov Model exclusively on authentic users' data.

3.8 Simplicity Versus Sophistication

With due respect to sophisticated methods for Continuous Authentication which can extract subtle features and insights from any given touch stroke dataset, the usefulness of certain basic and simpler techniques cannot be undermined. We compare some of these ideas below.

Leingang [26] demonstrated the simplicity and better performance of the classical J48 Decision Tree method over other sophisticated techniques such as LSTM-RNN, Random Tree, Bayesian Network; J48 was applied on a raw version of the HMOG dataset (used without any augmentation or feature extraction applied).

Ku [27] observed that accuracy and the touch stroke pattern are correlated and concluded that pattern length also might be an important factor to be explored.

Keeping in view the huge volume of computation involved in the application of various ML techniques and with the limited processing and memory power of CPUs in mobile devices, Gunn [28] created a simulation wherein the touch stroke data from a mobile device is sent to a cloud-based server; various ML functional libraries being available on this server, viz. Random Forests, Support Vector Machine, Long

Short-Term Memory–Recurrent Neural Networks, etc. Performance of SVM was noticed to surpass those of the others.

Choi [29] experimented with a One-Class Probabilistic Randomized One Hidden Layer Network (OC-RMPNet) which does not require iterative learning; it was applied on both the HMOG and Touchalytics datasets and was shown to exhibit acceptable performance.

Toward more sophisticated works in recent times, Chang [30] applied Deep Learning Networks such as Kernel Deep Regression Network (KDRN) which learns from pre-extracted features applied on the Touchalytics dataset. Features which were otherwise hidden during direct observation could be extracted through KDRNs.

In the proposal of Lee [31], a One-Class Support Vector Machine (OCSVM) was applied on 6-digit touch keystroke features together with data from built-in motion sensors. Two features, namely, –grot (game rotation) and sizeDn (size when a key is pressed) were found to be the most influencing features.

3.9 Most Recent Methods

Among the most recent works, Syed [36] extracts 14 features through an app and extensively preprocesses the same. These preprocessed data are treated with Support Vector Machine, Logistic Regression, Naive Bayes, Random Forest, and Multilayer Perceptron independently. Random Forest performed best than all the other techniques. In another case, Acien [35] applied a Support Vector Machine (with Radial Basis Function) on the UMDAA-02 dataset. The dataset contains 7 modes (including behavioral and biometric) of data collected in a semi-controlled environment.

3.10 Building Datasets

A few databases were created as by products of the various developments in this field of research as illustrated in Table 1.

For instance, the MOBIKEY password database was contributed by Antal [13]. A popular dataset contributed by Frank [32] is the Touchalytics database. The massive HMOG (Hand, Movement, Orientation and Grasp) dataset containing 96 features classified into (i) grasp resistance and (ii) grasp stability, respectively, was built by Sitova [33]. El-Abed [34] collected touch stroke data from mobile devices in portrait and landscape orientations to build the publicly available RHU benchmark dataset. The most recent contribution in the area of datasets was by Acien [35], who have contributed the UMDAA-02 dataset which was built under natural conditions.

Table 1 Prominent datasets for touchstroke dynamics

Dataset	Contributor	Features
MOBIKEY	Antal [13]	No. of samples: 9844 passwords
		3 types of passwords (easy, logical strong, strong)
		No. of volunteers: 54
		Data collection device: 13 nos. of Nexus 7 tablets
		No. of features: 7 {Key, Downtime, Uptime, Pressure, FingerArea, x, y-coordinates, and acceleration measured along x, y, z axes}
		No. of secondary features: 72–82 depending on the type of password
Touchalytics	Frank [32]	No. of volunteers: 41
		Data collection device: 4 smartphones with similar specifications
		No. of features: 7 {event code (e.g., FingerUp, FingerDown, FingerMove, MultiTouch), the absolute event time (in ms), the device orientation, x,y-coordinates, pressure, area of the screen, FingerOrientation, and ScreenOrientation.}
HMOG	Sitova [33]	No. of volunteers: 100
		Data collection device: 10 nos. of Samsung Galaxy S4
		No. of features: 96
RHU	El-Abed [34]	No. of features: 5
		Data collection device: Nexus 5 touchscreen mobile phone and Samsung Galaxy Note 10.1 tablet
UMDAA-02	Acien [35]	No. of volunteers: 48
		No. of features: 41
		Data collection device: a number of smartphones distributed by the researchers to the users (detailed hardware specifications not known)

4 Conclusions

In this paper, we have introduced the various fundamental concepts and aspects pertaining to Continuous Authentication (CA) techniques used in the security domain of mobile touchscreen devices.

As discussed at length, CA techniques have become quite necessary, keeping in view the fact that nowadays mobile devices have become an indispensable part of our lives. In fact, other than its use as a communication device or personal digital assistant, we also use it to store important personal data; not to mention the occasions where financial transactions are carried out.

A contemporary and novel classification of the methods is done which is based on the most significant works carried out in this domain. We classify nine different categories of methods which include general methods, password-based methods, typo-inclusive methods, behavior-adaptive methods, device-independent methods, ensemble methods, non-ML methods, simple versus sophisticated methods, and the most recent methods.

A number of touchstroke datasets resulted as spin-offs from the discussed research works; a few of which eventually became benchmarks in this singular domain of research. We familiarize the reader with these datasets along with their nature and differences. These datasets include MOBIKEY, HMOG, Touchalytics, RHU, and UMDAA-02.

The trends that have emerged in recent times have been elaborated in this paper.

References

1. Lella, A., Lipsman, A.: The US Mobile App Report, comSCORE, 21 Aug 2014
2. Lai, J., Zhang, D.: ExtendedThumb: a target acquisition approach for one-handed interaction with touch-screen mobile phones. IEEE Trans. Human-Mach. Syst. **45**, 362–370 (2015)
3. Abdulhakim, A.A., Abdul, M.A.: Touch gesture authentication framework for touchscreen mobile devices. J. Theor. Appl. Inf. Technol. **62**, 493–498 (2014)
4. Alariki, A.A., Manaf, A.B.A., Khan, S.: A study of touching behavior for authentication in touch screen smart devices. In: 2016 International Conference on Intelligent Systems Engineering (ICISE) (2016)
5. Al-Obaidi, N.M., Al-Jarrah, M.M.: Statistical keystroke dynamics system on mobile devices for experimental data collection and user authentication. In: 2016 9th International Conference on Developments in eSystems Engineering (DeSE), pp. 123–129 (2016)
6. Corpus, K.R., Gonzales, R.J.D., Morada, A.S., Vea, L.A.: Mobile user identification through authentication using keystroke dynamics and accelerometer biometrics. In: Proceedings of the International Workshop on Mobile Software Engineering and Systems—MOBILESoft 16, pp. 11–12 (2016)
7. Prabhakar, M., Priya, B., Narayanan, A.: Framework for authenticating smartphone users based on touch dynamics. J. Eng. Appl. Sci. **13**, 4604–4608 (2018). https://doi.org/10.3923/jeasci.2018.4604.4608
8. Temper, M., Tjoa, S.: The applicability of fuzzy rough classifier for continuous person authentication. In: International Conference on Software Security and Assurance, pp. 17–23 (2016)
9. Meng, W., Li, W., Wong, D.: Enhancing touch behavioral authentication via cost-based intelligent mechanism on smartphones. Multimed. Tools Appl. **77**(23), 30167–30185 (2018). https://doi.org/10.1007/s11042-018-6094-2
10. Meng, W., Wang, Y., Wong, D., Wen, S., Xiang, Y.: TouchWB: touch behavioral user authentication based on web browsing on smartphones. J. Netw. Comput. Appl. **117**, 1–9 (2018). https://doi.org/10.1016/j.jnca.2018.05.010
11. Roh, J.-H., Lee, S.-H., Kim, S.: Keystroke dynamics for authentication in smartphone. In: International Conference on Information and Communication Technology Convergence (ICTC), pp. 1155–1159 (2016)
12. Putri, A.N., Asnar, Y.D.W., Akbar, S.: A continuous fusion authentication for Android based on keystroke dynamics and touch gesture. In: International Conference on Data and Software Engineering (ICoDSE) (2016)
13. Antal, M., Nemes, L.: The MOBIKEY keystroke dynamics password database: benchmark results. In: Advances in Intelligent Systems and Computing Software Engineering Perspectives and Application in Intelligent Systems, pp. 35–46 (2016)
14. Kambourakis, G., Damopoulos, D., Papamartzivanos, D., Pavlidakis, E.: Introducing touchstroke: keystroke-based authentication system for smartphones. Secur. Commun. Netw. **9**(6), 542–554 (2014)

15. Salem, A., Zaidan, D., Swidan, A., Saifan, R.: Analysis of strong password using keystroke dynamics authentication in touch screen devices. In: Cybersecurity and Cyberforensics Conference, pp. 15–21 (2016)
16. Takahashi, H., Ogura, K., Bista, B., Takata, T.: A user authentication scheme using keystrokes for smartphones while moving. In: ISITA 2016, pp. 310–314. Monterey, California, USA (2016)
17. Zhang, H., Yan, C., Zhao, P., Wang, M.: Model construction and authentication algorithm of virtual keystroke dynamics for smart phone users. In: IEEE International Conference on Systems, Man, and Cybernetics, pp. 000171–000175. Budapest, Hungary (2016)
18. Zhou, L., Kang, Y., Zhang, D., Lai, J.: Harmonized authentication based on ThumbStroke dynamics on touch screen mobile phones. Decis. Support Syst. **92**, 14–24 (2016). https://doi.org/10.1016/j.dss.2016.09.007
19. Alshanketi, F., Traore, I., Ahmed, A.A.: Improving performance and usability in mobile keystroke dynamic biometric authentication. In: 2016 IEEE Security and Privacy Workshops (SPW), pp. 66–73 (2016)
20. Palaskar, N., Syed, Z., Banerjee, S., Tang, C.: Empirical techniques to detect and mitigate the effects of irrevocably evolving user profiles in touch-based authentication systems. In: IEEE 17th International Symposium on High Assurance Systems Engineering (HASE), pp. 9–16 (2016)
21. Lu, L., Liu, Y.: Safeguard: user reauthentication on smartphones via behavioral biometrics. IEEE Trans. Comput. Soc. Syst. **2**(3), 53–64 (2015)
22. Wang, X., Yu, T., Mengshoel, O., Tague, P.: Towards continuous and passive authentication across mobile devices: an empirical study. In: WiSec '17, Boston, MA, USA, pp. 35-45 (2017)
23. Kumar, R., Kundu, P.P., Phoha, V.V.: Continuous authentication using one-class classifiers and their fusion. In: 2018 IEEE 4th International Conference on Identity, Security, and Behavior Analysis (ISBA), pp. 1–8. Singapore (2018). https://doi.org/10.1109/ISBA.2018.8311467
24. Gautam, P., Dawadi, P.R.: Keystroke biometric system for touch screen text input on android devices optimization of equal error rate based on medians vector proximity. In: 2017 11th International Conference on Software, Knowledge, Information Management and Applications (SKIMA), pp. 1–7. Malabe (2017). https://doi.org/10.1109/SKIMA.2017.8294136
25. Roy, A., Halevi, T., Memon, N.: An HMM-based behavior modeling approach for continuous mobile authentication. In: Proceedings of IEEE International Conference on Acoustics, Speech and Signal Processing (ICASSP), pp. 3789–3793. IEEE (2014)
26. Leingang, W., Gunn, D., Kim, J., Yuan, X., Roy, K.: Active authentication using touch dynamics. IEEE Southeastcon. IEEE. https://www.researchgate.net/publication/328179313_Active_Authentication_Using_Touch_Dynamics. Accessed 22 May 2019
27. Ku, Y., Park, L., Shin, S., Kwon, T.: A guided approach to behavioral authentication. In: ACM SIGSAC conference on computer and communications security (CCS '18), pp. 2237–2239. Toronto, ON, Canada (2018). https://doi.org/10.1145/3243734.3278488
28. Gunn, D.J., Roy, K., Bryant, K.: Simulated cloud authentication based on touch dynamics with SVM. In: 2018 IEEE Symposium Series on Computational Intelligence (SSCI), pp. 639–644. Bangalore, India (2018). https://doi.org/10.1109/SSCI.2018.8628762
29. Choi, S., Chang, I., Teoh, A.: One-class random Maxout probabilistic network for mobile touchstroke authentication. In: 24th International Conference on Pattern Recognition (ICPR), pp. 3359–3364. Beijing, China (2019)
30. Chang, I., Low, C., Choi, S., Beng Jin Teoh, A.: Kernel deep regression network for touch-stroke dynamics authentication. IEEE Signal Process. Lett. **25**(7), 1–1 (2018). https://doi.org/10.1109/lsp.2018.2846050
31. Lee, H., Hwang, J.Y., Kim, D.I., Lee, S., Lee, S.H., Shin, J.S.: Understanding keystroke dynamics for smartphone users authentication and keystroke dynamics on smartphones built-in motion sensors. Secur. Commun. Netw. **2018**, 1–10 (2018)
32. Frank, M., Biedert, R., Ma, E., Martinovic, I., Song, D.: Touchalytics: on the applicability of touchscreen input as a behavioral biometric for continuous authentication. IEEE Trans. Inf. Forens. Secur. **8**(1), 136–148 (2013)

33. Sitova, Z. et al.: HMOG: new behavioral biometric features for continuous authentication of smartphone users. IEEE Trans. Inf. Forens. Secur. **11**(5), 877–892 (2016). https://doi.org/10.1109/tifs.2015.2506542
34. El-Abed, M., Dafer, M., Rosenberger, C.: RHU keystroke touchscreen benchmark. In: 2018 International Conference on Cyberworlds (CW) (Oct 2018)
35. Acien, A., Morales, A., Vera-Rodriguez, R., Fierrez, J.: MultiLock: mobile active authentication based on multiple biometric and behavioral patterns (2019). https://arxiv.org/abs/1901.10312. Accessed 18 May 2019
36. Syed, Z., Helmick, J., Banerjee, S., Cukic, B.: Touch gesture-based authentication on mobile devices: the effects of user posture, device size, configuration, and inter-session variability. J. Syst. Softw. **149**, 158–173, 2019. https://doi.org/10.1016/j.jss.2018.11.017

Hermit: A Novel Approach for Dynamic Load Balancing in Cloud Computing

Subasish Mohapatra, Subhadarshini Mohanty, Arunima Hota, Prashanta Kumar Patra, and Jijnasee Dash

Abstract Cloud computing is the recent trend in computing paradigms. It is widely being used by different communities due to its abundant opportunities as resources or services. The cloud hosts a variety of web applications and provides services on a pay-per-use basis. It has many computing issues related to server management, response time, content delivery, etc. As the number of users is raised alarmingly in the cloud, so to balance their need the load balancing has turned to be a major issue in cloud computing. It is one of the challenging tasks to schedule the load in this distributed environment. Hence, it is one of the prominent areas of research and development. In this paper, we have analyzed different metaheuristic algorithms and proposed a novel meta-heuristic algorithm for load balancing in the cloud computing environment. We have evaluated the performance of the algorithm with other competitive schemes such as Tabu Search, Simulated Annealing, Particle Swarm Optimization, and Genetic Algorithm, etc., in different measures like waiting time and turnaround time. The proposed approach outperforms existing schemes.

Keywords Cloud computing · Load balancing · Task scheduling · Hermit algorithm · Waiting time

---

S. Mohapatra (✉) · S. Mohanty · A. Hota · P. K. Patra · J. Dash
Department of Computer Science and Engineering, College of Engineering and Technology, Bhubaneswar, India
e-mail: subasish.mohapatra@gmail.com

S. Mohanty
e-mail: sdmohantycse@cet.edu.in

A. Hota
e-mail: arunimahota123@gmail.com

P. K. Patra
e-mail: pkpatra@cet.edu.in

J. Dash
e-mail: jijnasee.nit08@gmail.com

D. Mishra et al. (eds.), *Intelligent and Cloud Computing*,
Smart Innovation, Systems and Technologies 194,
https://doi.org/10.1007/978-981-15-5971-6_30

1 Introduction

Cloud computing is widely accepted through a vast range of users for providing a solution to the massive length of computational problems. The cloud surroundings include heterogeneous computing sources alongside processors, bandwidth, and reminiscence. By using the usage of the virtualized sources, the cloud computing surroundings can create a brand new utility state of affairs to offer distributed services within the shape of infrastructure, platform, and software on-demand basis. The advantages of cloud computing consist of location independence, availability, reliability, and optimized cost [1]. To gain the above goals, the task needs to be scheduled properly among various resources. To provide optimal facilities like larger memory, higher bandwidth, and computational power, the VM migration is done by shifting the live VMs on execution from one physical machine to another machine [2, 3]. The VM migration technique helps in remapping the digital machine and physical resources dynamically with flexible allocation and reallocation of resources [4]. Task scheduling is a subset of load balancing, which is now one of the promising issues in the cloud computing environment. It ensures that each computing service is sent effectively and reasonably. Load balancing presents a better satisfactory service by optimizing resource utilization and response time [5]. Load balancing allows scalability, maximizing the performance, and minimizing the response time by using optimally utilizing the available sources and minimizing resource consumption [6, 7]. According to VM allocation techniques, load balancing algorithms are categorized into static and dynamic. Static algorithms don't depend on the present conditions of the system and also inefficient while the customer demands are changing over time. Dynamic algorithms are online algorithms, where the VMs are allocated dynamically according to the loads at each time interval. These algorithms could dynamically configure the VM placement combining with the VM migration technique. Similarly based on scheduling strategies the scheduling process may execute in a central or distributed scheduler. The central load balancing algorithms in clouds are commonly supported by a centralized controller. The distributed algorithm gets rid of the overhead posed by the central algorithm scheduler and improves the reliability and scalability of the network. In the deterministic algorithm, the algorithms use the information about the properties of the nodes and the characteristics of the processes to be scheduled deterministically allocate process to nodes but in other hand, the probabilistic algorithm uses the information regarding the static attributes such as number of nodes, processing capability, and the network topology. It is very rare to find optimal solutions for load balancing algorithms. Task scheduling in cloud computing is an NP-complete problem. Therefore, most of the researchers of the proposed algorithms are specializing in finding an approximate solution for VM load balancing problems. For this reason, we classify the surveyed algorithms into three categories, which include heuristic, meta-heuristic, and hybrid. The goal of heuristic algorithms is finding a good solution for a particular problem [8]. Metaheuristic algorithms follow a set of uniform strategies to assemble and clear up problems. Combining the heuristic algorithm to meet the initial VM placement and the meta-heuristic algorithm to optimize the location of VMS at some stage in migration leads to the concept of the hybrid algorithm.

The organization of the paper is as follows. The next section explains the literature review. Then the subsequent sections are proposed model, simulation, result discussion. Finally, the conclusion and future works are outlined in the last section.

2 Literature Review

The objective of a scheduling algorithm in a distributed system is to spread the load on the processing nodes to boom the aid usage and minimizing the total task execution time. The basic scheduling model is depicted in Fig. 1. Various scheduling strategies along with their advantage and disadvantages are discussed below. The authors have proposed an adaptive resource allocation algorithm for the cloud system with preemptable tasks to increase the utilization of clouds [9]. This method fails to optimize cost and time. To overcome the problem, the authors have proposed an efficient scheduling algorithm and enhanced the efficiency of the cloud platform [10]. The performance of the given algorithms could also be increased by varying different cloud parameters [11]. A modified ant colony based algorithm has been proposed to overcome the complex problem of increased complexity [12]. A hybrid approach is proposed by utilizing the benefits of the genetic algorithm and the static algorithm. Completion time, cost, bandwidth, distance, and reliability are taken as QoS parameters. Static is applied after selection, crossover, and mutation to improve the local search ability [13]. The authors have proposed the GA-based scheduling set of rules, which specialize in load balancing. The fixed bit string representation is used as an encoding scheme for chromosomes. These algorithm aims to attain load stability while minimizing makespan. The preliminary populace is generated randomly. The pool of chromosomes undergoes a random single-factor crossover,

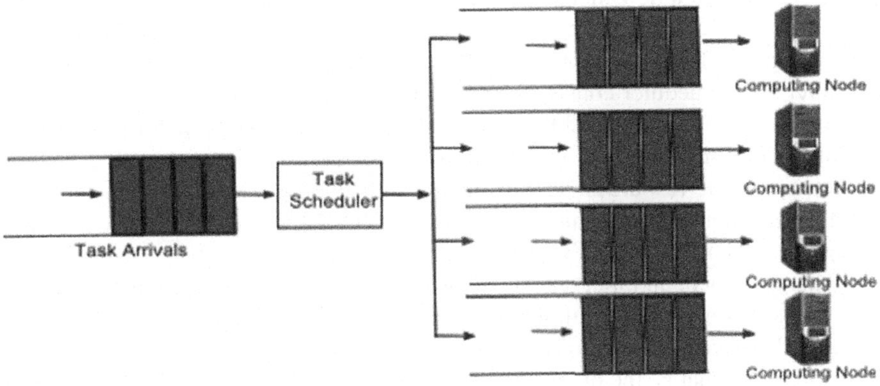

Fig. 1 Basic scheduling model

in which relying upon the crossover point, the portion lying on one side of the crossover site is exchanged with the other side [14]. The genetic algorithm based scheduling approach which uses an advanced version of max–min to generate an initial population has been proposed to optimized the makespan [15]. CloudSim is used to simulate the cloud environment. Job spanning time and load balancing Genetic Algorithm (JLGA) has been proposed in [16]. The objective is to reduce job completion time, reduce average job makespan, and achieve load balance. MATLAB is used to simulate the cloud environment. A new GA-based scheduling has been presented and it achieves better load balance and optimal response time. The processors are selected randomly. To improve the efficiency of GA, the Logarithmic Least Square Matrix based algorithm is used for calculating the priority of the processors [17]. The proposed algorithm also solves the problem of idle state and starvation. PSO-based algorithm has been proposed that considers three independent objectives [18]. They're minimizing the overall execution time of tasks, assembly the QoS necessities of tasks, and reduction of value. A particle is encoded by using a two-dimensional vector having a period of double the range of subtasks. Initial particles are generated using a greedy algorithm. Modification of particle evolving formula is done to suit the encoding scheme. Pareto domination principle is used to find a pareto-optimal solution. Different variants of continuous PSO, named integer PSO has been proposed in [19]. It outperforms the SPV while there is a huge distinction in the task lengths and processing speed of resources. It optimizes both makespan and cost. It is used for independent task scheduling. Pareto optimization is used. A novel task scheduling technique primarily based on PSO has been presented. SJFP (Shortest Job to Fastest Processor) algorithm is used to initialize the solution for PSO. Its objective is to lessen the makespan [20].

3 Proposed Model

We have proposed a new metaheuristic algorithm that balances the load in a cloud environment dynamically. The requests from different geographical channels are submitted to the data center controller. The controller forwards the load to the central scheduler. The scheduler consists of a load balancing algorithm to map the request to the virtual machines. A table is maintained to check the availability of virtual machines. They dynamically update the status of each machine after they acquire and release of the resources. The resources with the highest priority are informed to the data center controller and accordingly assign the resources to the competing requests. The working model is shown in Fig. 1.

In this approach, an input task queue was considered, where each task is accompanied (and defined) by the number of clock cycles as well as the length of its code. We should also have the knowledge of every virtual machine prior to the execution of the program, that is, the processor speed and the threshold value of the number of tasks that can be concurrently executed each virtual machine. In this heterogeneous environment, the job queue is handled by a master server and also the allocation of the incoming task to any virtual machine. Our load balancing model, henceforth known as the Hermit Model, consists of individual Hermit threads. Each Hermit is a lightweight process or thread running in each virtual machine comprised in the cloud system. The role of the Hermit threads is to regularly update the Central Master Server with the status of the virtual machines. For every task in the task queue, the hermits will calculate the probability for allocation in their virtual machines and notify the Master Server, which will determine the maximum probability. For every task, the hermits will calculate the probability for the best allocation, and the maximum probability is chosen. The probability will depend upon two things: the first parameter is a factor multiplied with the processor speed. The second parameter is the usage value of each virtual machine, which is a measure of how idle or busy a virtual machine is. Now we want that an idle machine should be preferred for allocation over a machine already busy. So the usage parameter is raised to a negative power that is the probability of successful allocation is inversely proportional to the usage of a virtual machine. The magnitude of the power of Usage is kept large enough to discourage allocation, no matter how high its processing power is. In other words, we can say that the machine which has the highest usage will be least preferred. Next, we calculate the likelihood for each incoming task. The virtual machine having maximum probability is chosen. Still, there is one check to be performed, which is that the virtual machine that has been chosen shall not have crossed its threshold value. Otherwise, if we do not perform this check we might overload a machine. The threshold value is known to us and is provided as input. If the machine has not reached the threshold value, the task is assigned to that virtual machine. All comparisons and calculations and the job of allocation is done by the Master Server.

3.1 Parameters Utilized in the Environment

In the algorithm, the following parameters have been introduced and used:

1. T_k: kth task
2. V_i: ith Virtual Machine
3. L_k: the load of T_k in terms of Kilo Lines of Code (KLOC)
4. Q: Q is the crossover factor, taken as 0.8
5. η^i: attractiveness of V_i determined by its processing speed
6. τ_i: Usage Value of V_i
7. $P_k(i)$.: Probability of assigning T_k to V_i
8. α, β: Curve fitting parameters for $P_k(i)$
9. $V^i{}_{curr}$: Current number of tasks running in V_i
10. $V^i{}_{thresh}$: Threshold value or the maximum number of tasks that can run in $V^i{}_{thresh}$
11. ρ_i: Decay constant for Usage value τ_i of V_i
12. ϕ_i: Factor to determine ρ_i for V_i

3.2 Terminologies

$$\begin{cases} Q\,L_k, & \text{if task } T_k \text{ is allocated to } V_i \\ 0, & \text{if no task is being allocated} \end{cases} \tag{1}$$

Where τ_i is the Usage Value of V_i, the ith Virtual Machine.

$$\eta^i = \text{processor speed}(V_i) \tag{2}$$

Where the processor speed is provided at the beginning of execution as input and the processor speed has been kept constant.

$$\rho_i = \Phi * (\text{Processor speed } V_i) \tag{3}$$

such that the max ρ_i lies in the range (0.5,1.5).

This is because it is desired that the decay constant be such that the usage is decreased by about 10% every clock cycle. The decay constant is taken to be proportional to the processor speed.

$$= \frac{(\tau_i)^\alpha \times (\eta_i)^\beta}{\sum_{i=1}^{N} (\tau_i)^\alpha \times (\eta_i)^\beta} \tag{4}$$

where $\alpha = -3$, $\beta = 1.2$

When we want to assign a task, we need to first calculate the probability of the task assignment for a particular virtual machine with the help of the Hermit Thread running in it. Then the maximum probability is chosen, and if the ith Virtual Machine has the maximum probability and its threshold has not been reached, then that machine is allotted with the task.

$$\tau_i = (1 - \rho_i)\tau_i + \Delta\tau_i^k \tag{5}$$

This is the usage update formula, where the virtual machine usage is updated as and when the task gets completed. The method will be called periodically and usage will be updated.

3.3 Hermit Algorithm

1. The observer thread is initialized to run on the master server
2. N 'Hermit' threads are created for the N number of Virtual Machines in the cloud system
3. For each of the N virtual machine V_i, the threads are initialized with

   a. V^i_{thresh}, η_i are set according to V_i specifications

   b. Usage τ_i is set to a constant value 0.5
   c. V^i_{curr} is set to 0

4. The task queue T is obtained.
5. For each task T_k, while task queue T is not empty $\rho_k(i)$ is calculated.
6. The virtual machine V_i for which the value of $\rho_k(i)$ is maximum is determined.
7. If $V^i_{curr} < V^i_{thres}$, then the task is assigned to V_i and V^i_{curr} is incremented by 1.
8. Else if, $V^i_{curr} = V^i_{thres}$ the virtual machine V_j with the next highest probability, $\rho_k(j)$ is selected for task allotment and V_j^{curr} is incremented.
9. Else if all Virtual Machines have reached their threshold values, then task assignment is halted for 1000 ms and control is shifted to step 5.
10. For all VMs V_i, for the Usage τ_i the Hermit thread is regularly updated every 1 ms. Every time a task gets completed in V_i, V^i_{curr} is decremented by 1.
11. If Task Queue T is not found empty, control is shifted to step 4.
12. Stop.

4 Simulation Environment

Performance evaluations are made in a simulated cloud environment called cloud analyst. It is a graphical user interface that runs on CloudSim architecture. It facilitates configuring the attributes for simulation and modeling. We have tested the Hermit algorithm individually for load balancing after that we have compared the performance with other competitive algorithms for evaluation of its efficiency. The Hermit Model of load balancing in virtual machines was coded and implemented in Java. The concept of multithreading was used in order to simulate the parallel functioning of virtual machines in a cloud environment. The Java implementation of Hermit Model consists of three classes—Central, Hermit, and Task. The class Central represents the main thread running in the Master Server in the cloud service. The Central class extends the predefined Java class Observer in order to observe other threads that are in this case, the individual Hermit threads. The Hermit class represents the individual Hermit threads running in the virtual machines. Hermit class extends the predefined java class Observable which enables the status of the thread to be observed by the Central thread. Task class defines the input to the system, that is, the parameters a task must carry. The Task class consists of attributes like length of code (in KLOC), CPU cycles required for execution, burst time, etc. After accounting the number of virtual machines in the environment as input, the java class Random was used to generate random CPU speeds of individual virtual machines, the number of tasks, the code length, and CPU cycles of each task. The tasks are independent in nature. Our simulation is executed in Windows 7 Ultimate Operating System on a 2.66 GHz. CPU with 4 GB of RAM and 500 GB of HDD of Intel Core i5 processor. Java programming language is used for implementation.

4.1 Result Discussion

Initially, we illustrated the process of the proposed algorithm and then compared the performance with other competitive schemes (Fig. 2).

The number of tasks (randomly chosen) was 88 and the tasks had the following configuration (Table 1):

After using the Hermit model while assigning tasks to the virtual machines, the orders of assignment of tasks were as follows (Table 2).

Similarly, we have carried out the simulation for varying the number of tasks and virtual machines. Comparisons are made with other competitive algorithms such as SA, TS, GA, PSO, etc., and the makespan result for different approaches in varying the task and virtual machines are shown in Table 3.

The above result can be represented graphically as follows:

Figures 3 and 4 represent the makespan comparison of different methods in varying the iterations. It is observed that as the number of input tasks increases the proposed algorithm performs better than other algorithms. As the cloud environment is dynamic and the scheduling should be fast, the algorithm is better suited in comparison to other metaheuristic algorithms.

5 Conclusion and Future Work

In this paper, we have studied various models and algorithms, which are evolutionary approaches for balancing the load in a cloud computing environment. Each method has certain merits and demerits and is suitable for a particular design and topology. It is not necessary that an algorithm will always be better than the other heuristic approach in all cases. After doing a complete study, we propose a model for efficiently balancing load in a cloud computing environment. In fact, this proposed model outperforms other competitive schemes. Our goal was to develop a dynamic approach that suitably takes into account all the parameters of virtual machines. The Hermit Model of load balancing is not based on any assumptions and can be made to fit better to a cloud system with different values curve fitting parameters like alpha and beta. In future, this proposed algorithm can be integrated with other metaheuristic algorithms for increasing the efficiency in different measures like average waiting time, virtual machine cost, data transfer cost, etc. The Hermit Model can provide a large number of future research opportunities.

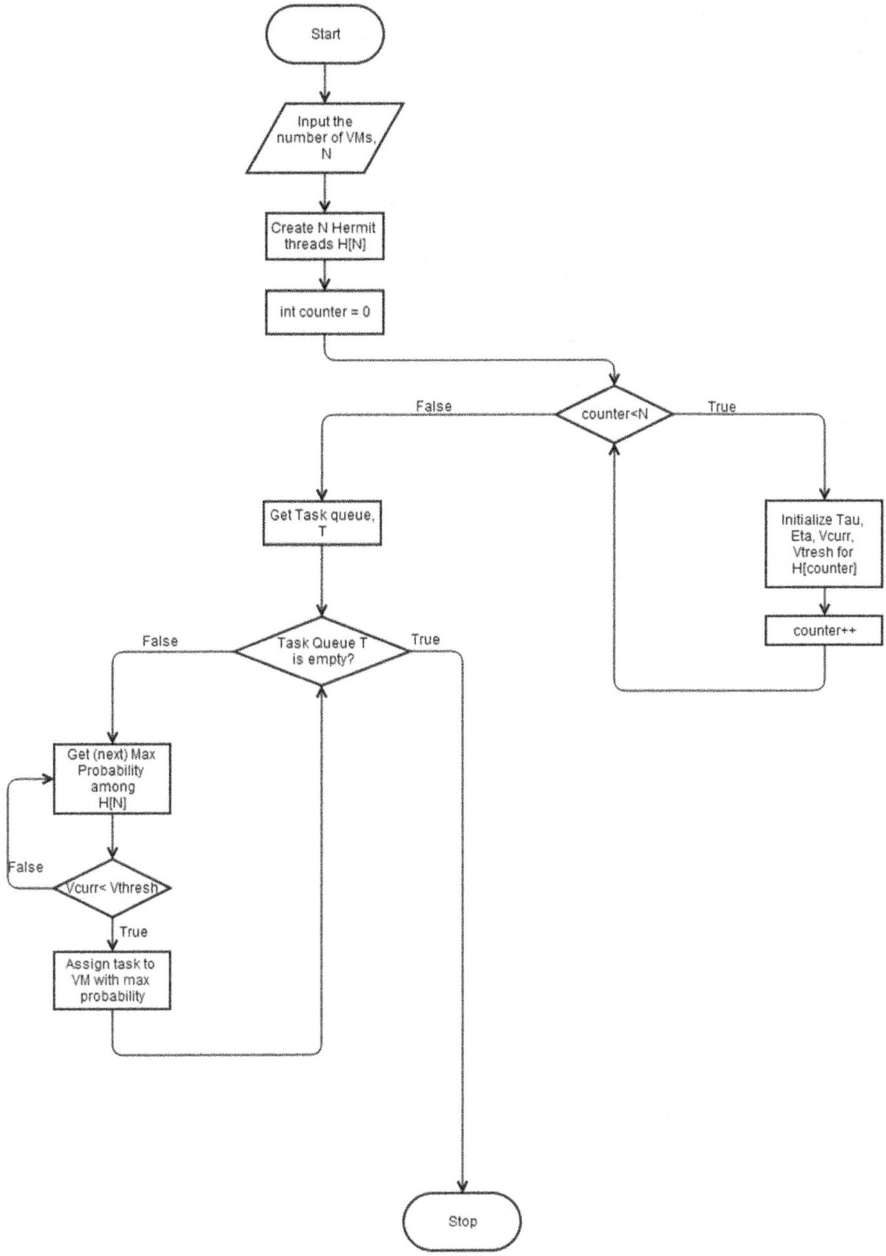

Fig. 2 Flowchart of Hermit model

Table 1 Task configuration

Task Id	Clock cycles	KLOC	Task Id	Clock cycles	KLOC	Task Id	Clock cycles	KLOC
1	296,544	5	31	405,041	9	60	698,430	3
2	238,770	7	32	288,741	3	61	123,382	5
3	627,519	9	33	120,347	6	62	745,194	4
4	150,458	7	34	132,828	4	63	446,633	3
5	343,777	7	35	414,874	1	64	594,330	0
6	385,708	3	36	301,771	7	65	759,082	5
7	915,375	2	37	618,590	0	66	376,528	1
8	549,191	8	38	117,355	8	67	558,839	9
9	295,865	3	39	810,827	6	68	947,148	10
10	268,870	4	40	613,364	9	69	837,127	7
11	508,523	1	41	231,835	0	70	825,837	7
12	101,662	6	42	996,818	1	71	272,997	5
13	590,657	6	43	139,898	6	72	100,890	5
14	313,502	4	44	328,192	2	73	410,102	4
15	124,314	1	45	420,085	0	74	374,526	5
16	578,066	1	46	413,909	5	75	196,330	3
17	107,364	5	47	582,523	6	76	140,828	8
18	590,016	3	48	630,766	8	77	170,474	7
19	687,847	10	49	172,374	10	78	833,807	2
20	318,809	1	50	306,834	4	79	831,291	10
21	938,692	7	51	164,958	7	80	397,575	1
22	955,820	1	52	610,717	5	81	928,020	7
23	227,842	3	53	579,886	10	82	967,290	6
24	614,695	6	54	772,427	3	83	505,772	2
25	523,201	10	55	598,258	8	84	183,034	6
26	122,631	10	56	426,438	8	85	962,240	1
27	101,039	9	57	104,274	1	86	766,094	8
28	144,204	1	58	828,042	9	87	311,245	10
29	704,508	4	59	770,160	9	88	260,520	8
30	929,106	1						

Table 2 Task assigned to VMs

Task Id	Assigned VM	Task Id	Assigned VM	Task Id	Assigned VM
1	2	31	1	60	5
2	5	32	5	61	2
3	1	33	2	62	2
4	4	34	0	63	5
5	3	35	2	64	1
6	0	36	2	65	1
7	2	37	3	66	3
8	5	38	1	67	2
9	1	39	0	68	3
10	2	40	2	69	2
11	4	41	5	70	2
12	0	42	4	71	5
13	4	43	1	72	5
14	2	44	3	73	4
15	1	45	0	74	2
16	3	46	2	75	0
17	5	47	1	76	1
18	1	48	2	77	5
19	2	49	5	78	2
20	3	50	5	79	2
21	4	51	2	80	3
22	3	52	0	81	4
23	5	53	0	82	1
24	1	54	2	83	1
25	2	55	5	84	5
26	0	56	1	85	2
27	3	57	1	86	5
28	5	58	5	87	1
29	5	59	2	88	1
30	4				

Table 3 Makespan calculation of different metaeuristics in varying iterations

Iteration	(Task, VMs)	SA	TS	GA	PSO	Hermit
100	(50,10)	135.87	127.44	99.66	85.14	82.66
	(50,20)	96.23	87.45	70.22	65.69	62.86
	(50,30)	70.27	65.93	55.56	48.22	40.11
	(100,10)	300.85	249.23	188.35	165.66	160.94
	(100,20)	188.72	145.12	110.17	105.64	102.58
	(100,30)	135.63	122.32	101.26	75.69	72.35
	(500,10)	1823.6	1622.5	1115.3	966.23	924.44
	(500,20)	800.76	700.73	645.45	591.28	493.76
	(500,30)	712.21	600.53	449.33	365.56	261.11
500	(50,10)	120.87	100.04	92.23	86.13	80.53
	(50,20)	82.21	75.32	67.28	55.32	50.61
	(50,30)	62.23	58.97	52.16	45.22	41.11
	(100,10)	260.26	214.32	175.55	158.27	152.94
	(100,20)	155.59	130.27	104.62	96.61	90.66
	(100,30)	112.03	100.31	96.26	77.54	68.37
	(500,10)	1702.2	1200.0	1098.2	954.13	915.33
	(500,20)	750.39	699.90	631.76	538.28	490.76
	(500,30)	629.53	555.32	428.33	312.78	221.10

Fig. 3 Makespan of different algorithms for 100 iterations

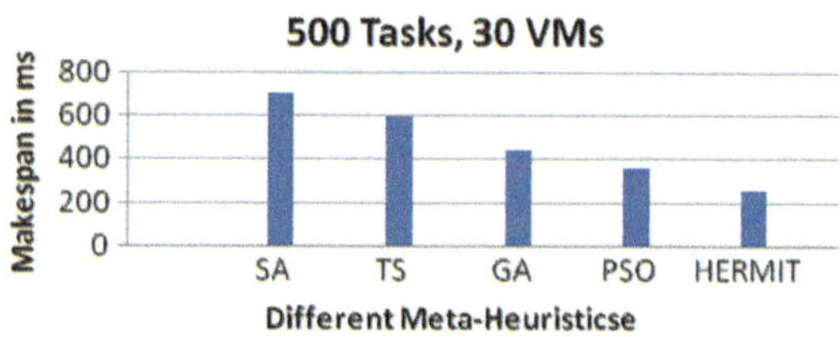

Fig. 4 Makespan of different algorithms for 500 iterations

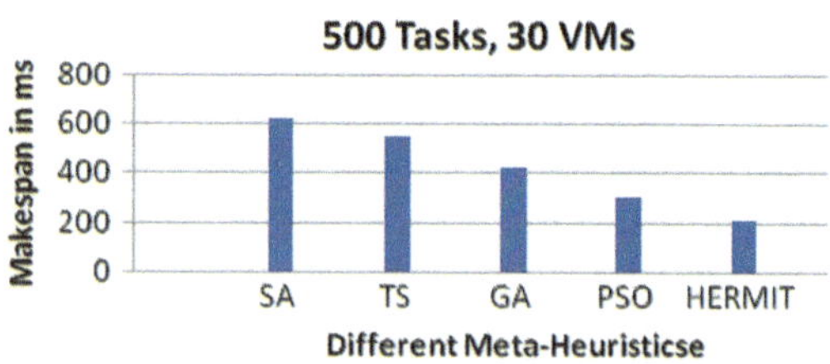

References

1. Rimal, B.P., Choi, E., Lumb, I.: A taxonomy and survey of cloud computing systems. In: Proceedings of the Fifth International Joint Conference on INC, IMS and IDC, pp. 44–51. IEEE (2009)
2. Jin, H., Gao, W., Wu, S., Shi, X., Wu, X., Zhou, F.: Optimizing the live migration of virtual machine by CPU scheduling. J. Netw. Comput. Appl. **34**(4), 1088–1096 (2011)
3. Mann, Z.A.: Allocation of virtual machines in cloud data centers- A survey of problem model and optimization algorithms. ACM Comput. Surv. (CSUR) **48**(1), 11 (2015)
4. Jun, C.: Ipv6 virtual machine live migration framework for cloud computing. Energy Procedia **13**, 5753–5757 (2011)
5. Patel, R., Patel, S.: Survey on resource allocation strategies in cloud computing. Int. J. Eng. Res. Technol. **2**(2) (2013)
6. Kansal, N.J., Chana, I.: Cloud load balancing techniques: a step towards green computing. Int. J. Comput. Sci. **9**(1), 238–246 (2012)
7. Randles, M., Lamb, D., Taleb-Bendiab, A.: A comparative study into distributed load balancing algorithms for cloud computing. In: Proceedings of the 24th IEEE International Conference on Advanced Information Networking and Applications Workshops (WAINA), pp. 551–556 (2010)
8. Talbi, E.G.: Metaheuristics: From Design to Implementation. 74. John Wiley & Sons (2005)
9. Moharana, S.S., Ramesh, R.D., Powar, D.: Analysis of load balancers in cloud computing. Int. J. Comput. Sci. Eng. **2**(2), 101–108 (2013)
10. Sharma, T., Banga, V.K.: Efficient and enhanced algorithm in cloud computing. Int. J. Soft Comput. Eng. (IJSCE) **3**(1) (2013)
11. Parveen, S., Neha, K.: Study of optimal path finding techniques. Int. J. Adv. Technol. 4(2) (2013)
12. Kumar, A., Bawa, S.: Generalized ant colony optimizer: swarm-based meta-heuristic algorithm for cloud services execution computing. **101**(11), 1609–1632 (2019)
13. Matos, J.G., Marques, C.K., Liberalino, C.H.: Genetic and static algorithm for task scheduling in cloud computing. Int. J. Cloud Comput. **8**(1), 1–9 (2019)
14. Singh, S., Kalra, M.: Scheduling of independent tasks in cloud computing using modified genetic algorithm. In: Proceedings of the International Conference on Computational Intelligence and Communication Networks (CICN),pp. 565–569. IEEE (2014)
15. Wang, T., Liu, Z., Chen, Y., Xu, Y., Dai, X.: Load balancing task scheduling based on genetic algorithm in cloud computing. In: Proceedings of the 12th International Conference on Dependable, Autonomic and Secure Computing (DASC), pp. 146–152. IEEE (2014)
16. Pilavare, M.S., Desai, A.: A novel approach towards improving performance of load balancing using genetic algorithm in cloud computing. In: Proceedings of the International Conference on Innovations in Information, Embedded and Communication Systems (ICIIECS), pp. 1–4 (2015)
17. Feng, M., Wang, X., Zhang, Y., Li, J.: Multi-objective particle swarm optimization for resource allocation in cloud computing. In: Proceedings of the 2nd International Conference on Cloud Computing and Intelligence Systems, pp. 1161–1165. IEEE (2012)
18. Beegom, A.A., Rajasree, M.S.: A particle swarm optimization based pareto optimal task scheduling in cloud computing. In: Proceedings of the International Conference in Swarm Intelligence. Springer International Publishing, pp. 79–86 (2014)
19. Abdi, S., Motamedi, S.A., Sharifian, S.: Task scheduling using Modified PSO Algorithm in cloud computing environment. In: Proceedings of the International Conference on Machine Learning, Electrical and Mechanical Engineering, pp. 38–41 (2014)
20. Saxena, D., Saxena, S.: Highly advanced cloudlet scheduling algorithm based on particle swarm optimization. In: Proceedings of the Eighth International Conference on Contemporary Computing (IC3), pp. 111–116. IEEE (2015)

Development of Hybrid Extreme Learning Machine for Classification of Brain MRIs

Pranati Satapathy, Sateesh Kumar Pradhan, and Sarbeswara Hota

Abstract The current biomedical research focuses on automated classification of brain magnetic resonance images into normal or pathological. This paper presents an Extreme Learning Machine model hybridized with one of the recently developed swarm intelligence based optimization algorithm, i.e., Crow Search Algorithm (CSA) for the classification of brain MRIs. This work is broadly divided into three different processes, i.e., Extraction of features from the images, Reduction of features, and classification task. For feature extraction from MR images, Discrete Wavelet Transformation (DWT) is used. Some of the features are not useful for the classification model, so Principal Component Analysis (PCA) is used for reducing the features, thus decreases the computational complexity and memory requirements of the classification model. CSA is used for the optimal determination of the weights between the input layer and the hidden layer of the ELM. The proposed CSA-ELM model is used for the classification of the processed images. Different performance metrics, i.e., classification accuracy, specificity, sensitivity are considered for measuring the performances of the proposed model.

Keywords Magnetic resonance imaging · Crow search algorithm · Extreme learning machine · Discrete wavelet transformation · Principal component analysis

P. Satapathy · S. K. Pradhan
Department of Computer Science and Applications, Utkal University, Bhubaneswar, Odisha, India
e-mail: satapathy.pranati@gmail.com

S. K. Pradhan
e-mail: sateesh.cs@utkaluniversity.ac.in

S. Hota (✉)
Department of Computer Application, Siksha 'O' Anusandhan Deemed to be University, Bhubaneswar, Odisha, India
e-mail: sarbeswarahota@soa.ac.in

D. Mishra et al. (eds.), *Intelligent and Cloud Computing*,
Smart Innovation, Systems and Technologies 194,
https://doi.org/10.1007/978-981-15-5971-6_31

1 Introduction

Human brain suffers from various diseases. The different medical imaging techniques and analysis tools help the physicians and radiologists in identifying the affected tissues and in the diagnosis of diseases [1]. Magnetic Resonance Imaging (MRI) is one of the imaging techniques providing finer images of the human brain. Brain MRI generates informative images that facilitate further investigation [2]. The objective of biomedical research in brain MRIs is to automatically classify the MR images according to distinguished diseases with more accuracy and less human interference. So the software-based brain MRI analysis and classification is essential for the medical community and researchers have suggested various methods in this direction [3]. The image classification process is broadly categorized as dataset collection, data preprocessing including feature extraction and feature reduction, and then classification [4, 5].

From the review of related papers on brain image classification, it is found that variants of Artificial Neural Network (ANN) techniques have been used for classification of brain MRIs. Before applying classification algorithms, the feature extraction and feature reduction processes are performed. In [6], neural network model was used for brain image classification and it produced a better result. Kharrat et al. [7] proposed GA-SVM model to classify brain MRIs. The spatial Gray-Level Dependence method (SGLDM) was used for feature extraction from normal and tumor regions. Genetic Algorithm (GA) was used for feature selection and SVM was used for classification purpose. Zhang et al. [8] employed the neural network based model for the classification of brain MR images. In the preprocessing task, DWT and PCA were applied for the feature extraction and feature reduction tasks. The weights of the BPNN model were optimized by the scaled conjugate gradient (SCG) method. The authors in [9] applied the modified sine cosine (MCSA) method with Extreme Learning Machine (ELM) for the brain image classification. In [10], an improved Particle Swarm Optimization (PSO) algorithm is hybridized with ELM for the brain image classification. ELM is proposed in [11] to overcome the limitations of the ANN approach.

The literature study suggests that swarm intelligence based algorithms are used for the optimal determination of hidden weights and biases in ELM. In this work, the Crow Search Algorithm (CSA) is used with ELM for the optimal determination of hidden weights and biases.

In this paper, the three image datasets are preprocessed with 2-D DWT and PCA. The goal of this paper is to develop the CSA-ELM-based brain MR image classification model. The paper layout is as follows. Section 2 deals with the methods used in this work. The simulation study is explained in Sect. 3. Section 4 deals with the conclusion and future scope of this work.

2 Methodologies

The methodologies used in data preprocessing and classification are described in this section. For data preprocessing, DWT and PCA are used. ELM algorithm and CSA are used for the classification task.

2.1 Data Preprocessing

An image is decomposed into several sub-bands, i.e., LH1, HL1, HH1, and LL1 recursively using 2-D DWT. The first three sub-bands represent detail images and the fourth sub-band corresponds to an approximate image. These four sub-bands are again decomposed into the second level of detail and approximation images. This process is repeatedly executed to achieve the desired level of resolution. The coefficients obtained in this process are useful features. For transforming the feature space to a minimal feature space, PCA is the most suitable technique [12]. The features are selected with respect to variance.

2.2 Crow Search Algorithm (CSA)

On analyzing the living mechanism of crows, Askarzadeh proposed a new meta-heuristic optimization algorithm known as CSA [13]. A lot of work has been developed based on the CSA. The authors in [14] used the CSA for the image segmentation of brain MRI images. They used CSA as the optimization algorithm. Gupta et al. in [15] proposed an optimized version of CSA for the diagnosis of Parkinson's disease. The authors in [16] proposed a modified CSA for the load dispatch problem. The literature study reveals that CSA has been applied in various domain. Crows are considered as one of the most intelligent animals in the world. CSA is inspired from the following behaviors of the crows.

- Crows live in the groups.
- Crows remember the hiding places of their excess food.
- A crow may follow another one to steal the foods of others.
- Crows can move randomly in order to mislead the thieves for protecting their food source.

The mathematical formulation of the CSA is described as follows:
The population of N crows is represented using Eq. (1).

$$C^k = \left\{c_1^k, c_2^k, \dots c_N^k\right\},\ k = \{0,\ 1,\ 2,\ \dots\ \text{TotGen}\} \tag{1}$$

TotGen represents the total number of generations. The new population C^{k+1} is generated using two conditions, i.e., awareness of the crow about the other and unawareness of the crow about the other crow.

From these above two states, one state is selected based on the awareness probability factor APF_i^k. The new crow position is updated using Eq. (2).

$$c_i^{k+1} = \begin{cases} c_i^k + r_i \times fl \times \left(m_j^k - c_i^k\right), r_j \geq \text{APF}_i^k \\ \text{random position, otherwise} \end{cases} \quad (2)$$

Here r_i and r_j are random numbers between 0 and 1. The parameter fl controls the flight length. m_j^k represents the best position obtained by crow j at generation k. The basic steps of CSA are shown in the flowchart in Fig. 1.

Fig. 1 Flowchart of CSA

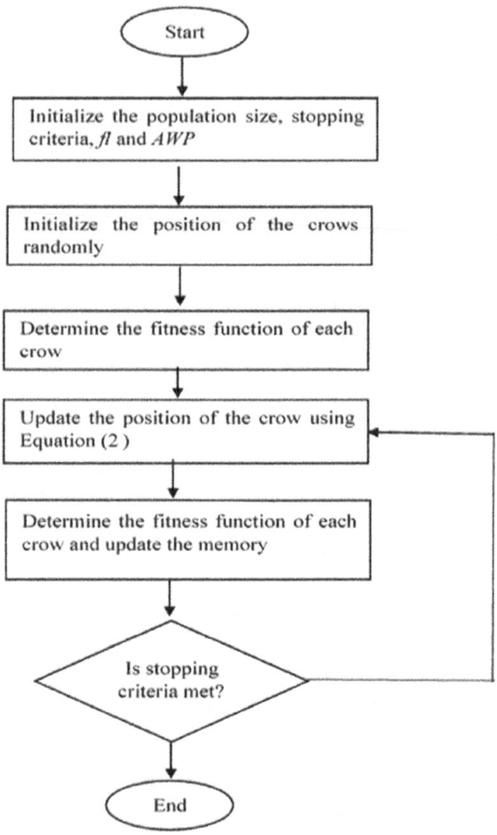

Fig. 2 Proposed CSA-ELM model

2.3 Proposed ELM-CSA Model

This proposed CSA-ELM model consists of four different methodologies embedded together. First of all, 2D DWT is used for extracting features from the brain MR images. For feature reduction, PCA is used. ELM is used for the classification purpose. For the optimal determination of the hidden layer weights and biases, CSA is used as the optimization algorithm. Figure 2 describes the workflow of CSA-ELM model.

ELM is one of the popular learning method used for Single Hidden Layer Feed Forward Neural Network (SLFN) [17]. It avoids the limitations of gradient descent learning [18]. In order to avoid the limitations of the randomness of the hidden layer parameters, evolutionary and swarm intelligence based optimization algorithms, i.e., GA, PSO, ABC, etc., are used. In this work, CSA is used to train the ELM with an aim to increase the classification performance.

3 Simulation Study

This section deals with the datasets description and data preprocessing and classification task.

3.1 Dataset Description and Preprocessing

The datasets considered in this work are taken from the Medical school of Harvard University. The three image datasets are considered, i.e., Glioma, Multiple sclerosis, and Alzheimer for simulation purpose. In this study, 2-D DWT is used for feature extraction from the collected images up to 3-level decomposition. To reduce the feature vector, PCA is used in this work. Table 1 describes the features after applying PCA.

3.2 Classification

For the classification of brain MR images into normal or diseased, the CSA-ELM method is used. For training and testing the model, the fivefold cross-validation

Table 1 Feature reduction using PCA

Name of the datasets	No. of original features	No. of reduced features
Alzheimer	1296	34
Glioma	1296	24
Multiple Sclerosis	1296	39

method is used. During the implementation of ELM, 60 number of hidden nodes and sigmoid activation function are considered. For determination of the input weights and hidden biases, CSA algorithm is used as an optimization algorithm. The population size is 50 and 100 number of iterations are considered. The values of fl and APF are set to 2 and 0.2, respectively. The classification performance of the proposed model is analyzed using Accuracy, Specificity, and Precision as defined in Eqs. (3–5) with TP, TN, FP, and FN.

$$\text{Accuracy} = \frac{\text{TP} + \text{TN}}{\text{TP} + \text{FP} + \text{TN} + \text{FN}} \tag{3}$$

$$\text{Specificity} = \frac{\text{TN}}{\text{TN} + \text{FP}} \tag{4}$$

$$\text{Precision} = \frac{\text{TP}}{\text{TP} + \text{FP}} \tag{5}$$

In this simulation study, the values of the above three measures are shown in Table 2. The graphical representation of accuracy is shown in Fig. 3 for the three datasets.

The performance results are depicted in Table 2 and Fig. 3. The experiments are performed for the accuracy, specificity, and precision values. From the simulation

Table 2 Performance measures for the three datasets using different models

Name of the dataset	Name of the model	Accuracy (in %)	Specificity	Precision
Alzheimer	Basic ELM	88.57	0.90	0.8666
	PCA-ELM	91.43	0.90	0.875
	PCA-CSA-ELM	**94.29**	**0.95**	**0.933**
Glioma	Basic ELM	79.07	0.9	0.8889
	PCA-ELM	81.4	0.65	0.7586
	PCA-CSA-ELM	**85.37**	**0.947**	**0.944**
Multiple Sclerosis	Basic ELM	77.27	0.6667	0.5833
	PCA-ELM	86.36	0.80	0.70
	PCA-CSA-ELM	**90.91**	**0.86**	**0.77**

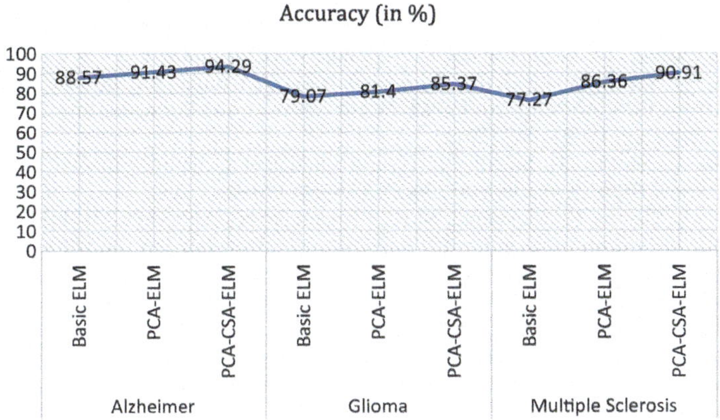

Fig. 3 Graphical representation of accuracy of the three different datasets

results, it is concluded that the CSA-ELM model with feature reduction using PCA outperformed the basic ELM model for the classification of brain MR images in this work.

4 Conclusion

An efficient classification model is proposed in this paper for the brain MRIs. 2-D DWT is used for extracting features from the brain MR images. To get useful features, PCA is used for reducing the feature set. The results of feature reduction are shown in Table 1. For the determination of the initial weights and hidden biases of ELM, CSA is applied. The performance of the CSA-ELM model with PCA as the feature reduction technique is evaluated using three of the metrics, i.e., accuracy, specificity and precision. It is found in this study that the CSA-ELM model with feature reduction using PCA outperformed the basic ELM model for the classification of brain MR images.

References

1. Jiang, J., Trundle, P., Ren, J.: Medical image analysis with artificial neural networks. Comput. Med. Imag. Graph. **34**(8), 617–631 (2010)
2. Chaplot, S., Patnaik, L., Jagannathan, N.: Classification of magnetic resonance brain images using wavelets as input to support vector machine and neural network. Biomed. Signal Process. Control **1**(1), 86–92 (2006)
3. Mohan, G., Subashini, M.: MRI based medical image analysis: survey on brain tumor grade classification. Biomed. Signal Process. Control **39**, 139–161 (2018)

4. Shree, N., Kumar, T.: Identification and classification of brain tumor MRI images with feature extraction using DWT and probabilistic neural network. Brain Inf. **5**, 23–30 (2018)
5. Kumar, V., Sachdeva, J., Gupta, I., Khandelwal, N., Ahuja, C.: Classification of brain tumors using PCA-ANN. In: 2011 IEEE World Congress on Information and Communication Technologies, pp. 1079–1083 (2011)
6. Dahshan, E., Hosny, T., Salem, A.: Hybrid intelligent techniques for MRI brain images classification. Digit. Signal Process. **20**(2), 433–441 (2010)
7. Kharrat, A., Gasmi, K., Messaoud, M.B., Benamrane, N., Abid, M.: A hybrid approach for automatic classification of brain MRI using genetic algorithm and support vector machine. Leonardo J. Sci. **17**, 71–82 (2010)
8. Zhang, Y., Dong, Z., Wu, L., Wang, S.: A hybrid method for MRI brain image classification. Expert Syst. Appl. **38**(8), 10049–10053 (2011)
9. Nayak, D.R., Dash, R., Majhi, B., Wang, S.: Combining extreme learning machine with modified sine cosine algorithm for detection of pathological brain. Comput. Electr. Eng. **68**, 366–380 (2018)
10. Nayak, D.R., Dash, R., Majhi, B.: Discrete ripplet-II transform and modified PSO based improved evolutionary extreme learning machine for pathological brain detection. Neurocomputing **282**, 232–247 (2018)
11. Huang, G.B., Zhu, Q., Siew, C.: Extreme Learning machine: theory and applications. Neurocomputing **70**, 489–501 (2006)
12. Wold, S., Esbensen, K., Geladi, P.: Principal component analysis. Chemom. Intell. Lab. Syst. **2**(1–3), 37–52 (1987)
13. Askarzadeh, A.: A novel metaheuristic method for solving constrained engineering optimization problems: crow search algorithm. Comput. Struct. **169**, 1–12 (2016)
14. Oliva, D., Hinojosa, S., Cuevas, E., Pajares, G., Avalos, O., Gálvez, J.: Cross entropy based thresholding for magnetic resonance brain images using crow search algorithm. Expert Syst. Appl. **79**, 164–180 (2017)
15. Gupta, D., Sundaram, S., Khanna, A., Hassanien, A.E., De Albuquerque, V.H.C.: Improved diagnosis of Parkinson's disease using optimized crow search algorithm. Comput. Electr. Eng. **68**, 412–424 (2018)
16. Mohammadi, F., Abdi, H.: A modified crow search algorithm (MCSA) for solving economic load dispatch problem. Appl. Soft Comput. **71**, 51–65 (2018)
17. Huang, G.B., Zhu, Q.Y., Siew, C.K.: Extreme learning machine: a new learning scheme of feed forward neural networks. Neural Netw. **2**, 985–990 (2004)
18. Huang, G.B., Zhou, H., Ding, X., Zhang, R.: Extreme learning machine for regression and multiclass classification. IEEE Trans. Syst. Man Cybern. **42**(2), 513–529 (2012)

Delay and Disruption Tolerant Networks: A Brief Survey

Satya Ranjan Das, Koushik Sinha, Nandini Mukherjee, and Bhabani P. Sinha

Abstract Challenged networks represent a very special class of networks characterized by widely varying network conditions such as intermittent connectivity, a heterogeneous mix of resource-constrained nodes, long and variable message communication time, bidirectional data-rate asymmetries and high failure rate of nodes. In such deployment scenarios, connectivity is not well served either by the standard Internet architecture and protocols or by popular mobile ad hoc network (MANET) and Wireless Sensor Network (WSN) protocols. Such networks are characterized by the Delay/Disruption Tolerant Networking (DTN) model that uses automatic store-and-forward mechanisms to provide assured data delivery under such extreme operating conditions. In this paper, we present a brief survey on DTN architecture, routing, potential applications and future research challenges. We envision that the research outcomes on this fairly recent topic will have a profound impact on various applications including interplanetary Internet, deep space explorations and future Mars missions, Artic Observing Network (AON) to explore biological, physical, and chemical processes/changes in polar regions.

Keywords Challenged networks · Delay/disruption tolerant networks · Interplanetary Internet · Arctic observing network · Bundle protocol

S. Ranjan Das (✉) · B. P. Sinha
Department of Computer Science and Engineering, Siksha 'O' Anusandhan Deemed to be University, Bhubaneswar, Odisha, India
e-mail: satyadas@soa.ac.in

N. Mukherjee
Department of Computer Science and Engineering, Jadavpur University, Kolkata, India

K. Sinha
Department of Computer Science, Southern Illinois University, Carbondale, USA

D. Mishra et al. (eds.), *Intelligent and Cloud Computing*,
Smart Innovation, Systems and Technologies 194,
https://doi.org/10.1007/978-981-15-5971-6_32

1 Introduction

Internet connectivity is mainly based on wired links along with the essential wireless technologies like satellite and short-range mobile links. Emerging wireless network technologies, unlike Internet experience features like link disruptions for long durations, high error rates, long and variable delays and large bidirectional data-rate asymmetries. Examples of wireless networks in such challenged environments include (1) military wireless networks that connect satellites, aircrafts, troops in the battlefield, under-water and land-based sensors etc., (2) interplanetary networks that connect earth with other space-stations and planets, (3) special-purpose wireless networks that connect mobile devices on earth. Development of an appropriate model of such networks along with suitable protocol for communication in such challenged environment constitute an important research of current interest.

A Delay and Disruption Tolerant Network (DTN) is such a network model which is a coordinated mixture of smaller special-purpose networks, including the Internet. DTNs provide compatibility within and between the networks by absorbing long delays and disruptions and by inter-network communication protocol translation. DTN was originally developed for InterPlanetary Network (IPN), which is a concept of providing Internet-like services across space-stations and planets, supporting deep space exploration. But it is envisioned that DTN can have broader domain of applications such as public-service applications, and scientific, military, commercial applications where link delay or disruption is a prime concern. The following section highlights on the various aspects of DTN architecture.

2 DTN Architecture

In interplanetary as well as space wireless mobile communicating devices on earth, the number of moving communicating objects with limited operational power are increasing at a rapid rate. Links can be disturbed by obstacles when communicating nodes are in motion and links may even be shut down when nodes must maintain secrecy or conserve power. Due to these events, connectivity disruption occurs which is called as intermittent connectivity that leads to network partitioning. On the Internet, intermittent connectivity leads to data loss in the form of dropped packets or delayed packets or eventually ends the session with a failed application. DTNs, in contrast, are designed so as to manage delay and disruptions, and hence, provide communication between intermittently connected nodes. Two types of intermittent connectivity may be there

1. opportunistic connectivity, where although with a long delay communicating nodes in a network including sender and receiver may contact at an unscheduled time as and when the wireless mobile devices come within the communication range or come nearer to an information kiosk and
2. scheduled connectivity, which may need time synchronization.

The DTN architecture is so designed as to tackle this problem of intermittent connectivity in an effective way [5]. DTN architecture basically adopts variable-length messages for communication and a naming convention syntax that supports a broad range of addressing and naming conventions to increase flexibility. To support end-to-end reliability, it uses a store-and-forward mechanism over multiple communication paths and for long durations and hence each DTN node is associated with a storage component. The DTN architecture provides security mechanisms that protect the infrastructure from unauthorized access. The DTN network uses a bundle protocol on top of a "convergence layer," that is itself on top of other lower layers. The DTN Bundle Protocol [DTNBP] designs the message formats (called bundles) transferred between DTN bundle agents in the bundle communications which leads to the DTN store-and-forward overlay network [10].

2.1 Store-And-Forward Message Switching Mechanism

The mechanism of DTNs to deal with the problems associated with long or variable delay, intermittent connectivity, high error rates, and asymmetric data rates is by using a classical method, known as store-and-forward message switching. The complete message or parts of the message are forwarded from the storage of sourcenode to subsequent nodes until it reaches its destination node. Store-and-forwarding methods are not node-to-node relays but rather star relays; a central storage device at the center of all links is contacted by both the source and destination separately. Each node contains a storage place (such as hard disk), called as persistent storage which can hold messages for indefinite period. While waiting for next-hop routing-table lookup, Internet routers store the incoming packets for very short durations using buffers and memory chips. Unlike Internet routers which use buffers and memory chips for keeping the incoming packets temporarily, DTN routers use persistent storage of their DTN node queues for the following reasons:

1. retransmission of dropped or delayed messages can be done in case of any error in message communication.
2. a node in a communicating pair may be more reliable and quick than the other nodes while sending or receiving data.
3. unavailability of next hop communication link for a span of time.

This message switching mechanism provides important information such as retransmission bandwidth, intermediate storage space, and message size to DTN nodes in the network by transferring the complete message in a single transfer (Fig. 1).

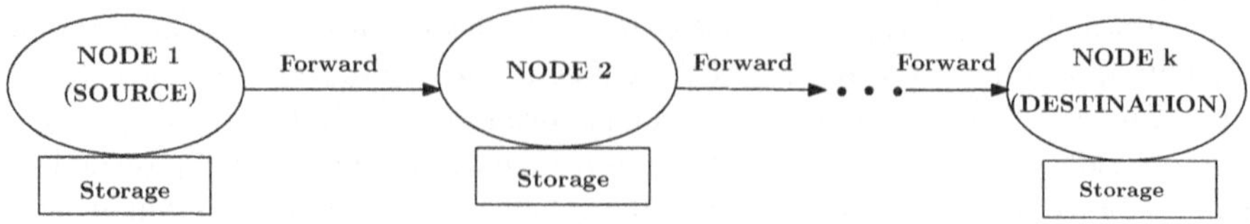

Fig. 1 Store-and-forward message switching model

3 Routing in DTN

All communication networks must have the ability for end-to-end data routing. Delay and disruption tolerant networks (DTNs) are identified by their intermittent connectivity, leading to non-continuous end-to-end paths. Well-known existing routing protocols fail to set up routes in such challenging conditions. Whenever it is impossible or extremely difficult to set up immediate end-to-end routes, DTN routing protocols can effectively communicate messages using the store-and-forward strategy where the message is stored in each hop and incrementally moves over the network to finally reach the destination. Three major DTN forwarding approaches are there such as (i) based on Knowledge, (ii) based on replication (epidemic). and (iii) based on history (probabilistic). DTN forwarding mainly concerns with the number of copies that an algorithm should distribute into the network.

3.1 Bundle Protocols

DTN uses the store-and-forward message switching framework using a new overlay transmission protocol, called the bundle protocol, just above the lower layer protocols such as physical, data link, and network layer. Under specific situations like long network delays and disruptions, the bundle protocol combines with the lower layer protocols to communicate in the same or different set of network protocols. The whole bundle or bundle fragments are stored and then forwarded among multiple nodes. For all DTN nodes, there is a common bundle protocol but each DTN node has separate lower layer protocol based on the communication environment. Figure 2 (top) shows the bundle protocol layer overlay and (bottom) comparison between Internet protocol stack (left) and DTN protocol stack (right).

Bundles are composed of the following three components: (1) a bundle header that contains one or more DTN blocks, (2) user data of a source application, that also contains control information (such as ways of data acquisition and processing) given by the source application for the destination application, and (3) a bundle trailer (optional), that contains zero or more DTN blocks. Bundles have distinct features such as (1) can be arbitrarily long, (2) can be fragmented and then reassembled at the destination, (3) does not alter the Internet protocol data, (4) can enhance the level of object-data encapsulation implemented by the TCP/IP layer protocols.

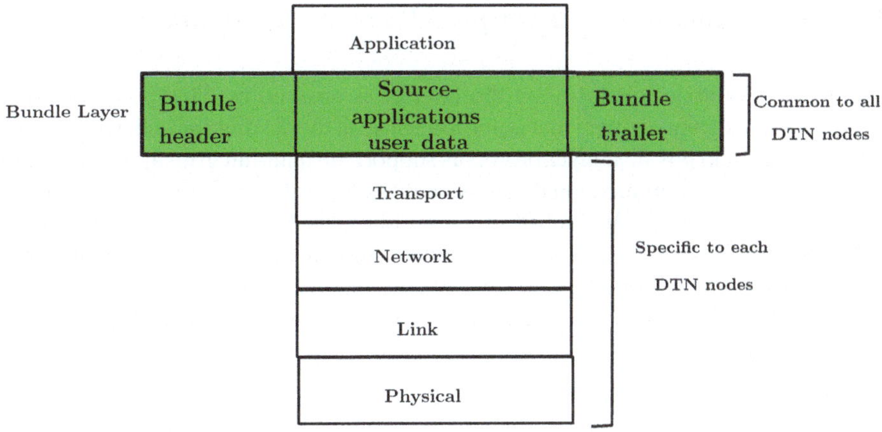

Fig. 2 Layers of DTN Node

Based on the type of service selected, there are optional acknowledgements from the receiving node. The convergence layer protocol stack may be conversational, like TCP, but minimally conversational lower layer protocols may be suitable on intermittently connected links with long delays. The bundle protocol which supports bundle exchanges is just above the convergence layer protocol stack and there is a convergence layer interface that separates between the lower layer protocols and the bundle layer protocol.

At any instance, a specific node may be treated as a source node or destination node, or bundle-forwarder:

- As a source or destination node: A DTN node can act like a source or destination node by sending or receiving bundles to or from other DTN nodes, but do not act as a bundle-forwarder. The bundle protocol in a DTN node needs persistent storage to work over long-delay links. Support for custody transfers is not mandatory.
- As a bundle-forwarder: A DTN node can act as bundle-forwarder between two or more other nodes in one of the following scenarios:
  - Bundle forwarding equivalent to routing: Bundles are forwarded among multiple nodes where each node operates on the same Lower layer protocols like the forwarding node does and needs persistent storage in case of long-delay links. Custody transfers are optional for a node.
  - Bundle forwarding equivalent to gateway: Bundles are forwarded among multiple nodes, where each node implements different Lower layer protocols and hence implemented similarly by the forwarding node. The node must need persistent storage and optionally supports custody transfers.

3.2 Delay Isolation via Transport-Protocol Termination

The end-to-end node reliability on the Internet is ensured by TCP protocol using retransmission of any segment that is not confirmed at the destination end, whereas the network, link, and physical layer protocols support the data-integrity applications. In a DTN, reliable communication depends on these lower layer protocols. The bundle protocol agents behave like substitutes for source-to-destination communication. Thus, the bundle protocol isolates the conversational lower layer protocols from long delays. The bundle protocol layer itself supports end-to-end data exchange. Though may be fragmented, usually bundles are sent as independent single units to subsequent DTN nodes.

3.3 Custody Transfers

In DTN protocol stack, both transport and bundle layers allow source-to-destination retransmission of lost data. But the associated reliability can be supported only at the bundle layer where source-to-destination retransmission is done by using custody

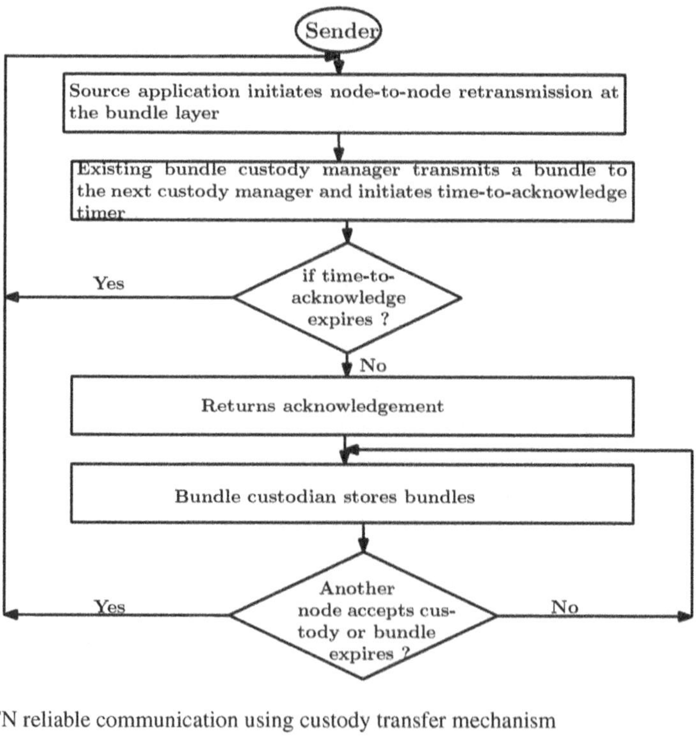

Fig. 3 DTN reliable communication using custody transfer mechanism

transfers. These transfers are prepared by bundle protocol agents between consecutive nodes, initiated at the source application. In Fig. 3, the basic flow of custody transfer mechanism is described.

The bundle protocol agent moves forward retransmission points continuously toward the destination using underlying reliable protocol along with custody transfers. The progressive retransmission points minimally reduce (1) the total time to send a bundle to destination, (2) the number of highly expected retransmission hops, and (3) the extra amount of network load because of retransmission. Thus, a DTN network gains in terms of total delay in message routing due to minimal hop-by-hop retransmission, as opposed to end-to-end retransmission in presence of links with long delays or lossy links.

4 Research Prospects

Delay and disruption tolerant network (DTN) is an emerging wireless networking paradigm that is still in preliminary stage of research till date. Survey works reported [1, 2] in DTNs, give a fair idea about the broad variety of classical technologies for open research in the field of DTNs. In the next subsection, we discuss about few challenging topics on DTNs which need further investigation.

4.1 Research Problems Related to Routing in DTN

Following are some existing problems that cover the different DTN sub-sectors.

1. To implement the bundle mechanism in an efficient manner, elimination of bundle replicas from intermediate nodes is needed when the bundle reaches the destination. This is an important issue which needs further research.
2. Control mechanism for communication route set up between DTN nodes is an aspect which needs detailed investigation.
3. Security aspects of the bundle layer protocol need to be investigated since there is no standardized security model, so far.
4. Bundles with short life spans such as stored unexpired bundles may be discarded at early stages to minimize the buffer occupancy, which can be considered as a promising congestion control strategy.
5. More investigations are expected in future on the application of erasure coding which needs a lot of processing power since bundles are to be delivered within a specific time interval.
6. During node congestion because of buffer overflow, bundle dropping rate highly increases and this leads to unnecessary consumption of bandwidth to retransmit the dropped bundles. As soon as the bandwidth occupancy of a receiving node for accepting incoming bundles crosses some specific value, the probability of

incoming bundle dropping will increase. However, more effective solution to this problem is called for. One solution to this issue is that a receiving node may inform the forwarder about the probability of incoming bundle dropping so that the forwarder may temporarily stop forwarding the bundle.

4.2 Future Research Challenges

1. Characteristic features of DTN are environment specific. Hence, it is really a challenge to design a generalized DTN model for a broad variety of applications.
2. Extensive research is needed for developing efficient, smart network learning mechanisms with fast and reliable communication whenever sufficient information about the network is not available.

5 Conclusion

Many real-life applications involving wireless network are faced with the problem of intermittently connected networks that operates in extreme environments which cause frequent disruptions or long delays in communication. This has led to the development of the potential network model termed as a Delay and Disruption Tolerant Network (DTN) in the literature. Significant research contributions have been made to provide suitable responses to DTN-based problems in some particular conditions. In this paper, we have presented a survey to discuss about valuable contributions on various aspects of DTN including architecture, buffer management, message routing, flow and congestion management, and cooperative mechanisms. Most of the researches are based on custom-made simulations in some specific scenarios. It would have been prudent if some benchmark problem situations on DTNs are formulated on which any designed protocol for fast and reliable communication may be tested for performance evaluation on a common testbed.

References

1. Zhang, Z.: Routing in Intermittently Connected Mobile Ad Hoc Networks and Delay Tolerant Networks: Overview and Challenges. IEEE Communications Surveys and Tutorials, Vol. 8, Issue No. 1, pp. 24–37 (2006)
2. Z. Zhang and Q. Zhang, Delay/Disruption Tolerant Mobile Ad Hoc Networks: Latest Developments, Wiley InterScience, Wireless Communications and Mobile Computing, pp. 1219–1232 (2007)
3. M. Narang et al., UAV-assisted edge infrastructure for challenged networks, 2017 IEEE Conference on Computer Communications Workshops (INFOCOM WKSHPS), Atlanta, GA, 2017, pp. 60–65.https://doi.org/10.1109/INFCOMW.2017.8116353.

4. Alhilal, A., Braud, T., Hui, P.: The Sky is NOT the Limit Anymore: Future Architecture of the Interplanetary Internet. IEEE Aerospace and Electronic Systems Magazine **34**(8), 22–32 (2019). https://doi.org/10.1109/MAES.2019.2927897. Aug
5. V. Cerf, et al., "Delay-Tolerant Network Architecture, IETF RFC 4838, Informational," April 2007. http://www.ietf.org/rfc/rfc4838.txt
6. Voyiatzis, A.G.: A survey of delay- and disruption-tolerant networking applications. J. nternet Eng. **5**, 331–344 (2012). June
7. Psaras, I., Wood, L., Tafazolli, R.: Delay-/Disruption-Tolerant Networking: State of the Art and Future Challenges (2009)
8. M. Khabbaz, C. Assi, W. Fawaz.: Disruption-Tolerant Networking: A Comprehensive Survey on Recent Developments and Persisting Challenges. Communications Surveys and Tutorials, IEEE. 14. 1 - 34. 10.1109/SURV.2011.041911.00093. (2012)
9. Caini, C., Cruickshank, H., Farrell, S., Marchese, M.: Delay- and disruption-tolerant networking (DTN): an alternative solution for future satellite networking applications. Proc. IEEE **99**(11), 1980–1997 (2011). https://doi.org/10.1109/JPROC.2011.2158378. Nov
10. F. Warthman "Delay tolerant networks (DTNs): A tutorial." (2012) http://ipnsig.org/wp-content/uploads/2012/07/DTN_Tutorial_v2.04.pdf
11. The Internet Research Task Force's Delay-Tolerant Networking Research Group (DTN-RG), http://www.dtnrg.org
12. The InterPlanetary Networking (IPN) project, described on the InterPlanetary Networking Special Interest Group (IPNSIG) site, http://www.ipnsig.org
13. Sabogal, D., George, A.D., "Towards resilient spaceflight systems with virtualization,", : IEEE Aerospace Conference. Big Sky, MT **2018**, 1–8 (2018). https://doi.org/10.1109/AERO.2018.8396689
14. Hong, H., El-Ganainy, T., Hsu, C., Harras, K.A., Hefeeda, M.: Disseminating Multilayer Multimedia Content Over Challenged Networks. IEEE Transactions on Multimedia **20**(2), 345–360 (2018). https://doi.org/10.1109/TMM.2017.2744183. Feb

Service-Oriented Software Engineering

Test Case Generation Based on Search-Based Testing

Rashmi Rekha Sahoo, Mitrabinda Ray, and Gayatri Nayak

Abstract Maximum coverage with minimum testing time is the main objective of a test case generation activity which leads to a multi-objective problem. Search-Based Testing (SBT) technique is a demanding research area for test case generation. Researchers have applied various metaheuristic (searching) algorithms to generate efficient and effective test cases in many research works. Out of these existing search-based algorithms, Genetic Algorithm (GA) and Particle Swarm Optimization (PSO) are the most widely used algorithms in automatic test case generation. In this paper, a test case generation approach is proposed using Cuckoo Search (CS) algorithm. CS has a controlling feature, Lévy flights, which makes it more efficient in searching the best candidate solution. It helps to generate efficient test cases in terms of code coverage and execution time. In our proposed method, test cases are generated based on path coverage criteria. Fitness of a test case is evaluated using branch distance and approximation level combined functions. The result is compared with PSO and with its variant Adaptive PSO (APSO). The experimental result shows that both the algorithms give nearly equal to the same result. Though the results are nearly equal, the implementation of CS is simple as it requires only one parameter to be tuned.

Keywords Search-based testing · Test case generation · Cuckoo search · PSO · APSO

R. R. Sahoo (✉) · M. Ray · G. Nayak
Department of Computer Science and Engineering, Siksha 'O' Anusandhan (Deemed to be University), Bhubaneswar, Odisha, India
e-mail: rashmirekhasahoo@soa.ac.in

M. Ray
e-mail: mitrabindaray@soa.ac.in

G. Nayak
e-mail: gayatrinayak@soa.ac.in

D. Mishra et al. (eds.), *Intelligent and Cloud Computing*,
Smart Innovation, Systems and Technologies 194,
https://doi.org/10.1007/978-981-15-5971-6_33

1 Introduction

Software testing phase plays a vital role in producing quality software products. The manual testing process is a labour-intensive and time-consuming task which consumes more than half of the time and resources required for the entire software development life cycle [1]. Hence, several methods have been proposed like random testing, symbolic execution, combinatorial testing, model-based testing, search-based testing, etc. [2]. In a software life cycle, many phases have the multi-objective problem. So searching techniques have been introduced to the software engineering field and are called Search-Based Software Engineering (SBSE) [3]. SBT is the subarea of SBSE [3]. The desired output of a test case generation activity is to produce a minimum number of test cases that should give maximum coverage in minimum time which is a multi-objective problem. So SBT gains popularity in the software testing research field. Several searching/metaheuristic algorithms have been applied till now for test case generation: GA, PSO, Ant Colony Optimization (ACO), Bee Colony Optimization (BCO), Simulated Annealing, Hill Climbing, etc. [2, 3]. Still, there is a requirement to reduce test case generation time. In this paper, the cuckoo search algorithm is used to generate test cases for path coverage. CS algorithm follows the reproduction strategy of the obligate brood parasitism of some cuckoo species and the lévy flight behaviour of some birds like fruit flies [4]. Obligate brood parasite cuckoos lay their eggs in the nest of other species, and the host bird unknowingly raises the broods of the cuckoo. If the host birds notice that the eggs are not their own, they simply throw the eggs away or leave the nest and build a new nest somewhere else. CS becomes a robust algorithm by possessing the lévy flights which maintain exploration and exploitation at the time of searching a candidate solution [4]. The fraction of nests (P_a) is the important parameter to be tuned to make the searching process efficient; other parameters like step size and lévy exponent are there but they are always set with the same constant value in almost all applications [5–9]. The experimental result is compared with the PSO-based test case generation. Though original PSO fails to balance exploration and exploitation, APSO has the inertia weight attribute which balances both [10]. As both the algorithms are claimed as simple and robust algorithms [2–4, 10], we compare the results of both the algorithms and trys to give a new direction for test case generation. In searching techniques, search process is guided through an objective/fitness function [11]. In this work, Branch Distance (BD) [12] and Approximation Level (AL) [13] combined fitness functions are used for path coverage. BD determines the distance (difference) between conditional nodes' input data to the desired data [12]. AL determines how far the control is from the target branch/path by counting the number of target nodes not covered [12, 13]. The combination of BD and AL are widely used in SBT named Combined Fitness Function (CFF) [13].

Fitness (test data) = Normalized Branch Distance + Approximation Level [12, 13]

Normalized BD = $1-(1.001)^{-\text{distance}}$ [14].

The objective of this paper is to automate test case generation using cuckoo search and to generate efficient test cases that provide maximum path coverage in minimum time and iterations. The novelty of this paper is the use of CFF in cuckoo search to cover a target path which is most critical in the probability of coverage point of view. In this work, Triangle Classification Problem is taken as a program under test and it is very difficult to automatically generate a test case for the equilateral triangle according to the probability of coverage.

The paper is organized as follows: in Sect. 2, related work on cuckoo search and PSO-based test case generation is given. Section 3 presents the proposed work on test case generation using CS. In Sect. 4, materials and methods explain the experimental setup. In Sect. 5, the experimental result is discussed. The conclusion and future work are presented in Sect. 6.

2 Related Work

Srivastava et al. [15] propose a cuckoo search-based approach to cover the most error-prone paths along with covering all transitions and all paths of a program. State machine is the input and the optimal test sequence is the output of the proposed approach. The CS-based approach outperforms GA and ACO. Sharma et al. [5] propose a framework for test case generation with the intent to implement various benchmark programs. But, in their paper, they have not done any implementation work. Khari and Kumar [6] generate optimized test suits of java programs by proposing a cuckoo search-based approach and compare the result with hill climbing and firefly algorithms. Their approach outperforms both the algorithms in terms of the number of test cases generated, number of iterations and time. Kumar and Muthukumarvel [7] propose a PSO- and CS-based approach for software quality estimation. They generate test cases using PSO and prioritize the test cases using improved CS. The results are evaluated using software metrics which yields an efficient result. Srivastava et al. [8] design a cuckoo search-based model for test effort estimation. CS optimizes the parameters used in estimating the effort through a use case diagram. They experimentally show that their approach gives better results than the other metaheuristic approach used so far for test effort estimation. Panda and Dash [9] generate test data for object-oriented programming using cuckoo search. They use the state machine diagram for path coverage and compare the results with GA. They get a better result with their proposed approach in terms of the number of iterations and the percentage of coverage. Sahoo and Ray [16] improvised the branch distance-based fitness named Improved Combined Fitness (ICF) and used it in PSO to generate test cases to cover a critical path. ICF gives better results in terms of percentage of coverage, execution time and the number of iterations as compared to existing branch distance functions.

3 Proposed Test Case Generation Approach

Figure 1 shows the proposed approach to generate test cases using the cuckoo search algorithm. Our objective is to automate test case generation activity and to achieve maximum coverage in minimum time. The proposed approach is implemented on a triangle classification problem. First, a Control Flow Graph (CFG) is derived from the Source program. Then the test adequacy criteria (path coverage) [17] is defined. Then, the BD-based fitness, i.e., CFF is a computed based on the path coverage. CFF guides CS algorithm to search for the best solution (test data). The algorithm executes the instrumented code and generates the test data and the test data is refined using CS algorithm and finally, the best solution (test data) is generated.

Algorithm: Test Case Generation using Cuckoo Search

Input: Instrumented Program
Output: Optimized test case

1. Initialize a random population of n test data (host nests), X_i.
2. Get a test data randomly by lévy flight behaviour, *i*.
3. Calculate its fitness (CFF), F_i.
4. Choose a test data randomly among the n test data say *j* and evaluate its fitness (CFF), F_j.
5. If $F_i > F_j$, then replace *i* by new solution *j* else let *i* be the solution.
6. Leave a fraction of P_a of the worst nest (test data) by building new ones at new locations using lévy flights.
7. Keep the current optimum test data (nest). Go to Step (2) if maximum iteration is not reached.
8. Get the optimum test data.

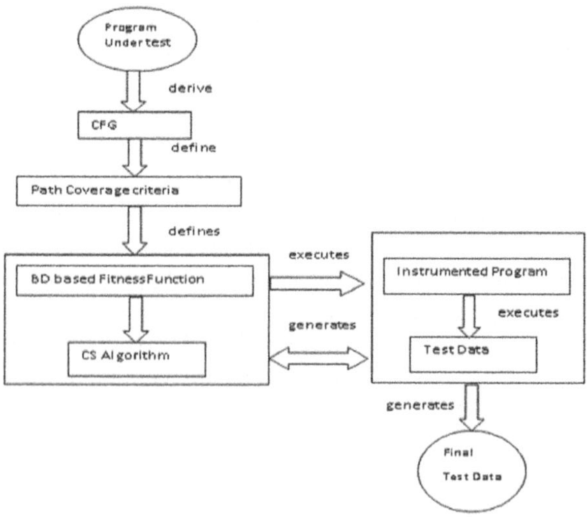

Fig. 1 Schematic diagram of the proposed approach

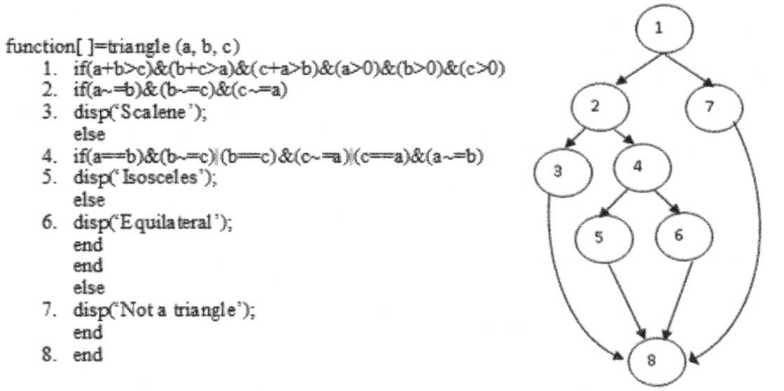

Fig. 2 **a** Source code of TCP in MATLAB [2] **b** CFG of Fig. 2a [2]

Case Study: Triangle Classification Problem (TCP)

CS algorithm is implemented on TCP. Though it is a small program, to get 100% path coverage for this problem is quite difficult, because according to probability theory [2], the probability of getting a combination of an equilateral triangle is least in comparison to the other three types (Isosceles, scalene and invalid triangles). So we consider the path leading to the equilateral triangle as a critical path or target path. Three integer values of three sides of a triangle are given as input to the TCP and it determines the type of triangle. Figure 2a shows the MATLAB code of the TCP program and Fig. 2b shows the corresponding CFG of Fig. 2a.

Path 1: Node1–Node7–Node8 (Not a triangle)

Path 2: Node1–Node2–Node3–Node8 (Scalene)

Path 3: Node1–Node2–Node4–Node 5–Node8 (Isosceles)

Path 4: Node1–Node2–Node4–Node 6–Node8 (Equilateral)

Figure 2b shows four linearly independent paths. Path 4 is considered as a critical path. There are 3 conditional nodes in Fig. 2a. The Conditional Statements (CSs) and their corresponding BDs according to Korel's branch distance function [12] are shown below.

CS1 = (a + b > c) & (b + c > a) &(c + a > b) & (a > 0) & (b > 0) &(c > 0)

BD1 = −2*(a + b + c)

CS2 = (a ~= b) & (b ~= c) &(c ~= a)

BD2 = min (min (abs (a−b), abs (b−c), abs(c−a))

CS3 = (a == b) & (b ~= c) || (b == c) &(c ~= a) ||(c == a) & (a ~= b)

BD3 = −abs (b−c)−abs(c–a)−abs (a−b)

AL of path 1 is 2 as it is far from the target by two nodes as shown in Fig. 2b. Similarly, AL of path 2 is 1 and path 3 and 4 is zero.

Table 1 Random population

Test Case	A	B	C
TC1	3	4	5
TC2	2	3	5
TC3	2	4	3
TC4	3	3	5
TC5	1	3	3

Implementation of the Algorithm

Random Population Initialization

Host nests are represented as test data of test cases. Each nest is 3 dimensional, representing three sides of the triangle. 5 test cases are randomly generated as shown in Table 1.

Iteration 1:

Assume that, by applying lévy flight [4], randomly chosen test data is test case 1, (3, 4, 5) which leads to a scalene and randomly choose another test data, let it be TC4, (3, 3, 5) leads to isosceles without applying lévy flights and calculate its CFF and compare it. According to Korel's branch distance function [12]:

- BD (TC1) = −2*((3 + 4 + 5) = −24
- Normalized Branch distance = $1-(1.001)^{24}$ = 1–1.024 = −0.024
- CFF of TC1, Fitness (TC1) = BD + AL = −0.024 + 2 = 1.976
- BD (TC4) = BD1 + BD2 + BD3, BD1 (TC4) = −2*(3 + 3 + 5) = −22

BD2(TC4) = min(abs(3–3), abs(3–5), abs(5–3)) = min(0,2,2) = 0
BD3 (TC4) = −abs (3–5)−abs (5–3)−abs (3–5) = −2–2−2 = −6
BD (TC4) = −22 + 0–6 = −28, Fitness (TC4) = −0.028 + 1 = 0.972

As Fitness (TC4) < Fitness (TC1), TC4 will be replaced with TC1. So discard TC1 and pass TC4 to next generation and mark TC4 as checked.

Iteration 2:

Select a test case randomly from the remaining unchecked test cases. Let it be TC2, repeat the process in iteration1; TC2 (2, 3, 5) leads to an invalid triangle:

BD (TC2) = −2*(2 + 3 + 5) = −20, CFF of (TC2) = −0.020 + 2 = 1.986.

No replacement is required as Fitness (TC4) < Fitness (TC2). So discard TC2 and pass TC4 to the next generation. Mark TC2 as checked.

Iteration 3:

Choose randomly another unchecked test case, let it be TC3.

TC3 (2, 4, 3) covers path 2. BD = BD1 + BD2

BD = −18 + 1 = −17, Fitness (TC3) = −0.017 + 2 = 1.983; since TC3 > TC4, discard it.

Iteration 4:

Similarly, let the randomly chosen test case is TC5 and its fitness be −0.018 which is less than TC4. So discard TC4 and pass TC5 to the next generation.

Iteration 5:

On reiterating the process, all the test cases are checked. Using lévy flights generate a new test case, let it be TC6 (5, 5, 5) which leads to an equilateral triangle and find its fitness value. Fitness (TC6) = 0. Hence the TC6 is the optimal test case.

4 Materials and Methods

MATLAB is used as an implementation platform. TCP is taken for experimental analysis. The experiment is done on 1000 populations and with a range [1, 20] for 100 iterations on TCP. For CS algorithm, only one parameter needs to be tuned, i.e., P_a: 0.25 [4–7].

5 Experimental Results

Table 2 shows the average results of 3 runs with 100 iterations. It shows the average number of test cases generated, the minimum number of iteration required to cover the target path and the average execution time. The results of PSO, APSO and CS are nearly equal. All these three algorithms are able to achieve the target path by giving 100% path coverage. All three algorithms achieve the target within 5 iterations. Though APSO and CS both achieve the target path in 3 iterations only, the execution time of CS is more than APSO. But the implementation of CS is simpler than PSO as it requires only Pa to be set [4]. CS convergence graph is shown in Fig. 3.

Table 2 Experimental results

Population: 1000 Range: [1–20]	Average number of test cases generated for				Iterations# to achieve target path	Average execution time in Seconds
Algorithms	Not a Triangle	Scalene	Isosceles	Equilateral		
PSO with CFF [15]	468	428.667	101.667	2.769	4	2.883164
APSO with CFF [15]	454	425.667	117	3.333	3	2.769037
CS with CFF	457.667	432	108	2.3333	3	4.402865

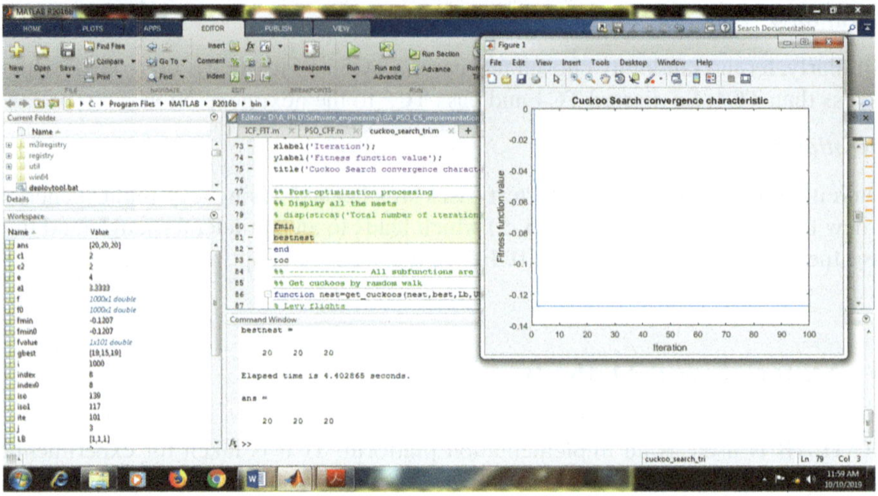

Fig. 3 CS Convergence characteristics

6 Conclusion and Future Work

Many research works are going on to fully automate the testing process. Test case generation is the most important activity of the software testing process. In this paper, an automatic test case generation is done using a cuckoo search algorithm. TCP problem is taken for evaluating the performance of the algorithm. The experimental result is compared with the PSO-based test case generation. It is observed that both the searching techniques give almost the same result and give 100% path coverage in the TCP problem. Though the results are nearly equal, CS algorithm is simpler than PSO as it requires few parameters and mainly the parameter (P_a) plays a vital role in efficient searching. It can be said that only one parameter (P_a) needs to be tuned. Our future work is to implement the algorithms in more complex programs and to predict software defects using searching techniques.

References

1. Maragathavalli, P.: Search-based software test data generation using evolutionary computation. Int. J. Comput. Sci. Inf. Technol. **3**(1), 213–223 (2011)
2. Sahoo, R.R., Ray, M.:Metaheuristic techniques for test case generation: a review. J. Inf. Technol. Res. **11**(1), 158–171 (2018)
3. Harman, M., Jia, Y., Zhang, Y.:Achievements, open problems and challenges for search based software testing. In: Proceedings of IEEE 8th International Conference on Software Testing, Verification and Validation, pp 1–12 (2015)
4. Yang, X., Deb, S.: Cuckoo search via levy flights. In: Proceedings of the Nabic—World Congress on Nature & Biologically Inspired Computing, pp. 210–214 (2009)

5. Sharma, S., Rizvi, S.A.M., Sharma, V.: A framework for optimization of software test cases generation using cuckoo search algorithm. In: 9th International Conference on Cloud Computing, Data Science & Engineering (Confluence), pp. 282–286. IEEE (2019)
6. Khari, M., Kumar, P.: An effective meta-heuristic cuckoo search algorithm for test suite optimization. Informatica **41**(3), 363–377 (2017)
7. Kumar, K.S., Muthukumarvel, A.: Optimal test suite selection using improved cuckoo search algorithm based on extensive testing constraints. Int. J. Appl. Eng. Res. **12**(9), 1920–1928 (2017)
8. Srivastava, P.R., Varshney, A., Nama, P., Yang, X.S.: Software test effort estimation: a model based on cuckoo search. Int. J. Bio-Inspired Comput. **4**(5), 278–285 (2012)
9. Panda, M., Dash, S.: Automatic test suite generation for object oriented programs using metaheuristic cuckoo search algorithm. Int. J. Control Theory Appl. **10**(18) (2017)
10. Roshan, R., Porwal, R., Sharma, C.M.: Review of search based techniques in software testing. Int. J. Comput. Appl. **51**(6), 42–45 (2012)
11. Baresel, A., Sthamer, H., Schmidt, M.: Fitness function design to improve evolutionary structural testing. In: Proceeding of the Genetic and Evolutionary Computation Conference, pp. 1329–1336 (2002)
12. Korel, B.: Automated software test data generation. IEEE Trans. Softw. Eng. **16**(8), 870–879 (1990)
13. Chen, Y., Zhong, Y.., Shi, T., Liu, J.: Comparison of two fitness functions for GA-based path-oriented test data generation. In: Fifth International Conference on Natural Computation, IEEE Computer Society, pp. 177–181 (2009)
14. Wegener, J., Baresel, A., Sthamer, H.: Evolutionary test environment for automatic structural testing. Inf. Softw. Technol. **43**, 841–854 (2001)
15. Srivastava, P.R., Singh, A.K., Kumhar, H., Jain, M.: Optimal test sequence generation in state based testing using cuckoo search. Int. J. Appl. Evol. Comput. (IJAEC) **3**(3), 17–32 (2012)
16. Sahoo, R.R., Ray, M.: PSO based test case generation for critical path using improved combined fitness function. J. King Saud Univ.-Comput. Inf. Sci. (2019). https://doi.org/10.1016/j.jksuci.2019.09.010
17. Mall, R.: Fundamentals of software engineering. 5th edn. PHI Learning Pvt. Ltd (2018)

A Hybrid Approach to Achieve High MC/DC at Design Phase

Gayatri Nayak, Mitrabinda Ray, and Rashmi Rekha Sahoo

Abstract In model-based testing, test cases are generated from the various type of UML diagrams. The objective is to cover the functional requirements present in the software under test. A single UML diagram is not capable to cover all functional requirements. To fulfill this, objective research is going on to generate test cases using UML diagram. In this paper, we propose a hybrid method that combines class diagram and activity diagram together to generate test cases using concolic testing and measure Modified Condition/ Decision Coverage (MC/DC) score using our proposed algorithm Test Suite Evaluator (TSE). We know that MC/DC criterion plays a vital role in regulated domains such as aerospace and safety critical domains, software quality assurance, as per DO-178B(C) certification by NASA. Our experimental result shows that this hybrid approach achieves significantly high MC/DC percentage over existing approaches in which test cases are generated based on either class diagram or activity diagram.

Keywords Concolic testing · Activity diagram · Class diagram · Modified condition/ decision coverage (MC/DC)

1 Introduction

Software Testing is defined as the verification of different Program Under Test (PUT). Testing is a very important phase of SDLC (Software Development Life Cycle) [2]. Software Testing is an activity to verify the software or its constraint in such a way

G. Nayak (✉) · M. Ray · R. R. Sahoo
Department of Computer Science and Engineering, Siksha 'O' Anusandhan (Deemed to Be) University, Bhubaneswar, Odisha, India
e-mail: gayatrinayak@soa.ac.in

M. Ray
e-mail: mitrabindaray@soa.ac.in

R. R. Sahoo
e-mail: rashmirekhasahoo@soa.ac.in

D. Mishra et al. (eds.), *Intelligent and Cloud Computing*, Smart Innovation, Systems and Technologies 194,
https://doi.org/10.1007/978-981-15-5971-6_34

that the software system is intended to be bug-free. The objective is to finding out different types of errors and missing requirements. Testing activities consist of design and execution of test suite. Testing may be manual or automatic throughout the SDLC. In manual testing, software tester acts as an end-user and check the software system to detect any unpredicted actions or error. Automation testing is used to examine and find the difference between actual results and expected results. This testing process is achieved using test scripts or any tool which can be applied for this testing. Software testing strategies are classified into two categories such as black box testing and white box testing [5]. In black box testing the design or implementation of the software system is unknown for the tester. Black Box testing process is either functional or non-functional. In case of white box testing the design and implementation part of the software system is being tested and is also well known by the tester.

There exist various types of coverage criteria in white box testing such as statement coverage, branch coverage, MC/DC coverage, and multiple condition coverage. To improve the quality of the avionics software, researchers have been working in automatic testing process that supports MC/DC criteria [4]. MC/DC is the second strongest coverage criteria but researchers are using this concept because it needs minimum $n+1$ number of test cases, whereas MCC requires 2^n test case for testing a system. MC/DC criteria improve the condition and decision coverage criteria. Here, each condition in a decision independently affects the outcome of the decision. Multiple condition coverage criteria are the strongest coverage criteria among all criteria. Here, every probable results of every condition in a decision are taken into consideration.

In model-based testing, the tester starts the testing process from the design phase, so that the efficiency with respect to cost and time is enhanced [1]. Visual representation of a system is represented by using Unified Modeling Language (UML) known as UML diagram. UML diagram also represents different actions and class of a system. There exist different types of UML diagrams. In this paper, we consider only the class diagram and activity diagram. To represent a static configuration of a software system and their needs like inheritance and association, etc., a class diagram is used [9]. The attributes and various methods of class diagram do not depend on user interaction with class. The activity diagram is used to represent the flow of the process [6]. It shows the step by step process of the operation performed in the system. But, activity diagram is unaware of all the attributes of the system and their constraints. Thus, to cover maximum number of attributes and their conditional constraints, we have merged both the diagram to increase MC/DC%.

The objective of this work is to generate effective test cases at design phase of SDLC and to improve MC/DC%. These test cases are generated from the combined Java source code which is obtained from both activity diagram and class diagram.

2 Related Work

Chilenski et al. [3] presents MC/DC coverage criteria for testing. They also explain about the advantages of MC/DC coverage criteria related to other coverage criteria.

Mall et al. [11] presents the concept of use case and sequence diagram to generate test cases. They also explain about the Use Case Dependency Graph that is derived from the use case diagram and Concurrent control flow graph that is derived from control flow graph to produce test cases. Using this concept integration testing is performed at design phase in model-based testing.

Mingsong et al. [7] proposed a method to generate test cases automatically from activity diagram. In this method, the test cases are generated randomly. After generating test cases, one can find the execution flow of the process. Then apply the suitable coverage criteria in that test cases which are generated in that flow of execution. Hence, the number of test cases is reduced and test adequacy criteria can be obtained.

Mohapatra et al. [10] have taken a set of UML diagram to generate a test sequence. They explain about the application of MC/DC criteria to a collaboration diagram. To generate test cases from a Java program, they apply concolic testing. Then COPECA tool is applied in the generated test cases to measure MC/DC%.

Parada et al. [8] proposed a method to generate test case from the UML activity diagram and sequence diagram. From the class diagram, structural code was generated. Java code includes various types of method signatures, and concept of constructor method with attribute initialization passed by parameter.

3 Proposed Work

Different types of UML diagrams have been used by the researchers to generate test cases such as *use case diagram, class diagram, sequence diagram, activity diagram, collaboration diagram, state Chart diagram* etc.

But, in this proposed approach, we have combined class diagram and activity diagram to generate test cases. These two diagrams are constructed using Argo UML2.x as shown in Fig. 1 and it supports many advanced features than UML1.x. We design UML activity diagram and class diagram of *Railway Ticket Reservation System (RTRS)* using some advanced features such as guard conditions, actions followed by condition, etc. Then, we combine the extracted Java codes got from these two diagrams to generate skeletal Java code. This skeletal Java code is then instrumented by the code instrumentor to generate Java code written as per jCUTE syntax. After obtaining Java code, we do concolic testing using jCUTE. This concolic tester generates test cases in the form of a test suite. Then, this test suite is supplied to Test Suite Evaluator (TSE) as an input along with Java code to calculate MC/DC%

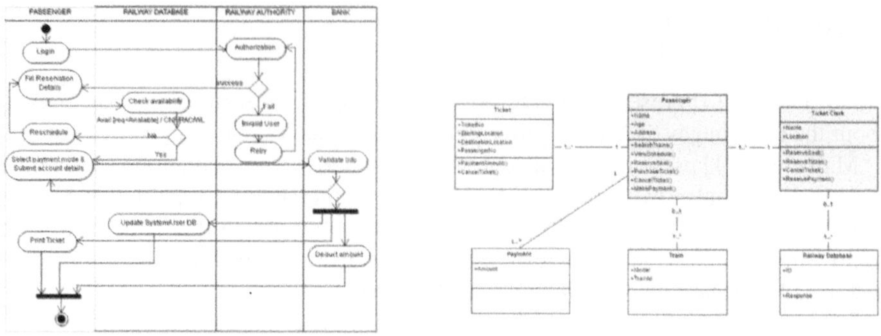

(a) Activity Diagram of RTRS (b) Class Diagram of RTRS

Fig. 1 UML diagrams of railway reservation system

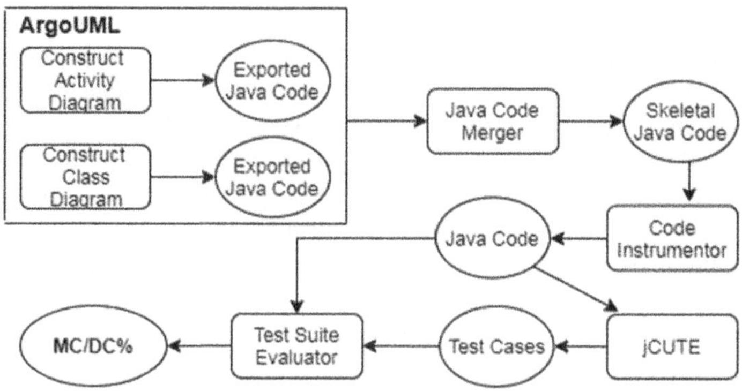

Fig. 2 Architecture of our proposed hybrid model

using Eq. (1). In this equation, contributing clauses help in finding MC/DC pairs. The proposed model of our approach is shown in Fig. 2.

$$MC/DC(P, TS) = \frac{No\_of\_Contributing\_Clauses}{Total\_No\_of\_Clauses} * 100 \quad (1)$$

Test Suite Evaluator (TSE) module is described in algorithm 1 that takes two inputs such as original Java code and its test cases.

Algorithm 1 Test Suite Evaluator

Input: Java Program(J), Test Suite(TS)
Output: MC/DC%

Create a predicate_container $\leftarrow$ $FindPredicate(J)$
Create Extended Truth Table for each predicate
I $\leftarrow$ Find Active Clauses for MC/DC pairs
N $\leftarrow$ Total Number of Clauses
I $\leftarrow$ Remove_duplicate-tests (I)
Compute MC/DC% using Eq. 1
Exit

3.1 Implementation

In this section, we describe the implementation details and analyze their results. During our experiment, in the first stage we construct activity diagram and class diagram as shown in Fig. 1. Then, we generate the Java code of the individual diagram and merge them together to a single Java file and named it as skeletal Java code.

This skeletal Java code is supplied as input to code instrumentor to generate pure Java code to be tested by jCUTE. Now, jCUTE takes this Java code as input and produces test cases and many other information as output as shown in Fig. 3.

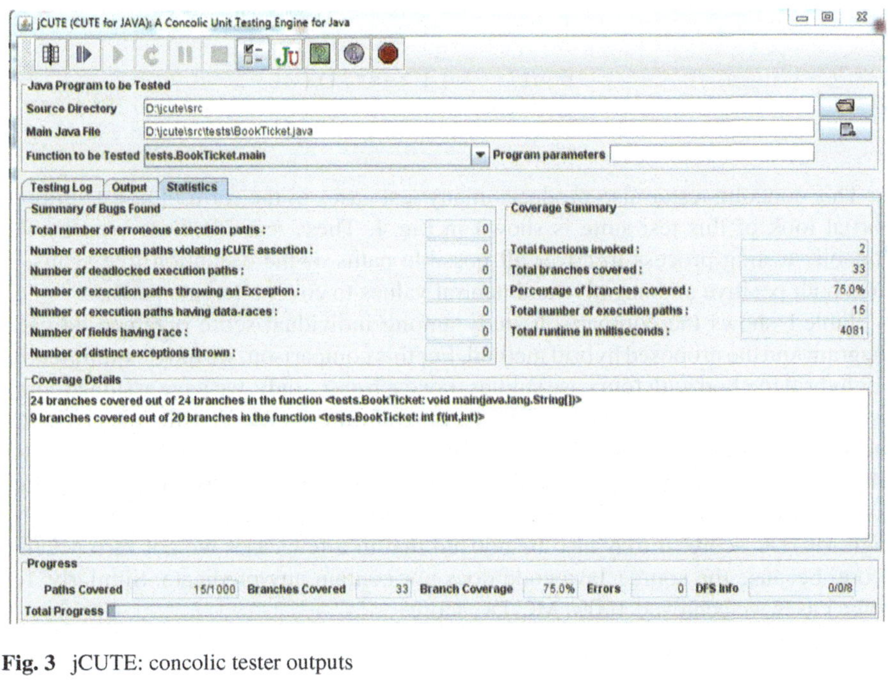

Fig. 3 jCUTE: concolic tester outputs

Fig. 4 Partial screen shoot of sample test cases generated

```
public void test1(){
    i=0;
    input = new Object[5];
    input[i++] = new Integer(-904098110);
    input[i++] = new Integer(388126733);
    input[i++] = new Integer(-2007583436);
    input[i++] = new Integer(1320244315);
    input[i++] = new Integer(-1154309457);
    i=0;
    cute.Cute.input = this;
    tests.BookTicket.main(null);
}
public void test2(){
    i=0;
    input = new Object[5];
    input[i++] = new Integer(1);
    input[i++] = new Integer(388126733);
    input[i++] = new Integer(-2007583436);
    input[i++] = new Integer(1320244315);
    input[i++] = new Integer(-1154309457);
    i=0;
    cute.Cute.input = this;
    tests.BookTicket.main(null);
}

public void test3(){
    i=0;
```

This concolic tester also produces many test cases in the form of a test suite. A partial look of this test suite is shown in Fig. 4. These test cases are used by the concolic testing process to cover all possible paths of the execution tree. Thus, it fires both positive and negative and neutral values to cover alternate paths.

Table 1 shows the comparison study among individual score obtained by each diagram and the proposed hybrid method. For this comparison, we have experimented our hybrid method with four case studies. For each case study, we have assumed some use cases whose corresponding activity diagram are considered for experimental work. From Table 1, it may be observed that our hybrid method achieved better MC/DC score as compared to the individual diagram. Because, class diagram supplies some extra information such as attributes and its constraints that helps in generating high MC/DC score. It may also be noticed that in a few cases we got zero MC/Dc score, because the source Java code does not contain any predicate. Similarly, for some cases we achieved 100% MC/DC score.

Table 1 Comparative study of obtained results

Sl No.	Case Study	UseCase	Activity Diagram		Class Diagram		Hybrid Approach	
			Test cases	MC/DC%	Test cases	MC/DC%	Test cases	MC/DC%
1	RTRS	Book Ticket	11	66.66	7	33.33	14	76.1
2	RTRS	Cancel Ticket	14	75.0	10	50.0	22	97.7
3	RTRS	Refund Money	12	72.2	7	33.0	18	85.7
4	e-shopping	Login	6	50.0	4	33	6	75.0
5	e-shopping	Place Order	6	75.0	4	50.0	6	75.0
6	e-shopping	Manage Product	4	50.0	3	33.3	7	76.1
7	e-shopping	view item	2	0	4	0	4	0
8	ATM	Ckeck Balance	5	50.0	4	33.33	7	63.6
9	ATM	Withdraw	6	75.0	5	50.0	12	83.3
10	ATM	Deposit	5	50.0	2	33	5	50.0
11	ATM	Fund Transfer	7	75.0	6	50.0	12	90.0
12	Elevator	Move up/Down	2	75.0	4	50.0	6	100
13	Elevator	Open/Close door	3	33.3	2	50.0	5	50.0

4 Conclusion

The proposed hybrid method generates high MC/DC% using activity and class diagrams together. We have experimented our proposed method using some well-known example that is *Railway Ticket Reservation System*, *ATM*, and several other case studies. The proposed method validates that the structural representation (class diagram) of any system provides more information about the key attributes such as visibility constraints and value access constraints. Hence, this leads to increase in number of test cases and MC/DC score as well.

In future, we extend our proposed approach to compute MC/DC% with some different UML diagrams.

References

1. Barisal, S.K., Behera, S.S., Godboley, S., Mohapatra, D.P.: Validating object-oriented software at design phase by achieving mc/dc. International Journal of System Assurance Engineering and Management **10**(4), 811–823 (2019)
2. Bhatt, D.: A survey of effective and efficient software testing technique and analysis. Iconic Research and Engineering Journals (IREJOURNALS) (2017)

3. Chilenski, J.J., Miller, S.P.: Applicability of modified condition/decision coverage to software testing. Software Engineering Journal **9**(5), 193–200 (1994)
4. Ghani, K., Clark, J.A.: Automatic test data generation for multiple condition and mcdc coverage. In: 2009 Fourth International Conference on Software Engineering Advances. pp. 152–157. IEEE (2009)
5. Khan, M.E., Khan, F., et al.: A comparative study of white box, black box and grey box testing techniques. Int. J. Adv. Comput. Sci. Appl **3**(6) (2012)
6. Kundu, D., Samanta, D.: A novel approach to generate test cases from uml activity diagrams. Journal of Object Technology **8**(3), 65–83 (2009)
7. Mingsong, C., Xiaokang, Q., Xuandong, L.: Automatic test case generation for uml activity diagrams. In: Proceedings of the 2006 international workshop on Automation of software test. pp. 2–8. ACM (2006)
8. Parada, A.G., Siegert, E., De Brisolara, L.B.: Generating java code from uml class and sequence diagrams. In: 2011 Brazilian Symposium on Computing System Engineering. pp. 99–101. IEEE (2011)
9. Shah, S.A.A., Shahzad, R.K., Bukhari, S.S.A., Humayun, M.: Automated test case generation using uml class & sequence diagram. British Journal of Applied Science & Technology **15**(3), (2016)
10. Swain, S.K., Mohapatra, D.P.: Test case generation from behavioral uml models. International Journal of computer applications **6**(8), 5–11 (2010)
11. Swain, S.K., Mohapatra, D.P., Mall, R.: Test case generation based on use case and sequence diagram. International Journal of Software Engineering **3**(2), 21–52 (2010)

Test Case Prioritization Using OR and XOR Gate Operations

Soumen Nayak, Chiranjeev Kumar, Sachin Tripathi, Lambodar Jena, and Bichitrananda Patra

Abstract In this paper, the efficiency of test case prioritization technique is enhanced by using the two gate operations. OR and XOR gate boolean operations are used for prioritizing the test cases. The time cost of regression testing is reduced to a large extent. Though regression testing improves the quality of the software, it consumes a lot of testing time. For the study, the dataset used by Nayak et al. has been chosen. For reducing the regression time cost, two basic operations are done. Initially pairing of test cases has been done using XOR operations. OR operations are used between them to enhance the fault detection rate. Finally pairing of test cases between the result and the left test cases in the test suite is done to detect residual non-checked error in the software. This process is continued till the resultant is all 1's. The experimental result of the proposed approach gives better performance and scalability than that of non-prioritized prioritization techniques and existing techniques. The fault detection rate of each technique as well as the proposed technique is measured by using average percentage of faults detected (APFD) metric.

Keywords Regression testing · APFD · Basic gate operations · Test case prioritization

---

S. Nayak (✉) · C. Kumar · S. Tripathi
Department of Computer Science and Engineering, IIT (ISM), Dhanbad 826004, India
e-mail: soumennayak@soa.ac.in; soumen.nayak@gmail.com

C. Kumar
e-mail: kumar.c.cse@ismdhanbad.ac.in

S. Tripathi
e-mail: var_1285@yahoo.com

S. Nayak · L. Jena · B. Patra
Department of Computer Science and Engineering, Siksha 'O' Anusandhan (Deemed to be University), Bhubaneswar, India
e-mail: lambodarjena@soa.ac.in; jlambodar@gmail.com

D. Mishra et al. (eds.), *Intelligent and Cloud Computing*,
Smart Innovation, Systems and Technologies 194,
https://doi.org/10.1007/978-981-15-5971-6_35

1 Introduction

Fault detection may be a necessity for maintaining the software during any stage of software development. Therefore, the software is checked again to validate that the software is working fine and there is no adverse effect on the already working portion of the software. A system under test (SUT) is said to regress if a changed or modified component fails or a new module, when appended with unmodified modules, causes failures in the unchanged portions by generating side-effects or feature interactions. According to Craig [1], "*Testing is a concurrent lifecycle process of engineering, using and maintaining testware (i.e. testing artifacts) in order to measure and improve the quality of the software being tested.*"

Regression testing [2] is a maintenance level activity that confirms that a recent code change has not adversely affected the existing functionalities. It comprises a full or partial selection of test cases that are re-executed to check that all the features and components still complies with the specified requirements. The adversity in the software might arise due to enhancements added in the subsequent versions of the software, error corrections, code optimization, and deletion of existing features. Therefore regression testing is required and it consumes maximum testing time. It can be performed using the following techniques:

1. Retest all—Here all the test cases available in the test suite are executed one by one. It consumes huge chunks of time and resources.
2. Regression test selection (RTS)—It is better to choose a part of test suite to re-execute rather than executing them all. The tests in RTS can be divided into two types—obsolete test cases and reusable test cases. The later tests can be used in succeeding regression cycles.
3. Regression test prioritization (RTP)—The test cases are given priority and selected based on some testing goals that can reduce the suite usage and saves time and effort.
4. Regression test minimization (RTM)—In this technique, the redundant tests are deleted permanently from the test suite in terms of features or codes exercised. Therefore, it reduces testing resources significantly.

The test suite sometimes grows exponentially due to the new test cases that are developed and added to test the new modules. The new modules are incorporated with the existing software in the subsequent versions depending on the requirement of the clients or customers. The reduction process could be understood properly if the tests are prioritized in some order. The purpose of RTP is to reduce the set of test cases based on some non-arbitrary rational criteria.

Regression test selection and regression test minimization have certain disadvantages because some empirical evidence shows that the fault detecting capacity of the test suite is severely degraded [3, 4]. Though there are safe regression selection techniques available [5] where fault detection capacities in the subset are not compromised with the original test suite but the safety conditions do not hold always [6]. Regression test prioritization techniques [7, 8] schedule the execution of test cases in an order that enhances the effectiveness in satisfying some performance goals. RTP technique eliminates the drawbacks of RTS and RTM since it never discards any test cases. The advantage with RTP is that the testing objective will be met very quickly thus reducing the time cost in regression testing.

In this manuscript, the test case prioritization is performed using OR and XOR boolean gate expressions. The test suite for the study has been taken from the literature [9]. In the fault matrix, the test case that detects the faults in the software is taken as 1 and the faults it could not detect are considered as 0. The test case is selected that contains maximum 1's in the row of the fault matrix. Then XOR operation is being performed between the test cases and the selected test case. The test case that gives maximum 1's in the resultant is selected. Between them OR operation is performed and the fault detection array subset is updated. For undetected error, the process is repeated until there are all 1's in the result. That means all the errors have been detected successfully.

The rest of the paper is organized as follows. Section 2 deals with related work. Section 3 explains the problem statements. Section 4 emphasizes on proposed work. Section 5 deals with experimentation and explanation of the steps. Section 6 contains the discussion part. And Sect. 7 contains conclusion and future works.

2 Related Work

Test case prioritization reduces the testing resource requirements to a large extent. Prior research on this topic by the researcher and academicians paved the way for further research. Rothermel et al. [10] proposed many methods to prioritize test cases based on test case execution information. The result shows that each of the technique enhances the fault detection rate. Elbaum et al. [11] have incorporated varying test costs fault severities into test case prioritization. Here, they have added a few more techniques along with the techniques mentioned in [10] and also proposed another metrics $APFD_c$ for measuring the performance of the mentioned techniques.

Kavita et al. [12] have proposed test prioritization techniques based on two factors that are based on severity of faults. The results so obtained are promising. Tyagi and Malhotra [13] have considered three factors for prioritization of test cases. The result obtained is better than that obtained by Kavita et al. Nayak et al. [9] has considered two factors, namely: fault rate and severity value of fault. The result so obtained is better than [12]. Muthusamy et al. [14] proposed a new metric-based priority based on four factors.

Farooq and Nadeem [15] have proposed a method to prioritize the test cases based on mutation testing. Here the researcher has taken the test cases that have the ability to kill the mutants. The most efficient test cases have been assigned the higher priority. Hettiarachchi and Do [16] investigated to know the effect of fuzzy expert system and the test case prioritization effectiveness. They have evaluated the result using three applications and found that the TCP technique along with requirement and risks improves the fault detection capability.

Various meta-heuristic algorithms can be used to solve TCP issues. Singh et al. [17] proposed a technique using ant colony optimization algorithm to prioritize the test case. Khatibsyarbini et al. [18] proposed firefly algorithm to solve the time-constrained TCP problem. Li et al. [19] have solved the sequencing problem in regression testing by using five meta-heuristic algorithms. They found that genetic algorithm gives a better result than rest, though greedy algorithm performs better in multimodal landscape.

3 Problem Statement

In this section, the problem statements are mentioned that are used in the manuscript.

Definition 1 *Test Case Prioritization Problem*

The TCP problem definition [10] can be elaborated as follows:

Given: T, a test suite, PeT, the permutation set of T and f is the function from PeT → the real numbers, ReN.

Problem definition: To discover T' Є PeT such that for all T", T" Є PeT and T" ≠ T' [f(T') ≥ f (T")].

PeT → possible set of sorting ordering of T and f → objective function (when put on any arrangement returns a particular value).

Definition 2 *Test Case Selection Problem*

The RTS problem is defined in the literature [6] as follows:

Given: The program, Pg, the modified version of Pg, Pg', and a test suite T.

Problem definition: Search a subset of T, T', with which to test Pg'.

4 Proposed Work

The proposed method prioritizes the tests with the goal to detect the faults present in the program as soon as possible following some testing criteria. The higher priority test cases are scheduled to execute earlier than the lower ones because the higher ones detect maximum faults in less time.

A. Proposed Method

The proposed method is given in an algorithmic form below:

Algorithm: Proposed TCP Algorithm

Input: T → test suite and FM → Fault Matrix
Output: T' → Ordered test suite
Begin

1. Initially T' → empty
2. In the fault matrix, the test case detecting fault(s) is considered as 1's and the faults that are undetected are considered as 0's
3. Choose the rows in the fault matrix that contain maximum number of 1's.
4. If (TCi == no. of 1)
5. Select any test case randomly.
6. Compare using XOR operation, the selected rows with all rows.
7. Select the rows that give maximum number of 1's.
8. Implement OR operation between them // will show how many faults are exposed.
9. Update the T' and F subsets. // Test case selection
10. Compare the results with remaining test case.
11. Repeat the steps 4–10, till the resultant or operation gives all 1's // All 1's shows all the faults are exposed.
12. Set T' → T

    End

B. Evaluation of Test Suite Effectiveness

The performance is computed by using a software metrics, average percentage of faults detected (APFD). Higher the value, quicker fault detection. Mathematical representation of the metric is as follows:

$$\text{APFD} = 1 - \frac{(\text{TF1} + \text{TF2} + \text{TF3} + \cdots + \text{TF}m)}{n * m} + \frac{1}{2 * n} \tag{1}$$

where, $\text{TF}_i \rightarrow$ exact location of the first test in T that discover fault, f
$m \rightarrow$ number of faults contained in SUT that is exposed under T.
$n \rightarrow$ total number of test cases in the T.

5 Experiment and Discussion

The test suite has been taken from the work of Nayak et al. [9]. The fault matrix consists of test case and the corresponding faults that can be detected. The information about the total number of faults each test case exposes is also mentioned. It is given in Table 1.

In the fault matrix given in Table 1, the test case detecting fault(s) is considered as 1's and the faults that are undetected are considered as 0's in Table 2.

Following the steps in the algorithm, we obtained the subset that can detect all the faults are as follows: {TC8, TC6, TC2, TC7}. After this selection step when all 1's are detected, this becomes the stopping criteria of the algorithm. Then the test cases are prioritized in the sequence given below:

TC8 > TC6 > TC2 > TC7 > TC1 > TC3 > TC4 > TC5 > TC9 > TC10.

Table 1 Test cases (TCs) as $TC_1 \ldots TC_{10}$ and $FaL_1 \ldots FaL_{10}$ as faults, 'V' illustrate the faults located by the T

TCs/Faults	TC1	TC2	TC3	TC4	TC5	TC6	TC7	TC8	TC9	TC10
FaL1								V	V	
FaL2		V	V		V					
FaL3				V		V				V
FaL4		V	V							
FaL5								V		
FaL6								V	V	
FaL7				V	V		V			
FaL8	V					V				
FaL9				V		V				V
FaL10	V							V		
Total FaL Located	2	2	2	3	2	3	1	4	2	2

Table 2. 1's in fault detection place and 0's in non-detection place

TCs/Faults	TC1	TC2	TC3	TC4	TC5	TC6	TC7	TC8	TC9	TC10
FaL1	0	0	0	0	0	0	0	1	1	0
FaL2	0	1	1	0	1	0	0	0	0	0
FaL3	0	0	0	1	0	1	0	0	0	1
FaL4	0	1	1	0	0	0	0	0	0	0
FaL5	0	0	0	0	0	0	0	1	0	0
FaL6	0	0	0	0	0	0	0	1	1	0
FaL7	0	0	0	1	1	0	1	0	0	0
FaL8	1	0	0	0	0	1	0	0	0	0
FaL9	0	0	0	1	0	1	0	0	0	1
FaL10	1	0	0	0	0	0	0	1	0	0

B. Discussion

The analysis of the proposed approach and the non-prioritized approach is done in this portion of the paper. The order of each of the technique is given below in Table 3.

The APFD score of the three techniques is found by the area under the curve if we plot T fraction in the X-axis and fault detected percentage in the y-axis, the result obtained is in Table 4 as follows (Fig. 1).

Since the APFD % of the proposed method is better than the two, so the method is robust enough to detect faults sooner than the two.

Table 3 Ordering of non-prioritized and prioritized techniques

Non-prioritized order	Nayak et al. [9]	Proposed order
TC1	TC8	TC8
TC2	TC4	TC6
TC3	TC9	TC2
TC4	TC6	TC7
TC5	TC2	TC1
TC6	TC1	TC3
TC7	TC5	TC4
TC8	TC10	TC5
TC9	TC3	TC9
TC10	TC7	TC10

Table 4 Techniques with their APFD%

Techniques	APFD %
Non-prioritized order	63
Previous work [9]	81
Proposed work	85

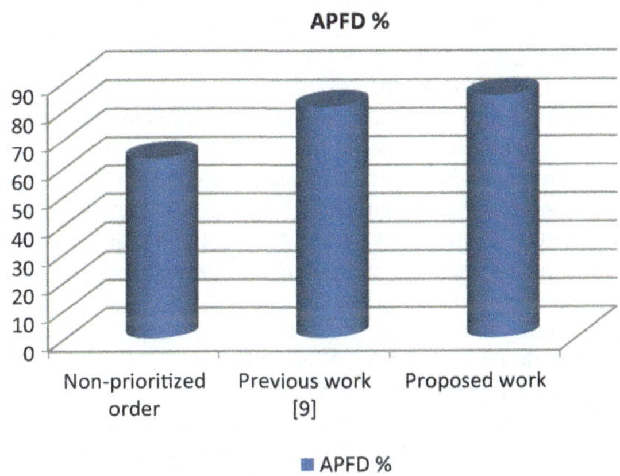

Fig. 1 APFD % of different prioritization techniques

6 Conclusion and Future Work

In this paper, a new prioritization technique is proposed taking OR and XOR gate into consideration. The results are analyzed using APFD metrics and the APFD % is found to be 85% for our proposed method which is better than the existing method. The higher the APFD score, the sooner the faults are detected. The testing resources will be saved and the quality will be improved. Further, the real-world application will be taken for our study purpose to check its effectiveness.

References

1. Craig, R.D., Jaskiel, S.P.: Systematic software testing. Artech House Publishers (2002)
2. Software Engg. Terminology, I.S.G.: IEEE standards collection. IEEE std 610.12-1990 (1990)
3. Wong, W.E., Horgan, J.R., London, S., Mathur, A.P.: Effect of test set minimization on fault detection effectiveness. Softw. Pract. Exp. **28**(4), 347–369 (1998)
4. Wong, W.E., Horgan, J.R., Mathur, A.P., Pasquini, A.: Test set size minimization and fault detection effectiveness: a case study in a space application. In: Proceedings of the 21st Annual International Computer Software and Applications Conference, pp. 522–528 (1997)
5. Rothermel, G., Harrold, M.J.: A safe efficient regression test selection technique. ACM Trans. Softw. Eng. Methodol. **6**(2), 173–210 (1997)
6. Rothermel, G, Harrold, M.J.: Analyzing regression test selection techniques. IEEE Trans. Softw. Eng. **22**(8), 529–551 (1996)
7. Rothermel, G., Untch, R., Chu, C., Harrold, M.: Test case prioritization: an empirical study. In: Proceedings of International Conference Software Maintenance, pp. 179–188 (1999)
8. Wong, W., Horgan, J., London, S., Agarwal, H.: A study of effective regression testing in practice. In: Proceedings of Eighth International Symposium Software Reliability Engineering, pp. 230–238 (1997)
9. Nayak, S., Kumar, C., Tripathi, S.: Effectiveness of prioritization of test cases based on faults. In: 3rd International Conference RAIT (2016)
10. Rothermel, G., Untch, R., Chu, C., Harrold, M.: Prioritizing test cases for regression testing. IEEE Trans. Softw. Eng. **27**(10), 929–948 (2001)
11. Elbaum, S., Malishevsky, A., Rothermel, G.: Test case prioritization: a family of empirical studies. IEEE Trans. Softw. Eng. **28**(2), 159–182 (2002)
12. Kavitha, R., Sureshkumar, N.: Test case prioritization for regression testing based on severity of fault. Int. J. Comput. Sci. Eng. (IJCSE) **2**(5), 1462–1466 (2010)
13. Tyagi, M., Malhotra, S.: An approach for test case prioritization based on three factors. Int. J. IT CS **4**, 79–86 (2015)
14. Muthusamy, T., Seetharaman, K.: Efficiency of test case prioritization technique based on practical priority factor. Int. J. Soft Comput. **10**(2), 183–188 (2015)
15. Farooq, F., Nadeem, A.: A fault based approach to test case prioritization. In: International Conference on Frontiers of Information Technology (FIT). IEEE (2018)
16. Hettiarachchi., C., Do., H.: A systematic requirements and risks-based test case prioritization using a fuzzy expert system. In: 19th International Conf. Software Quality, Reliability and Security. IEEE (2019)
17. Singh, Y., Kaur, A., Suri, B.: Test case prioritization using ant colony optimization. ACM SIGSOFT Softw. Eng. Notes **35**(4), 1–7 (2010)
18. Khatibsyarbini, M., Isa, M., Jawawi, D., Hamed, H., Suffian, M.: Test case prioritization using firefly algorithm for software testing. IEEE Access **7**, 132360–132373 (2019)
19. Li, Z., Harman, M., Hierons: Search algorithms for regression test case prioritization. IEEE Trans. Softw. Eng. **33**(4), 225–237 (2007)

Optimal Geospatial Query Placement in Cloud

Jaydeep Das, Sourav Kanti Addya, Soumya K. Ghosh, and Rajkumar Buyya

Abstract Computing resources requirements are increasing with the massive generation of geospatial queries. These queries extract information from a large volume of spatial data. Placement of geospatial queries in virtual machines with minimum resource and energy wastage is a big challenge. Getting query results from mobile locations within a specific time duration is also a major concern. In this work, a bi-objective optimization problem has been formulated to minimize the energy consumption of cloud servers and service processing time. To solve the problem, a crow search based bio-inspired heuristic has been proposed. The proposed algorithm has been compared with traditional First Fit and Best Fit algorithms through simulation, and the obtained results are significantly better than the traditional techniques.

Keywords Geospatial query · Cloud computing · Energy efficiency · Query placement · Optimization

---

J. Das (✉)
Advanced Technology Development Centre, Indian Institute of Technology Kharagpur, Kharagpur, West Bengal 721302, India
e-mail: jaydeep@iitkgp.ac.in

S. K. Addya · S. K. Ghosh
Department of Computer Science and Engineering, Indian Institute of Technology Kharagpur, Kharagpur, West Bengal 721302, India
e-mail: souravkaddya@cse.iitkgp.ac.in

S. K. Ghosh
e-mail: skg@cse.iitkgp.ac.in

R. Buyya
School of Computing and Information Systems, University of Melbourne, Melbourne, VIC 3010, Australia
e-mail: rbuyya@unimelb.edu.au

D. Mishra et al. (eds.), *Intelligent and Cloud Computing*,
Smart Innovation, Systems and Technologies 194,
https://doi.org/10.1007/978-981-15-5971-6_37

1 Introduction

Geospatial Query (GQ) processing is essential in the applications of geographical information systems (GIS), multimedia information systems (MIS), location-based services (LBS), etc. In GQ, the *location* is an essential attribute. To access GIS data, a user can generate GQ from their mobile devices either from a static location or mobile location. In the best case, if the query resolved in a nearby cloud data center (DC), which leads to less communication cost and propagation delay. These metrics will not change for GQs, which are accessed from a static location. Whereas GQs are coming from mobile location users, metrics will change very frequently. GQs extract information from a large volume of spatial data [1]. Therefore, the distribution of GQs in dynamic cloud DCs is challenging [2]. Cloud computing offers a shared pool of huge resources like CPU core, RAM, storage, bandwidth, etc. Virtualization made physical resources available to the user by offering an illusion of a dedicated system. This technique creates multiple virtual machines (VMs) with different configurations of CPU cores, RAM, storage, bandwidth, etc. As per users requirement, the cloud service provider (CSP) offers them required VMs.

GQ load distribution means shifting of GQs to the heavy load VMs to low load VMs. Make a stable cloud environment with equal GQ distribution. While shifting the GQ, keep in mind that the query should be resolved within its timespan and the energy consumption will be minimum. In the literature, very fewer number of works published in GQ placement in the cloud. Lee et al. [3] proposed a spatial indexing technique that works over HBase for big geospatial data. They considered containedIn, intersects, within Distance types of GQ and used GeoLife and SFTaxi Trajectories data points for experiments. In [4], Bai et al. proposed an indexing technique over HBase distributed database using kNN and window query processing algorithms to process huge data objects. A learning technique to resolve GQ in the cloud is discussed in [5]. Akdogan et al. [6] proposed the Voronoi diagram for the efficient processing of a varied range of GQs. The authors have deployed MapReduce-based approach to resolve the reverse nearest neighbor (RNN), maximum reverse nearest neighbor (MaxRNN), and k-nearest neighbor (kNN) queries. GQ resolution on the cloud using geospatial service chains has been discussed in [7].

On the other hand, many algorithms are proposed for load balancing in the cloud. Load balancing is a key factor for GQ placement in cloud servers. Kumar and Sharma [8] proposed an algorithm that is based on a proactive prediction based approach. It predicts future loads based on the past load history and distributed the loads from heavy load VMs to underloaded VMs. Calheiros et al. [9] proposed a workload prediction using the ARIMA model, which helps to get feedback from previous workloads and update the model accordingly. It helps to assign VMs dynamically and maintain the user QoS by reducing the rejection rate and response time. Garg et al. [10] proposed a mechanism to dynamic cloud resource allocation by assigning maximum workload to a DC as per SLA. To maximize the utilization of the cloud DC, they integrated noninteractive and transactional applications. It also reduces the penalty due to less SLA violation as a maximum number of the applications are run in DCs. [11] proposed a framework which schedules tasks decreasing overall energy of cloud data center. Geospatial query resolution is done using cloudlets in [12], and using fog computing in [13].

From the above literature survey, it is observed that most of the works are done separately for GQ processing and load distribution in the cloud environment. These two scenarios have been merged. In this work, GQs are not only resolved in the user-specified time span but also minimize the overall power consumption of the cloud environment due to an efficient GQ load balancing algorithm among the VMs. In this paper, it is assumed that the scheduling of GQs is highly heterogeneous. This work focus on the minimization of GQ resolving time span and energy consumption. The key contributions of this paper are:

- Optimal GQ load distribution with minimal timespan and energy consumption.
- Minimize the service time of the GQ.

The rest of the paper is organized as follows. Categories of the geospatial queries and modeling of processing those queries in the cloud platform are explained in Sect. 2. In Sect. 3, the problem-solution approach using crow search heuristic has been discussed. Performance analysis of the proposed scheme is presented in Sect. 4. Conclusion and future scope of the work are drawn in the last section.

2 System Model

Users generate the GQs through the user interface of web-enabled electronic gadgets, i.e., mobile, laptop, computer, etc. It will be submitted to the Cloud broker. The cloud broker will map the GQ with the existing query types. It is also needed to identify the requirements of geospatial data and geospatial services. There are three types of servers, i.e., Processing Server, Data Server, and Map Server available in the DC. The data server keeps geospatial data. The processing server processes the geospatial data. The map server helps to generate the map on processed geospatial data. After identification, the broker assigns a VM to execute GQ. When the execution is over, it generates the GQ results (GQR), which projected to the user interface. A pictorial view of the system model has shown in Fig. 1.

Assignment of appropriate VM to a GQ is a key feature to distribute the GQ load. Before the assignment of GQ, it is needed to know GQ types, which helps to select VM's specifications during VM to server mapping. The types of GQs [1] are mentioned below.

- **Filter Query**—This type of query[14] filters a particular geometry which presents in another geometry.
- **Within Distance**—It measures whether one geometry or object is present within a particular euclidean distance of another geometry or not.
- **Nearest Neighbor (NN)**—It measures whether geometries is the nearest neighbor of a particular geometry or not.

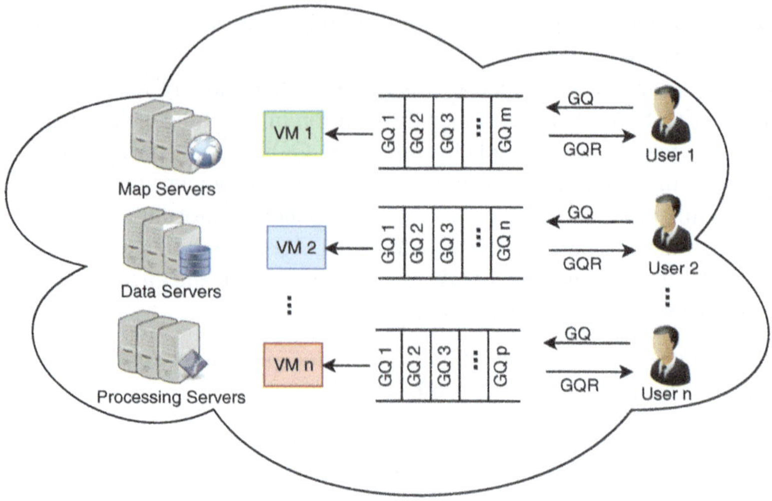

Fig. 1 Geospatial query processing model in cloud

- **Geospatial Join Query**—It compares one layer of geometry with the layers of the other geometries. Geospatial index type (that is, R-tree or Quadtree) must be the same on the geometry column of all the tables involved in the join operation.

2.1 *Service Configuration*

Cloud DC receives GQs from end-users. To assign GQs in VMs, it has to meet the resource requirements such as CPU, memory, etc. The VM assigns to the GQ if it meets the resource requirements of GQ. The selection of VMs with a number of GQs will be based on its capacity constraints. If requested VM specification is not available, then CSP may assign a higher configured VM. A cloud DC is configured with a large number of servers. VMs are also assigned to the nearby cloud DC of GQ. The nearby cloud DC should fulfill the resource requirement of GQ. Therefore, the optimal mapping of VM to the cloud server is important for maintaining the QoS of GQ services. Scheduling of the GQs is required to place into VMs. After the completion of GQ scheduling, an algorithm is needed for optimal selection of VM among all available VMs.

2.2 *VM to Cloud Server Mapping*

A suitable mapping of GQs to VMs is required while optimizing overall GQ processing time and power consumption of the system. A geospatial query set

$GQ = \{GQ_1, GQ_2, \ldots, GQ_p\}$ consists of 'p' number of geospatial queries. A VM set $\{vm_1, vm_2, \ldots, vm_q\}$ consists of 'q' VMs. Each GQ has two dimensional (CPU and memory) resource requests. A crow matrix is generated for the GQ requests set. The main focus of the GQ placement (GQP) algorithm is to the continuous optimization of processing time and power consumption. Next, two parameters are represented mathematically.

2.2.1 Processing Time Calculation:

The service time of the GQ processing can be defined as follows:

$$T_s = T_w + T_p \tag{1}$$

where, T_s is Service Time, T_w is Waiting Time in Queue, and T_p is Processing Time in VM. The queue discipline, m/m/1 [15], is considered . All GQ requests will be in a FIFO manner. The arrival rate of GQs is λ and μ will be the service rate. The steady-state probability of 'p' numbers of GQs can be calculated. Thus, the expected number of GQ request in the queue can be assumed, by which the T_w in the queue can also be computed. Similarly, the calculation of T_s can be done.

Now, to process large number of GQs, it needs more VMs. Increment of the number of VMs consumes more energy. If the number of VMs increases, then the energy consumed will be more. This is an NP-Hard problem. Trade-off between the number of VMs generation and overall energy consumption has been done. The relation between processing time in VM (T_p) and energy consumption

$$T_p \propto E_c \tag{2}$$

2.2.2 Power Consumption Calculation:

The total power consumption can be defined as below [15]

$$pwr_i = (pwr_i^{\max} - pwr_i^{\min}) * utz_i + pwr_i^{\min} \tag{3}$$

$pwr_i^{\max}$ and $pwr_i^{\min}$ are the maximum and minimum average power consumed while maximum and minimum utilization is occurred, respectively. utz_i is utilization of vm_i.

2.3 *Problem Formulation*

In this paper, the GQ placement algorithm is modeled as a bi-objective optimization problem. The objective is to minimize the energy consumption of cloud servers and processing time of GQ. The objective function of the GQP algorithm is mentioned below

$$\text{Minimize} \sum_{c=1}^{n} E_c \quad \text{and} \quad \text{Minimize} \sum_{p=1}^{m} T_p \tag{4}$$

Subject to the three constraints are mentioned below

- ***Assignment constraint***: It assures that the GQs are placed in such VMs where each GQ dimensions are matched with the VM dimensions.
- ***Capacity constraint***: It assures that the total VM requirements of the GQ set should be less than or equal to the total available VMs.
- ***Placement constraint***: It assures to the assignment of a GQ to only one VM which meets the resource requirements along with all dimensions.

3 Crow Search Algorithm for Geospatial Query Placement

To solve the aforementioned bi-objective optimization problem, the crow search algorithm (CSA) [16], has been chosen. CSA is a well-known algorithm that is used to solve many optimization problems. Unlike genetic algorithm (GA), ant colony optimization (ACO), particle swarm optimization (PSO), chaotic ant swarm (CAS),

Algorithm 1: Crow search algorithm for GQ placement (GQP).

```
Input: GQs from different users
Result: Assign GQs in appropriate VM
Evaluate the position of the GQs.
Initialize the memory of the GQs.
while iter < iter_max do
    Generation of a new assignment of GQs in set GQS''.
    for i = 1 to n do
        Randomly select a VM vm_j that gq_i follows.
        Define the awareness probability AP.
        if r_j ≥ AP_j^iter then
            Check the availability of the resources at VM vm_j.
            Update the position of gq_i in an array GQS'.
        else
            Generate a random VM vm_k with resources.
            Update the position of gq_i in the array GQS'.
        end
    end
    if FL < 1 then
        Set GQS'' = GQS'.
    else
        Perform 'x' random shuffles
        Update GQS''.
    end
    if f(GQS'') is better than f(GQS) then
        Set GQS = GQS''.
    end
end
```

CSA also makes use of a population of seekers to explore the search space. In CSA, the number of adjustable parameters is less (flight length (FL), and awareness probability (AP)) compare to the other optimization algorithms. As adjustable parameters are very difficult to manage. GQP is such an optimization problem where CSA can be used to find its optimal solutions. For GQ placement, GQs are considered as crows, and the best VM for GQ placement is equivalent to an optimal food source. A suitable VM selection is needed as all VMs are uniformly eligible for placing the GQs. As crows search for optimal food sources, similarly, the GQs are searching for appropriate VM for processing. The proposed crow search based algorithm for geospatial query placement into VM is described in Algorithm 1.

4 Performance Evaluation

All experiments are performed in CloudSim 4.0 simulator. The considered number of hosts 100 and each host has 16 processing elements. The number of VMs is varying from 100 to 500, and five types of VMs are considered. These are *Micro VM*(1 Core, 1 GB RAM), *Small VM*(1 Core, 2 GB RAM), *Medium VM*(2 Core, 2 GB RAM), *Large VM*(2 Core, 8 GB RAM), and *xLarge VM*(1 Core, 1 GB RAM). The number of cloudlets (here GQs) is considered within the range of 200−1200. The value of random shuffle is 20, and $iter_{\max}$ is 1000 for keeping the less complexity of the algorithm.

GQs are processed on a first-come, first-serve basis. The number of GQs against processing time graph has been displayed in Fig. 2. Also, the comparison has been done among GQP algorithm with existing Best Fit (BF), First Fit (FF), and Random allocation algorithms in the context of the number of used VM with respect to the number of geospatial queries which are depicted in Fig. 3. It has been observed that

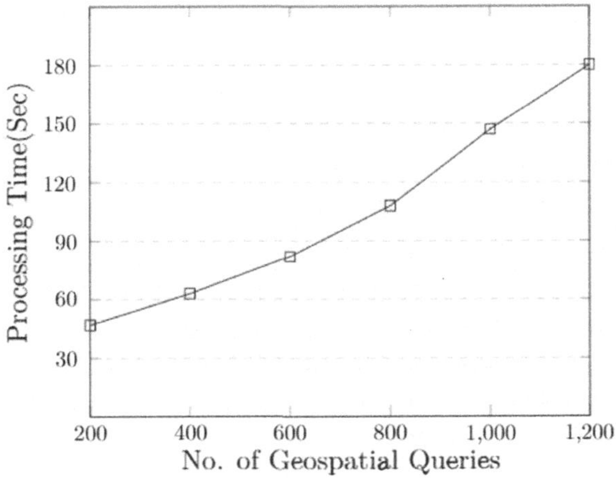

Fig. 2 Processing time versus number of geospatial queries

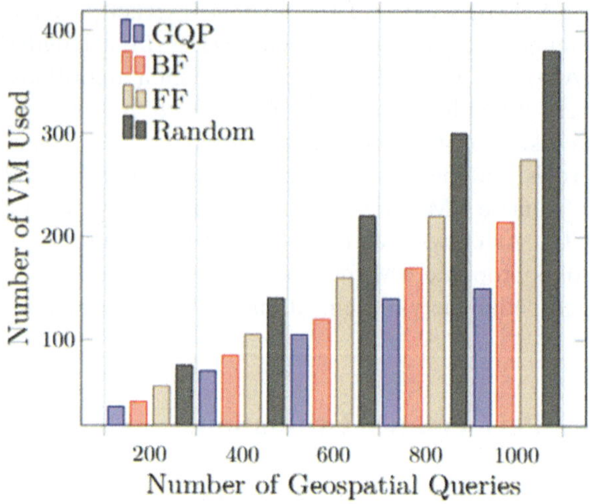

Fig. 3 VMs used by number of GQs for different placement strategies

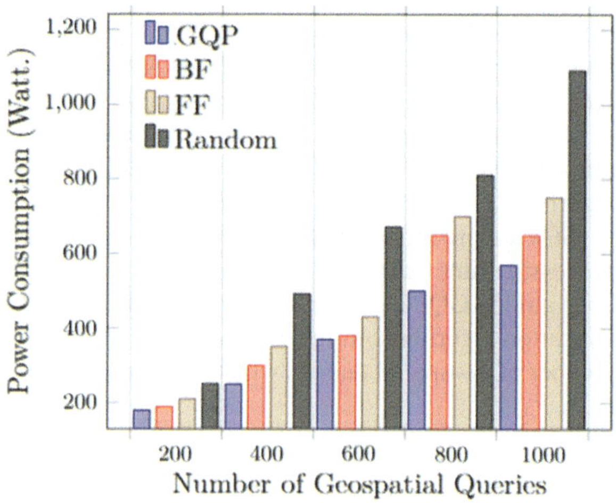

Fig. 4 Variation of power consumption with the number of GQs for different placement strategies

the number of used VMs are increased with the increment of the number of GQs. The number of VMs are used in GQP is lesser than the other three existing algorithms. In FF, first GQ is placed in the first VM. Next GQ will check whether it will fit in the first VM or not. If not, then it moves to the later VM. This moves toward nonoptimal solutions. In the case of FF, the algorithm checks all the VMs capacity and then decides where to move the GQs. This is also moved toward nonoptimal solutions. Figure 4 shows the overall power consumption against the different types of GQs placement strategies. As less number of VMs are used for resolving GQs in the GQP algorithm, this leads to the minimal power consumption in GQP compared to the other three algorithms.

5 Conclusions and Future Work

The processing of GQ from a mobile location in the cloud server is a challenging task. Mainly, GIS data resides in the spatial database in huge volumes. To process a large number of GQ leads to high propagation delay and response time. Also, it causes higher energy consumption in cloud servers. In this work, an optimization problem formulated and solved it using crow search based heuristic. The obtained results are significantly outperforming than traditional FF and BF. An extension of this work, GQ processing in a multi-cloud environment will be done.

References

1. Shekhar, S., Chawla, S.: Spatial Databases: A Tour. Prentice Hall Upper Saddle River, NJ (2003)
2. Yang, C., Huang, Q.: Spatial Cloud Computing: A Practical Approach. CRC Press (2013)
3. Lee, K., Ganti, R.K., Srivatsa, M., Liu, L.: Efficient spatial query processing for big data. In: Proceedings of International Conference on Advances in Geographic Information Systems (ACM SIGSPATIAL), pp. 469–472. ACM (2014)
4. Bai, J.W., Wang, J.Z., Huang, J.L.: Spatial query processing on distributed databases. In: Advances in Intelligent Systems and Applications, vol. 1, pp. 251–260. Springer, Berlin (2013)
5. Das, J., Dasgupta, A., Ghosh, S.K., Buyya, R.: A learning technique for vm allocation to resolve geospatial queries. In: Recent Findings in Intelligent Computing Techniques, vol. 1, pp. 577–584. Springer, Berlin (2019)
6. Akdogan, A., Demiryurek, U., Banaei-Kashani, F., Shahabi, C.: Voronoi-based geospatial query processing with mapreduce. In: Proceedings of International Conference on Cloud Computing Technology and Science (CloudCom), pp. 9–16. IEEE (2010)
7. Das, J., Dasgupta, A., Ghosh, S.K., Buyya, R.: A geospatial orchestration framework on cloud for processing user queries. In: Proceedings of International Conference on Cloud Computing in Emerging Markets (CCEM), pp. 1–8. IEEE (2016)
8. Kumar, M., Sharma, S.: Deadline constrained based dynamic load balancing algorithm with elasticity in cloud environment. Comput. Electr. Eng. **69**, 395–411 (2018)
9. Calheiros, R.N., Masoumi, E., Ranjan, R., Buyya, R.: Workload prediction using arima model and its impact on cloud applications qos. IEEE Trans. Cloud Comput. **3**(4), 449–458 (2015)
10. Garg, S.K., Toosi, A.N., Gopalaiyengar, S.K., Buyya, R.: Sla-based virtual machine management for heterogeneous workloads in a cloud datacenter. J. Netw. Comput. Appl. **45**, 108–120 (2014)
11. Primas, B., Garraghan, P., McKee, D., Summers, J., Xu, J.: A framework and task allocation analysis for infrastructure independent energy-efficient scheduling in cloud data centers. In: Proceedings of International Conference on Cloud Computing Technology and Science (CloudCom), pp. 178–185. IEEE (2017)
12. Das, J., Mukherjee, A., Ghosh, S.K., Buyya, R.: Geo-cloudlet: time and power efficient geospatial query resolution using cloudlet. In: Proceedings of 11th International Conference on Advanced Computing (ICoAC), pp. 180–187. IEEE (2019)
13. Das, J., Mukherjee, A., Ghosh, S.K., Buyya, R.: Spatio-fog: a green and timeliness-oriented fog computing model for geospatial query resolution. Simul. Model. Practice Theory **100**, 102043 (2020)
14. Güting, R.H.: An introduction to spatial database systems. VLDB J. Int. J. Very Large Data Bases **3**(4), 357–399 (1994)

15. Satpathy, A., Addya, S.K., Turuk, A.K., Majhi, B., Sahoo, G.: Crow search based virtual machine placement strategy in cloud data centers with live migration. Comput. Electr. Eng. **69**, 334–350 (2018)
16. Askarzadeh, A.: A novel metaheuristic method for solving constrained engineering optimization problems: crow search algorithm. Comput. Struct. **169**, 1–12 (2016)

Measuring MC/DC Coverage and Boolean Fault Severity of Object-Oriented Programs Using Concolic Testing

Swadhin Kumar Barisal, Pushkar Kishore, Anurag Kumar, Bibhudatta Sahoo, and Durga Prasad Mohapatra

Abstract Fault-Based Testing targets to detect a certain set of faults in a program. There are different faults that may be found in the source code of a given program. This paper considers boolean faults only. From the literature survey, it is found that very few works have been done on targeting MC/DC (Modified Condition / Decision Coverage) score and boolean fault severity together. This motivates to present a novel technique to measure MC/DC Coverage and boolean fault severity of object-oriented programs using concolic testing. This work uses concolic testing for generating test cases for some standard Java programs. This paper proposes an algorithm that uses these test cases to measure test metrics such as MC/DC percentage and fault severity level named FEP (Fault Exposing Potential) of each Java program. FEP is measured for each predicate or boolean expression present in a program and it represents the severity level of boolean faults present in a program. This approach is validated by experimenting on twenty moderate size Java programs.

Keywords Concolic testing · MC/DC · Test cases · Boolean faults · Fault-based testing

---

S. K. Barisal (✉) · P. Kishore · A. Kumar · B. Sahoo · D. P. Mohapatra
National Institute of Technology, Rourkela, Odisha, India
e-mail: swadhinbarisal@gmail.com

P. Kishore
e-mail: 518cs1002@nitrkl.ac.in

A. Kumar
e-mail: 218CS3320@nitrkl.ac.in

B. Sahoo
e-mail: bdsahu@nitrkl.ac.in

D. P. Mohapatra
e-mail: durga@nitrkl.ac.in

S. K. Barisal
Siksha 'O' Anusandhan (Deemed to be) University, Odisha, Bhubaneswar, India

D. Mishra et al. (eds.), *Intelligent and Cloud Computing*,
Smart Innovation, Systems and Technologies 194,
https://doi.org/10.1007/978-981-15-5971-6_38

1 Introduction

Whenever some alterations are done with the conditional expressions of a program, fault-based testing plays a vital role in demonstrating that some faults present in the program are incorrect. Alternate expressions have got numerous classes. For making programs potentially correct, program expressions are replaced with the alternate ones. For doing fault-based testing, test cases help us to differentiate the original program from all its alternatives by identifying mutants. Whenever software is changed, regression testing is used for re-validating the changed software. There are many approaches to improve the steps for doing regression testing. Based on some principles such as coverage score or mutation score, test case prioritization [5] approach is used to find out test cases with the highest priority.

In regression testing, new test cases are added to the existing test cases, which leads to an increase in the number of test cases and ultimately increasing the regression testing cost.

There exist many testing techniques for effectively testing the software. But, concolic testing has become familiar as it tests the input program thoroughly and tries to cover all execution paths. Koushik Sen et al. [9] defined concolic testing as a hybrid testing method that does randomized execution and symbolic execution together. With the flow of execution, it is expected to explore all paths while testing. Constraint solvers are used to select the most probable path whenever applied on the symbolic input values. Input values are also collected with a constraint solver to give high path coverage. The advantageous aspect of this testing is that it covers all possible paths while executing the program and is still cost-effective too.

The remaining portion of this paper is written as follows: Sect. 2 highlights some basic concepts for understanding this work. Section 3 presents the schematic model of the proposed work. Section 4 discusses the implementation work and the result analysis. Section 5 compares proposed work with some existing works. Section 6 contains conclusions of proposed work with some future directions to the work, that readers can extend.

2 Basic Concepts

This section contains some definitions required for understanding the proposed approach.

MC/DC criteria [1, 4, 11], make sure that each condition should affect the decision's outcome independently. Thereby, a thoughtful strategy should be there to select the test case. A predicate with n conditions, there should be at least $n+1$ test cases [8].

In MC/DC testing method, we found a total count of $n+1$ contributing test cases, where n represents the number of clauses in the given predicate. But, whenever we are using MCC (Multiple Condition Coverage) then, 2^n number of test cases are generated. Hence, we choose MC/DC over MCC criteria. While in case of sequential computation of test cases, MC/DC testing is used to replace MCC testing.

$$MC/DC\_\text{Score}(J,\ TS) = \frac{\text{Number\_of\_independent\_conditions}}{\text{Number\_ of\_ Total\_conditions}} \tag{1}$$

2.1 Fault-Based Testing

The faults may potentially help us for evaluating, as well as designing the test suites to measure their effectiveness in code coverage. In functional and structural testing, fault knowledge is commonly used for testing. To do fault-based testing [3], the fault model helps us in identifying potential faults in a program whichever is tested. Also, this model helps us to create or evaluate the test suites' performance based on the hypothetically identified faults.

Basic idea of fault-based testing is that we have to select those test cases which can distinguish between the program to be tested from the alternative one containing hypothetical faults. The program that needs to be tested is modified into hypothetically faulty programs. We can evaluate the thoroughness of a test suite by seeding the faults. Thorough testing is a part of test adequacy criterion. Also for augmenting a test suite, test cases can be selected with fault seeding. Finally, we calculate the number of faults killed in the program to measure the mutation score.

2.2 Fault Classes

The existing testing techniques generate test cases for boolean expressions present in a program [6, 10] . Boolean expressions are nothing but a logical expression that carries logical predicates inside a program. Usually, a boolean predicate P carries n conditions. We know that for n conditions, it is possible to have $2^n - 1$ tests that can be utilized to detect boolean faults and this is denoted as P_0. Then, we aim to distinguish P from the faulty rest of the expressions, that is $P \bigoplus P_0 = TRUE$. In case of MCC criterion, we require 2^n tests but, if n grows to large value then this requires an exponential number of tests. If we cover only $m < 2^n$ tests, then $2^n - m$ combinations are left uncovered. The fault coverage by m tests is computed using Eq. 2, while the rest of the cases remains uncovered. Thus, the tester should focus on these remaining uncovered combinations for testing and FEP is computed as per Eq. 2. In this equation, m represents covered tests. We have considered several boolean faults as listed in Table 1 for this work. This table represents fault names and their meanings.

$$\text{Fault\_Coverage}(n,\ m) = FEP(n,\ m) = 1 - \frac{(2^n - m)}{2^n} \tag{2}$$

Table 1 Types of fault considered for MC/DC testing

Sl No.	Fault symbol	Fault name	Fault meaning	Example
1	F1	Variable negation fault	Variable is negated	$(x_1'+(x_2.x_3)).x_4$
2	F2	Expression negation fault	Expression is negated	$(x_1+(x_2.x_3))'.x_4$
3	F3	Term omission fault	A term is negated	$(x_1.x_4$
4	F4	Variable omission fault	A term is omitted	$(x_2.x_3)).x_4$
5	F5	Variable reference fault	Variable is replaced	$(x_1+(x_2.x_3)).x_3$
6	F6	Operator reference fault	Operator is replaced	$(x_1.(x_2.x_3)).x_4$
7	F7	Clause conjunction fault	A term x_1 replaced by $x_1.x_4$	$(x_1+(x_2.x_3)).x_1.x_4$
8	F8	Clause disjunction fault	A term x_1 replaced by $x_1 + x_4$	$(x_1+(x_2.x_3)).(x_1 + x_4)$
9	F9	Stuck-at-zero fault	A condition is stuch at zero	$(x_2.x_3).x_4$
10	F10	Stuck-at-one fault	A condition is stuch at one	x_4

2.3 *Concolic Testing*

According to Koushik Sen et al. [9], concolic testing is defined as a hybrid testing method. This method uses symbolic execution and randomized execution together. Concolic testing usually finds more number of paths compared to other testing criteria. It uses constraint solvers with symbolic input values to select alternate possible path. The constraint solvers are used to negate the condition under evaluation, then it uses concrete values for executing the conditional statements.

3 Proposed Approach

This section presents the proposed method to measure boolean fault finding rate and MC/DC score of each test suite. Figure 1 shows the block diagram of the proposed approach. The detailed description of the proposed approach is given below

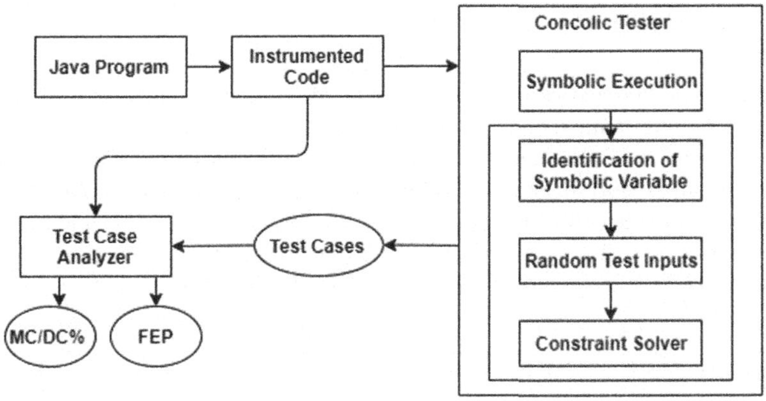

Fig. 1 Modular diagram of the proposed approach

3.1 *Overview*

This approach uses concolic testing for generating test cases. The concolic tester takes an instrumented Java program as input for generating test cases. Then, the proposed algorithm *Test Cases Analyzer (TCA)* is used to measure FEP and MC/DC%.

3.2 *Detailed Description*

Our approach targets to measure MC/DC coverage and fault severity achieved by the test cases obtained from concolic tester. This concolic tester tool is called jCUTE, which is an open source tool and available at Internet.[1] It takes modified Java program and produces various outputs like test cases, branch coverage percentage, number of paths visited, number of errors found, time taken, etc. These programs are taken from the GitHub repository.[2] Initially, the Java code is transformed into an instrumented Java code. This translation targets to rewrite the code as per the concolic executor syntax. The obtained test cases and the source code are given as input to the Test Case Analyzer (TCA) algorithm to compute MC/DC% and FEP value. This algorithm gives various outputs such as number of functions invoked, number of predicates covered, total number of conditions (C), total execution time, number of independent conditions (I) covered, etc. Then, it computes MC/DC score using Eq. 1, that uses I and C values. Then, it computes FEP using Eq. 2. The proposed Test Case Analyzer (TCA) is explained in the next subsection.

[1] https://osl.cs.illinois.edu/software/jcute/.

[2] https://github.com/osl/jcute/tree/master/src/tests.

3.3 TCA Algorithm

The purpose of this algorithm is to compute MC/DC percentage and FEP. It has two major modules named concolic tester and test case analyzer (TCA). Again, each module contains few submodules. Concolic tester has submodules like symbolic execution, random execution, and constraint solver. TCA has submodules like predicate reader and test case reader. The *predicate reader* module is used to find predicates from the input code. The *test case reader* module is used to read the test cases from the test file. Then, it computes MC/DC score using Eq. 1, and FEP values using Eq. 2.

Algorithm 1 Test Case Analyzer

Input: Java Program(J), Test Cases(TC_s)
Output: MC/DC%, FEP
1: P $\leftarrow$ $add(PredicateReader(J))$
2: For each $Predicatep \in P$ do
3: I $\leftarrow$ $I \cup$ Find_Independent_TestPair(p)
4: $N \leftarrow Total\_Tests(p)$
5: End For
6: I $\leftarrow$ Remove_duplicate (I)
7: M $\leftarrow$ Count(I)
8: Compute MC/DC% using Eq. 1
9: Compute Fault_Coverage (FEP) using Eq. 2
10: Exit

Algorithm 1 demonstrates stepwise description to show the working of test case analyzer. Step 1 exports all predicates. Steps 2 to 8 are used to compute MC/DC score. Step 9 is used to compute FEP score. These two results are used to measure the effectiveness of the testing process.

4 Implementation and Results

To implement the proposed approach, we have considered Java programming. This approach has a concolic tester to generate test cases. Another module called Test Case Analyzer (TCA) is used to measure MC/DC coverage percentage and fault exposing potential (FEP).

Implementation: Our approach is implemented using Java Language. The concolic tester takes Java program as input and generates test cases. Concolic testing is a hybrid process, where symbolic execution replaces actual variables with symbolic notations and then fires random inputs to these symbolic variables for execution. After symbolic execution, the constraint solver is used to negate the condition to visit the alternate path, until all paths in the program execution tree are visited. Then, the second module is executed that is Test Cases Analyzer (TCA) for measuring FEP and and MC/DC score of twenty programs and the results are shown in Table 2.

Table 2 Results of Test Case Analyzer

Sl. No	Program name	No. of predi-cates	No. of test cases	Total No. of condi-tions	Idependent conditions	FEP	MC/DC score
1	GradeCal.java	6	5	12	4	0.097	0.33
2	Waterfall.java	2	8	9	5	0.048	0.55
3	Wildlife.java	4	6	12	2	0.0	0.16
4	NS2.java	18	13	29	2	0.073	0.06
5	Regression.java	2	2	12	3	0.036	0.25
6	StringBuffer1.java	5	9	14	6	0.015	0.42
7	SwitchTest2.java	6	14	16	10	0.0	0.62
8	AssertTest.java	0	1	0	0	0	0.0
9	AssertTest1.java	3	6	7	7	0.488	1.0
10	CAssume2.java	5	14	11	10	0.0	0.90
11	Comparision1.java	17	28	43	41	1.367	0.95
12	Condition.java	4	8	9	7	37.5	0.77
13	Condition1 .java	1	5	3	3	0.976	1.0
14	ConditionDemo1.java	6	9	16	16	0.024	1.02
15	Demo1.java	3	1	8	0	0.0	0.0
16	AssetTest2.java	7	8	21	21	0.001	1.0
17	FruitBasket .java	12	9	25	4	0.0	0.16
18	IfSample.java	6	13	14	14	0.085	1.0
19	InsertionSort.java	7	8	14	4	0.024	0.28
20	SortingAlgos.java	26	9	50	25	0.0	0.50

5 Comparison Study

Yu et al. [10] proposed a method to compare MC/DC test criteria and MUMCUT test criteria, and also has done analysis on fault-detection ability. This analysis helps to prioritize test cases.

Hsu et al. [5] proposed "Enhanced Additional Greedy Algorithm" for prioritization purpose. They tried to determine the fault severity level for test cases by experimenting on several programs. But in the proposed approach, we measure MC/DC score and FEP score to establish a correlation between them.

Fraser et al. [2] proposed an algorithm for logic satisfiability and compared with the performance of MUMCUT and minimal-MUMCUT strategies. They claimed that their approach solves boolean expressions satisfiability in less number of test cases compared to MUMCUT. Thus, They have generated the test cases, but not went for prioritizing them. On the other hand, we have generated test cases from boolean expressions and prioritized them using concolic testing.

Kapoor et al. [7] presented different techniques to handle issues such as (i) finding necessary and sufficient conditions for detecting different class of faults, (ii) introduce a hierarchical structure of faults to show the relationship between Operator Reference Fault (ORF), Associative Shift Fault (ASF), and other kinds of faults (iii) Presented a formula for building fault hierarchy. So, all these techniques are going to help readers to generate effective test cases.

Gargantini et al. [3] proposed a method to derive test cases directly from a boolean expression based on possible faults on that expression. Particularly, they have considered boolean expressions that satisfy the Disjunctive Normal Form (DNF) only. Thus, they have targeted to discover a particular set of test cases and their possible faults.

Kaminski et al. [6] proposed three improvements to logic-based testing, that are (i) improvement for Relational Operator Replacement (ROR) in mutation testing, (ii) presented the power of ROR operator for logic base test criteria such as MC/DC, (iii) proved theoretically that minimal-MUMCUT finds more faults than MC/DC.

6 Conclusions

Fault-based testing has been proved to be an effective testing approach. The proposed method is validated for boolean faults using Java programs, because even though some programs achieved 100% MC/DC still they are not fully free from threats as indicated by their corresponding FEP values. Thus, we conclude that achieving full coverage is not enough to claim a system is trust-worthy. Therefore, researchers have to give attention on fault-based testing along with coverage-based testing.

For future work, readers can extend this work to prioritize test cases by taking fault classes and influence value of each fault with respect to the fault hierarchy.

References

1. Elbaum, S., Rothermel, G., Kanduri, S., Malishevsky, A.G.: Selecting a cost-effective test case prioritization technique. Softw. Qual. J. **12**(3), 185–210 (2004)
2. Fraser, G., Gargantini, A.: Generating minimal fault detecting test suites for boolean expressions. In: 2010 Third International Conference on Software Testing, Verification, and Validation Workshops, pp. 37–45. IEEE (2010)
3. Gargantini, A., Fraser, G.: Generating minimal fault detecting test suites for general boolean specifications. Inf. Softw. Technol. **53**(11), 1263–1273 (2011)
4. Holloway, C.M.: Towards understanding the DO-178C/ED-12C assurance case. In: 7th IET International Conference on System Safety, Incorporating the Cyber Security Conference 2012, pp. 1–6. IET (2012)
5. Hsu, Y.C., Peng, K.L., Huang, C.Y.: A study of applying severity-weighted greedy algorithm to software test case prioritization during testing. In: 2014 IEEE International Conference on Industrial Engineering and Engineering Management (IEEM), pp. 1086–1090. IEEE (2014)
6. Kaminski, G., Ammann, P., Offutt, J.: Improving logic-based testing. J. Syst. Softw. **86**(8), 2002–2012 (2013)
7. Kapoor, K., Bowen, J.P.: Test conditions for fault classes in boolean specifications. ACM Trans. Softw. Eng. Methodology (TOSEM) **16**(3), 10 (2007)

8. Khatibsyarbini, M., Isa, M.A., Jawawi, D.N., Tumeng, R.: Test case prioritization approaches in regression testing: a systematic literature review. Inf. Softw. Technol. (2017)
9. Sen, K., Agha, G.: CUTE and jCUTE: Concolic unit testing and explicit path model-checking tools. In: International Conference on Computer Aided Verification, pp. 419–423. Springer, Berlin (2006)
10. Yu, Y.T., Lau, M.F.: A comparison of MC/DC, MUMCUT and several other coverage criteria for logical decisions. J. Syst. Softw. **79**(5), 577–590 (2006)
11. Zamli, K.Z., Al-Sewari, A.A., Hassin, M.H.M.: On test case generation satisfying the MC/DC criterion. Int. J. Adv. Soft Comput. Appl. **5**(3) (2013)

Asynchronous Testing in Web Applications

Sonali Pradhan and Mitrabinda Ray

Abstract Web applications are fast progressive and more interactive to the world-wide use of computers. It is important to understand testing with web applications and the modelling process of web applications. A finite state machine is widely used to conduct state-based testing of web application. However, the state-space explosion is a common problem by using many input choices provided to the users. It has the ability where without any specific order the values can be entered in the web application. This combined to create state explosion issue. To solve the problem, this paper suggests a reduction analysis by reducing the number of transitions with giving constraints on the inputs by compressing Finite State Machine (FSM).

Keywords Web application · FSM · Fsmweb · Clustering · State explosion

1 Introduction

Web application plays an important role in the growth of the software market [1]. The users run the web application interface as clients, send data and get responses by the server via the Internet. It is a combination of recent technologies, different languages and programming models. A web site is a collection of web pages and collection of HTML elements [1]. The pages are linked together to retrieve information according to the user requirements. By the client request, a static web page that is an HTML file having a specific Uniform Resource Locator (URL) address is sent to a server and then a client views the required information. A dynamic web page is a collection of HTML content where without refreshing the whole page, only the required portion

S. Pradhan (✉)
Department of Computer Science and Engineering, C V Raman Global University, Bhubaneswar, India
e-mail: sonalipradhan7@gmail.com

M. Ray
Department of Computer Science and Engineering, Siksha 'O' Anusandhan, Deemed to be University, Bhubaneswar, India
e-mail: mitrabinda@soa.ac.in

D. Mishra et al. (eds.), *Intelligent and Cloud Computing*,
Smart Innovation, Systems and Technologies 194,
https://doi.org/10.1007/978-981-15-5971-6_39

of the page is going to be refreshed to get the updated data which is send to the user. A Web application uses different web browsers and web technology across the web, using HTML pages over the internet. Any system that has huge number of states, create state space explosion issue and the size of the state space grows exponentially in the system.

In our approach, finite state machines are used. Using finite state machines the behavior of web model can be studied in web applications. The heterogeneousness nature of web applications is handled by these models and they are not depending on another in the implementation. To generate black box tests from FSMs [2–4], various methods have been proposed. Modeling a complete web application with FSMs is a complex part due to the number of choices provided to the users in the web pages. Such as selection elements can have many choices and many inputs for text fields. The data also can be entered in any order. These are the main causes, which leads to state space explosion issue. The FSMWeb model is a hierarchical collection of aggregate Finite State Machines by compressing transitions [5]. Bottom level FSMs are derived from individual web page and the logical web pages are the parts of web pages. The full web application is represented as top level FSM. At the lower level FSMs, black box testing is performed with combination of test sequences.

Related work on web application testing using FSMs are discussed in Sect. 2. There is a short summary given in Sect. 3 for the FSMWeb method and aggregate FSM and the transitions covered in our case study. An analysis of using compressed FSMs is discussed in Sect. 4. Section 5 describes the conclusion of our work.

2 Backgrounds

Some work concerning with Finite State Machine (FSM) for generating test cases are discussed here. Using FSM, Huang [6] suggests covering each edge/transition. Howden [7] recommends covering the states with complete trips through the FSM, with no looping, and Pimont and Rault [8] suggest covering edge pairs. A spanning tree is generated from the FSM and then test sequences are generated traversing the tree [2]. They assume both specification FSM and design FSM. This type of association detects certain types of design faults. Fujiwara et al. [3] referred to Chow's approach. Taking the term W-method, Wp method is developed known as partial W-method. Then Chow's method is considered to be better instead of Pimont and Rault's method.

To test object oriented programs [9, 10] and designs [11, 12], FSMs have also been used. FSMs are derived from the source using symbolic execution [9, 10]. Turner and Robson [12] extract FSMs from the design of classes. Offutt and Abdurazik [11] derive tests from state chart diagrams [13]. UML state chart is an advance approach of finite state machines. Offutt et al. [4, 14] developed an FSM model, they defined several testing criteria on the transitions such as Modified Condition or Decision Coverage named as MCDC coverage criteria.

Conallen [15] extends UML which models the web application architecture. In the extension, the web pages of a web site are modeled using UML class diagram. Both the client and server pages are separated by each class. By the diagram respective different behavior of the page is captured. Manola [16] creates an XMLbased Web Object Mode using OMG's Object Management Architecture [13]. Li et al. [17] combine and extend the two research [15, 16] discussed above which create a modeling method using UML diagrams that express the web application design. They captured and tested the behavior of software using FSM models.

Kung et al. [10] propose that using Object-Oriented program the test cases can be automatically generated for web applications. This type of testing is black-box in nature. But concerned with scalability, it is not validated. Using decision tables [19], Di Lucca et al. [18] have defined a testing strategy which is not validated for web applications. A theoretical approach white box strategy is suggested where slicing of the programs can be supported in regression testing [20]. Large web applications cannot be scaled by this approach. Kung [21] presents an agent-based framework. This black-box technique automatically generates test cases using simple AI methodologies which is also not validated or implemented. Wu and Offutt [22, 23] propose a web application modeling technique known as atomic sectioned web application. That is based on regular expressions using dynamic web applications. The big web application can be scaled by this white-box approach. It has been preliminarily validated. The atomic sections and FSMs give some difference. In atomic sections the structure of the web applications is discussed where as FSMs model [24] describes the behavior of the web model.

3 Conversion of Web Application to FSM

Testing web applications using FSMWeb method is introduced in paper [2]. In phase 1, working of web model is divided into four stages. (1) First the web application is partitioned into clusters; (2) Then logical web pages are defined; (3) For each cluster FSMs are built; (4) The entire web application is built using FSM. In phase 2 test cases are generated from the model explained by Phase 1.

Cluster is the collections of web pages and software modules. The user level functions are implemented by the logical web pages. In step one; the web application is partitioned into clusters. That is known as the highest level of abstraction. The lower level abstraction implements user level function. A single considerable function is associated with individual cluster with individual web pages. Individual web pages are associated with software modules at the lowest level of abstraction. The web pages are modeled as multiple Logical Web Pages (LWP) which facilitates testing of the modules. A LWP accepts data from the user through an form. The form is associated with HTML elements. Logical web pages are completely described as sets of inputs and their actions. A LWP inputs are atomic in nature. This means that data entered into a text field is considered to be one specific user input. There is no

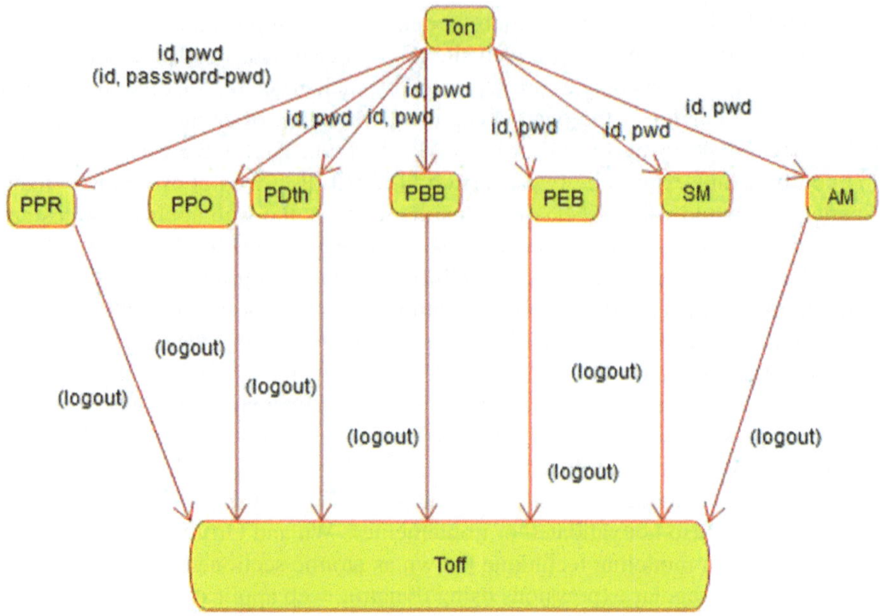

Fig. 1 Top-level aggregate FSM for APB

specific order to enter the inputs in the form according to the requirement given in the web pages.

Lower level FSMs are combined together to create Aggregate FSMs (AFSMs). Each node in an Aggregate FSM called as AFSM. AFSM represents a lower level FSM. The transitions represent information in between lower levels FSMs. During phase 2, test sequences are generated using FSMWeb method. A case study is considered here to depict the FSMWeb method. This web application is Airtel Payment Bank (APB) shown in Fig. 1. The application manages data associated with Payment prepaid (PPR), Payment Postpaid (PPO), Payment DTH (PDth), Payment Broadband (PBB), Payment Electricity Bill (PEB), Send Money (SM) and Add Money (AM). The top-level aggregate FSM for APB is shown in Fig. 1. It represents a three layered FSM. The top-level abstraction implements functions that are identified by the user. Clusters are individual web pages associated with modules. Each module performs a single function. APB services are partitioned with different services according to the requirement and that is available for each user type in the application. The login and logout pages are globally available to the user. The login and logout pages are T_{on} as *Turn on* and T_{off} as *Turn off* states, respectively. At the lower level FSM, the defined states are PPR, PPO, PDth, PBB, PEB, SM and AM.

To generate test cases, the FSMs and aggregate FSMs are considered. Association of individual FSM creates aggregate FSMs. In the first step, test paths are generated for each FSM based on various coverage criteria. The transitions cover set for the top-level aggregate FSM in APB is shown in Fig. 2. They test cases are given below

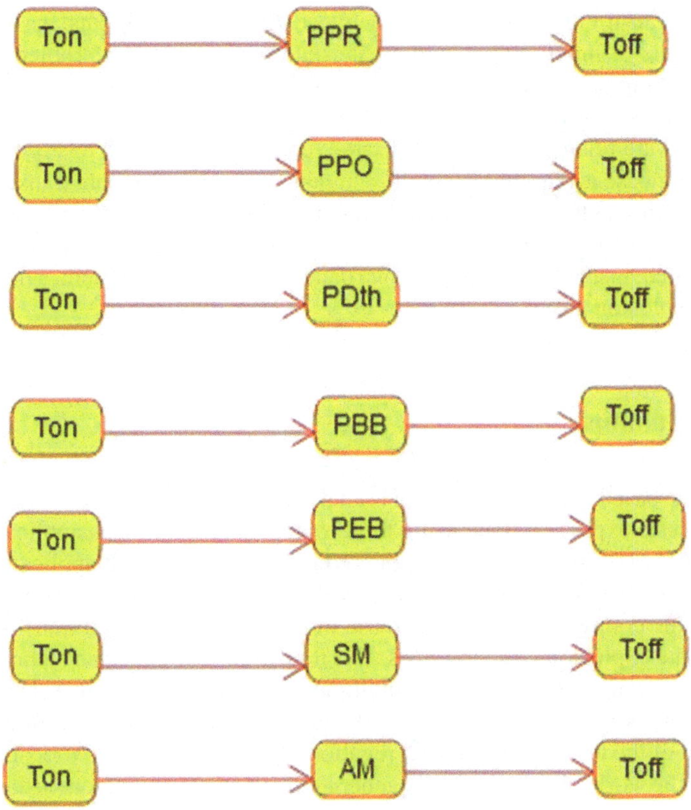

Fig. 2 Transitions cover for APB

for Fig. 2. The test suite is the collection of test cases. In Fig. 2 we find seven test cases and combination of the test cases make a test suite.

4 Reduction Analysis

The focus here is to reduce state explosion problem [25] and on the number of transitions and states using reduction analysis. Here we discuss the savings gained by FSM web modeling and compression techniques. Each link is traversed and each form is exercised until a node is reached either already visited or newly visited. In the web page, HTML elements of the form are extracted to get the inputs in the web application. If a web page has more than one form then each form is treated as its own logical web page.

The inputs chosen from a database and by clustering, we can actually reduce the total cost of the generating paths. The number of tests depends on the input values and number of transitions considered by each test. Adding more inputs, associate with large numbers of input options will create state space explosion problem. A single form related inputs, becoming single transition by the compression technique. The actual savings occur in that case. Usually N input variables find N database access. If the inputs are related in a form and same record is maintained to store them then only one database is required to get those inputs. It would save more transitions to be generated. It is a better way to find individual values for unrelated inputs. For related inputs the best example is login to user name and password. As these two inputs are clearly related, instead of two, find to one database access is sufficient. Thus, reduces the state explosion issue.

5 Conclusions

A technique based on web application is analyzed in this paper which helps to reduce the state space explosion issue using finite state machines. A web application has a greater number of input fields to be entered by the users. A web application allows number of inputs choices and without any specific order for the given values to be entered. This combines to create increase in number of states and transitions. That leads to state explosion issue in FSMs. The transitions compression technique helps us to reduce unnecessary access of multiple database lookup which decreases the generation of extra transitions. Testing effectiveness depends upon the inputs chosen, generation of the testing path and aggregation of the paths. Future work suggests for extending the work with dynamically generated web pages and increases the potential to handle it.

References

1. Offutt, J.: Quality attributes of web software applications. IEEE Softw.: Spec. Issue Softw. Eng. Internet Softw. **19**(2), 25–32 (2002)
2. Chow, T.: Testing software design modeled by finite-state machines. IEEE Trans. Softw. Eng. SE-4 (3), 178–187 (1978)
3. Fujiwara, S., Bochmann, G., Khendek, F., Amalou, M., Ghedasmi, A.: Test selection based on finite state models. IEEE Trans. Softw. Eng. **17**(6), 591–603 (1991)
4. Offutt, J., Liu, S., Abdurazik, A., Ammann, P.: Generating test data from state based specifications. J. Softw. Test. Verification Reliab. **13**(1), 25–53 (2003)
5. Andrews, A.A., Offutt, J., Alexander, R.T.: Testing web applications by modeling with FSMs. Softw. Syst. Model. **4**(1), 326–345 (2004)
6. Huang, J.C.: An approach to program testing. ACM Comput. Surv. **7**(3), 113–128 (1975)
7. Howden, W.E.: Methodology for the generation of program test data. IEEE Trans. Comput. **24**(5), 554–560 (1975)
8. Pimont, S., Rault, J.C.: A software reliability assessment based on a structural and behavioral analysis of programs. In: Proceedings of the 2nd International Conference on Software Engineering (ICSE 1976), San Francisco, CA, USA, pp. 486–491 (1976)

9. Gao, J.Z., Kung, D., Hsia, P., Toyoshima, Y., Chen, C.: Object state testing for object-oriented programs. In: Proceedings of the 19th Annual International Computer Software and Applications Conference (COMPSAC'95), Dallas, TX, USA, pp. 232–238 (1995)
10. Kung, D., Suchak, N., Gao, J., Hsia, P., Toyoshima, Y., Chen, C.: On object state testing. In: Proceedings of the 18th Annual International Computer Software and Applications Conference (COMPSAC'94), Los Alamitos, CA, USA, pp. 222–227 (1994)
11. Offutt, J., Abdurazik, A.: Generating tests from UML specifications. In: Proceedings of the Second International Conference on the Unified Modeling Language (UML'99), Fort Collins, CO, USA, pp. 416–429 (1999)
12. Turner, C.D., Robson, D.J.: The state-based testing of object-oriented programs. In: Proceedings of the Conference on Software Maintenance (ICSM 1993), Montreal, Quebec, Canada, pp. 302–310 (1993)
13. The Object Management Group Unified Modeling Language Specification, vol. 1.3 (1999)
14. Offutt, J., Xiong, Y., Liu, S.: Criteria for generating specification-based tests. In: Proceedings of the Fifth International Conference on Engineering of Complex Computer Systems (ICECCS'99), Las Vegas, NV, USA, pp. 119–131 (1999)
15. Conallen, J.: Modeling web application architectures with UML. Commun. ACM **42**(10), 63–70 (1999)
16. Manola, F.: Technologies for a web object model. Internet Comput. 60–68 (1999)
17. Li, J., Chen, J., Chen, P.: Modeling web application architecture with UML. In: Proceedings of the 36th International Conference on Technology of Object-Oriented Languages and Systems (TOOLS-Asia'00), Xi'an, China, pp. 265–274 (2000)
18. Di Lucca, G.A., Fasolino, A.R., Faralli, F., De Carlini, U.: Testing web applications. In: Proceedings of the 18th International Conference on Software Maintenance (ICSM), Montreal, Canada, pp. 313–319 (2002)
19. Binder, R.V.: Testing Object-Oriented Systems: Models, Patterns, and Tools. Addison-Wesley (2000)
20. Xu, L., Xu, B., Chen, Z., Jiang, J., Chen, H.: Regression testing for web applications based on slicing. In: Proceedings of the 27th Annual International Computer Software and Applications Conference (COMPSAC'03), p. 652–656 (2003)
21. Kung, D.: An agent-based framework for testing web applications. In: Proceedings of the 28th Annual International Computer Software and Applications Conference (COMPSAC'04), vol. 2, Hong Kong, China, pp. 174–177 (2004)
22. Wu, Y., Offutt, J.: Modeling and testing web-based applications, Technical report ISE-TR-02–08, Department of Information and Software Engineering, George Mason University, Fairfax, VA, July 2002. https://www.cs.gmu.edu/~tr_admin/2002.html
23. Wu, Y., Offutt, J.: Modeling and testing of dynamic aspects of web applications, under minor revision (2009)
24. Pradhan, S., Ray, M., Patnaik, S.: Coverage criteria for state-based testing: a systematic review. Int. J. Inf. Technol. Project Manag. (IJITPM), **10**(1), 1–20 (2019)
25. Pradhan, S., Ray, M., Patnaik, S.: Clustering of web application and testing of asynchronous communication. Int. J. Ambient. Comput. Intell. (IJACI) **10**(3), 33–59 (2019)

Architecture Based Reliability Analysis

Sampa ChauPattnaik and Mitrabinda Ray

Abstract The rapid growth of the software products implies the reuse of software in the software development process. In component-based system (CBS), the architecture of the software is analyzed for reliability analysis. Reliability estimation is based on the interactions between the components, individual component's average execution time, reliability, transition probabilities, and transition reliability between any two components. A path-based approach for reliability assessment in architectural level is proposed.

Keywords Reliability estimation · Component dependency graph · State-based model · Path-based model · Operational profile

1 Introduction

Software means execution of a program. For this, some input values are given and output values for the given input values occur by executing some statements or instructions given in a program. A failure occurs if the actual output value deviates from the expected output. The defined failure is different from application to application. Software reliability is defined as the failure-free probability of software within a specified period of time under specified conditions. Some terms associated with software reliability are: fault, failure, etc. A fault means logic or instruction is not correct. The incorrect logic execution will cause a failure. Hence, faults and failures are correlated.

S. ChauPattnaik (✉) · M. Ray
Department of Computer Science and Engineering, Siksha 'O' Anusandan (Deemed to be) University, Bhubaneswar, India
e-mail: sampachaupattnaik.soajan17@gmail.com

M. Ray
e-mail: mitrabindaray@soa.ac.in

D. Mishra et al. (eds.), *Intelligent and Cloud Computing*,
Smart Innovation, Systems and Technologies 194,
https://doi.org/10.1007/978-981-15-5971-6_40

Software Development Life Cycle (SDLC) consists of various phases such as requirement phase, design phase, coding phase, testing phase, etc. In the design phase or early phase of the software development life cycle, no failure data are available. A predictive model is needed to estimate the reliability of the software [1, 2]. The researchers developed several reliability growth models [3–6] used in the testing phase, where reliability assessment is based on the analysis of failure data. Software reliability prediction during the design phase is significant for the developer to find the flaws in software design. Due to the lack of information about the operational profile [7] and failure information of the components, at the design time, the reliability prediction depends upon the software architecture. Software reliability assessment in the design phase is based on the architecture of the software. The software reliability analysis at the architectural level focuses on how architecture meets its quality requirements. For this, it requires sequence diagrams that show the behavioral structure of the system.

The outline of this paper is represented as follows: Sect. 2 describes the software reliability analysis in architectural level; Sect. 3 describes the literature survey on both state-based models and path-based models. In Sect. 4, we proposed method for path-based reliability estimation for the ATM case study in architectural level and Sect. 5 presents the conclusion.

.

2 Background Study

In this section, we give an overview of the software reliability analysis at the architectural level. The software reliability estimation is either done as the design level or at the code level. Software Reliability Analysis occurs in the following way: (i) Black Box Reliability Analysis: The software's internal information are not considered, (ii) Software Metric Based Reliability Analysis: It is based on the static analysis like lines of code, number of statements, complexity, etc., of the given software and its development process and conditions. (iii) Architectural-based Reliability analysis. Architectural-based reliability analysis is also called component-based reliability analysis [8]. Architectural-based analysis is based on three ways. These are. State-Based model, Path-Based model, and Additive Model (Fig. 1).

State-based models: This types of models [8] estimate software reliability analytically. A Markov property is assumed the transfer of control between components, that is, the software architecture represented with a discrete-time Markov chain (DTMC), continuous-time Markov chain (CTMC), or semi-Markov process (SMP). Several models are explained in the literature. Those models are Littlewood Model [4], Cheung Model [3], Lapri Model [9], Kubat Model [5],and Gokhale Model [10].

Path-based Model: For estimation of the application reliability this model considers the possible execution paths. Different paths are obtained by the testing sequence of components either experimentally or algorithmically. Each path reliability is calculated by the product of reliabilities of each component along that path

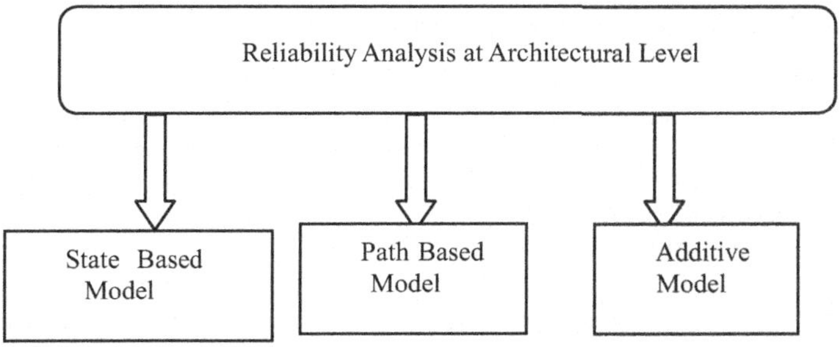

Fig. 1 Different architectural reliability analysis models

and the system reliability is computed by the average of path reliabilities over all paths. Different path-based models are described in the past 28 years. The software architecture is represented as Component Dependency Graph (CDG), Component Call Graph, or Markov Structure. Different models are Shooman Model [6], Krishnamurthy Model [11], Yacoub Model [12], Hamlet [13], Zhang [14], Hus's model [15], etc.

Additive models: The architecture of the software is not considered in additive models. The system reliability estimation is based on the components failure data. It considers the software reliability growth model in nonhomogeneous Poisson process (NHPP). Everett model [16], Xie model [17] are some of the additive models.

2.1 Comparison Between State-Based Model and Path-Based Models

In this section, we explain the basic difference between the state-based and path-based model.

Table 1 shows the uncertainty parameter for state-based models as well as path-based models. Since we stated that both the above models are similar, some common parameters like operational profile, module reliability, and component reliability are used. Apart from some parameters like use case scenarios and transition reliability, which we find in path-based models but not in state-based models. The application reliability is estimated either by solving the composite method or by the hierarchical method.

3 Related Work on Path-Based and State-Based Models

In this section, the different state-based models and path-based models used in different applications to estimate the software reliability is represented. The

Table 1 State-based model versus path-based model

State-based model	Path-based model
It combines the software architecture with the failure behavior to obtain the system reliability. It represents the three types of failure models such as the probability of failure, constant failure rate, and time-dependent failure intensity	It represents the failure behavior of components using the probability of failure or reliability
The uncertainty factors influencing this model are Time between Failure, Component Failure Rate, Module Reliability, Transition Probability components, etc.	The uncertainty factors affecting this model are Individual Component Reliability, Number of Component Executions, Probability of a Scenarios, Transition Reliability, Average Execution Time of a component, etc.
It is analytically describes the infinite number of paths or loops obtained in an intermediate graph called control flow graph	It considers the finite number of component executions traces, which correspond to system test cases

researchers have proposed several works on component-based reliability models that are represented either by state-based or path-based approaches. Cheung Model [3] described the reliability estimation by considering the DTMC system architecture with known component reliability of the system. The reliability estimation of the application is determined with respect to user environment. Yacoub Model [12] explained a path-based model for the system reliability which is based on the execution of the scenarios. A CDG (Component Dependency Graph) is used to estimate the system reliability. For reliability estimation, parameters like average execution time of the components, probability of the scenarios, individual component reliability etc., are calculated from the operational profile. Hus's model [18] is based on path reliability to estimate the overall system reliability used in the early phase of software development. It explains the three structure of the program, i.e., sequence, branch, and loop. The sensitivity analysis occurs in this model to determine the critical components and paths for resource allocation. A new approach SVM (Support Vector Machine) is proposed [19] for reliability estimation in component-based software systems to optimize the reliability evaluation in terms of the structure complexity. A disjoint path analysis is obtained [20] for network reliability analysis to identify the critical links between the components by considering the probability between the components. Component-based software reliability analysis is based on software architecture. Usage profile [21] model is used in architecture-based software reliability analysis to find the difference between components' testing profile and practical profile. It is very difficult to estimate accurate reliability of a system for CBS. The researchers have developed several mathematical approaches for reliability estimation. Soft computing methods [22] can be used to estimate the component-based reliability system.

4 Proposed Work for Path-Based Reliability Estimation

The path-based approach is represented here, which is based on execution scenarios [2, 12] for large-scale systems. The sequence diagrams are used to model the scenarios that specify in a time sequence manner. The component execution probabilities assigned to scenarios in a profile is similar to the operational profile. We illustrate a simple component-based application for ATM (Automatic Teller Machine) application system. The architecture consists of the following components Deposit, Check balance, Withdraw, Transaction, Log in, Bad Pin, and Print Receipt.

The analysis specifies a set of scenarios in the execution of the application,.such as enterPassword(), verifyAccount(), withDrawcash(), checkBalance(), printReceipt(), requestPassword(), etc. With the help of scenarios, a probabilistic model is used for the reliability analysis in component-based systems. To deduce the probabilistic model, we use a graph called as Component Dependency graph (CDG) which is similar to Control flow Graph (CFG). The CDG shows the relationship between components and determines the probable execution paths.

4.1 Component Dependency Graph (CDG)

The above scenarios can be represented as an intermediate graph called CDG [12] to estimate the whole system reliability. A CDG consists of the following: A node (n) (represented as component name CN, Average Execution time AET), edges (e) (interaction between two components which can be considered as the transition probability, PTr), and a Start node (b) and end node (E). Thus a CDG template as <n,e,b,E>.

An ATM bank system [23] is employed to validate the architecture-based reliability model. This system consists of 11 components. We conducted a path-based analysis of this software to obtain the reliability of each component and the overall system reliability. Figure 2 explains the sequence diagram of the ATM system with different scenarios. The corresponding component dependency graph for the sequence diagram is described in Figs. 3 and 4.

The algorithm explains the path-based model. For the procedure component number, component names C_i, transition reliability TR_{ij}, transition probability tp_{ij} between components are input values. The transition probabilities between two components and component reliabilities are described in Tables 2 and 3. The algorithms consider the number of paths from the starting component (B) to ending component (E). Then calculate the reliability along the execution path [15] and calculate the average of the total path which is considered as the system reliability. In path-based reliability approach [23], we obtained the following path list. They are as follows: Path1 = CR1-CR2-CR11.

Path2 = CR1-CR2-CR 3-CR5-CR11

Path3 = CR1-CR2-CR4-CR5-CR11

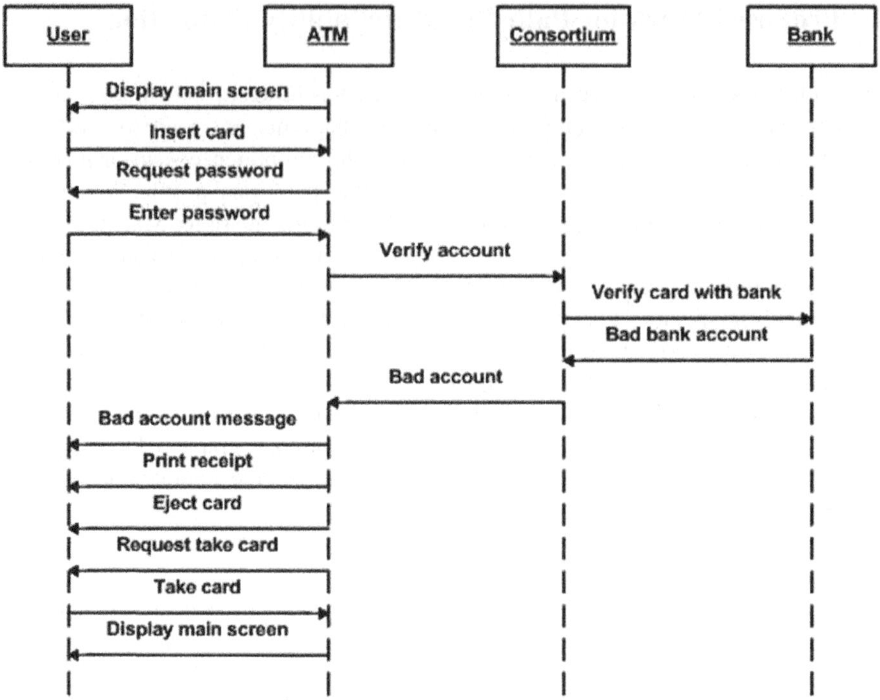

Fig. 2 ATM system sequence diagram [20]

Path4 = CR1-CR2-CR4-CR8-CR6-CR11
Path5 = CR1-CR2-CR4-CR6-CR11
Path6 = CR1-CR2-CR3-CR6-CR11
Path7 = CR1-CR2-CR3-CR8-CR6-CR11
Path 8 = CR1-CR2-CR3-CR6-CR7-CR6-CR11
Path 9 = CR1-CR2-CR3-CR6-CR7-CR6-CR11.

So, the total system reliability is 0.557. Whereas, in Cheung's composite method [3] (state-based model) the system reliability is 0.560. Here, we calculate the system reliability using both path-based as well as state-based approach.

4.2 Comparative Analysis

In this section, we compare the reliability estimation of state-based method using Cheung's model with the path-based approach from the case study given in Sect. 4.1. Table 4 shows a comparative result with actual reliability evaluated from the experiment.

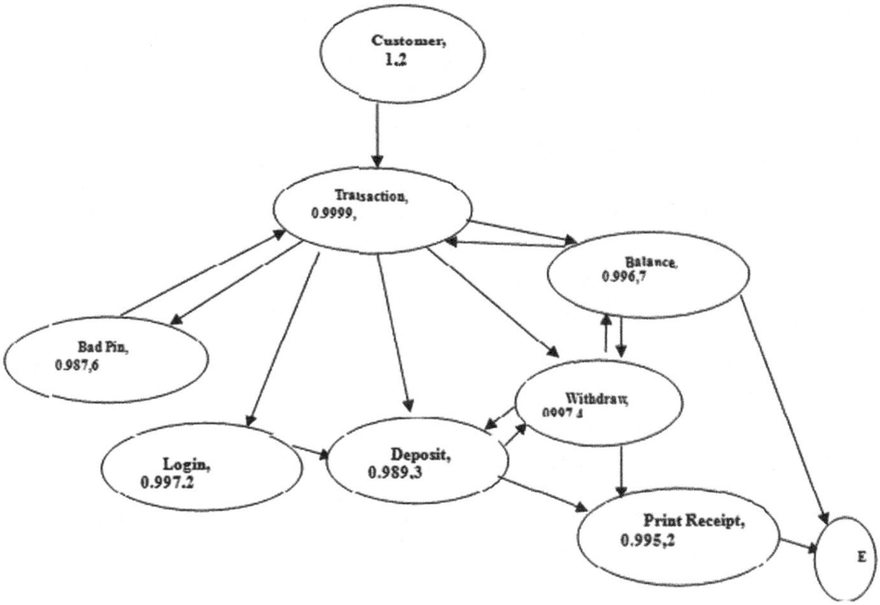

Fig. 3 A CDG for the ATM system

Procedure Path based Reliability Analysis of Scenario
    Input CDG //Component Dependency Graph
    Output Re_{appl} *// Application Reliability*
Initialization
    $Re_{appl}=0$, $Re_{temp}=1$, P=1
Method
push $<C_i, CR_i>$, Re_{temp} into Stack
while Stack <> Empty do
    pop $<C_i, CR_i>$, Re_{temp}
    while (P)
        $\forall < C_j, CR_j > \in$ child (C_i)
            push (C_i, CR_i, $Re_{temp}=Re_{temp}*CR_i*tp_{ij}*TR_{ij}$);
            P++;
        end
    Re_{appl} += Re_{temp}
end

Fig. 4 Path-based reliability analysis algorithm

Table 2 Transition probability (tp) between components

$tp_{1,2} = 1.0$				
$tp_{2,3} = 1.0$	$tp_{2,4} = 0.999$	$tp_{2,11} = 0.001$		
$tp_{3,5} = 0.227$	$tp_{3,6} = 0.669$	$tp_{3,8} = 0.104$		
$tp_{4,5} = 0.227$	$tp_{4,6} = 0.669$	$tp_{4,8} = 0.104$		
$tp_{5,2} = 0.048$	$tp_{5,6} = 0.951$	$tp_{5,11} = 0.001$		
$tp_{6,3} = 0.4239$	$tp_{6,4} = 0.4239$	$tp_{6,7} = 0.1$	$tp_{6,9} = 0.4149$	$tp_{6,11} = 0.0612$
$tp_{7,6} = 1.0$				
$tp_{8,6} = 1.0$				
$tp_{9,6} = 0.01$	$tp_{9,10} = 0.99$			
$tp_{10,6} = 1.0$				

Table 3 Component reliability of the ATM system

Component #	Reliability
CR1	1.0
CR2	0.982
CR3	0.97
CR4	0.96
CR5	1.0
CR6	0.996
CR7	0.99
CR8	1.0
CR9	1.0
CR10	0.8999
CR11	1.00

Table 4 Comparison of results

Actual reliability of the ATM system	Using Cheung's composite		Proposed path-based	
	Estimated reliability	Error	Estimated reliability	Error
0.526	0.560	3.4%	0.567	4.1%

From the above table, we get two values, i.e., using state-based method system reliability estimation is 0.560 and in path-based method system reliability estimation is 0.577. We observed the difference between these two methods and get the result as 3.4% and 4.1%. This implies that the result obtained by both the models give possibly accurate estimations compared to the actual reliability.

5 Conclusion and Future Research

This paper addresses the architecture-based reliability models. We classify the basics of system architecture and failure behavior. The reliability predictions of both the state-based and path-based approaches are the same and the state-based models analytically explain an infinite number of states, whereas path-based models restrict the number of paths to ones observed experimentally during the testing. The state-based approach is better compared to path-based approach as we can apply various failure models in state-based reliability estimation approach. The failure model helps to decide the impact of individual component reliability on the overall system reliability. This phenomenon will not be found in path-based approaches.

The proposed method has applied an effective path-based reliability analysis into the ATM system case study. But the above path-based method has the following limitations. (i) The path selection is done by considering the starting component to reaching end component with different paths. This causes different path reliability values for the given system. (ii) The different software architecture may possibly have different paths and any path can be selected by the tester, this leads to the uncertainty for the complex path selection. Hence, uncertainty assessment for software architecture based reliability estimation is a topic of future scope.

References

1. Pham, H., Pham, M. S.: Software reliability models for critical applications. United States. https://doi.org/10.2172/10105800. https://www.osti.gov/servlets/purl/10105800
2. Pradhan, S., Ray, M., Patnaik, S.: Coverage criteria for state-based testing: a systematic review. Int. J. Inf. Technol. Project Manag. (IJITPM) **10**(1), 1–20 (2019)
3. Cheung, R.C.: A user-oriented software reliability model. IEEE Trans. Soft. Eng. **6**(2), 118–125 (1980)
4. Littlewood B.: Software reliability model for modular program structure. IEEE Trans. Reliab. 241–246 (1979)
5. Kubat, P.: Assessing reliability of modular software. Oper. Res. Lett. 35–41 (1989)
6. Shooman, M.L.: Structural models for software reliability prediction. In: Proceedings of the 2nd International Conference on Software Engineering, pp. 268–280. IEEE Computer Society Press (1976)
7. Musa, J.D.: Operational profiles in software-reliability engineering. IEEE Softw. **10**(2), 14–32 (1993)
8. Goševa-Popstojanova, K., Trivedi, K.S.: Architecture-based approach to reliability assessment of software systems. Perform. Eval. **45**(2–3), 179–204 (2001)
9. Laprie, J. C.: Depndability evaluation of software systems in operation. IEEE Trans. Softw. Eng. **10**(6), 701–714 (1984)
10. Gokhale, S. S., Trivedi K. S.: Reliability prediction and sensitivity analysis based on software architecture. *13th International Symposium on Software Reliability Engineering, Proceedings*. IEEE. 64–75 (2002)
11. Krishnamurthy, S., Mathur, A.P.: On the estimation of reliability of a software system using reliabilities of its components. In: Proceedings The Eighth International Symposium on Software Reliability Engineering, pp. 146–155. IEEE (1997)

12. Yacoub, S., Cukic, B., Ammar, H.H.: A scenario-based reliability analysis approach for component-based software. IEEE Trans. Reliab. **53**(4), 465–480 (2004)
13. Hamlet, D., Mason, D., Woitm, D.: Theory of software reliability based on components. In: Proceedings of the 23rd International Conference on Software Engineering. ICSE 2001, pp. 361–370. IEEE (2001)
14. Zhang, F., et al.: A novel model for component-based software reliability analysis. In: 11th IEEE High Assurance Systems Engineering Symposium, pp. 303–309. IEEE (2008)
15. Hsu, C.J., Huang, C.Y.: An adaptive reliability analysis using path testing for complex component-based software systems. IEEE Trans. Reliab. **60**(1), 158–170 (2011)
16. Everett, W.W.: Software component reliability analysis. In: Proceedings 1999 IEEE Symposium on Application-Specific Systems and Software Engineering and Technology. ASSET'99 (Cat. No. PR00122), pp. 204–211. IEEE (1999)
17. Xie, M., Wohlin, C.: An additive reliability model for the analysis of modular software failure data. In: Proceedings of Sixth International Symposium on Software Reliability Engineering. ISSRE'95, pp. 188–194. IEEE (1995)
18. Hou, C., et al.: A scenario-based reliability analysis approach for component-based software. IEICE Trans. Inf. Syst. **98**(3), 617–626 (2015)
19. Tyagi, K., Sharma, A.: A rule-based approach for estimating the reliability of component-based systems. Adv. Eng. Softw. **54**, 24–29 (2012)
20. Inoue, T.: Reliability analysis for disjoint paths. IEEE Trans. Reliab. **68**(3), 985–998 (2019)
21. Vasilache, S., Tanaka, J.: Synthesis of state machines from multiple interrelated scenarios using dependency diagrams. In: 8th World Multiconference on Systemics, Cybernetics and Informatics, pp. 49–54. SCI (2004)
22. Peng, N.: A SVM reliability evaluation model for component-based software systems. In: 2013 2nd International Symposium on Instrumentation and Measurement, Sensor Network and Automation (IMSNA), pp. 704–708. IEEE (2013)
23. Wang, W.L., Pan, D., Chen, M.H.: Architecture-based software reliability modeling. J. Syst. Softw. **79**(1), 132–146 (2006)

A Study on Decision Making by Estimating Preferences Using Utility Function and Indifference Curve

Suprava Devi, Mitali Madhusmita Nayak, and Srikanta Patnaik

Abstract To get the desired goal, in various fields, decision making is being employed, which has been gaining attention in current decades, for which efficient evaluation of individual preferences should be done. Preferences are vital elements of decision making to estimate the effects regarding individual behavior and their social status. The main objective of the paper is to present various quantitative techniques regarding preference relation in a decision making process. This paper shows how the decision making process can be enhanced through preferences and utility function. The indifference curve is also incorporated for decision making involving uncertainty. This study presents the existing studies by investigating the relationship between the choices, preferences and indifference curve in the decision making process.

Keywords Decision making · Choices · Preferences · Rationality · Local nonsatiation · Convexity · Utility function · Indifference curve

1 Introduction

In day to day life everyone comes across a lot of choices and options, for which decisions have to be taken, depending upon the situation they face or the actions they need to choose. People are continuously engaged in decision making processes in today's world which may be in personal or professional areas. Many of the decisions are common and have taken together under time and pressure [1, 2]. So, decision making process is a complex and hard task to do.

S. Devi (✉) · S. Patnaik
Department of Computer Science and Engineering, Siksha 'O' Anusandhan Deemed to be University, Bhubaneswar 751030, Odisha, India
e-mail: Suprava_devi@yahoo.in

M. M. Nayak
Department of Mathematics, Siksha 'O' Anusandhan Deemed to be University, Bhubaneswar 751030, Odisha, India

D. Mishra et al. (eds.), *Intelligent and Cloud Computing*,
Smart Innovation, Systems and Technologies 194,
https://doi.org/10.1007/978-981-15-5971-6_41

Decision-making has been considered as the suitable method for choosing a suitable alternative and to understand how alternative values are compared while making a selection is the main objective for decision making process. Individual decision making is very challenging because the decision maker has to be considered many dissimilar alternatives and contradictory issues while making decision [3].

Decision making is the process of recognizing and selecting options which depends upon the values and preferences of the decision maker. Decision making denotes that there are alternative choices that are taken for granted to be considered, and in such situations we have to recognize majority of possible alternatives and from this we have to select the best one which fulfill our target [4], from all the alternative choices available.

Correct representation of the decision maker's preferences is the basic step of decision making. People always have a set of preferences and values, depending on the type of product liked or disliked by them. Preferences can be measured by measuring the satisfaction of a person with a particular product which is compared with the opportunity cost of that product because whenever people buy one product; they sacrifice the chance to buy a contending product. Preferences determine the types of products that people will buy within their budget, and these preferences will help in fulfilling the demand of customer. From the consumer behavior we assume that in general from the all choices bundles the consumer has distinct preference from which the consumer has to accept the most favored bundle that are accessible [5]. This helps in building a model that how a consumer experience about deal with one product besides another through which forecast of consumer behavior can be possible.

Preferences are generally considered as primitive to decision-making which influence behavior in many aspects. Impatient individuals hesitate to spend money, time, or resources at the present to get benefits in the future. The individuals who make consistent intertemporal decisions have a constant rate of discounting [6]. The discounting factor represented by a function called utility function which calculates the individual consumer preferences [7].

Depending upon the individual problem the utility function has various shapes. Utility function represented as a curve called indifference curve which have different forms [8]. When it is convex upwards, it is known as risk-averse utility function (where risk is avoided), when it is concave upwards, it is known as risk-prone utility function (where risk is taken), or when it is linear, the curve is a straight line (risk neutral). The last one is comparatively simple one because it reduces the risk of a decision maker having without any consistent preference [9].

This article help the decision maker carry their thoughts in the direction of risk into the study of business related problems which involves uncertainty. Preference theory has immense possible assessment for the decision maker who needs to get better consistency of their decisions and also predict the consumer behavior across various fields. It enlightens how efficient people make decisions which maximize their utility in the long run and minimizes regret.

2 Choices, Preferences and Rationality

Decision making indicates that there are number of choices which are need to be considered and in such situations we not only recognize the various alternatives but also to select the one that has the highest efficiency and also satisfy the decision maker objectives. Rationality is the foundation of individual behavior. For the reason of choice we have preference relation over the different choices we are faced with [10]. Preferences having some own conditions in order to accomplish the basic necessities are classified as rational.

2.1 Choices

Regarding the products and services the decisions that consumer make regarded as consumer choice. In order to study the consumer behavior, we acknowledge that the decisions made by the consumers that to purchase which commodities or consume that commodity over time.

Choices contain giving up something to achieve others. Evaluating different choices with benefits helps in decision making. Consumer choice referred to the bundle which maximizes consumer utility along with their preferences and budget line. This also depends upon how the consumer measures two commodities. An individual's choices reveal his/her preferences.

Let us consider that certain choice set A $\subset$ P, suppose that the individual choose the element of A that he prefers most. By simplifying this, the individual choice can be presented by the choice rule as,

$C\ (A; \succsim) = \{x \in A|\, x \succsim y \text{ for all } y \in A\}$

As per the set of items in A individual akin to as much as any of the other options [11].

2.2 Preferences

Preferences denotes to certain features that the consumer wants to be present in a commodity or service by making it preferable which may by the level of happiness, degree of satisfaction, utility from the product, etc. Preferences are the main factors that influence consumer demand [12, 13].

Suppose the consumer selects from among C commodities and the commodity space is given by X $\subset$ RC+.

For example, if $x \succcurlyeq y$, we could also write $y \preceq x$, where $\preceq$ is the "no better than" relation. If $x \succcurlyeq y$ and $y \succcurlyeq x$, then the consumer is "indifferent between x and y," represented as $x \sim y$. The indifference curve I_y is defined as the set of all bundles that are indifferent to y. That is, $I_y = \{x \in X|y \sim x\}$.

Suppose the preference relation $\succsim$ on set A:
when $\precsim$ represent the duplicate of $\succsim$ when it is described as
$p \precsim q \Leftrightarrow q \succsim p$
when $\succ$ represent the asymmetric element of $\succsim$ it is described as
$p \succ q \Leftrightarrow$ [[$p \succsim q$ but not $q \succsim p$]
when $\sim$ represent the symmetric element of $\succsim$ it is designated as
$p \sim q \Leftrightarrow$ [$p \succsim q$ and $q \succsim p$]
that means when $p \succ q$ at the time of decision making 'p' is strictly preferred to 'q' and when $p \sim q$ at the time of decision making, there is no difference between p and q [14].

2.3 *Rationality*

Rationality denotes that people who are well-informed about the possible outcomes are capable of making reasonable and consistent choices which maximizes utility.

Rules of rationality can relate to preferences where whole objects are taken into consideration. Four axioms of rationality are the basics that the decision makers require for the rational causes. A preference relation is rational if it obeys the four axioms.

1. Completeness

   $\forall$A, B $\in$ X: A $\succcurlyeq$ B or B $\succcurlyeq$ A or both at the same time

2. Transitivity

   $\forall$A, B, C $\in$ X: A $\succcurlyeq$ B and B $\succcurlyeq$ C, then A $\succcurlyeq$ C

3. Reflexiveness

   $\forall$A $\in$ X: A $\succcurlyeq$ A

4. Continuity

   $\forall$A, B, C $\in$ X: if A $\succcurlyeq$ B $\succcurlyeq$ C, then $\exists\ \lambda \in [0, 1]$ such that B ~ λ A + $(1-\lambda)$ C.

Some other axioms which are studied and added by Samuelson as preferable because people are inherently insensible that always want more what they like is called monotonicity.

3 Properties of Preferences

3.1 *Monotonicity*

Monotonicity refers to the relation of preference '$\succcurlyeq$' which is monotone.

If A $\succ$ B for any A and B such that $Al > Bl$ for all $l = 1, \ldots, K$.

which is said to be strongly monotone if $Al \geq Bl$ for all $l = 1, \ldots, K$ and $A_i > B_i$ for some $i \in \{1, \ldots, K\}$ denotes that A $\succ$ B.

Both monotonicity and strong monotonicity tions implies "extra is superior". Monotonicity states that if P is strictly preferred to Q then each and every element of P is greater than the consequent element of Q. Strong monotonicity states that if every element of P is partially as large as the consequent element of Q and partially one element of P is strictly larger, then P is strictly preferred to Q.

(Weak) $\forall$P, Q $\in$ Z: if P $\geq$ Q then $p \succcurlyeq$ Q

(Strong) $\forall$P, Q $\in$ Z: if P $>$ Q then P $\succ$ Q

For the given consumption bundle, the consumer prefers the bundle that have more of at least one good and which is not fewer in any other good. Consumer preferences are monotonic, which means that consumption of more goods means more satisfaction [15, 16].

Following example explained the distinction between monotonicity and strong monotonicity. Suppose bundle P = (2, 2) and bundle Q = (2, 4). If $\succcurlyeq$ implies strongly monotone, then Q $\succ$ P. On the other hand, if $\succcurlyeq$ implies monotone, then in this case Q is not strictly preferred to P.

Monotonicity implies local nonsatiation, but the converse isn't true.

If $\succcurlyeq$ is strongly monotone, then it is monotone.

If $\succcurlyeq$ is monotone, then it is locally nonsatiated.

3.2 *Local Nonsatiation*

The local nonsatiation property of consumer preferences states that for very bundle of commodities there is always another bundle of commodities which is randomly close to it is also preferred to it [17].

A preference relation $\succcurlyeq$ is said locally nonsatiated if for all $x \in X$ and $\varepsilon > 0$, there present a point y such that $\|y - x\| < \varepsilon$ and $y \succ -x$. This shows that, for every x, for all time there is a point close to it that the consumer strictly prefers to x. if some commodities are not good then the consumers would sometimes prefer less of them is referred by local nonsatiation. This does not mean that all commodities are bad if preferences are non-satiated. The look of indifference curve depends upon the implications of monotonicity or local nonsatiation. When $I_x = \{y \in X | \, y \sim x\}$ then the slope is thin and downward sloping which is look like Fig. 1.

Figure 2 shows the indifference curves were thick because the point x and the neighborhood of x where all points are indifferent to x. But here violation of local-nonsatiation (or monotonicity) occurs because in this region there is no strictly preferred point.

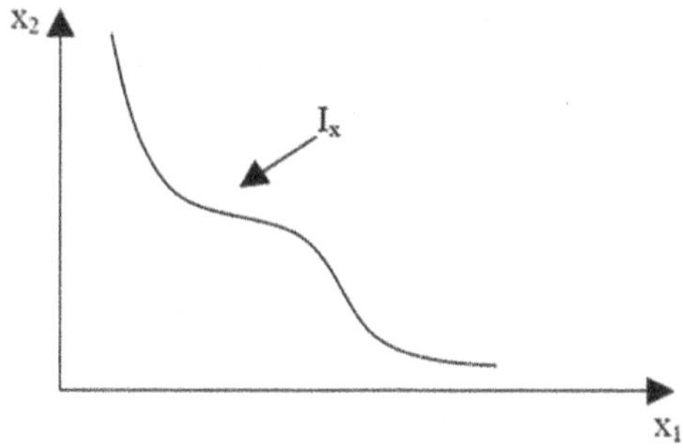

Fig. 1 Thin indifferent curve

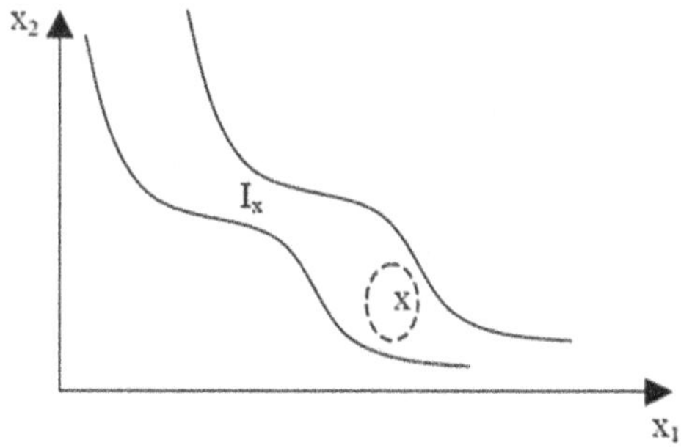

Fig. 2 Thick indifference curve

3.3 Convexity

Preferences that are good and organized are convex in nature because majority of people consumed commodities together. Mostly consumer desires to deal with some of one commodity for some of the other and lastly ends with consuming both the commodities rather than focusing on only one commodity [18]. Convexity of preferences means peoples ordering of different outcomes, usually with reference to quantity of various commodities consumed which are strictly preferred **averages to extremes** which mean that consumer tastes automatically satisfy weak convexity.

Suppose A, B, and C are three consumption bundles and on the consumption set X the preference relation $\succcurlyeq$ is called **convex** if

A, B, C $\in$ X where B $\succcurlyeq$ A and C $\succcurlyeq$ A, and for every $\theta \in [0, 1]$:

θ B +(1−θ) C $\succcurlyeq$ A

that means any two bundle is at least as good as the third bundle.

The preference relation $\succcurlyeq$ is called **strictly convex** if

A, B, C $\in$ X where B $\succcurlyeq$ A and C $\succcurlyeq$ A, and B $\neq$ C, and for every $\theta \in [0, 1]$:

θ B +(1−θ) C $\succcurlyeq$ A

which means for any two different bundles each bundle regarded as at least as good as the third bundle [17, 19].

4 Utility Function

Utility means 'usefulness'. In consumer theory utility is the capability of a commodity to fulfill human wants. The quality of a commodity is measured in utils which satisfy human wants. A mathematical technique or tool which is used to evaluate preferences is called a utility function [20]. Utility function computes preferences over a set of commodities and services. Utility is the level of satisfaction that consumers get for choosing and consuming a commodity or service.

A function is considered as a utility function U (x) which assigns a numerical value to every consumption bundle $x \in X$ depending on how it categorizes, or generally U: $X \rightarrow R$, or $U \subseteq X \times R$, where R denotes to the set of real numbers. For any x and y a utility function U(x) that represent a preference relation $\succcurlyeq$ can be represented as,

$\forall x \in X, \forall y \in X, U(x) \geq U(y)$ if and only if $x \succcurlyeq y$

i.e. function U(x) allocates a numerical value to x which is slightly larger than the value it allocates to y where x is somewhat as good as y.

Suppose in the given figure the indifference curve I_x can be identified by the distance beside the line $x2 = x1$ travelled before intersecting I_x. In this case each I_x will intersect the line once because indifference curves are downward sloping and a distinct number attached with it. Because of preferences are convex, i.e. if $x \succ y$, then I_x lies over and right of I_y and that's why I_x posses a higher number than I_y. The number which is associated with I_x is known as utility of x [21] (Fig. 3).

Utility function the same axioms as preferences because they are simply numerical representation of them. Like monotonicity axioms, the utility function is increasing i.e. for the more you have the commodity the larger are the utility.

4.1 Utility Is an Ordinal Concept

To measure the preferences of a consumer on an ordinal scale the function which is used is called ordinal utility function. This function shows which option is better than the other so consumer decision-making process under conditions of certainty use the term ordinal utility (Fig. 4).

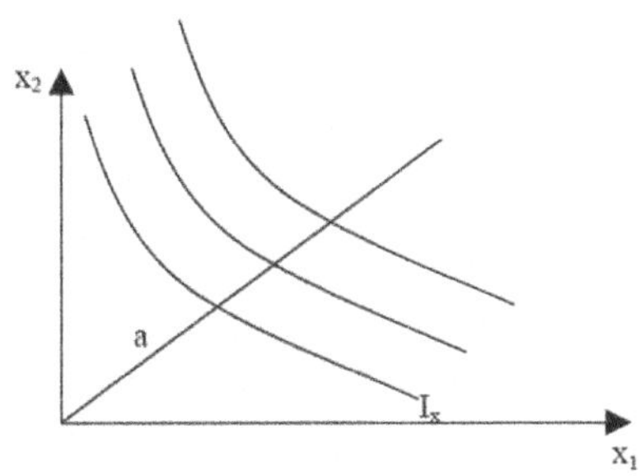

Fig. 3 Levels of indifference curves

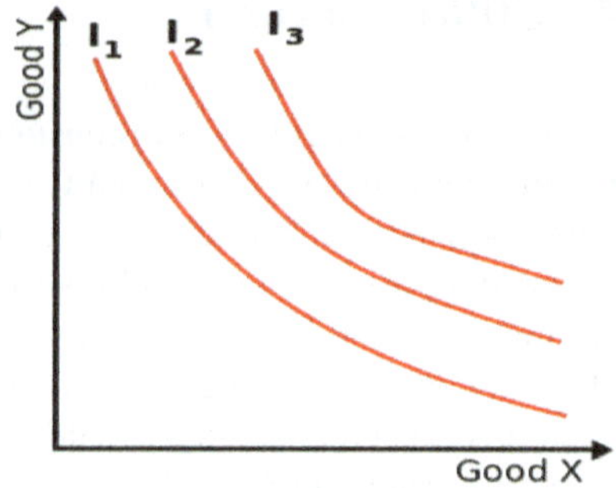

Fig. 4 An indifference curve

If U (x) correspond to $\succcurlyeq$ and f () is considered as monotonically growing function, then $V(x) = \mathrm{f}(\mathrm{U}(x))$. For Cobb-Douglas utility function $u(x) = x_1^a x_2^{1-a}$, it is difficult to evaluate because $x1$ and $x2$ are multiplied together with different exponents. But, for monotonically increasing function $\mathrm{f}(z) = \log(z)$,

$$V(x) = \log[x_1^a x_2^{1-a}] = a \log x_1 + (1 - a) \log x_2.$$

4.2 *Basic Property of Utility Function*

Same preferences can be shown by many utility functions. Here V () represents the same preferences as U (). The indifference curves are convex if preferences are convex. For example, the Cobb-Douglas utility function,

$$u(x_1, x_2) = x_1^{\frac{1}{4}} x_2^{\frac{1}{4}}$$

The following figure shows the three-dimensional (3D) graph of the above function.

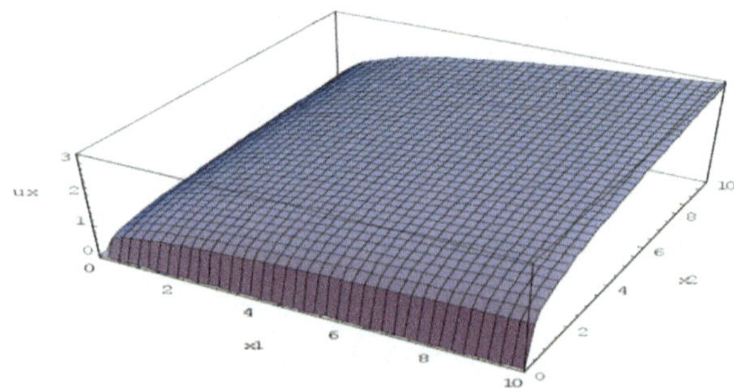

Fig. 5 Function u (x)

Fig. 6 Level sets of u (x)

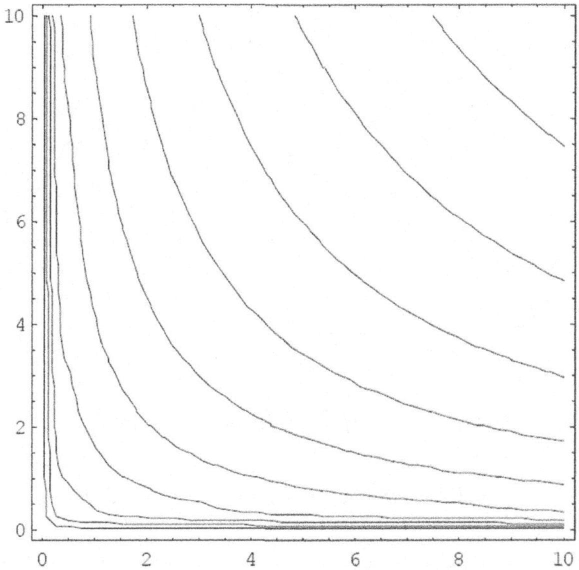

Figures 5 and 6 shows different levels of I_x for a variety of utilities and the utility function are convex in nature.

5 Indifference Curve

Indifference curve is a graph which shows various combinations of two commodities which give a consumer equal level of satisfaction and utility. The axes of the graph represent one commodity each (e.g. good A and good B).This makes the consumer indifferent about the commodities he consumed which mean each point on the **indifference curve** designates that a consumer is **indifferent** between the two points and all points present on the indifference curve give him the same utility. It is a downward sloping curve which is convex to the origin. The following graph shows a combination of two commodities that the consumer consumes.

Figure 7 shows the indifference curve U having two bundles of commodities A and B which provide the same level of satisfaction to the consumer. So the consumer gains the same level of satisfaction at any point on the same curve.

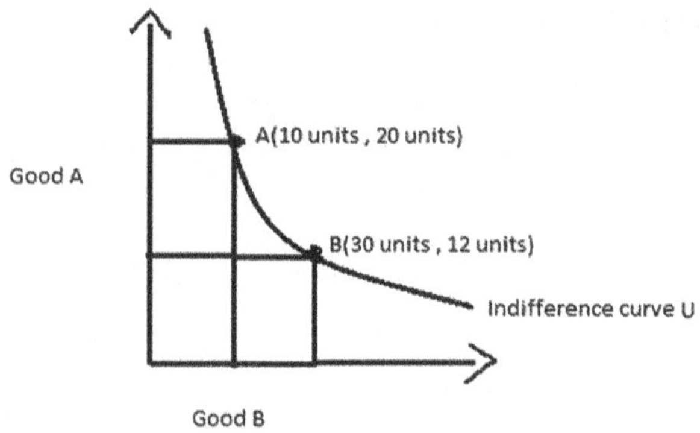

Fig. 7 Indifference curve

5.1 *Properties of Indifference Curve*

5.1.1 Indifference Curves Are Always Thin

If the indifference curve is thick, it violates the **non**-satiation property of preferences, which means the point B in the above figure has more quantity than point A, which contradicts, so it does not exist.

5.1.2 Indifference Curves Cannot Intersect Each Other

Here, the consumer can not differentiate between points (A and D) and (B and C). According to monotonicity, here the consumer strictly prefers A to B, and also C to D. By positioning all this we get,

$A \succ B \sim C \succ D \sim A$

So, the curve must present to the northeast of the other by the rule of monotonicity (Fig. 8).

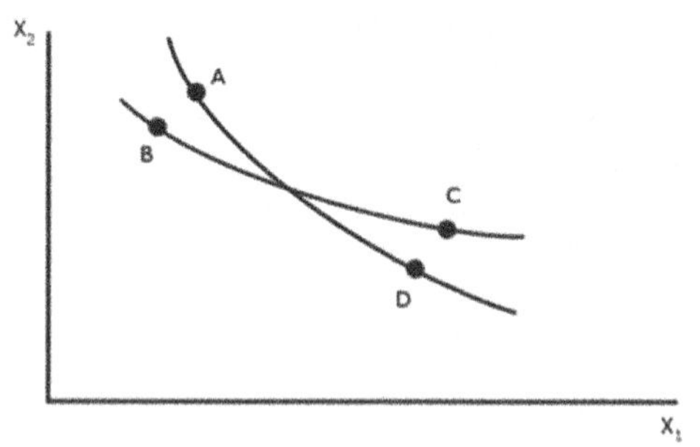

Fig. 8 Indifference curves that cross

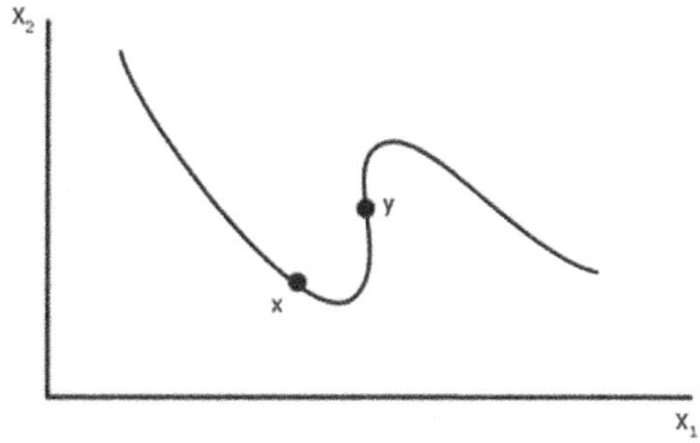

Fig. 9 An upward sloping indifference curve

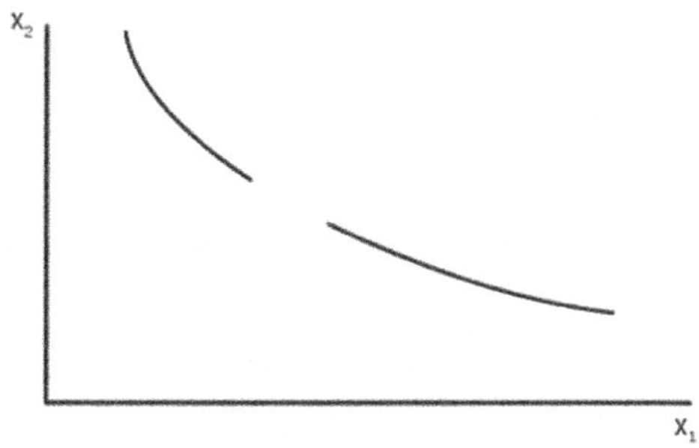

Fig. 10 Gap in the indifference curve

5.1.3 Indifference Curves Slope Downward

Downward sloping of **indifference curves** must **occur which is** from left to right that means the **curve** is sloped negatively. Suppose the curve is not slope downward then the points x and y present on the same curve.

For which, $y_i \geq x_i$ for all i, and $y_i > x_i$ for some i, as in the corresponding Fig. 9. This opposes the rule of monotonicity, which states that y is strictly prefers to x.

5.1.4 Indifference Curves Are Unbroken

As shown in Fig. 10, there are no gaps present in the indifference curve. This means that preferences are complete, transitive and continuous which implies that the utility functions are continuous.

5.1.5 Indifference Curves Are Convex to the Origin if Preferences Are Convex

Assume that the two points x and y lie on the same curve. Following the rule of convexity, $t\,x + (1 - t)\,y$ must lies on a bigger indifference curve, where $t \in [0, 1]$. The bigger curve must present to the right of the previous indifference curve according

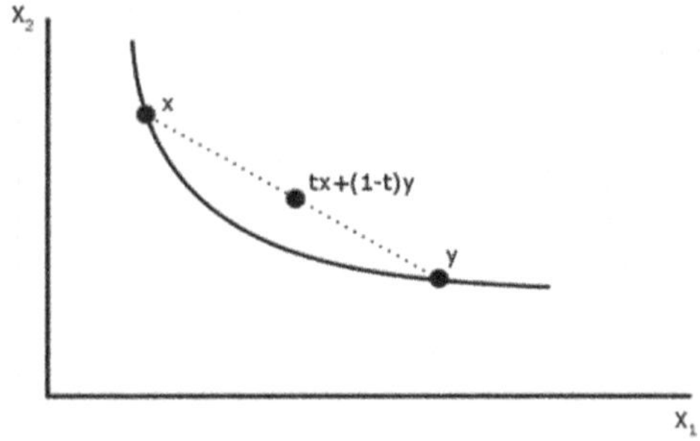

Fig. 11 Convex indifference curves

to the assumption of monotonicity. So as shown in the figure the indifference curve is convex (Fig. 11).

5.2 *Marginal Rate of Substitution*

The concept of indifference curve lies on the properties of marginal rate of substitution. Marginal Rate of Substitution of Good X_1 for Good X_2 is $(MRSX_1X_2) = \Delta X_2/\Delta X_1$. The marginal rate of substitution (MRS) is calculated as the consumer sacrifices some part of one commodity in exchange for another commodity for maintaining the utility level same. MRS is calculated as,

$$\mathrm{MRS} = \frac{dx_2}{dx_1}$$

where dX_1 is the number of commodity that will be sacrificed to get dX_2. Marginal rate of substitution is the slope of the indifference curve. The MRS is decreasing along with the indifference curve (Fig. 12).

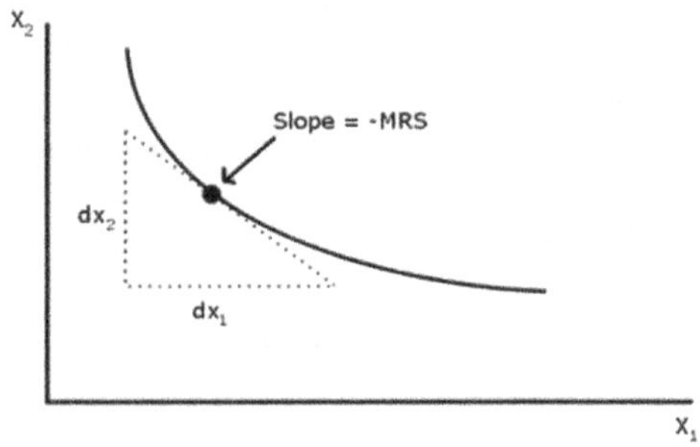

Fig. 12 Graph of MRS

6 Summary and Conclusions

There is a requirement to build up and employ a decision-making process which is both scientific and understanding of individual differences and perspectives. Fundamentally decisions are imprecise so individual identify this deficiency and will make a longer time attempting to locate the optimal choice. This can be considered as a variety of ranges of decisions and every decision can be observed as different levels of preferences. This article presents the theory, extensions, and some applications of preferences in decision making. We try to find to give the decision maker with an approach and interface that makes it as easy and straightforward as possible to state his/her preferences. The applications included in the paper help outs in real-world decision making. Choice portrays how people make decisions which have wide application areas. People's preferences with respect to commodities they purchase and the time they need are described by indifference curves. The paper presents that many outcomes and behaviors related preference parameters.

References

1. Gong, M., Simpson, A., Koh, L., Hua Tan, K.: Inside out: The interrelationships of sustainable performance metrics and its effect on business decision making: Theory and practice. Resour. Conserv. Recycl. **128**, 155–166 (2018)
2. De Couck, M., Caers, R., Musch, L., Fliegauf, J., Giangreco, A., Gidron, Y.: How breathing can help you make better decisions: two studies on the effects of breathing patterns on heart rate variability and decision-making in business cases. Int. J. Psychophysiol. **139**, 1–9 (2019)
3. Ben-Akiva, M., McFadden, D., Train, K.: Foundations of stated preference elicitation: consumer behavior and choice-based conjoint analysis. Found. Trends® Econ. **10**(1–2), 1–144 (2019)
4. Fulop, J.: Introduction to decision making methods. In: BDEI-3 Workshop, Washington, pp. 1–15 (2005)
5. Donnelly, R., Ruiz, F.R., Blei, D., Athey, S.: Counterfactual inference for consumer choice across many product categories (2019). arXiv preprint. arXiv:1906.02635
6. Echenique, F.: New developments in revealed preference theory: decisions under risk, uncertainty, and intertemporal choice (2019). arXiv preprint arXiv:1908.07561
7. Sharma, Aran, Pillai, Rajnandini: Customers' decision-making styles and their preference for sales strategies: Conceptual examination and an empirical study. J. Pers. Selling Sales Manag. **16**(1), 21–33 (1996)
8. Shahtaheri, Y., Flint, M.M., Jesús, M.: A multi-objective reliability-based decision support system for incorporating decision maker utilities in the design of infrastructure. Adv. Eng. Inf. **42**, 100939 (2019)
9. Sheng, C.L.: A general utility function for decision-making. Math. Model. **5**(4), 265–274 (1984)
10. Danaf, M., Becker, F., Song, X., Atasoy, B., Ben-Akiva, M.: Online discrete choice models: applications in personalized recommendations. Decis. Support Syst. **119**, 35–45 (2019)
11. Levin, J., Milgrom, P.: Introduction to choice theory. http://web.stanford.edu/~jdlevin/Econ20202 (2004)
12. Warren, C., Peter McGraw, A., Van Boven, L.: Values and preferences: defining preference construction. Wiley Interdiscip. Rev. Cogn. Sci. **2**(2), 193–205 (2011)

13. Athey, S., Blei, D., Donnelly, R., Ruiz, F., Schmidt, T.: Estimating heterogeneous consumer preferences for restaurants and travel time using mobile location data. In: AEA Papers and Proceedings, vol. 108, pp. 64–67 (2018)
14. Awad, I., Lateefeh, H.A., Hallam, A., El-Jafari, M.: Econometric analysis of consumer preferences and willingness-to-pay for organic tomatoes in Palestine: choice experiment method. (2019)
15. Mizrachi, D., Salaz, A.M., Kurbanoglu, S., Boustany, J., ARFIS Research Group.: Academic reading format preferences and behaviors among university students worldwide: a comparative survey analysis. PloS one **13**(5), e0197444 (2018)
16. Franke, Ulrik, Ciccozzi, Federico: Characterization of trade-off preferences between non-functional properties. Inf. Syst. **74**, 86–102 (2018)
17. Mas-Colell, A., Whinston, M.D., Green, J.R.: Microeconomic Theory, vol. 1. Oxford University Press, New York (1995)
18. Oguego, C.L., Augusto, J.C., Muñoz, A., Springett, M.: Using argumentation to manage users' preferences. Future Gener. Comput. Syst. **81**, 235–243 (2018)
19. Board, S.: Preferences and utility. UCLA, Oct (2009)
20. Brabant, Q., Couceiro, M., Dubois, D., Prade, H., Rico, A.: Extracting decision rules from qualitative data via Sugeno utility functionals. In: International Conference on Information Processing and Management of Uncertainty in Knowledge-Based Systems, pp. 253–265. Springer, Cham (2018)
21. Zhao, W.-X., Ho, H.-P., Yu, J., Chang, C.-T.: The roles of aspirations, coefficients and utility functions in multiple objective decision making. Comput. Ind. Eng. (2019)

Binary Field Point Multiplication Implementation in FPGA Hardware

Suman Sau, Paresh Baidya, Rourab Paul, and Swagata Mandal

Abstract In 1980, Victor Miller and Neal Koblitz proposed Elliptic Curve Cryptography (ECC) that is popularly known as asymmetric key crypto algorithm which is rooted in the ECDLP on finite fields known as Elliptic Curve Discrete Logarithm Problem. The main advantage of ECC, over the other public key cryptography (like RSA), is that it gives equivalent security performance level but using the smaller key size length. In this paper, we proposed an ECC hardware implementation based on FPGA. Also, hardware implementation of Elliptic Curve Digital Signature Algorithm (ECDSA) in FPGA and its comparison with other implementation is presented. The ECC hardware implementations are categorized into two main implementing technologies, and they are FPGA and ASIC implementations. We first discuss elliptic curves standards, point multiplication over binary fields GF(2^m), and the use of prime fields on the hardware implementation. The proposed method evaluates the performance analysis of ECC implementation in an FPGA.

Keywords ECC · FPGA · ECDSA · Cryptography · Koblitz curve

S. Sau (✉) · P. Baidya · R. Paul
Siksha O Anusandhan (Deemed to be University), Bhubaneswar, Odisha, India
e-mail: sumansau@soa.ac.in

P. Baidya
e-mail: pareshbaidya@soa.ac.in

R. Paul
e-mail: rourabpaul@soa.ac.in

S. Mandal
Jalpaiguri Government Engineering College, Jalpaiguri, India
e-mail: swaga89@gmail.com

D. Mishra et al. (eds.), *Intelligent and Cloud Computing*,
Smart Innovation, Systems and Technologies 194,
https://doi.org/10.1007/978-981-15-5971-6_42

1 Introduction

Applications which are used for safety and security like medical service, aviation safety, power distribution, army device are very critical application in terms of data security and confidentiality. In these types of applications, we have to keep the confidentiality and authenticity of data at the highest priority. Data security and confidentiality can be preserved using the cryptography algorithm which is mainly divided into two types of algorithms mainly called symmetric and asymmetric key cryptography. Asymmetric key cryptography which is also called a public key cryptography uses different secret keys to data encrypt and decrypt the confidential data. ECC is a class of asymmetric key cryptography. There is a demand for hardware implementation of ECC algorithms for the abovementioned critical implementation. In the ECC algorithm, the most expensive operation is binary point field multiplication which we have proposed here as the efficient implementation of binary point field multiplications using FPGA. In this paper, we have proposed a binary point multiplication prescribed by NIST for public key ECC cryptography and ECDSA algorithm for authentication.

The paper is organized as follows, Sect. 3 gives a theoretical background of ECC as well as binary multiplication. Section 3 briefs the implementation process in FPGA hardware followed by the performance analysis and conclusion.

2 Related Work

The use of ECC in different applications, e.g., authentication (ECDSA), banking transactions, digital authentication system mobile security, sensor networks, and many privacy secure system applications are increasing day by day. ECC is broadly used in different network topologies like TLS, SSL, [4], WTLS, WAP [5]. Different ECC standards are proposed by different country and organizations (like IEEE, ANSI, etc).

From the standardization point (ECC curve domain parameters) of view, they are as follows:

- American National Standard X9.62 2005 recommended for Public Key Cryptography for the Financial Services Industry [6]
- Wireless Application Forum, WAP WTLS: Wireless Application Protocol Wireless Transport Layer Security Specification, Feb. 1999
- Public Key Cryptography for the Financial Services Industry: Key Agreement and Key Transport Using Elliptic Curve Cryptography, American National Standard X9.63-2001, 2001 [6]
- Institute of Electrical and Electronics Engineers. Specifications for Public Key Cryptography, IEEE Standard 1363–2000, Aug. 2000 [6]
- Specifications for Public Key Cryptography—Amendment 1: Additional Techniques, IEEE Standard 1363A-2004, Oct. 2004 [6]

- NIST FIPS (2000) Standard
- ECC Brainpool. "ECC Brainpool standard curves and curve generation." October 2005
- Committee on National Security Systems. "National information assurance policy on the use of public standards for the secure sharing of information among national security systems." 1 October 2012 [6]
- ANSSI FRP256V1 (2011) (Standard of France and other EU country)
- Chinese State Cryptography Administration (SCA) published the national public key cryptographic algorithm based on ECC , known as SM2 [7]
Radio Frequency Identification (RFID) tags, cell phones, smartcard industries (MasterCard, Digital Signature Trust Co, DataKey, and MIPS Technologies, etc.) has shown their extreme interest in ECC due to its small key size with respect to other asymmetric crypto algorithms, low area, and fast implementation on hardware. A hardware design can provide significant enhancements of security for safeguarding private key and rest information contrast to software solutions. That is why a demand for the hardware execution of the ECC increases its power and performance efficiency with respect to software implementations based on microcontrollers. Mathematically there are different options for ECC implementations. Figure 1 has shown the different available ECC implementetion techniques. Color boxes denote the category in which our used ECC design falls.

3 Theoretical Background

3.1 Elliptic Curve Cryptography (ECC)

Public key cryptography and symmetric block cipher scheme provides high security of digital data transaction. Public key cryptography uses two non-identical keys: one is public key another is private key but in symmetric block cipher scheme, it uses only one key to encryption and decryption of the data. Where storing of the symmetric key is infeasible there public key cryptography is efficient to secure the data. Among all the classical cryptography methods RSA cryptography is most popular and its hardness depends on the factorization of a number but the key size is large. Elliptic Curve Cryptography (ECC) provides the same level security as RSA and consumes the computational power and resources of the device because of its smaller key size. Elliptic Curve Discrete Logarithm Problem is defined as let F be an elliptic curve over the finite field Kp and $p = q^n$ where q is prime. Given P, Q $\in F(K_p)$, a scalar point k satisfies the equation Q = kP, if it exists. Here the public key is Q and the private key is k. The hardness of the ECDLP is to solve the equation because there is no mathematical proof to solve the equation. This intractibility increases the security level of the elliptic curve cryptography. This paper introduces two public key algorithms: one is Elliptic Curve Diffie–Hellman (ECDH) which is used for key exchange protocol and the other one is Elliptic Curve Digital Signature (ECDSA) which is used for digital signing.

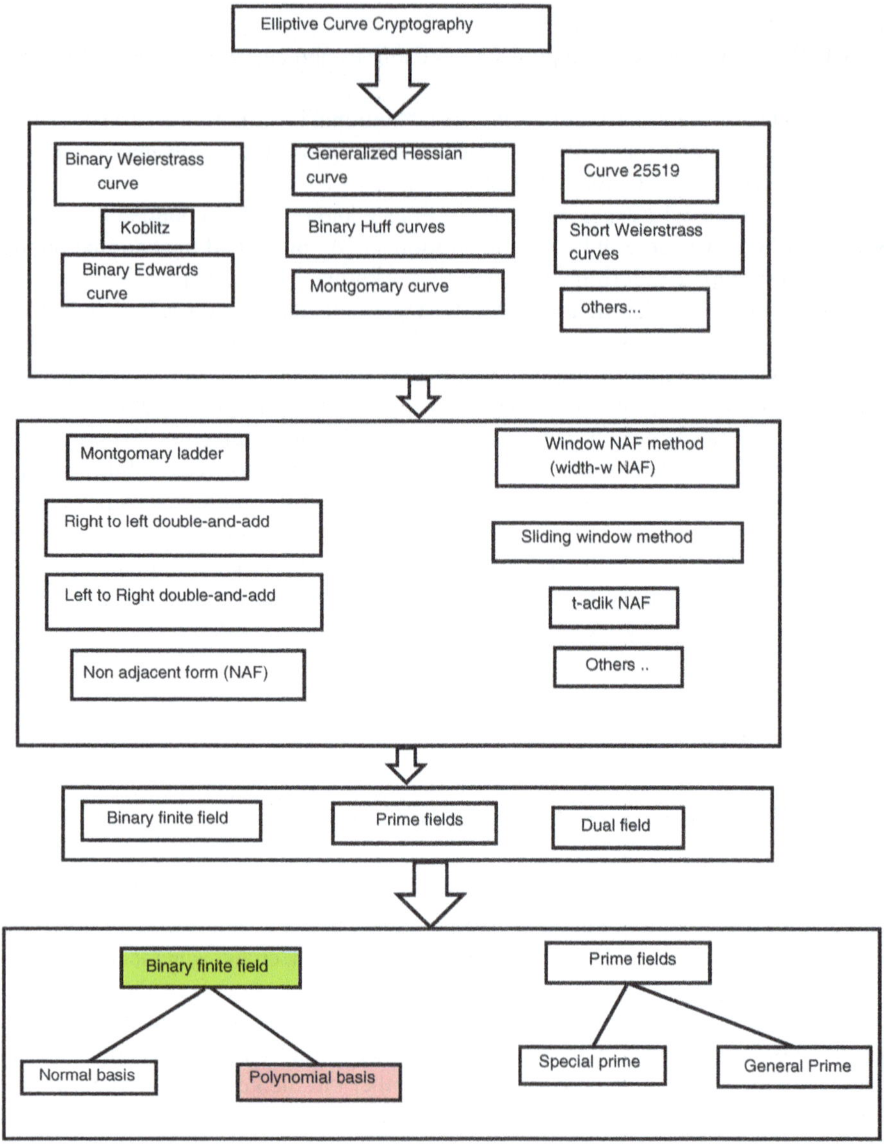

Fig. 1 ECC classification and design strategy

3.2 ECDSA Digital Signature

A digital signature algorithm is used to verify the authenticity of a digital content or documents. Digital signature confirms the receiver that the message is delivered by the authentic sender. The ECDSA consists of three steps: key generation, signature generation, and signature verification.
The ECDSA Algorithm:
Key generation:
1. First generate a random number as a private key.
2. Then compute Q(x, y)=d.G(x, y). Here Q is the public key.
Signature generation:
1. Choose a random number k from Zq.
2. Compute $R = (r_x, r_y) = k.G$
3. Calculate $r = r_x$ mod p.
4. Calculate $s = k^{-1}(H(m) + r.d)$ mod p.
This gives the result r and s.

Signature verification:
1. First compute t=(s-1)mod q.
2. Compute$v_1 = (H(m).t)$ mod q and $v_2 = (r.w)$mod q.
3. Calculate $(f_1, f_2) = (v_1.G + v_2.Q)$ mod q.
If the value of $f_2 = r$, then the verification is successful.

4 ECC Hardware Implementation on FPGAs

ECC design and implementation in an embedded system using FPGA are categorized based on the choosing of the finite field curves into three groups as shown in Fig 1. Major ECC implementations on FPGAs are on prime and binary fields, respectively. Another design type is the implementation with dual-field property.

4.1 Implementations of the Point Multiplication Based on FPGA in Binary Fields

Standard Curves overs the prime fields recommended by the NIST FIPSs are P-192, P-224, P-256, P-384, and P-521. Also, the Standard Curves over the binary fields recommended by the NIST FIPSs are K-163, B-163, K-233, B-233, K-283, B-283, K409, B-409, K-571, and B-571. There are two main advantages using GF binary field curve mathematics over the GF prime field hardware implementations which are as follows:

Table 1 ECC on Koblitz curves

Curve basis	Field	Board	Area	Frequency (Mhz)	Time (us)
Polynomial [17]	233	Stratix II(EP2S180F1020C3)	38056 ALM	181.06	8.09
Polynomial [12]	233	Virtex V4 (XC4VFX12)	2431	155.376	604
Gaussian Normal [13]	233	Stratix V	16421 ALM	246.1	6.8
Polynomial [14] [single]	233	Virtex2 (XC2V4000)	14091 slices	51.70	8.72
Polynomial [14] [parallel]	233	Virtex 2 (XC2V4000)	15916 slices	51.70	7.22
Polynomial [17]	163	Stratix II	20525 ALM	187.48	4.91
Polynomial [18]	163	Stratix II	13472 ALMs	155.5	26
Polynomial [20]	163	Virtex (XC5-110T)	2708 slices+5 BRAM	222.65	55
Gaussian [19]	163	Virtex(XC-2V2000)	6494 Slices+6 BRAMs	128	–
Polynomial [Proposed]	233	Kintex 7	22344	100	11.55

- The bit addition is performed using mod 2 and hence represented in hardware by simple XOR gated, and no Carry chain is required.
- The bit multiplications are represented in hardware using AND gates.

First FPGA-based(Altera Flex 10K FPGA) point multiplication implemented for Koblitz curve K-163 is proposed in [8] which computed in 45.6 ms. There are other implementations of FPGA-based point multiplication for the Koblitz curves which are presented in [14, 17]. Contributions of this FPGA implementations are as follows:

- Point operation parallelization and interleaving is proposed in [16]
- parallelization of scalable pointwise multiplication that can support all 5 NIST Koblitz curves without reconfiguring structures is proposed in [14]
- Hybrid 233 bit Karatsuba multipliers are used for ECC point multiplication, which obtain the maximum slice utilization of the FPGA [10] (we are using this one)

Table 1 shows the FPGA-based hardware implementation of different Kobilz curve ECCs. A high-speed ECC-based on FPGA for Koblitz curves is proposed in [15]. Physical attack resists ECC multiplier design which is proposed in [10], and here a masked Karatsuba Multiplier is proposed to prevent the side channel DPA attack. In Fig. 2 a generic N-bit masked multiplier is shown. The multiplicands A and B are masked with M_a and M_b, respectively. The input to the masked multiplier is the

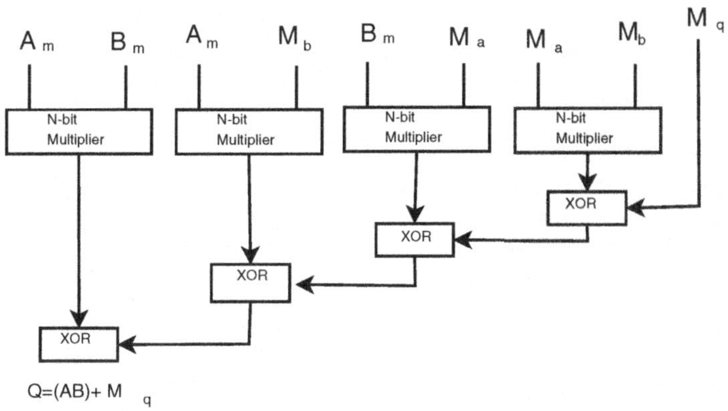

Fig. 2 Generic N-bit masked multiplier

masked value A_m and B_m, and the masks M_a, M_b, and M_q. The output Q is the product of unmasked multiplicands and the mask M_q.

5 Conclusion

In this paper, an FPGA-based hardware implementation of point multiplication (K-233) on the binary field is proposed. Performance table on the basis of area and critical time with similar kinds of previous implementation are presented in Table 1. We achieve better performance with respect to the existing results.

References

1. P.Moreira, J.Christiansen and K.Wyllie," The GBTx Link Interface Asic," v1.7 Draft, Oct.2011
2. Miller, V.S., Use of elliptic curve in cryptography, Advances in Cryptology, in: Proceedings of the Crypto'85, 1986, pp. 417-426
3. Koblitz, N.: Elliptic curve cryptosystems. Math. Comput. **48**, 203–209 (1987)
4. Dierks, T., Rescorla, E., The Transport Layer Security (TLS) Protocol Version 1.2, August 2008
5. WAP WTLS,, Wireless Application Protocol Wireless Transport Layer Security Specification Wireless Application Protocol Forum, February 1999
6. https://www.cnss.gov/Assets/pdf/minorUpdate1_Oct12012.pdf
7. Xianghong Hu, Xin Zheng , Shengshi Zhang , Weijun Li, Shuting Cai and Xiaoming Xiong,A High-Performance Elliptic Curve Cryptographic Processor of SM2 over GF(p)
8. K Itoh, M Takenaka, N Torii and S Okada, Implementation of Elliptic Curve Cryptographic Coprocessor over $GF(2^m)$ on an FPGA, in Proceedings of the International Workshop on Cryptographic Hardware and Embedded Systems (CHES), Worcester, Lecture Notes in Computer Science, Vol. 1965, (Springer-Verlag), 2000, pp. 25-40

9. Hankerson, D., Rodriguez-Henriquez, F., Ahmadi, O.: Parallel Formulations of Scalar Multiplication on Koblitz Curves. Journal of Universal Computer Science **14**(3), 481–504 (2008)
10. http://cse.iitkgp.ac.in/ debdeep/osscrypto/eccpweb/pubs/vdat2007.pdf
11. K Jarvinen and J Skytta, High-Speed Elliptic Curve Cryptography Accelerator for Koblitz Curves,16th International Symposium on Field-Programmable Custom Computing Machines, 2008, pp. 109-118
12. Cinnati Loi, K.C., Ko, S.B.: High performance scalable elliptic curve cryptosystem processor for Koblitz curves. Microprocess. Microsyst. **37**, 394–406 (2013)
13. C. Realpe-Munoz, P. and Velasco-Medina, J., High-performance elliptic curve cryptoprocessors over GF(2m) on Koblitz curves, Analog Integr. Circ. Sig. Process, Vol. 85, 2015, pp. 129-138
14. Ahmadi, O., Hankerson, D., Rodriguez-Henriquez, F.: Parallel Formulations of Scalar Multiplication on Koblitz Curves. Journal of Universal Computer Science **14**(3), 481–504 (2008)
15. Jarvinen, K.: Optimized FPGA-based elliptic curve cryptography processor for high-speed applications. Integration, the VLSI journal **44**, 270–279 (2011)
16. Jarvinen, K., Skytta, J.: Fast point multiplication on Koblitz curves: Parallelization method and implementations. Microprocess. Microsyst. **33**, 106–116 (2009)
17. Jarvinen, K. and Skytta, J., High-Speed Elliptic Curve Cryptography Accelerator for Koblitz Curves, in Proceedings of the 16th International Symposium on Field-Programmable Custom Computing Machines, 2008, pp. 109-118
18. Jarvinen, K., Skytta, J.: On Parallelization of High-Speed Processors for Elliptic Curve Cryptography, IEEE Trans. Very Large Scale Integr. Syst. **16**(9), 1162–1175 (2008)
19. Dimitrov, V.S., Jarvinen, K., Jacobson, M.J., Chan, W.F. and Huang, Z., FPGA Implementation of Point Multiplication on Koblitz Curves Using Kleinian Integers, in Proceedings of the International Workshop on Cryptographic Hardware and Embedded Systems (CHES), Worcester, Lecture Notes in Computer Science, Vol. 1965, (Springer-Verlag), 2006, pp. 445-459
20. Cinnati Loi, K.C., Ko, S.B.: Parallelization of Scalable Elliptic Curve Cryptosystem Processors in GF(2m). Microprocess. Microsyst. **45**, 10–22 (2016)

Automatic Text Summarization for Odia Language: A Novel Approach

Sagarika Pattnaik and Ajit Kumar Nayak

Abstract In the era of Artificial Intelligence, Natural Language Processing (NLP) has become an integral part in the process of development. The methodology aims at making the machine or the computer more interactive with humans saving time. Auto text summarization is one such subfield of NLP that has gained importance in the present scenario as it helps to achieve the task faster. This paper focuses on the morphological aspect of Odia language in the computational scenario. An overview of text summarization process with its progress in Indian language has been discussed. It also suggests a simple statistical method for auto text summarization. The result obtained can be considered significant.

Keywords NLP · Text summarization · TF-IDF · F score

1 Introduction

Auto Text summarization is the process of making an abridged version of a text document, without losing its information content. The overall process is automated with the help of a computer. Text summarization approaches can be categorized into extractive, i.e., shallow or knowledge-poor approach and abstractive, i.e., deep or knowledge-rich approach. Knowledge-rich approaches require deep linguistic knowledge and are harder to implement than shallow approaches. Most of the research have been carried out for extractive methods due to their simplicity and cost-effectiveness. Statistical, linguistic, and machine learning approaches have made the summarizers efficient in different aspects. Different summarization tools are also available for English and other European languages like Resoomer, Auto Summarizer, Tools 4 Noobs, and SUMMA [1–3].

S. Pattnaik (✉)
Department of CSE, ITER, SOA (Deemed to be University), Bhubaneswar, India
e-mail: sagarika.pari@gmail.com

A. K. Nayak
Department of CS&IT, ITER, SOA (Deemed to be University), Bhubaneswar, India
e-mail: ajitnayak2000@gmail.com

D. Mishra et al. (eds.), *Intelligent and Cloud Computing*,
Smart Innovation, Systems and Technologies 194,
https://doi.org/10.1007/978-981-15-5971-6_43

This paper focuses on the extractive method of text summarization and tries to give a computational framework to Odia language considering its intrinsic features [4]. Indian languages particularly Odia are not computationally advanced in comparison to European languages like English, and it has become an important criterion for being motivated to carry out the task. The application of a summarizer in various tasks like research-patents, tele-health supply chain, story lines of events, extracts from news articles, etc. has encouraged to perform the experiment.

The rest of the paper is organized as follows:

Section 2 presents some of the relevant works on the present task. Section 3 provides a morphological view of Odia language. Section 4 describes the data set used. Section 5 discusses the proposed model for Odia text summarization. In Sect. 6, evaluation process is discussed. Finally in Sect. 7, the paper is concluded with a future direction for further research.

2 Literature Survey

Statistical extractive text summarization models are the basis for pioneering works. Luhn [5], who is considered as the father of text summarization proposed a model for English documents based on the count of word frequency and their relative position in the sentence. Another statistical model has also been developed by Edmundson considering additional features like key words, cue words, Title and heading words, and sentence location [6]. Hans et al. [7] have suggested a single document automatic text summarization using Term Frequency-Inverse Document Frequency (TF-IDF) for English text documents. The result of this research produces 67% of accuracy. Unlike other artificial intelligence which needs machine learning, this automatic summarization experiment does not need any machine learning due to the use of existing libraries such as NLTK and Text Blob. Statistical-based extractive text summarization is further carried out by several researchers, considering different parameters like TF-IDF in combination with linguistic features, etc. [7, 8] and has got an acceptable result.

Other techniques like machine learning approaches that require a voluminous corpus have also been applied to European languages to get higher results. Kupiec et al. [9] has applied Bayesian classification function and claimed of getting a desired result. Other machine learning approaches applied to text documents are SVM, CRF [10], Hidden Markov [11], Neural Network [12, 8], and Naïve Bayes [10].

Though meager in comparison to European languages, a few noticeable works have been done in Indian languages. P. Shah and N. P. Desai [13] have given a brief summary of automatic extractive text summarization techniques adopted for various foreign and Indian languages. Techniques adopted are Sentence scoring method, combination of Random Indexing and Page Rank, Bayesian theory based, Clustering, SVM, Stochastic based method using TF-IDF score. The general compression ratio is maintained within the range of 20 to 25%.

Summarization models based on statistical methods like considering the count of occurrence of thematic words [14] and TF-IDF value of unigram and bigram terms [15] have been proposed and have achieved an appreciable result. Another summarizer based on SVM a machine learning technique has been applied to Hindi text document from the news domain [13]. They have achieved 72% accuracy with 50% compression ratio and 60% accuracy with 25% compression ratio.

Similarly summarizers have been developed for Bengali text documents basing on different techniques like statistical sentence scoring technique [16, 17] and K means clustering [18] are reported. An efficient summarizer for the Tamil language has also been proposed [19] that uses graph theoretic scoring technique. Word analysis and sentence analysis have been performed during the process. The text ranking algorithm used is not domain-specific and does not need annotated corpora. Rouge evaluation tool has been used for performance measure and has got a score of 0.4723.

From the literature survey, it has been analyzed that summarization models have achieved their heights in languages like English and the reasons are manifold like availability of voluminous corpus, availability of NLP tools compatible with the language, etc. But for Indian languages unavailability of adequate amount of machine compatible corpus especially for Odia language has been the reason for slow growth.

3 Overview of Odia Language

Odia is a morphologically rich language [4]. It has an agglutinative nature. The use of infix is rare in Odia. There are no capital letters to distinguish the words as a noun. The structure of the sentence is in the form of *subject object verb (sov)*. Instead of prepositions as in English language, Odia language uses postposition. Unlike Hindi, in Odia language most of the postpositions are attached to the main word (Noun/pronoun) like Marathi. Unlike Hindi and English the pronouns in Odia do not distinguish between genders.

These are some of the features that make computation of the language complex.

In computational linguistics, representing a language is a major problem and to create a computational model for resource-poor language like Odia is a difficult task.

4 Data Set

Documents belonging to varied domains like news, stories, etc. are taken as input for the model. It comprises of Odia texts, on an average size of 30 lines. The documents are normalized before being analyzed by the model. Different linguistic features are taken into consideration during the normalization process. The reference gold standard summaries are framed by expert linguists.

5 Proposed Model

The model is based on the extractive method applying statistical computation on the input data set.

The sequences of steps that are followed for the process are

1. Preprocessing
2. Sentence extraction
3. Context ordering
4. Final output summary

5.1 Preprocessing

Preprocessing of the input text corpus comprises sentence segmentation, tokenization, stemming, and removal of stop words. We have developed our own tool for the mentioned preprocessing steps.

5.1.1 Sentence Segmentation

It is the process of dividing a running text into sentences. On the basis of punctuation mark, the process is carried out.

5.1.2 Tokenization

Each sentence is broken down into individual components called tokens. The existence of spaces is considered as token delimiter.

5.1.3 Stemming

On the basis of similarity among tokens, i.e., number of characters similar among the compared words, stemming is carried out.

5.1.4 Stop Word Removal

There are words in a document, whose presence does not have any impact in the calculation of sentence score. These words are removed during the preprocessing step.

5.2 Sentence Extraction Process

The model uses Term Frequency-Inverse document (TD-IDF) [20] a simple statistical method for extracting the output summary. The terms with the highest TF-IDF score are the terms that best characterize the topic of the document and the TF-IDF value increases proportionally to the number of times a word appears in the document. It is offset by the frequency of the word in the corpus. This helps to adjust for the fact that some words appear more frequently in general. Each sentence is scored using the TF-IDF values calculated. The sentences are then ranked according to the score value. Finally, the sentences are then sorted in descending order and we get a summary based on the most important sentences found in a document. The number of sentences to be extracted depends on the compression rate which is kept at 50%. This simple statistical technique is novel as far as the application on Odia text document is concerned.

Algorithm 1: Summarization Process

1. `Input text document`
2. `Preprocessing`
   (i) `Sentence segmentation`
   (ii) `Tokenization`
   (iii) `Stop word removal`
   (iv) `Stemming`
3. `Calculate TF_IDF of the tokens according to Eq. 1:`

$$TF - IDF = TF * IDF \tag{1}$$

$$TF = \frac{\text{frequency of a word or a term in a document}}{\text{Total words in a document}} \tag{2}$$

$$IDF = \log \frac{\text{total number of documents}}{1 + \text{number of documents containing the term}} \tag{3}$$

4. `Calculate the sentence score` sc_i `of each sentence in the document according to Eq. 4.`

$$Sc_i = \sum TF\_IDF \tag{4}$$

5. `Rank the sentences according to their sentence score values.`
6. `Select the required number of high scoring sentences in consideration with the compression ratio.`

5.3 Context Ordering

5.3.1 Information Ordering

Information ordering is of the following types:

Chronological Ordering
It is the method of ordering sentences by the date of the document.
Example: For summarizing news articles.

Coherence
This method chooses ordering on the basis of similarity among neighboring sentences.

Topical Ordering
In this method, ordering is done according to the topic in the source document.

5.4 Sentence Realization

It includes

- Compression and simplification of sentences
- Checking further for coherence
- Keeping longer or more descriptive phrases before short, reduced, or abbreviated forms

6 Evaluation

The output summary is validated by comparing it with expert summaries created by linguists.

F score is taken as the metric for evaluation.

Figure 1 shows the resultant F scores of 10 documents and their average value.

Figures 2 and 3 show an example of input text document and its output summary.

$$\text{Precision}(p) = \frac{\text{Number of sentences similar in system generated summary and ideal summary}}{\text{Number of sentences in system generated summary}} \quad (5)$$

$$\text{Recall}(r) = \frac{\text{Number of sentences similar in system generated summary and ideal summary}}{\text{Number of sentences in ideal summary}} \quad (6)$$

$$F\text{score} = \frac{2.p.r}{p + r} \quad (7)$$

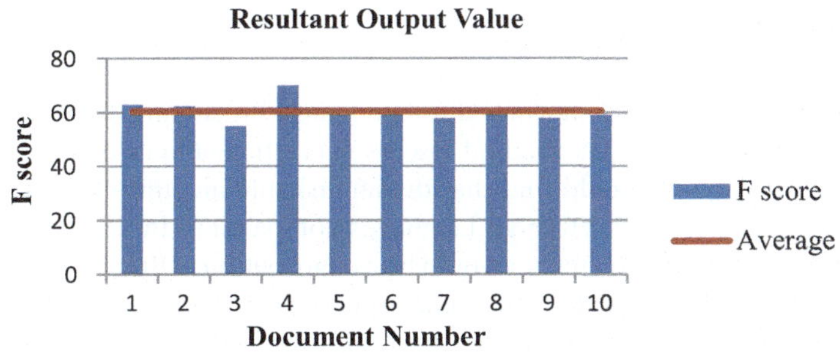

Fig. 1 Resultant F score values of documents

ଭାରତୀୟ କ୍ରିକେଟ ଦଳର ପୂର୍ବତନ ଅଧିନାୟକ ସୌରଭ ଗାଙ୍ଗୁଲି ଆଜି ଗୋପବନ୍ଧୁ ଭବନ ପରିଦର୍ଶନରେ ଆସିଥିଲେ । ସ୍ଥାନୀୟ ବାରବାଟୀ ଷ୍ଟାଡିୟମରେ ଅନୁଷ୍ଠିତ ପୂର୍ବାଞ୍ଚଳ ରଣଜୀ ଦିନିକିଆ କ୍ରିକେଟର ଆଜି ଓଡିଶା-ବଙ୍ଗଳା ମ୍ୟାଚରେ 'ମହାରାଜା' ଭାଗ ନେଇଥିଲେ । ଖେଳ ଶେଷ ହେବା ପରେ କୋଲକାତା ଅଭିମୁଖେ ଯାତ୍ରା କରିବା ପୂର୍ବରୁ ସୌରଭ ଗୋପବନ୍ଧୁ ଭବନ ଆସି ଉତ୍କଳମଣି ପଣ୍ଡିତ ଗୋପବନ୍ଧୁ ଦାସଙ୍କ ପ୍ରତିମୂର୍ତ୍ତିରେ ମାଲ୍ୟାର୍ପଣ କରିଥିଲେ । ଲୋକସେବକ ମଣ୍ଡଳର ଆଜୀବନ ସଦସ୍ୟ ନିରଞ୍ଜନ ରଥ, 'ସମାଜ' ସମ୍ପାଦକ ଶରତ ମିଶ୍ର ଓ ମୁଖ୍ୟ ପରିଚାଳକ ହେମନ୍ତ କୁମାର ସାହୁ କୋଲକାତା ନାଇଟ୍ ରାଇଡର୍ସ ଅଧିନାୟକ ସୌରଭଙ୍କୁ ସ୍ୱାଗତ କରିବା ସହ 'ସମାଜ' ମାଧ୍ୟମରେ କିପରି ଦୁଃଖି ଦରିଦ୍ରଙ୍କୁ ସହାୟତା ପ୍ରଦାନ କରାଯାଉଛି ସେ ସମ୍ପର୍କରେ ଉଲ୍ଲେଖ କରିଥିଲେ । ସେହିପରି ସ୍ୱାଧୀନ ଭାରତ ସହ ସ୍ୱତନ୍ତ୍ର ଓଡିଶା ରାଜ୍ୟ ଗଠନରେ ରାଜ୍ୟର ସର୍ବବୃହତ ସମ୍ବାଦପତ୍ର 'ସମାଜ'ର ଭୂମିକା ସମ୍ପର୍କରେ ଅବଗତ କରାଇଥିଲେ । ସୌରଭଙ୍କ ସହ ଓଡିଶା କ୍ରିକେଟ ଆସୋସିଏସନ କୋଷାଧ୍ୟକ୍ଷ ସତ୍ୟ ମହାନ୍ତି, ଉପସଭାପତି ନବରଞ୍ଜନ ପଟ୍ଟନାୟକ, ଯୁଗ୍ମ ସମ୍ପାଦକ ସଞ୍ଜୀବ ଦତ୍ତ, ଓଡିଶା କ୍ରିକେଟ ଏକାଡେମୀ ନିର୍ଦ୍ଦେଶକ ଉପେନ୍ଦ୍ର ଦାସ, କୋଚ ତଥା ଓଡିଶା ଦିନିକିଆ ରଣଜୀ ଦଳର ମ୍ୟାନେଜର ମାନସ ରାୟ ଉପସ୍ଥିତ ଥିଲେ ।

Transliterated text

bhāratiẏa kriketa dalara purbatana adhināẏaka saurabha gāṅguli āji gopabandhu bhabana paridarsanare āsithile. sthāniẏa bārabāṭi sṭāḍiẏamare anusṭhita purbāñchaḷa raṇaji dinikiā kriketara āji oḍisā-baṅgaḷā miāchare 'mahārājā' bhāga neithile. kheḷa sesa hebā pare kolakātā abhimukhe jātrā karibā purbaru saurabha gopabandhu bhabana āsi utkaḷamaṇi paṇḍita gopabandhu dāsaṅka pratimurttire māliārpaṇa karithile. lokasebaka maṇḍaḷara ājibana sadasẏa nirañjana ratha, 'samāja' sampādaka sarata misra o mukhẏa parichāḷaka hemanta kumāra sāhu kolakātā nāiṭ rāiḍarsa adhināẏaka saurabhaṅku sbāgata karibā saha 'samāja' mādhẏamare kipari duḥkhi daridraṅku sahāẏatā pradāna karājāuchhi se samparkare ullekha karithile. sehipari sbādhina bhārata saha sbatantra oḍisā rājẏa gaṭhanare rājẏara sarbabruhata sambādapatra 'samāja'ra bhumikā samparkare abagata karāithile. saurabhaṅka saha oḍisā kriketa āsosiesana kosādhẏakhia satẏa mahānti, upasabhāpati nabarañjana paṭṭanāẏaka, jugma sampādaka sañjiba datta, oḍisā kriketa ekāḍemi nirdesaka upendra dāsa, kocha tathā oḍisā dinikiā raṇaji daḷara miānejara mānasa rāẏa upasthita thile.

Fig. 2 Input text document

ଭାରତୀୟ କ୍ରିକେଟ ଦଳର ପୂର୍ବତନ ଅଧିନାୟକ ସୌରଭ ଗାଙ୍ଗୁଲି ଆଜି ଗୋପବନ୍ଧୁ ଭବନ ପରିଦର୍ଶନରେ ଆସିଥିଲେ । ସ୍ଥାନୀୟ ବାରବାଟୀ ଷ୍ଟାଡିୟମରେ ଅନୁଷ୍ଠିତ ପୂର୍ବାଞ୍ଚଳ ରଣଜୀ ଦିନିକିଆ କ୍ରିକେଟର ଆଜି ଓଡିଶା-ବଙ୍ଗଳା ମ୍ୟାଚରେ 'ମହାରାଜା' ଭାଗ ନେଇଥିଲେ । ସୌରଭଙ୍କ ସହ ଓଡିଶା କ୍ରିକେଟ ଆସୋସିଏସନ କୋଷାଧ୍ୟକ୍ଷ ସତ୍ୟ ମହାନ୍ତି, ଉପସଭାପତି ନବରଞ୍ଜନ ପଟ୍ଟନାୟକ, ଯୁଗ୍ମ ସମ୍ପାଦକ ସଞ୍ଜୀବ ଦତ୍ତ, ଓଡିଶା କ୍ରିକେଟ ଏକାଡେମୀ ନିର୍ଦ୍ଦେଶକ ଉପେନ୍ଦ୍ର ଦାସ, କୋଚ ତଥା ଓଡିଶା ଦିନିକିଆ ରଣଜୀ ଦଳର ମ୍ୟାନେଜର ମାନସ ରାୟ ଉପସ୍ଥିତ ଥିଲେ ।

Transliterated text

bhāratiẏa kriketa daḷara purbatana adhināẏaka saurabha gāṅguli āji gopabandhu bhabana paridarsanare āsithile. sthāniẏa bārabāṭi sṭāḍiẏamare anusṭhita purbāñchaḷa raṇaji dinikiā kriketara āji oḍisā-baṅgaḷā miāchare 'mahārājā' bhāga neithile. saurabhaṅka saha oḍisā kriketa āsosiesana kosādhẏakhia satẏa mahānti, upasabhāpati nabarañjana paṭṭanāẏaka, jugma sampādaka sañjiba datta, oḍisā kriketa ekāḍemi nirdesaka upendra dāsa, kocha tathā oḍisā dinikiā raṇaji daḷara miānejara mānasa rāẏa upasthita thile.

Fig. 3 Output summary

7 Conclusion

Text summarization is a vital area in Natural Language Processing. It has a prominent role in our ongoing works. The method discussed is efficient in its own aspect. From the literature survey, we could find that distinguishable amount of work has been done for European and other foreign languages compared to Indian languages. So, this proposed summarization model would be a contribution to the society, especially for Odia language. But there are still some aspects the model lacks and needs to be focused like applying better redundancy removal process and consideration of more linguistic features that can aid in increasing the efficiency of the summarizer. Our future research will concentrate on the development of more efficient summarizers that does not contain the aforementioned flaws and on the process of development of corpus.

References

1. https://www.resoomer.com/en
2. https://www.tools4noobs.com/summarize
3. https://www.pypi.org/project/summa
4. Pradhan K.C., Hota B.K., Pradhan B.: Saraswat Byabaharika Odia Byakarana. Styanarayan Book Store, Fifth Edition (2006)
5. Luhn, H.P.: The automatic creation of literature abstracts. IBM J. Res. Dev. **2**(2), 159–165 (1958)
6. Edmundson, Harold P.: New methods in automatic extracting. J. ACM (JACM) **16**(2), 264–285 (1969)
7. Qaroush, A., et al.: An efficient single document arabic text summarization using a combination of statistical and semantic features. J. King Saud Univ.-Comp. Inform. Sci. (2019)
8. Narayan, S., Shay, B.C., Mirella L.: Ranking sentences for extractive summarization with reinforcement learning. arXiv preprint arXiv:1802.08636 (2018)
9. Kupiec, J., Pedersen, J., Chen, F.: A Trainable document summarizer. Dans les actes de ACM Special Interest Group on Information Retrieval (SIGIR), pp. 68–73 (1995)
10. Neto, J.L., Alex, A.F., Celso, A.A.K.: Automatic text summarization using a machine learning approach. Brazilian Symposium on Artificial Intelligence, Springer, Berlin, Heidelberg (2002)
11. Conroy, J.M., Dianne, P. O.: Text summarization via hidden markov models. Proceedings of the 24th annual international ACM SIGIR conference on Research and development in information retrieval, ACM (2001)
12. Khosrow, K.: Automatic text summarization with neural networks. In Proceedings of Second International Conference on Intelligent Systems, IEEE, Texas, USA, June, pp. 40–44.3 (2004)
13. Prachi, S., Nikita, P. D.: A survey of automatic text summarization techniques for Indian and foreign languages. International Conference on Electrical, Electronics, and Optimization Techniques (ICEEOT), pp. 4598–4601, IEEE (2016)
14. Kumar, K.V., Yadav, D.: An improvised extractive approach to hindi text summarization. In Information Systems Design and Intelligent Applications, pp. 291–300, Springer, New Delhi (2015)
15. Vijay, S., et al.: Extractive text summarisation in Hindi. 2017 International Conference on Asian Language Processing (IALP). IEEE (2017)
16. Sarkar, K.: Bengali text summarization by sentence extraction. arXiv preprint arXiv:1201.2240 (2012)

17. Abujar, S., et al.: A heuristic approach of text summarization for Bengali documentation. 2017 8th International Conference on Computing, Communication and Networking Technologies (ICCCNT). IEEE (2017)
18. Akter, S., et al.: An extractive text summarization technique for Bengali document (s) using K- means clustering algorithm. 2017 IEEE International Conference on Imaging, Vision & Pattern Recognition (icIVPR). IEEE (2017)
19. Kumar, S., Ram, V.S., Devi, S.L.: Text extraction for an agglutinative language, Proceedings of Journal: Language in India, pp. 56–59 (2011)
20. Salton, G., et al.: Automatic text structuring and summarization. Inf. Process. Manage. **33**(2), 193–207 (1997)

Insight into Diverse Keyphrase Extraction Techniques from Text Documents

Upasana Parida, Mamata Nayak, and Ajit Ku. Nayak

Abstract Keyphrase refers to a set of terms which best present the document content in a brief way. As the world is already approached toward digital documents, it is more crucial to find out a particular document accurately and efficiently. This can be achieved by keyphrase extraction as it describes the core information of documents which will be helpful to find target documents for the users. The keyphrase extraction is one of the challenging research areas in the natural language processing field because of the involvement of the voluminous unstructured and unorganized data. This paper surveys various keyphrase extraction technologies and their respective families. A systematic review of this paper provides an overview of the existing technologies with their pros and cons, and explores the direction for future development and research in this field.

Keywords Keyphrase extraction · Information retrieval · Unsupervised method · Semi-supervised method · Supervised method

1 Introduction

While working with text, extracting keyphrase is one of the major tasks as they are the smallest unit of the text that can summarize the content and pin down the most relevant information. Different terminologies are used for referring the words containing most relevant information of the document such as keywords, keyphrases, key terms, or key segments. This document refers to the relevant term as keyphrase.

U. Parida (✉)
Computer Science and Engineering, SOA Deemed to be University, Bhubaneswar, India
e-mail: upasanaparida@gmail.com

M. Nayak · A. Ku. Nayak
Computer Science and Information Technology, SOA Deemed to be University, Bhubaneswar, India
e-mail: mamatanayak@soa.ac.in

A. Ku. Nayak
e-mail: ajitnayak@soa.ac.in

D. Mishra et al. (eds.), *Intelligent and Cloud Computing*,
Smart Innovation, Systems and Technologies 194,
https://doi.org/10.1007/978-981-15-5971-6_44

The idea behind it is to reduce the bag of words into handful of words in such a way that the text representation dimensionality reduces without compromising with the core concept. Keyphrase extraction can be integrated successfully in the following fields: [1, 2].

- **Information Retrieval**:

Retrieving relevant information and discarding irrelevant information from huge structured or unstructured data is called information retrieval. Vector Space Model and Boolean Model are some of the models used for information retrieval. Here keyphrase extraction technique can be integrated to provide a better result.

- **Indexing:**

The terms used for defining the topic of the document are called Indexing. The motive of indexing is to set up vocabulary and can be used to extract keyphrases. Index can be in any form: phrase, words, or numeric.

- **Text Summarization:**

The process of creating summary from the text having the main points is called text summarization. This process could be used in keyphrase extraction to convert the large document into small relevant text.

- **Text Classification:**

The task which provides a predefined category to the text is called text classification. For example, gender can be classified as male or female.

- **Bio Text Mining:**

Bio text mining refers to the study of the procedure text mining applied to biomedical text mining field [3].

- **Question Answering System:**

Question answering systems (QASs) reply question's answers which are required in languages. The research work in QASs started in 1960s and afterward, a huge number of QASs have been developed.

The phases involved in various methods of keyphrase extraction [4] are broadly described as follows:

- Candidate generation: All the possible candidate keyphrases are detected from the text.
- Property calculation: Determination of properties and statistics required for the purpose of scoring.
- Ranking: Finding the rank for each candidate keyphrase and top candidates are finally selected for representing the text (Fig. 1).

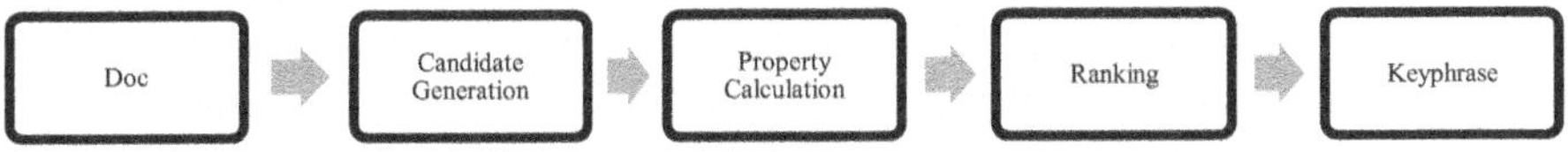

Fig. 1 Block diagram of keyphrase extraction method

This paper is organized as follows: the first section, an overview of different approaches of keyphrase extraction methods; the second section describes related notable existing works; the third section is analyzing and concluding about different discussed approaches and the fourth section describes the future scope for research in this field.

2 Overview of Different Approaches of Keyphrase Extraction

Keyphrase extraction techniques can be categorized into unsupervised, semi-supervised, and supervised [5]. Unsupervised methods further can be subdivided based on the approach, such as linguistic, statistics, graph-based approach, or machine learning based or hybrid approach (Fig. 2).

A. **Unsupervised Methods**

Unsupervised methods have a lot of advantages over supervised methods as they are independent of training data. The existing unsupervised approach can be divided into linguistic, statistical, graph-based, machine learning based, and hybrid method.

A1. **Linguistic Approach**:

Linguistic approaches are based on the rules derived from the linguistic features of the terms, sentences, and graphs. This method provides accuracy but with a disadvantage

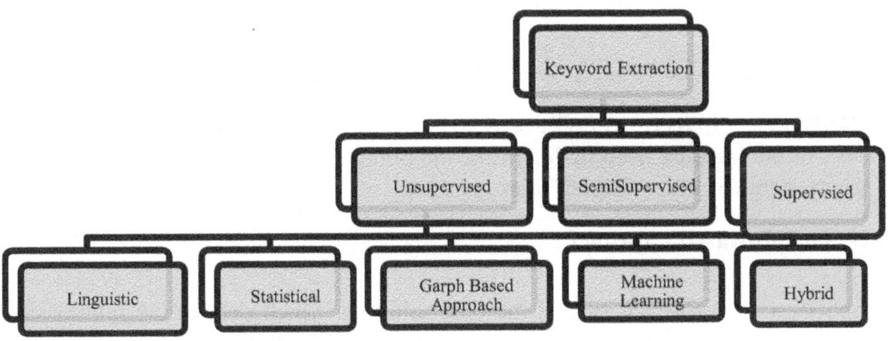

Fig. 2 Hierarchy of different keyphrase extraction

of being computationally intensive and requires domain and language expertise. Lexical analysis, syntactic analysis [31], semantic [6, 32], and disclosure analysis are some of the linguistic approaches of keyword extraction which are very complex in nature [7]. Electronic dictionary, tree tagger, WordNet, n-grams are some of the resources used for the purpose of lexical analysis [8].

- **Tree Tagger**—This tool is used to provide the text with part of speech and lemma information.
- **WordNet**—This is one of the English dictionary based lexical database that is designed by the influence of psycholinguistic theory of human lexical memory.

  Similarly noun phrases, chunks are used as syntactic analysis resources [9].

A2. **Statistical Approach:**

Statistical approaches are simple methods with no prior training data requirement. In addition to this, it is language and domain independent. Comparing with linguistic approach, it provides less accuracy but a large volume of dataset makes it possible to perform statistical analysis and provides good result [10]. This approach generally learns precious statistical information from large corpus to extract keyphrases. Some of the statistics methods term frequency-inverse document frequency (TF-IDF) and Patricia (PAT) tree are used to find out the statistics of the word from the document to identify keyphrases.

- **TF-IDF-**This TF*IDF is a technique for information retrieval that measures a term's frequency (TF) and its inverse document frequency (IDF). The TF is simply the frequency of terms in a document. The IDF is a well-judged weight of a term in the text collection. It is first introduced in 1972 and is extremely popular then onward in natural language processing. If d represents a document, $|D|$ represents all considered document (size of corpus) and w represents a word in the document. In a document D and a word w, the expression for computing TF-IDF is as given in Eq. (1):

$$TF * IDF = f_{w,d} * \frac{\log|D|}{f_{w,d}} \tag{1}$$

- **Patricia (PAT) tree**—A PAT tree is no other a digital tree where each independent bits of the keys are used to determine the branch of the tree. So Patricia trees are represented as binary digital trees [11, 12].

A3. **Graph-Based approach:**

Graph-Based approach is one of the best solutions which defines the relationship and structural information in an efficient way [13, 14]. In this approach, documents are considered as graphs with the terms as vertices and their relationship represented by the edges. This approach works based on the idea of voting or recommendation. This means, when one vertex connects to another vertex, by default it is casting a vote to that vertex. TextRank and PageRank are some of the graphical approach automatic keyword extraction algorithm. TextRank is used for text ranking where else PageRank is used to compute the weight of the web pages [15].

A4. **Machine Learning Approaches:**

The machine learning approach works either supervised or unsupervised, but in reality supervised approach is preferred between these two approaches. The supervised method develops a model which gets trained by a set of keyphrases. The learning data set requires manual intervention which is a very tedious and time-consuming process. This approach includes various machine learning models like Super Vector Model (SVM), C4.5, Naïve Bayes, etc [16].

- **SVM**—SVM is a supervised algorithm associated with a learning algorithm, and is commonly adopted for classification and regression challenges.
- **Naïve Bayes**—Naive Bayes are a collection of classification algorithms based on Bayes' Theorem. Bayes' Theorem states that if *A* and *B* are two events, then mathematically they can be represented by the equation below:

$$P(A|B) = \frac{P(B|A) * P(A)}{P(B)} \tag{2}$$

A5. **Hybrid Approach:**

Either combining two or more approaches discussed above or by using heuristics (such as position and length), a hybrid approach can be modeled for unsupervised keyphrase extraction. It will provide the benefits of both methods for extracting keyphrases.

B. **Semi-supervised Methods:**

Semi-supervised methods aim to construct a function that is perfectly smooth to the intrinsic structure of the phrases. The assumption is that semantically connected phrases are most likely to have relevant scores, the function to be estimated is needed to provide the title phrases a higher score and also locally smooth on the constructed hyper-graph [2, 5, 17].

C. **Supervised Methods:**

Unlike unsupervised methods, supervised methods are domain dependent and require labeled training data. The shortcoming of the supervised methods is that the result generally is found to be biased.

3 Related Prominent Existing Works on Keyphrase Extraction

This section reviews some of the noteworthy keyword extraction works of different approaches.

(A) Krulwich and Burkey [6] (1996) used heuristics methods such as the position, the length, the layout features of the terms, html, and similar tags, the text formatting, etc. for selecting keyphrases from a document. As the heuristics are syntactic ones, the phrases are present in section headers and use the acronyms.

(B) Pasquier [18] (2010) used sentence clustering and Latent Dirichlet Allocation (LDA) for keyphrase extraction of a single document. The algorithm is based on clustering of sentences of the documents for emphasizing the sections of text that are semantically correlated.

(C) Turney [3] (2003) used statistical relation among keyphrases to enhance the coherence of the detected keyphrases.

(D) Erkan and Ramdev [14] (2004) introduced graph-based technique for extracting important sentences. This is based on the concept of eigenvector centrality in a graph representation of sentences. LexRank is very insensitive to the noise in the data.

(E) Mihalcea and Tarau [12] (2004) proposed the model TextRank. The TextRank algorithm is followed by the PageRank algorithm of Google. TextRank can also be used for sentence extraction instead of keywords. It does not require core lexical knowledge, domain knowledge, or language knowledge. TextRank does not rely on the local context but recursively drawn from the entire text.

(F) Litvak and Last [15] (2008) have given a comparison between supervised and unsupervised approaches for keywords extraction. The methods are established on the graph-based syntactic representation of text and web documents.

(G) The output of the HITS algorithm on a set of summarized documents achieved comparatively to supervised methods. The authors advised based on the simple degree-based rankings from the first iteration of HITS, rather than running it to its convergence, which should be considered.

(H) Cai& Cao move forward [16] (2017) SVM rank model which is used for ranking keywords that is compared by some traditional algorithms such as term frequency and inverse document frequency(TF-IDF), Latent Dirichlet allocation(LDA) and Text Rank. This experiment improved precision and recall of SVM ranking for keyword extraction 6% and 5%, respectively.

(I) Palshikar [17] (2007) introduced a hybrid structural and statistical technique for detection of keyphrases from an individual document. The undirected co-occurrence network, using a dissimilarity measure between two terms, computed from the frequency of their co-occurrence in the preprocessed and lemmatized document, as the edge weight, was shown to be appropriate for the centrality measures based approach for keyword extraction.

(J) Decong Li et al. [5] (2010) introduced a semi-supervised base by using a commonly believed concept that the heading of a document is generally determined to show the core concept of the document and thus it is natural to have close semantics to the title of the document.

(K) Turney proposed GenEx [19] [2002], one of the supervised learning approaches in classification problem. A GenEx is an amalgam of two independent modules, Genitor and Extractor. Extractor is the real automatic keyphrase extraction system. The most important key feature of GenExis is the capability to maintain its performance throughout various domains.

(L) Frank et al. introduce Keyphrase Extraction Algorithm (KEA) [20] (1999) which is one of the popular supervised keyphrase extraction approaches. It emphasizes on selecting and weighting candidate and scoring process of the term. KEA utilized two features firstly occurrence and secondly TF xIDF for candidate keyphrases as the classification scores together with Naïve Bayes as the machine learning algorithm. KEA developer also tried his best to increase the precision and recall value with the help of integration with the degree of domain dependence.

4 Conclusion

From the discussion above, a survey on different approaches by various researchers for the intention of keyphrase extraction was performed and the details of various properties selection criteria commonly adopted to score the candidate keyphrases based on their significance of the target text. Some of the important work done in this field have been listed chronologically. Table 1 provides some noteworthy inferences drawn from the discussed research works.

5 Future Work

Although the existing keyphrase extraction techniques are quite beneficial, still need exists to develop more advanced unsupervised, domain-independent, and language-independent techniques. The review done in this paper highlights the positive and negative part of some renowned keyphrase extraction techniques. Based on these pros and cons of different approaches, future research can be extended until we find a satisfactory result in this field.

Table 1 Summarization of different keyphrase extraction techniques

References	Technique	Advantages	Disadvantages
8, 13	a. Semantic b. Syntactic	a. High-quality results that better match the user's expectations but less computational cost. b. Provides satisfactory result on the readers assigned keywords	a. Complex schema matching. b. Used information from single document
3	Statistical	Not domain dependent	More time consuming for feature calculation
15, 16, 17	a. LexRank b. TextRank c. HIITs	a. Insensitive to noisy data b. Require core lexical knowledge, domain knowledge, or language knowledge. c. Works good for both with training data and no training data	a. Unable to deal with multi-documents. b. Not suitable for a large volume of texts. c. Unable to generate graphs from large units (subgraphs)
18	SVMRank	Better result than TextRank and TF-IDF	Automatic artificial candidate keyphrase ranking not exist
19	Hybrid approach	Combination of structural and statistical approaches	Does not consider the word position only depends on the frequency
12	Semi-Supervised	Very effective and robust, recall is high	Precision is low
20, 9	a. GenEx b. KEA	a. Simple Implementation. b. Works good for short text	a. Complex Computation. b. Unsuitable for large text

References

1. Zhang, C.: Automatic keyword extraction from documents using conditional random fields. J. Comput. Inform. Syst. **3**, 1169–1180 (2008)
2. Li, D., Li, S.: Hypergraph-based inductive learning for generating implicit key phrases. In Proceedings of the 20th international conference companion on World Wide Web, pp. 77–78. ACM (2011)
3. Turney, P. D.: Coherent keyphrase extraction via web mining. arXiv preprint cs/0308033. (2003)
4. Wang, Q., Sheng, V.S., & Wu, X.: Document-specific keyphrase candidate search and ranking. Exp. Syst. Appl. **97**, 163–176(2018)
5. Li, D., Li, S., Li, W., Wang, W., Qu, W.: A semi-supervised key phrase extraction approach: learning from title phrases through a document semantic network. In Proceedings of the ACL 2010 conference short papers, pp. 296–300. Association for Computational Linguistics (2010)
6. Krulwich, B., Burkey, C.: Learning user information interests through extraction of semantically significant phrases. In Proceedings of the AAAI Spring Symposium on Machine Learning in Information Access, pp. 100–112. Menlo Park: AAAI Press (1996)
7. Hulth, A.: Improved automatic keyword extraction given more linguistic knowledge. In Proceedings of the 2003 conference on Empirical methods in natural language processing, pp. 216–223. Association for Computational Linguistics (2003)

8. Barzilay, R., Elhadad M.: Using lexical chains for text summarization. Advances in automatic text summarization, pp. 111–121 (1999)
9. Wan, X., Xiao, J.: Single document keyphrase extraction using neighborhood knowledge. In AAAI, vol. 8, pp. 855–860 (2008)
10. Luthra, S., Arora, D., Mittal, K., & Chhabra, A.: A statistical approach of keyword extraction for efficient retrieval. Int. J. Comput. Appl. **168**(7) (2017)
11. Chien, L. F.: PAT-tree-based keyword extraction for Chinese information retrieval. In ACM SIGIR Forum, vol. 31, pp. 50–58. ACM (1997)
12. Mihalcea, R., Tarau, P.: Textrank: Bringing order into text. In Proceedings of the 2004 Conference on Empirical Methods in Natural Language Processing, pp. 404–411 (2004)
13. Chang, J. Y., Kim, I. M.: Analysis and evaluation of current graph-based text mining researches. Adv. Sci. Technol. Lett. **42**, 100–103 (2013)
14. Erkan, G., Radev, D.R.: Lexrank: Graph-based lexical centrality as salience in text summarization. J. Art. Intell. Res. **22**, 457–479 (2004)
15. Litvak, M., Last, M.: Graph-based keyword extraction for single-document summarization. In Proceedings of the Workshop on Multi-source Multilingual Information Extraction and Summarization, pp. 7–24. Association for Computational Linguistics (2008)
16. Cai, X., Cao, S.: A Keyword Extraction Method Based on Learning to Rank. In 13th International Conference on Semantics, Knowledge and Grids (SKG).pp. 194–197. IEEE (2017)
17. Palshikar, G. K.: Keyword extraction from a single document using centrality measures. In International Conference on Pattern Recognition and Machine Intelligence, pp. 503–510. Springer, Berlin, Heidelberg (2007)
18. Pasquier, C.: Task 5: Single document keyphrase extraction using sentence clustering and Latent Dirichlet Allocation. In Proceedings of the 5th International Workshop on Semantic Evaluation, pp. 154–157. Association for Computational Linguistics (2010)
19. Abulaish, M., Anwar, T.: A supervised learning approach for automatic keyphrase extraction. Int. J. Innov. Comput. Inform. Cont. **8**(11), 7579–7601 (2012)
20. Witten, I.H., Paynter, G.W., Frank, E., Gutwin, C., Nevill-Manning, C.G.: Kea: Practical automated keyphrase extraction. In Design and Usability of Digital Libraries: Case Studies in the Asia Pacific, pp. 129–152. IGI Global (2005)

Review on Usage of Hidden Markov Model in Natural Language Processing

Amrita Anandika, Smita Prava Mishra, and Madhusmita Das

Abstract Hidden Markov Model (HMM) is a probabilistic graphical model, which allows us to calculate a sequence of unknown or unobserved variables from a set of observed variables. Predicting weather conditions (hidden) on the basis of types of clothes worn by someone (observed) is a simple example of HMM. The Markov process assumption is based on a simple fact that the future is dependent only on the present not on the past. HMM are useful as they are being able to model stochastic processes. Again human spoken language is a probabilistic language. Hence using a stochastic modeling technique, i.e., HMM, we can be able to achieve remarkable results in relatively limited scopes.

Keywords Hidden Markov model · Stochastic process · Markov assumption · Probabilistic model

1 Introduction

Natural Language Processing (NLP) is an area of computer science concerned with human and computer interaction. The main goal of NLP is the generation of natural language and making a computer to understand that natural language. HMM is one of the first developed models used in the field of NLP. It is the most favorable among all other machine learning approaches because it is domain independent as well as language independent. Hidden Markov Model (HMM) is a statistical or probabilistic model developed from Markov chain. A Markov chain is a mathematical process having a set of different states and probabilities for variables. It tells about the probabilities of sequences of any random variable or state. A Markov chain is widely recognized as Markov process. Like Markov chain, HMM also has a collection of

A. Anandika (✉)
Department of CSE, S'O'A University, Bhubaneswar, Odisha, India
e-mail: amrita.anandika@gmail.com

S. P. Mishra · M. Das
Department of CS & IT, S'O'A University, Bhubaneswar, Odisha, India

D. Mishra et al. (eds.), *Intelligent and Cloud Computing*,
Smart Innovation, Systems and Technologies 194,
https://doi.org/10.1007/978-981-15-5971-6_45

dissimilar or distinct states and probabilities for a variable. It tries to predict the future state of the variable by using probabilities that are based only on current and previous state like tomorrow's weather condition can be predicted by observing today's weather [1]. The key difference between Markov chain and HMM is that in HMM states are hidden or unobserved, meaning that the states are not directly visible to an observer. HMM is used for machine learning and data mining tasks.

2 Preliminaries

Hidden Markov model considers both observed states (words that we see in the input) and hidden states (part of speech tags). A hidden Markov model consists of five elements such as

$S =$ A finite number of states ($s_1 \ldots s_n$).

$\Pi =$ Start probability. It defines starting probability p(s) at state s.

$A =$ Transition probability p($s_i | s_j$). It defines the probability of a process moving from state s_i to states$_j$.

$O = o_1, o_2 \ldots o_p$ a sequence of P observations.

$B =$ Emission probability. It defines the probability of output signal produced by a state [2].

We can define HMM as a three tuple representation as stated below.

$$\lambda = (A,\ B,\ \pi) \quad (1)$$

Again HMM has two assumptions.

1. Markov Assumption: Probability of a particular state is dependent only on its previous state. It can be formulated as shown in Eq. (2):

$$\mathrm{P}(s_i | s_1 \ldots s_{i-1}) = \mathrm{P}\ (s_i | s_{i-1}) \quad (2)$$

2. Output Independence: Probability of output observation o_i depends on the state that produces the observation, i.e.,s_i, neither on any other states nor any other observations [1] as stated in Eq. (3):

$$\mathrm{P}(o_i | s_1 \ldots s_i \ldots, s_p, o_1, \ldots, o_i, \ldots o_p) = \mathrm{P}(o_i | s_i) \quad (3)$$

HMM is widely used in different NLP research areas. Some of them are discussed below.

3 Applications of HMM in NLP

HMMs can be applied in those areas where the goal is to extract a data sequence that is not observed immediately. HMM is widely used in various fields of NLP. Some of them are listed below:

- Speech recognition
- Spell-checking
- Part of speech tagging
- Named entity recognition
- Chunking

3.1 Speech Recognition

Speech recognition systems are comprised of a collection of statistical models, which represent a range of sound of the language that needs to be identified. In general, speech has a sequential structure and it can be determined as a sequence of spectral vectors across the audio frequency range. To construct such models HMM provides a natural framework [3]. Speech recognition has two elements such as Feature extraction and Feature matching.

As shown in Fig. 1, voice sample with no noise is provided as input to the front-end module and feature vector is obtained as output. In feature matching, the feature vector obtained from an unknown voice sample is applied in opposition to an acoustic model. Once the feature vector is extracted, an acoustic model is created. The output

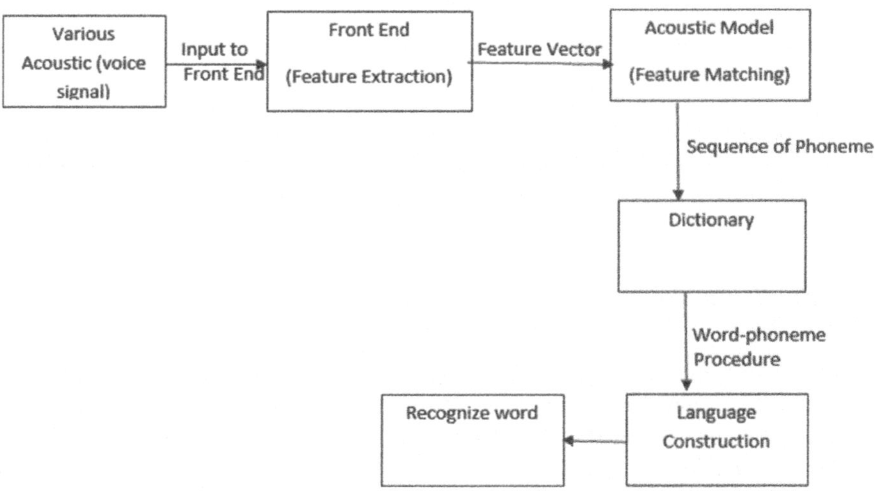

Fig. 1 Diagram of speech recognition process

of front-end acts as input to the acoustic model. The output of acoustic model is measured as a known word. Please note that the first paragraph of a section or subsection is not indented.

Representation of an Acoustic model: In speech recognition, phoneme is the basic unit of sound. Phoneme is the minimal unit required to distinguish among different meanings of words. In HMM for speech recognition, each phoneme is assigned to a unique HMM. When an unknown word arrives, it will be applied to all HMM models and the HMM with maximum score will be considered as recognized word [4]. Different types of HMM used for speech recognition task are mentioned below.

1. Context-Independent Phoneme HMMs
2. Context-Dependent Triphone HMM, and
3. Whole-Word HMM

Among all triphones, HMM is widely used in speech recognition.

Front-End: Function of Front-End is to parameterize an input signal like audio, to a sequence of output features.

Vector Quantization: It has the function of mapping feature vectors into symbols. It is also known as acoustic modeling. These symbols represent HMM state.

HMM model creation: HMMs are created for every basic sound unit, i.e., phoneme. Then all HMMs are linked together for representing the vocabulary that is under consideration. Bi-gram model is used for representing the relationship between the words in a given sentence.

Training: In the training phase, a large amount of voice data is supplied to HMM model. By using this, HMM adjusts its probability distribution and transition matrix.

Recognition: There are five steps for automatic speech recognition method [5]. They are

1. Feature Extraction
2. Acoustic Model
3. Lexicon
4. Language Model, and
5. Decoder: Here Viterbi algorithm is used to combine all these to get a word sequence from speech by using dynamic programming.

3.2 Spell-checking

Sometimes prepared text data in NLP, statistical language modeling, and information retrieval system contains typing error and grammatical errors, which decrease its information value. Therefore spell-checking is a must required phenomenon. Spell-checking can be defined as a task of identifying and marking incorrectly or wrongly spelled words in a text file that is written in a natural language. Errors related to misspelled words can be divided into two classes. They are non-word errors and real-word errors.

Generally, human typing leads to non-word errors, which arise due to three major reasons such as Typographic errors, Cognitive errors, and Phonetic errors [6]. Spell-checking is defined as an operation of detecting the best possible correct word sequence in a given list of incorrect words. HMM and Viterbi algorithm are the most efficient methods for classification of the sequenced data. In HMM, words warped with errors in any sentence are considered as a sequence of observation o. For this o, possible corrections w are considered as hidden states. Then Viterbi algorithm assigns the most probable corrected word sequence to the given list of possible incorrect words [7]. Basic elements of HMM are represented below.

1. Observation matrix P($o|w$)
2. State transition matrix P

The heuristic procedure for probability P($o|wj$) with $(w|t)o$ observation, correspond to the state w_j is estimated as shown below,

$$\mathrm{P}(o|w_j) = C - j\,x\,C_i = 0\,(C - i) \tag{4}$$

Here j = Order of words offered by the spell-checking module.
C = Number of suggested corrections by the spell-checking module.
$x\,c_i = 0\,(C - i)$ = Normalization constant.

3.3 Part of Speech Tagging

Part of speech (POS) tagging is a method of identifying a word in a text or document as related to an exact part of speech that is derived from its definition and from its circumstance or finding its relationship with neighboring and correlated words in a sentence, or in a passage. In a basic form, it is the identification of words as nouns, verbs, adjectives, adverbs, etc. In numerous applications like Question Answering, Speech Recognition, Machine Translation, POS tagging is required. POS-tagging algorithms are categorized into two unique groups such as

1. Rule-Based POS taggers
2. Stochastic POS taggers

Rule-based POS taggers
Automatic POS tagging is a part of NLP field, where statistical methods are doing well in comparison to rule-based methods. Rule-based approaches use contextual information for allocating tags to any unknown word.

Stochastic POS taggers
Stochastic tagger consists of different approaches used for solving the problems in part of speech tagging. A simple stochastic tagger rephrases words according to the probability of a word that occurs with a certain tag. In other words, the tag which

appears with a word in a regular interval within the training set will be defined as an ambiguous occurrence of that word. A substitute for the word frequency approach is the N-gram approach. It states that the most appropriate tag for any given word can be found by considering the probability of its occurrence with n previous tags. N-gram approach is used to calculate the probability of occurrence of a given sequence of tags [8].When stochastic tagger incorporates previously mentioned two approaches with tag sequence probability and word frequency measurement then, it is defined as Hidden Markov Model.

When we are doing POS tagging, our goal is to discover the sequence of tags S as given in Eq. 5.

$$S = \arg\max_S^{\arg\max} p(S|W) \tag{5}$$

Here, we find the sequence of POS tags with the highest probability given on a sequence of words. As tag emissions are overlooked in HMM, Bayes' rule is applied to change the probability into an equation as given below,

$$S = {}_S^{\arg\max} p(S|w) = {}_S^{\arg\max} \frac{p(w|S)p(S)}{p(W)} = {}_S^{\arg\max} p(W|S)p(S) \tag{6}$$

Here p*(W)* is a constant. Any changes in the sequence S will not change the probability P*(W)*. Hence eliminating it will not make any variation in the final sequence S which maximizes the probability. The probability P(W|T) is the probability of getting a sequence of words given on the sequence of tags [9]. HMM gives us probabilities but we need the actual sequence of tags. Hence, we use Viterbi algorithm, which can give us the tag sequence having the highest probability of being correct given on a sequence of words.

3.4 Named Entity Recognition

NER is an Information Extraction process concerned with extracting Named Entities (NE) from raw data and categorizing these named entities into some predefined classes like name of a person, location, organization, city, country, and time and date expression, etc. Named entity can be defined as a word or a phrase which distinctly identifies an item from a set of items having similar attributes. Several machine learning approaches are used for NER, but Hidden Markov Model (HMM) based machine learning approach is widely used to identify NEs, because using HMM building a NER system is easy. It is language independent and dynamic in nature. In HMM a join probability is assigned to paired observation and label sequence. The probability of the next state occurrence depends only on its previous state [10]. According to Viterbi algorithm, HMM is used to discover the maximum possible tag sequence in the state space of the probable tag distribution that is created according to the state transition probabilities [11]. This algorithm helps in finding the optimal tags within linear time. The main aim of this algorithm is to consider only the most probable sequences among all state sequences. Viterbi algorithm has the following HMM parameters:

- Set of States P where $|P| = M$. M represents the number of states.
- Observations Q $|Q| = N$. Where N represents the total number of output alphabets.
- Transition probability X.
- Emission probability Y.
- Probability π.

According to Viterbi algorithm for each state P, we can define HMM as stated in equation.

$$\eta_T(P) = \max_{P_1, \ldots, P_{T-1}} \mathrm{P}\{X1 = P_1, \ldots, X_{T-1} = P_{T-1}, X_T = P, Q_1, \ldots, Q_T\} \tag{7}$$

As per Markov property, if the maximum possible path which ends with state P at time T is equal to certain P^* at time $T - 1$, then P^* is the value of the last state of the maximum possible path that ends at time $T - 1$. From this following recursion can be defined as stated in Eq. (8).

$$\eta_T(P) = \max_{P} X_{P,R}\beta_R(Q_T)\eta_{T-1}(P) \tag{8}$$

Here, $\beta_R(Q_T)$ represents the probability to have Q_T when the Markov state is R [12]. NER system based on HMM is easy to understand and implement and also easy to analyze. Sequence labeling difficulty can be solved efficiently by HMM.

3.5 *Chunking*

Chunking is a process of extracting meaningful short phrases or chunks from a sentence. It is also known as shallow parsing. Chunking can also be defined as the identification of noun phrases. Although part of speech tagging defines words as nouns, verbs, adjectives, etc. it does not give any idea regarding the structure of a sentence. The purpose of a shallow parser is to divide a text into segments that correspond to certain syntactic units [13]. In maximum cases, shallow parsing is considered to be a tagging problem. According to statistical point of view, tagging can be solved as a maximization problem.

Let A = set of output tags. B = input vocabulary of an application. Suppose an input sentence,$A_1 \ldots A_t$ is given. The process involves discovering the state sequence having maximum probability on the model [14]. Meaning that the output sequence tags $A = A_1 \ldots A_t$. This process can be formulated as shown:

$$A = \underset{A}{\arg\max}\, p(A|B) = \underset{A}{\arg\max}\, p(A).p(B|A) \tag{9}$$

This maximization process is independent of input sequence. This formula is widely used for solving POS tagging efficiently.

3.6 Machine Translation

Machine Translation (MT) is defined as a process of automatic conversion of one natural language into another language by preserving the actual meaning of the input text [15]. HMM is widely used in various MT areas like line alignment, word and text alignment, time alignment, etc. [16]. All of them being sequence alignment problems that can be easily harnessed from HMM. Line alignment is adjusting multiple lines to start with the same margin of a page having equal spacing. Word alignment is a process of breaking down an instruction into small blocks. Text alignment is defined as the arrangement of text in association with margin.

4 Conclusion

Hidden Markov Model method is an effective, simple, and refined way to model the sequence families. It is domain and platform independent. HMM consists of different efficient dynamic programming algorithms to detect the most appropriate label sequence and significant probability of that label sequence. It also gives higher evaluation performance among all machine learning approaches. NLP being a sequence labeling problem may be appropriately mapped to HMM. HMM along with Viterbi algorithm is capable to address many of the NLP sub-problems like speech recognition, spell-checking, POS tagging, Named entity recognition, machine translation, etc. Further research may be carried out for various Indian and non-Indian languages which are yet to explore the potential of this useful tool like HMM.

References

1. Sudha, M., Nusrat, J., Deepti, C.: Named entity recognition using hidden markov model (HMM). Int. J. Nat. Lang. Comput. **1**(4) 68–73 (2012)
2. Chopra, D., Morwal, S.: Named entity recognition in english using hidden markov model. In: Int. J. Comput. Sci. Appl. **3**(1), 293–297 (2013)
3. Rupali S., Chavan, Ganesh Sable, S.: An overview of speech recognition using HMM. Int. J. Comput. Sci. Mob. Comput. **2**(4), 233–238 (2013)
4. Najkar N., Razzazi F., Sameti H.: An evolutionary decoding method for HMM-based continuous speech recognition systems using particle swarm optimization. Int. J. Patt. Anal. Appl. 327–339 (2014)

5. Najkar N., Razzazi F., Sameti H.: A novel approach to HMM-based speech recognition systems using particle swarm optimization. In: International Journal on Mathematical and Computer Modeling, 11–12 (2010)
6. Daniel, H., Jan, S., Jozef J.: Unsupervised spelling correction for the slovak text. Int. J. Inform. Commun. Technol. Ser. **11**(5), 345–351 (2013)
7. Jayalatharachchi E., Wasala, A., Weerasinghe, R.: Data-driven spell checking: The synergy of two algorithms for spelling error detection and correction. In: International Conference on Advances in ICT for Emerging Regions (ICTer), pp. 7–13, IEEE, Colombo (2012)
8. Nakagawa T., Kudo, T., Matsumoto, Y.: Unknown word guessing and part-of- speech tagging using support vector machines. In: Proceedings of the Sixth Natural Language Processing Pacific Rim Symposium (2001)
9. Padro, M., Padro, L.: Developing Competitive HMM POS Taggers Using Small Training Corpora. Estal, (2004)
10. Ekbal, A., Bandyopadhyay, S.: A hidden markov model based named entity recognition system: bengali and hindi as case studies. In: Ghosh A., De R.K., Pal S.K. (eds.) Pattern Recognition and Machine Intelligence. Lecture Notes in Computer Science, vol. 4815, Springer, Berlin, Heidelberg (2007)
11. Chopra D., Joshi N., Mathur I.: Named entity recognition in Hindi using hidden markov model. In: IEEE Second International Conference on Computational Intelligence & Communication Technology, pp. 581–586, (2016)
12. Chopra D., Morwal S.: Named entity recognition in Punjabi using Hidden Markov Mod-el. Int. J. Comput. Sci. Eng. Technol. **3**(7) (2012)
13. Song, M., Song, I.Y., Hu, X., Allen, R.B.: Integrating text chunking with mixture hidden markov models for effective biomedical information extraction. In: Sunderam, V.S., van Albada, G.D., Sloot, P.M.A., Dongarra, J.J. (eds.) Computational Science—ICCS, Springer, Berlin, Heidelberg (2005)
14. Kudo T., Matsumoto, Y.: Use of support vector learning for chunk identification. In: Proceedings of CoNLL 2000 and LLL 2000, Saarbrucken, Germany, pp. 142–144 (2000)
15. Yonggang, D., William, B.: HMM word and phrase alignment for statistical ma-chine translation. In: IEEE Transactions on Audio, Speech & Language Processing (2006)
16. Fatemeh, M., Abolghasem, S.: Improvement of time alignment of the speech signals to be used in voice conversion. In: Int. J. Speech Technol. **21**(9), 79–87 (2018)

Performance of ELM Using Max-Min Document Frequency-Based Feature Selection in Multilabeled Text Classification

Santosh Kumar Behera and Rajashree Dash

Abstract In text classification feature selection is used to reduce the feature space to improve the classification accuracy. In this paper, we propose a method max-min document frequency-based feature selection and we applied Extreme Learning Machine (ELM) model to improvise the text classification performance. For this text classification, we used the multilabel Reuters dataset which consists of 10788 number of documents. In this experiment, the ELM model performs better using max-min document frequency-based feature selection in terms of precision, recall, and F-measure as is compared to the ELM model using full feature space without using any feature selection technique.

Keywords Text categorization · Feature selection · Document frequency · ELM

1 Introduction

In today's world of fast-growing electronic text documents on the web, it has become important to deal with a new technique for a better result. Searching for information in such a large amount of data is now a big challenge. Based on the content present in the text document, Text classification (TC) assigns the predefined categories to the text document. TC is popular in many different applications such as spam filtering [2], sms filtering [3], topic detection [4], document categorization [5], bioinformatics [6], sentiment analysis [7], etc. High dimensional dataset can also be reduced through proper feature selection, in some cases, principle component analysis (PCA) is also being used [13].

S. K. Behera (✉) · R. Dash
Department of Computer Science and Engineering, Siksha 'O' Anusandhan
Deemed to be University, Bhubaneswar Odisha, India
e-mail: santoshbehera@soa.ac.in

R. Dash
e-mail: rajashree_dash@yahoo.co.in

D. Mishra et al. (eds.), *Intelligent and Cloud Computing*,
Smart Innovation, Systems and Technologies 194,
https://doi.org/10.1007/978-981-15-5971-6_46

Records or document are represented as a vector in the Text classification process. The features are generally represented as term frequency (tf) or term frequency-inverse document frequency (tf-idf) score in vector space. As there is a huge number of text documents, the number of feature spaces is also large. With a large number of features, it will degrade the classification accuracy. The corresponding model will also take a significant amount of time to get the result. So we need to apply feature extraction and feature selection techniques to minimize the feature space to get better accuracy and also help the model to compute the result in less time. Feature extraction is the process of transforming raw data into modeling features, and feature selection is the process of filtering the unnecessary features and giving importance to the feature which will classify the relevant text document.

Text classification mainly consists of 5 steps: Preprocessing, Feature Extraction and Selection, Learning the model, Evaluation.

2 Related Work

This examination is identified with the work on feature selection and extreme learning machine.

2.1 *Extreme Learning Machine*

Extreme learning machine (ELM), a learning algorithm proposed by Huang [11], is a single hidden layer feed forward network (SLFNs) which is generally used for classification and regression. It is a faster learning algorithm which provides better performance with less human interaction. ELM is also used in many application areas like classification and regression [14, 15]. Randomly, the input weights, biases, and hidden nodes are assigned to ELM. The output is computed through inverse operation.

Given n distinct samples $(x_i, t_i) \in R^k X R^m (i = 1, 2, 3....n)$, the single layer feed forward neural network with p hidden nodes with activation function $g(x)$ is represented by

$$O_j = \sum_{i=1}^{p} \beta_i g(w_i \cdot x_j + b_i), \; j = 1, 2...n \tag{1}$$

The weight vector $w_i = [w_i 1, w_i 2, ...w_i k]^T$ is the connection between hidden nodes to input nodes. The weight vector $\beta_i = [\beta_i 1, \beta_i 2....\beta_i p]^T$ is the connection between hidden nodes to output nodes and b_i is the bias of the ith hidden nodes. The above equation can be written as

$$H\beta = O \tag{2}$$

where

$$\begin{bmatrix} g(a_1.x_1 + b_1) & \dots & g(a_n.x_1 + b_n) \\ \vdots & \vdots & \vdots \\ g(a_1.x_n + b_1) & \dots & g(a_n.x_n + b_n) \end{bmatrix}_{nXp}, \beta = \begin{bmatrix} \beta_1^T \\ \vdots \\ \beta_n^n \end{bmatrix}_{pXm}, O = \begin{bmatrix} O_1^T \\ \vdots \\ O_n^n \end{bmatrix}_{nXm}$$

Here, H is the output matrix. The β can be written as $\beta = H^{\dagger}T$. Here, $H^{\dagger}$ is the Moore–Penrose inverse of H.

2.2 *Feature Selection with ELM Model*

ELM with large feature space will increase the number of input nodes which will decrease the performance of the model. So, the feature selection technique is applied to minimize the feature space which in turn decreases the number of input nodes.

In a study [8], the authors have proposed a k-means clustering-based feature selection in which they combined the feature of Bi-Normal Separation (BNS) and they represented it using cosine-similarity to reduce the feature space to model the Extreme Learning Machine (ELM) and Multilayer Extreme Learning Machine (ML-ELM).

In another study [1], the authors have proposed a new clustering-based feature selection technique. The standard k-means clustering technique in addition to tf-idf and WordNet helps to form an efficient and optimized feature space to model the ELM.

In another survey [9], the authors have applied an improved Markov Boundary-based algorithm to reduce the feature space. Then they enhanced the performance of the ELM by extracting features and by using genetic-based weight assignments.

In a survey [10], the high dimensional data is reduced to select the distinguished features, which improves the classification accuracy. They experimented Normalized Difference Measure (NDM), which takes the relative document frequencies for feature selection They compared with other technique like chi square(CHI), odds ratio(OR), Gini index(GINI), balanced accuracy measure(ACC2), etc.

3 Data Collection

A corpus is a collection of many unlabeled text documents which can be categorized into many some predefined categories. In this study, we have used the standard benchmark Reuters dataset. The Reuters dataset is a multi-class and multi-labeled dataset [12, 16]. It has more than one class and each document belongs to more than one class. It has 90 classes, 7769 training documents, and 3019 testing documents. It is the ModApte (R(90)) subset of the Reuters-21578 benchmark.

3.1 Text Preprocessing

A text can be a sequence of characters, words, phrases, paragraphs, etc. This test preprocessing step can diminish the noise and improve the accuracy of the model. Each file in the corpus follows the following steps:

—Removing hyphens, digits, punctuation marks, special characters, and numbers.
—Removing stop words.
—Tokenization and token normalization.
—Stemming and lemmatization.

Stemming helps in removing prefix or suffix of a given word to get the root word but not always this gives good results. So we apply lemmatization which performs morphological analysis of a given word which gives the base form of all its inflectional forms.

4 Proposed Model

The proposed max-min document frequency-based feature selection with ELM is shown in Fig. 1. The various phases of max-min document frequency-based feature selection with ELM are given below:

- Reuters-21578 ModeApte(R(90)) dataset is collected. The entire corpus consists of *10788* files; after Processing, the corpus is divided into training and testing datasets. The training dataset consists of *7769* files and the testing dataset consist of *3019* files.
- In the Tokenization phase, the entire text is converted to lower case, then we remove numbers and punctuation [<, >, !.@, ., "," , ∗, (,),' , 1, 2, ...] . We remove leading and trailing white spaces and split the entire text into smaller pieces called tokens.
- In filtering the stop words, we remove the common words such as *is, are, i, we, the, will, and be* which do not have any important meaning and removing these words will not put any impact in text classification . From nltk.corpus, we have imported the English stop-word list.
- Stemming is the process of chopping a word into its base or root word. This is an important phase because the same word can be in a different form in different files. For example *works, worked, and working* all indicate its root word *work*. Here, we used the standard Porter stemmer algorithm for the process of stemming.

Lemmatization reduces from an inflectional form of the word to the base form by using lexical knowledge of the word. Like the word *studying* in stemmer, it will be studi but by using a lemmatizer it will be *studying*. Similarly, the word *cries,crys* will become *cri* in stemmer but *cry* in a lemmatizer.

- In this phase of feature extraction, we convert the textual data into some numeric score which helps to extract the the distinguishing features from the text. They are many methods like *countvectorizer*, *tf-idfvectorizer*, *bag-of-words*, etc. In this experiment, the tf-idf vectorizer concept is applied. The *term frequency (tf)* is the frequency of each word in the corpus. It will increase when the number of words in a document increases. The *inverse document frequency(idf)* is used to give importance to the rare words in the document.

$$tf - idf : W_{ij} = tf_{ij} * log_2 \frac{N}{n} \tag{3}$$

Now we have a complete feature matrix with *tf-idf* score, next we need to select the important feature using the good feature selection technique.

- The *tf-idf* feature matrix is used in the feature selection technique. Here, we have used the max-min document frequency-based feature selection technique. The vocabulary is built for the entire corpus with its corresponding *tf-idf* score. Then we check the relation of a term across all documents in the corpus. If a term appears frequently in the Corpus, i.e. across many documents, then we ignore that term which is represented as *Max-df*. If *Max-df* = *0.50*, then it shows that in more than *50%* of the document the term appears. It shows these terms are the common words but stop words are not removed from the vocabulary list as well its score from the *tf-idf* matrix. Similarly, if *Min-df* = *0.03*, it shows that in less than *3%* of document the term appears. So we remove these terms because very rare terms are also not useful in the classification process. In this way, the vocabulary list with its corresponding *tf-idf* score matrix is reduced. We keep only those terms with their *tf-idf* score which pass the *Max-df* and *Min-df* threshold values.
- The optimized feature matrix acts as the input neurons $x_i....x_n$ to the ELM model. The random input weights $w_i....w_n$ between -1 and 1 are chosen and the bias value b_i is also chosen. Then the activation function sigmoid is applied. The number of hidden neurons is chosen randomly. In general, the number of hidden neurons is chosen carefully after many iterative processes. The output weight is calculated by using pseudo inverse to the output of the activation function. Then, the actual output is calculated. The actual output is normalized and transformed to 0 or 1 by using some threshold value.
- The predicted label is now compared with the input label and then *precision, recall*, and *F-measure* are calculated.

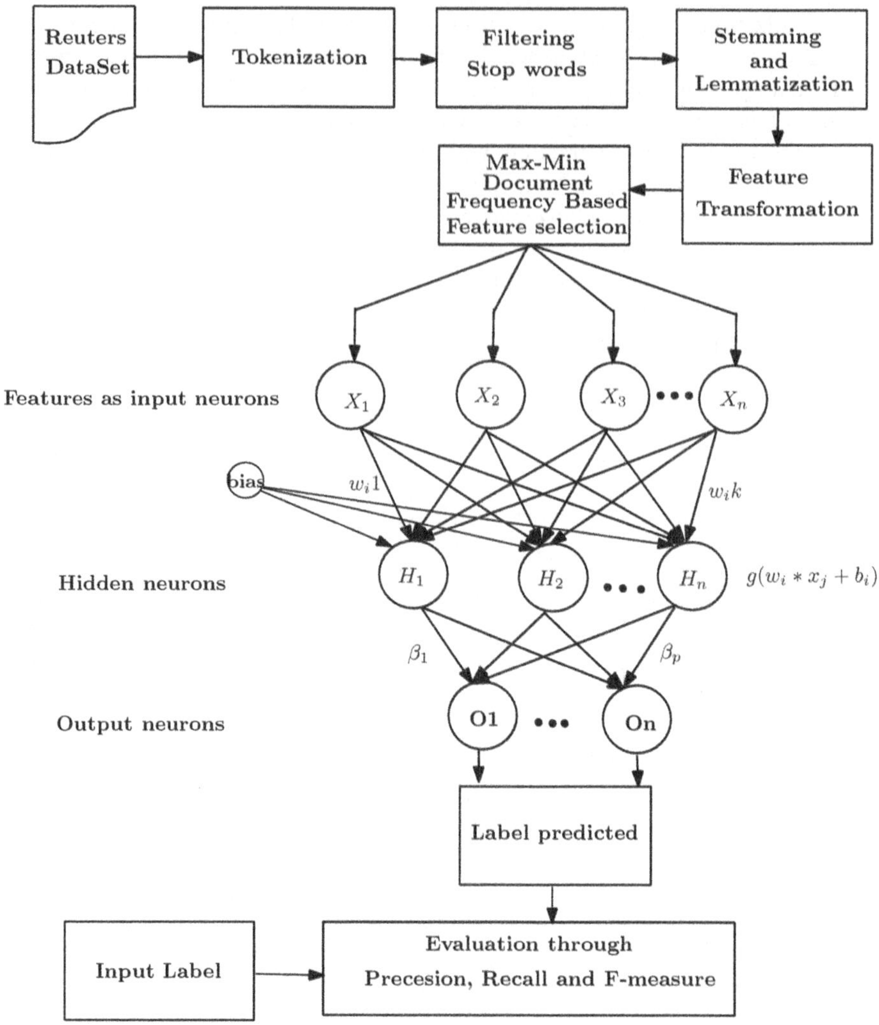

Fig. 1 Max-Min document frequency-based feature selection with ELM

5 Result and Analysis

The classification process is evaluated by using Max-Min document frequency-based feature selection with and without using any feature selection technique. We have applied the evaluation metrics like precision, recall, and F-measure (Tables 1, 2, 3, 4).

Table 1 Precision table

Min-df-Max-df	0.1–0.2	0.1–0.3	0.1–0.4	0.1–0.5	0.1–0.6	0.1–0.7	0.1–0.8	0.1–0.9	0.1–1.0
Feature selected	56	56	56	58	60	61	61	61	61
Precision score	0.86	0.90	0.91	0.90	0.90	0.90	0.90	0.90	0.90

Table 2 Recall table

Min-df-Max-df	0.1–0.2	0.1–0.3	0.1–0.4	0.1–0.5	0.1–0.6	0.1–0.7	0.1–0.8	0.1–0.9	0.1–1.0
Feature selected	56	56	56	58	60	61	61	61	61
Recall score	0.42	0.46	0.49	0.50	0.50	0.50	0.50	0.50	0.50

Precision measures how much true positive labeled data has been retrieved among total retrieved data, which is given in Eq. 4.

Recall measures how many relevant data have been retrieved among the total number of retrieved data, which is given in Eq. 5.

F-measure takes both precision and recall and calculates the harmonic mean. The balance between precision and recall is represented in F-measure which is given in Eq. 6.

$$p = \frac{t_p}{t_p + f_p} \tag{4}$$

$$r = \frac{t_p}{t_p + f_n} \tag{5}$$

$$F = \frac{2pr}{p + r} \tag{6}$$

The number of features is auto-selected based on the threshold values of *Max-df* and *Min-df*. We compare our Max-Min document frequency experimental results with and without any feature selection technique. Initially, the *Min-df* = *0.1* and *Max-df* = *0.2*, which results in *precision* = *0.86*, *recall* = *0.42*, and *F-measure* = *0.56*. This result shows we are removing a feature that appears in less than *10%* of document, and similarly removing features that appear in more than *20%* of the document.

Table 3 F-measure table

Min-df-Max-df	0.1–0.2	0.1–0.3	0.1–0.4	0.1–0.5	0.1–0.6	0.1–0.7	0.1–0.8	0.1–0.9	0.1–1.0
Feature selected	56	56	56	58	60	61	61	61	61
F-measure	0.56	0.60	0.63	0.64	0.64	0.64	0.64	0.64	0.64

Table 4 Precision, Recall, and F-measure without any feature selection

Without any feature selection, number of features getting selected = 20682		
Precision	Recall	F-measure
0.88	0.45	0.56

When we increase the threshold value of *Max-df* with fixed *Min-df*, our experiment shows that with *Max-df = 0.5* the *precision* is *0.90*, *recall* is *0.50*, and *F-measure* is *0.64* after onwards mostly the number of *features, precision, recall,* and the measure are constant. It shows that by removing features which appear in more than *40%* of document, new features are not added. When we compare the feature selection technique with normal feature matrix, which we get without applying any feature selection technique, it shows that the number of *features = 20682*, with *precision = 0.88, recall = 0.45*, and *F-measure = 0.56*. It indicates that with large number of features also it does not help to increase the *precision, recall, and F-measure.* So the Max-Min document frequency-based feature selection performs better with less number of features in terms of *precision, recall, and F-measure.*

6 Conclusion

In this experiment, the Max-Min document frequency-based feature selection with ELM is compared with only ELM without any feature selection technique. This result shows the Max-Min document frequency-based feature selection with ELM performs better in terms of *precision, recall, and F-measure.* As future work, the parameters like *number of hidden neurons, activation function, Min-df, and Max-df* need to be tuned with different threshold values. Further study can be performed with different feature selection techniques and classification algorithms.

References

1. Roul, R. K., Gugnani, S., & Kalpeshbhai, S. M. : Clustering based feature selection using extreme learning machines for text classification. In 2015 Annual IEEE India Conference (INDICON) (pp. 1-6). IEEE (2015)
2. Guzella, T.S., Caminhas, W.M.: A review of machine learning approaches to spam filtering. Expert Systems with Applications **36**(7), 10206–10222 (2009)
3. Idris, I., Selamat, A.: Improved email spam detection model with negative selection algorithm and particle swarm optimization. Applied Soft Computing **22**, 11–27 (2014)
4. Zeng, J., Zhang, S.: Variable space hidden Markov model for topic detection and analysis. Knowledge-Based Systems **20**(7), 607–613 (2007)
5. Jiang, L., Li, C., Wang, S., Zhang, L.: Deep feature weighting for naive Bayes and its application to text classification. Engineering Applications of Artificial Intelligence **52**, 26–39 (2016)
6. Saeys, Y., Inza, I., & Larraaga, P. : A review of feature selection techniques in bioinformatics. bioinformatics, 23(19), 2507-2517 (2007)
7. Medhat, W., Hassan, A., Korashy, H.: Sentiment analysis algorithms and applications: A survey. Ain Shams engineering journal **5**(4), 1093–1113 (2014)
8. Roul, R. K., & Sahay, S. K. : K-means and wordnet based feature selection combined with extreme learning machines for text classification. In International Conference on Distributed Computing and Internet Technology (pp. 103-112). Springer, Cham (2016)
9. Yin, Y., Zhao, Y., Zhang, B., Li, C., Guo, S.: Enhancing ELM by Markov Boundary based feature selection. Neurocomputing **261**, 57–69 (2017)
10. Rehman, A., Javed, K., Babri, H.A.: Feature selection based on a normalized difference measure for text classification. Information Processing & Management **53**(2), 473–489 (2017)
11. Huang, G.B., Zhu, Q.Y., Siew, C.K.: Extreme learning machine: theory and applications. Neurocomputing **70**(1–3), 489–501 (2017)
12. Asuncion, A., & Newman, D. : UCI machine learning repository (2007)
13. Dash, R., Dash, R., Mishra, D.: A hybridized rough-PCA approach of attribute reduction for high dimensional data set. European Journal of Scientific Research **44**(1), 29–38 (2010)
14. Zheng, W., Qian, Y., Lu, H.: Text categorization based on regularization extreme learning machine. Neural Computing and Applications **22**(3–4), 447–456 (2013)
15. Li, M., Xiao, P., & Zhang, J. : Text classification based on ensemble extreme learning machine. arXiv preprint arXiv:1805.06525 (2018)
16. Thaoma, M. :The Reuters Dataset.https://martin-thoma.com/nlp-reuters/. (2017)

Issues and Challenges Related to Cloud-Based Traffic Management System: A Brief Survey

Sarita Mahapatra, Krishna Chandra Rath, and Srikanta Pattnaik

Abstract In the present scenario, increasing traffic volume is a prime concern worldwide. Many urban areas suffer from acute traffic congestion due to various reasons like insufficient road capacity, etc. Congestions adversely affect major sectors like economy, environment, and health. Smart solutions to this timely topical problem may be provided by using Intelligent Transportation Systems (ITS) technologies. ITS urban traffic management mostly deals with the collection and processing of a huge amount of geographically distributed information to manage distributed infrastructure and components of road traffic. Due to the distributed aspect of the problem, there is a need to develop a novel, scalable Cloud-based ITS platform. The Cloud-based ITS hosts the collection of software services which leads to the development of the Cloud-based Traffic Management System (CTMS). In this paper, we did a survey on distinct aspects of CTMS such as adaptive intersection control algorithm integrated with microscopic prediction mechanism and microscopic traffic simulation tool tightly integrated with the ITS-cloud.

Keywords Traffic management · Cloud-based traffic management · Cloud-based ITS

S. Mahapatra (✉)
Department of Computer Science and Information Technology, Siksha 'O' Anusandhan Deemed to Be University, Bhubaneswar, Odhisha and Research Scholar, Utkal University, Bhubaneswar, India
e-mail: saritamahapatra4@gmail.com

K. C. Rath
P.G. Department of Geography, Utkal University, Bhubaneswar, Bhubaneswar, India

S. Pattnaik
Department of Computer Science and Engineering, Siksha 'O' Anusandhan Deemed To Be University, Bhubaneswar, India

D. Mishra et al. (eds.), *Intelligent and Cloud Computing*,
Smart Innovation, Systems and Technologies 194,
https://doi.org/10.1007/978-981-15-5971-6_47

1 Introduction

Traffic congestion in the last decade has increased exponentially in India. Congestion and the related slow urban mobility can have a major negative effect on both the economy and the quality of life. During peak hours in four major cities in India such as Kolkata, Delhi, Mumbai, and Bengaluru, traffic congestion costs the economy Rs 1.47 lakh crore annually as per the study made in [25].

To meet the traffic congestion challenge, infrastructure development may not be a viable option. Instead, there is a need for the development of smart solutions such as a Traffic Management System (TMS) which is inherently associated with Intelligent Transportation System (ITS). TMS and ITS provide effective solutions to improve incident response time, reduce road traffic congestion and get overall satisfaction during travel. A few distinct services of TMS include (1) traffic prediction that enables early detection of traffic issues, (2) vehicle routing to reduce commuter journey, (3) effective parking management, and (4) interaction between prediction services and routing to improve traffic flow control.

From the last few years, some novel methods and techniques in TMS have been proposed by various researchers across the globe using intelligent technologies like wireless communications, data mining, D2D communication, cloud computing, and so on. Using new sources of traffic data, different transportation applications has been created. For real-time road traffic control and decision-making, these applications using big data can be used. To exploit these positives of using big data in TMS applications, conventional models experience some technical challenges like (1) management and storage of data, (2) analysis and processing of data, (3) problems in adapting to uncertain changes, and (4) managing data from heterogeneous sources. These technical intricacies led to the development of an adaptable distributed framework for TMS which efficiently stores, processes, and analyzes bit intelligent traffic data. To respond to real-time big data challenges in TMS, a distributed cloud computing framework can be used as a cost-effective and scalable solution.

In this paper, we attempt to study some prominent research works on a distributed framework for intelligent traffic management system. The characteristic features of a distributed intelligent traffic management system are (1) ease in implementing in a broad range of applications, (2) quick processing and analysis of huge amount of traffic data, (3) elasticity, high availability, and fault tolerance, and (4) low-latency data storage. The remaining of the paper is structured as follows: we started with architectures in road traffic management, modeling, and simulation. Then, we reviewed major related works to traffic management systems (TMSs) and then cloud-based traffic management systems (CTMSs), then we identified critical issues and challenges related to CTMS.

2 Architectures for Road-Traffic Management and Monitoring

Management and monitoring of road traffic is one of the prime functionalities of the ITS. Whenever multiple traffic flow directions intersect, to guarantee safety in situations, random control over the traffic flow is essential. To deal with conflicting areas, the flow of vehicles from different directions can be scheduled using a traffic control system where the control algorithm ensures proper management and avoids unwanted delays. Use of traffic signals and signages present in concerned areas such as intersections and signaled pedestrian crossings is the most prominent approach in traffic management. In the another approach, using wireless technologies or variable message signs, a traffic controller monitors individual vehicles and can respond autonomously [1]. This idea provides finer control over the traffic (1) enabling vehicles on an optimal route, (2) setting variable speed limits, and (3) in guiding the vehicles with traffic lights' status in future.

2.1 Traffic Modeling and Simulation

In this section, we discuss the current research in traffic simulation and modeling along with the developed traffic management schemes. TMSs utilize traffic modeling to predict future situations [8, 23] and to control objective-based flow of control. The modeling strategies are also widely used to evaluate performance of TMSs [6, 7, 9–11]. Basically traffic modeling can be viewed in four interrelated levels, as shown in Fig. 1.

3 Distributed-Framework-Based Traffic Management System

This section discusses a sound background on distributed processing systems and investigates on application of distributed technologies in traffic management. Traffic management and vehicle control techniques discussed in the previous section can be used to prepare a database to prepare an effective, adaptive, and robust traffic management system which is assigned with data collection from sensors distributed geographically and monitoring traffic lights at different intersections. In advanced traffic management systems, it can be assumed that D2D wireless communication nodes send commands to vehicles. To provide the needed coverage, D2D nodes also have to be distributed geographically. All components of the distributed framework such as D2D node, sensor, intersection and controller must be associated with some functionality to perform its function which leads to the design of distributed adaptive TMS. With the huge increase in users along with the need of high processing power

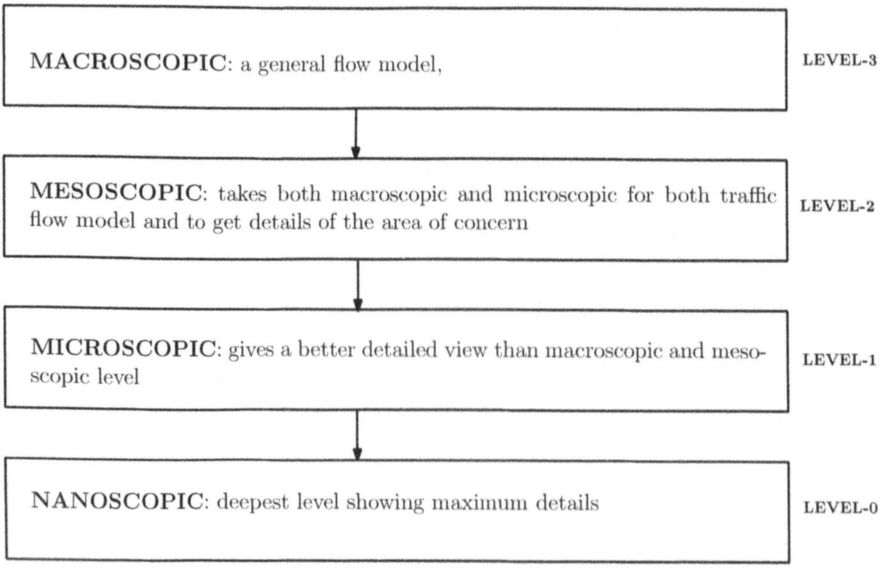

Fig. 1 Levels of traffic modeling (increasing level of depth from top to bottom)

and scalability, two popular and widespread distributed computing environments came into the picture such as grid computing and cloud computing. A grid computing system was proposed in [16] which provides a scalable distributed computing framework to respond to large, computationally complex problems.

The cloud has been viewed as a user-friendly version [22] or extension or a superset of the grid [21]. Cloud computing with the main focus of platform transparency and scalability, primarily provides a universal execution platform for different applications. Both cloud and grid computing systems are service-oriented architectures (SoA) [18]. Grid computing implements the software as a service (SaaS) approach. Cloud computing implements services at three distinct levels: SaaS, platform as a service (PaaS), and infrastructure as a service (IaaS) [19]. SaaS provides support to develop and access software components. PaaS provides users access to a big platform to implement their services. In case of IaaS, access to hardware infrastructures is given with managed access to cloud resources which enables to implement a broad range of software.

Table 1 gives a comparison of the distinct features of cloud and grid frameworks [20]. The complex cloud services, usually implemented as a single atomic unit, are the primitive parts of the cloud or grid system. Both cloud and grid computing are different in terms of service allocation and use. In most of the grid-based systems, a service is interpreted as a straight access to computational power which is nothing but running a parallel client application on each node [19], which is a step better than a standard cluster computer. Both cloud and grid computing principles can be used to design a Cloud-based ITS distributed processing platform. Cloud-based ITS

Table 1 Feature comparison between cloud and grid systems

Feature	Grid	Cloud
Architecture	SoA	SaaS, PaaS, IaaS
Sharing resources	Shared resources	Assigned resources are not shared
Control approach	Decentralized control	Centralized control
Heterogeneous resources	Combination of heterogeneous resources	Combination of heterogeneous resources
Usability	Hard to manage	User-friendly
Virtualization	Virtualization of data and computing resources	Virtualization of hardware and software resources

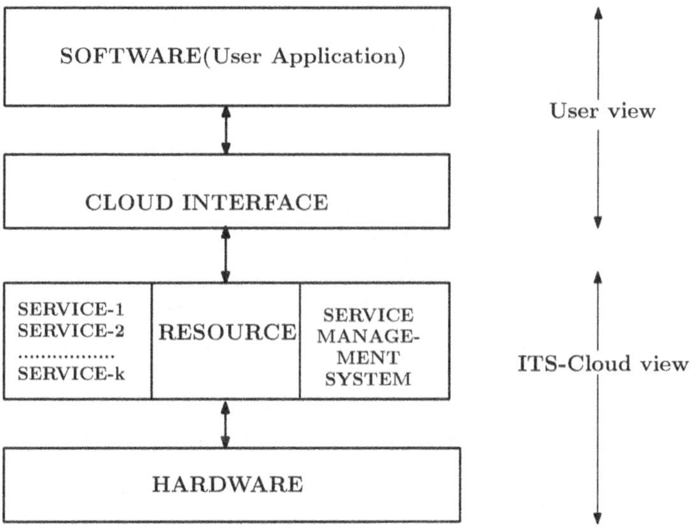

Fig. 2 Cloud-based ITS-Architecture

is developed to host the different CTMS and ITS applications. These application-specific designs need to adapt to suitable mechanisms from both cloud as well as grid computing frameworks. Cloud-based ITS follows the general distributed processing paradigm available in both cloud and grid systems along with service discovery, flexible messaging system, access, and life cycle management (Fig. 2). Features related to cloud computing included in ITS-cloud are

1. On-demand allocated dynamic services to be used only by single user.
2. Use of a multi-dynamic service allocation mechanism which supports portability, duplication, dynamic allocation, and availability to many users.
3. Service containers for dynamic and multi-dynamic service connectivity and life cycle management.

4 Design of Cloud-Based ITS Framework

Cloud-based ITS is a cloud-based distributed computing framework with the primary objective to host all related applications which form the Cloud-based Traffic Management System (CTMS) and to monitor their inter-communications. The Cloud-based ITS framework has the following essential components along with a cloud-interface [1]:

1. Services: These are the prime components of the service-oriented architecture (SoA) model. Services are the segregated cloud segments required to do assigned functions in the system, and using a controlled interaction between software services, the cloud-based ITS model operates effectively.
2. Service Management System: The Service Management System (SMS) is responsible for the acquisition of existing services and allocation of new services to be associated with suitable resources.
3. Resources: These are considered as service containers and required to initiate and manage other services for the whole life cycle of the service.

5 Limitations and Future Enhancements

Though CTMS is considered as a complete operational traffic management system, several improvements can be thought of which are presented in this section in the form of limitations and possible future enhancements related to the traffic management methods and CTMS in urban traffic-related situations.

A few future enhancements in the context of traffic management methods include

1. Emergency vehicle modeling: The main challenge for both CTMS and traffic simulator would be emergency vehicle modeling and handling by minimizing the duration required for the emergency vehicle to reach the incident location.
2. Automatic tuning methods: To adapt to the behavior of traffic management at different conditions, parameter control may be done using suitable automatic tuning methods. This leads to the prospect of future investigations on different tuning parameters which affect traffic management.
3. ITS vehicle lanes: As per [24], traffic performance can be studied by using vehicle lens which can make traffic simulators and models more effective to support various traffic conditions.

The limitations and future improvements over CTMS include

1. Dynamic adaptation: Cloud-based traffic management system should dynamically adapt to changes because when, where, and which vehicle leaves or joins the cloud cannot be predicted.
2. Cloud collaboration: It is a key challenge for cloud collaboration and efficient architecture flexibility to share available resources.

3. Communication: Due to the need for reliable communication, establishing cloud infrastructure is a challenging task for the implementation of cloud-based vehicular traffic management.

6 Conclusion

In this paper, we briefly address the issues of distributed processing and managing road traffic information from various sources. The traffic management performance was found to be further improved using distributed data management mechanisms such as Cloud-based ITS platform. Use of Cloud-based ITS is deployed for various traffic-related applications, and hence enhances the features like dependability, adaptability, and scalability of the urban traffic management system.

Cloud-based Traffic Management System (CTMS) which uses various methods and algorithms was developed in Cloud-based ITS as a software service component. Underlying features of ITS-cloud such as reliability, robust data management, scalability, self-configuration, and capability provide smart solutions to traffic management. CTMS is also used efficiently in providing support for dynamic routing, intersection control, and structured traffic management in large geographic regions. CTMS flexibly configures itself, once provided with basic data on the road traffic network. In this paper, we also discuss on ITS mechanisms such as adaptive intersection control which helps in reducing travel time and overall energy consumption.

References

1. Jaworski, P.: Cloud Computing Based Adaptive Traffic Control and Management (2013). https://pdfs.semanticscholar.org/e7b8/876e81881a89bf00a1f827955183e8b4ceaf.pdf
2. Jaworski, P., Tim E., Jonathan M., Keith J.B.: Cloud computing concept for Intelligent Transportation Systems. In: 2011 14th International IEEE Conference on Intelligent Transportation Systems (ITSC) (2011): 391–936. https://ieeexplore.ieee.org/stamp/stamp.jsp?tp=&arnumber=6083087&tag=1
3. Rahimi, M. M. and Hakimpour, F.: Towards A Cloud Based Smart Traffic Management Framework, Int. Arch. Photogramm. Remote Sens. Spatial Inf. Sci., XLII-4/W4, 447-453, https://doi.org/10.5194/isprs-archives-XLII-4-W4-447-2017. https://www.int-arch-photogramm-remote-sens-spatial-inf-sci.net/XLII-4-W4/447/2017/
4. Iftikhar Ahmad, Rafidah Md Noor, Ihsan Ali, Muhammad Imran and Athanasios Vasilakos, "Characterizing the role of vehicularcloud computing in road traffic management, International Journal of DistributedSensor Networks (2017), Vol. 13(5), https://doi.org/10.1177/155014771770872https://journals.sagepub.com/doi/pdf/10.1177/1550147717708728
5. Henry Liu, Jun-Seok Oh, and Will Recker. Adaptive signal control system with online performance measure for a single intersection. Transportation Research Record: Journal of the Transportation Research Board, pages 131138, 2002
6. Seung-Bae Cools, Carlos Gershenson, and Bart DHooghe. Self-organizing traffic lights: A realistic simulation. Advances in Applied Self-organizing Systems, pages 4150, 2008
7. McKenney, Dave: White, Tony: Distributed and adaptive traffic signal control within a realistic traffic simulation. Engineering Applications of Artificial Intelligence **26**(1), 574583 (2013)

8. D.A. Roozemond. Using intelligent agents for urban traffic control systems. In Proceedings of the international conference on artificial intelligence in transportation systems and science, pages 6979. Citeseer, 1999
9. M. Wiering.Multi-agent reinforcement learning for traffic light control. http://igitur-archive.library.uu.nl/math/2007-0330-200425/wiering00multi.pdf, 2000
10. Ioannis Papamichail and Katerina Kampitaki. Integrated ramp metering and variable speed limit control of motorway traffic flow. In 17th IFAC World Congress, pages 14084 14089,2008
11. Victor Firoiu and Marty Borden. A study of active queue management for congestion con- trol. In Nineteenth Annual Joint Conference of the IEEE Computer and Communications Societies, pages 1435 1444 vol.3, 2000
12. Rakha, Hesham, Aerde, Van: Michel: Comparison of Simulation Modules of TRANSYT and INTEGRATION Models. Transportation Research Record: Journal of the Transportation Research Board **1566**(1), 17 (1996)
13. Wilco Burghout, H.N. Koutsopoulos, and I. Andreasson. Hybrid mesoscopic-microscopic traffic simulation. Transportation Research Record: Journal of the Transportation Research Board, 1934(1):218 255, 2005
14. Emmanuel Bourrel, Rue Maurice Audin, and Vaulx-en-velin Cdex. Mixing Micro and Macro Representations of Traffic Flow: a Hybrid Model Based on the LWR Theory. Transportation Research Record: Journal of the Transportation Research Board, 1852(1):193 200, 2003
15. William D Gropp, Dinesh K Kaushik, David E Keyes, and Barry F Smith. High-performance parallel implicit CFD. Parallel Computing, 27(4):337362, March 2001
16. I. Foster and C. Kesselman. The grid: blueprint for a new computing infrastructure. 2004
17. J Shiers. The worldwide LHC computing grid (worldwide LCG). Computer physics communications, 177(1-2):219 233, 2007
18. Latha Srinivasan and Jem Treadwell Hp: An Overview of Service-oriented Architecture. HP Software Global Business Unit publication, Web Services and Grid Computing Service-Oriented Architecture (2005)
19. Ian Foster, Yong Zhao, Ioan Raicu, and Shiyong Lu. Cloud Computing and Grid Comput- ing 360-Degree Compared. In 2008 Grid Computing Environments Workshop, pages 110,November 2008
20. L.M. Vaquero, L. Rodero-Merino, Juan Caceres, and Maik Lindner. A break in the clouds: towards a cloud definition. ACM SIGCOMM Computer Communication Review, 39(1):50 55, 2008
21. G.V. McEvoy and B. Schulze. Using clouds to address grid limitations. In Proceedings of the 6th international workshop on Middleware for grid computing, 2008
22. Jeremy Geelan. Twenty-One Experts Define Cloud Computing. Cloud Expo, online article: http://cloudcomputing.sys-con.com/node/612375, 2009
23. Venkatesh, M., Srinivas, V.: Autonomous Traffic Control System Using Agent Based Technology. International Journal of Advancements in Technology **2**(3), 438445 (2011)
24. Bart van Arem, Cornelie J. G. van Driel, and Ruben Visser. The Impact of Cooperative Adaptive Cruise Control on Traffic-Flow Characteristics. IEEE Transactions on Intelligent Transportation Systems, 7(4):429436, December 2006
25. https://timesofindia.indiatimes.com/india/traffic-congestion-costs-four-major-indian-cities-rs-1-5-lakh-crore-a-year/articleshowprint/63918040.cms

A Survey on Clustering Algorithms Based on Bioinspired Optimization Techniques

Srikanta Kumar Sahoo and Priyabrata Pattanaik

Abstract In this growing informative world, in every domain, we can get a large amount of raw data, so it is a huge task to find proper and valid information from it. For this task, it is required to categorize data into different groups of similar behaviors. Over the years many authors have provided different techniques for clustering. Again, environment, domain, and applications are changing rapidly in different organizations, keeping this in mind many researchers are still modifying or developing new clustering algorithms. Now it is important to select or develop a proper clustering algorithm suitable for us. In this paper, we have tried to present some recent clustering algorithms. We have mainly focused on bioinspired optimization algorithms for the clustering problem. Later in the paper, we have also given a comparative study. This can help in selecting the proper clustering algorithm for the required domain.

Keywords Optimization-based clustering · Bioinspired clustering · Data clustering · Cluster center · SICD

1 Introduction

Nowadays, organizations like engineering, science, education, military, business, finance, etc., need automation. In the automation systems, strong algorithms are required for decision-making and processing. Hence, artificial intelligence, machine learning, data mining techniques become very important. Clustering is an important technique used in data mining for data analysis. In clustering, data elements having similar behaviours are collected together to form a group called a cluster. Behaviors of data elements of one cluster are somehow different from the data elements of other clusters.

S. K. Sahoo (✉) · P. Pattanaik
Institute of Technical Education and Research, Siksha 'O' Anusandhan (Deemed to be University), Bhubaneswar, Odisha 751030, India
e-mail: srikantasahoo@soa.ac.in

P. Pattanaik
e-mail: priyabratapattanaik@soa.ac.in

D. Mishra et al. (eds.), *Intelligent and Cloud Computing*, Smart Innovation, Systems and Technologies 194,
https://doi.org/10.1007/978-981-15-5971-6_48

In literature, there are different clustering algorithms proposed, which can be mainly divided into three categories: partition-based, density-based, and hierarchical clustering. In hierarchical clustering, once the data element assigned to a cluster, it cannot be reallocated. In density-based clustering, the nearby data elements are grouped to form a cluster, which may not give proper clustering when the size and dimension of the data set increases. In partition-based clustering, data elements are assigned to the cluster heads according to the similarity in behavior. So inter-cluster similarity is more and intra-cluster similarity is less. The partition-based clustering is simple and has less execution time. In this paper, we concentrate on partition-based clustering.

It has been observed that the classical clustering algorithms (including partition-based: *k*-mean, *k*-mode, *k*-medoids, PAM, CLARA, etc.) are somehow unable to classify the data sets into proper groups. This is due to the high dimensionality of data sets. Sometimes the size of data sets also matters. So, one of the ways to properly classify data sets into clusters is to use an optimization algorithm to maximize the inter-cluster similarity and minimize the intra-cluster similarity. Hence, the clustering problem can be considered as an optimization problem. In literature, it has been found that many optimization techniques are used for clustering. Among different optimization techniques, the nature-inspired meta-heuristic optimization algorithms are widely used for clustering applications, as they give global optimality with better time complexity.

In the remainder of the sections, we have presented a review of some existing methods, followed by a brief description of some optimization-based clustering algorithms, finally, a comparative study of these algorithms is presented.

2 Literature Review

In the early days, the classical clustering methods like *k*-mean and fuzzy *c*-mean (FCM) are found to be effective in image segmentation problems [1]. With emerge of soft-computing optimization algorithms are found to be more efficient for optimal clustering. Later it is also observed that most of the cases k-mean and FCM are used for problem objectives which then optimized using optimization algorithms. With this, many hybrid clustering algorithms using optimization techniques have been developed.

A list of these optimization algorithms and their application for the clustering problem in recent times is shown in Table 1. And later in the paper, we have described some of them along with a comparative study.

Table 1 Recent optimization algorithms

Algorithm name	Algorithm name
Particle swarm optimization [2]	Grasshopper optimization [13]
Ant colony based optimization [3]	Class topper optimization [14]
Lion optimization algorithm [4]	Elephant search optimization [15]
Bee colony optimization [5]	Wolf pack search algorithm [16]
Biogeography-based optimization [6]	Dolphin partner optimization [17]
Teaching-learning optimization [7]	Group counseling optimization [18]
Whale optimization algorithm [8]	Firey algorithm [19]
Clown fish optimization [9]	Hunting search [20]
Cuckoo search optimization [10]	Krill Herd [21]
Crow search optimization [11]	Fruit fly optimization [22]
Bat optimization [12]	Dolphin echolocation [23]
Genetic algorithm [24]	

3 Some Recent Optimization-Based Clustering Algorithms

3.1 Particle Swarm Optimization-Based

Particle Swarm Optimization is based on the searching behavior of swarms like a group of birds. It is a population-based meta-heuristic search method. The particles (birds) move in the direction of food destination by changing their position and velocity. Each position of the particle is a candidate solution (pbest) and the particle having the best position is the global solution (gbest). The particles move until the optimal solution (gbest) is found. In PSO, for each iteration of the algorithm velocity and positions are updated based on an objective function. The velocity and position can be updated using the following equations [2]:

$$V_i^{k+1} = \omega V_i^k + c1 * r1 * \left(P_i^k - X_i^k\right) + c2 * r2 * \left(G_i^k - X_i^k\right) \quad (1)$$

$$X_i^{k+1} = X_i^k + V_i^{k+1} \quad (2)$$

where V_i, P_i, X_i, G_i are velocity, pbest position, current position, and gbest positions, respectively.

PSO based clustering algorithm:

Step 1: Randomly initialize the cluster centers (Particle).
Step 2: Repeat steps 3–7 for a maximum number of iteration
Step 3: For each particle, repeat steps 4–7

Step 4: For each data vector, repeat steps 5–7
Step 5: Find the distance of the data vector to each cluster center and assign the data vector to the cluster whose distance is minimum.
Step 6: Calculate fitness value and update pbest and gbest.
Step 7: Update cluster centers using velocity (1) and position formula (2).

In [2], kernel density estimation technique along with PSO is used for clustering. Here the authors initialize the swarm; find each particle's personal best position and personal dense position using KDE (local maxima); then finally updates the velocity and position of particles. The author has given their modified position and fitness updating equation. This method overcomes the drawbacks of traditional PSO based clustering such as premature convergence to local optima and problem in setting learning coefficients.

3.2 *Ant Colony Optimization Based*

Ant colony optimization is based on the movement of ants from one destination to another. This movement depends on the pheromone intensity in the path and distance. At certain time t the probability of movement from source s to destination d by an ant k is calculated as follows [3]:

$$P_{sd}^{k}(t) = \frac{[\tau_{sd}(t)]^{\alpha}[\eta_{sd}]^{\beta}}{\sum_{k \in \text{allowed}(k)} [\tau_{sk}(t)]^{\alpha}[\eta_{sk}]^{\beta}} \quad \text{if } d \in \text{allowed}(k) \tag{3}$$

Here τ_{sd} is the intensity of pheromone in the path, $\eta_{sd} = 1/\text{distance}_{sd}$, α and β are constants that tell the influence of pheromone. After a certain time t', the intensity of pheromone is updated as [3]:

$$\tau_{sd}\left(t + t'\right) = (1 - \rho)\tau_{sd}(t) + \Delta\tau_{sd} \tag{4}$$

Here ρ is the rate of evaporation and $\Delta\tau_{sd}$ is the total sum of pheromone submitted by all ants. The process repeats for some maximum number of iterations to give an optimal solution.

In [3], the author has used the conventional clustering technique LEACH to divide into clusters with one cluster head each, and they used the ACO technique to move the mobile sinks to cluster heads and collect data directly from them. This reduces the energy and lifetime of battery-operated devices in the household.

3.3 Bee Colony Optimization Based

Bee colony optimization is based on the style of the food collection process by honey bees. In bee colony optimization, initially, all the honey bees search for food source independently, when a bee finds food source it comes back with some nectar amount to its hive to inform others. The bee having better nectar amount become the scout bee and some of the other bees follow it. The process repeated for some max number of times to get to an optimal solution.

A modified bee colony optimization used for clustering proposed in [5]. Here the data point allocation is done in two passes (forward and backward) of search space. Then the un-allocated data points given a chance for reallocation. The algorithm is as follows:

BCO based clustering algorithm:

Step 1: Initialize the number of clusters and randomly choose the cluster heads.
Step 2: Repeat steps 3–6 for some maximum number of iterations
Step 3: Assign data points to clusters according to the probability of loyalty and generate partial solutions.
Step 4: Analyze the partial solutions generated in step 3 according to SICD value and recruit the best among partial solutions.
Step 5: Assign the un-allocated data points to different clusters according to the probability of loyalty.
Step 6: Determine the global best solution using SICD value and update.

Here, in the algorithm in step 3, the data points assigned to a specific cluster head by the probability of trust of the data point with the cluster. And the selection of the best solution in step 4 is made by computing SICD (sum of inter-cluster distance) value using the following [5]:

$$\text{SICD}(n, r) = \sum_{i=1}^{n} \sum_{j=1}^{r} d(o_j, c_i) \tag{5}$$

Here, n is the number of clusters, r is the number of objects in the cluster, and $d(o_j, c_i)$ is the Euclidian distance between jth object to ith cluster. The author has also represented a hybrid approach of clustering by combining MBCO and k-means techniques. This also gives better performance compared to some existing classical algorithms.

3.4 Whale Optimization Algorithm Based

The Whale Optimization Algorithm is based on the hunting behavior of humpback whales. The whales dive down the water around 12 m and create bubbles around the subject. And then move upwards in the water to attack the subject. This is called bubble-net attacking. The bubble-net attacking is as follows:

The whales shrink and encircle the subject by considering the present best position as the position of the target subject and other whales update their positions referring to it. This can be done by the following equations [8]:

$$D = \left|C \cdot X^*(t) - X(t)\right| \quad (6)$$

$$X(t+1) = X^*(t) - A \cdot D \quad (7)$$

where X^* is the current best whale position and X is the position of a searching whale. Value of A and C are computed as $A = 2 \cdot a \cdot r - a$ and $C = 2 \cdot r$ with a varying from 2 to 0 and r a random value between 0 and 1.

The whales can also attack using a spiral movement. The position of the whale, in this case, varies in a helix shape. This can be done by the following equation [8]:

$$X(t+1) = D \cdot \mathrm{e}^{bl} \cos(2\pi l) + X^*(t) \quad (8)$$

In a bubble-net attack to find the optimal solution, the whale moves randomly to find target subjects. This random movement can be done by the equations [8]:

$$D = |C \cdot X_{\text{rand}} - X| \quad (9)$$

$$X(t+1) = X_{\text{rand}} - A \cdot D \quad (10)$$

Whale optimization-based clustering algorithm is as follows:

Step 1: Randomly initialize the cluster centers (Particle).
Step 2: Repeat steps 3–8 for a maximum number of iteration
Step 3: For each particle, repeat steps 4–6
Step 4: For each data vector, repeat steps 5–6
Step 5: Find the distance of the data vector to each cluster center and assign the data vector to the cluster whose distance is minimum.
Step 6: Calculate fitness value and update best search agent X^*.
Step 7: For each search agent X repeat step 8
Step 8: Update the position of search agent X, either by shrink or spiral move using (7) or (8). If $A > 1$ then randomly choose another target using (10).

3.5 Class Topper Optimization Algorithm Based

The Class Topper Optimization algorithm [14] works according to the learning behavior of students in a class. The performance index (PI) of each student can be calculated by considering exam marks of different subjects. All students in a section learn from the section topper (ST) and update their PI to become the ST. Similarly, all section toppers learn from class topper (CT) to become the next class topper. This process repeats for $E_{\max}$ number of examinations, and in each examination for each section and across all sections PI updated. Finally, the ST with the best PI becomes CT. This algorithm is used for clustering applications, by considering each student as a search agent and each section as a cluster. Here at the student-level, learning from ST is performed using Eqs. (11) and (12), again learning in section level is performed using Eqs. (13) and (14), where ST learns from CT [14].

$$I^{(S,E+1)} = I_{\mathrm{WF}} I^{(S,E)} + cn_2\left(\mathrm{ST}^{(\mathrm{SE},E)}_{\text{centroids}} - S^{(S,E)}_{\text{centroid}}\right) \tag{11}$$

$$S^{(S,E+1)}_{\text{centroid}} = S^{(S,E)}_{\text{centroid}} + I^{(S,E+1)} \tag{12}$$

$$J^{(\mathrm{SE},E+1)} = I_{\mathrm{WF}} J^{(\mathrm{SE},E)} + cn_1(\mathrm{CT}^{(E)}_{\text{centroids}} - \mathrm{ST}^{(\mathrm{SE},E)}_{\text{centroid}}) \tag{13}$$

$$\mathrm{ST}^{(\mathrm{SE},E+1)}_{\text{centroid}} = \mathrm{ST}^{(\mathrm{SE},E)}_{\text{centroid}} + J^{(\mathrm{SE},E+1)} \tag{14}$$

Here E is examination, S is student, SE is section, ST is section topper, CT is the class topper, I_{WF} is inertia weight factor, c is acceleration coefficient, n_1, n_2 are random values between 0 and 1.

The clustering algorithm is as follows:

Step 1: Repeat steps 2–11 for $E_{\max}$ number of examinations.
Step 2: Repeat steps 3–10 for $\mathrm{SE}_{\max}$ number of sections
Step 3: Update improvements of ST by learning from CT using (13) and (14).
Step 4: Repeat steps 5–9 for $S_{\max}$ number of students
Step 5: Update the improvements of S by learning from ST using (11) and (12).
Step 6: Randomly initialize the cluster centers in the first iteration/examination.
Step 7: Assign data points to different clusters as per the distance.
Step 8: Find new cluster centers by taking mean.
Step 9: Calculate SICD value as the performance index.
Step 10: If new cluster centers have better SICD value then update cluster centers or update section topper (ST).
Step 11: Update class topper (CT) as best among the STs.
Step 12: Set optimal solution as CT.

This clustering algorithm's performance was compared with some of the well-known optimization algorithms for clustering and found that with the same level of time complexity it gives better results.

4 Comparative Study

Table 2 depicts the comparative study of the algorithms discussed in Sect. 3.

Table 2 Optimization algorithms comparative study

References	Technique used	Data set used	Evaluation parameters	Advantages	Limitations
Alswaitti et al. [2]	PSO, kernel density estimation (KDE)	Iris, wine, breast cancer, glass, etc. from UCI repository	Classification accuracy (CA), standard deviation, Dunn index (DI)	Better accuracy, cluster compactness, and repeatability	Execution time is more for large datasets
Wang et al. [3]	LEACH, ACO	50–200 sensor nodes with initial energy 0.5 J distributed over 10,000 m^2 area	Energy consumption, residual energy and network lifetime	Better performance than LEACH and mobile-P	It does not perform well when the network size increases
Das et al. [5]	MBCO, *k*-means	Iris, wine, glass, cancer, etc. from UCI repository	SICD value, classification error, time complexity	Better clustering accuracy with almost similar time complexity	The convergence rate is low with higher dimension data sets
Nasiri and Khiyabani [8]	WOA	ART, iris, wine, CMC, thyroid, etc.	Mean, standard deviation, rank	Intra-cluster distance and standard deviation found to be better than some well-known algorithms	It does not work for local search. Performance can be further improved
Das et al. [14]	CTO	Iris, cancer, wine, CMC, HV from UCI repository.	SICD value, average percentage of error, time complexity	Parallel search enables it to perform better for high dimensional data sets	Fails to perform well for nonspherical data

5 Conclusion

In this paper, we have made a study of different optimization algorithms used for clustering, with position updating equations. In real life there are many classification problems which need to minimize/maximize specific parameters (objective function); hence clustering algorithm with optimization is a helpful tool. Once the objective function for the particular problem is defined, one can use these algorithms. According to the requirement, these objective functions may change. Here we have studied some recently used optimization-based clustering algorithms, also made a comparative study that can help many aspirants. Apart from this, we have listed twenty-two such algorithms that give the aspirants a choice to choose for a particular application. Our future work remains to design an algorithm which can be useful for anomaly detection in social media or similar type of applications.

References

1. Binu, D.: Cluster analysis using optimization algorithms with newly designed objective functions. Expert Syst. Appl. **42**, 5848–5859 (2015)
2. Alswaitti, M., Albughdadi, M., Isa, N.: Density-based particle swarm optimization algorithm for data clustering. Expert Syst. Appl. **91**, 170–186 (2018)
3. Wang, J., Cao, J., Li, B., Lee, S., Sherratt, R.: Bio-inspired ant colony optimization based clustering algorithm with mobile sinks for applications in consumer home automation networks. IEEE Trans. Consum. Electron. **61**(4), 438–444 (2015)
4. Yazdani, M., Jolai, F.: Lion optimization algorithm (LOA): a nature-inspired metaheuristic algorithm. J. Comput. Des. Eng. **3**, 24–36 (2016)
5. Das, P., Das, D., Dey, S.: A modified Bee Colony Optimization (MBCO) and its hybridization with *k*-means for an application to data clustering. Appl. Soft Comput. **70**, 590–603 (2018)
6. Simon, D.: Biogeography-based optimization. IEEE Trans. Evol. Comput. **12**(6), 702–713 (2008)
7. Shahsamandi, P., Sadi-nezhad, S.: An improved teaching-learning based optimization approach for fuzzy clustering. In: Third International Conference on Advanced Information Technologies and Applications, pp. 43–50 (2014)
8. Nasiri, J., Khiyabani, F.: A whale optimization algorithm (WOA) approach for clustering. Cogent Math. Stat. ISSN: 2574–2558 (2018)
9. Christ, J., Subramanian, R.: Clown fish queuing and switching optimization algorithm for brain tumor segmentation. Biomed. Res. **27**(1), 65–69 (2016)
10. Lakshmi, K., Visalakshi, N., Shanthi, S., Parvathavarthini, S.: Clustering categorical data using *K*-modes based on cuckoo search optimization algorithm ICTACT. J. Soft Comput. **8**(1), 1561–1566 (2017)
11. Parvathavarthini, S., Karthikeyani Visalakshi, K., Shanthi, S., Madhan Mohan, J.: Crow search optimization based fuzzy *C*-means clustering for optimal centroid initialization. Taga. J. Graphic. Technol. **14**, 3034–3045 (2018)
12. Vellaichamy, V., Kalimuthu, V.: Hybrid collaborative movie recommender system using clustering and bat optimization. Int. J. Intell. Eng. Syst. **10**(1), 38–47 (2017)
13. Łukasik, S., Kowalski, P., Charytanowicz, M., Kulczycki, P.: Data clustering with grasshopper optimization algorithm. In: Proceedings of the Federated Conference on Computer Science and Information Systems, vol. 11, pp. 71–74 (2017)

14. Das, P., Das, D., Dey, S.: A new class topper optimization algorithm with an application to data clustering. IEEE Trans. Emerg. Topics Comput. **99**, 1 (2018)
15. Deb, S., Fong, S., Tian, Z.: Elephant search algorithm for optimization problems. In: Tenth International Conference on Digital Information Management (ICDIM), IEEE, pp. 249–255 (2015)
16. Yang, C., Tu, X., Chen, J.: Algorithm of marriage in honey bees optimization based on the wolf pack search. In: The 2007 International Conference on Intelligent Pervasive Computing, IPC. IEEE, pp. 462–467 (2007)
17. Shiqin, Y., Jianjun, J., Guangxing, Y.: A dolphin partner optimization. GCIS'09 WRI Global Congr. Intell. Syst. IEEE **1**, 124–128 (2009)
18. Eita, M., Fahmy, M.: Group counseling optimization. Appl. Soft Comput. **22**, 585–604 (2014)
19. Yang, X.: Firefly algorithm, stochastic test functions and design optimization. Int. J. Bio-Inspired Comput. **2**(2), 78–84 (2010)
20. Oftadeh, R., Mahjoob, M., Shariatpanahi, M.: A novel metaheuristic optimization algorithm inspired by group hunting of animals: Hunting search. Comput. Math Appl. **60**(7), 2087–2098 (2010)
21. Gandomi, A.H., Alavi, A.H.: Krill herd: a new bio-inspired optimization algorithm. Commun. Nonlinear Sci. Numer. Simul. **17**(12), 4831–4845 (2012)
22. Pan, W.-T.: A new fruit fly optimization algorithm: taking the financial distress model as an example. Knowl. Based Syst. **26**, 69–74 (2012)
23. Kaveh, A., Farhoudi, N.: A new optimization method: dolphin echolocation. Adv. Eng. Softw. **59**, 53–70 (2013)
24. Halder, A., Pramanik, S., Kar, A.: Dynamic image segmentation using fuzzy *C*-means based genetic algorithm. Int. J. Comput. Appl. **28**(6), 15–20 (2011)

Automated Question Answering System: An Overview of Solution Strategies

Debashish Mohapatra and Ajit Kumar Nayak

Abstract All the answers cannot be found at one place even when anyone searches it on the web, so a system is needed that can give accurate answers to the questions asked. This is where the question answer system comes into place. Question Answering (QA) system can be defined as an approach in the computer science field which deals with the building of the systems that can automatically provide answers of any type and at any point of time asked by humans in natural language. The basic idea of developing such a system is to have man-machine interaction. In this paper, few methods in question answering system to retrieve an answer whenever a question is asked.

Keywords QA system · Natural language processing · Surface pattern · Pattern matching

1 Introduction

When something is searched on the internet, no one gets an accurate result. The answers vary because the answers are in an ordered list, e.g., "*Where is the birthplace of Mahatma Gandhi?*" [1] The answer should be, "Porbandar, Gujarat". Instead of getting such an accurate answer, the user gets a variety of answers. This creates confusion as which answer is the correct one. In order to solve this issue or problem question answering system is developed.

D. Mohapatra (✉)
Department of Computer Science and Engineering, Institute of Technical Education and Research, Siksha 'O' Anusandhan (Deemed to be University), Bhubaneswar, Odisha 751030, India
e-mail: mohapatra.debashish171@gmail.com

A. K. Nayak
Department of Computer Science and Information Technology, Institute of Technical Education and Research, Siksha 'O' Anusandhan (Deemed to be University), Bhubaneswar, Odisha 751030, India
e-mail: ajitnayak@soa.ac.in

D. Mishra et al. (eds.), *Intelligent and Cloud Computing*,
Smart Innovation, Systems and Technologies 194,
https://doi.org/10.1007/978-981-15-5971-6_49

Question Answering System (QAS) can be defined as to build a system which can answer all the questions accurately asked by the humans in natural language or a computer program which gets the correct answer from a structured database of a huge amount of knowledge and information which is collected from an unstructured collection of natural language documents, texts, data, etc. There are few areas from where the documents are collected in natural languages which are as follows: any set of Wikipedia pages, a set of web pages on particular topics, news reports, a local collection of old documents, etc. Question Answering system deals with a variety of questions such as facts, lists, how, when, where, which, cross-lingual, etc. There are two types of domains in which the question answering system broadly works: (a) Closed domain and (b) Open domain. Closed domain [2] can be defined as domain-specific questions on that particular topic, e.g., Questions on trees, food, animals, etc. This leads to a limited number of questions as it is specific to only one domain. Open domain is more generic and contains any type of questions from any domain. This domain has more variety and diverse questions to be asked from.

The rest of the paper is organized as follows: Sect. 2 discusses about the architecture of the question answering system. Section 3 is about Processes of question answering system. Section 4 is explaining about solution strategies of question answering system. Section 5 describes Challenges in question answering system. Section 6 explains the types of question answering system. Section 7 is the conclusion.

2 Architecture of Question Answering System

The diagram represents the architecture of the question answering system (Fig. 1).

General steps that are followed by any question answering system:

When a user passes any type of question to the question answering system. The question is passed to the question processing block. The question processing block [3], has three parts which are as follows: question analysis, question classifier, and question reformulation. The overall function of the question answering block is to process the question in such a way which leads to the representation of the information needed. The output of the question answer processing block is to form a set of queries to further process it in the document processing module to get the information required. The queries are then processed to the document processing block. The document processing block has three parts which are as follows: Information retrieval, Paragraph Filtering, Paragraph Ordering. The role of document processing block paragraph ordering is to rank the documents the ordering of relevant documents fetched. The only motivation is to convert the documents into paragraphs to make the system faster. The relevant documents are then sent to the answer processing block. The answer processing block has three parts which are as follows: answer identification, answer extraction, answer validation. The role of this block is to identify, extract, and validate the actual answer needed to be provided to the user.

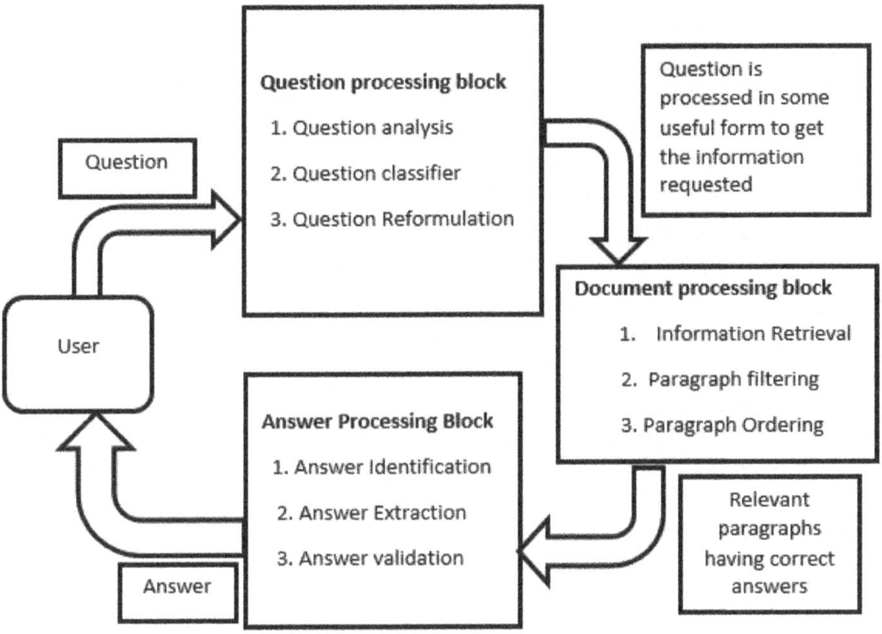

Fig. 1 Architecture of question answering system

3 Types of Processing

Following are the different types of processing [3], required in question answering system which are described as follows.

3.1 Question Processing

The role of the question processing block is to analyze, classify the type of question, and reform it in such a way that it can help to fetch the right answers.

Question Analyzer: The main job is to focus on the actual portion of the question to get to the actual answer. For example, "What is the longest river in the world?" so we need to focus on the "longest river" and "world". This contains what and where question to get the answer tailored. This makes it easier to get the answer from the resources, be it already available, or need to be searched out.

Question classifier: The duty of classifier is to find out what type of question it is. For example, Why, when, what, how, etc., it basically acts as a guidance to the answer to be found for the question asked by the user. The question only is not enough for finding the accurate answer.

Question Reformation: The main objective is to expand the keywords found in the question so that the answers can be found accurately. After the question type and what to focus is found, then a list of keywords is passed on for further searching of answers.

3.2 Document Processing

The document processing block is responsible for converting relevant documents fetched into paragraphs to make it easier for the system to get the right or correct answer.

Paragraph Filtering: The paragraph filtering is basically filtering the documents that consist of the keywords from the question asked after the reformation. If we have a large amount of documents to search from, then we can discard the paragraphs which do not have the keywords associated with it.

Paragraph Ordering: In this, we order the paragraph according to the possibility of getting the accurate results. This can be done in three ways:

1. The number of keywords found in a sequence increases the probability of getting the accurate answer.
2. The keywords distant from the accurate answers result in the next ordering as it has less chances of getting the answer.
3. Last but not the least; the keywords are missing in a paragraph. These paragraphs are not considered to have the results we want or simply being discarded.

Information Retrieval: Whenever a question is asked a lot of data is found on the web related to it. The goal is to find relevant documents and ranks then according to the possibility of getting the right result.

3.3 Answer Processing

The answer processing block identifies, extract, and validate the answers from the given data that has been processed from the document processing.

Answer Identification: It is not always necessary that we can get an answer within the question itself. So it is better to depend on a few things such as name, organization. etc. This works when we have to find out the results from other documents.

Answer Extraction: The paragraphs that we have are used to find answers by using the parser. The best way to get the answer is to rank the paragraphs in sequence or rank then to the probability of getting the right answers. If we don't get the answers in the first place then the system will automatically select the second-best possible answer.

Answer Validation: Confidence in the answer can be done in two way. One way to have relevant resource or knowledge that can give the correct answer type. If the specific data type is used, then we can use the web to validate the answer.

4 Question Answer Solution Strategies

There are many strategies to solve the question answering system better but out of those the following have been explained because these approaches cover major questions that can be solved (Fig. 2).

Linguistic Approach: In this, a lot of natural language knowledge for processing the data and for this were all need a linguistic person, so that we can solve the natural language [4] issues in any forms such as grammatical. New technology helps the machine to learn itself from everyday usage and being more prominent and efficient. Thus, reduces effort after a specific period of time. In contrast to the new method, old technology highly relied on AI and then merging with natural language. But the data collected will always be in different forms of lang uage even though it is taken from any source. The source can be highly demanded on the internet and unstructured data. A lot of processing is done on the user's query such as parsing, POS tagging, and so on. The only role of such processing is to have a prediction of the output. The major challenge is in which language the question is asked and to check whether the machine knows it or not. Then how the question can be broken and made the question more simplified. This makes easy for the machine to do work and easier prediction for the accurate output.

Statistical Approach: This approach depends on existing data [5], whose answers are already there. The machine just has to guess the right answer every time. The only job is to make the system efficient enough to answer the existing questions. This is the trending stuff going on in theweb-based applications.

Pattern Matching Approach: In this, the question finds out the pattern or the questions hidden [6] in the questions only, and such series of data is then processed further to find out the answers. Suppose a question like when is the next world cup? Then, this question is broken into pieces and pattern is made with the existing one.

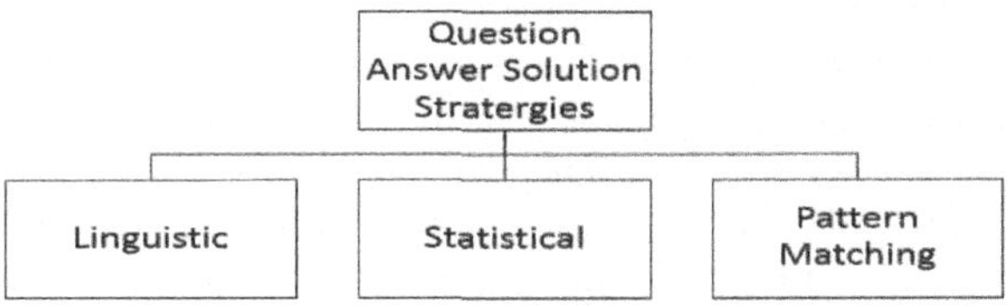

Fig. 2 Figure shows different question answer solution strategies

5 Challenges in Question Answering System

There are many challenges to overcome in building a question answering system out of which following are explained as they are the most common issues to be found to be solved.

Lexical Gap: The same words may have different meaning [7], in different languages. This causes difficulty to guess out the meaning for the machine. Such as Pooja is buying thing for puja. Here Pooja is a person and puja is an event that happens in Indian families. In such a situation, the machine is not capable of discriminating the words which are almost having different meaning but same representation. This creates a gap between such words and errors may occur. The system can answer any query asked by humans if it is trained continuously and bridging the lexical gap over a period of time. String Normalization and String similarity Normalization allow matching of string in slightly different forms and conversation in lowers cases.

Ambiguity: This is the phenomenon where one-word sentence has a different meaning. This is the difficulty that has to be solved. These can be lexical, semantic, and syntactical. This is the flip side of ambiguity. The method is basically used to overcome the lexical gap [8]. If the system is not trained enough then it leads to loss of precision in the system. We use ambiguous and nonambiguous data. Nonambiguousness is the process of selecting one type of meaning for most of the phrases. Corpus based data are basically uses prediction, frequency of words used, from unstructured data.

Multilingualism: There are many languages and every person use it on the basis of its comfort level. There are many languages to one can express but having a large corpus [9], in which one sentence is having the same meaning in different languages. The much easier method is to have QA that have a large potential to process the data. This challenge will always be there as it is a very vast field to understand.

6 Types of Automated Question Answering System

There are many types of question answering systems. These can be broadly divided into three categories which are as follows:

Voice: The best example of automated question answering system using voice is Google Duplex [10]. This system is basically calls on behalf of the user. The system is so accurate that it can talk like a real human being. The calls get received when the receiver is absent. The call gets recorded and can be fully saved in the cell phone of the user. The user does not have to worry about the important calls. This makes the user's life easier, and the technology is best for busy people who get multiple calls. This technology is developed by Google, which is a multinational company which provides various serious and leading technologies. Android is developed by Google and it is the most widely used technology in smartphones these days. The call recording can only be done when the user gives permission to it otherwise it is sent to the non-recording line. The non-recorded call stays with Google server and the best part is that it can be used as evidence if there is any case.

Image: In image processing [11], the text or image is captured by the machine or the system. The image is scanned and all the text and image are extracted for it. The data that is taken will be in structured or unstructured form and that will be stored in the database. Socratic is an android application that scans the image of the mathematics problems and then gives the answers accordingly. The system is basically good for students. The company has only aimed to make education easier and available to all for everyone through digital sources. The application takes the image of the unsolved mathematics problem and scans it. The image is then processed and then takes all the text from it. The system matches with the problem and then identifies the solution to it.

Text: Hike [12] (Chat Application), that is highly popular for its features. This is an Indian product and used by youths of India, mostly. It has got an AI assistant popularly known as Natasha, who gives basic answers like fake calls, jokes, thoughts, weather condition, news, and so on. The best part is that it has a hidden mode that keeps the chats private, and 128-bit encryption. It can be used for calling over the internet, video calls, texting, and a lot more fun stuff.

7 Conclusion

The survey paper concludes about the comprehensive overview of the question answering system, its challenges, and a few applications that are broadly being used. The QA system is highly problem-specific. Broadly, the QA system uses techniques based on linguistic approach, statistical approach, and pattern-based approach. The QA system is divided into three categories which are in the form of voice, text, and image.

References

1. Dwevedi, S.K., Shingh, V.: Research and reviews in question answering system. In: International Conference on Computational Intelligence: Modeling Techniques and Applications (CIMTA) (2013)
2. https://en.wikipedia.org/wiki/Natural_language_processing
3. Allam, A.M.N., Haggag, M.H.: The question answering system: A survey. Int. J. Res. Rev. Inf. Sci. (IJRRIS) **2**(3) (September 2012)
4. Clark, P., Thompson, J., Porter, B.: A knowledge-based approach to question answering. In: Proceedings of AAAI'99 Fall Symposium on Question-Answering Systems (1999)
5. Ittycheriah, A., Franz, M., Zhu, W.J., Ratnaparkhi, A., Mammone, R.J.: IBM's statistical question answering system. In: Proceeding of the Text Retrieval Conference TREC-9
6. Zhang, D., Lee, W.S.: Web based pattern matching and mining approach to question answering. In: Proceeding of the 11th Text Retrieval Conference (2002)
7. https://www.daytranslations.com/blog/lexical-english-language/

8. Anjali, M.K., Babu Anto, P.: Ambiguities in natural language processing. Int. J. Innov. Res. Comput. Commun. Eng. **2**(5) (October)
9. https://en.wikipedia.org/wiki/Multilingualism
10. https://www.cnet.com/how-to/what-is-google-duplex/
11. https://en.wikipedia.org/wiki/Socratic.org
12. https://en.wikipedia.org/wiki/Hike_Messenger

Study on Google Firebase for Real-Time Web Messaging

Rahul Patnaik, Rajesh Pradhan, Smita Rath, Chandan Mishra, and Lagnajeet Mohanty

Abstract In today's fast-paced life real-time communication with everyone is essential as well as appreciated by everyone. This study introduces a real-time database server called the Google Firebase API and it features through a web messaging app. A real-time database is the type of database that stores and fetches the data stored in it very quickly. But firebase's this feature is just tip of the iceberg, it's much more than that it provides developers with a variety of features that not just creates a secure system but also develop communication- based applications with ease. This article will cover how firebase helps backend as well as storage purpose of an application and reduces the burden from the technical point of view.

Keywords Google firebase API · Authentication · Cloud messaging · Real-time database

1 Introduction

Gone are the days when people used to write down every important detail into the logs. Over the years, technology has replaced the pen and paper by bringing virtual database i.e. cloud based. Every important detail that need to jotted down

R. Patnaik · R. Pradhan · S. Rath (✉) · C. Mishra · L. Mohanty
Department of Computer Science and Engineering, Institute of Technical Education and Research, Siksha 'O' Anusandhan (Deemed to be University), Bhubaneswar, Odisha 751030, India
e-mail: smitarath@soa.ac.in

R. Patnaik
e-mail: patnaik.rahul2503@gmail.com

R. Pradhan
e-mail: raj4568111@gmail.com

C. Mishra
e-mail: mishrachandan1997.cm@gmail.com

L. Mohanty
e-mail: lagnajeet2000@gmail.com

D. Mishra et al. (eds.), *Intelligent and Cloud Computing*,
Smart Innovation, Systems and Technologies 194,
https://doi.org/10.1007/978-981-15-5971-6_50

into the books and done into memory of a cloud server. A cloud service has some characteristics that differentiate it from the conventional web hosting it is versatile, a client can have as many or as few service as they would like at any particular time, and the service is operated by the provider in full. The provider gives a real-time database which can be private or public. Private cloud services are used by companies or firmware's whereas public cloud services are accessible by all the users on the internet. Public cloud applications, usually by a minute or hour, are sold on demand. Only the CPU cycles, space and bandwidth that consumers use account for. Cloud computing eliminates all the hassles that come with storing your own data, and it is accessible from anywhere. The flexibility is more expanded by the customization of cloud-based application by modification of power, storage, and bandwidth as users' needs change.

Khedkar and Thube [1] described about firebase and its features, also compared firebase with mongo Singh [2] explained about the real time database storage, authentication, hosting and database. Hari Shankar Singh, Uma Shankar Singh [3] implemented website development through firebase and describe its essential features to make a real time models. Chatterjee et al. [4] implemented real time android application using firebase that send text information about videos, text and images over internet. Wadkar and Patil [5] differentiated and focused advantages and limitations of firebase infrastructure over traditional database. Ozsoyoilu and Snodgrass [6] explained about temporal and real time database in real-time applications. Lahudkar et al. [7] implemented firebase to explain NoSQL database. Srivastava et al. [8] implemented web messaging through firebase across various platforms.

1.1 Types of Services

- SaaS (Software as Service):- A software distribution system in which applications are hosted by a third party provider and made available over the web to subscribers.
- PaaS (Platform as a Service):- A system in which a third-party service hosts platforms and software for application development on its own network and makes them accessible over the network to subscribers.
- IaaS (Infrastructure as a service):- A system in which databases, storage and other virtualized computing resources are managed by a third-party company and made available to developers on the cloud.
- BaaS (Backend as a service):- A model in which developers outsource all the behind the scene aspects of a web application so that they only have to maintain the frontend.

A cross platform application ranging from iOS to android has its own database. The basic idea of the database is to fetch the sorted data in a given time. Any reactive application which immediately reflects in the interface which is used by people uses a real time database. The data of the application is stored a JSON file and can be accessed by an individual across platforms.

2 Firebase

Firebase is a BaaS (Backend as a service) providing company operated by Google. Basically, its's a real time database i.e. it is a NoSQL database. It takes information in one large JSON tree. Simple information is stored in records while hierarchically complex data is stored. While most of the databases use HTTP to sync the data, firebase uses WebSocket is much faster than the traditional HTTP. All the data can be synced from a single WebSocket depending upon the Client's network Speed. While being fast, it is also very secure because it supports OAuth2 for authentication via third-party applications.

2.1 Features of Firebase

Once the firebase API is enabled into the application, its features can be used with small simple lines of codes.

- Analytics: This service allows the designer of the software to know how the user is using it. The SDK enabled captures every detail, events, properties as well as any custom-data. Dashboard shows summarised report of the application usage ranging from number of visitors to most used feature of the application.
- Authentication: The foremost important thing an application requires is an authorised entry of the users in the application. The auth feature of the API allows users to sign into the app through Gmail, twitter, Facebook, etc. The auth variable has unique id which is very unique for every user and auth variable value gets defined in security and firebase rules. Authenticated user experience can be customized by the developer.
- Storage: Firebase also has a storage facility that can store content such as images, videos and audio directly from the SDK server. Uploading and downloading control is with the user while developer has all the details of the files that is being circulated in the application.
- Firebase cloud messaging: cross-platform interactive application which allows developer to build app across various platform like android, iOS, web-based allowing transfer of messages at no cost. An FCM (firebase cloud messaging) implementation includes an app server interacting with FCM via HTTP or XMPP protocol (Fig. 1).
- Hosting: this feature of firebase is used to provide fast and secure static hosting for the application. Firebase hosting is a web application service hosting for programmers in the production class. For Hosting, with a single command, you

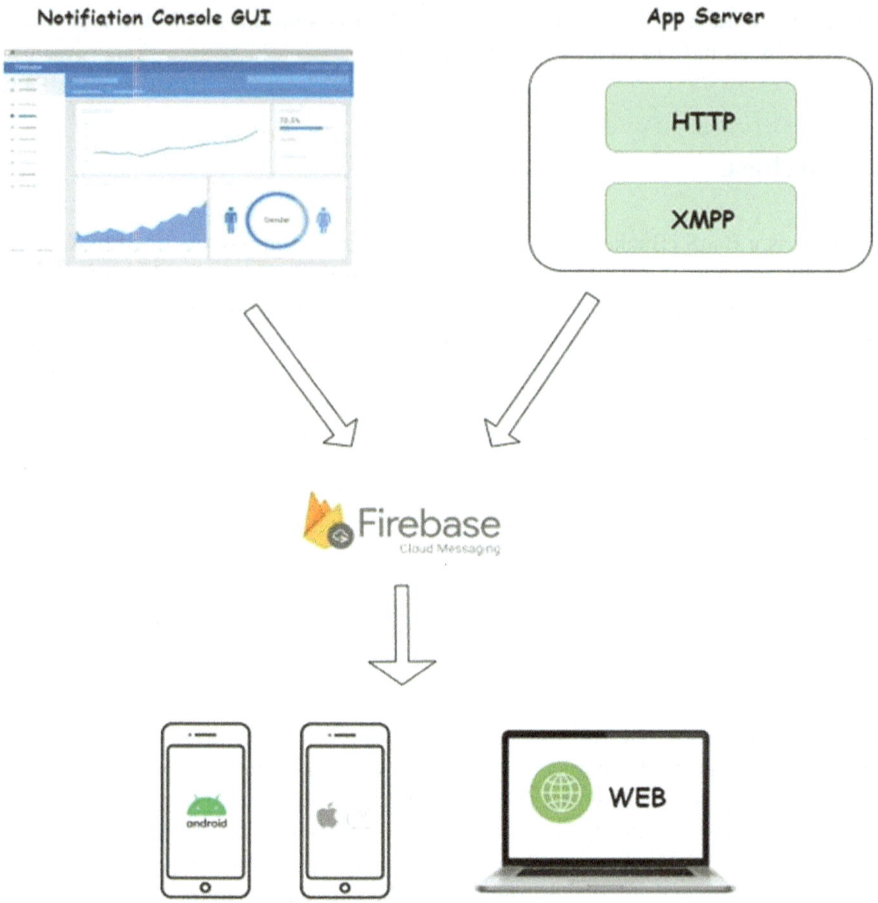

Fig. 1 Firebase cloud messaging framework

can submit web services and static content quickly and effectively to a Content Delivery Network (CDN). It requires custom domain support, Regional CDN and SSL Certificate Auto Provided.

- Real time database: firebase doesn't require any SQL queries to see/fetch the data that is why firebase is known to have NoSQL database. Even if the system goes offline database securely stores the data in its servers.
- Crash reporting: any error that occurs in the application during its run or even when the app is offline it comes as a crash to the firebase. Every crash is reported to the developer and necessary actions are taken in order for the better experience. Custom events can be created to catch the steps that lead to the crash. In that case the same crash can be avoided very easily by the application.
- App indexing: when the application is already installed in the user's device, if he/she searches for some content and that matches the related content of the

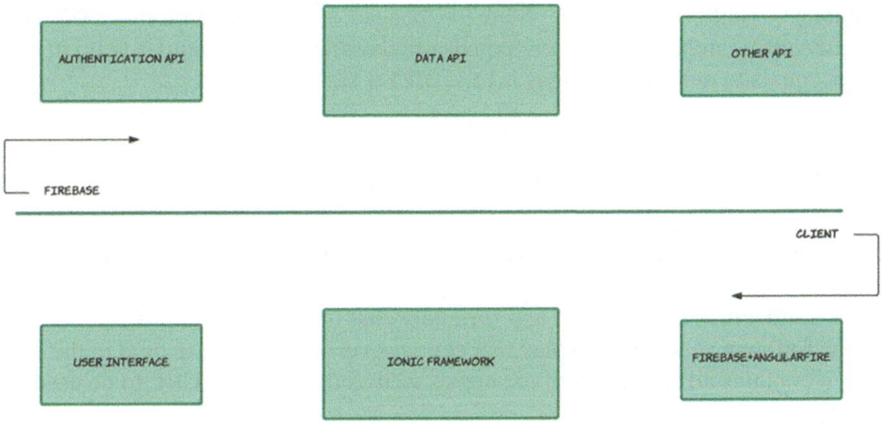

Fig. 2 Interaction between client and firebase network

application; the application will come live from the search results itself. But if the application isn't installed an install card shows up in search results (Fig. 2).

2.2 *Adding Firebase to Project*

First, enable firebase for the project i.e. to add firebase to the project. Steps included to do so are:

1. Create a project.
2. In the project dialog, select the project.
3. Billing plan dialog box would appear (firebase has pay as you go i.e. one has to pay to before he/she uses the service of firebase).

2.3 *Enable Google Auth*

Allow users to sign into the app using their Google account, enable Google Auth. In Firebase console click on the authentication section.

- Set up sign in method and go to the edit option of google. Enable it and give a name to the project. In this case, whitelist few client IDs such that they can use the app otherwise they would find authentication error.
- Add Firebase Auth object private Firebase Authm Auth; mAuth = FirebaseAuth.getInstance();
- To check registration status private Firebase Auth. Auth State Listenerm Auth Listener; @Override protected void on Create (Bundle savedInstanceState){mAuthListener = newFirebaseAuth.AuthStateListener(){@Override

public void on Auth State Changed (@NonNullFirebaseAuthfirebaseAuth){FirebaseUser user equals firebaseAuth.getCurrentUser(); if(user not equals to null){//User signs in}else{//User signs out}}};}.

2.4 *Initialize Firebase SDK*

On the project overview page, click on the web icon (“</>”) to add firebase to the app.

First add a nickname to your app (Friendly chat is used in our case).

The following code will display that contains version of firebase used in the app. Plus; it contains all the details of the app as well as the database URL to be used by your app.

<script src = “https:www.gstatic.com/firebasejs/7.2.2/firebase-app.js”></script>

//this code contains version of firebase and file type of firebase app

Copy the SDK to project’s library and paste the initialization code snippet in the footer. firebase.initializeApp(firebaseConfig);

//this will initialise the firebase in the project.

2.5 *Initialize Firebase Storage*

In the firebase console click on storage it will prompt for secure rules. Set cloud storage location.

2.6 *Chat Module*

To Configure the firebase in app use GCP console using the following command

Firebase login–no localhost

No local host is used if on remote shell

After completion of authorization, a link will be provided in GCP console a link which will direct to the log in page.

This is the verification code enter it in the Cloud shell prompt and to set up the firebase project use the code

Firebase use–add

Give an alias name as it helps in managing multiple apps at a time

Following images in Fig. 3 shows the chat module between 2 people as well as the details in the database storage details.

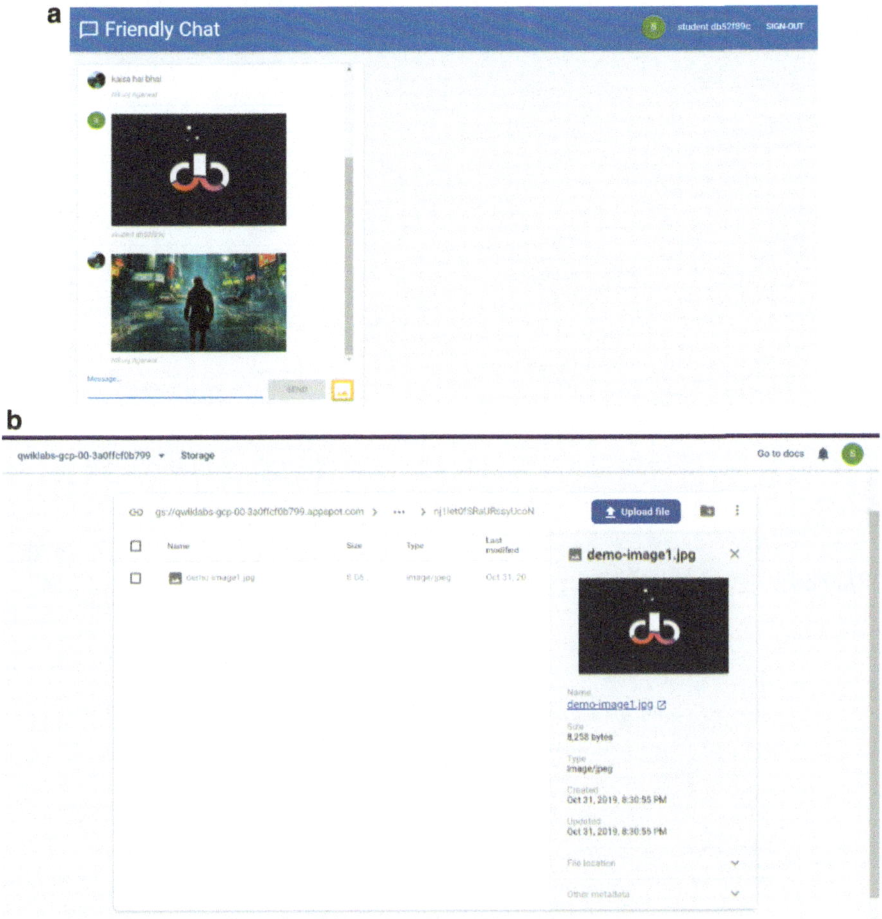

Fig. 3 **a** Chat **b** image shared details in the storage section **c** database details for the image sent **d** messages tab details for text message

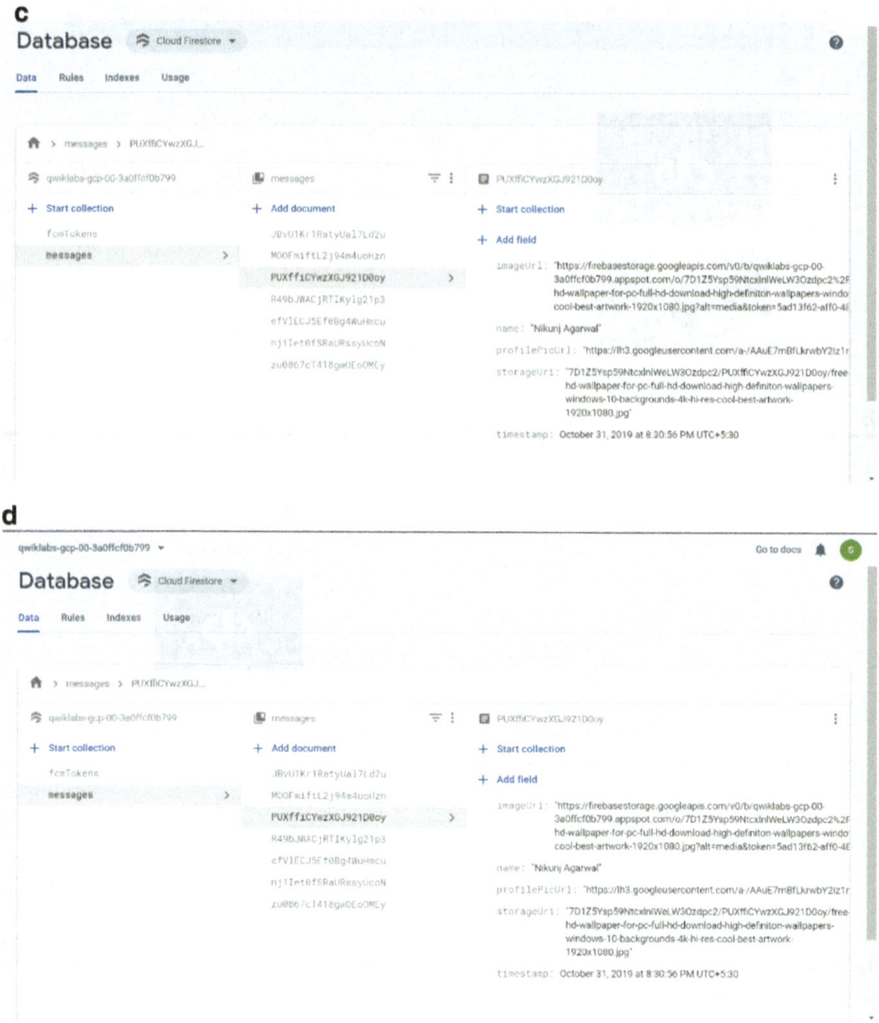

Fig. 3 (continued)

3 Conclusion

This paper looked at Google Firebase and its capabilities, which are very important in the development of an application in real time. Developing a web or mobile application could be an exhaustive process as it involves a lot of time and cost. Firebase reduces the workload from the developer's shoulders by handling major chunk of process of the application. Most of the work like authentication and storage was handled by firebase in our project. What's the cherry on top is that firebase will keep on upgrading its features as there are many more functionalities as well fail-proof techniques to be introduced in the future making future-proof framework and the usage of storage will always be there.

References

1. Khedkar, S., Thube, S.: Real time databases for applications. Int. Res. J. Eng. Technol. (IRJET) **4**(6), 2078–2082 (2017)
2. Singh, N.: Study of Google firebase API for android. Int. J. Innov. Res. Comput. Commun. Eng. **4**(9), 16738–16743 (2016)
3. Singh, S.H., Singh, S.U..: Study on Google firebase for website development (the real time database). Int. J. Eng. Technol. Sci. Res. IJETSR **4**(3), 364–367 (2017)
4. Chatterjee, N., Chakraborty, S., Decosta, A., Nath, A.: Real-time communication application based on android using Google firebase. Int. J. Adv. Res. Comput. Sci. Manage. Stud. **6**(4), 74–79 (2018)
5. Wadhar, M.C., Patil, P.P.: Traditional infrastructure versus firebase infrastructure. Int. J. Trend Sci. Res. Dev. **2**(4), 2050–2053 (2018)
6. Ozsoyoglu, G., Snodgrass, R.T.: Temporal and real-time databases: a survey. IEEE Trans. Knowl. Data Eng. **7**(4), 513–532 (1995)
7. Lahudkar, P., Sawale, S., Deshmane, V., Bharambe, K.: NoSQL database-google's firebase: a review. Int. J. Innov. Res. Sci. Eng. Technol. **7**(3) (2018)
8. https://firebase.google.com/docs/cloud-messaging/js/case-studies

Dealing with Sybil Attack in VANET

Binod Kumar Pattanayak, Omkar Pattnaik, and Sasmita Pani

Abstract With the increment in human populace and financial advancement, the populace of private vehicles on the street has likewise expanded. This has brought about expanded mishap likelihood and fatalities on the streets. Vehicular Ad hoc Networks (VANETs) have indicated guarantee of cutting down street mishaps and fatalities thereof by empowering correspondence between vehicles. VANETs likewise enable the street administrators to control and screen vehicles for rash driving and snappy help. Be that as it may, these systems face difficulties of Design, Routing, Communication, and Security. Utilization of remote vehicle for correspondence has left these systems powerless against various sort of security assaults. One of the significant helplessness in such systems is caused when a malignant vehicle or RSU can gain different characters. The assault is named as Sybil Attack. In VANETs, a noxious vehicle may send wrong messages identified with traffic, mishap ahead, street shut, and so on. This may compel the engine driver to take an alternate course, hence making him inclined to an untoward occurrence. In this paper, we have used different methodologies like PKC, PVM, and RTM techniques to detect Sybil attack in VANET.

Keywords RSU (Road side Unit) · Public key cryptography (PKC) · Position verification method (PVM) · Resource testing method (RTM)

B. K. Pattanayak
Department of Computer Science and Engineering, Institute of Technical Education and Research, Siksha 'O' Anusandhan (Deemed to be University), Bhubaneswar, Odisha 751030, India
e-mail: binodpattanayak@soa.ac.in

O. Pattnaik (✉) · S. Pani
Department of Computer Science and Engineering, Government College of Engineering, Keonjhar, Odisha 758002, India
e-mail: omkar29in@gmail.com

S. Pani
e-mail: susmitap20@gmail.com

D. Mishra et al. (eds.), *Intelligent and Cloud Computing*,
Smart Innovation, Systems and Technologies 194,
https://doi.org/10.1007/978-981-15-5971-6_51

1 Introduction

The research community nowadays focus on the concept of Mobile ad hoc network (MANET) which is implemented in road and transport communication where vehicles act as nodes and it forms a Vehicular Ad hoc Network (VANET). It is also regarded as a subset of MANET. VANET was created in October 2002, by the Federal Communications Commission (FCC). The main objective of this concept is to provide safety on roads and secure driving. To create a VANET environment, some parameters are necessarily required like: On-Board Unit (OBU) and a Road Side Unit (RSU) that must be installed along the roads. Dedicated Short Range Communications (DSRC) protocol is used to communicate between OBU and RSU. Again this protocol is modified later and form a new protocol called Wireless Access for Vehicular Environment (WAVE). It is based on IEEE 802.11p group. The architecture of VANET supports vehicle to vehicle (V2V), as well as vehicle to infrastructure (V2I) interchanges associated with RSU and OBU.

Security is necessarily required for VANET due to its wireless communications. So, there are different types of attacks possible in VANET. **Sybil attack** is one of the harmful attacks among them. In this attack, multiple identities are created by the attacker, thereby trying to forge or steal new identities from the neighboring vehicles. The attacker performs various malicious operations with the help of Sybil attack. An illusion is created by the attacker by sending messages to different vehicles from different identities. So, there occurs a jamming in the network which enforces the driver of vehicles to divert from the original route.

Generally, it affects two major areas like Routing and Voting and Reputation Systems as well. It may appear as a malicious node at more than one location at a time in geographical routing protocols [1]. Also, it may disrupt the cluster head and disrupt the valuable information in cluster-based routing protocols [2]. But in the second case, a voting process is performed and the vehicle's exact location is monitored and the misbehaving node is identified. When a malicious vehicle performs Sybil attack, all messages come from a particular location that circulates the false information in the entire network. So, the network may be unable to communicate with the neighboring nodes.

The rest of the paper is organized as follows. Section 2 describes the attack model along with the assumptions. Section 3 covers the way Sybil Attack Works. Section 4 includes the methods of protection from the Sybil attack in VANET. Section 5 concludes the paper.

2 Attack Model and Assumptions

In this part, the Sybil attack model is defined, thereby presenting some assumptions on the system and the attacker's probable actions.

2.1 Attack Model

In wireless networks, the new neighboring nodes are discovered by the mobile nodes to whom the beacon packets are periodically broadcasted in the network. In VANET, the vehicle is to be considered as a node. Sybil attack is based on the idea of multiple identities being produced by a malicious node [5]. Also, the identities and positions are determined by the beacon packets. A malicious node has the potential to create multiple identities without the knowledge of legitimate nodes. A malicious driver may have additional identity information because either he has borrowed identity from another driver or might be stealing identity from neighbors. The main objective of Sybil attack detection is that only one identity is assigned to each physical node [3].

In this paper, we assume a malicious node to be a physical node which produces multiple identities and correspondingly, the Sybil nodes are the fabricated identities produced by the malicious node.

Assumptions

The assumptions are focused here on the basis of Sybil attack detection and the attacker's actions.

2.2 Attack Detection

In this paper, we have assumed the following assumptions about the detection of Sybil attack in VANET. (i) The Sybil attack is launched by the individual drivers (vehicles) which are the basic threat to the system. We also assume that the Sybil attack may be created by multiple malicious vehicles. (ii) The RSUs are deployed separately along the road and to maintain the authentication of infrastructure, an Electronic License Plate (ELP) is issued to the whole network [7]. (iii) We focus on the same radio module called Received Signal Strength Indicator (RSSI) such as DSRC which is based on Radio Frequency (RF) communication strategy [6]. (iv). Finally, we assume that all the vehicles must be equipped with GPS devices along with digital maps since the accurate position is determined by GPS technology. But in real-world scenario, we cannot rely on roadside base stations present in RSU for the whole network.

Attackers' Actions

The following are the possible actions conducted by the attacker.

1. Compromised RSBs: The attacker may be successful in compromising the RSBs (part of RSU) which are a semi-trusted parties. The detection of compromised RSBs can be cancelled by the DMV (Department of Motor Vehicle). But, there is still a chance for the attacker to retrieve the information stored in the RSBs. So, the attacker uses its technique to recover the amount of information quickly from the RSBs.

2. Announce false messages: It is also regarded *as* False data injection. The false message can be signed by a vehicle before it is broadcasted in the network. The message can also be signed by a pseudonym that is certified by CA. So, it is impossible to detect the message. When the number of vehicles is more than the number of attackers in a network, the voting scheme is used. But, if the false identities are generated sufficiently by the attacker, then the Sybil attack occurs.
3. Sybil attack: It refers to a situation wherein the attacker may pretend to be multiple vehicles by virtue of using multiple pseudonyms. When multiple pseudonyms are used to sign the message by a vehicle, the Sybil attack is formed in VANET. The privacy of a vehicle is compromised when the vehicle is identified from a group of pseudonyms it uses. Therefore, RSBs and vehicles tend to sign multiple messages in the attacker's pseudonyms [4].

3 How Sybil Attack Works

The concept of Sybil attack was first introduced by Douseur [8]. This is a threat to the functionality of VANET. The attacker uses multiple identities in order to send messages to other nodes that belong to the network. The identities of other nodes may be spoofed by a node called malicious node that is also regarded as the Sybil attacker. Sybil nodes are those nodes whose identities are spoofed by the Sybil attacker. An illusion is created in the network wherein message is sent to the vehicles that an accident or a traffic jam has already taken place, and thus other vehicles are forced to change their routes and tend to move along the attacker's route. Also, the false information may be injected by the Sybil attacker into the network. The attacker also tries to disrupt the warning message sent by the actual vehicle in the network and fulfill its target. So, this may put the life of passengers in danger [9].

The different types of possible Sybil attacks are given in Fig. 1.

In Fig. 1, it is clear that Sybil attack classification is carried out on the basis of the type of communication, identity, and their participation in the network [10]. These features are explained clearly in the following paragraphs.

1. **Communication category**: When a Sybil node receives a radio message from an honest node, the message may be listened by a malicious node present in the

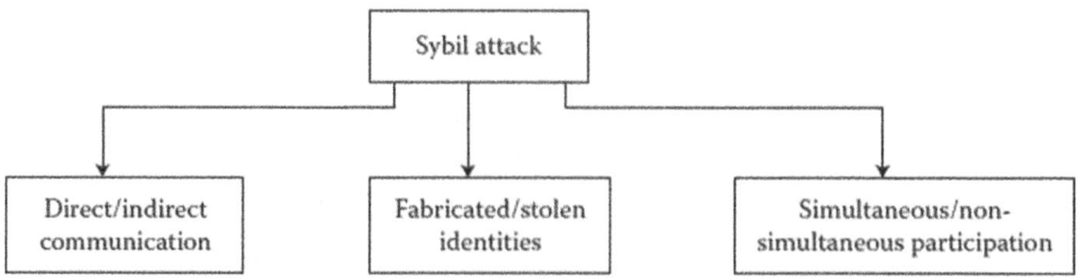

Fig. 1 Different forms of Sybil attack

network. Along the same path, the malicious devices actually send messages but not the Sybil nodes. In direct communication, the legitimate nodes communicate with all the Sybil nodes which are created by malicious devices. But in indirect communication, the malicious node helps the legitimate nodes to enter the area of Sybil nodes.

2. **Identity category**: In this category, a new Sybil identity is created by an attacker and that is generated randomly with a 32-bit integer which is a fabricated identity or the identities of the neighboring nodes are stolen by the attacker.
3. **Participation category**: The malicious nodes create multiple Sybil identities which can participate simultaneously in the attack. In this category, at a time, one identity is used for each node, and thus an observation takes place in the network that many times a particular node may join or leave the network.

4 Protection Methods for Sybil Attack in VANET

In VANET, the protection mechanism can be classified into three different types that are: (i) Public Key Cryptography method, (ii) Methods based on position verification and (iii) Resource Testing method.

4.1 Public Key Cryptography Method

Public key cryptography is a standard method to detect Sybil attack and also, it provides authentication mechanism [11]. The concept of a symmetric cryptography is used to develop the security solution which is based on the combination of digital certificates with signatures.

In this technique, a CA (central authority) is present, that is responsible for issuing the certificates, thereby maintaining the hierarchy. A secure channel is used to communicate between CAs and monitor the issued certificates which are used by every signed message. In VANET, the signed information sent by vehicles may be verified by the receiver as demonstrated in Fig. 2. In this technique, it is verified if a message is valid or invalid. The messages which are attached with certificates are to be considered as valid ones and the messages having no certificates are ignored. So, the Sybil attacks can be prevented. At a time, only one certificate is assigned to each node, and to maintain better privacy, it is necessary to change the certificates from time to time. But in VANET, due to rapid changes in mobility, it is difficult to implement PKI. But to maintain privacy, the key pair (Public/Private) must be authenticated by the CA. But, it may not be useful to check the identity of a vehicle [12].

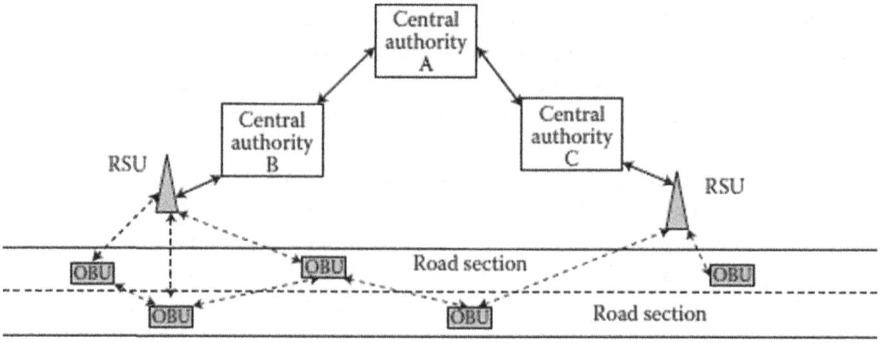

Fig. 2 Hierarchy of central authority

4.2 Position Verification Method

In a vehicular environment, the concept of sensors, GPS has also an important role in creating a node (vehicle). It may also be possible to tamper the sensors on the part of an attacker and send false information to other vehicles present in the network. To achieve effective moving of vehicles in the network, accurate positions of the vehicle are essentially required. So it is an important task for all the nodes to maintain an accurate position in the network. This is very much important for geographic routing. In this routing scheme, accurate positions of all nodes are essential because intermediate nodes are responsible for transmitting messages to the destination based on their locations.

The method of verification of a secure position was first addressed by the authors in [13], called as **verifiable multilateration** approach. To fix the position of the vehicle, a trustworthy network is built by the base stations which measures the sending and receiving information sequentially in time. In this approach, a node neither delays the reply time nor advances the time. So, the nearer distance is fixed for communication.

The second proposal is given by authors in [14], where the concept of threshold values is used to determine the correct position of the node by receiving information from other nodes. In this technique, nodes rely on GPS rather than base station scheme to determine their positions. Beacon messages are used for sending and receiving location information of nodes and also for verifying it through specified checks.

There are four types of check strategies based on the above concept which is given below

- Acceptance Range Threshold;
- Mobility Grade Threshold;
- Maximum Density Threshold;
- Map-based Technique.

In **Acceptance Range Threshold**, the verified node obtains a maximum threshold value which determines their communication ranges in the network. So, it helps to detect wrong positions of nodes.

In **Mobility Grade Threshold**, the last node position is known, and thus, if any false information about changing of position is detected, then an extreme speed is defined for a hub. In order to communicate among the nodes, signal messages are used.

The idea about **Maximum Density Threshold** is to accommodate a specific region where a maximum number of vehicles can be physically present.

In the Map-based technique, the correct location of the vehicle is to be determined.

In order to detect Sybil attack based on the position, the parameters like collision avoidance, resource availability, emergency alert, and traffic conditions can be useful. A VANET may be affected by inserted bogus packets, dropping packets, replying packets, and modifying existing packets which are created by the malicious attacker [14, 22].

There are two efficient schemes for localization like spectrum-based and spectrum free techniques. In spectrum-based technique, in order to locate the vehicle's position, the estimated distance between a receiver and transmitter must be calculated. Some of the techniques are: Time of Arrival (TOA), Angle of Arrival (AOA), and Received Signal Strength Indicator (RSSI). But distance free localization technique helps in gathering other position information. In order to achieve more accuracy in localization, distance-based method must be adopted [15].

According to Yan et al. a new method has been developed for Sybil attack in which vehicles can listen to GPS coordinates of other vehicles and also visualize the neighbor's vehicle with the help of radar transmission range. It is also called as a virtual vehicle eye. Therefore, it is easy to track the actual location of the neighboring vehicle and malicious vehicles are separated from others [16].

The limitation of this method is that when an existing vehicle position is acquired by a target vehicle, then they are in the radar network of the established vehicle [18, 23, 24]. Such a scenario is depicted in Fig. 3.

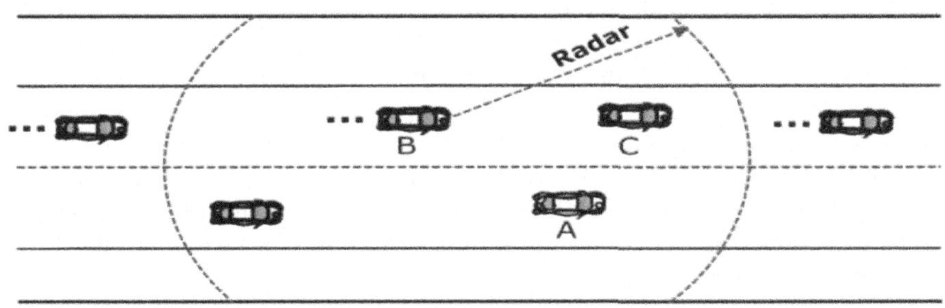

Fig. 3 A potential Sybil attack

As shown in Fig. 3, A gets *C*'s position to be *Lc*. *A* claims to have to victimized node *B* that its position is Lc, and its ID is IDa. *B* recognizes a vehicle that is located at *Lc* and then concludes that it is the location of *A* [14].

4.3 Resource Testing Method

In this technique, assumption is generated in the user's mind for the limited resources of each physical entity. The concept of radio resource testing specifies that a message is broadcasted by a node to all its neighbor nodes and try to listen to the response message by a channel which is selected randomly.

The nodes which do not respond to the broadcasted message positively in the same channel are assumed as Sybil entities and the Sybil attack is detected. Otherwise the neighbor node responds to the same channel [17, 25].

The resources which can be shared between malicious vehicles and Sybil entities are memory, IP, and computational recourses. So it is possible to generate for Sybil attack conditions.

According to the researcher's opinion, this approach may not be suitable for VANET. The malignant vehicle can produce numerous integrity which is not match with the identity of the vehicle present in the network. So they are registered separately in other list [15].

The main objective of the resource testing method is to detect the Sybil attack and check the fake identities from the attacker. But in the practical scenario, to obtain sufficient fake identities for an attacker is not a difficult task. So this method may not be implemented in VANET [19, 23].

5 Conclusion and Future Scope

In this paper, we have analyzed the detection, as well as a protection mechanism, for Sybil attack in vehicular communication. Also, we have assumed the attacker's action in different scenarios. In this case, we prefer the position verification method as an efficient technique to protect from Sybil attack in real-world environment. Extensive work is still required in the future. Also, we focus on the acceptance range threshold technique of position verification method to detect the wrong position of more number of Sybil nodes in VANET. In future, the performance of our method can be examined using NS2 and traffic simulators called SUMO and MOVE for real-world environment.

References

1. Karlof, C., Wagner, D.: Secure routing in wireless sensor networks: attacks and counter measures. Ad hoc Netw. J. (Elsevier) **1**, 293–315 (2003)
2. Sood, M., Vasudeva, A.: Perspectives of Sybil attack in routing protocols of mobile ad hoc network. Comput. Netw. Commun. (NetCom) **131**, 3–13 (2013)
3. Ali Akbar Pouyan, Mahdiyeh Alimohammadi**, "**Sybil Attack Detection in Vehicular Networks", *Computer Science and Information Technology* 2(4): 197–202, 2014
4. Yu, B., Xu, C.Z., Xiao, B.: Detecting Sybil attacks in VANETs. J. Parallel Distrib. Comput. (Elsevier) (9 February, 2013)
5. Newsome, J., Shi, E., Song, D., Perrig, A.: The Sybil attack in sensor networks: analysis and defenses. In: Third International Symposium on Information Processing in Sensor Networks, pp. 259–268. IPSN 2004 (2004)
6. http://grouper.ieee.org/groups/scc32/dsrc/index.html
7. Hubaux, J.P., Capkun, S., Luo, J.: The security and privacy of smart vehicles. IEEE Secur. Privacy Mag. **2**(3), 49–55 (2004)
8. Douceur, J.R.: The Sybil attack. In: Proceedings of the International Workshop on Peer to Peer Systems, pp. 251–260 (March 2002)
9. Grover, J., Gaur, M.S.: Sybil attack in VANETs detection and prevention. Secur. self-organ. Netw. Printed in the United States of America on acid-free paper-10 9 8 7 6 5 4 3 2 1 (2011)
10. Newsome, J., Shi, E., Song, D., Perrig, A.: The Sybil attack in sensor networks: analysis and defenses. In: Proceedings of International Symposium on Information Processing in Sensor Networks, pp. 259–268 (April 2004)
11. Raya, M., Hubaux, J.P.: Securing vehicular ad hoc networks. J. Comput. Secur. 39–68 (April 2007)
12. Khalili, A., Katz, J., Arbaugh, W.A.: Towards secure key distribution in truly ad-hoc networks. In: Proceedings of IEEE Workshop on security and Assurance in Ad-Hoc Networks (2003)
13. Martucci, L.A., Kohlweiss, M., Anderson, C., Panchenko, A.: Self-certified Sybil-free pseudonyms. In: WiSec'08: Proceedings of the First ACM Conference on Wireless Network Security, pp. 154–159. New York, NY, USA, ACM Press (2008)
14. Yan, G., Olariu, S., Weigle, M.C.: Providing VANET security through active position detection. Comput. Commun. **31**(12), 2883–2897 (2008)
15. Isaac, T., Zeadally, S., Camara, J.S.: Security attacks and solutions for vehicular ad hoc networks. Commun. IET **4**(7) (June 2010)
16. Samara, G., Al-Salihy, W.A., Sures, R.: Security analysis of vehicular ad hoc networks (VANET). In: IEEE Second International Conference on Network Applications, Protocols and Services (2010)
17. Zhou, T., Choudhury, R.R., Ning, P., Chakrabarty, K.: P^2 DAP-Sybil attacks detection in vehicular ad hoc networks. IEEE J. Sel. Areas Commun. **29**(3) (March 2011)
18. Ibrahim, K.: Data aggregation and dissemination in vehicular ad-hoc networks. Doctoral dissertation, Old Dominion University, Norfolk, Virginia (2011)
19. Levine, B.N., Shields, C., Margolin, N.B.: A survey of solutions to the Sybil attack. MA, University of Massachusetts, Amherst (2006)
20. Saini, M., Kumar, K., Bhatnager, K.V.: Efficient and feasible methods to detect Sybil attack in VANET. Int. J. Eng. Res. Technol. **6**, 431–440 (2013)
21. Hao, Y., Tang, J., Cheng, Y.: Cooperative Sybil attack detection for position based applications in privacy preserved VANETs. In: IEEE Communications Society in the IEEE Globecom 2011 Proceedings
22. Pathre, A., Agrawal, C., Jain, A.: Identification of malicious vehicle in VANET environment. J. Glob. Res. Comput. Sci. **4**, 30–34 (2013)

23. Rahbari, M., Jamali, M.A.J.: Efficient detection of Sybil attack based on cryptography in VANET. Int. J. Netw. Secur. Appl. (IJNSA) **3**, 185–195 (2011)
24. Feng, X., Li, C., Chen, D., Tang, J.: A method for defensing against multi-source Sybil attacks in VANET. Peer-to-Peer Netw. Appl. (Springer) **10**, 305–314 (2017)
25. Grover, J., Gaur, M.S., Laxmi, V.: Multivariate verification for Sybil attack detection in VANET. Online https://doi.org/10.1515/comp-2015-0006 (December 2015)

Role of Intelligent Techniques in Natural Language Processing: An Empirical Study

Bishwa Ranjan Das, Dilip Singh, Prakash Chandra Bhoi, and Debahuti Mishra

Abstract Nowadays intelligent computing with natural language processing (NLP) is growing rapidly world wide. This paper describes the use of intelligent computing strategies in the field of NLP. Various intelligent techniques such as Artificial Neural Network, Support Vector Machine, Conditional Random Field, Hidden Markov Assumption, and Bayes' Rule and with some Stochastic Process are used in sub fields of NLP such as Machine Translation, Sentiment Analysis, Information Retrieval, Information Extraction, Question Answering and Named Entity Recognition are used in to develop a reliable and efficient system with high accuracy. In this application area, both training and testing datasets with authorized corpus are required to test the proposed model. Most of the exact experimental values are recorded to handle the real-time situation for further study.

Keywords Stochastic · Corpus · Natural language generation · Support vector machine · Natural language understanding · Natural language processing

---

B. R. Das · P. C. Bhoi
Department of Computer Science and Information Technology, Institute of Technical Education and Research, Siksha 'O' Anusandhan (Deemed to be University), Bhubaneswar, Odisha 751030, India
e-mail: biswadas.bulu@gmail.com

P. C. Bhoi
e-mail: prakashbhoi@gmail.com

D. Singh (✉) · D. Mishra
Department of Computer Science and Engineering, Institute of Technical Education and Research, Siksha 'O' Anusandhan (Deemed to be University), Bhubaneswar, Odisha 751030, India
e-mail: dillipsingh0306@gmail.com

D. Mishra
e-mail: debahutimishra@soa.ac.in

D. Mishra et al. (eds.), *Intelligent and Cloud Computing*,
Smart Innovation, Systems and Technologies 194,
https://doi.org/10.1007/978-981-15-5971-6_52

1 Introduction

Natural Language Processing (NLP) is a subarea of artificial intelligence focusing interaction between human and computer language to process and analyze large amounts of data to make the computer understand. Natural Language Understanding (NLU) and Natural Language Generation (NLG) are the two major categories of NLP [1]. NLP is mainly based on statistics and grammar (rule-based). Using hand-coding, many language processing systems were designed by devising heuristic rules for *stemming*. Most of the NLP research works are heavily relied on machine learning techniques because of the *statistical revolution* in the late 1980s and mid-1990s [2].

In NLP, different techniques of machine learning algorithms are applied. A large set of *features* are taken as input by these algorithms that are generated from the input data. The common handwritten rules were produced by *decision trees* algorithm. Based on attaching the weights, i.e., *real valued* to input feature for each, *soft, probabilistic decisions* are made by the *statistical models* that are focused on the research. The relative certainty of many different possible answers rather than only one is expressed by such models producing more reliable results [3, 4].

The body of the paper contains the introduction followed by various research areas in natural language processing in Sect. 2. All the intelligent techniques are elaborated thoroughly in Sect. 3, roles and methods described in Sect. 4, and finally conclusions with future work are discussed in Sect. 5.

2 Research Area in NLP [5]

There are many research areas in natural language processing such as *phonetics* and *phonology* which explains how to represent sound in writing for the formation of phonemes in a particular human language [5]. *Morphology* is another research area that analyze, identify, and describe the structure and formation of words. There are some syntax rules and principles that are given to individual languages for a given sentence structure [6, 7]. *Lexical Analysis* is another area that converts a sequence of characters to a sequence of tokens carrying an objective to divide the text into paragraphs, sentences, and words [8]. *Syntactic Analysis* analyzes the grammatical structure of a sentence transforming each word into a structure. If the rules of the language are violated then those word sequences would get rejected [9]. *Semantic Analysis* checks for the semantical correctness to find the absolute meaning where the object of the task domain has to be mapped with the syntactic structure [9]. *Discourse Integration* is the denotation of an individual sentence that may depend on the sentence which precedes it and might influence the meaning of the sentence that follows it [10], and last but not the least *Pragmatic Analysis* is done to understand the language that will focus on the structure set of a text to extract actual meaning [11].

3 Intelligent Techniques

Natural Language Processing follows many techniques which are discussed below.

3.1 Artificial Neural Network

The neural networks are made up of artificial neurons that are similar to the functioning of the neurons as information processing units in the human brain [12, 13]. The three-stage system of the human nervous system is shown below in Fig. 1.

The central system of the brain is represented in a neural net that continuously receives information for making accurate decisions. Left to right represents the signals through the system and right to left represents feedback [14].

3.1.1 Single Layer Feed-Forward Network

The inputs are directly connected to the outputs in a single layer Feed-forward Network via a series of weights [15]. Every input and output is connected by the synaptic links. In each neuron node, sum of the products of the weights is calculated. The neuron takes the higher threshold values as its activated value is represented in Fig. 2.

$$0y_j = f(\text{net}_j) = 1 \text{ if net}_j \geq 0$$
$$0 \text{ if net}_j \leq 0 \text{ where net}_j = \sum x_i w_{ij}$$

3.1.2 Multi-layer Feed-Forward Network

Multi-layer feed-forward network comprises of input, hidden, and output layer [16]. Hidden neurons are the computational units of the hidden layer [3, 4]. In $(1 - m1 - m2 - n)$, where input neurons are denoted as 1, $m1$ belongs to the first hidden layer, $m2$ belongs to the second hidden layer, and the output n in the output layer, all in a multi-layer feed-forward network as shown below in Fig. 3.

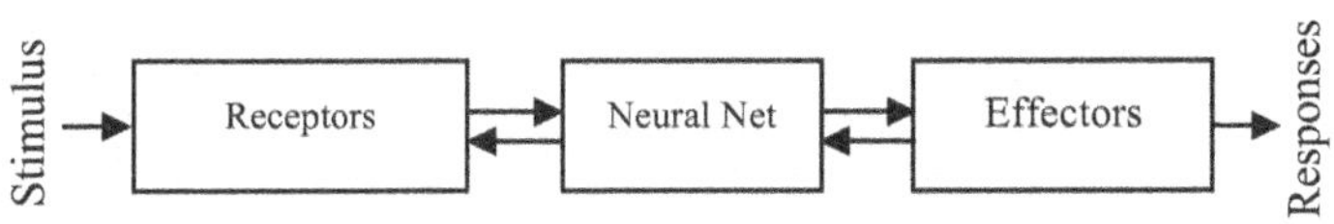

Fig. 1 Neuronfunctions

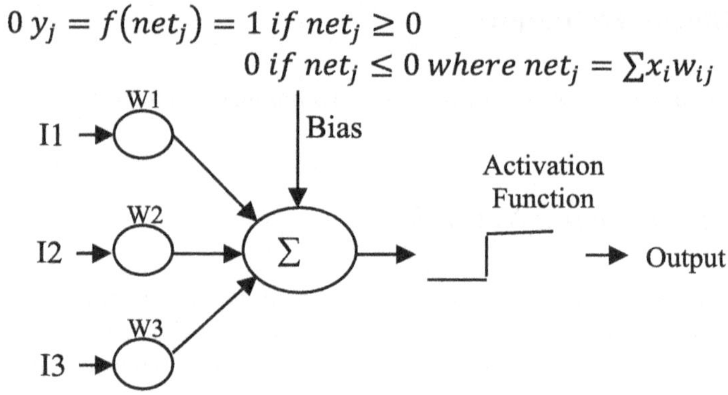

Fig. 2 Neural network of a single layer

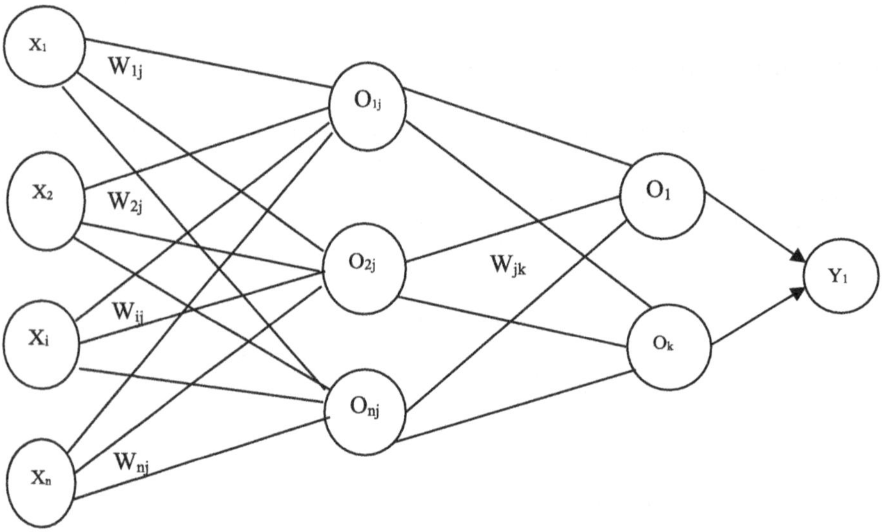

Fig. 3 Feed forward network of Multi-layer

$x_1, x_2, x_3, \ldots, x_n$ are the input

$w_{k1}, w_{k2}, w_{k3}, \ldots, w_{kn}$ are synaptic weight

$$U_k = \sum_{j=1}^{m} w_{kj} \times x_j \tag{1}$$

$$Y_k = \varphi(U_k + b_k) \tag{2}$$

where

$$U_k = \text{linear combiner output.}$$

$$\varphi() = \text{is the activation function.}$$

$$Y_k = \text{output signal of neuron apply an affine transformation to the output.}$$

$$U_k = \text{of the linear combiner in the model.}$$

$$V_k = U_k + b_k \tag{3}$$

where

$$b_k \text{ is bias value value which may} + \text{ve or} - \text{ve.}$$

The relationship between activation potential V_k of a neuron k or the induced local field and the U_k that is the output of linear combiner which is modified. The origin never passes through the graph of V_k versus U_k. An external parameter of an artificial neuron k is the Bias b_k.

Using (1), (2), and (3) we may formulate the combinations as follows:

$$V_k = \sum w_{kj} x_j \tag{4}$$

$$Y_k = \varphi(V_k) \tag{5}$$

The role of the neural network came into view as a powerful machine learning tool in the areas of image recognition and speech processing. The data collection and the extraction of characteristics of the texts were considered by applying the modern tools of machine learning like support vector machine, artificial neural networks, random forests for the automatic classification of various abstracts and scientific articles. In order to work for the local and global textual semantics, a collective application of recurrent and convolutional neural network was proposed to enable a high-order label correlation with a computational complexity [17, 18].

3.2 *Support Vector Machine (SVM)*

Classification and regression problems are mostly carried out by Support Vector Machine (SVM) that is a supervised machine learning algorithm but is generally used for classification problems. It is also used in pattern analysis due to its better

Fig. 4 Textual data for classification

generalized performance [19]. Its gives high degree of accuracy in big number of features set, in NLP it is mostly used to categorize text. To define a simple problem, a 2-class problem in which the classes are separable by nonlinearity, this machine is used. Let D be the given dataset as $(X_1, y_1), (X_2, y_2), \ldots, (X_D, y_D)$ in which X_i is denoted as the set of training tuples associated with y_i where y_i is class labels. One of two values can be taken for each y_i, which is $+1$ or -1 (i.e. $y_i \in \{+1, -1\}$), as Fig. 4, says.

The equation $wx + b = 0$, can be written for a separating hyper-plane, where x is the vector input, w is the adjustable weight, and b is the bias. The training tuples are two-dimensional, e.g., $x = \{x_1, x_2\}$, where x_1 and x_2 are the attribute values of A_1 and A_2, respectively, for x. The training data along with the test data are separated into two classes by an optimal hyper-plane. It also maximizes the margin by a separating hyper-plane. The margin M and two different parallel lines is expressed as $wx + b = +1,\ M = 2/\|w\|$. The margin is maximized by $r = 1/\|w\|$and minimized by$\|w\| = \|w\|^2/2$, subject to$d_i(w \cdot x_i + b) \geq 1$, where $i = 1, 2, 3, \ldots, l$. A support vector is a training tuple that lands on either side of the margin. It can also carry out the nonlinear classification. All the feature vectors a represented in their dot products, then the optimization problem is written in usual form. SVM handles nonlinear hypotheses by simply substituting all the dot products of x_i and x_j in the dual form with a certain Kernel function $K(x_i, x_j)$. The polynomial kernel function with a degree d such that $K(x_i, x_j) = (x_i \times x_j + 1)^d$ is focused among different kinds of available Kernel function. From all the combinations of features up to d, the d degree of the optimal separating hyper-plane found by the polynomial kernel function. The set of functions can be considered as hypothesis space. The linearly separable case is nearly finished. A decision-making function $f(x)$ is given by the nonlinear SVM classifier as shown in (6), such that

$$f(x) = \sum_{i=1}^{m} w_i K(x, z_i) + b, \quad g(x) = \text{sign}(f(x)) \tag{6}$$

Let $g(x)$ is $+1$, x is denoted as class C_1, and for $+1$, x can be denoted as class C_2. The number of support vectors is m and kernel is K which is a high dimensional space and can be implicit map vectors that can be evaluated efficiently. $K(x, z_i) = (x, z_j)^d$ represents the polynomial kernel [20].

Support vector machines are studied for learning text classifiers that examine the actual properties of learning with text data. Sentiment analysis is one of the most widely studied research areas in the field of NLP. To resolve the NLP related issues like imbalanced training data and the difficulty of obtaining sufficient training data, two methods have been developed to help support vector machines deal with those unique features [21–23].

3.3 Conditional Random Field

Conditional random field (CRF) is an undirected graphical model to specify output values with the given input nodes that are calculated by the conditional probability [24, 25]. CRF is a model to calculate the conditional probability. CRF is defined as $P(S|O)$, the conditional probability distribution where $S$is on the observed sequence random variables set O, and finding the maximum probability $S*$ through the training to make $S^* = \text{argmax} P(S|O)'$.

The following equation is based on the undirected graphical model

$$\exp\left(\sum_j \lambda_j t_j(y_{i-1}, y_i, x, i) + \sum_k \mu_k s_k(y_i, x, i)\right) \quad (7)$$

where $s_k(y_i, x, i)$ is a state feature function, $t_j(y_{i-1}, y_i, x, i)$ is a transition feature function, λ_j and μ_k are parameters to be estimated from training data. At the time of defining feature function, the characteristic of the training data is constructed to express some observation of the empirical distribution as a set of real valued features $b(x, i)$ [26, 27].

$$b(x, i) = \begin{cases} 1 \text{ if the observation at position i is the word anything} \\ 0 \qquad\qquad \text{otherwise} \end{cases} \quad (8)$$

$$t_j(y_{i-1}, y_i, x, i) = \begin{cases} b(x, i) \text{ if } y_{i-1} = \text{IN and } y_i = \text{NNP} \\ 0 \qquad\qquad \text{otherwise} \end{cases} \quad (9)$$

$$p(y|x, \lambda) = \frac{1}{z(x)} \exp\left(\sum \lambda_j f_j(y, x)\right) \quad (10)$$

where $z(x)$ is the normalize factor [6].

4 Role and Methods of SVM in NLP

There are many intelligent techniques available in the field of deep learning and machine learning. SVM is one of the high giving accuracy techniques used for classification and regression. Here it is used for classification of pattern such as noun and verb of a given text. Support Vectors are basically coordinate features of the n-dimensional space consisting of a numeric value from which a class will be predicted. Takes a huge number of coordinate values and separates it by the most valuable hyper-plane and maximizes it. SVM plays a major role to classify texts, words, and phrases in a systematic manner. Here one example is to classify noun, verb, adjective, adverb, preposition, pronoun, and many more lexical units of a given text or corpus. It is known as Part of Speech Tagging on the literature point of view. Morphological Analysis is another concept of language processing to find the root or base word using intelligent techniques like SVM, Artificial Neural Network, Conditional Random Field, etc. Most of the vital and important concept in NLP is Machine Translation, i.e., to translate one language to another language in terms of text and speech where Machine Learning and Deep Learning, a part of intelligent technique can be used. Named Entity Recognition is one of the important concepts which classify the person's name, place name, and organization name by any intelligent technique [28–30].

5 Conclusion and Future Work

Here we have basically described for Morphological Analysis with intelligent technique, i.e., SVM, neural network, conditional random field. It provides a very good interface to interact with this system to find the exact root word with all its grammatical structure including person, number, and gender for Odia language. Sometimes the ambiguous words are very difficult to find the exact root word. Further modification is needed to handle the ambiguous words in the near future. The technology that is applied to develop a robust morphological analyzer is applicable for some other Indian languages and can be used in different application areas like Machine Translation, Question Answering, Part of Speech Tagging, Named Entity Recognition, Information Extraction, and Information Retrieval.

References

1. Juraffsky, D., Martin, J.H.: Speech and Language Processing (2011)
2. Dash, N.S.: Indian scenario in language corpus generation. In: Dash, N.S., Probal, D., Pabitra, S. (eds.) Rainbow of Linguistics, vol. 1, pp. 129–162, Kolkata: T. Media Publication (2007)
3. Romanov, A., et al.: Application of natural language processing algorithms to the task of automatic classification of Russian scientific texts. Data Sci. J. **18**(37), 1–17. https://doi.org/10.5334/dsj-2019-037 (2019)
4. Kaur, J., Saini, J.R.: A study of text classification natural language processing algorithms for Indian languages. VNSGU J. Sci. Technol. **4**(1), 162–167 (2015)
5. Bharati, A., Chaitanya, V., Sangal, R.: Natural language processing a paninian perspective, department of computer science and engineering Indian institute of technology Kanpur, with contributions from K. V. Rama krishnamacharyulu Rashtriya Sanskrit Vidyapeetha, Tirupati, Prentice-Hall of India New Delhi (1994)

6. Singh, S., Gupta, K., Shrivastava, M., Bhattacharyya, P.: Morphological richness offsets resource demand—experiences in constructing a pos tagger for Hindi. In: Proceedings of the COLING/ACL 2006 Main Conference Poster Sessions, Sydney, Australia, pp. 779–786 (2006)
7. Jena, I., Chaudhury, S., Chaudhry, H., Dipti, M.: Developing oriya morphological analyzer using lt-toolbox. In: Information Systems for Indian languages Communications in Computer and Information Science, vol. 139, pp. 124–129 (2011)
8. Mohanty, S., et. al.: Object oriented design approach to orinet system: online lexical database for Oriya language. In: IEEE proceedings of LEC-2002. University of Hyderabad, Hyderabad, India (2002)
9. De Cao, D., Croce, D., Pennacchiotti, M., Basili, R.: Combining word sense and usage for modeling frame semantics. In: Proceedings of the Semantics in Text Processing Conference, pp. 85–101 (2008)
10. Machine Learning of Natural Language, Walter Daeleman, CNTS Language Technology Group Department of Linguistics, University of Antwerp, Belgium. walter.daelemans@ua.ac.be
11. Mahapatra, D., Mahal, K.: Adhunika Odia Byakarana, 5th edn. (2010)
12. Bobrow, D., Kaplan, R., Kay, M., Norman, D., Thompson, H., Winograd, T.: GUS, a frame-driven dialogue system. Artif. Intell. **8**, 155–173 (1977)
13. Das, B.R., Patnaik, S., Dash, N.S.: Development of Odia language corpus from modern news paper texts: some problems and issues. In: Proceedings of the International Conference On Intelligent Computing, Communication and Devices (ICCD 2014), pp. 18–19 Apr 2014, SOA University, Bhubaneswar, India, Springer Book Series on AISC, pp. 88–94 (2015)
14. Goldberg, Y.: A primer on neural network models for natural language processing. J. Artif. Intell. Res. **57**, 345–420. Computer Science Department Bar-Ilan University, Israel (2016)
15. Haykin, S.: Neural Networks, A Comprehensive Foundation. PHI (1998)
16. Norvig, R.: Artificial Intelligence, A Modern Approach, Pearson Prentice Hall
17. http://esl.fis.edu/grammar/langdiff/phono.htm
18. Chen, G., Ye, D., Xing, Z., Chen, J., Cambria, E.: Ensemble application of convolutional and recurrent neural networks for multi-label text categorization. In: B: 2017 International Joint Conference on Neural Networks (IJCNN), pp. 2377–2383. IEEE (2017)
19. Hearst, M.A.: Support vector machine. IEEE Intell. Syst. **13**(4), pp. 18–28 (1998)
20. Mokanaranganet, T., et al.: Tamil Morphological Analyzer Using Support Vector Machines, pp. 15–23. Springer International Publishing Switzerland (2016)
21. Li, Y., Bontcheva, K., Cunningham, H.: Adapting SVM for Natural Language Learning: A Case Study Involving Information Extraction. Department of Computer Science, the University of Sheffield, UK
22. Preety, S.D.: Sentiment analysis using SVM and Naive Bayes algorithm. Int. Inst. Technol. Bus. (Jhundpur, Sonipat, Haryana) IJCSMC **4**(9), 212–219 (September, 2015)
23. Joachims, T.: Text categorization with support vector machines: learning with many relevant features. University at Dortmund Informatik LS8, Baroper Str. 301 44221 Dortmund, Germany
24. Wallach, H.M.: Conditional random fields: an introduction. Technical Reports (CIS) (2004). Available at: http://works.bepress.com/hanna_wallach/1
25. Lafferty, J., McCallum, A., Pereira, F.C.: Conditional random fields: probabilistic models for segmenting and labeling sequence data. In: Proceedings of the 18th ICML'01, pp. 282–289 (2001)
26. Ankurparikh.: Part-of-speech tagging using neural network. In: Proceedings of ICON-2009: 7th International Conference on Natural Language Processing, Report No: IIIT/TR/2009/232 (2009)
27. Schmid, H.: Part-of-speech tagging with neural networks. In: COLING '94 Proceedings of the 15th Conference on Computational Linguistics, vol. 1, pp. 172–176
28. Patel, C., Gali, K.: Part-of-speech tagging for Gujarati using conditional random fields. In: Proceedings of the IJCNLP-08 Workshop on NLP for Less Privileged Languages, pp. 117–122. Hyderabad, India (2008)

29. Ranjan, P., Basu, H.V.S.S.A.: Part of speech tagging and local word grouping techniques for natural language parsing in Hindi. In: Proceedings of the 1st International Conference on Natural Language Processing (ICON 2003), Mysore (2003)
30. Saharia, N., Das, D., Sharma, U., Kalita, J.: Part of speech tagger for assamese text. In: Proceedings of the ACL-IJCNLP 2009 Conference Short Papers, pp. 33–36. Suntec, Singapore (2009)

Feature Selection and Classification for Microarray Data Using ACO-FLANN Framework

Pradeep Kumar Mallick, Sandeep Kumar Satapathy, Shruti Mishra, Amiya Ranjan Panda, and Debahuti Mishra

Abstract Classification is a data mining technique used to predict group membership for data instances. The main objective of a classifier is to discover the hidden class level of the unknown data. The classifier performance depends upon the data size, number of classes, and dimension of feature spaces. The classifier accuracy has been improved by applying optimization techniques. There are many optimization techniques that are developed for this purpose such as Particle Swarm Optimization (PSO), ACO, ABC, DE, MLP, FLANN, PSO-FLANN, etc. In this proposed work, a new model is proposed for the classification of microarray data. Artificial Neural Network(ANN) uses Ant Colony Optimization(ACO) to tune the parameters of ANN. Principal Component Analysis (PCA) is used for dimensionality reduction. Here in the first step, the reduced dataset is optimized by using Ant Colony Optimization(ACO) and after that, in the second step, the optimized dataset is trained to Functional Link Artificial Neural Network (FLANN). This model is called ACO-FLANN. The proposed model(ACO-FLANN) has been compared with PSO-FLANN. The

P. K. Mallick · A. R. Panda
School of Computer Engineering, Kalinga Institute of Industrial Technology (KIIT) Deemed-to-be University, Bhubaneswar, Odisha, India
e-mail: pradeepmallick84@gmail.com

A. R. Panda
e-mail: amiya872@gmail.com

S. K. Satapathy (✉)
School of Computer Science and Engineering, VIT University, Chennai, Tamilnadu, India
e-mail: sandeepkumar04@gmail.com

S. Mishra
Department of Computer Science and Engineering, VIT University, Amaravati, Andhra Pradesh, India
e-mail: shruti_m2129@yahoo.co.in

D. Mishra
Department of Computer Science and Engineering, Institute of Technical Education and Research, Siksha 'O' Anusandhan (Deemed to be University), Bhubaneswar, Odisha 751030, India
e-mail: mishradebahuti@gmail.com

D. Mishra et al. (eds.), *Intelligent and Cloud Computing*,
Smart Innovation, Systems and Technologies 194,
https://doi.org/10.1007/978-981-15-5971-6_53

simulation shows that the proposed classification technique is superior and faster than PSO-FLANN.

Keywords FLANN · PSO · PSO-FLANN · ANN

1 Introduction

Now a days the microarray data sets contain a large number of gene expressions. The Classification performance depends up on the number of features of gene expression data sets. So the major concern is to reduce the dimension of the data set. In this proposed work, principal component analysis(PCA) is used for feature selection and dimensionality reduction for gene expression data set.

Here in this paper, Ant Colony Optimization (ACO) is used for the optimization of the data set and Functional Link Artificial Neural Network (FLANN) is used for training the optimized data set for classification. The paper is organized as follows: Sect. 2 describes the literature survey, Sect. 3 describes the proposed model and algorithm, Sect. 4 describes Result and Simulation, and finally, the last Sect. 5 describes conclusion and future work.

2 Literature Review

David Martens et al. has proposed a new ACO algorithm, i.e., Ant-Miner+. The proposed Ant-Miner technique can handle both binary and multiclass classification problems and generates rule lists consisting of propositional and interval rules. Ant-Miner+ accuracy is superior than the accuracy obtained by the other Ant-Miner versions and competitive or better than the results achieved by the compared classification techniques [1].

Ahmed Nejmedine Machraoui et al. has applied a hybrid segmentation algorithm for classification and detection of breast cancer cells, i.e., ACO-Otsu segmentation method. This method gives more performance than standard Ostu method [2].

In this paper, an unsupervised gene selection method called **MGSACO**, which incorporates the ant colony optimization algorithm into the filter approach, by minimizing the redundancy between genes and maximizing the relevance of genes is discussed. Moreover, **a new fitness function** is applied in the proposed method which does not need any learning model to evaluate the subsets of selected genes [3].

Debahuti Mishra et al. *has* proposed a hybrid method **called Rough ACO,** where ACO is hybridized with Rough Set Theory. The proposed method is successfully applied for dimensionality reduction of a feature set by choosing a subset of the original features that contains most of the essential information from a gene expression data [4].

In this paper the ant colony optimization (ACO) algorithm is introduced to select genes relevant to cancers first, then the multi-layer perceptions (MLP) neural network and support vector machine (SVM) classifiers are used for cancer classification. Experimental results show that selecting genes by using ACO algorithm can improve the accuracy of BP and SVM classifiers [5].

Hualong Yu et al. has discussed a simple modified ant colony optimization (ACO) algorithm to select tumor-related marker genes, and support vector machine (SVM) is used as a classifier to evaluate the performance of the extracted gene subset [6].

In this paper, Ant colony optimization (ACO) based classification is used for the analysis of gene expression data. Here cAnt-Miner is employed, a variation of the classical Ant-Miner classifier, to interpret numerical gene expression data. Experimental results on well-known gene expression datasets show that the ant-based approach is capable of extracting a compact rule base and provides good classification performance [7].

In this paper, Lei Shi et al. has discussed **an Ant Colony Optimization and rough set based ensemble approach** for improving the generalization ability and efficiency of the ensemble for biomedical classification. Ant Colony Optimization and rough set theory are incorporated to select a subset of all the trained component classifiers for aggregation. Experimental results show that compared with the existing methods, it not only decreases the size of the ensemble, but also obtains higher prediction performance [8].

Jyotsna Bansal et al. has taken three different datasets named Leukemia, Lung Cancer, and Prostate from the UCI machine learning repository and applied efficient **association based ant colony optimization** for improving the classification accuracy. Finally, ACO mechanism has been applied on the final dataset to find the classification accuracy [9].

Large sampled datasets have thousands of samples and less number of attributes, so it takes long training and testing time. Ant Colony Optimization (ACO) is a metaheuristic approach to get optimal solution for computationally intractable problems. In the first module of this paper, modified ACO is applied to select the most promising set of features from large attributed datasets. In the second module, competitive sample selection is applied to get only necessary and unique samples from the large sampled dataset. The results show that two objectives are fulfilled (i) Number of samples can be reduced and (ii) Most accurate class can be predicted [10].

Satchidananda Dehuri et al. has discussed a new algorithm, i.e., an improved particle swarm optimization (IPSO) which is used to train the functional link artificial neural network (FLANN) for classification and named it ISO-FLANN. In contrast to MLP, FLANN has less architectural complexity, easier to train and more insight may be gained in the classification problem [11].

Satchidananda Dehuri et al. has discussed another algorithm, i.e., HFLANN proved to be better than the best results found by its competitor like RBF and FLANN with backpropagation learning. The architectural complexity is low, whereas training time is little bit costly as compared to FLANN [12].

Pournamasi Parhi et al. *has* proposed PSO-FLANN framework of Feature Selection and Classification for Microarray Data. Here PSO is used to optimize the data.

Functional Link Artificial Neural Network (FLANN) is used to tune the parameters of FLANN. PSO-FLANN is compared with DA and gives 80% accuracy.

Sashikala Mishra Et al. has proposed a Classification technique called BAT-FLANN. Here bat algorithm is used to update the weights of a Functional Link Artificial Neural Network (FLANN) classifier. Here the proposed model has been compared with FLANN and PSO-FLANN [13] and the simulation shows that the proposed classification technique is superior and faster than FLANN and PSO-FLANN [14].

After this literature reviewed, we found that using Ant Colony Optimization (ACO) and Functional Link Artificial Neural Network (FLANN) there is no work done till now. So it is a promising area for research.

3 Proposed Work

In this work, a new model is proposed for the classification of microarray data. Functional Link Artificial Neural Network (FLANN) uses Ant Colony Optimization (ACO) to tune the parameters of FLANN (Fig. 1).

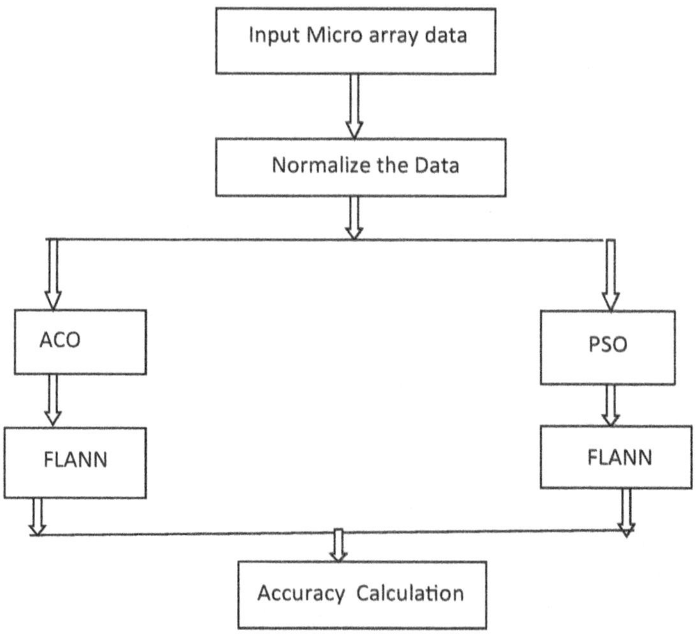

Fig. 1 ACO-FLANN architecture

The algorithm is as follows:

Step 1. ***Read the Data set (input).***

Step 2. ***Distribute the data set in two variable s training and Testing. Training contained 80% data and testing content 20 % data.***

Step.3 ***Normalize the Training data***

norm_data = (inputs - min(min(inputs))) ./ (max(max(inputs)) - min(min(inputs)));

Step.5 ***Input the normalized data to ACO for optimization.***

Step 6. ***Create the ACO model***

Step 6. ***Initialize the ACO parameters.***

Maxit= 3 [Maximum Number of iteration]
nAnt=40 [Number of Ants (population size)]
tauo= 0.1 [initial Pheromone]
alpha=1 [Pheromone exponential weight]
beta=0.02; [Heuristic exponential Weight]
rho = 0.1 ;[evaporation rate]

Step7. Move the ants

```
for it=1:MaxIt

  % Move Ants
  for k=1:nAnt
    ant(k).Tour=[];

    for l=1:nVar

      P=tau(:,l).^alpha.*eta(:,l).^beta;

      P=P/sum(P);

      j=RouletteWheelSelection(P);

      ant(k).Tour=[ant(k).Tour j];

    end

    ant(k).x=N(ant(k).Tour);

    [ant(k).Cost, ant(k).Sol]=CostFunction(ant(k).x);

    if ant(k).Cost<BestSol.Cost
      BestSol=ant(k);
    end

  end
```

Step7. Update Pheromones

```
for k=1:nAnt

   tour=ant(k).Tour;

   for l=1:nVar

      tau(tour(l),l)=tau(tour(l),l)+Q/ant(k).Cost;
     nd
    end
```

Step 8. Calculate Evaporation

```
tau=(1-rho)*tau;
```

Step 9. Store Best Cost

```
BestCost(it)=BestSol.Cost;
```

Step 10. Plot Best cost figure

Step11. Initialize the weights of FLANN

```
for j=1:length(inputs)
   x=xx(j,:);
   for i=1:size(x,2)
      X=[1 x(i) sin(pi.*x(i)) cos(pi.*(x(i)))]
      XX(:,i)=sum(X)
   end
   jj(j,:)=XX;
end

su_inp=[]
for k=1:length(inputs)
   su_inp(k)=sum(jj(k,:))
end
```

Step 12. Train the Network

```
P= [su_inp];
T = [targets1];
net = linearlayer;
net.trainParam.goal= 0.5;
net = train(net,su_inp,targets1)
[Xs,Xi,Ai,Ts] = preparets(net,su_inp,targets1);
net = train(net,Xs,Ts,Xi,Ai);
net1 = newlind(P,T);
Y = sim(net1,P);
perf = perform(net,Ts,Y)
```

Step 13. Calculating accuracy

Step 14. Plot the mean square error graph

Step 15. Plot Validation graph

Step 16. Plot Error histogram graph

4 Result and Simulation

This section describes the simulation of our proposed model, performance evaluation, and experiments carried to explore the properties. There are three different data sets such as breast cancer, leukemia, and lung cancer taken for evaluation purpose. All the three data sets features has been reduced by using PCA which is mentioned in the Table 1. After implementation by using our proposed model, we got higher accuracy performance that PSO-FLANN. The detail accuracy, mean square error, validation graph, and error histogram is mentioned below in Figs. 2, 3, 4, 5, and 6.

Table 1 Classification accuracy result table

Data sets	Size of data sets	Size of data sets after reduction using PCA	Accuracy achieved by PSO-FLANN (%)	Accuracy achieved by ACO-FLANN (%)
Leukemia	72 * 7129	72 * 2	78	81.7
Lung cancer	197 * 581	197 * 8	85.3	90
Breast cancer	98 * 1213	98 * 97	86.3	91

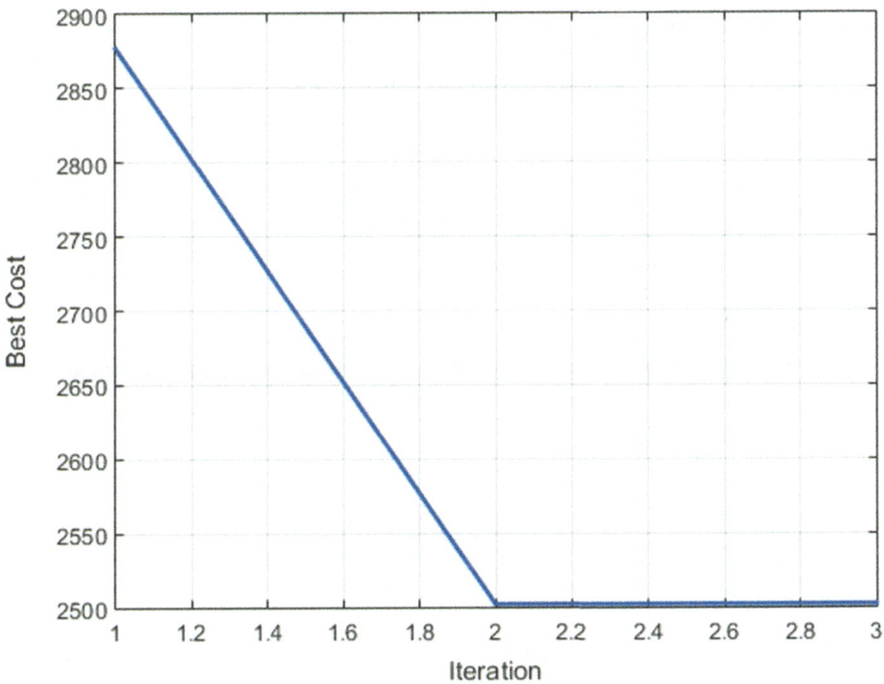

Fig. 2 Best cost

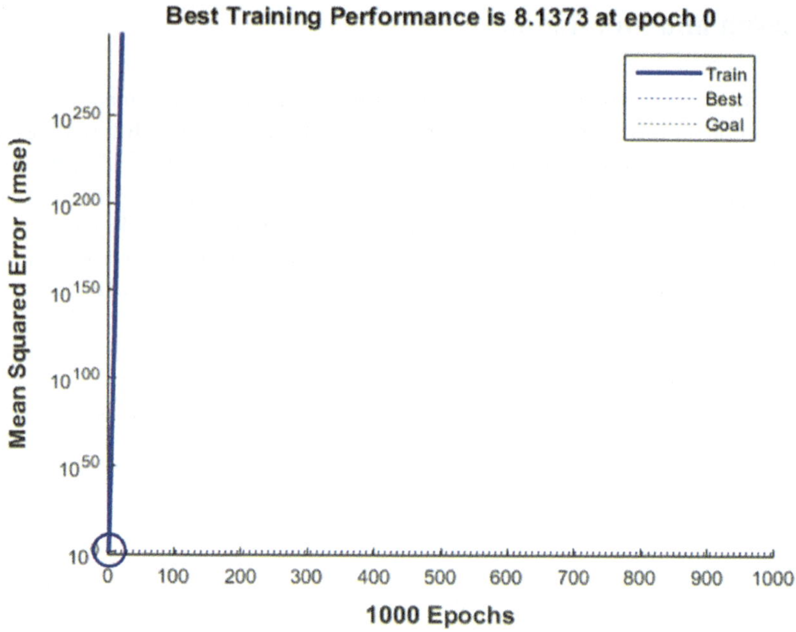

Fig. 3 Training performance

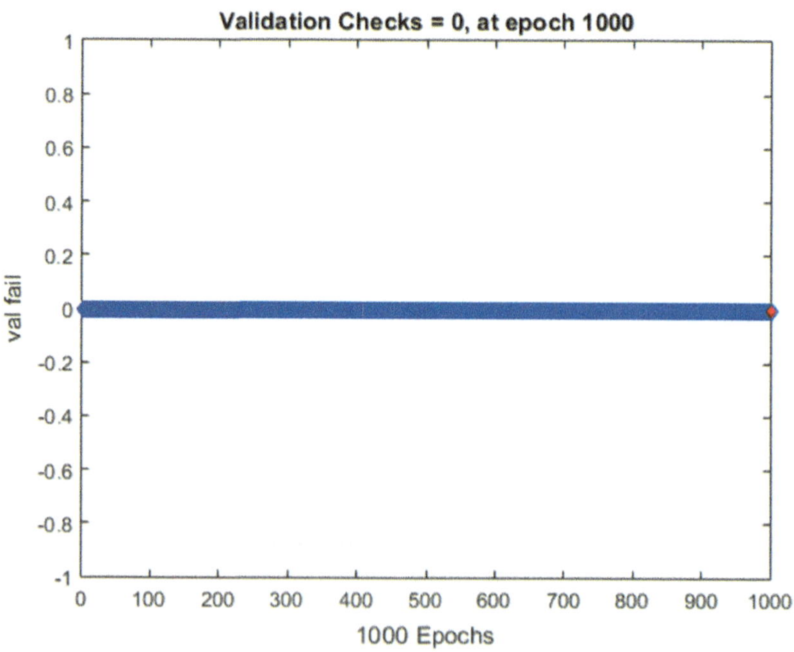

Fig. 4 Validation

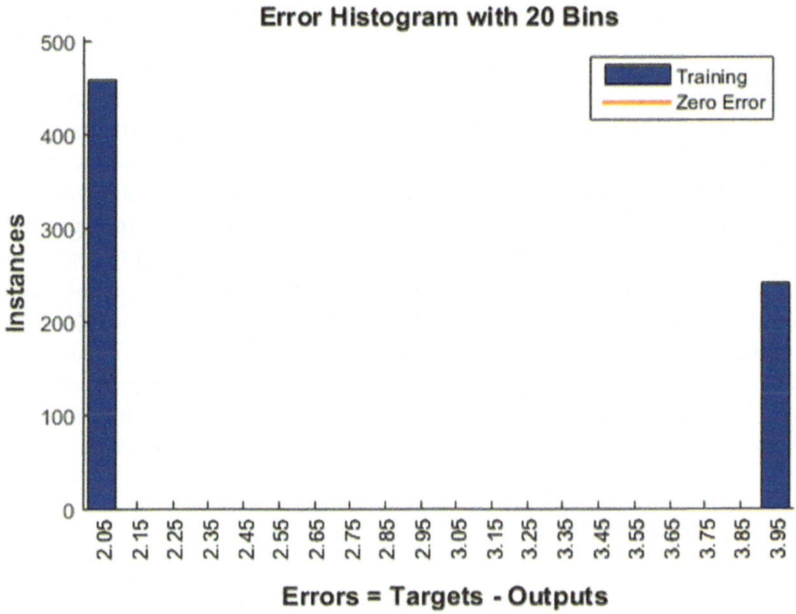

Fig. 5 Error histogram

Fig. 6 FLANN training

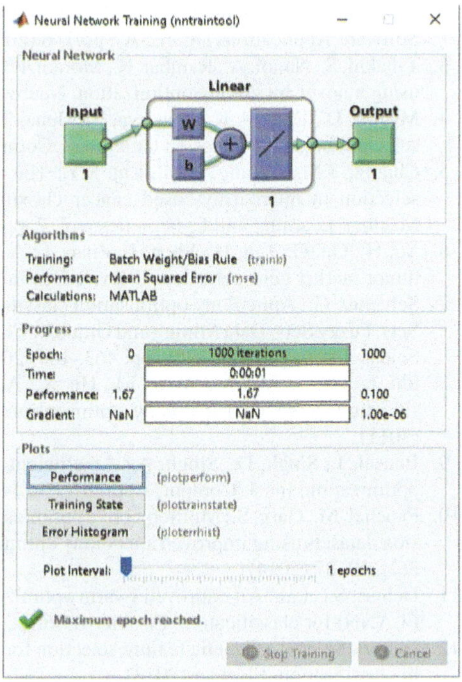

5 Conclusion

The predominant objective of a classifier is to discover the hidden classification level of the unknown data. The classifier performance depends upon the data size, wide variety of classes, and dimension of characteristic spaces. The classifier accuracy has been expanded by way of making use of optimization techniques. Several optimization techniques such as Particle Swarm Optimization (PSO), ACO, ABC, DE, MLP, FLANN, PSO-FLANN, etc., are developed for optimization purpose. In this proposed work, a new framework is proposed for the classification of microarray records. Functional Link Artificial Neural Network (FLANN) uses Ant Colony Optimization (ACO) to tune the parameters. This model is called ACO-FLANN. The proposed model (ACO-FLANN) has been compared with various datasets. Simulation suggests that the proposed classification method is advanced and more faster than BAT-FLANN and PSO-FLANN and gives good accuracy.

References

1. Martens, D., De Backer, M., Haesen, R., Vanthienen, J., Snoeck, M., Baesens, B.: Classification with ant colony optimization. IEEE Trans. Evol. Comput. **11**(5), 651–664 (2007)
2. Machraoui, A.N., Cherni, M.A., Sayadi, M.: Ant colony optimization algorithm for breast cancer cells classification. In: IEEE International Conference on Electrical Engineering and Software Applications (ICEESA), pp.1–6 (2014)
3. Tabakhi, S., Najafi, A., Ranjbar, R., Moradi, P.: Gene selection for microarray data classification using a novel ant colony optimization. Neurocomputing **168**, 1024–1036 (2015)
4. Mishra, D., Rath, A.K., Acharya, M., Jena, T.: Rough ACO: a hybridized model for feature selection in gene expression data. Int. J. Comput. Commun. Technol. **1**(1), 85–98 (2009)
5. Chiang, Y.M., Chiang, H.M., Lin, S.Y.: The application of ant colony optimization for gene selection in microarray-based cancer classification. In: IEEE International Conference on Machine Learning and Cybernetics, vol. 7, pp. 4001–4006 (2008)
6. Yu, H., Gu, G., Liu, H., Shen, J., Zhao, J.: A modified ant colony optimization algorithm for tumor marker gene selection. Elsevier Genomics, Proteomics Bioinf. **7**(4), 200–208 (2009)
7. Schaefer, G.: Ant colony optimisation classification for gene expression data analysis. Rough Sets, Fuzzy Sets, Data Mining and Granular Computing of the series Lecture Notes in Computer Science, Springer, vol. 5908, pp. 463–469 (2009)
8. Shi, L., Xi, L., Ma, X., Weng, M., Hu, X.: A novel ensemble algorithm for biomedical classification based on ant colony optimization. Appl. Soft Comput. Elsevier **11**, 5674–5683 (2011)
9. Bansal, J., Singh, D., Singh, A.: An efficient medical data classification based on ant colony optimization. Int. J. Comput. Appl. **87**(10), 14–18 (2014)
10. Panchal, M., Garg, S.: Multiobjective optimization for classifying samples of large gene expression datasets using improved ant colony optimization and competitive sample selection. IJCSC **5**(2), 79–87 (2014)
11. Dehuri, S., et.al.: An improved swarm optimized functional link artificial neural network (ISO-FLANN) for classification. Elsevier J. Softw. Syst. (2010–2012)
12. Dehuri, S., et.al.: Genetic feature selection for optimal functional link artificial neural network in classification. Elsevier (2013)

13. Parhi, P., Mishra, D., Mishra, S., Shaw, K.: A novel PSO-FLANN frame work of feature selection and classification for microarray data. Elsevier Procedia Eng. **38**, 1644–1649 (2012)
14. Mishra, Sashikala, Shaw, Kailash, Mishra, Debahuti: A new meta-heuristic bat inspired classification approach for microarray data. Elsevier Procedia Technol. **4**, 802–806 (2012)

An Ensemble Approach for Automated Classification of Brain MRIs

Sarada Prasanna Pati, Sarbeswara Hota, and Debahuti Mishra

Abstract Classification of brain Magnetic Resonance Imaging (MRI) is finding widespread use in clinical practice as it greatly helps the radiologists and neurologists to diagnose the brain abnormalities with more accuracy and precision. Several machine learning models are been proposed in recent times to effectively classify the brain MRIs. Ensemble of classification models produces better classification performance compared to individual models. In this work, we have proposed and validated a Particle Swarm Optimization (PSO) based linear ensemble model for the classification of brain MRIs. The ensemble model combines the outcomes of three heterogeneous base classifiers, namely Artificial Neural Network (ANN), Least-squares Support Vector Machines (LS-SVM), *k*-Nearest Neighbors (*k*-NN). PSO is applied to optimally decide the individual contribution or weightage of the three base models towards the final results of the ensemble model as a whole. Discrete wavelet transform (DWT) is applied in the first stage for feature extraction and Principal Component Analysis (PCA) is applied for reducing the dimensionality of the extracted feature vectors before classification.

Keywords Brain MRI · Classification · Classifier ensemble · Particle swarm optimization

---

S. P. Pati (✉) · D. Mishra
Department of Computer Science and Engineering, Institute of Technical Education and Research, Siksha 'O' Anusandhan (Deemed to be University), Bhubaneswar, Odisha 751030, India
e-mail: saradapati@soa.ac.in

D. Mishra
e-mail: debahutimishra@soa.ac.in

S. Hota
Department of Computer Application, Institute of Technical Education and Research, Siksha 'O' Anusandhan (Deemed to be University), Bhubaneswar, Odisha 751030, India
e-mail: sarbeswarahota@soa.ac.in

D. Mishra et al. (eds.), *Intelligent and Cloud Computing*,
Smart Innovation, Systems and Technologies 194,
https://doi.org/10.1007/978-981-15-5971-6_54

1 Introduction

Critical brain diseases are often considered as serious health hazards, which may involve disabilities of some form or even threat to life. Early and accurate detection of such diseases is vital for proper treatment planning and possible recovery. In recent years, Magnetic Resonance Imaging (MRI) has emerged as one of the popular choices to study the human brain. MRI is basically a noninvasive technique that provides features like superior soft tissue differentiation, high spatial resolution, and better contrast [1]. MRI can detect a variety of conditions of the brain such as cysts, tumors, bleeding, swelling, developmental and structural abnormalities, infections, inflammatory conditions, or problems with the blood vessels, etc. The advances in brain MR imaging technology produce high quality multi-spectral images that contain large volumes of rich and sensitive data about the soft tissue anatomy of the human brain [2]. Analysis of these large and complex MRI datasets has become a tedious and complex task for clinicians, who have to manually extract important information. This manual analysis is often time-consuming and prone to errors due to various inter- or intra-operator variability studies and the inferences are also non-reproducible [3]. These difficulties in brain MRI data analysis required inventions in computerized methods to improve disease diagnosis with high accuracy and fast speed. This is where the digital image processing techniques along with machine learning aid further diagnosis, treatment planning, prior and post-surgical procedures. Automated image classification systems for brain MR images to detect normal or pathological brain conditions are of great demand as they could highly enhance the clinical diagnostics of brain related diseases by providing high diagnostic accuracy at a very fast speed [4, 5].

Image classification refers to the process of automated grouping of images or predicting the type or category (predefined) to which an image belongs based on distinctive visual elements in the image called the image features. Machine learning in amalgamation with image processing techniques have widely been used in developing image classification models which can automate the process of recognizing images [4–7]. Literature on automated medical image processing suggests that Artificial Neural Network (ANN) [5, 8], Support Vector Machines (SVM) [6, 9], *k*-Nearest Neighbors (*k*-NN) [5, 10], Classification and Regression Tree (CART) [10], Random forest [10], Convolutional Neural Network (CNN) [11], Probabilistic Neural Network (PNN) [7, 12], and Fuzzy *c*-Means Clustering (FCM) [13] are some of the commonly used techniques used in developing medical image classification models.

Ensemble learning has been one of the most active areas of interest for many researchers in recent years. The primary objective behind constructing ensemble models is to combine the outputs of several weak learners to build a strong learning model, and thereby increasing the prediction performance and accuracy of the model. The major finding that encourages the use of ensemble technique in supervised learning is that the ensemble models often give better accuracy compared to the individual methods that make them up [14, 15].

In this work, we propose a multi-classifier ensemble model for classifying the MR brain images as normal or diseased. Discrete wavelet transform (DWT) is applied in the first stage for feature extraction from MRIs and Principal component analysis (PCA) is applied subsequently to reduce the dimension of the feature vectors. The ensemble model combines the outcomes of three heterogeneous base classifiers, namely ANN, *k*-NN, and Least-square SVM (LS-SVM). The three base classifiers are first independently trained on the MR image datasets. In the next phase, Particle Swarm Optimization (PSO) is applied to optimally decide the individual contribution or weightage of the three base models toward the final results of the ensemble model as a whole.

The rest of the paper is organized as follows. The proposed approach is presented in Sect. 2. The simulation study and result analysis are presented in Sects. 3 and 4, respectively. The conclusion is given in Sect. 5.

2 Proposed Approach

The image classification models are mostly developed as supervised learning models that learn to identify different classes of images by analyzing the feature vectors of the example images during the training phase. A medical image classification system in general consists of three basic steps, namely, feature extraction, feature reduction, and classification. The proposed system also comprises of same three steps as depicted in Fig. 1.

An efficient feature extraction method is considered to be the key toward building an efficient classifier model [16]. Feature extraction refers to the extraction of the

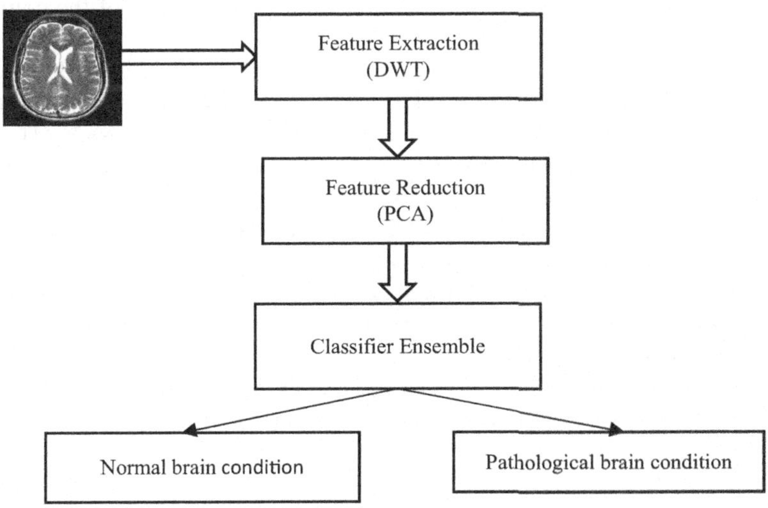

Fig. 1 Framework for brain MRI classification

relevant features from the original image or a given region of interest within an image to prepare the feature vector which will subsequently be the input to the classifier to classify the image to one of the target classes. The proposed system employs 2-Dimensional DWT for feature extraction and thus use extracted wavelet coefficients as feature vectors. The wavelets represent localized frequency information for the function in a signal which can be used for classification.

Selection of an efficient technique for feature reduction also has a significant impact on the final outcome realized in a classification task [5, 17]. High dimension of feature vectors may lead to design complexity of the model and may also result in over-fitting/underfitting of data. Therefore, it becomes necessary to reduce the dimension of the feature vectors in order to lower the computational complexity and increase the discriminative power and generalization ability of the model so that it can make more accurate prediction. PCA is applied for reducing the dimensionality of the extracted feature vectors in this work.

The effectiveness of an ensemble model greatly depends on its design and the way the individual outputs of the base models are integrated. Each individual model may offer good performance alone but by aggregating the output of multiple individual models the error committed by the individual models on the same data gets averaged out and the chance of over-fitting also is reduced. This, in turn, lowers the variance, results in better generalization ability for the model, and improves the prediction performance. The different ensemble models for classification differ in terms of how they are built out of the multiple base classifiers and the way the individual outputs are combined or processed to obtain the final output of the overall model. Some of the popular ensemble techniques include bagging, boosting, and stacking [14, 15]. Weighted combination model of multiple classifier systems is yet another approach for constructing ensemble models. The basic idea here is to find an optimal combination of weights that could be assigned to the output of base classifiers based on their classification accuracy such that the whole performance of the combined system is optimized.

A weight optimized linear ensemble model is finally proposed in the last step for classifying the brain images (Fig. 2). The ensemble model combines the results of three heterogeneous base classifiers, namely ANN, k-NN, and LS-SVM. The three base classifiers are first independently trained on the MR image datasets. In the next phase, PSO is applied to optimally decide the individual contribution or weightage of each of the three base models toward the final results of the ensemble model as a whole by considering the mean square error (MSE) of the overall ensemble model as the fitness function. Here weights assigned to each base classifier are considered as free parameters and their combination values are optimized using PSO to get the best classification performance.

The linear combination of the three base classifiers representing the weighted sum of the individual classification models can be written as Eq. (1).

$$\tilde{y} = \sum_{i=1}^{k} w_i y_i \tag{1}$$

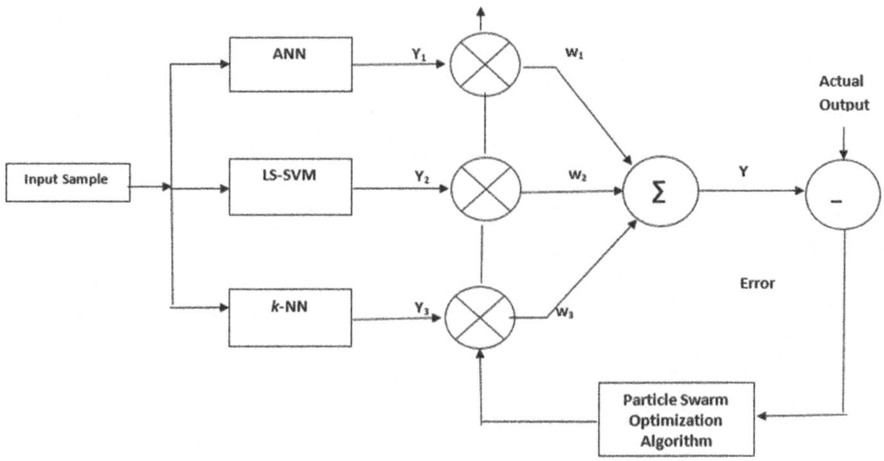

Fig. 2 Proposed model of classifier ensemble

Here, $\tilde{y}$ is the final predicted output of the ensemble system and y_i is the output of each individual model. k(=3) is the number of classifier. Further, w_i represents the weight assigned to the output of the base classifiers such that $\sum_{i=1}^{k} w_i = 1$.

3 Simulation Studies

The proposed classifier ensemble model is experimented on the two brain image datasets for two different brain diseases, namely, Alzheimer and Glioma collected from [18], and supported and maintained by the Medical School of Harvard University. These datasets consist of axial, T2-weighted, 256 × 256 pixel MR brain images of both normal and diseased brain conditions. The Alzheimer and Glioma dataset contains 101 and 123 number of brain MRI samples, respectively. Two-dimensional DWT (2D-DWT) is applied in the first stage for feature extraction which decomposes the MR images to get the wavelet coefficients that would work as the feature vectors for classification at the next level. Decomposition up to level three of the Haar wavelet has been used in this work to extract 32 × 32 features for each brain MR image. PCA is applied subsequently to reduce the dimension of the feature vectors. PCA exploits the correlation among the features in the original dataset and performs an orthogonal transformation on the features to reduce them to a smaller set of linearly uncorrelated variables called principal components [19]. The principal components represent the reduced set of features that retains most of the variations in the original data and possess enough ability to discriminate between the target classes. As per the standard procedure, the input feature vectors were normalized before applying PCA. The threshold for variance taken for selecting the principal components is 95%. The dimensionality of the feature vectors is thus reduced from 1024 to 18 and 12 for Alzheimer and Glioma datasets, respectively, after applying PCA. After the feature extraction and dimensionality reduction of the feature vectors, the final step is to classify the images with respect to their predetermined class labels (normal/diseased). The feature vectors are then input to the three base classifiers independently to find the predicted output.

The image samples are divided into 70:30 ratio for training and testing purposes. The network architecture of ANN has a great influence on the classification performance and hence are chosen carefully [20], after a series of trials. The best network architecture for ANN that exhibits the least MSE value, obtained in our case, has one hidden layer with 10 hidden layer nodes. The transfer function employed is "sigmoid". The main objective of choosing LS-SVM over SVM is to incur less computational cost. For LS-SVM, the Radial Basis Function (RBF) kernel is used. To get the best classification performance in case of LS-SVM, the kernel and regularization parameters (γ and C) need proper tuning. After careful evaluation, the γ and C were picked. For k-NN, the value of k is considered to be five.

To compare and validate the performance of the proposed method, another two established bioinspired metaheuristics, namely, Artificial Bee Colony (ABC) and Ant Colony Optimization (ACO) were also applied in another two sets of experiments to optimize the weighted combination of the output of the base classifiers to get the final output.

4 Result Analysis

The classification performance of the proposed model for the two image datasets is measured in terms of Accuracy, Sensitivity, Specificity, and Precision. Tables 1, 2, 3, and 4 represent a comparative study of the classification performance exhibited by each of the classification methods in terms of Accuracy, Sensitivity, Specificity, and Precision, respectively. From these empirical results, it is evident that the proposed ensemble model outperforms all the individual classification models used as base classifiers. The classification accuracy achieved in case of Alzheimer and Glioma are 95.4% and 96.24%, respectively. Furthermore, it is also observed

Table 1 Accuracy (in percentage)

Method	Alzheimer	Glioma
ANN	81.18	84.21
LS-SVM	89.1	91.33
k-NN	86.13	87.96
Ensemble-ACO	94.05	93.23
Ensemble-ABC	93.56	94.73
Ensemble-PSO	95.04	96.24

Table 2 Sensitivity (in percentage)

Method	Alzheimer	Glioma
ANN	85	89.04
LS-SVM	91.66	93.5
k-NN	91.22	93.05
Ensemble-ACO	95.08	92.59
Ensemble-ABC	95	96.1
Ensemble-PSO	96.66	97.4

Table 3 Specificity (in percentage)

Method	Alzheimer	Glioma
ANN	75.6	78.33
LS-SVM	85.36	89.28
k-NN	79.54	81.96
Ensemble-ACO	92.5	94.23
Ensemble-ABC	90.24	92.85
Ensemble-PSO	92.68	94.64

Table 4 Precision (in percentage)

Method	Alzheimer	Glioma
ANN	83.6	83.33
LS-SVM	90.16	92.3
k-NN	85.24	85.89
Ensemble-ACO	95.08	96.15
Ensemble-ABC	93.44	94.87
Ensemble-PSO	95.08	96.15

that the proposed PSO based linear ensemble model gives superior results compared to the other two linear models using ABC and ACO in terms of all performance measures, i.e., Accuracy, Sensitivity, Specificity, and Precision.

5 Conclusion

In this work, we proposed an ensemble-based model for the classification of brain MRIs in order to detect pathological brain conditions. The ensemble model combines the results of three heterogeneous base classifiers, namely ANN, *k*-NN, and LS-SVM. The three base classifiers are first independently trained on the MR image datasets. The proposed method used PSO for optimizing the contribution of the three individual base classifiers to the overall performance of the ensemble model. The model was validated through simulation studies and it was observed that this model exhibits better classification accuracy compared to all three individual base models, as well as the ABC and ACO based ensemble model.

References

1. Chen, R., Yang, S., Wang, C.: MRI brain tissue classification using unsupervised optimized extenics-based methods. Comput. Electr. Eng. **58**, 489–501 (2017)
2. Ortiz, A., Gorriz, J., Ramirez, J., Salas-Gonzalez, D., Llamas-Elvira, J.: Two fully-unsupervised methods for MR brain image segmentation using SOM-based strategies. Appl. Soft Comput. J. **13**, 2668–2682 (2013)
3. Hemanth, D., Vijila, C., Selvakumar, A., Anitha, J.: Performance improved iteration-free artificial neural networks for abnormal magnetic resonance brain image classification. Neurocomputing **130**, 98–107 (2014)
4. Mohan, G., Subashini, M.: MRI based medical image analysis: survey on brain tumor grade classification. Biomed. Signal Process. Control **39**, 139–161 (2018)
5. El-Dahshan, E.S.A., Hosny, T., Salem, A.B.M.: Hybrid intelligent techniques for MRI brain images classification. Digit. Signal Process. **20**, 433–441 (2010)
6. Chaplot, S., Patnaik, L., Jagannathan, N.: Classification of magnetic resonance brain images using wavelets as input to support vector machine and neural network. Biomed. Signal Process. Control **1**, 86–92 (2006)
7. Varuna Shree, N., Kumar, T.: Identification and classification of brain tumor MRI images with feature extraction using DWT and probabilistic neural network. Brain Inf. **5**, 23–30 (2018)
8. Zhang, Y., Dong, Z., Wua, L., Wang, S.: A hybrid method for MRI brain image classification. Expert Syst. Appl. **38**, 10049–10053 (2011)
9. Amin, J., Sharif, M., Yasmin, M., Fernandes, S.: A distinctive approach in brain tumor detection and classification using MRI. Pattern Recognit. Lett. 1–10 (2017)
10. Gupta, T., Gandhi, T., Gupta, R., Panigrahi, B.: Classification of patients with tumor using MR FLAIR images. Pattern Recognit. Lett. 1–6 (2017)
11. Gao, X., Hui, R., Tian, Z.: Classification of CT brain images based on deep learning networks. Comput. Methods Programs Biomed. **138**, 49–56 (2017)
12. Saritha, M., Paul Joseph, K., Mathew, A.: Classification of MRI brain images using combined wavelet entropy based spider web plots and probabilistic neural network. Pattern Recogn. Lett. **34**, 2151–2156M (2013)
13. Maitra, M., Chatterjee, A.: A Slantlet transform based intelligent system for magnetic resonance brain image classification. Biomed. Signal Process. Control **1**, 299–306 (2006)
14. Rokach, L.: Ensemble-based classifiers. Artif. Intell. Rev. **33**(1–2), 1–39 (2010)
15. Dietterich, T.: Ensemble methods in machine learning. Mult. Classif. Syst. Lect. Notes Comput. Sci. **1857**, 1–15. Springer, Heidelberg (2000)
16. Jui, S.L., Zhang, S., Xiong, W., Yu, F., Fu, M., Wang, D., Hassanien, A., Xiao, K.: Brain MRI tumor segmentation with 3D intracranial structure deformation features. IEEE Intell. Syst. **31**, 66–76 (2016)
17. Zhang, Y., Wang, S., Ji, G., Dong, Z.: An MR brain images classifier system via particle swarm optimization and kernel support vector machine. Hindawi Publ. Corp. Sci. World J. Article ID 130134 (2013)
18. http://www.med.harvard.edu/AANLIB
19. Shlens, J.: A Tutorial on Principal Component Analysis. arXiv:14041100,2014
20. Benyelloul, K., Aourag, H.: Bulk modulus prediction of austenitic stainless steel using a hybrid GA–ANN as a data mining tools. Comput. Mater. Sci. **77**, 330–334 (2013)

An Analysis of Various Demand Response Techniques in Smart Grid

Nilima R. Das, Mamata Nayak, S. C. Rai, and Ajit K. Nayak

Abstract Recently a lot of research has been made on different scheduling methods applied in Demand Response programs in residential grid networks. Even if they are different in their methodologies their main aim is to reduce the consumers' expense of energy and demand during peak hours in order to reduce the production costs and increase the supply of electricity. The scheduling methods provide schedules that help to guide the consumption pattern of the users so that the peak load and consumption during peak hours can be reduced. Some of the recently developed methods are discussed and analyzed here based on their properties and limitations that provide a better understanding of the concept of Demand Response and the future research scope in this area.

Keywords Demand side energy management · Demand response · Peak hour consumption

1 Introduction

1.1 Smart Grid and DSM

Smart Grid is an up-and-coming concept in electrical grid systems. It claims the advancement of technology and information system with highly advanced electrical infrastructures that can increase the reliability, security, and efficiency. Reliability is one of the most important challenges for a grid system. Reliable operation of the grid requires an absolute balance between supply and demand [1]. However, maintaining balance is a very challenging task as it is difficult to control the demand of the consumers. This can only be possible with the implementation of the smart grid. The

N. R. Das (✉) · M. Nayak · A. K. Nayak
Faculty of Engineering and Technology, SOA (Deemed to Be University), Bhubaneswar, India
e-mail: nilimadas@soa.ac.in

S. C. Rai
Department of IT, Silicon Institute of Technology, Bhubaneswar, India

D. Mishra et al. (eds.), *Intelligent and Cloud Computing*,
Smart Innovation, Systems and Technologies 194,
https://doi.org/10.1007/978-981-15-5971-6_55

smart grid can improve the energy efficiency in producing and using electricity. To meet high demands the utilities generally depend on more fossil fuel-based plants, but uncertainty in generation, forces them to keep a higher surplus of resources which can increase the cost of electricity. The other way to maintain the balance between demand and supply is to use new techniques that give more importance to user involvement. In order to maintain a proper balance between the demand and supply, the smart grid requires the integration of Demand Side Management (DSM) techniques. DSM incorporates energy conservation and energy efficiency programs, Demand Response (DR) programs, and residential or commercial demand management programs. These programs when implemented increases the efficient use of the available energy instead of building new production and transmission framework or infrastructures. The main objective of a DSM program is to reduce or change the time of consumption [1]. The DSM techniques are broadly classified into two categories: Energy Efficiency Programs and Demand response programs. Energy efficiency programs encourage consumers to use energy-efficient appliances which can avoid wastage of energy and help in conserving more energy. DR programs are used to scale down the energy demand during peak hours which can turn down the risk of blackouts and voltage fluctuations and prevents further expenditure on new power plants. Some systems may also use energy storage devices to moderate the effect of high demand periods.

1.2 Demand Response

DR motivates the consumers to change their consumption pattern so that it can match with the available supply. The DR scheme employs intelligent techniques to increase the efficient use of energy at the demand side. DR can effectively maintain a balance between the demand and available supply. With DR programs it is easy for the utilities to handle or manage the consumer loads. The utility implements load control and management programs by following bill saving schemes or incentives for the consumers to motivate them to change their consumption patterns. Changing usage pattern reduces the peak time load which is the main objective of DR.

Benefits of DR Several benefits of demand response programs are the sources of motivations for this study.

- DR helps in reducing demand during peak hours. By reducing the demand DR can remove inconveniences and increase the availability of electricity with an upgraded level of service. It also leads to the virtual elimination of blackouts.
- Reduction in demand reduces the operational cost of the utilities.
- DR can help improve overall power plant utilization rates by stimulating energy consumption during off-peak hours and flattening the demand curve.
- For distributors of electricity, DR can provide an alternative to heavy investment in transmission infrastructure by avoiding high penalties during peaks that exceed contracted levels of power.

DR can be a more cost-effective alternative than adding generation capabilities to meet the peak and occasional demand spikes because it causes a reduction in peak demand and avoids system emergencies. For all these advantages, DR models are required to be implemented at the user side. Motivating global consumers to participate in DR schemes in residential sector can bring in more energy savings in the electrical market.

2 Recent Advancements in DR

Recently a lot of research has been made on different scheduling methods useful for DR programs in residential grid networks. Some of the methods are discussed in this section and grouped according to the programming paradigms they have used.

2.1 Linear Programming Method (LP)

The article [2], proposes a DR technique that finds optimal operation schedule for the household appliances with an objective to reduce the financial expenditure of an Energy Consumer (EC). It uses a linear programming method to calculate a deterministic scheduling solution without considering the uncertainties and a stochastic scheduling strategy to figure out the uncertainties. The authors in [3], present a Home Energy Management System (HEMS) model that employs a heuristic method to supervise the operation of household appliances. It provides an algorithm that takes into consideration multiple users and their load priority and finds the optimal load schedule for them.

2.2 Integer Linear Programming Method (ILP)

The authors in [4], present a security constrained unit commitment (SCUC) model that concurrently schedules the energy produced from generating units, spinning reserve, and reserve contributed by DR resources. The projected model is described as a two-phase stochastic mixed-integer programming (SMIP) model. In [5], the authors have proposed an Integer Linear Programming (ILP) based consumption scheduling system to minimize the peak time load and generate optimal power and operation time for user appliances. The article [6], defines a DR problem that comprises of the functions relating to electricity consumption, generation, and storage. It tries to reduce the system's Peak to Average Ratio (PAR). In [7], the authors present a solution for the DR problem based on a closed loop method that predicts the electricity demand of each EC, combines the demand from all ECs in a neighborhood, and sends the collective load information to the Utility Company (UC). The UC can use this load information for energy management and TOU price decisions. Based on the TOU pricing presented by the UC, the residential ECs adjust their loads to minimize their electricity payment with a consideration to their comfort level. In [8], the authors present the energy management of an EC that consumes energy from the grid, as well as energy produced from its own solar energy production unit. The EC tries to optimize its energy use with an aim to reduce its payment. The work done in [9], presents a Home Energy Management System (HEMS) for Smart Homes (SH) with a variety of load profiles set up with photovoltaic (PV) generation, energy storage, and DC demand. The HEMS optimizes the use of local renewable energy and minimizes energy waste caused due to AC and DC conversions and charging and discharging of the storage device. The authors in [10], describe a DR model for a residential community where loads are aggregated and directly dispatched by Load Aggregator (LA). The optimal scheduling is done by the LA.

2.3 Nonlinear Programming Method (NP)

In [11], a technique is proposed to optimize the DR programs based on TOU pricing. The objective of the method is to minimize the electricity costs and maximize load factor. The proposed method motivates the ECs to shift and change their demands according to the price set by the utility. The main objectives of this program are peak-shaving, valley filling, and load-curve smoothing.

2.4 Dynamic Programming Method (DP)

The authors in [12], propose a dynamic pricing scheme that can provide incentives to the ECs for getting an aggregate load profile suitable for UCs. It presents a distributed algorithm implemented as a cooperative game played among the ECs that reduces the total cost and PAR of the system. In [13], dynamic energy management method has been used for residential DR. The dynamic energy management method finds an optimized schedule for controllable appliances and PHEVs using dynamic programming method.

2.5 Convex Programming Method (CP)

In [14], the DR problem is represented through a convex optimization method which is able to achieve near optimal performance with the predecided energy storage capacity. In [15], a RTP scheme is used to reduce the PAR, maximize the payoff of the ECs and the UCs profit while solving the DR problem. It tries to compensate the wholesale market cost based on the real energy consumption by the ECs. In [16], the authors propose models of expenses to UCs emerging from client request uncertainties and models of client satisfaction with the distinction between the declared demand and real utilization. In [17], the DR problem is described as a noncooperative game. The optimization process calculates the schedule one day before the actual implementation. The method proposed in [18], uses a day-ahead DR mechanism synchronized by an autonomous central unit; the ECs are concerned in reducing their financial expenditure by producing or storing energy instead of only buying it. In [19], an immediate load billing scheme has been introduced to motivate the ECs to change their consumption patterns. The method is described through an aggregating game. A distributed control approach has also been considered to solve the same game. In [20], the authors use a game-theoretical approach that provides a decision-making scheme for electricity supplier based on real-time pricing technique. The authors in [21], define a multi objective optimization problem that helps the ECs to minimize their energy cost and the inconveniences caused to them. The energy consumption is optimized to trim down the cost and PAR of the system. The authors in [22], suggest a Mixed Integer Nonlinear programming model using the distributed approach for demand side energy management with a dynamic pricing scheme. The article [23], provides a model of DR problem using a game between a UCs and ECs. A distributed algorithm is developed to determine the NE. The work [24], incorporates a local level DR, integrating the welfare of multiple UCs and a large number of ECs. The system finds optimal production and consumption of electricity for a locality. In [25], a DR model is defined based on a dynamic pricing technique termed as day-ahead hourly pricing (DAHP). In DAHP the prices of electricity for every hour are decided one day before the actual implementation avoiding the price uncertainty. In [26], a real-time information based DR model has been designed that can work with sophisticated communication networks in SG. The method tries to reduce PAR of the entire system. In [27], the authors have provided a distributive DR system based on Bayesian game theory, which can be implemented by an autonomous system. It is assumed that each player has statistical information about the behavior of its opponents so that it can maximize its utility based on that information.

2.6 Heuristic Method (HM)

The article [28], provides a DR model that finds out the consumption model in the presence of dynamic price. A probabilistic Bayesian behavior model is suggested that can learn the energy consumption patterns of shiftable appliances in response to the broadcast prices and a multiple linear regression model is proposed that can forecast the hourly energy consumption of curtailable appliances. The optimization model uses a Genetic Algorithm to find optimal consumption pattern for the EC. The work done in [29], presents an energy management technique that can schedule the time varying appliances of an EC based on their precedence. The comfort levels of the appliances are calculated from the information like the time of operation, their power requirements, and priorities which is used by the system to find an optimal schedule for their operation.

3 A Comparative Analysis

A comparative analysis has been made for the discussed articles and the results of the analysis have been presented in Table 1. The comparison is made on the basis of the following points: (1) Programming paradigm used (2) Objective of the work, (3) Whether it is implemented in a centralized or distributed manner, (4) Pricing tariff used, (5) Area of application, and (6) Limitations of the work.

4 Conclusion

Most of the works studied here use conventional programming techniques like LP and NP methods to resolve the DSM problems. However, it is very difficult to find solutions for a problem involving a large number of controllable devices which have several computation patterns and heuristics using these programming techniques. The computational time taken by the LP methods is also very large for complex problems. Similarly, NP methods require complex computations to get the solution. The DP methods require a large volume of memory if the problem size is very large. On the other hand, heuristic-based algorithms can provide near optimal solution with less computational complexity and time. The heuristic-based methods like Genetic Algorithm, Ant Colony Optimization, and Particle Swarm Optimization can search very large spaces of candidate solutions and find globally optimal solution in polynomial time. As distributed and coordinated intelligence is required at all levels of the electric grid like generation, transmission, and distribution, the computational intelligence mechanisms that include artificial and bioinspired intelligence paradigms can easily learn and adjust to new situations, generalize and abstract the existing situations, and find an association between different situations which, in turn, help to develop effective and robust algorithms for grid management. Some of the works in the literature use a heuristic approach for optimization. However, they have not considered the advantages of using storage systems or Reusable Energy Sources (RES) in the optimization process. So, there is a need for developing some heuristic optimization algorithms that can combine the use of storage systems and RES to support a robust DR system.

Table 1 Comparative analysis of the DR methods

References	Method used	Goal	Approach followed	Pricing tariff	Application area	Limitations
[2, 3]	LP	Reduce peak load, Minimize PAR and daily electricity payments	Single user, Multiple users Centralized, Distributed	Fixed, RTP, TOU, CPP	Residential commercial	Applied for limited number of appliances, storage system not used, users have to share their total consumption information with the UC that limits the practical implementation of the system, pricing method needs advanced communication system between UC and EC
[4–10]	ILP, MILP	Reduce load demand by controlling selected appliances, Minimize peak hourly load	Single user, Multiple users Centralized, Distributed	Fixed, TOUP, RTP	Residential	Computational complexity due to ILP, storage system not used, sharing of total consumption information with the UC that limits the practical implementation of the system
[11]	NP	Reduce total load and maximize load factor calculated as average daily load/peak load	Multiple users Distributed	TOUP	Residential	Computational complexity, storage not used

(continued)

Table 1 (continued)

References	Method used	Goal	Approach followed	Pricing tariff	Application area	Limitations
[12, 13]	DP	Reduce peak demand by reducing consumption	Multiple users Distributed	Multi-TOUP and multi-CPP	Residential	Storage not used, needs advanced automated environment to implement complex price structures, storage not used
[14–27]	CP	Minimize average electricity cost and PAR	Single user, Multiple users Centralized, Distributed	TOUP, RTP, Dynamic	Residential	Needs advanced communication and automation system between UC and EC due to complex pricing method, storage not used. Some has used shared storage that may cause synchronization problem
[28, 29]	HM	forecast hourly energy consumption of curtailable appliances, reduce PAR, total energy cost	Single user, Multiple users Distributed	IBR, RTP, Dynamic	Residential	Needs advanced communication and automation system between UC and EC due to distributed approach and price broadcasting, storage not used

References

1. Mohsenian-Rad, A., Wong, V., Jatskevich, J., Schober, R., Leon-Garcia, L.: Autonomous demand-side management based on game-theoretic energy consumption scheduling for the future smart grid. IEEE Trans. Smart Grid **1**(3), 320–331 (2010)
2. Chen, X., Wei, T., Hu, S.: Uncertainty-aware household appliance scheduling considering dynamic electricity pricing in smart home. IEEE Trans. Smart Grid **4**(2), 932–941 (2013)
3. Abushnaf, J., Rassau, A., Lodzimierz, W.: Impact of dynamic energy pricing schemes on a novel multi-user home energy management system. Electric Power Syst. Res. **125**, 124–132 (2015)
4. Parvania, M., Fotuhi-Firuzabad, M.: Demand response scheduling by stochastic SCUC. IEEE Trans. Smart Grid **1**(1), 89–98 (2010)
5. Zhu, Z., Tang, J., Lambotharan, S., Chin, W.H., Fan, Z.: An integer linear programming and game theory based optimization for demand-side management in smart grid. In: 2011 IEEE GLOBECOM Work-shops (GC Wkshps), pp. 1205–1210. IEEE, USA (2011)
6. Hubert, T., Grijalva, S.: Modeling for residential electricity optimization in dynamic pricing environments. IEEE Trans. Smart Grid **3**(4), 2224–2231 (2012)
7. Ozturk, Y., Kumar, D.S., Kumar, S., Lee, G.: An intelligent home energy management system to improve demand response. IEEE Trans. Smart Grid **4**(2), 694–701 (2013)
8. Adika, C., Wang, L.: Autonomous appliance scheduling for household energy management. IEEE Trans. Smart Grid **5**(2), 673–682 (2014)
9. Zhao, C., Dong, S., Li, F., Song, Y.: Optimal home energy management system with mixed types of loads. CSEE J. Power Energy Syst. **1**(4), 29–37 (2015)
10. Nan, S., Zhou, M., Li, G.: Optimal residential community demand response scheduling in smart grid. Appl. Energy **210**, 1280–1289 (2018)
11. Farhadi, P., Taheri, B.: Smart meter tariff pricing using load demand response model. In: Smart Grid and Cities Congress and Fair (ICSG), 5th International Conference 2017, pp. 135–139. IEEE, Istanbul (2017)
12. Caron, S., Kesidis, G.: Incentive-based energy consumption scheduling algorithms for the smart grid. In: Smart Grid Communications (SmartGridComm), First IEEE International Conference 2010, pp. 391–396. IEEE, USA (2010)
13. Muratori, M., Rizzoni, G.: Residential demand response: dynamic energy management and time-varying electricity pricing. IEEE Trans. Power Syst. **31**(2), 1108–1117 (2016)
14. Huang, L., Walrand, J., Ramchandran, K.: Optimal demand response with energy storage management. In: IEEE Third International Conference on Smart Grid Communications (SmartGridComm) 2012, pp. 61–66. IEEE, Taiwan (2012)
15. Ping Qian, L., Zhang, Y., Huang, J., Wu, Y.: Demand response management via real-time electricity price control in smart grids. IEEE J. Sel. Areas Commun. **31**(7), 1268–1280 (2013)
16. Yang, P., Tang, G., Nehorai, A.: A game-theoretic approach for optimal time-of-use electricity pricing. IEEE Trans. Power Syst. **28**(2), 884–892 (2013)
17. Atzeni, I., Garcia, L., Scutari, G., Palomar, D., Fonollosa, J.R.: Non-cooperative and cooperative optimization of distributed energy generation and storage in the demand-side of the smart grid. IEEE Trans. Signal Process. **61**(10), 2454–2472 (2013)
18. Atzeni, I., Ordóñez, L., Scutari, G., Palomar, D., Fonollosa, J.R.: Demand-side management via distributed energy generation and storage optimization. IEEE Trans. Smart Grid **4**(2), 866–876 (2013)
19. Chen, H., Li, Y., Louie, R.H., Vucetic, B.: Autonomous demand side management based on energy consumption scheduling and instantaneous load billing: an aggregative game approach. IEEE Trans. Smart Grid **5**(4), 1744–1754 (2014)
20. Bu, S., Richard Yu, F.: A game-theoretical scheme in the smart grid with demand-side management: towards a smart cyber-physical power infrastructure. IEEE Trans. Emerging Top. Comput. **1**(1), 22–32 (2013)

21. Liu, Y., Yuen, C., Huang, S., Hassan, N., Wang, X., Xie, S.: Peak-to-average ratio constrained demand-side management with consumer's preference in residential smart grid. IEEE J. Sel. Top. Signal Process. **8**(6), 1084–1097 (2014)
22. Barbato, A., Capone, A., Chen, L., Martignon, F., Paris, S.: A distributed demand-side management framework for the smart grid. Comput. Commun. **57**, 13–24 (2015)
23. Kamyab, F., Amini, M., Sheykhha, S., Hasanpour, M., Jalali, M.M.: Demand response program in smart grid using supply function bidding mechanism. IEEE Trans. Smart Grid **7**(3), 1277–1284 (2016)
24. Maharjan, S., Zhu, Q., Zhang, Y., Gjessing, S., Basar, T.: Demand response management in the smart grid in a large population regime. IEEE Trans. Smart Grid **7**(1), 189–199 (2016)
25. Jia, L., Tong, L.: Dynamic pricing and distributed energy management for demand response. IEEE Trans. Smart Grid **7**(2), 1128–1136 (2016)
26. Ye, F., Qian, Y., Hu, R.Q.: A real-time information based demand-side management system in smart grid. IEEE Trans. Parallel Distrib. Syst. **27**(2), 329–339 (2016)
27. Ininahazwe, H., Muriithi, C.M., Kamau, S.: Optimal demand-side management for smart micro grid with storage. J. Power and Energy Eng. **6**(02), 38 (2018)
28. Mnatsakanyan, A., Kennedy, S.: A novel demand response model with an application for a virtual power plant. IEEE Trans. Smart Grid **6**(1), 230–237 (2015)
29. Khan, A., Javaid, N., Khan, M.I.: Time and device based priority induced comfort management in smart home within the consumer budget limitation. Sustain. Cities Soc. **41**, 538–555 (2018)

Energy Efficient Chrip Signal Using Stockwell Transform

Subhashree Sahoo, Sushree Bibhuprada B. Priyadarshini, Amiya Bhusan Bagjadab, and Santosh Kumar Majhi

Abstract In this document, we are studying a technique of time-frequency analysis called Synchrosqueezing Stockwell Transform (SSST) to satisfy the processing and interpretation of high resolution signals. The SSST technique squeezes and reconstructs the outcomes of complicated spectra coefficients along the direction of frequency. In order to concentrate the energy distribution on the time-frequency spectrum throughout the actual instantaneous frequency of the signal, and to improve the corresponding time-frequency resolution is the aim. The experimental results performed on chirp signals depict that the proposed method can correctly give a better energy concentration around the instantaneous frequency with less smearing of energy.

Keywords Chirp signal · Instantaneous properties · synchrosqueezed transform (SST) · *S*-Transform · Synchrosqueezing stockwell transform (SSST)

1 Introduction

Efficient strategies for analysis and parameter estimation of chirp signals are getting increasingly vital in sonar, geology, biomedical and radio detection, and ranging applications. Analyzing Instantaneous properties (e.g., instantaneous frequencies,

S. Sahoo
Pondicherry University, Kalapet, Pondicherry, India
e-mail: Ssubhashree46@gmail.com

S. B. B. Priyadarshini
Institute of Technical Education and Research, Siksha 'O' Anusandhan (Deemed to be University), Bhubaneswar, Odisha 751030, India
e-mail: bimalabibhuprada@gmail.com

A. B. Bagjadab (✉) · S. K. Majhi
Veer Surendra Sai University of Technology, Burla, India
e-mail: amiya7bhusan72@gmail.com

S. K. Majhi
e-mail: smajhi_cse@vssut.ac.in

D. Mishra et al. (eds.), *Intelligent and Cloud Computing*,
Smart Innovation, Systems and Technologies 194,
https://doi.org/10.1007/978-981-15-5971-6_56

instantaneous amplitudes, and instantaneous phases), the native properties of signals have additionally been a crucial topic for over twenty years. A chirp may be a signal with increasing frequency (up-chirp) or decreasing frequency (down-chirp) over time.

Signal processing involves analyzing, synthesizing, and modifying signals that are widely described as features that convey information about the conduct or characteristics of certain creations such as noise and pictures. Keep in mind that no data can be transmitted by a purely monochromatic signal $a\cos(\omega_0 t + \theta)$. A modulation is required for this purpose, and one of the feasible methodologies is to implement modulation of amplitude. Suppose $m(t)$ be a positive function that corresponds to the transmitted data. We tend to obtain the signal $x(t)$ by multiplying the carrier frequency signal $\cos\omega_0 t$ through $m(t)$. The Fourier Transform is based on the traditional technique of characterizing signal characteristics. It is possible to determine the energy distribution of the signal by evaluating its spectrum. Fourier spectrum assessment worldwide describes standardized harmonic elements, and therefore, requires numerous extra harmonic elements to simulate nonstationary information. The instantaneous frequency could be a helpful idea for describing non-monochromatic signal [1–7]. It is defined because of the temporal derivative of the oscillations phase θ. Distinction to Fourier frequency the instantaneous frequency is mostly time-dependent frequency. It is defined as

$$v(t) = \frac{1}{2\pi}\frac{\mathrm{d}\phi}{\mathrm{d}t} \tag{1}$$

i.e., basically as a temporary stage oscillation derivative. One would have the Instant angular rate without the factor $1/2\pi$. The instantaneous frequency of a sinusoidal signal is constant and equals the expected oscillation frequency [7]. In the context of frequency noise and phase noise, the thought of the instantaneous frequency is particularly necessary; however, it is applied in conjunction with chirped optical pulses with a time-dependent instantaneous frequency.

Fourier spectrum of oscillating signal doesn't give probability distribution of instantaneous frequencies. The relation in between Instantaneous frequency and Fourier frequency is considerably a lot of refined. Let $m(t)$ be a positive function that matches the transmitted data. The signal is obtained by multiplying the frequency signal of the carrier $\cos\omega_0 t$ by $m(t)$. $x(t) = m(t)\cos\varphi(t)$ And it is frequently noted that $m(t)$ is the $x(t)$ signal's instantaneous amplitude. On the contrary, the need to modulate phase or frequency needs the Instant Phase definition. By reasoning on the amplitude comparable to the earlier one, the signal is frequently noted. $x(t) = a\cos\omega_0(t)$ has a steady amplitude and an instantaneous phase has an instantaneous amplitude. Let $x(t)$ be a real signal and the associated canon pair $[\theta(t), \varphi(t)]$. The feature $a(t)$ appearing in this couple is $x(t)$ stage $\pi(t)$ instantaneous amplitude. This is the Instant Phase. The Instant Frequency is the time derivative of $f(t)$.

2 Related Work

The short-term Fourier Transform (STFT) and Continuous Wavelet Transform (CWT) are used intensively to evaluate and process multi-component signals, that is, modulated wave superposition. The Synchrosqueezing could be a post-processing technique by reassigning the coefficients in scale or frequency to circumvent the uncertainty relationships intrinsic in these linear transformations. Synchrosqueezing is an algorithm for the assessment of time-frequency signals. It is intended to decompose signals with time-varying oscillatory characteristics into their constituent parts. The objective of Synchrosqueezing Transform (SST) is to provide a more concentrated display of multi-component signals in the time-frequency plane. SST also offers a signal decomposition technique that allows distinct waves to be separated and demodulated. In the context of audio signals, SST was initially implemented. To analyze signals such as Short-Time Fourier Transform (STFT), Continuous Wavelet Transform (CWT), there are many time-frequency transforms [1]. The Short-Time Fourier Transform, as well as the Wavelet Transform, are the main methods to concur recently split a signal into time and frequency components. Synchrosqueezing is linked to the reassignment algorithm time-frequency class [8]. It is used to estimate the Instantaneous frequency of signal from its time-frequency representation. Time-frequency reassignment method was originally introduced from the study of STFT, which studies the Instantaneous frequency around their center frequency in the spectrogram diagram [6]. Synchrosqueezing Transform is a powerful tool for analyzing the signal. There are various methods of Synchrosqueezing transform (SST) which has been approached such as Fourier synchrosqueezed transform (FSST) and Wavelet synchrosqueezed transform (WSST). Synchrosqueezing Transform (SST) is an extension of the Wavelet Transforms that breaks down a signal into components that are well separated in the time-frequency plane enabling these components to be reconstructed [2, 3]. Synchrosqueezing is an algorithm for the evaluation of time-frequency signals. It is intended to decompose signals with time-varying oscillatory characteristics into their constituent parts. The aim of Synchrosqueezing Transform (SST) is to provide a more concentrated, time-frequency plane representation of multi-component signals. SST also offers a signal decomposition technique that allows distinct waves to be separated and demodulated [5]. In the context of audio signals, SST was initially implemented. To analyze signals such as Short-Time Fourier Transform (STFT), Continuous Wavelet Transform (CWT), there are many time-frequency transforms. Figures 1 and 2 show the linear and hyperbolic scenarios, respectively.

Synchrosqueezing Transform is a powerful tool for analyzing the signal. There are various methods of Synchrosqueezing transform (SST) which has been approached such as Fourier synchrosqueezed transform (FSST) and Wavelet Synchrosquee Transform (WSST). Synchrosqueezing Transform (SST) represents an extension of the Wavelet Transform that decomposes a signal into components that are well segregated in the time-frequency plane allowing the reconstruction of such components. [6, 9] SST is a well-defined time-frequency representation that allows the

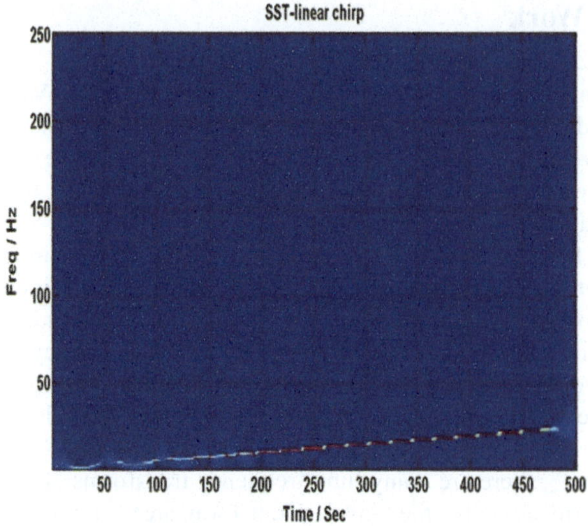

Fig. 1 synchrosqueezed transform for linear chirp signal

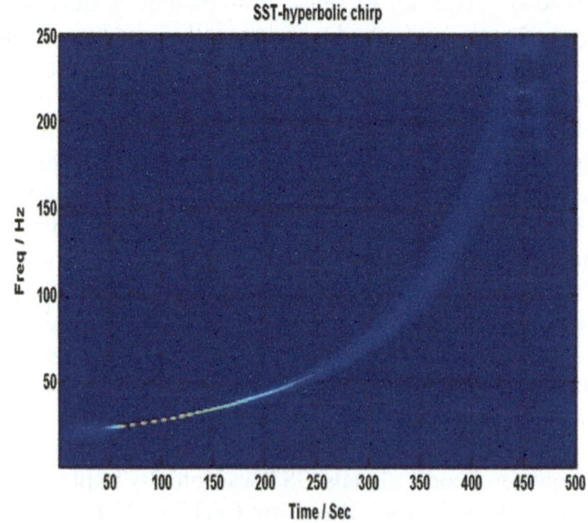

Fig. 2 synchrosqueezed transform for hyperbolic chirp signal

identification of instantaneous frequencies. Signals $f(t)$ have the general form as follows:

$$f(t) = \sum_{k=1}^{k} f_k(t) + e(t) \tag{2}$$

where each component of

$$f_k(t) = A_k(t)(\cos 2\pi \varphi_k(t)) \tag{3}$$

is a Fourier like oscillatory mode with time-varying amplitude and frequency.

$$e(t) = \text{measurement error}$$

The SST was so originally introduced within the context of Wavelet analysis, and it's not been entirely custom-made to the STFT. For instance, proposes an SST-like decomposition supported the STFT that comes with an approximation result, however, this remodel doesn't enable the reconstruction of the modes that are nevertheless one amongst the most characteristics of the original SST. A lot of problematic, the assumptions on the modes are the world, whereas the STFT and therefore the CWT are native transforms [2, 8]. Time-frequency depictions provide a significant tool for time-series signal analysis. They will give insight into the complex structure of a "multi-layered" signal that consists of many components, like the distinct phonemes during voice auditory communication, or a measuring device signal and its delayed echo. There are many types of time-frequency (TF) analysis methods; either "linear" or "quadratic" approaches belong to the overwhelming majority [5, 10].

As such, we are proposing a generalized strategy to transform Synchrosqueezing to address diffusions in both time and frequency aspects. We have shown that the wavelet diffusion only happens at the frequency dimension for the signal with a steady frequency. Based on this observation, to promote synchrosqueezing, the initial signal with instantaneous frequency varying in time is mapped to another analytical signal with steady frequency. In order to achieve a time-frequency representation with a focused wavelet ridge, a time-scale domain restore operation is then provided.

3 Stockwell Transform

The *S*-Transform presented in the current correspondence is the associated expansion of the Continuous Wavelet Transform (CWT) ideas and is based on a moving and ascendable Gaussian window locating. The *S*-Transform's advantages are because of the reality that the modulating sinusoids get installed on the time axis with significance, while the locating ascendable Gaussian window dilates and interprets [7].

For its indigenous spectral component properties, the S-transform could be a time-frequency illustration far-famous. One drawback to the S-transform is that it shares with the Continuous Wavelet Transform, the short-time Fourier Transform, and the redundant illustration of the time-frequency region, and therefore, the consumption of computing resources characteristic [5, 9–13]. It's a Stockwell time-frequency distribution that combines STFT and WT's excellent characteristics. S-Transform can be defined as below

$$S(\tau, f, p) = \int_{-\infty}^{+\infty} h(t) * \omega(\tau - t, f, p) \mathrm{e}^{-j2\pi f t} \mathrm{d}t \tag{4}$$

where "p" denotes the set of parameters that estimate the shape and characteristics of "w" and it denotes the window function of the S-Transform, i.e.

$$\omega(t, f, p) = \frac{\|f\|}{\sqrt{2\pi\rho}} \mathrm{e}^{\frac{-t^2 f^2}{2p^2}} \tag{5}$$

Another way to defined S-Transform is the

$$S_x(t, f) = \int_{-\infty}^{+\infty} x(\tau) \|f\| \mathrm{e}^{-\pi} (t - \tau)^2 f^2 \mathrm{e}^{-j2\pi\tau} \mathrm{d}\tau \tag{6}$$

It is one of the most simple and reversible figures current in the mapping of the integer. S-Transform is well-known for its indigenous spectral section characteristics as a time-frequency distribution. The S-Transform's primary property is that it unambiguously aggregates a time-frequency space frequency-dependent resolution.

S-transform delivers better signal clarity compared to Gabor transform, but the clarity is worse than Wigner's S-Transform distribution function works well in frequency domain assessment when the frequency of input is small. S-transform has greater clarity in the time domain when the input frequency is high.

4 Problem Statement

The following problems associated with the S-Transform analysis of the signals. Some of them are discussed below

- SST is a very efficient algorithm for giving better energy concentration along the direction of Instantaneous frequency of signal.
- S-Transform provides the time-frequency representation of a signal. But the spectrogram analysis of S-Transform shows that there is heavy smearing of energy in the time-frequency domain of the signal.

- S-Transform representation of the signal results in the bad quality representation of the signal around its Instantaneous Frequency.
- Energy of the signal is not conservative around Instantaneous Frequency.

5 Proposed Approach

In our work, we have proposed a Synchrosqueezed Stockwell transform approach to overcome the problems associated with Stockwell Transform. Further, the above signal is needed for the analysis of S-Transform based Synchrosqueezing Transform. SSST provides better signal representation than S-Transform analysis. The SST was so originally introduced within the context of Wavelet analysis, and it's not been entirely custom-made to the STFT. For instance, proposes an SST-like decomposition supported the STFT that comes with an approximation result, however, this remodel doesn't enable the reconstruction of the modes that are nevertheless one amongst the most characteristics of the original SST. A lot of problematic, the assumptions on the modes are the world, whereas the STFT, and therefore, the CWT are native transforms. Time-frequency depictions afford a powerful tool for time-series signal analysis. They will provide insight into the complex structure of a "multi-layered" signal that consists of many components, like the different phonemes during voice auditory communication, or a measuring device signal and its delayed echo. There are various categories of time-frequency (TF) analysis algorithms; either "linear" or "quadratic" approaches belong to the overwhelming majority.

As such, we are proposing a generalized strategy to transform Synchrosqueezing to address diffusions in both time and frequency aspects. It has been illustrated that the wavelet diffusion only happens at the frequency dimension for the signal with a steady frequency. Based on this observation, to promote synchronous squeezing, the initial signal with instantaneous frequency varying in time is mapped to another analytical signal with steady frequency. In order to achieve a time-frequency representation with a focused wavelet ridge, a time-scale domain restore operation is then provided [7]. The S-transform presented in this correspondence is the associated expansion of the Continuous Wavelet Transform (CWT) ideas and is based on a moving and ascending Gaussian window locating them. It is shown to have some intriguing features that are absent from the ongoing Transform wavelet. It offers frequency-dependent resolution while preserving an instant connection with the Fourier spectrum, the S-Transform is exclusive therein. The S-Transform's advantages are due to the reality that the modulating sinusoids are installed on the time axis with significance, while the locating ascendable Gaussian window dilates and interprets [11]. For its indigenous spectral component characteristics, the S-transform could be a time-frequency illustration far-famous.

There are various methods of synchrosqueezed transform (SST) which have been approached such as FSST, WSST. Here we study SSST for analyzing the signals. SSST is proffered to meet the high resolution signal processing and interpretation. Here we have studied the new algorithm SSST to improve the signal quality. It is already known that SST is a very efficient algorithm for giving better energy concentration along the direction of Instantaneous frequency of signal. *S*-Transform provides the time-frequency representation of a signal. But the spectrogram analysis of *S*-Transform shows that there is heavy smearing of energy in the time-frequency domain of the signal. To reduce the smearing of energy and to provide a better representation of the signal we introduce a new transform, i.e., Synchrosqueezed *S*-Transform which yields less smearing of energy and gives a well concentrated signal along its Instantaneous frequency in time domain. From Fig. 1, we can conclude that SST of Linear chirp signal gives a better representation of the signal around its instantaneous frequency in its time-frequency domain. But when the same signal is applied in the *S*-Transform (Fig. 3), it results in the smearing of energy around Instantaneous frequency. While the same signal yields better representation around its instantaneous frequency with less smearing of energy in Synchrosqueezed Stockwell Transform (Fig. 5).

Similarly, in Fig. 2, we can conclude that SST of Hyperbolic chirp signal gives a better representation of the signal around its instantaneous frequency in its time-frequency domain. But when the same signal is applied in the *S*-Transform (Fig. 4), it results in the smearing of energy around Instantaneous frequency. While the same signal yields better representation around its instantaneous frequency with less smearing of energy in Synchrosqueezed Stockwell Transform (Figs. 5 and 6).

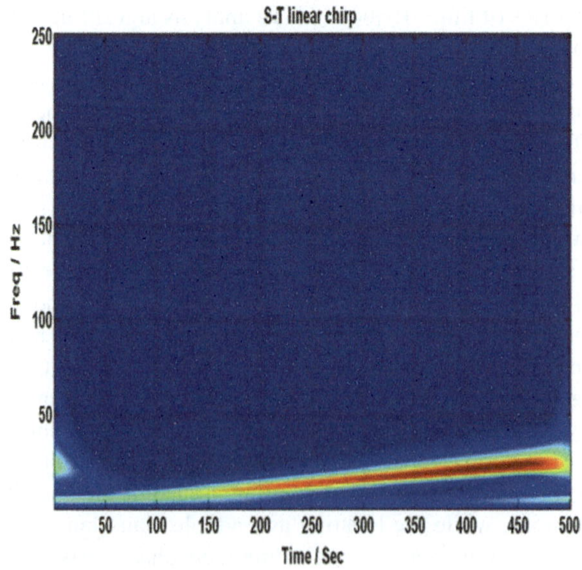

Fig. 3 Stockwell transform for linear chirp signal

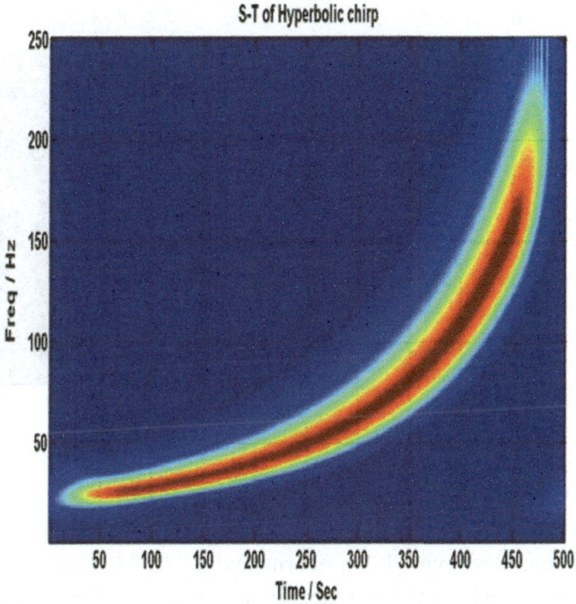

Fig. 4 Stockwell transform for hyperbolic chirp signal

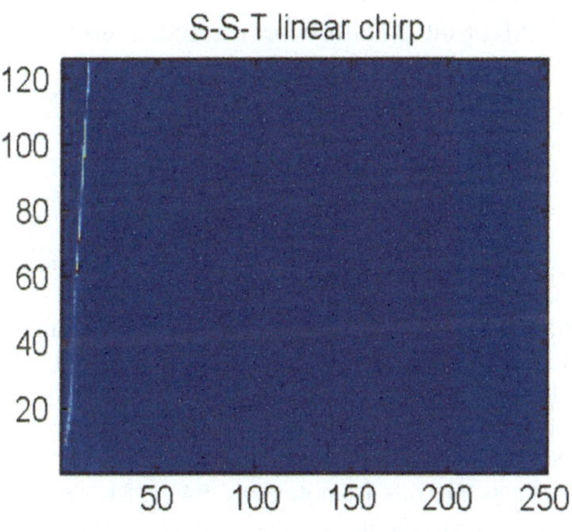

Fig. 5 Synchrosqueezed Stockwell transform for linear chirp signal

6 Conclusion

In this article, we have discussed the proposed research context S-Transform based Synchrosqueezing Transform (SSST) investigated some of the factors that are responsible for the quality degradation and lower signal resolution around its instantaneous frequency of the signal in time-frequency domain. We have discussed our proposed

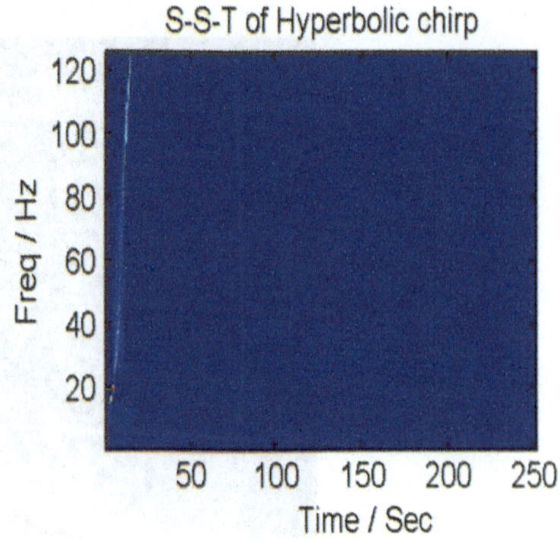

Fig. 6 Synchrosqueezed Stockwell Transform for Hyperbolic chirp signal

approach S-Transform based Synchrosqueezing Transform (SSST) able to overcome all the problems associated with S-Transform analysis of the Chirp signal. This allows us to conclude that SSST provides better signal resolution than *S*-Transform. These signals should deal with the above transforms to get better energy concentration around Instantaneous frequency, better resolution, and less smearing of energies. So the quality of the signal remains intact during the analysis of SSST than *S*-Transform analysis. Furthermore, information can be extracted from these chirp signals.

References

1. Boashash, B.: Estimating and interpreting the instantaneous frequency of a signal Algorithms and applications. Proc. IEEE **80**(4), 540–568 (1992)
2. Daubechies, I., Jianfeng, L., HT, W.: Synchrosqueezed wavelet transforms: an empirical mode decomposition-like tool. Appl. Comput. Harmon. Anal. **30**(2), 243–261 (2011)
3. Herrera, R.H., Han, J., van der Baan, M.: Applications of the synchrosqueezing transform in seismic time-frequency analysis. Geophysics **79**(3) (2014)
4. Huang, Z., et al.: Synchrosqueezing *S*-transform and its application in seismic spectral decomposition. IEEE Trans. Geosci. Remote Sens. **54**(2), 817–825 (2016)
5. Jabloun, M., et al.: Estimation of the instantaneous amplitude and frequency of non-stationary short-time signals. Sig. Process. **88**(7), 1636–1655 (2008)
6. Daubechies, I., Wang, Y.G., Wu, H.T.: ConceFT: concentration of frequency and time via a multitapered synchrosqueezed transform. Phil. Trans. R. Soc. A (2016)
7. Li, C., Liang, M.: A generalized synchrosqueezing transform for enhancing signal time–frequency representation. Sig. Process. **92**(9), 2264–2274 (2012)
8. Auger, F., et al.: Time-frequency reassignment and synchrosqueezing: an overview. IEEE Signal Process. Mag. **30**(6), 32–41 (2013)
9. Daubechies, I., Lu, J., Wu, H.T.: Synchrosqueezed wavelet transforms: An empirical mode decomposition-like tool. Appl. Comput. Harmon. Anal. **30**(2), 243–261 (2011)
10. Stockwell, R.G., Mansinha, L., Lowe, P.R.: Localization of the complex spectrum: the *S* transform. IEEE Trans. Signal Process. **44**(4), 998–1001 (1996)

11. Stockwell, R.G., Mansinha, L., Lowe, R.P.: Localization of the complex spectrum: the *S* transform. IEEE Trans. Signal Process. **44**(4), 998–1001 (1996)
12. Picinbono, B.: Instantaneous frequency of a signal, time-frequency analysis: concepts and methods 37–60
13. Picinbono, B.: On instantaneous amplitude and phase of signals. IEEE Trans. Signal Process. **45**(3), 552–560 (1997)

Harmonic and Reactive Power Compensation Using Hybrid Shunt Active Power Filter with Fuzzy Controller

Alok Kumar Mishra, Prakash Kumar Ray, Akshaya Kumar Patra, Ranjan Kumar Mallick, Soumya Ranjan Das, and Ramachandra Agrawal

Abstract This paper presents a Fuzzy controlled Hybrid Shunt Active Power Filter (HSAPF) used for reactive power and harmonic compensation. Comparing Passive Filter with Active Filter, the former is being bulky and its design is complex, and the later is not cost-effective, hence a hybrid structure of shunt active and shunt passive filter is considered. MATLAB/Simulink models of HSAPF are developed to get a low distorted and in phase source current waveform at the input side. The compensation process for shunt active filter is different from predictable methods such as $(p - q)$ theory or $(i_d - i_q)$ theory, in which only the source current is to be sensed. The proposed system is tested at both steady state and transient condition, and its performance is then estimated in terms of various parameters like Total Harmonic Distortion (THD), Input Power Factor (IPF), percentage overshoot for Fuzzy Logic Controller (FLC) used for reference current estimation and all the obtained results are compared with Proportional Integral Derivative (PID) Controller used for reference current estimation.

A. K. Mishra (✉) · A. K. Patra · R. K. Mallick · R. Agrawal
EEE Department, ITER, S'O'A (Deemed to be University), Bhubaneswar 751030, India
e-mail: malok2010@gmail.com

A. K. Patra
e-mail: hiakp@yahoo.com

R. K. Mallick
e-mail: ranjanmallick@soa.ac.in

R. Agrawal
e-mail: ramachandra1agrawal@gmail.com

P. K. Ray
EE Department, College of Engineering and Technology, Bhubaneswar 751003, India
e-mail: pkrayiiit@gmail.com

S. R. Das
EE Department, IIIT, International Institute of Information Technology, Bhubaneswar 751003, India
e-mail: srdas1984@gmail.com

D. Mishra et al. (eds.), *Intelligent and Cloud Computing*,
Smart Innovation, Systems and Technologies 194,
https://doi.org/10.1007/978-981-15-5971-6_57

Keywords PID · FLC · Reactive power · Harmonic compensation · THD · Power quality

1 Introduction

Nowadays in almost all applications, power electronics-based loads are being used such as UPS, SMPS, ASD, etc. All of these power electronics-based equipment draw the current from the supply which is of nonlinear in nature and consists of lots of harmonic [1]. Due to this nonlinear current, the utility system has to supply a quite large amount of reactive power from the supply, as the current drawn from the supply is not at all sinusoidal and the source voltage and current are not in phase, again the Point of Common Coupling (PCC) gets affected due to the same [2–4]. To mitigate these harmonic issues traditionally passive filters are being used [5]. Fixed compensation with resonance problem is the demerits of Passive filters, which are being overcome by a power electronic-based device called active filter [6]. Active filters are known to be costly for higher rating, hence a low rated SAPF and a low-cost Shunt Passive Filter (SPF) called HSAPF is formed for a three-phase three wire system [7]. To estimate the reference current for the SAPF, many control techniques such as instantaneous power theory (p-q) or synchronous reference frame theory $(i_d - i_q)$ are available in the literature [8]. The present paper is focused on a novel control method, to estimate the current references which do not require the load reactive volt-ampere [9]. To control the inverter dc voltage of the capacitor and to extract the current references, Fuzzy controller is applied. Various simulations are performed and results are analyzed both for PID and FLC of the proposed HSAPF system [10].

2 HSAPF System

SAPF provides a better solution for the harmonic mitigation. However, shunt active filtering is costly. So, in many applications even these days passive filters are being used for providing a cost-effective solution for harmonic mitigation. To construct the SPF resistors, inductors, and capacitors are used which makes them massive. Hence, HSAPF is created by low rated and low-cost SAPF and SPF, respectively, which in turn, removes the demerits of pure passive and pure active filters, while the advantages of both remain intact. In this paper, a HSAPF is formed by the use of a low rated three-phase voltage source PWM converter and a shunt connected LC passive filter, for designing the passive filters the reactive power requirement is taken as 2 KVAR which is divided as 1000 VAR for single tuned 5th harmonic filter and 500 VAR each for single tuned 7th harmonic filter and 11th and 13th double tuned filter as shown in Fig. 1.

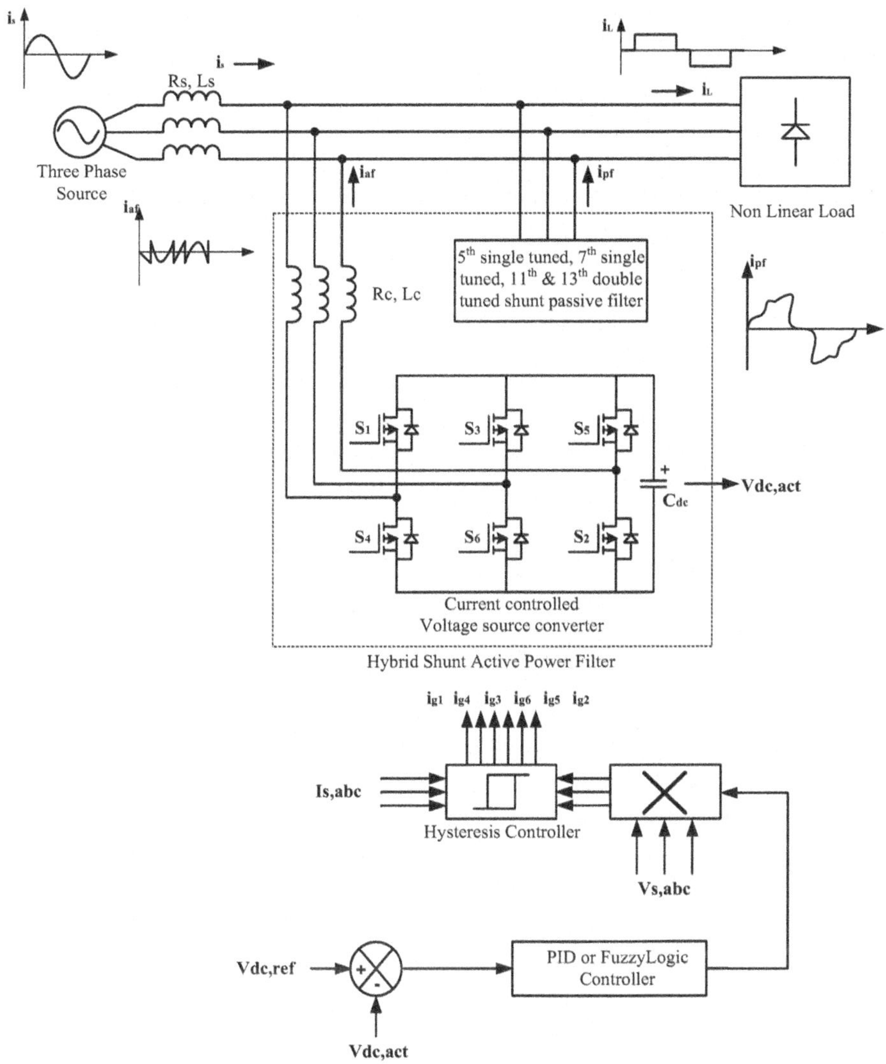

Fig. 1 HSAPF system

3 Control Algorithms

At PCC equal magnitude with opposite polarity harmonic current has to be injected in order to accomplish harmonic compensation, which in turn, the distribution systems power quality gets improved by canceling the original distortion.

The instantaneous source current can be represented as (from Fig. 1)

$$i_s(t) = i_L(t) - i_{af}(t) \tag{1}$$

The source voltage is represented as

$$V_s(t) = V_m \sin \omega t \tag{2}$$

In case if the load is nonlinear, the load current is having harmonic component, as well as fundamental component, that can be represented as

$$\begin{aligned} i_L(t) &= \sum_{n=1}^{\infty} I_n \sin(n\omega t + \varphi_n) \\ &= I_1 \sin(\omega t + \varphi_1) + \left(\sum_{n=2}^{\infty} I_n \sin(n\omega t + \varphi_n) \right) \end{aligned} \tag{3}$$

The instantaneous load power can be calculated as

$$\begin{aligned} p_L(t) &= v_s(t) * i_L(t) \\ &= V_m I_1 \sin^2 \omega t * \cos \varphi_1 + V_m I_1 \sin \omega t * \cos \omega t * \sin \varphi_1 \\ &\quad + V_m \sin \omega t * \left(\sum_{n=2}^{\infty} I_n \sin(n\omega t + \varphi_n) \right) \end{aligned} \tag{4}$$

$$p_L(t) = p_f(t) + p_r(t) + p_h(t). \tag{5}$$

This load power is having harmonic, reactive, and active power. From (4), the real power drawn from the load is represented as

$$p_f(t) = V_m I_1 \sin^2 \omega t * \cos \varphi_1 = v_s(t) * i_s(t). \tag{6}$$

If the active filter gives the harmonic and reactive power, then the source current $i_s(t)$ will be pure sinusoidal and in phase with the source voltage. The three-phase source currents after compensation can be expressed as

$$i_{sa}^*(t) = \frac{p_f(t)}{v_s(t)} = I_1 \cos \varphi_1 \sin \omega t = I_{max} \sin \omega t. \tag{7}$$

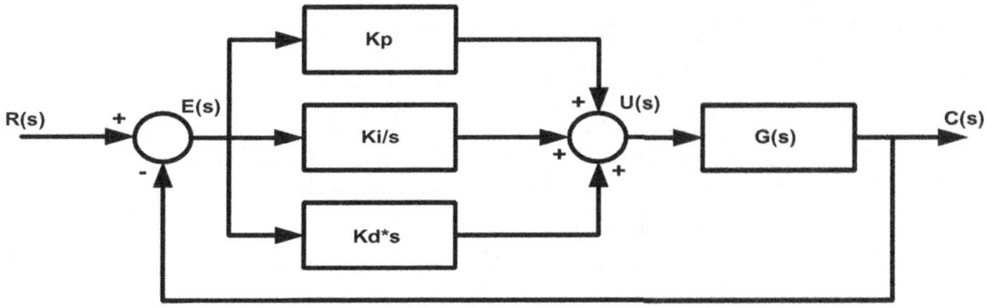

Fig. 2 PID controller basic structure

where $I_{max} = I_1 \cos \varphi_1$.

Similarly,

$$i_{sb}^{*}(t) = I_{max} \sin(\omega t - 120°). \quad (8)$$

$$i_{sc}^{*}(t) = I_{max} \sin(\omega t + 120°). \quad (9)$$

Fuzzy or PID Controller is used to regulate the inverter DC-link capacitor voltage and to calculate approximately this I_{max}(current references crest value).

4 PID Controller for Reference Current Estimation

Figure 2 represents a PID control system. The Proportional, Integral, and Derivative actions are applied to *e*(*t*) (error signal) in a PID controller, again which is applied to the plant model. Mathematically PID controller is represented in (10).

$$u(t) = K_p e(t) + K_i \int_0^t e(t) + K_d \frac{de(t)}{dt} \quad (10)$$

5 FLC for Reference Current Estimation

Proposed FLC used for Reference Current Estimation is shown in Fig. 3.

Any object has a degree of membership which varies from 0 to 1 in the fuzzy set theory, which makes fuzzy logic to handle all situations in a natural way. The shape of membership functions can also be trapezoidal or Gaussian, etc., depending on the applications and can be symmetrical or asymmetrical. The power converter voltage

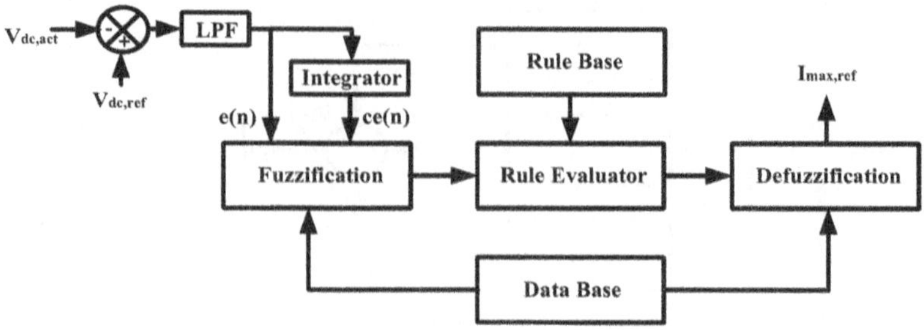

Fig. 3 FLC for reference current estimation

error is defined in linguistic variables such as Positive Big Error (PBE), Negative Big Error (NBE), Positive Medium Error (PME), Negative Medium Error (NME), Positive Small Error (PSE), Negative Small Error (NSE), Zero Error (ZEE), where each variable is defined by a gradually varying triangular membership function. For error (*e*) and change in error (*de*), seven fuzzy sets are chosen. For a particular application depending on the input resolution, the number of fuzzy levels is selected which is not fixed. Higher will be the input resolution more has to be the number of fuzzy levels. Each input is assigned a membership function μ which is not quantized in the classical sense to each fuzzy set. To diminish the difficulty in calculation here, triangular fuzzy set values are used. We may also use bell-shaped or trapezoidal fuzzy set depending on the application as shown in Fig. 4. For n number of variable, the fuzzy rules are n-dimensional. The sum of rules is known as rule *R*. FIS editor edits the input and output variables, which are *e, de,* and output. After editing, we have to design the membership function for each variable. The final step involves writing rules in rule editor using the rule given in Table 1.

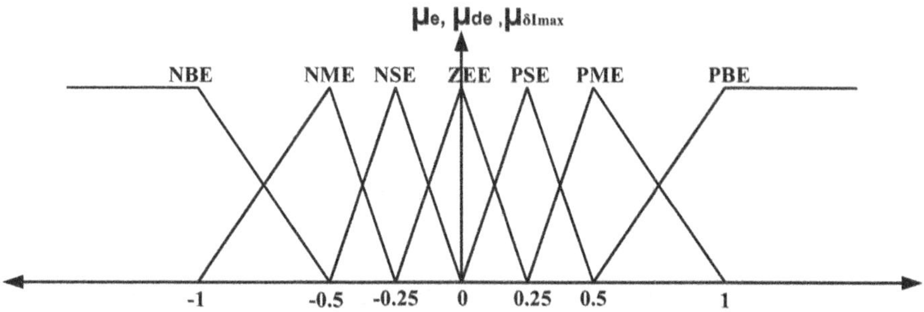

Fig. 4 Membership function used in fuzzification

Table 1 Fuzzy rules for closed loop control of voltage source converter

*e**de*	NBE	NME	NSE	ZEE	PSE	PME	PBE
NBE	NBE	NBE	NBE	NBE	NME	NSE	ZEE
NME	NBE	NBE	NBE	NME	NSE	ZEE	PSE
NSE	NBE	NBE	NME	NSE	ZEE	PSE	PME
ZEE	NBE	NME	NSE	ZEE	PSE	PME	PBE
PSE	NME	NSE	ZEE	PSE	PME	PBE	PBE
PME	NSE	ZEE	PSE	PME	PBE	PBE	PBE
PBE	ZEE	PSE	PME	PME	PBE	PBE	PBE

6 Results and Discussion

To investigate the performance of the proposed system a simulink model of a HSAPF System for reactive power and harmonic compensation is developed as shown in Fig. 5. For nonlinear load selection diode bridge rectifier with *R* load on its dc side is taken for which filtering is required. The designed parameter used in the simulation is given in Table 2. The system performance is investigated in four steps. Step: 1 When HSAPF is not connected to the system. Step: 2 When SPF is connected alone to the system. Step: 3 When HSAPF is connected to the system with a PID controller used for reference current estimation. Step: 4 when HSAPF is coupled to the distribution system and to estimate the reference current FLC is used. The load

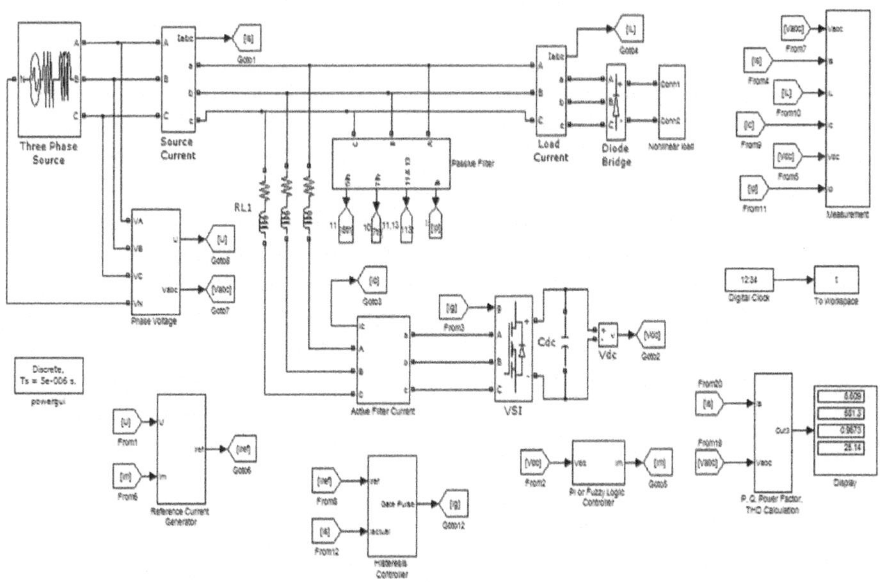

Fig. 5 Simulink model of HSAPF system

Table 2 Designed parameter of HSAPF system

Parameters	Values
Supply voltage (V_{rms})	230 V
Supply frequency (f)	50 Hz
Source impedances	
Source resistance (R_s)	0.1 Ω
Source inductance (L_s)	0.5 Mh
Nonlinear load	
Diode rectifier	6-diode
Load resistance (R_L)	40 Ω
Shunt active power filter	
Power converter	6-MOSFETs/6-diodes
Filter resistance (R_c)	0.4 Ω
Filter inductance (L_c)	1.35 mH
DC-link capacitance (C_{dc})	2000 μF
Reference capacitor voltage ($V_{dc,\ ref}$)	680 V
Shunt passive power filter	
5th harmonic filter	$K_{var} = 1$, Quality factor $Q = 50$
7th harmonic filter	$K_{var} = 0.5$, Quality factor $Q = 50$
11th and 13th harmonic filter	$K_{var} = 0.5$, Quality factor $Q = 50$
PI controller	$K_P = 0.35$, $K_I = 15$

or source current harmonic spectrum along with various waveforms is shown in the Fig. 6a–f, when HSAPF is not connected to the system. The load or source current is having a THD of 25.14%. From Fig. 7a–f, it is clearly visible that the reduction of THD from 25.14 to 17.5% when SPF is connected to the system. To further improve the performance of the system active filter is inserted into the existing system makes the system hybrid. Figures 8a–f and 9a–f gives various waveforms of HSAPF system with PID and FLC is used for reference current estimation, respectively. To test the validity of the PID and FLC, a load change is applied at $t = 0.3$ s and $t = 0.5$ s. From Figs. 8a–f and 9a–f it is clear that Fuzzy Logic Controlled HSAPF system gives better performances compared to PID Controlled HSAPF system. Various performance parameters obtained in the simulation are given in Table 3 for comparison.

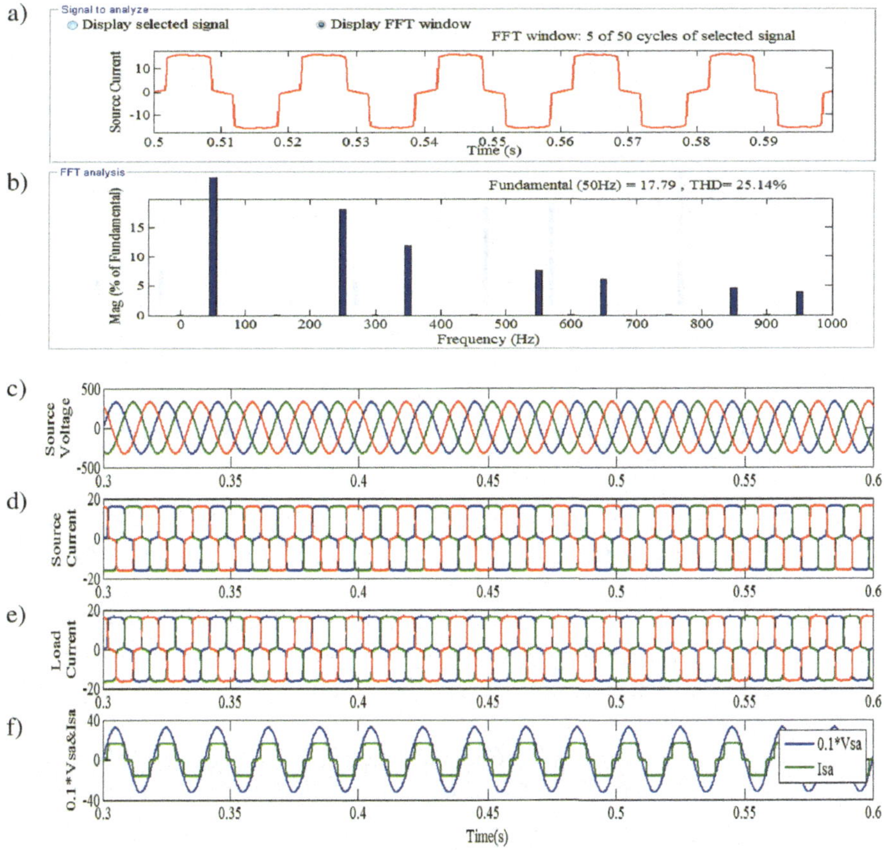

Fig. 6 Simulation results of the proposed system without HSAPF

7 Conclusion

The design, modeling, and simulation of HSAPF system for reactive power and harmonic compensation purposes considered in MATLAB/Simulink environment. The outcomes of the simulation work provided low THD of the source current with upgraded AC mains IPF and reduced supplied mains reactive power. Comparing with Fuzzy and PID Controller used for reference current estimation, FLC gives better result in terms of THD of supply current, IPF, compensation of reactive power and reduced percentage overshoot of capacitor voltage as well as regulated capacitor voltage with reduced overshoot when load disturbance occur. The prototype of the proposed HSAPF can be developed, which can be aimed as future work.

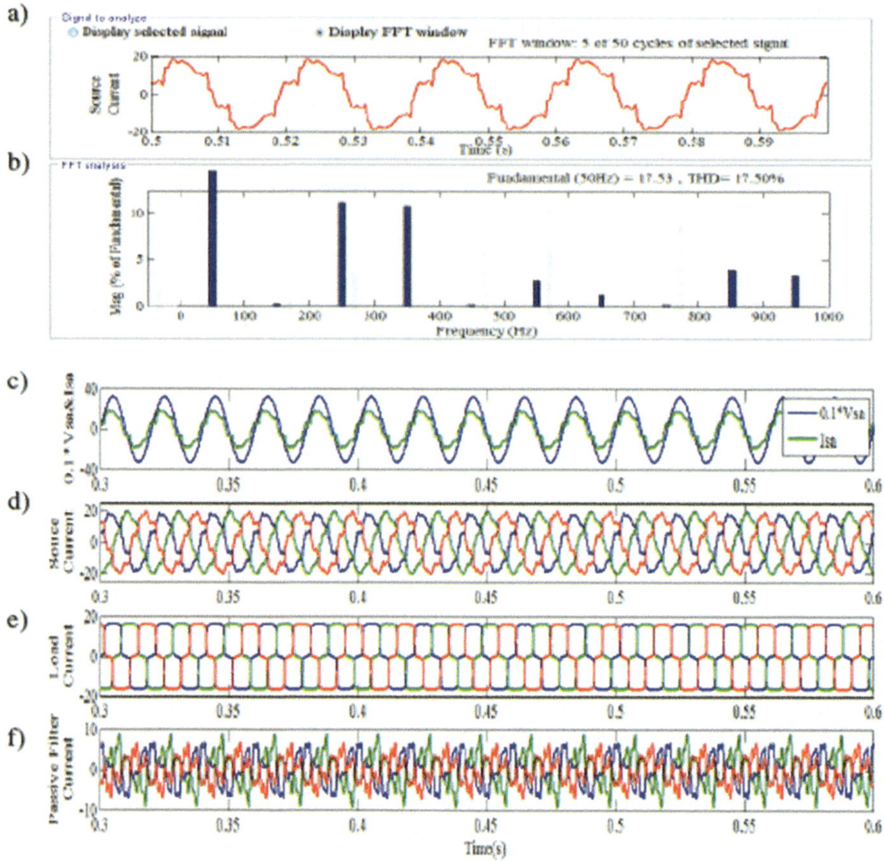

Fig. 7 Simulation results of the proposed system when SPF connected alone

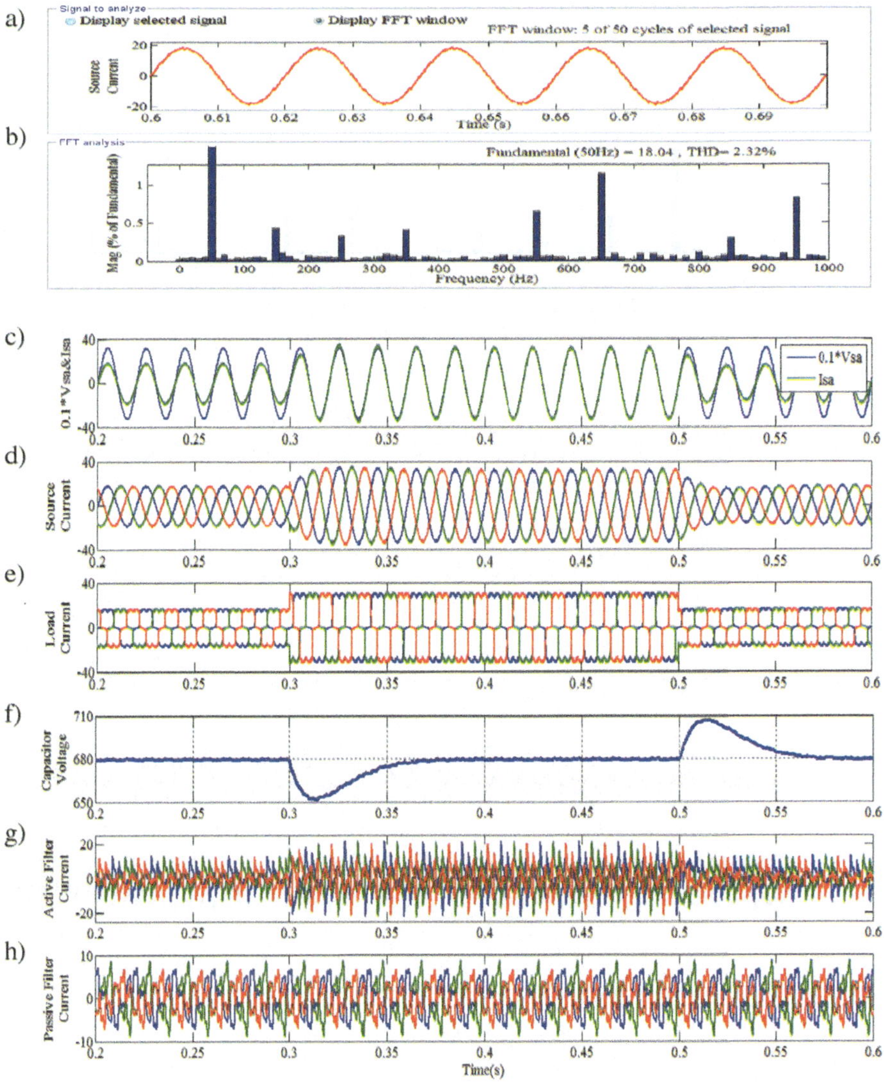

Fig. 8 Simulation results of PID controlled HSAPF system

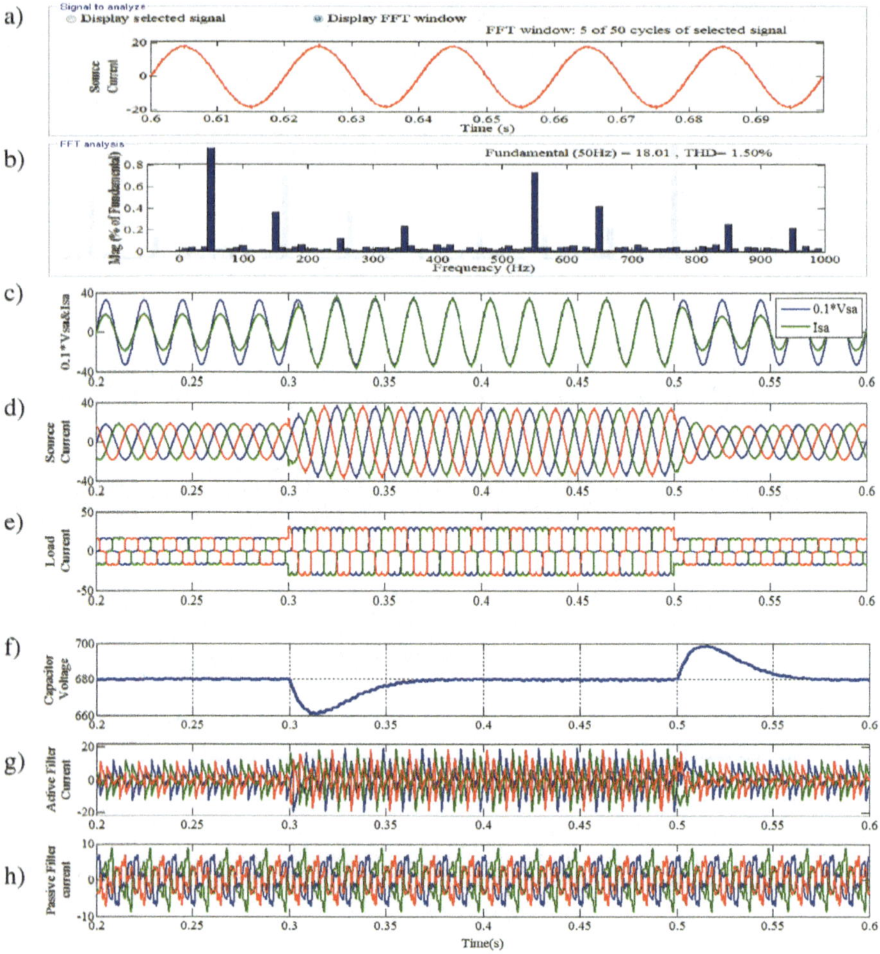

Fig. 9 Simulation results of fuzzy controlled HSAPF system

Table 3 Performance parameters of the system under different condition

Performance parameter / Different condition	THD (%)	Active power (K_w)	Reactive power (VAR)	Power factor	% Overshoot of V_{dc}
Without HSAPF	25.14	8.609	581.30	0.9673	–
SPF connected alone	17.50	8.372	69.30	0.9679	–
PID controlled HSAPF	02.32	8.730	14.22	0.9925	4.41
Fuzzy controlled HSAPF	01.50	8.271	2.644	0.9983	2.94

References

1. Haugan, P.T.S., Tedeschi, E.: Reactive and harmonic compensation using the conservative power theory. In: International Conference on Ecological Vehicles and Renewable Energies. Elsevier (2015)
2. Thirumoorthi, P., Yadaiah, N.: Design of current source hybrid power filter for harmonic current compensation. Simul. Model. Pract. Theor. **52** (2015) (Elsevier)
3. Thirumoorthi, P., Yadaiha, N.: Design of current source hybrid power filter for harmonic current compensation. Simul. Model. Pract. Theor. **52** (2015)
4. Mishra, A.K., Pathak, M.K., Das, S.: Isolated converter topologies for power factor correction—A comparison. In: IEEE International Conference on Energy, Automation and Signal 2011 Dec 28, pp. 1–6
5. Lam, S., Wong, M.C., Choi, W.H., Cui, X.X., Mei, H.M., Liu, Z.J.: Design and performance of an adaptive low-DC-voltage-controlled LC-hybrid active power filter with a neutral inductor in three-phase four-wire power systems. IEEE Trans. Ind. Electr. **61** (2014)
6. Rahmani, S., Hamadi, A., Haddad, K.A., Dessaint, L.A.: A combination of shunt hybrid power filter and thyristor-controlled reactor for power quality. IEEE Trans. Ind. Electr. **61** (2014)
7. Das, S.R., Ray, P.K., Mohanty, A.: Comparative assessment for power quality improvement using hybrid power filters. In: Recent Findings in Intelligent Computing Techniques, pp. 257–265. Springer, Singapore (2018)
8. Das, S.R., Ray, P.K., Mohanty, A.: Fuzzy sliding mode based series hybrid active power filter for power quality enhancement. Adv. Fuzzy Syst. (2018)
9. Das, S.R., Ray, P.K., Mohanty, A., Panigrahi, T.K.: Application of artificial intelligence techniques for improvement of power quality using hybrid filters. In: Computational Intelligence in Data Mining, pp. 719–729. Springer, Singapore (2020)
10. Ray, P.K., Das, S.R., Mohanty, A.: Fuzzy-controller-designed-PV-based custom power device for power quality enhancement. IEEE Trans. Energy Convers. **34**(1), 405–14 (2018)

Optimal Bidding of Market Participants in Restructured Power Market Adopting MBF Method

Ramachandra Agrawal, Purba Sengupta, Anasuya Roy Choudhury, Debashis Sitikantha, Ilyas Ahmed, and Manoj Kumar Debnath

Abstract Presently, the generating firms throughout the world experience the critical problem of maximizing their own benefit while trading power to the market controller. Strategic bidding can be adopted for this purpose. For profit maximization, winning the auction in the hourly competitive power market to retail partially or entirely the total power demanded of a specific hour is of vital importance. A very recent optimization method acknowledged as Mouth Brooding Fish (MBF) has been suggested to get rid of the strategic bidding issue of the generating firms. The proposed optimization technique is adopted to find the bidding strategies in modern dynamic electric power market along with the power producing firms and bulk power consuming customers altering the MCP. The small power demanding market participants are in collective demand form. Normally, the hourly trading duration information of the rivals in the electric power market of the auction process with winning objective is unknown and an independent participant has to form his own strategy by predicting the other participant's parameters out of the previous market history. The outcomes of the proposed MBF method evidence faster convergence and enhanced reliability to that obtained using TLBO and PSO methods. It is tested on an IEEE-30 Bus system incorporating 6 Power producing firms and 2 bulk power consumers. A new participant's entry effect on the present market bidding behavior is also discussed in this work.

Keywords Strategic bidding problem · Market clearing price · Power producer · Market operator · Hour ahead electric market · Mouth brooding fish concept

---

R. Agrawal · P. Sengupta · A. R. Choudhury · D. Sitikantha · M. K. Debnath (✉)
Electrical Engineering Department, ITER, SOA Deemed to be University, Bhubaneswar 751030, Odisha, India
e-mail: mkd@gmail.com

I. Ahmed
Electrical Engineering Department, ABIT, Biju Pattnaik University of Technology, Rourkela, Odisha, India

D. Mishra et al. (eds.), *Intelligent and Cloud Computing*,
Smart Innovation, Systems and Technologies 194,
https://doi.org/10.1007/978-981-15-5971-6_58

1 Introduction

To set up a financially productive power platform, the presence of an organized spot advertise that is offer based, security compelled, monetary transfer having nodal estimation is desired to be embraced like the case of Hourly markets of Australia as depicted in [1]. Diverse offering procedures and the aggressive deregulated power markets are displayed by various famous analysts in [2, 3]. Electricity markets such as in Norway, the retailers present govern their restraining infrastructures in the business sectors [4], yet it is unwanted as making undesirable trading conditions. Step savvy offering techniques for the players of the market were produced in [5], but the power consumer part was not taken up. Here, as in [6], both the power producer and customer side offering are dealt alongside the fixed offering methodology in the power market. In these cases, the opponent creating organization's offering data is vague bringing about less revenue earned from this market, as discussed enough by David in [7]. In [8], a fantastic work with another power creating unit indistinguishable to the first generating set utilizing PSO, suggested in [9]. In [10], it is revealed the offering methodology in a discounted power market. A nearby uni-modular inquiry through LUS related to TLBO strategy connected to AGC with multiple sources of control organize is accounted for [11], and the same technique is used to enhance the results acquired in [12, 13].

Presently, electrical power sectors are under restructuring which results in individual organizations which eventually leads to the creation of a competitive electricity marketplace by avoiding the monopolies [14, 15]. A literature survey on competitive electricity markets is reported in [16]. Game theory technique [17] and Nash equilibrium in [7], are amalgamated to resolve the profit maximization problem of the generating firms in [18–20].

Different pricing methods are compared by Yuan in [21], which are being adopted in ubiquitous deregulated markets. The wholesale electricity market considering the allied market power and the core competition among the participants have been investigated by Ashok et al. [22]. Liu, et al. have broadly discovered the deregulation consequence in the whole Asian market [23]. The MBF was fundamentally proposed in [24]. In this work, the mathematical models convoluted are nonlinear in nature like the real-time problems and classical methods can't be accepted to resolve the problem. Hence, a heuristic method such as the recent MBF method has been used in this work, where, every participant (the upper level or the lower level) can independently optimize its own profit function in every step, but these are affected by others decision in a successive way. The rest segments are put in the way as being explained.

Segment II comprises the detailed scientific analysis of the offering issues of the electrical energy organizations and the customers. Segment III demonstrates the recommended planning of calculation using MBF technique. Segment IV introduces the use of MBF method to take care of the offering issues detailed in portion II. Segment V discusses the outcomes found for the ideal offering issue with IEEE 30-Bus transmission framework. The impact of the involvement of another power supplying organization into the current power trading market. Segment VI incorporates the conspectus of this work along with its possible future extension where constant significance essential for the functional power markets are highlighted.

2 Problem Formulation

Here, we have developed the mathematical objective functions for the generating firms and the market power purchasers where the competitive power market is based on auction method and a day ahead market. Every generating firm will deliberately maximize its self-benefit by bidding. Usually, each generating firm sets an hourly available volume and the equivalent power generating prices to the next day market. For this mathematical model associated with bidding, as well as an energy output dispatch for power producers, are offered as follows. The power production cost function follows a quadratic function of the power output.

In a pool type of power exchanging market, "*m*" power supplying units (called as IPPs) are taking part with an amassed interest through their offerings. A power provider offers bid based power and is being submitted to the Market Operator (MO). Here, fixed offer sell off concepts are embraced. The *i*th provider of power has a structure as

$$G_i(P_i) = (a_i + b_i P_i) \tag{1}$$

where $G_i(P_i)$ is *i*th power provider's function that is produced with a_i and b_i as the bidding variables.

Also, Large or the major customers bid as:

$$W_j(L_j) = (c_j - d_j L_j) \tag{2}$$

where $W_j(L_j)$ is *j*th demanded power of a major customer. Presently, the MO with the free market activity from everyone will plan for the total power market activity structure and also decide the point of convergence, *R* as the MCP. It likewise determines the net amount cleared amid the frame of time alongside the production and requests of power by each of the market players. In this way, we can persuade the ability to produce power from

$$a_i + b_i P_i = R \tag{3}$$

$$c_j - d_j L_j = R \tag{4}$$

Along with the conditions

$$\sum_{i=1}^{m} P_i = Q(R) + \sum_{j=1}^{n} L_j \quad (5)$$

Here

$$P_{i\min} \leq P_i \leq P_{i\max} \text{ and } i = 1, 2, 3 \ldots m \quad (6)$$

and

$$L_{j\min} \leq Lj \leq L_{j\max} \quad j = 1, 2, 3 \ldots n \quad (7)$$

The net customer demanded power is $Q(R)$, which is much less power and is R value dependent and the price of the power demand elastic variation factor (K) with the expression

$$Q(R) = Q_0 - KR \quad (8)$$

Excluding conditions, solution of Eqs. (1)–(4), will be as stated below

$$\left[R = \frac{Q_0 + \sum_{i=1}^{m} \frac{a_i}{b_i} + \sum_{j=1}^{n} \frac{c_j}{d_j}}{K + \sum_{i=1}^{m} \frac{1}{b_i} + \sum_{i=1}^{m} \frac{1}{d_j}} \right] \quad (9)$$

Once R is found, the ith unit's power production and jth large power demanding customer are to be found with the help of the expressions presented below

$$\left(P_i = \frac{R - a_i}{b_i} \right) \quad (10)$$

Here $i =$ 1, 2, 3…m

$$\left(L_j = \frac{c_j - R}{d_j} \right) \quad (11)$$

Here $j =$ 1, 2, 3…n.

Based on the conditions of equalities, we will get the production and requests of power by the market players separately. Presently, if the breaking points of the curve are veered off from the predefined values as in conditions (6) and (7), then at that point they are kept inside the cut off points. Anyway, we accept that the energy yields as acquired from condition (10), is practical to be traded by the specific producing organization as any transient requirement is not being considered like incline limits like up and down slope, shut time and the start-up expenses, and so on. The cost parameters involved function is given as

$$C_i(P_i) = e_i + f_i P_i^2 \tag{12}$$

The maximization of benefit by the ith unit is as
Maximize

$$F(a_i, b_i) = RP_i - C(P_i) \tag{13}$$

Meeting the conditions: Presented by Eqs. (5)–(8).
So, likewise the jth has the demand form

$$B_j(L_j) = g_j - h_j L_j^2 \tag{14}$$

and the consumer maximizes the following fitness function

$$B(c_j, d_j) = B_j(L_j) - RL_j \tag{15}$$

We need to discover the consumer payment function parameters, to obtain the maximum value of B (c_j, d_j) with the requirements stated by the Eqs. (5)–(8).

This work recommends MBF technique to take care of the offering issues of each power provider and the same for the major customers with the optimum condition. Now, two parameters thought to be settled based on the given information and the other two are found by the adopted MBF method. When the yields were tallied with the work reported earlier and the customary PSO as well, MBF outcomes are substantially excellent. If the sellers of power could realize the offering pattern of others then the cost cleared could be moved with the end goal to make a higher advantage out of it. The technique prescribed here may in like manner be used viably to find the perfect strategies of offering for different number of merchants, buyers, and various power features as the standards of the significant number of business areas over the globe changes.

3 Solution Algorithm

3.1 Particle Swarm Optimization (PSO)

Restricting the social learning aspects in the PSO [9], causes fast convergence. But each particle learns from the global the best position even if the current position is far away from the desired global maximum or minimum, which may cause other swarms to fly to this region and get stacked in a local maximum or minimum and is quite undesirable. The PSO technique is empirically far better and is implementable with much ease as the literature survey evidences. But it might get trapped in the local extremes in complex, multi-modal objective equations.

Although there exist different modified PSOs still while solving problems with a large no of local optima, then it might be trapped at the local extremes and is the biggest limitation of the PSO. In normal PSO method, a particle updates its position and velocity out of its and other's best experience simultaneously. It is more probable to get trapped in case the search space is corresponding to fitness functions having multi-modal complexities with a large number of local optima.

3.2 Teaching Learning Based Optimization (TLBO)

The usual teaching-learning process of a classroom where one learns from another for improving one's knowledge level has been depicted here. This algorithm proposes two ways of learning, i.e., from the teacher and via interaction with other co-learners. Here a group of learners is treated as the population and the different subjects resemble different design parameters of the objective function to be optimized. The fitness value of the objective function is the results of the student while the best fitness value in the whole search space is treated as the teacher. From the fitness values, it is concluded that the best teacher is superior to other teachers. A good teacher results in a better mean for the learners. Additionally, mutual interaction among the Learners also helps in obtaining better results. Based on the mentioned process, a mathematical model is developed to optimize nonlinear continuous function. The teacher is the most knowledgeable fellow in the human society, so the best student may be treated as a teacher. The teacher transfers his/her knowledge among students, which in turn, increases the level of the entire class to obtain good grades.

3.3 MBF (Mouth Brooding Fish) Algorithm

In the ocean, numerous species play the numbers diversion and communicate a huge number of eggs (having no idea to parental basis) with the expectation that some little rate (a modest number) will really get by to adulthood; individuals from the cichlids family give considerable vitality in raising their adolescents. Instead of using vitality and assets as tremendous quantities of eggs, these angles generate generally few posterities, however, offer them assurance until such time that they are completely useful. When the guardians leave their offspring, life is to be sure intense. Notwithstanding, in light of the fact that they are practically more develop, bigger, to some degree more experienced and quicker swimmers, they have a superior possibility of existence. While numerous submerged animals have procedures to ensure themselves from damage, for example, disguise, not all have techniques for securing their young ones, as well. Mouth brooders have the capacity to fare thee well and ensure their posterity, to a great extent because of an exceptionally strange system. Mouth brooders secure their babies by utilizing the mouth as a safe house. African cichlids, which are maternal mouth brooders, are a case of fishes that keeps on securing their babies irrespective of the eggs have been brought forth. Once discharged, the cichlid employments unmistakable social prompts when risk is available to let her young realize the time has come to swim back and look for insurance. When they develop, what's more, the mother's safe house has insufficient space for every one of them, at the season of peril, a portion of the feeble cichlids are abandoned and need to look up with the peril and nature alone.

All together to pick the most ideal qualities for the control variables, the issues and the outcomes of tuning variables. In spite of the fact that we should expect the control variables don't change to make the MBF calculation and are a populace (*n*Fish) oriented method, i.e., the number of fishes will experience the task of settling a problem in MBF calculation. As it is unmistakable picking some of the populace under 5 isn't rationale, despite the fact that it relies upon a client necessity and could be a number greater than 2. MBF calculation technique comprises of 4 primary parts with the end goal to locate the most ideal outcomes for the issues. The primary developments of every cichlid. The suggested calculation utilizes the developments with the principle factor, along these lines, utilizing all consequences for cichlids development could help the MBF calculation to locate the most ideal response for the issue.

1. The principle developments of every cichlid
2. The extra developments of left ones
3. Hybrid with roulette wheel determination
4. Shark assault or impacts of risk development

1. The impact of *SP* capacity of the mother with the impact of developments is characterized as pursues:

$$\text{1st. Effect} = \text{SP} \times \text{Cichlids. Movements} \tag{16}$$

The very purpose is figured as pursues

$$\text{SP} = \text{SP} \times \text{SP}_{\text{damp}} \tag{17}$$

Here, SP_{damp} is the SP of mother's aging effect and lies somewhere in the range from 0.85 to 0.95.

2. The next issue of developments is to overcome during the past iterations which are unique in relation to be within 1 and 2.

$$\text{2nd. Effect} = \text{Dis} \times (\text{Cichlids. Best} - \text{Cichlids. Position}) \tag{18}$$

3. The third factor is the inclination of being equal to advance toward the most ideal position of entire cichlids group.

$$3\text{rd. Effect} = \text{Dis} \times (\text{Cichlids. Best} - \text{Cichlids. Position}) \tag{19}$$

4. The last factor is the developments of the nature patterns or power. Nature dependably has a tendency to the best conceivable answer in spite of the fact that these patterns are moderate yet have their consequences for all animals. In MBF in excess of 60% distinction in every cell, will be chosen.

$$\text{New N.F.P} = 10 \times \text{SP} \times \text{Nature Force. Position(Selected Cells)} \tag{20}$$

The fourth impact may be governed through the measure of scattering (Dis):

$$4\text{th. Effect} = \text{Dis} \times (\text{New N.F.P} - \text{Nature Force. Position}) \tag{21}$$

where, Nature Force Position is the most desired place of the previous cycle. The entirety of aggregate impacts spoke to the fundamental developments of every cichlid. The impacts of nature drift just go to an activity when the entire assembly drift (i.e., the best expense of current iteration under 15% superior to the best expense of the last emphasis) isn't fit as a fiddle. For greater effortlessness in depicting this idea, the general advances are appeared in Algorithm (1). In the fundamental developments of cichlids, every youngster moves to encompass scattering as pursues:

$$\text{ASDP} = 0.1 \times (Var_{\text{Max}} - Var_{\text{Min}}), \text{ASDN} = -\text{ASDP} \tag{22}$$

The most extreme cut off points of the issue's variety individually, from that point forward, by including the computed development one can locate another situation for cichlids.

$$\text{Cichlids. Movements} = -\text{Cichlids. Movements} \tag{23}$$

In MBF calculation,

$$nm = 0.04 \times n\text{Fish} \times \text{SP } e - 0.431 \tag{24}$$

These left out cichlids with the end goal to make due from risk need to move by ultra-extra encompassing scattering points of confinement and the developments thereof.

$$NCC = [n\text{Var} \times P\text{dis}] \tag{25}$$

The position acquired from subtracted UASDP or UASDN by the left out cichlids from the chose cells.

$$\text{Left Cichlids. Position} = \text{UASDP} \pm \text{Cichlids}'(\text{Selected Cells}) \tag{26}$$

MBF calculation for the 4% of the populace utilizes extra developments named shark assault impact.

$$n\text{shark} = 0.04 \times n\text{Fish} \quad (27)$$

Shark assault influences 4% of cichlids populace on position and developments as pursues

$$\text{Cichlids. New Position} = \text{Shark Attack} \times \text{Cichlids. Position} \quad (28)$$

There are three distinct algorithms for MBF method and are quite lengthy and are omitted to avoid undue space spoil. Hence, only the dealing equations with the proper concept has been explained here.

4 Test Result and Discussion

The execution of MBF, which is the suggested technique, the IEEE-30 bus network is utilized, which includes six generating units, depicted as independent generating firms and major customers as loads. The initial input data are presented in Table 1. Q_0 is 300 and K is 5 for the gross load. The parameters, i.e., b_i and d_j of the generators and the major customers, the suggested and earlier techniques, and the optimal parameters of the strategies of bidding are shown in Table 2. The MBF based MCP is superior to the conventional TLBO and PSO techniques. Even if the major customers incur less benefit but the gross benefit is evidently higher than the works reported and are put in Table 3.

Figure 1 exhibits the MBF convergence characteristics corresponding to TLBO and PSO techniques with the same number of iterations. Benefits of six generating firms are represented in Fig. 2. However, Fig. 3, shows the benefits of major customers

Table 1 Generators and large consumers data

Generator	e	f	P_{min} (MW)	P_{max} (MW)
1	6.0	0.0112	40	160
2	5.25	0.0525	30	130
3	3.0	0.1375	20	90
4	9.75	0.0253	20	120
5	9.0	0.075	20	*100
6	9.0	0.075	20	100
***Large consumer	g	h	L_{min} (MW)	L_{max} (MW)
1	30	0.04	0	200
2	25	0.03	0	150

Table 2 Bidding strategies for IPPs

Generator	MBF	TLBO	PSO
S. No.	b_i	b_i	b_i
1	0.0976	0.0864	0.1064
2	0.2687	0.4344	0.4967
3	0.6693	0.4228	1.3009
4	0.2103	0.1375	0.2395
5	0.2850	0.1500	0.7096
6	0.2850	0.1500	0.7096
Large consumer	d_j	d_j	d_j
1	0.0800	0.0800	0.3784
2	0.0600	0.0600	0.2838

Table 3 MCP and individual profits ($)

Generator	Proposed MBF		TLBO [12]		PSO [8]	
S. No.	Power	Profit	Power	Profit	Power	Profit
1	160.00	1678.3	160.00	1678.3	160.00	1370.1
2	107.38	728.8	85.1094	728.8	105.8371	588.12
3	25.07	372.7	36.1438	372.7	48.5923	324.80
4	47.69	432.0	62.0521	432.0	120.00	428.90
5	37.83	287.1	61.8753	287.1	49.0859	180.71
6	37.83	287.1	61.8753	287.1	49.0859	180.71
Large consumers	Load	Profit	Load	Profit	Load	Profit
1	127.73	858.3	146.4839	858.3	170.4639	1162.32
2	86.98	376.2	111.9785	376.2	143.9518	621.73
MCP	19.7814		18.1492		16.4729	
Total profit	5109.10		5027.72		4857.4	

are less while the gross profit of all the participators is much more compared to all earlier methods reported. Figure 4 indicates that the benefits of each vary as a new power generating unit enters in the market with identical characteristics to the first unit already present. Currently, the power transactions among different participant has changed, which is the cause of the profit variation as shown in the Fig. 4. However, the total benefit is enhanced as shown in Fig. 5. Table 4 depicts the MCP and profit of all the participants.

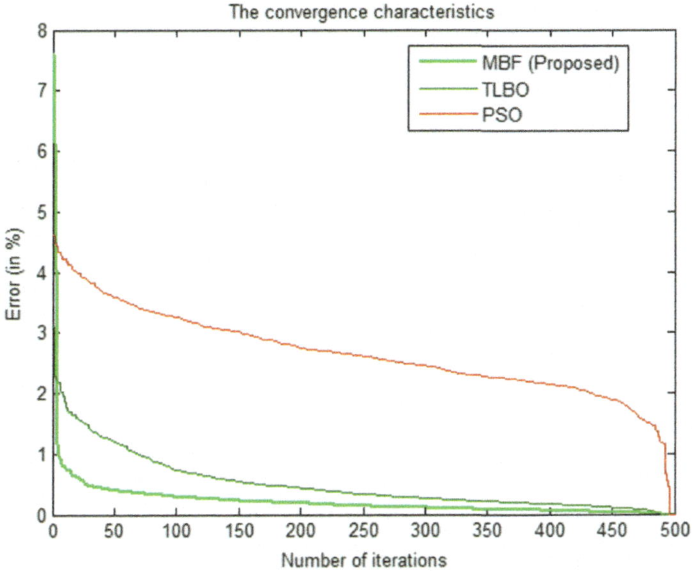

Fig. 1 Convergence characteristics of proposed CLPSO and PSO

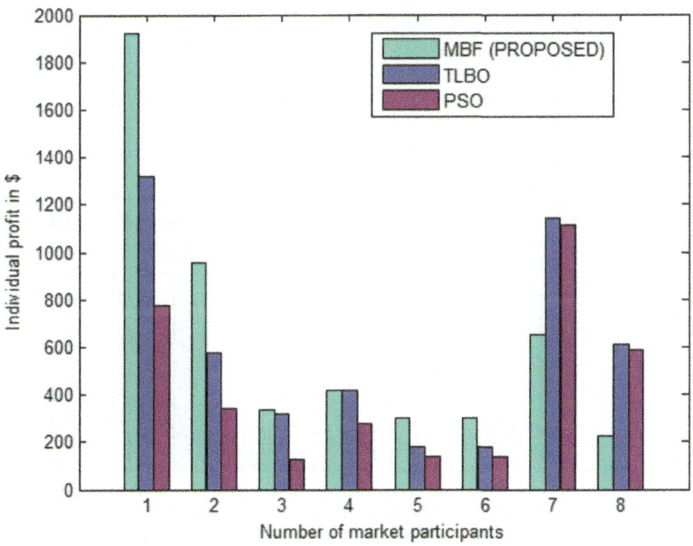

Fig. 2 Comparison of individual profit of IPPs

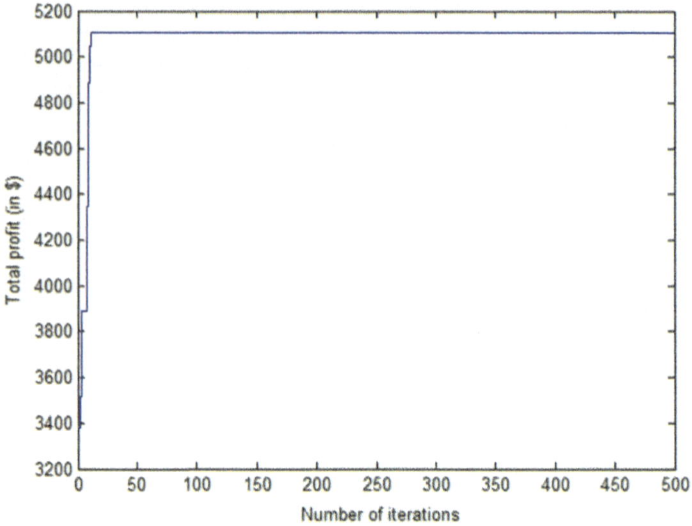

Fig. 3 Total profit maximization using proposed CLPSO

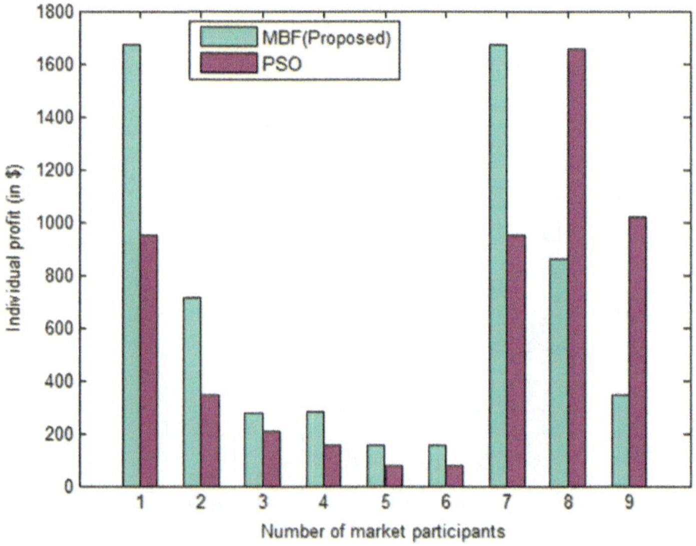

Fig. 4 Comparison of individual profits

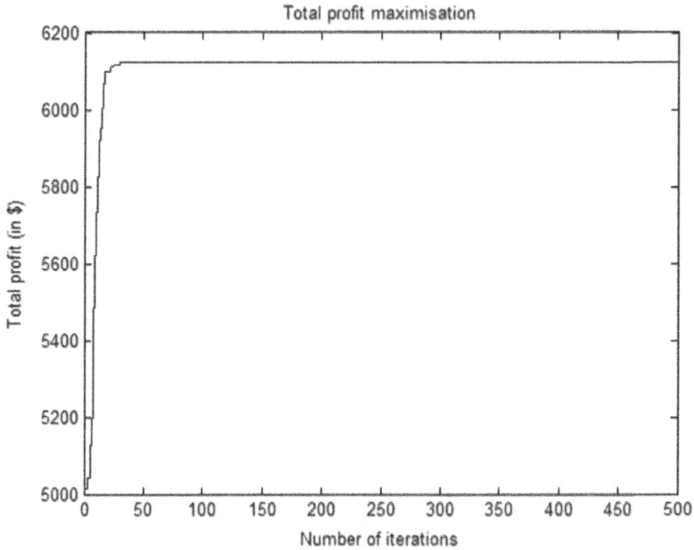

Fig. 5 Total profit maximization using MBF

Table 4 MCP and profit ($) of participants

Generator	Proposed MBF method		TLBO [12]		PSO [8]	
	Power	Profit	Power	Profit	Power	Profit
1	160.0000	1670.3	122.4546	1684.5	160.0000	948.5
2	31.4349	712.6	40.0000	551.1	80.7450	342.3
3	20.0000	274.5	30.0000	420.1	39.0117	209.3
4	20.0000	280.5	22.5751	243.9	78.5590	156.3
5	61.5417	154.6	34.4765	329	31.5215	74.5
6	61.5417	154.6	33.3129	320.8	31.5215	74.5
7	160.0000	1670.3	122.4546	1684.5	160.0000	948.5
Large consumer	Load	Profit	Load	Profit	Load	Profit
1	147.1093	859.0	100.0000	487.3	200	1654.42
2	147.1093	346.5	110.9099	60.5	150	1015.81
MCP	21.7821		21.1272		13.7282	
Total Profit	6123.0		5781.53		5424.00	

5 Conclusion

This Paper effectually executes the MBF technique for strategically bidding for two large consumers and six IPPs in a day-ahead electricity market during any specific hour of a trading period which may be iterated for 24 times. The operating unit constraints (minimum up and down times, up and down ramp limits, etc.) are not reflected. The Goal of an individual market contestant is to maximize its own benefit. This is done by calculating the MCP by implementing the procedures. The MBF provided better net profit as compared to PSO method as reported in Indicon-2010 [8]. It is also perceived to perform better in the event of a new generator entry into the market. The efficacy of the proposed MBF method is evaluated on IEEE 30-bus system successfully without the mentioned constraints and compared with PSO for hourly load demand.

Future research may be executed by utilizing the proposed technique for the 24-h trading duration by operating the generating units continuously, where the unassumed operating constraints may be considered to devise a practical market place.

References

1. Prabavathi, M., Gnanadass, R.: Energy bidding strategies for restructured electricity market. Int. J. Electr. Power Energy Syst. **64**, 956–966 (2015)
2. Zhang, D., Wang, Y., Luh, P.B.: Optimization based bidding strategies in the deregulated market. In: Proceedings of the 21st International Conference on Power Industry Computer Applications. Connecting Utilities. PICA 99. To the Millennium and Beyond (Cat. No. 99CH36351), pp. 63–68. IEEE (1999)
3. Wen, F.S., David, A.K.: Strategic bidding for electricity supply in a day-ahead energy market. Electr. Power Syst. Res. **59**(3), 197–206 (2001)
4. Fleten, S.-E., Pettersen, E.: Constructing bidding curves for a price-taking retailer in the Norwegian electricity market. IEEE Trans. Power Systems **20**(2), 701–708 (2005)
5. Ma, L., Fushuan, W., David, A.K.: A preliminary study on strategic bidding in electricity markets with step-wise bidding protocol. In: IEEE/PES Transmission and Distribution Conference and Exhibition, vol. 3, pp. 1960–1965. IEEE (2002)
6. Wen, F., David, A.K.: Optimal bidding strategies for competitive generators and large consumers. Int. J. Electr. Power Energy Syst. **23**(1), 37–43 (2001)
7. Wen, F., David, A.K.: Optimal bidding strategies and modeling of imperfect information among competitive generators. IEEE Trans. Power Syst. **16**(1), 15–21 (2001)
8. Kumar, J.V., Pasha, S.J., Vinod Kumar, D.M.: Strategic bidding in deregulated market using particle swarm optimization. In: 2010 Annual IEEE India Conference (INDICON), pp. 1–6. IEEE (2010)
9. Kennedy, J.: Particle swarm optimization. In: Encyclopedia of Machine Learning, pp. 760–766 (2010)
10. Shukla, U.K., Thampy, A.: Analysis of competition and market power in the wholesale electricity market in India. Energy Policy **39**(5), 2699–2710 (2011)
11. Sahu, B.K., Pati, T.K., Nayak,J.R., Panda, S., Kar, S.K.: A novel hybrid LUS–TLBO optimized fuzzy-PID controller for load frequency control of multi-source power system. Int. J. Electr. Power Energy Syst. **74**, 58–69 (2016)
12. Mallick, R.K., Agrawal, R., Hota, P.K.: Bidding strategies of Gencos and large consumers in competitive electricity market based on TLBO. In: 2016 IEEE 6th International Conference on Power Systems (ICPS), pp. 1–6. IEEE (2016)
13. Hota, P.K., Agrawal, R.C., Mallick, R.K.: Bidding strategies of IPP and large consumers in competitive electricity market based on CLPSO. In: MT090, ICAMT, Visakhapatnam, India, pp. 103–110 (2016)
14. Ferrero, R.W., Rivera, J.F., Shahidehpour, S.M.: Application of games with incomplete information for pricing electricity in deregulated power pools. IEEE Trans. Power Syst. **13**(1), 184–189 (1998)

15. David, A.K., Wen, F.: Strategic bidding in competitive electricity markets: a literature survey. In: 2000 Power Engineering Society Summer Meeting (Cat. No. 00CH37134), vol. 4, pp. 2168–2173. IEEE (2000)
16. Haghighat, H., Seifi, H., Kian, A.R.: Gaming analysis in joint energy and spinning reserve markets. IEEE Trans. Power Syst. **22**(4), 2074–2085 (2007)
17. Munz, U., Schumm, P., Wiesebrock, A., Allgower, F.: Motivation and learning progress through educational games. IEEE Trans. Ind. Electron. **54**(6), 3141–3144 (2007)
18. Weber, J.D., Overbye, T.J.: An individual welfare maximization algorithm for electricity markets. IEEE Trans. Power Syst. **17**, 590–596 (2002)
19. Pang, J.-S., Fukushima, M.: Quasi-variational inequalities, generalized Nash equilibria, and multi-leader-follower games. CMS **6**(3), 373–375 (2009)
20. Wu, Y.-K.: Comparison of pricing schemes of several deregulated electricity markets in the world. In: 2005 IEEE/PES Transmission & Distribution Conference & Exposition: Asia and Pacific, pp. 1–6. IEEE (2005)
21. Shukla, U.K., Thampy, A.: Analysis of competition and market power in the wholesale electricity market in India. Energy Policy **39**(5), 2699–2710 (2011)
22. Ni, Y., Zhong, J., Liu, H.: Deregulation of power systems in Asia: special consideration in developing countries. In: IEEE Power Engineering Society General Meeting, 2005, pp. 2876–2881. IEEE (2005)
23. Lai, M., Tong, X., Yang, H., Bing, P.: Dynamic bidding analysis in power market based on the supply function. Comput. Math. Appl. **58**(1), 25–38 (2009)
24. Jahani, E., Chizari, M.: Tackling global optimization problems with a novel algorithm—Mouth Brooding Fish algorithm. Appl. Soft Comput. **62**, 987–1002 (2018)

Stabilizing and Trajectory Tracking of Inverted Pendulum Based on Fuzzy Logic Control

Akshaya Kumar Patra, Alok Kumar Mishra, Ramachandra Agrawal, and Narayan Nahak

Abstract This manuscript presents a SIMULINK model of Inverted Pendulum (IP) and design of a Fuzzy Logic Controller (FLC) to control of Cart Position (CP), and Angular Position (AP) of the pendulum under uncertainties and disturbances. The FLC is a novel approach whose gains dynamically vary with respect to the error and change in the error signal. The validation of the improved control performance of FLC is established by comparative result investigation with other published control algorithms. The comparative results clearly reveal the better response of the proposed approach to control the system dynamics within the stable range with respect to accuracy, robustness, and ability to handle uncertainties.

Keywords Inverted pendulum · Angular displacement · Angular velocity · FLC

1 Introduction

The control of IP is a standard problem among all other problems in the control system engineering field owing to nonminimum phase, nonlinear and under-actuated characteristics as reported in the references [1–3]. Additionally, the IP system exhibits a significant number of industrial applications like self-balancing two-wheeled vehicles or a kind of sag-way, guided missiles, rockets, intelligent robots, and other crane

---

A. K. Patra (✉) · A. K. Mishra
Department of Electrical and Electronics Engineering, ITER, S'O'A University, Bhubaneswar 751030, India
e-mail: hiakp@yahoo.com

A. K. Mishra
e-mail: malok2010@gmail.com

R. Agrawal · N. Nahak
Department of Electrical Engineering, ITER, S'O'A University, Bhubaneswar 751030, India
e-mail: ramachandra1agrawal@gmail.com

N. Nahak
e-mail: narayannayak@soa.ac.in

D. Mishra et al. (eds.), *Intelligent and Cloud Computing*,
Smart Innovation, Systems and Technologies 194,
https://doi.org/10.1007/978-981-15-5971-6_59

models [4]. In this present study, an adaptive law of control approach is being tested and analyzed to choose the IP system as it possesses a significant relevance with the control dynamics. During the past three decades, so many control strategy techniques are suggested and tested for control of AP of pendulum within the stable range. Time-discrete and switching PID control strategy is implemented in IP problems with variable control gains based on the measured AP of the pendulum [5, 6]. However, the optimal gain parameter setting, a lesser range of robust control and need for change of gain setting with varying conditions are the major limitations to limit the real-time application of these controllers. Among other projected robust control algorithms implemented for limiting the AP of the pendulum are fuzzy [7], Linear Quadratic Regulator (LQR) [8–10], neurocontrol [11], backstepping control [12], passivity control [13], state feedback control [14], H-infinity ($H\infty$) control [15], Sliding Mode (SM) control [16], Fuzzy-Sliding Mode (FSM) control [17, 18], and BLQG control [19]. However, even if these control techniques are implemented effectively to control the AP the pendulum with better accuracy, still fail to handle various constraints and random change found in a trajectory of motion in the pendulum. These control techniques are not completely insensitive to the disturbances and the uncertainties of the model despite the improved performance. Hence, optimal control parameters setting for better performance and for avoiding slow response following process disturbance, the current work suggests an alternative novel technique implementing the fuzzy logic control approach.

The FLC approach concept leads to enhanced control performance with respect to robustness and delay of time compensation characteristics to counteract the negative impact of associated errors, disturbances, and uncertainties. The stabilizing control law $u(t)$ is developed based on the proposed approach [20–22]. Application of the FLC to control the AP of pendulum results to ensure a better robust controller in comparison to other contemporary well-accepted methods under both harmonized and incompatible uncertainties.

The highlights of this manuscript are as follows:

- Development of a SIMULINK model of an IP.
- Design of the FLC to control the AP of the pendulum within a stable range.
- Estimation of the control actions of the FLC under a huge deviation of process disturbance.
- Comparative investigation to certify the better response of the FLC.

This manuscript is structured as follows. Section 2 concisely illustrates the IP system with mathematical details reflecting its dynamic characteristics. Also, it clearly demonstrates the simulation execution of the system in MATLAB environment. A detailed presentation on how the control technique is formulated and how it is implemented for this problem is presented in Sect. 3. Comparative results of the proposed approach with other published control techniques and the related analysis are provided in Sect. 4. The concluding comments are summarized in the Sect. 5.

2 Problem Formulation and Modeling

2.1 System Overview

The closed-loop model of the IP is depicted in Fig. 1a. The applied horizontal force $F(t)$ and $v(t)$ are reflected as the process disturbance and the sensor noise, respectively, in this study. The controller receives information about the AP of the pendulum as an input to provide the optimal control force $u(t)$ and it balances the pendulum.

2.2 System Modeling

Figure 1b reflects the cart-pendulum model connected to a flexible cart rail with a free-swinging pole. The CP is being controlled by a DC motor. The nonlinear IP modeling through the Newton's Law based mathematical equations have been carried out. It is presumed that the vertical force does not affect the CP and the CP is disturbed by the horizontal force $F(t)$ based on the operation of the DC motor [23–25]. All the physical activities of the IP system are expressed mathematically and specified by Eqs. (1) and (2). All nomenclature and specifications for IP are shown in Tables 1 and 2, respectively. The SIMULINK model of the IP is established with respect to the Eqs. (1) and (2) as displayed in Fig. 1c. The model states and parameters and specifications are represented in Tables 1 and 2.

$$(M+m)\frac{\mathrm{d}^2x(t)}{\mathrm{d}t^2} - ml\frac{\mathrm{d}^2\theta(t)}{\mathrm{d}t^2}\cos\theta(t) + ml\left(\frac{\mathrm{d}\theta(t)}{\mathrm{d}t}\right)^2\sin\theta(t) + b\frac{\mathrm{d}x(t)}{\mathrm{d}t} = F(t) \tag{1}$$

$$\left(i + Ml^2\right)\frac{\mathrm{d}^2\theta(t)}{\mathrm{d}t^2} - mgl\,\sin\theta(t) = ml\frac{\mathrm{d}^2x(t)}{\mathrm{d}t^2}\cos\theta(t) \tag{2}$$

2.3 Response of IP Model

There are four roots in the IP system. One of them lies in right-hand side of the complex plane. As a result, the system becomes unstable. This needs the design of an adaptive controller for improving the stability of the system by means of shifting the roots into the left-hand side of the complex plane. The IP system SIMULINK model in the open-loop form is depicted in Fig. 1c. The IP system consists of two inputs and four outputs. The control force $u(t)$ and applied horizontal force $F(t)$ are the two inputs of the IP system. The CP and AP of the pendulum are two outputs of the IP system. An uncontrolled system dynamics such as AP of the pendulum, and

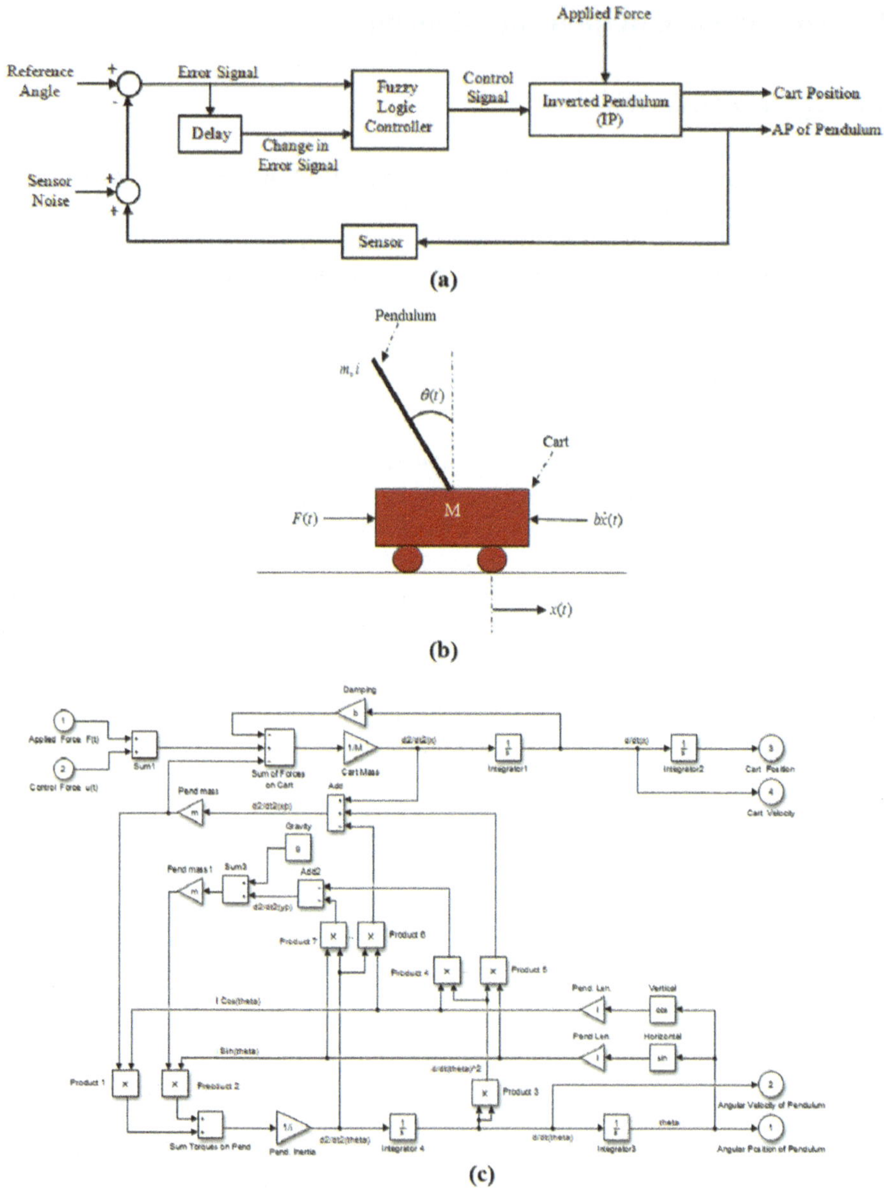

Fig. 1 **a** IP **b** Schematic model of the IP system; **c** Simulation model of the nonlinear IP system; **d** The horizontal force acting on the cart at the time of 0.1 s

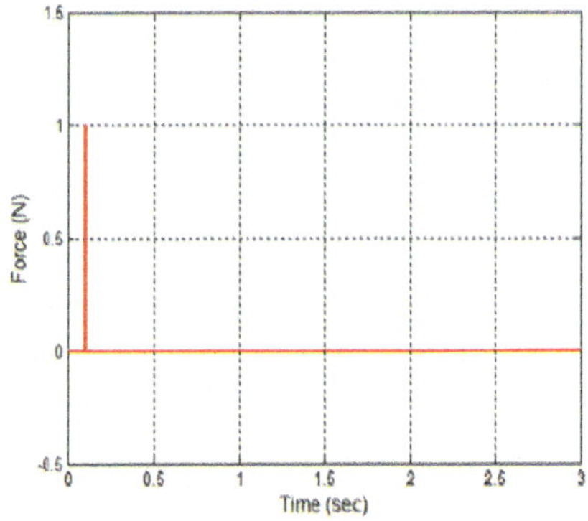

(d)

Fig. 1 (continued)

Table 1 IP model states and parameters

Symbol	Description
$F(t)$	Horizontal force acting on the cart.
$b\frac{\mathrm{d}x(t)}{\mathrm{d}t}$	Frictional force acting on the cart.
$x(t), \frac{\mathrm{d}x(t)}{\mathrm{d}t}, \frac{\mathrm{d}^2x(t)}{\mathrm{d}t^2}$	Cart position, cart velocity, cart acceleration, respectively.
$\theta(t), \frac{\mathrm{d}\theta(t)}{\mathrm{d}t}, \frac{\mathrm{d}^2\theta(t)}{\mathrm{d}t^2}$	AP, AV, angular acceleration of pendulum, respectively.

Table 2 IP model specification values

Symbol	Physical meaning	Value
M	Cart mass	0.5 kg
m	Pendulum mass	0.2 kg
i	Inertia	0.3 kg m^2
g	Gravitational acceleration	9.8 m/s^2
b	Frictional coefficient	0.1 N s/m
l	Pendulum length	0.3 m

CP are being observed owing to the application of 1 N impulsive horizontal force $F(t)$ on the cart at the time $t = 1.0$ s.

The uncontrolled system dynamics are illustrated in Fig. 2a, b. Figure 2a, b illustrates the unstable dynamics under various model uncertainties and disturbances. The unstable dynamics can be reduced by applying suitable control techniques. In this case, the AP of the pendulum is the most essential outcome that needs to be controlled within a stable range through suitable control techniques, and CP is analyzed in order to view the motion trajectory.

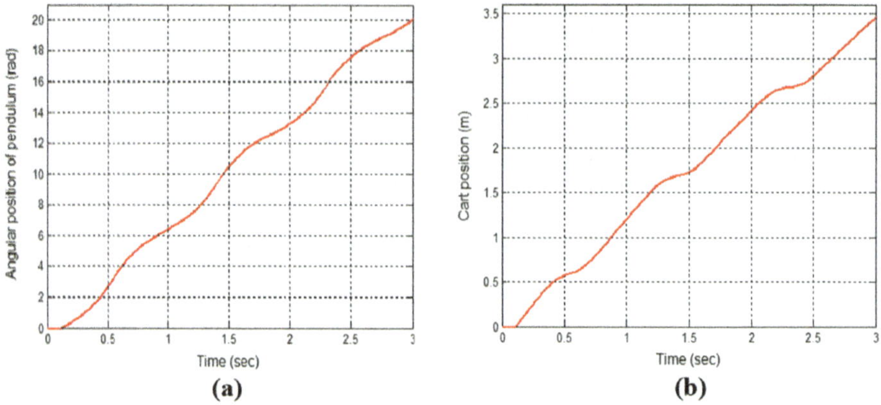

Fig. 2 **a** AP of a pendulum with the application of $F(t)$; **b** CP with the application of $F(t)$

3 Control Algorithm

The FLC algorithm is demonstrated in this section. The closed-loop system response with respect to robustness, accuracy, and stability is analyzed. The control specifications such as settling time t_s, steady-state error e_{ss}, maximum overshoot $O_{Max,}$ and maximum undershoot U_{Max} are also evaluated and examined with proper validation of the controller action.

3.1 FLC Design

The basic block diagram of the proposed fuzzy logic controller for control of AP and CP in IP system is shown in Fig. 1a. There are two separate meanings of fuzzy logic. Generally, fuzzy logic is the further application of multivalued logic and is also known as a logic system. We can also say fuzzy logic is the same as fuzzy sets theory, which relates to the collection of objects with unsharp boundaries and the membership is a point of degree. There is a specific object which degree of membership in a given set which can vary between the range 0–1 in fuzzy set theory. Fuzzy logic is based on sound quantitative and also deals with imprecise information and data. Mathematical theory the values of fuzzy variables are expressed by proper English language. Error in AP of the pendulum in IP system can be defined in linguistic variables like Negative Big (NB), Negative Medium (NM), Negative Small (NS), Zero (ZE), Positive Small (PS), Positive Medium (PM), Positive Big (PB), and each variable can be defined by varying triangular membership function. Seven fuzzy levels were chosen and were defined by a fuzzy set library whose values of the error signal are $e(t)$ and change in error signal is $de(t)$. The larger the number of fuzzy levels, the higher the input resolution. We know that a rule is n-dimensional and n is the number of variables

Table 3 Fuzzy rules for closed loop control of AP of pendulum

$e(t)\backslash de(t)$	NB	NM	NS	ZE	PS	PM	PB
NB	NB	NB	NB	NB	NM	NS	ZE
NM	NB	NB	NB	NB	NS	ZE	PS
NS	NB	NB	NM	NS	ZE	PS	PM
ZE	NB	NM	NS	ZE	PS	PM	PB
PS	NM	NS	ZE	PS	PM	PB	PB
PM	NS	ZE	PS	PM	PB	PB	PB
PB	ZE	PS	PM	PM	PB	PB	PB

included in the rule. The sum of rules is known as rule *R*. FIS editor edits the input and output variables, which are $e(t)$, $de(t)$, and output. After editing, the membership function for each variable is established [7]. The final step involves writing rules in rule editor using the rule given in Table 3.

4 Outcomes and Discussions

The response of the closed-loop IP with the proposed FLC is described in detail in this section. The proposed control approach (FLC) is compared with other popular control algorithms to justify its enhanced performance.

4.1 Performance Analysis of IP System with FLC

In this section, all physical activities of the closed-loop IP model with suggested FLC are examined under different conditions and the huge deviation of applied horizontal force. The closed-loop system dynamics with 1 N impulsive horizontal force $F(t)$ at the time of 1.0 s are displayed in Fig. 3a, b. The outcomes clearly specify the pendulum achieves the zero AP with less t_s and the cart also attains the balance position where the system is absolutely stable. To achieve the enhanced system response, the required control force $u(t)$ is generated by the suggested FLC and demonstrated in Fig. 4a.

4.2 Robustness of the FLC

Figure 4b illustrates the AP of the pendulum with suggested FLC under the huge deviation of the applied horizontal force $F(t)$. The time-domain outcomes under a huge deviation of applied forces show the enhanced performance of the closed-loop

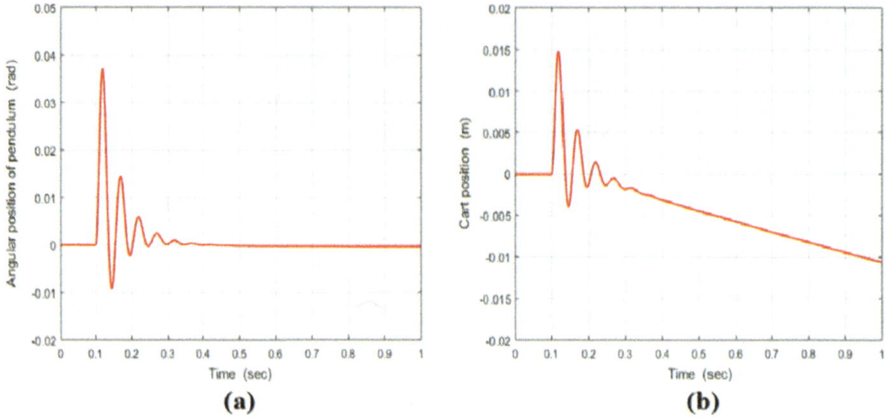

Fig. 3 **a** AP of the pendulum with the application of $F(t)$ based on FLC; **b** CP with the application of $F(t)$ based on FLC

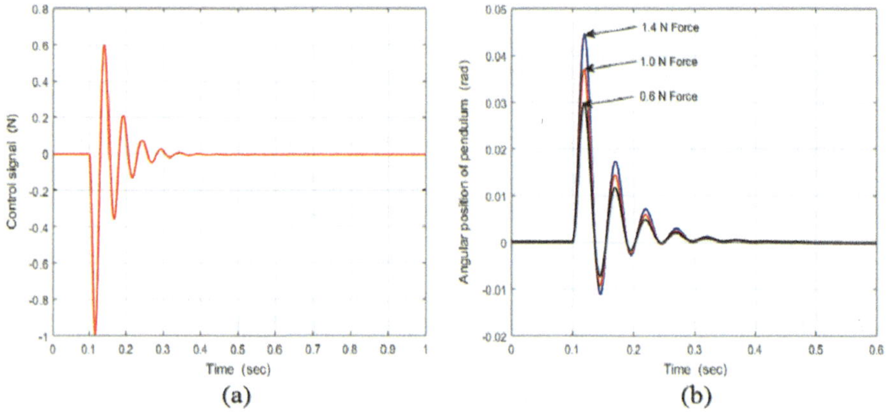

Fig. 4 **a** Controlled signal $u(t)$ of the FLC; **b** AP of the pendulum with a deviation of ± 40% applied $F(t)$ based on FLC

system with FLC. Overall in each case, the pendulum achieves finally zero AP with less t_s. As indicated by the outcomes, the suggested FLC robust performance under the huge deviation of applied horizontal forces is much better.

4.3 *Stability Investigation*

There are four roots in the IP system. One of them lies in the right-hand side of the complex plane as displayed in Fig. 5a. It signifies the system dynamics are unstable in nature. For enhancement of stability of the IP system, the FLC is developed and

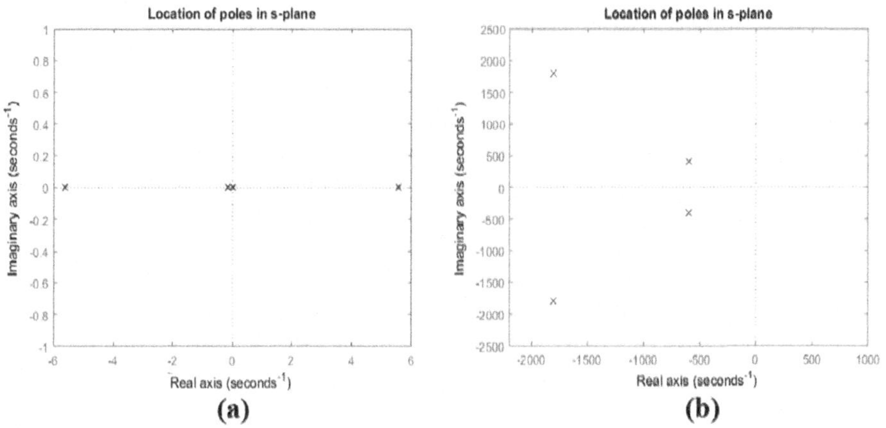

Fig. 5 **a** Location of system roots in the complex plane; **b** Location of closed-loop system roots in the complex plane

implemented in the IP system. As a result, four roots of the IP system are dragged toward the left-hand side of the complex plane as displayed in Fig. 5b. Consequently the system stability is enhanced due to the suggested FLC.

4.4 *Comparative Study*

The suggested FLC approach is compared with other popular control approaches such as PID, $H\infty$, FSM, and BLQG to justify its enhanced performance as a controller. Figure 3a illustrates the effect of the applied force in the AP of the pendulum with the proposed FLC approach. Table 4 presents a comparative analysis with respect to t_s(sec), O_{Max}(rad), U_{Max}(rad), noise (%), and e_{ss}(%). The effect of applied force in the AP of pendulum applying different control approaches as PID, $H\infty$, FSM, and BLQG is also demonstrated in Table 4 based on the references [5, 15, 18, 19],

Table 4 Comparative result analysis related to AP of the pendulum

Controller	PID [5]	$H\infty$ [15]	FSM [18]	BLQG [19]	FLC (Proposed)
Applied force (N)	1	1	1	1	1
t_s(sec)	2.8	2.1	1.8	1.9	0.37
O_{Max}(rad)	0.1	0.107	0.042	0.041	0.037
U_{Max}(rad)	0.01	0.045	0.151	0.01	0.008
Noise (%)	10	5	5	5	5
e_{ss}(%)	0	0	0	0	0

respectively. Similar working conditions are followed with the same level of sensor noise in all control techniques application for comparison.

The AP of pendulum under 1 N impulsive horizontal force is tested. The corresponding results are presented for the various control approaches along with the proposed FLC with respect to time domain specifications such as O_{Max}(rad), U_{Max}(rad), and t_s(sec). The results signify the better controllability of the FLC. The simulation results also demonstrate the high noise and chattering elimination capability with high robustness for the proposed approach. Overall, by looking at the above comparative analysis, the findings of suggested approach advantages are the higher accuracy and stability, more robustness, high noise, and chattering elimination capability, and better capability to handle uncertainty under various conditions and huge deviation of road disturbance.

5 Conclusions

The manuscript proposed a novel control strategy (FLC) to balance the pendulum. To justify its enhanced performance, it has been applied and tested to control the system dynamics of the IP system within the stable range. Initially, the IP system is modeled as a fourth-order state-space representation. Then the proposed control approach (FLC) is designed and implemented. The comparative results clearly reflect that the suggested FLC arrives at better performance than the other control approaches such as PID, $H\infty$, FSM, and BLQG with respect to stability, reliability, and robustness under various abnormal conditions and disturbances. The better performance of the suggested approach in terms of improved accuracy and stability, enhanced robustness, high noise, and chattering elimination capability, and better ability to control uncertainty justifies its real-time application.

References

1. Iqbal, J., Islam R, U., Syed, Z.A., Abdul, A.K., Ajwad, S.A.: Automating industrial tasks through mechatronic systems—a review of robotics in industrial perspective. Tehnic ki vjesnik—Tech. Gaz. **23**, 917–924 (2016)
2. Ajwad, S.A., Asim, N., Islam, R.U., Iqbal, J.: Role and review of educational robotic platforms in preparing engineers for industry. Maejo Int. J. Sci. Technol. **11**, 17–34 (2017)
3. Bettayeb, M., Boussalem, C., Mansouri, R., Al-Saggaf, U.: Stabilization of an inverted pendulum-cart system by fractional PI-state feedback. ISA Trans. **53**, 508–516 (2014)
4. Iqbal, J., Ullah, M., Khan, S.G., Khelifa, B., Ukovic, S.C.: Nonlinear control systems -a brief overview of historical and recent advances. Nonlinear Eng. **6**, 301–312 (2017)
5. Ghosh, A., Krishnan, T., Subudhi, B.: Brief paper—robust proportional-integral-derivative compensation of an inverted cart-pendulum system: an experimental study. IET Control Theor. Appl. **6**(8), 1145–1152 (2012)

6. Wang, C., Yin, G., Liu, C., Fu, W.: Design and simulation of inverted pendulum system based on the fractional PID controller. In: IEEE 11th Conference on Industrial Electronics and Applications (ICIEA). pp. 1760–1764 (2016)
7. Magana, M.E., Holzapfel, F.: Fuzzy-logic control of an inverted pendulum with vision feedback. IEEE Trans. Educ. **41**(2), 165–170 (1998)
8. Ozana, S., Pies, M., Slanina, Z., Hajovsky, R.: Design and implementation of LQR controller for inverted pendulum by use of REX control system. IEEE Int. Conf. Circ. Syst. **1**, 343–347 (2012)
9. Kumar, E.V., Jerome, J.: Robust LQR controller design for stabilizing and trajectory tracking of inverted pendulum. Proc. Eng. **64**, 169–178 (2013)
10. Prasad, L.B., Tyagi, B., Gupta, H.O.: Modelling and simulation for optimal control of nonlinear inverted pendulum dynamical system using PID controller and LQR. In: 6th Asia, Modelling Symposium (Ams), pp. 138–143 (2012)
11. Pasemann, F.: Evolving neurocontrollers for balancing an inverted pendulum. Netw. Comput. Neural Syst. **9**, 1–4 (1998)
12. Deng, L., Gao, S.: The design for the controller of the linear inverted pendulum based on backstepping. Int. Conf. Electron. Mech. Eng. Inf. Technol. (EMEIT). **6**, 2892–2895 (2011)
13. Jörgl, M., Schlacher, K., Gattringer, H.: Passivity based control of a cart with inverted pendulum. Appl. Mech. Mater. **332**, 339–344 (2013)
14. Žilić, T., Pavković, D.: Modeling and control of a pneumatically actuated inverted pendulum. ISA Trans. **48**, 327–335 (2009)
15. Lambrecht, P., Vander, G.: H-infinity control of an experimental inverted pendulum with dry friction. IEEE Contr. Syst. Mag. **13**(4), 44–50 (1988)
16. Wai, R.J., Chang, L.J.: Adaptive stabilizing and tracking control for a nonlinear inverted-pendulum system via sliding-mode technique. IEEE Trans. Ind. Electron. **53**, 674–692 (2006)
17. Tao, C.W., Taur, J., Chang, J.: Adaptive fuzzy switched swing-up and sliding control for the double-pendulum-and-cart system. IEEE Trans. Syst. Man Cybern. B Cybern. **40**(1), 241–252 (2010)
18. Chen, C.S., Chen, W.L.: Robust adaptive sliding-mode control using fuzzy modelling for an inverted-pendulum system. IEEE Trans. Ind. Electron. **45**(2), 297–306 (1998)
19. Patra, A.K., Rout, P.K.: Backstepping linear quadratic gaussian controller design for balancing an inverted pendulum. IETE J. Res. **1**, 1–15 (2019)
20. Khalil, H.K.: Nonlinear Systems, 3rd edn. Prentice Hall, Upper Saddle River, New Jersey (2002)
21. Patra, A.K., Rout, P.K.: Adaptive sliding mode Gaussian controller for artificial pancreas in TIDM patient. J. Process Control **58**, 23–27 (2017)
22. Patra, A.K., Rout, P.K.: Backstepping sliding mode Gaussian insulin injection control for blood glucose regulation in TIDM patient. J. Dyn. Sys. Meas. Control. **140**(9): 091006-091006-15 (2018)
23. Irfan, S., Mehmood, A., Razzaq, M.T., Iqbal, J.: Advanced sliding mode control techniques for inverted pendulum: modelling and simulation. Eng. Sci. Tech., Int. J. https://doi.org/10.1016/j.jestch.2018.06.010 (2018)
24. Ronquillo-Lomeli, G., Ríos-Moreno, G.J.: Nonnlinear identification of inverted pendulum system using Volterra polynomials. Mech. Based Des. Struct. Mach. **44**(1), 5–15 (2016)
25. Kajita, S., Kanehiro, F., Kaneko, K., Fujiwara, K.: Biped walking pattern generation by a simple three-dimensional inverted pendulum model. Adv. Robot. **17**(2), 131–147 (2003)

Fuzzy Logic Controller Design for Stabilizing and Trajectory Tracking of Vehicle Suspension System

Akshaya Kumar Patra, Alok Kumar Mishra, and Ramachandra Agrawal

Abstract The aim of this paper is to design a Fuzzy Logic Controller (FLC) in a Vehicle Suspension (VS) system for ride comfort improvement through absorption of shock that results from an uneven and rough road. The FLC is a novel approach whose gains dynamically vary with respect to the error and change in the error signal. The validation of the improved control performance of FLC is established by comparative result investigation with other published control algorithms. The comparative results clearly reveal the better response of the suggested approach to control the oscillation of the VS system within a stable range with respect to the accuracy, robustness and capability to control uncertainties.

Keywords Ride comfort · Road profile · Vehicle suspension system · FLC

1 Introduction

In recent times, many researchers in the field of Vehicle Suspension (VS) system dynamics have devoted to arriving at an optimal solution with conciliation between vehicle handling, ride comfort and stability. There is a need for a better approach today as the above problems are very much evident in modern vehicle cases. Specific to luxury cars, even if outstanding ride qualities are achieved, it is limited to acquiring adequate handling behavior. With reference to sports vehicles, although it is provided with very good handling capability, it fails to provide desirable ride quality. There are

A. K. Patra (✉) · A. K. Mishra
Department of EEE, ITER, Siksha 'O' Anusandhan (Deemed to be University), Bhubaneswar, India
e-mail: hiakp@yahoo.com

A. K. Mishra
e-mail: malok2010@gmail.com

R. Agrawal
Department of EE, ITER, Siksha 'O' Anusandhan (Deemed to be University), Bhubaneswar, India
e-mail: ramachandra1agrawal@gmail.com

D. Mishra et al. (eds.), *Intelligent and Cloud Computing*,
Smart Innovation, Systems and Technologies 194,
https://doi.org/10.1007/978-981-15-5971-6_60

many options in between for variations in the designing stage of the vehicle manufacturers to meet the customer needs. From the designing point of view, passenger comfort and vehicle control are the two primary objectives to be considered. Road disturbances such as bumps or potholes are to be handled not to sacrifice the passengers' comfort. At the control stage, these factors are generally considered either through keeping the vehicle body from rolling and pitching excessively, or by maintaining good contact between the tire and the road.

Nowadays, hydraulic dampers (shock absorbers) and springs are extensively used for vehicle suspensions. As a principle, these are charged with the job of absorbing bumps, minimizing the vehicle's body motions during acceleration, braking and turning of the vehicle, and keeping the tires in contact with the road. However, from the design point of view, these objectives are contradictory to each other to achieve all simultaneously at their optimum level.

The spring and the damper are the two essential components in a VS design. The spring design mostly depends on the weight of the vehicle. The damper design is based on the suspension's placement on the compromise curve and so it is essential to be perfectly chosen to make the optimal vehicle performance for any type of vehicle. For ideal performance, the damper should act such that passengers are isolated from low-frequency road disturbances and high-frequency road disturbances are absorbed. High damping is essential to achieve for the best isolation of passengers from low-frequency disturbances.

However, even though it is desirable to design a high damping system, it is on the other degrade the high frequency absorption rate. In other way, providing low damping for the damper offers adequate high-frequency absorption in terms of scarifying low-frequency isolation. To meet these contradictory objectives, it is essential to design and focus on automotive suspensions without compromising any of the factors mentioned above. As a solution to the above, three types of suspensions can be improved. These are passive, fully active, and semi-active types of suspensions. The spring and damper are the two basic components of the conventional passive suspension. Both the components are considered and fixed at the design stage. The suspension stores energy in the spring. Later, it dissipates energy through the damper.

It is very much needed for further research to develop robust control algorithms to enhance the performance of the VS system. This in turn enhances the vehicle capabilities to handle the aforementioned issues. An ideal design of the VS system needs to achieve many performance characteristics such as (1) control of body movement, (2) control of suspension movement, and force distribution. From the performance point of view, the VS should be able to isolate the body for comfort against the road impact and inertial disturbances. These are generally associated with cornering and braking or acceleration of the vehicle system [1]. As discussed above, many performance objectives are conflicting in nature, that to enhance one the other factor degrades. Considering all the objectives, as a goal for the designing of a suspension system is difficult to meet [2]. The minimization of vertical force to the passengers can be achieved by minimizing the vertical vehicle body acceleration of the suspension. Another factor which plays a vital role in passengers' comfort is the optimal contact between the wheels and the road, and this is essential in various driving circumstances in order to maximize safety factors [3]. Among a few designs in the past, the system presented in [4] based on unconstrained optimizations for the Passive Suspension (PS) system case is widely accepted and used. This successfully performs to achieve the desirability of low-suspension stiffness, reduced Unsprung Mass (UM), and an optimum damping ratio for better controllability. As the PS system performs satisfactorily to some extent, it is considered in many applications for the VS system. However, both the suspension spring and damper do not supply energy to the Suspension System (SS). They only control the motion of the vehicle body and wheels by limiting the suspension velocity. This is computed according to the rate specified by the designer. To overcome this issue, the Active Suspension (AS) system is considered as an efficient option for this application. The AS systems have the ability to dynamically respond to changes in the road profile. It is due to this fact that it can supply energy to produce relative motion between the body and the wheel. Sensors are provided in the suspension system to measure the parameters dynamically. The parameters such as body velocity, suspension displacement, and wheel velocity and wheel and body acceleration are sensed and computed for the controller as input parameters [5]. An AS can be thought of as the integration of the passive components to actuators that supply additional forces. These forces are computed by a control law based on the input data from the sensors fixed to the vehicle. For modeling the real-time dynamic conditions, the uncertainties due to system design and other external disturbances are needed to be considered for the controller. This motivates a better control design to increase the robustness and controllability under uncertainties and disturbances.

During the past three decades, so many control strategy techniques have been suggested and tested by absorbing the shocks due to a rough and bumpy road in the case of the VS system. Time-discrete and switching PID control strategy is implemented in VS problems with variable control gains based on the measured suspension variables [6, 7]. However, the optimal gain parameter setting, a lesser range of robust control, and need for change of gain setting with varying conditions are the major limitations to limit the real-time application of these controllers. Among other projected robust control algorithms applied for limiting the oscillation and velocity of the VS system are fuzzy logic control [8–13], fuzzy-PID control [14], genetic algorithm [15], neural networks [16], Neuro-Fuzzy (NF) control [17], Linear Quadratic Regulator (LQR) [18], H-infinity control [19], and Sliding Mode (SM) control [20–22]. However, even if these control techniques are implemented effectively by absorbing the shocks due to rough and bumpy roads in case of the VS system with enhanced accuracy and damping of oscillation, they still fail to handle various constraints and random changes found in a suspension environment. These approaches are not fully sensitive to the disturbances and the uncertainties of the model in spite of the improved performance. Hence, optimal control parameter setting for still better performance and for avoiding slow response following road disturbances (road impact), the current work suggests an alternative novel technique based on the fuzzy control concept. Application of the suggested approach to control the oscillation and velocity in a vehicle suspension system results in ensuring a better robust controller in comparison to other contemporary well-established approaches under both harmonized and incompatible uncertainties.

The highlights of this manuscript are as follows:

- Development of a SIMULINK model of a VS system.
- Design of a novel FLC to regulate the oscillation of the VS system within a stable range.
- Evaluation of the control actions of the FLC under a huge deviation of road disturbance.
- Comparative investigation to certify the better response of the FLC.

This manuscript is prepared as follows. Section 2 concisely illustrates the VS model with mathematical details reflecting on its dynamic characteristics of its suspension vibration process. Also, it clearly demonstrates the simulation execution of the system in MATLAB environment. A detailed presentation on how the control technique is formulated and how it is implemented for this problem is presented in Sect. 3. Comparative results of the proposed approach with other published control techniques and the related analysis are provided in Sect. 4. The concluding comments are summarized in Sect. 5.

2 Problem Formulation and Modeling

The problem formulation and modeling of the VS system are described in this section.

2.1 System Overview

The overall closed-loop model of the VS system is depicted in Fig. 1a. The road disturbance $w(t)$ and $v(t)$ are reflected as the process disturbance and the sensor noise, respectively, in this study. The controller receives information about the oscillation of the VS system as input to provide the optimal control force $u(t)$, and it is applied in between the UM and the Sprung Mass (SM) to reduce the relative motion between them.

2.2 Modeling of VS System

In the year 1958, the concept of the VS system emerged into limelight. It has been found in the later period that the vibration suppression capability is limited in case of traditional PS and semi-active suspension systems. This leads to a focus on an AS system with additional control forces as a better alternative to the above drawbacks to suppress the oscillations and be used in the modern vehicle industry. The major

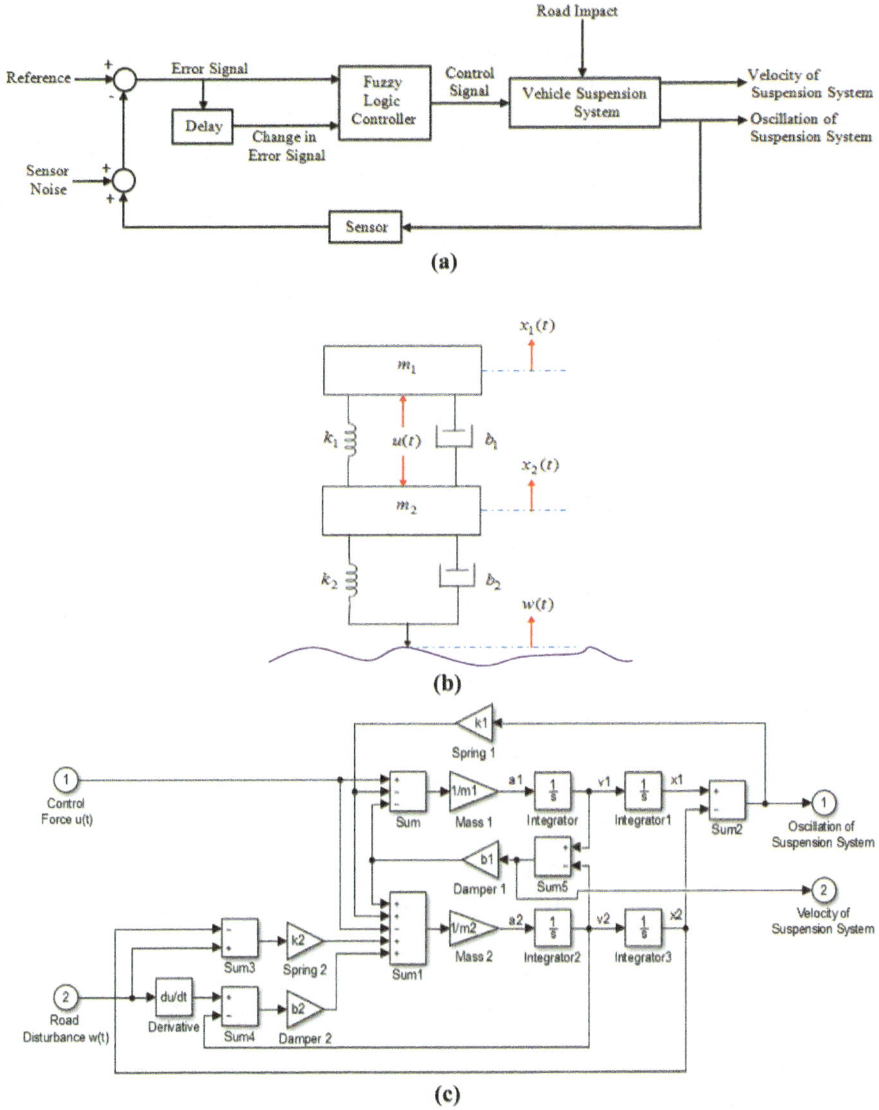

Fig. 1 **a** The VS model with FLC; **b** schematic model of the VS system; **c** simulation model of the nonlinear VS system

factor that differentiates the active suspension system performance-wise is the ability to inject energy into system dynamics via actuators unlike energy dissipations in case of conventional SS. To produce the desirable control force to handle the variety of road disturbances (road impact) in real-time applications, actuators are placed in between the UM and the SM in an AS system. The most challenging task of designing an

AS system is to enhance the ride comfort by absorbing the shocks due to rough and bumpy roads. The force actuator in the case of the AS system is capable to add and also dissipate energy from the system. This results in an increase in the ride comfort and vehicle handling due to the capability of the SS to regulate the vehicle altitude and to eliminate the adverse effects of braking and vehicle roll during cornering and braking. The schematic model of the VS system is demonstrated in Fig. 1b.

All the physical activities of the VS system are mathematically expressed and specified in Eqs. (1) and (2). All states and parameters of the model are represented in Table 1. All specifications of the VS model are documented in Table 2. The SIMULINK diagram of the VS model is established with respect to Eqs. (1) and (2) as displayed in Fig. 1c.

$$m_1\left(\frac{\mathrm{d}^2x_1(t)}{\mathrm{d}t^2}\right) = -b_1\left(\frac{\mathrm{d}x_1(t)}{\mathrm{d}t} - \frac{\mathrm{d}x_2(t)}{\mathrm{d}t}\right) - k_1(x_1(t) - x_2(t)) + u(t) \quad (1)$$

$$\begin{aligned} m_2\left(\frac{\mathrm{d}^2x_2(t)}{\mathrm{d}t^2}\right) &= b_1\left(\frac{\mathrm{d}x_1(t)}{\mathrm{d}t} - \frac{\mathrm{d}x_2(t)}{\mathrm{d}t}\right) + k_1(x_1(t) - x_2(t)) \\ &\quad + b_2\left(\frac{\mathrm{d}w(t)}{\mathrm{d}t} - \frac{\mathrm{d}x_2(t)}{\mathrm{d}t}\right) + k_2(w(t) - x_2(t)) - u(t) \end{aligned} \quad (2)$$

Table 1 VS model states and parameters

Symbol	Description
$w(t)$	Road impact acting on the UM
$x_1(t)$	Displacement of the SM in the vertical direction
$x_2(t)$	Displacement of the UM in the vertical direction
$\dot{x}_1(t)$	Velocity of the SM in the vertical direction
$\dot{x}_2(t)$	Velocity of the UM in the vertical direction
$[x_1(t) - x_2(t)]$	Oscillation of suspension system
$u(t)$	Applied control force between the SM and UM

Table 2 VS model specification values

Symbol	Quantity/meaning	Value
m_1	SM (mass of the vehicle supported by the suspension)	2500 kg
m_2	UM (mass of the axle and wheel)	320 kg
k_1	Spring constant of SS	80,000 N/m
k_2	Spring constant of wheel and tire	500,000 N/m
b_1	Damping coefficient of SS	350 N s/m
b_2	Damping coefficient of wheel and tire	15,020 N s/m

2.3 Performance Analysis of VS System

There are four poles in the VS system. One of them lies on the right-hand side of the complex plane. As a result, the system becomes unstable. This needs the design of an adaptive controller for improving the stability of the system by means of shifting the poles into the left-hand side of the complex plane. The VS system SIMULINK model in the open-loop form is depicted in Fig. 1c. The VS system has two inputs and two outputs. The control force $u(t)$ and road disturbance $w(t)$ are the two inputs of the VS system. The oscillation and velocity of the suspension are the two outputs of the VS system. An uncontrolled oscillation and velocity of the suspension system are observed owing to the application of 0.1 m impulsive road disturbance on the unsprung mass at the simulation time of 1.0 s. The irregular oscillation and velocity of the SS are illustrated in Fig. 2a, b. Figure 2a, b illustrates the unstable dynamics under various model uncertainties and disturbances. These unstable dynamics can be reduced by applying suitable control algorithms. In this case, the oscillation of the suspension system is the most essential outcome that needs to be controlled within a stable range through suitable control techniques, and the velocity of the suspension system is analyzed in order to view the motion trajectory.

3 Control Algorithm

The FLC control algorithm is demonstrated in this section. The closed-loop system response with respect to robustness, accuracy, and stability are analyzed. The control specifications such as settling time t_s, steady-state error e_{ss}, maximum overshoot

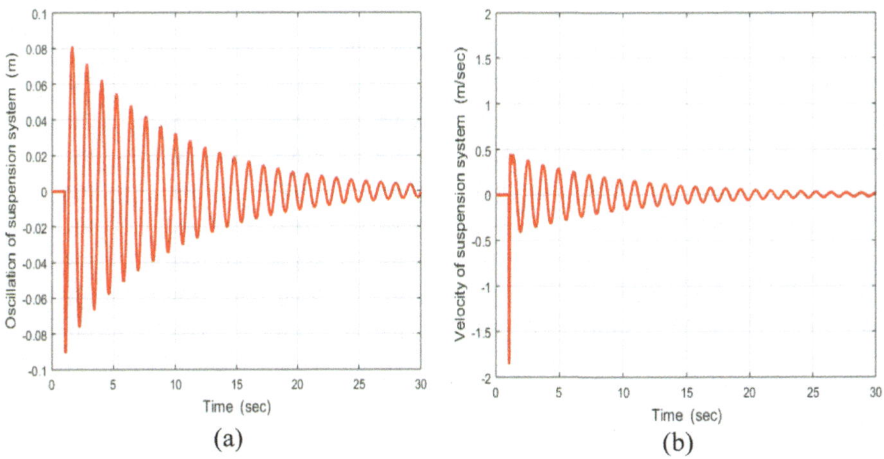

Fig. 2 **a** Response of oscillation of the SS with road impact; **b** response of velocity of the SS with road impact

O_{Max}, and maximum undershoot U_{Max} are also evaluated and examined with proper validation of the controller action.

3.1 FLC Design

The basic block diagram of the proposed fuzzy logic controller for control of oscillation in a VS system is shown in Fig. 1a. There are two separate meanings of fuzzy logic. Generally, fuzzy logic is the further application of multivalued logic and is also known as a logic system. We can also say fuzzy logic is same as the fuzzy set theory, that which relates to collection of objects with unsharp boundaries and the membership is a point of degree. There is a specific object whose degree of membership in a given set can vary between the 0 and 1 in fuzzy set theory. Fuzzy logic is based on sound quantitative and also deals with imprecise information and data. In mathematical theory, the values of fuzzy variable are expressed in English language. Errors in AP of a pendulum in IP system can be defined through linguistic variables like Negative Big (NB), Negative Medium (NM), Negative Small (NS), Zero (ZE), Positive Small (PS), Positive Medium (PM), and Positive Big (PB), and each variable can be defined by a varying triangular membership function. Seven fuzzy levels were chosen and were defined by fuzzy set library whose values of the error signal is $e(t)$, and change in error signal is $\mathrm{d}e(t)$. The larger the number of fuzzy levels, the higher the input resolution. We know that a rule is n dimensional and n is the number of variables included in the rule. The sum of rules is known as rule R. FIS editor edits the input and output variables, which are $e(t)$, $\mathrm{d}e(t)$, and the output. After editing, the membership function for each variable is established [11]. The final step involves writing rules in a rule editor using the rule given in Table 3.

Table 3 Fuzzy rules for closed-loop control of oscillation in the VS system

$e(t)\backslash \mathrm{d}e(t)$	NB	NM	NS	ZE	PS	PM	PB
NB	NB	NB	NB	NB	NM	NS	ZE
NM	NB	NB	NB	NB	NS	ZE	PS
NS	NB	NB	NM	NS	ZE	PS	PM
ZE	NB	NM	NS	ZE	PS	PM	PB
PS	NM	NS	ZE	PS	PM	PB	PB
PM	NS	ZE	PS	PM	PB	PB	PB
PB	ZE	PS	PM	PM	PB	PB	PB

4 Result and Discussions

The performance of the closed-loop VS system with FLC is described in detail in this section. The suggested control approach is compared with other popular control algorithms to justify its enhanced performance.

4.1 Performance Analysis of VS System with FLC

In this section, all physical activities of the closed-loop VS model with suggested FLC are examined under different conditions and the huge deviation of road disturbances. The time-domain response of the oscillation and the velocity of the suspension system with 0.1 m impulsive road disturbance at the simulation time of 1.0 s is displayed in Fig. 3a, b. The output results clearly specify that the suspension system attains zero oscillation and zero velocity with less settling time and also attains the balance position where the system is absolutely steady. To achieve the enhanced system response, the required control force $u(t)$ is generated by the suggested FLC and is demonstrated in Fig. 4.

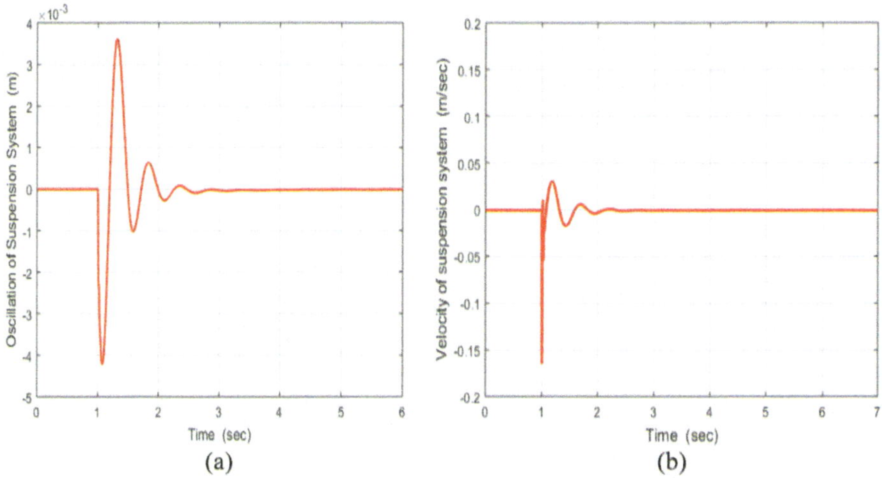

Fig. 3 **a** Response of oscillation of the SS with road impact based on FLC; **b** response of velocity of the SS with road impact based on FLC

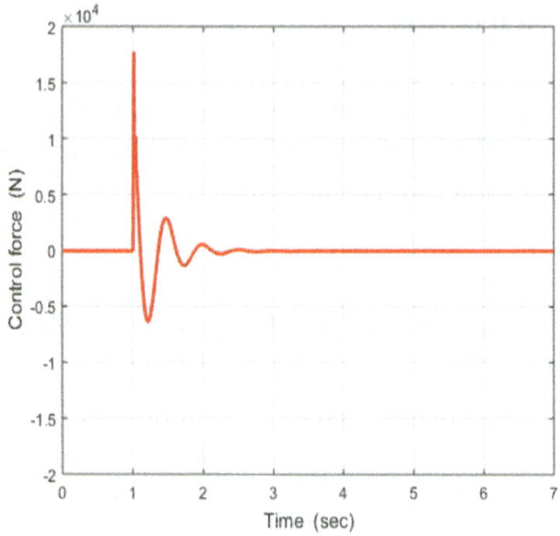

Fig. 4 Control force generated by FLC

4.2 Robustness of the FLC

Figure 5a, b illustrates the oscillation and the velocity of the VS system with suggested FLC under the huge deviation of road disturbance $w(t)$. The time domain outcomes under a huge deviation of road disturbances show the enhanced performance of the closed-loop VS model with FLC. Overall in each case, the VS system achieves finally zero oscillation and zero velocity with less settling time. As indicated by the

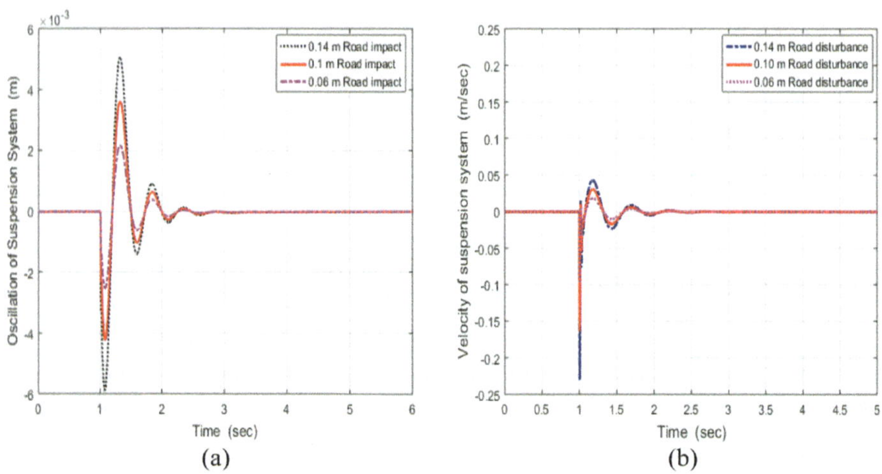

Fig. 5 **a** Oscillation of the SS with deviation of ±40% road impact based on FLC; **b** velocity of the SS with deviation of ±40% road impact based on FLC

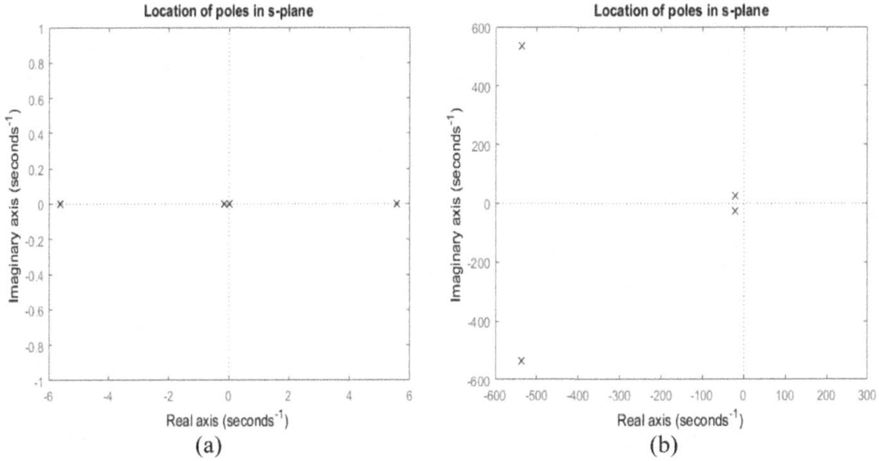

Fig. 6 **a** Position of system poles in a complex plane; **b** position of closed-loop system poles in the complex plane

results, the suggested controller's robust performance under the huge deviation of road disturbances compared to other published control techniques is much better.

4.3 Verification of Stability

There are four poles in the VS system. One of them lies on the right-hand side of the complex plane as displayed in Fig. 6a. It signifies the system dynamics are unstable in nature. For enhancement of stability of the VS system, the FLC is developed and implemented in the VS system. As a result, four poles of the VS system are dragged toward the left-hand side of the complex plane as displayed in Fig. 6b. Consequently, the system stability is enhanced due to the suggested FLC.

4.4 Comparative Study

The proposed FLC control approach is compared with other popular control approaches such as PID, NF, $H\infty$, and SM to justify its enhanced performance as a controller. Figure 3a illustrates the effect of road impact in the oscillation of the suspension system with the suggested FLC approach. Table 4 presents a comparative analysis with respect to t_s (s), O_{Max} (m), U_{Max} (m), noise (%), and e_{ss} (%). The effect of road disturbance in the oscillation of the suspension system applying different control approaches such as PID, NF, $H\infty$, and SM is also documented in Table 4 based on the references, and [6, 17, 19, 21], respectively. Similar working conditions

Table 4 Comparative result analysis related to oscillation of the suspension system

Controller	PID [6] [Ref. Fig. 5a]	NF [17] [Ref. Fig. 2a]	$H\infty$ [19] [Ref. Fig. 2a]	SM [21] [Ref. Fig. 4a]	Proposed FLC [Ref. Fig. 3a]
Road disturbance (m)	0.1	0.1	0.1	0.1	0.1
t_s (s)	4.2	3.7	3.1	3.0	2.5
O_{Max} (m)	0.0050	0.0047	0.0043	0.0040	0.0036
U_{Max} (m)	0.0042	0.0046	0.0041	0.0048	0.0041
Noise (%)	10	5	5	5	1
e_{ss} (%)	0	0	0	0	0

are followed with the same level of sensor noise in all control technique applications for comparison.

The oscillations of the suspension system under 0.1 m impulsive road disturbance are tested. The corresponding results are presented for the various control approaches along with the proposed FLC with respect to time-domain specifications such as O_{Max} (m), U_{Max} (m), and t_s (s). The results signify the better controllability of the FLC. The simulation results also demonstrate the high noise elimination capability with high robustness for the proposed approach. Overall, by looking at the above comparative analysis, the findings of suggested approach advantages are the higher accuracy and stability, more robustness, high noise elimination capability, and better capability to handle uncertainty under various conditions and huge deviation of road disturbance.

5 Conclusions

In this manuscript a novel control strategy, FLC, is proposed based on a self-tuned approach. To justify its enhanced performance, it has been applied and tested to control the oscillation and the velocity of the suspension system in the vehicle. In suggested FLC, a self-tuned algorithm is utilized to enhance the control performance. The comparative results clearly reflect that the suggested FLC arrives at better performance than the other control approaches such as PID, NF, $H\infty$, and SM with respect to stability, reliability, and robustness under various abnormal conditions and disturbances. The better response of the suggested approach with respect to improved accuracy and stability, enhanced robustness, high noise elimination capability, and better ability to control uncertainty justifies its real-time application.

References

1. Wright, P.G.: The application of active suspension to high performance road vehicles. In: Microprocessors in Fluid Engineering. IMechE Conference Publications (1984)
2. Fateh, M.M., Alavi, S.S.: Impedance control of an active suspension system. J. Mechatron. **19**, 134–140 (2009)
3. Lin, J.S., Kanellakopoulos, I.: Nonlinear design of active suspension. In: 34th IEEE Conference 1995, vol. 17, pp. 45–59 (1995)
4. Alleyen, A., Hedrick, J.K.: Nonlinear adaptive control of active suspensions. IEEE Trans. Control Syst. Technol. **3**(1), 845–860 (1995)
5. Esmailzadeh, E., Taghirad, H.D.: Active vehicle suspensions with optimal state feedback control. J. Mech. Sci. **200**(4), 1–18 (1997)
6. Kumar, M.S.: Development of active suspension system for automobiles using PID controller (2008)
7. Talib, M.H.A., Darns, I.Z.M.: Self-tuning PID controller for active suspension system with hydraulic actuator. In: 2013 IEEE Symposium on Computers & Informatics (ISCI), pp. 86–91, Apr 2013
8. Pekgökgöz, R.K., Gürel, M.A., Bilgehan, M., Kisa, M.: Active suspension of cars using fuzzy logic controller optimized by genetic algorithm. Int. J. Eng. Appl. Sci. (IJEAS) **2**(4), 27–37 (2010)
9. Li, H.: Reliable fuzzy control for active suspension systems with actuator delay and fault. IEEE Trans. Fuzzy Syst. **20**(2), 342–357 (2012)
10. Lin, Y.-J., Lu, T.-Q., Padovan, J.: Fuzzy logic control of vehicle suspension systems. Int. J. Veh. Des. **14**(5), 457–470 (1993)
11. Yoshimura, T., Nakaminami, K., Kurimoto, M.: Active suspension of passenger cars using linear and fuzzy logic controls. Control Eng. Pract. **1**, 41–47 (1998)
12. Yeh, E.C., Tsao, Y.J.: A fuzzy preview control scheme of active suspension for rough road. Int. J. Veh. Des. **15**(1), 166–180 (1994)
13. Yang, C.: Active Suspension System of Automobile Based on Fuzzy Control. Northeastern University (2007)
14. Yang, Q., Zhou, K., Zhang, W., Xing, X., Yuan, C.: Fuzzy-PID control on semi active air suspension. Trans. Chin. Soc. Agric. Mach. **39**(9), 24–29 (2008)
15. Du, H., Lam, J., Sze, K.Y.: Non-fragile output feedback H1 vehicle suspension control using genetic algorithm. Eng. Appl. Artif. Intell. **16**, 667–680 (2003)
16. Wang, Y.J.: Analysis of Vehicle Suspension Control Using Neural Networks. Northeastern University (2007)
17. Aldair, A.A., Wang, W.J.: A neurofuzzy controller for full vehicle active suspension systems. J. Vib. Control **18**(12), 1837–1854 (2012)
18. ElMadany, M.M., Abduljabbar, Z.S.: Linear quadratic Gaussian control of a quarter-car suspension. Veh. Syst. Dyn. **32**(6), 479–497 (1999)
19. Kaleemullah, M., Faris, W.F., Hasbullah, F.: Design of robust H∞, fuzzy and LQR controller for active suspension of a quarter car model. In: 2011 4th International Conference on Mechatronics (ICOM), pp. 1–6, May 2011. IEEE
20. Li, H., Yu, J., Hilton, C., Liu, H.: Adaptive sliding-mode control for nonlinear active suspension vehicle systems using T-S fuzzy approach. IEEE Trans. Ind. Electron. **60**(8), 3328–3338 (2013)
21. Chen, P.C., Huang, A.C.: Adaptive sliding mode control of nonautonomous active suspension systems with time-varying loading. J. Sound Vib. **282**, 1119–1135 (2005)
22. Patra, A.K., Rout, P.K.: Adaptive sliding mode Gaussian controller for artificial pancreas in TIDM patient. J. Process Control **59**, 13–27 (2017)

Fuzzy-Controlled Power Factor Correction Using Single Ended Primary Inductance Converter

Alok Kumar Mishra, Akshaya Kumar Patra, Ramachandra Agrawal, Shekharesh Barik, Manoj Kumar Debanath, Samarjeet Satapathy, and Jnana Ranjan Swain

Abstract This paper presents a comparative analysis between the two controllers, PI and fuzzy logic controller, used for power factor correction using DC to DC Single Ended Primary Inductance Converter (SEPIC) converter. MATLAB/Simulink models of SEPIC converter are developed to improve the power factor and output voltage regulation. Hysteresis Current Control (HCC) technique is used for input Power Factor Correction (PFC) and Proportional Integral Controller (PIC) or Fuzzy Logic Controller (FLC) is adopted for the regulation of the output voltage. The system is tested at both steady state and transient condition, and its performance is then estimated and compared in terms of various parameters like Total Harmonic Distortion (THD), Input Power Factor (IPF), Output voltage ripple for PIC and FLC.

Keywords FLC · PIC · SEPIC converter · HCC · Power factor correction circuits

A. K. Mishra (✉) · A. K. Patra
EEE Department, ITER, Siksha 'O' Anusandhan (Deemed to be University), Bhubaneswar 751030, India
e-mail: malok2010@gmail.com

A. K. Patra
e-mail: hiakp@yahoo.com

R. Agrawal · M. K. Debanath · S. Satapathy · J. R. Swain
EE Department, ITER, Siksha 'O' Anusandhan (Deemed to be University), Bhubaneswar 751030, India
e-mail: ramachandra1agrawal@gmail.com

M. K. Debanath
e-mail: mkd.odisha@gmail.com

S. Satapathy
e-mail: samarajeets832@gmail.com

J. R. Swain
e-mail: jnana5500@gmail.com

S. Barik
CSE Department, DRIEMS (Autonomous), Cuttack 754022, India
e-mail: shekharesh@gmail.com

D. Mishra et al. (eds.), *Intelligent and Cloud Computing*, Smart Innovation, Systems and Technologies 194,
https://doi.org/10.1007/978-981-15-5971-6_61

1 Introduction

According to the international standards and for the best power transfer and utilization, power factor correction has become a necessity. To get a DC output voltage, we use a rectifier and a parallel capacitor [1] as shown in Fig. 1a. The input voltage and current are shown in Fig. 1b. Figure 1b depicts a pulse-shaped input current with much harmonics content causing a very poor power factor. The IEEE and IEC are some international entities to define or standardize the permissible limits of the harmonic content in line current such as IEEE 519 and IEC 61000-3-2 [2]. The design of an AC to DC power converter to overcome these power quality issues like obtaining Unity Power Factor (UPF) at the AC input mains and a close regulation of the DC output voltage has been discussed in [3–7]. So, for small power uses in single-phase supplies, DC–DC converters are used which are switch based and controlled accordingly to ensure UPF at the mains' side. The main motive is to emulate a resistive circuit when seen from mains' side and improve the power factor. Theoretically, there exist three families of non-isolated power factor correction topologies which are Buck, Boost and Buck-Boost topologies [8, 9]. The Buck topology is generally used when low output voltage is required but has high-frequency commuted current at the input. Due to commuted nature it exhibits a discontinuous nature, and hence should be connected with a high-speed recovery circuit. The major drawback faced by Buck topology is the inclusion of a filter of high frequency range within the diode bridge and the power source. The filtering inductor causes a gradual variable current in the input end in the topology of a Boost converter. This topology provides high output voltage and leads to over-voltage stress at the switches [10]. The third family is Buck-Boost topology and finds its applications in power factor correction circuits. It includes Cuk and SEPIC converters [11, 12]. These converters differ from each other by the output voltage polarity. In case of the Cuk converter, the position of the free-wheeling diode and the inductor is reversed from that of the SEPIC converter to obtain reversed output voltage polarity. In this paper, we have used the SEPIC-type PFC circuit to improve the input-side power factor of the diode bridge rectifier, lower the THD factor with decreased ripple contained voltage at output [13–16]. The entire modelling and the design schemes of PFC, AC to DC power converters are carried out in the MATLAB/Simulink environment.

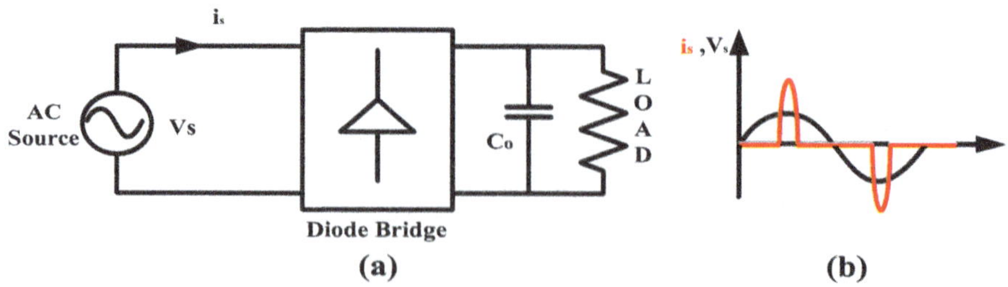

Fig. 1 **a** AC–DC rectifier, **b** source voltage and current waveform

2 Circuit Configuration and Design Equation

To improve the power factor at the input and to get a regulated voltage at the output, we have to interface the SEPIC converter in between the bridge rectifier and the load as shown in Fig. 2. Adopting this topology for a low-voltage DC bus provides a low ripple current at its input, which is rare in case of conventional converters of Buck-Boost type. With adequate coupling coefficient design in a SEPIC converter, the same magnetic core can accommodate the output and input inductors. When switch is on inductor L_1 and L_2 stores energy, inductor current increases linearly and output capacitor supply power to the load and diode is reverse biased applying KVL and KCL the equations are as described in (1)–(4). When switch is OFF applying KVL and KCL the equation are described in (5)–(8).

$$L_1 \frac{di_1}{dt} = v_{in} \tag{1}$$

$$L_2 \frac{di_2}{dt} = v_{C_1} \tag{2}$$

$$C_1 \frac{dv_{C_1}}{dt} = -iL_2 \tag{3}$$

$$C_o \frac{dv_{C_o}}{dt} = -\frac{v_{C_o}}{R} \tag{4}$$

$$L_1 \frac{di_1}{dt} = v_{in} - v_{C_1} - v_{C_o} \tag{5}$$

$$L_2 \frac{di_2}{dt} = -v_{C_1} \tag{6}$$

$$C_1 \frac{dv_{C_1}}{dt} = iL_1 \tag{7}$$

$$C_o \frac{dv_{C_o}}{dt} = iL_1 + iL_2 - \frac{v_{C_o}}{R} \tag{8}$$

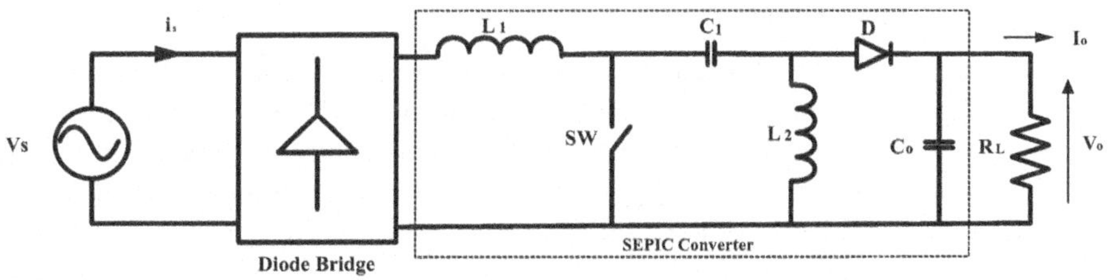

Fig. 2 SEPIC converter topology for power factor correction

Using the above equations we can find the values of circuit components which are described in (9)–(12), where v_s is the RMS source voltage, d is the duty cycle, ΔiL represents inductor current ripple, Δv_c, Δv_o are the voltage ripples of the output voltage and capacitor voltage, respectively, and f_s is the switching frequency. P_o is the output power, V_o is the output voltage, and I_o is the output current.

$$L_1 = \frac{v_s * d}{\Delta i L_1 * f_s} \quad (9)$$

$$L_2 = \frac{v_s * d}{2 * \Delta i L_2 * f_s} \quad (10)$$

$$C_1 = \frac{i L_2 * d}{\Delta v_{C_1} * f_s} \quad (11)$$

$$C_o = \frac{P_o}{4\pi * f_s * V_o * \Delta v_o} \quad (12)$$

3 Different Methods of Power Factor Correction

There are two objectives of power factor correction: 1. To get a regulated voltage at the output. 2. The wave shape of the current at the input must be a sine wave. To achieve the first objective, we have to use a feedback loop at the output. There are two methods to achieve the second objective. The first method is called "Multiplier Approach" and the second is called "Voltage Follower Approach". In Multiplier Approach, a feedback loop of input current is used to govern the DC to DC power converter to function as an input voltage programmed current sink as shown in Fig. 3. In this paper, Multiplier Approach is used. Multiplier Approach Control is further subdivided into four different methods for generating the gate pulse for the SEPIC converter, such as

1. Hysteresis current control,
2. Peak current control,
3. Average current control,
4. Borderline current control.

Comparing with the above four methods, Hysteresis Current Control (HCC) provides very low distorted waveforms of input current and does not need any ramp compensation, hence, here we will take the hysteresis control technique. In Fig. 4, two sinusoidal current references $I_{P,ref}$ and $I_{V,ref}$ are generated from I_{ref}, which shows the current peak and valley of the inductor, respectively. In this control technique, if the current of the inductor falls below the lower reference $I_{V,ref}$, then the switch is ON and if the current of inductor rises above the upper reference $I_{P,ref}$, the switch is

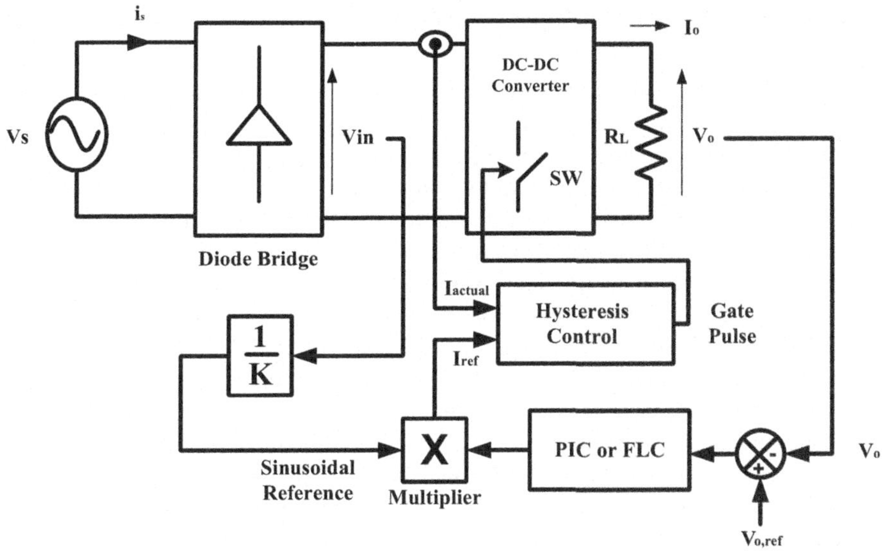

Fig. 3 Multiplier approach control technique

Fig. 4 Hysteresis current control

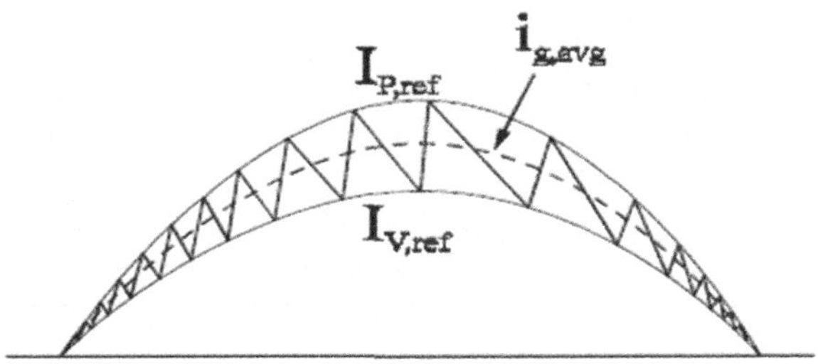

OFF. This control technique provides a variable frequency control structure which is the demerit of this method.

4 PIC for Output Voltage Regulation

The name Proportional-Integral controller itself says that this is a type of control loop mechanism adopted in continuous modulated operation of control. A PIC repeatedly calculates the error in between a set point (SP) value and measured process value (PV) and gets a correction accordingly in the PI base. In real-time operations, it accordingly provides an accurate and respective change of correction to a control function. For example, furnace temperature control that it applies a derivative term to effectively correct the error despite a huge change.

5 FLC for Output Voltage Regulation

The basic block diagram of the proposed FLC for output voltage regulation of the SEPIC converter is shown in Fig. 5.

Any object has a degree of membership which varies from 0 to 1 in the fuzzy set theory, which makes fuzzy logic to handle all situations in a natural way. The shape of membership functions can also be trapezoidal or Gaussian, etc. depending on the applications and can be symmetrical or asymmetrical. The SEPIC converter voltage error is defined in linguistic variables such as Positive Big Error (PBE), Negative Big Error (NBE), Positive Medium Error (PME), Negative Medium Error (NME), Positive Small Error (PSE), Negative Small Error (NSE), and Zero Error (ZEE), where each variable is defined by a gradually varying triangular membership function. For error (e) and change in error (de), seven fuzzy sets are chosen. For a particular application depending on the input resolution, the number of fuzzy levels is selected which is not fixed. Higher will be the input resolution, the more the number of fuzzy levels are. Each input is assigned a membership function μ which is not quantized in the classical sense to each fuzzy set. To diminish the difficulty in calculation, here triangular fuzzy set values are used; we may also use bell-shaped or trapezoidal fuzzy set depending on the application, as shown in Fig. 6. For n number of variables, the fuzzy rules are n dimensional. The sum of rules is known as rule

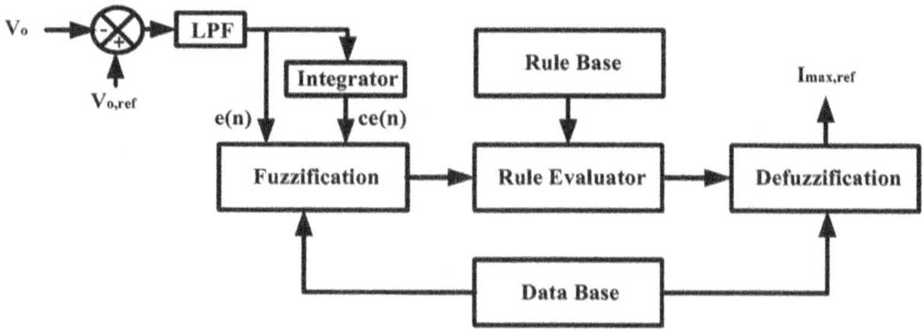

Fig. 5 Basic block diagram of FLC

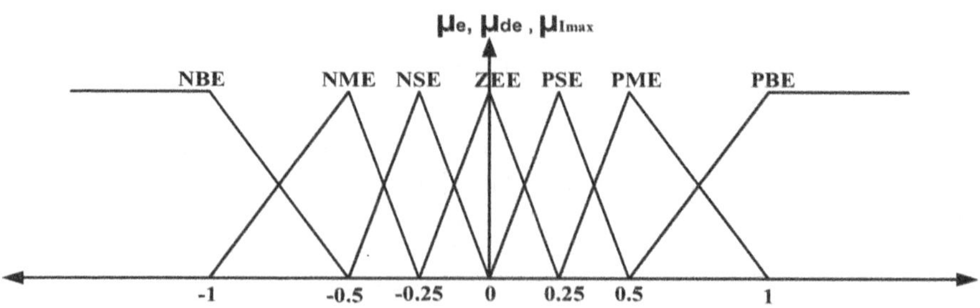

Fig. 6 Membership function used in fuzzification

Table 1 Fuzzy rules for closed-loop control of the SEPIC converter

*e**de*	NBE	NME	NSE	ZEE	PSE	PME	PBE
NBE	NBE	NBE	NBE	NBE	NME	NSE	ZEE
NME	NBE	NBE	NBE	NME	NSE	ZEE	PSE
NSE	NBE	NBE	NME	NSE	ZEE	PSE	PME
ZEE	NBE	NME	NSE	ZEE	PSE	PME	PBE
PSE	NME	NSE	ZEE	PSE	PME	PBE	PBE
PME	NSE	ZEE	PSE	PME	PBE	PBE	PBE
PBE	ZEE	PSE	PME	PME	PBE	PBE	PBE

R. FIS editor edits the input and output variables, which are *e*, d*e* and output. After editing, we have to design the membership function for each variable. The final step involves writing rules in the rule editor using the rule given in Table 1.

6 Results and Discussion

To investigate the performance of the proposed system, a Simulink model of a SEPIC Converter used for power factor correction is developed as shown in Fig. 7. The designed parameter used in the simulation is given in Table 2. The system performance is investigated in four steps. Step: 1 when the SEPIC converter is not connected to the system. Step: 2 when the SEPIC converter is connected to the system with

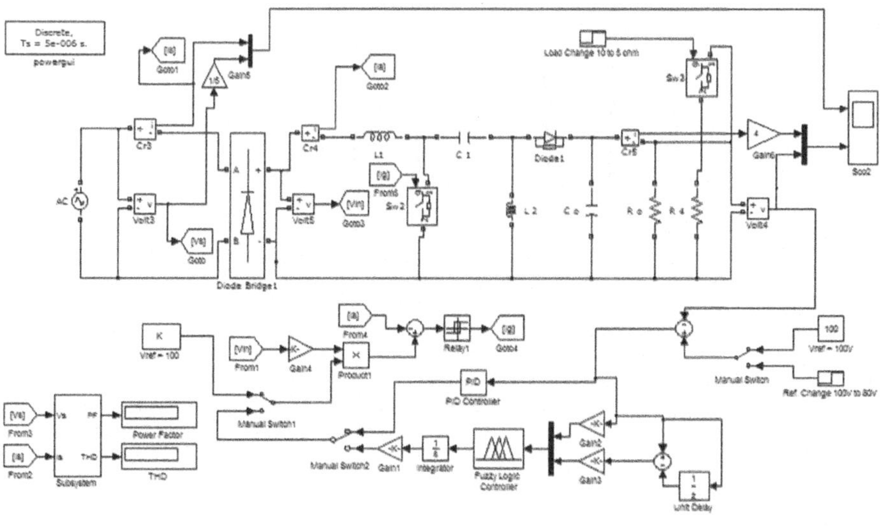

Fig. 7 Simulink model of the SEPIC converter

Table 2 Designed parameter of the SEPIC converter

Parameters	Values
Input voltage RMS value	120 V
Supply frequency	50 Hz
L_1, L_2	1 mH, 10 mH
C_1, C_o	10 mF, 10 mF
Load power	1 kW
Reference output voltage	100 V
Switching frequency	20 kHz
Voltage and current ripple	0.05
Duty cycle	0.45

an inner current loop only and without the output voltage feedback loop. Step: 3 when the SEPIC converter is connected to the system with PIC in the feedback loop. Step: 4 when the SEPIC converter is connected to the system with FLC in the feedback loop. To convert AC to DC, we generally connect a bridge rectifier and filter capacitor across the load as shown in Fig. 1; the simulation result of the same and its harmonic spectrum are shown in Fig. 8a–d. To upgrade the efficacy of the input current and output voltage, a SEPIC converter is used as discussed in Fig. 2. Various

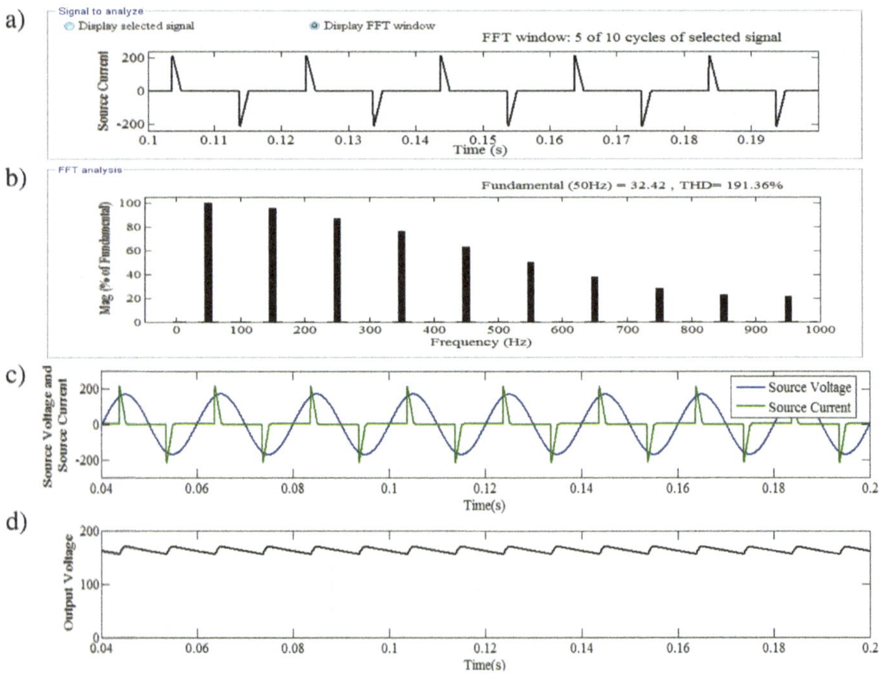

Fig. 8 Simulation results of the proposed system without the SEPIC converter

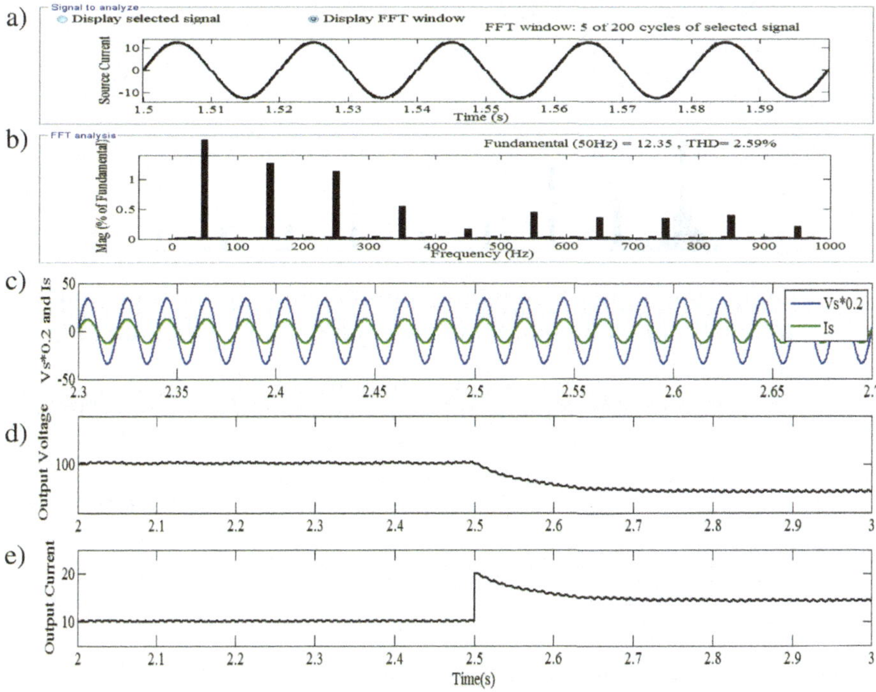

Fig. 9 Simulation results of the proposed system in an open loop

waveforms at steady state in an open loop are shown in Fig. 9a–e. From Fig. 9d, it is clear that when the load is increased at $t = 2.5$ s, load voltage is not a regulated one. To get a regulated output voltage and improved input current wave, PIC or FLC is used in the output feedback loop. The dynamic response of the system is shown in Figs. 10a–f and 11a–f, respectively, for PIC or FLC used in the outer loop. At $t = 1$ s, load resistance is decreased from 10 to 5 Ω and at $t = 2$ s, reference voltage is decreased from 100 to 80 V. With this dynamic change, the regulated output voltage is obtained which can be clearly seen from Figs. 10e and 11e, respectively, for PIC or FLC. Various performance parameters obtained in the simulation are as in Table 3 for comparison.

7 Conclusion

The design, modelling and simulation of a SEPIC converter used for power factor correction are considered in a MATLAB/Simulink environment. The outcomes of the simulation work provided low THD of the supply current with improved IPF and reduced output voltage ripple. Comparing with PIC and FLC used in the output feedback loop, FLC gives better results in terms of THD of supply current, IPF and

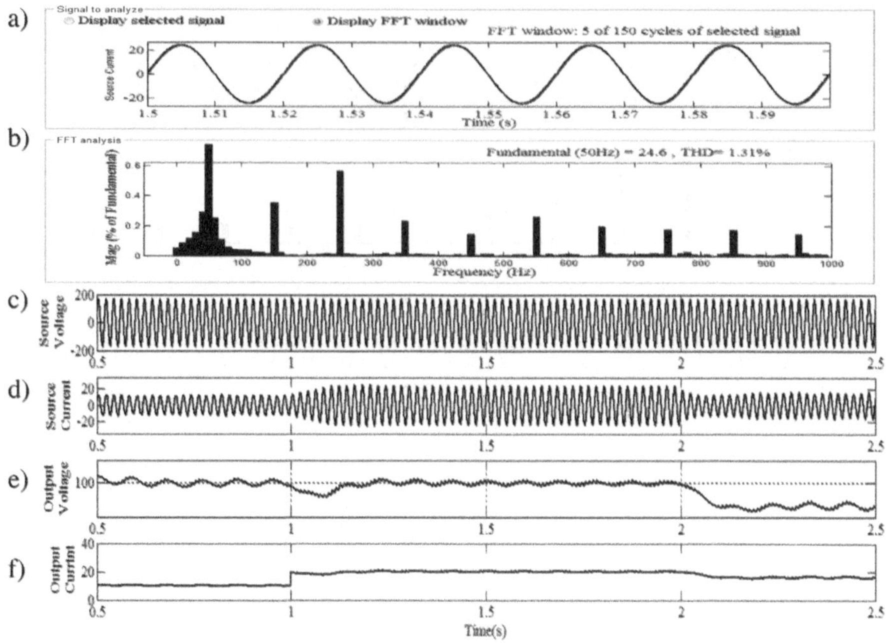

Fig. 10 Simulation results of PI-controlled SEPIC converter

output voltage ripple. During dynamic conditions, FLC gives a well-regulated voltage at the output with reduced ripple. The prototype of the proposed SEPIC converter can be developed, which would be attempted as a future work.

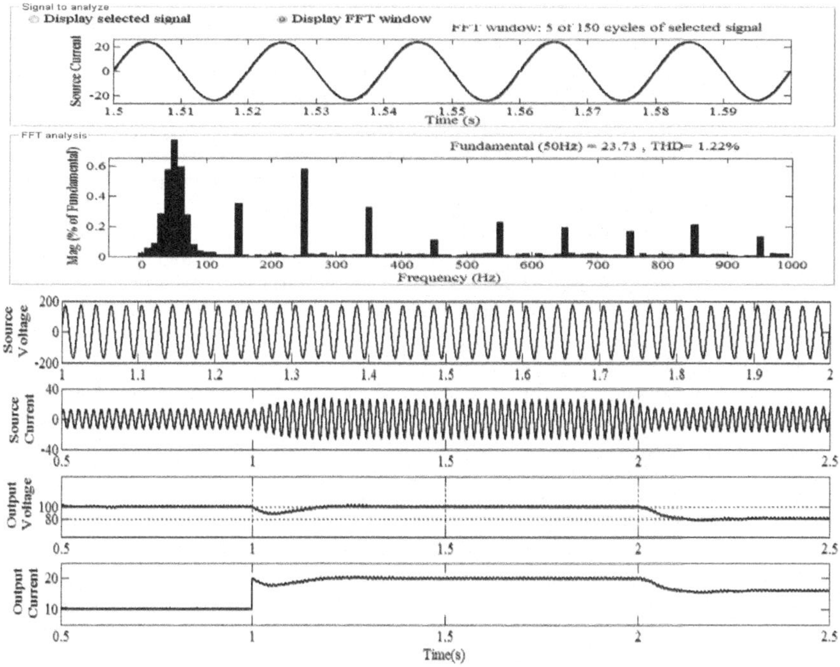

Fig. 11 Simulation results of fuzzy logic-controlled SEPIC converter

Table 3 Performance parameters of the system under different conditions

Performance parameter / Different condition	THD %	Power factor	Output voltage ripple %
Without SEPIC converter	191.36	0.4479	9.2
Open-loop SEPIC converter	2.59	0.9992	3.1
PI-controlled SEPIC converter	1.31	0.9995	6.0
Fuzzy logic-controlled SEPIC converter	1.22	0.9998	1.5

References

1. Kuiyuan, W.: The comparison and choice of several power factor correction methods. In: IEEE Vehicle Power and Propulsion Conference, pp. 1–5, Sept 2006. IEEE
2. 519–2014: IEEE Recommended Practice and Requirements for Harmonic Control in Electric Power Systems, pp. 1–29, 11 June 2014
3. Pressman, I.: Switching Power Supply Design. McGraw-Hill, New York (1991)
4. Singh, B., Singh, B.N., Chandra, A., Al-Haddad, K., Pandey, A., Kothari, D.P.: A review of single-phase improved power quality AC-DC converters. IEEE Trans. Ind. Electron. **50**(5), 962–981 (2003)
5. Batarseh, I.: Power Electronics Circuits, 3rd edn. Wiley

6. Richard, R.: The fundamentals of power factor correction. Int. J. Electr. Eng. Educ. **31**, 213–229 (1994)
7. Manias, S., Dziogas, P., Olivier, G.: An AC To DC converter with improved input power factor and high power density. IEEE Trans. Ind. Appl. **IA-22**(6), 1073–1081 (1986)
8. Dah, D., Ki, S.K.: Light-load efficiency improvement in buck-derived single-stage single-switch PFC converters. IEEE Trans. Power Electron. **28**(5), 2105–2110 (2013)
9. Liu, X., Xu, J., Chen, Z., et al.: Single-inductor dual-output buck–boost power factor correction converter. IEEE Trans. Ind. Electron. **62**(2), 943–952 (2015)
10. Kanaan, H.Y., Al-Haddad, K., Fnaiech, F.: Switching-function-based modeling and control of a SEPIC power factor correction circuit operating in continuous and discontinuous current modes. IEEE Int. Conf. Ind. Technol. **1**(8), 431–437 (2004)
11. Umamaheswari, M.G., Uma, G.: Analysis and design of reduced order linear quadratic regulator control for three phase power factor correction using Cuk rectifiers. Electr. Power Syst. Res. **1**(96), 1–8 (2013)
12. Mishra, A.K., Pathak, M.K., Das, S.: Isolated converter topologies for power factor correction—a comparison. In: IEEE International Conference on Energy, Automation and Signal, pp. 1–6, 28 Dec 2011
13. Mahdavi, M., Farzanehfard, H.: Bridgeless SEPIC PFC rectifier with reduced components and conduction losses. IEEE Trans. Ind. Electron. **58**(9), 4153–4160 (2011)
14. Umamaheswari, M.G., Uma, G., Isabella, L.A.: Analysis and design of digital predictive controller for PFC Cuk converter. J. Comput. Electron. **13**(1), 142–154 (2014)
15. Poorali, B., Adib, E.: Analysis of the integrated SEPIC-flyback converter as a single-stage single-switch power-factor-correction LED driver. IEEE Trans. Ind. Electron. **63**(6), 3562–3569 (2016)
16. Melo, P.F., Gules, R., Ribeiro, E.F., et al.: A modified SEPIC converter for high-power-factor rectifier and universal input voltage applications. IEEE Trans. Power Electron. **25**(2), 310–321 (2010)

Data Center Networks: The Evolving Scenario

Tapasmini Sahoo, Suraj Kumar Naik, and Kunal Kumar Das

Abstract Data centers (DCs) have emerged as an irreplaceable and essential infrastructure to support this ever-growing trend as a result of the exponential growth of Internet services. A DC typically houses a large number of computing and storage nodes, interconnected by a network that has been specially designed, namely DC. The DCN serves as a foundation for interaction and plays a key role in the optimization of DC operations. In this paper, a brief survey of various architectures of data center network (DCN) is presented based on their switching and transmission technology. This paper also depicts a comparison of different topologies of representative data center networks with respect to several properties to highlight their advantages and disadvantages.

Keywords DCN · Topology · DCN architecture · DC · Switching technique

1 Introduction

The ever-increasing number of applications and services needed from the Internet requires an infrastructure known as Data Center (DC) [1]. A volume of resources in computing and storage is required in DCs for serving the ever-growing services and applications. In response to these needs, information centers are being created in a centralized and controlled setting to accommodate a large number of networked servers. The Data Center Network (DCN) is a specially constructed network in which a huge number of servers capable of computation and storage are present interconnected using bandwidth-efficient networks. In addition to the connectivity, the DCNs

T. Sahoo (✉) · S. K. Naik · K. K. Das
Siksha 'O' Anusandhan (Deemed to be University), Bhubaneswar, Odisha 751030, India
e-mail: tapasminisahoo@soa.ac.in

S. K. Naik
e-mail: surajnaik@soa.ac.in

K. K. Das
e-mail: kunaldas@soa.ac.in

D. Mishra et al. (eds.), *Intelligent and Cloud Computing*,
Smart Innovation, Systems and Technologies 194,
https://doi.org/10.1007/978-981-15-5971-6_62

also allows operational optimizations of the data center to achieve a balance between the increasing price and the business continuity needed.

The architecture and principle of operation of DCN differ from traditional networks such as LAN and WAN. This paper presents a review of the operational challenges and requirements for the DCNs. Several proposals for the DCN architectures have been reported in the literature attempting to deliver consistently elevated capability across all servers [2, 3].

Initially, the DCNs employed electrically switched hierarchical architectures with moderate network sizes. However, such architectures do not scale up well to support larger number of servers, calling thereby improvization through the use of optical switching techniques. The optical system administration innovation is appropriate to address the difficulties, for example, oversubscribed system and higher data rate. Optical system components support on-request availability and limit where required in the system, accordingly allowing the development of slender, however, flexible interconnects for huge server pools [4]. Optical connections can bolster higher data rates over longer separations utilizing less power than copper links. In addition, optical switches are more energy efficient than the electrical ones, bringing about lower heat dissemination and less-expensive cooling cost.

The design methodology of a DCN is of premium significance as it impacts the overall operational efficiency. The design of a DCN, or its topology, straightforwardly reflects on its versatility, cost, adaptation to noncritical failure, nimbleness and power utilization [5]. Ordinary DCNs have been planned to utilize a tree-like topology. Typically, the DCN employs hierarchical topology where the tree's leaves (end-hubs or servers) are associated with Top-of-Rack (ToR) switches and these (ToR) switches are associated with accumulation switches which are thus associated with center switches at the root of the tree. This topology has before long appeared to suffer various downsides of scale, limit, unwavering quality, use and power spending plan [2]. Therefore, efforts have been made to address a few (or the majority) of the above issues in the tree-based DCN topology. These models can be classified as server-driven and switch-driven while being dependent on their framework, for example, electronic versus optical DCNs.

The main drawback in contemporary DCs is the measure of their energy efficiency. Typically during peak network load or busy hour, DCNs are tuned for optimum efficiency and thus are extremely hungry for energy. It has been observed that a typical DCN facility features a power density of over a hundred times that of an office building in the same locality. Thus, the DCN design structure and energy efficiency are the two reticular issues in a network. The present review addresses the popular DCN architectures, such as electrically switched DCN architecture, followed by the hybrid electro-optical architecture, full optical architecture and passive DCN optical architecture. Next, the paper presents some studies on the energy efficiency in DCNs.

However, it's challenging to get an efficient design for DCs, which should be fault tolerant, more scalable and should offer less latency. The various requirements for DCs are as follows.

DCs should support bandwidth-intensive applications. Many applications in the DC need to move huge amounts of data between host and the network which develops their bottleneck performance. Good support for typical traffic patterns needs to be provided by a well-designed network architecture.

DCs should have low-end commodity switches to achieve better performance to price ratio. In some architectures, non-programmable COTS switches are utilized at the low end rather than the top of the line ones, in view of the perception that the per-port cost of the low-end switches is a lot less expensive than that of the top of the line ones.

DCs should have graceful performance degradation means; it is essential that we plan our system architecture in such a way that the design should be fault tolerant and also debase nimbly.

2 Data Center Network

Data center is a gigantic group of PCs that is claimed and worked by utilizing an association. They are regularly structured to streamline both expense and execution. Information offices developed to turn into the most incredible approach to achieve huge scale computational wants with unreasonable execution the utilization of the littlest measure of assets.

The basic three-tier DCN architecture follows a tree structure composed of three layers of switches namely access, aggregate and core layers as shown in Fig. 1. The servers in the lowest layers are connected without delay to one of the part

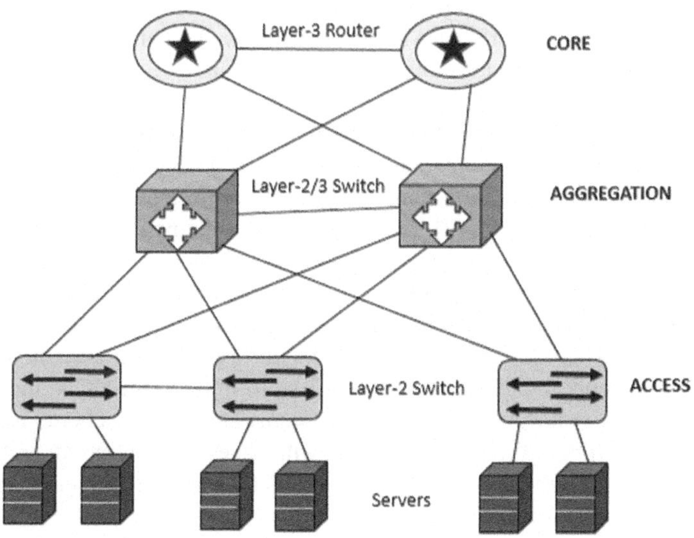

Fig. 1 Three-tier DCN architecture

layer switches. The combination layer switches interconnect together a couple of get admission to layer switches. All of the combination layer switches are related to each different by core layer switches. Core layer switches are also accountable for connecting the records' core to the Internet. The three tier architecture is the common community architecture used in information centers [6]. While the three-tier structure is unable to take care of the growing demand for cloud computing, the increased layers of the three-tier DCN are pretty oversubscribed. Moreover, scalability is another predominant difficulty in three-tier DCN. Major troubles faced by means of the three-tier structure include scalability, fault tolerance, strength efficiency, and cross-sectional bandwidth.

The DCN infrastructure interconnects servers in an information core and throughout data centers. It can be normally labeled based totally on two dimensions, particularly transmission technological know-how and scale. There are distinct forms of transmission in DCs, such as electrical, Wi-Fi and optical. All the transmission technologies are widely utilized in computer networks but in the context of DCN, they show their respective merits.

In electrical transmission-based DCs, electrical cables and specifically twisted pair cables are mostly used due to the ease in deployment. Yet, they habitually force a length limitation of around a hundred meters; consequently, they must be utilized intra insights focuses.

Optical transmission is viewed as one of the most encouraging advancements for DCs. Its more prominent data transmission, smaller size, lighter weight, and reduced power utilization than that of electrical links make it invaluable to utilize. Utilizing procedures like WDM, a hundred Gbps optical connections have been created, while electrical transmission innovations are generally confined to 40 Gbps. Optical fibers can work with electrical switches with handsets [7, 8]. Optical fibers can likewise work with the exception of photoelectric change through coordinating a light emission from any info port to any yield port. Optical transmission shows a high-performance index for long-distance transmission.

DCNs employ wireless technologies for providing greater bandwidth and inherent flexibility. More often than not in DCs, servers are thickly stuffed inside short ranges, causing the radio condition static to aside from incessant development of people and types of gear. In such a situation, remote systems are normally reasonable to outfit viewable pathway associations [9]. Wi-Fi framework sending should be possible over racks with limited directional reception apparatuses or staged exhibit radio wires, to such an extent that there will be shirking of sign obstruction.

All the above-discussed transmission technologies are well specified for intra-data center networks. Thus the DCN topologies for intra-DCNs are classified in accordance with the technologies used for transmission. In the intra-DCN, the electrical switched topology is nevertheless the mainstream. The optical scheme is projected as the technology for data center networking with excessive bandwidth and less cost. In the interim, Wi-Fi transmission is managed as a decision to improve the current system framework next to all the more cabling expenses.

Electrical Element-Based Topologies: Usually, intra-DC's system plans are mostly dependent on electronic switches. As indicated by the capacity of gadgets, electrical component-based topologies are commonly prepared in switch-driven or server-driven structure. Structure goals of the topology generally give emphasis on bisection bandwidth, hop distance, expected number of hops, cabling complexity, scalability, fault tolerance, etc.

Hybrid Topologies: These topologies of DCN are based on electrical/optical elements. Due to the regular request for expanding transmission capacity, optical switching is best suited for DCNs. The first venture of utilizing optical switching in DCs is to combine the electrical techniques of switching with optical switching.

3 DCN Topologies

The topology of a DC refers to the general construction of a DC. Communication networks also support DCs that assist in aggregating information and enabling them to flow through the system. This can be achieved in many ways, and some DC topology models have arisen. For example, many adopt a tree-based topology, including what is called a network with three layers of a "three-tier DC": access, aggregate and heart. Certain topologies of the DC include systems where one "main" server is connected to many other servers and where there are cross-linked or cross-indexed servers for various functionality styles. A "leaf-spine" solution, for instance, adds different components to a core "spine layer" that supports heavy directional traffic networks. DCN topologies are also categorized based on their switching techniques.

3.1 Electrical Switching-Based Topologies

The broad classification of network topologies based on electrical switching are of two kinds, switch-driven topologies and server-driven topologies. In switch-driven DCs, switches assume the essential liability in system development and information transmission. Systems in traditional DCs are regularly developed depending on a tree-based progressive topology with a few levels as per the system scale. A two-level DC structural design comprises edge-level switches and core-level switches to hold hundreds or thousands of servers. But for large-scale DCs, the three-level structural design might be increasingly fitting, where switches are organized in three stages: edge, aggregate and core.

Fat Tree Fat Tree is mostly used in intra-DCNs, which interconnects indistinguishable switches where the limit of the system pairs as we go upward the tree is as shown in Fig. 2. It receives a unique example of a Clos topology. Fat tree is a high transmission capacity low-latency topology. It is well organized to avail bisection bandwidth without utilizing costly core switches [10].

Figure 2a shows an example of a 4-ary Fat tree. A p-ary Fat tree is a three-layer topology. Each pod in it consists of $(p/2)^2$ servers and two layers of $p/2p$ port switches.

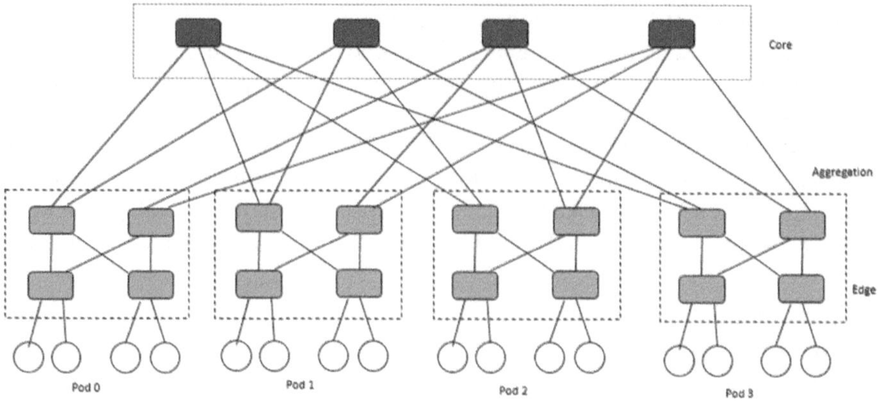

Fig. 2 Basic structure of Fat tree

Each edge switch connects to $p/2$ servers and $p/2$ aggregate switches. Each aggregate switch connects to $p/2$ edge and $p/2$ core switches. There will be $(p/2)^2$ core switches and each connects to p pods.

The benefit of Fat tree is that all switches are indistinguishable and modest ware items which can be utilized for all switches. Moreover, there are various equivalent cost ways between any two hosts, and can avail full bisection bandwidth in principle. Fat tree has an indistinguishable transmission capacity at any bisection.

Each layer has the equivalent totaled data transmission. Each port has the same speed as the end has. If the packets are distributed uniformly in an available link, then each device can transmit at line speed. Fat tree's high cabling difficulty is a detriment. A Fat tree constructed from 16-port GigE switches will consist of 16 pods, each with an edge layer and an aggregation layer with 8 switches each, as an example of this topology. Eight hosts are allocated each to the edge switches in each pod. The network serves 1024 servers, each with 8 hosts, consisting of 128 subnets with 64 identical cost paths in separate pods between any given pair of hosts.

Server-driven topologies: In switch-driven topologies, servers are just endpoints in the system. In any case, servers could assume a progressively significant job in an intra-DCN for the quick development of server equipment.

The significance of servers within the network matters a lot by setting the interconnection and routing instead of the switches present in the network. In such DCs, servers act as not only the handiest cease hosts, but additionally as communicating nodes in between them. The usual method for such topologies is to recursively create a hierarchical high degree community by a couple of low-level units. These topologies generally encompass hypercube systems which reduce path duration and improve bandwidth via wealthy inter-cellular network connections. The hypercube structure, however, requires complicated cabling that's difficult to maintain.

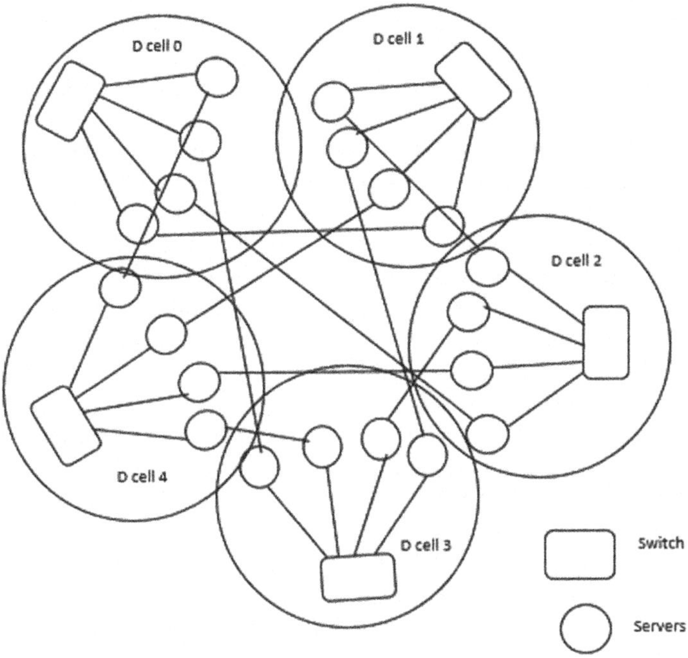

Fig. 3 Basic structure of D Cell topology

D Cell It is based on normal hierarchical server-centric topology [11] through a recursive production. It interfaces servers (inside the equivalent rack) to a neighborhood D cell that has a dreadful part higher network limit as contrasted and the tree-based topologies [12]. D cell does not make any bottleneck due to the appropriation of the traffic all through all connections. This structure utilizes servers outfitted with more than one system port and littler switches, to helpful asset the improvement of its recursively portrayed engineering. On this system topology, a server is joined to various others through smaller switches. Those hyperlinks are assumed to be bidirectional (Fig. 3).

3.2 Hybrid Topologies

As per the consistent expansion of transmission capacity request in DCs, optical switching is the best solution. Optical switching in DCs is applied in combination with electrical switching. For instance, optical switching is utilized for conglomerating traffic among racks. Since optical gadgets accompany different qualities and specifications, optical systems in DCs are frequently recognized by the use of optical switches.

OSA (Optical Switching Architecture) It investigates the achievability of a totally optical-based system among ToR switches. It includes optical transceivers within ToR switches with different fibers for send and receive data packets; the multiplexers (Mux) multiplex several fiber optical signals to one fiber's output; the 1 × 4 Wavelength Selective Switch (WSS) forwards the signal to 4 DWDM channels or wavelengths; the circulators are used to achieve bidirectional transmission over a single fiber; the 16 × 16 bipartite MEMS optical switch grants associations from any info port to any one of the yield ports; the signals of different wavelengths are combined into one fiber by coupling; the de-multiplexers (DeMux) just do the reverse function of a Mux, thus resulting multiple wavelengths to different fibers from a single optical signal. In view of the assessed traffic request', the WSS and MEMS switches are reconfigured [13] (Fig. 4).

OSA, in its existing form, has constraints. The challenge is to scale OSA from a container size to a larger DC consisting of tens to hundreds of thousands of servers. This includes nontrivial efforts in the management of structural design and control. The first obstacle is the port density of MEMS. While the MEMS 1000-port is hypothetically possible, there are 320 ports in the largest MEMS starting today. One characteristic approach to expand the port thickness is by means of interconnecting numerous little MEMS switches. Notwithstanding, this represents an extra prerequisite for quick organized circuit switching. Furthermore, a bigger system size requires

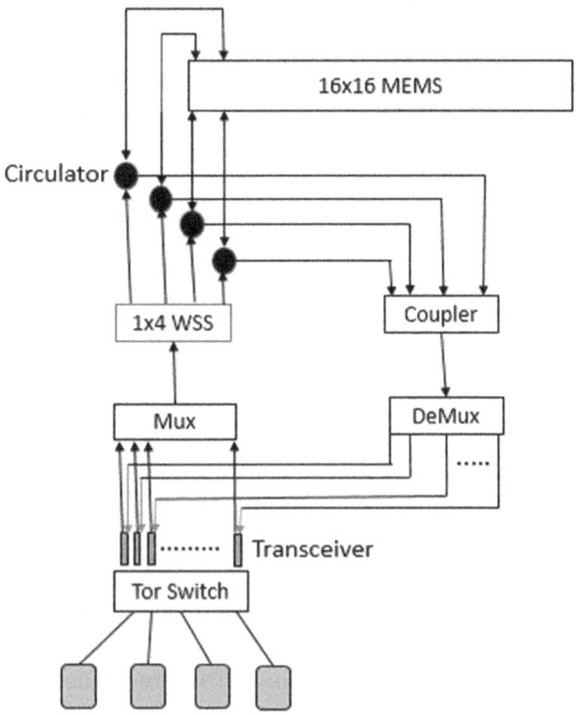

Fig. 4 Basic structure of OSA

more control and traffic management. Additionally, the increase in the end-to-end delay is another considerable issue.

Helios: A Hybrid Electrical/Optical Switch It is a multilayered tree of pod switches and core switches on two scales. Core switches can be either electrical package switches or optical circuit switches; one type of switch's advantages compensate for the other type of weaknesses [14]. The inter-pod communication is handled by circuit switching. In the congested portion of inter-pod communication, packet switching helps to convey all-to-all bandwidth. The mix switching results in power consumption, complexity reduction and increased performance for given traffic.

In Fig. 5, each unit has various hosts (named "H") associated with the pod switch by short copper joins. In connection with the central switching unit, the pod switch contains various optical handsets (named "T"). Half of the uplinks from each device in this system are connected with packet switches, each requiring an optical handset as well. Before associating with a single optical circuit switch, the other portion of uplinks from each switch passes through an optical multiplexer (named "M").

The hybrid DC switch architecture essentially incorporates the advantages of both technologies and uses as a fully equipped packet-switched network with the same measures of performance but at low cost and complexity and less power consumption. This determines the traffic subset best suited for circuit switching and dynamically reconfigures the topology of the network at runtime based on changing patterns. It does not require any end-host modification, rather it demands switch software modification.

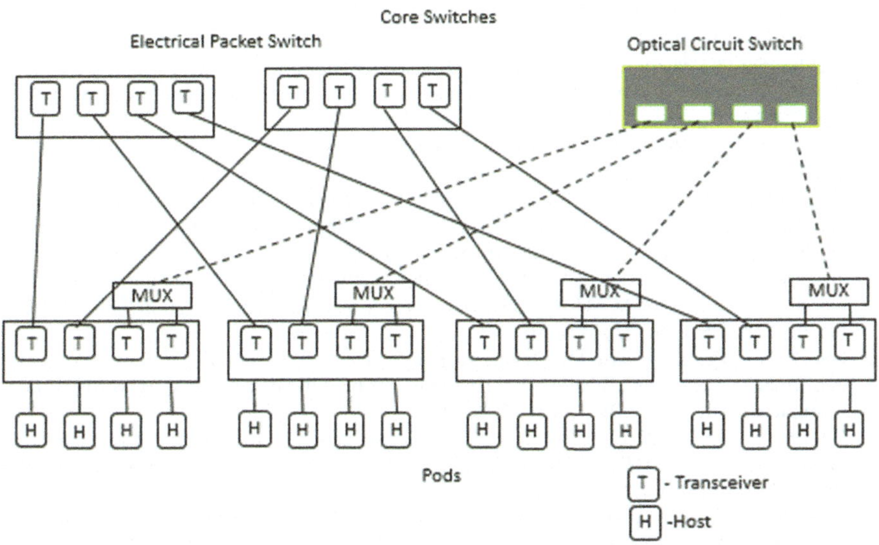

Fig. 5 Basic structure of Helios topology

4 Conclusion

We addressed some representative topologies of networks of DCs in this survey. We discussed the properties of these topologies, the techniques for improving their efficiency and techniques of fault tolerance. For future possible DC networks, scientists continue to suggest potentially new topologies with an appreciable balanced value of efficiency, cost and sustainability.

References

1. Cisco global cloud index: forecast and methodology, 2014–2019. In: White Paper, Cisco, San Jose, CA, USA, Oct 2015. http://www.cisco.com/c/en/us/solutions/collateral/serviceprovider/global-cloudindex-gci/Cloud_Index_White_Paper.pdf
2. Greenberg, A., Hamilton, J., Maltz, D.A., Patel, P.: The cost of a cloud: research problems in data center networks. SIGCOMM Comput. Commun. Rev. **39**(1), 68–73 (2009)
3. Armbrust, M., Fox, A., Griffith, R., Joseph, A.D., Katz, R.H., Konwinski, A., Lee, G., Patterson, D.A., Rabkin, A., Stoica, I., Zaharia, M.: Above the clouds: A Berkeley view of cloud computing. Technical Report UCB/EECS-2009-28, Feb 2009. EECS Department, University of California, Berkeley. http://www.eecs.berkeley.edu/Pubs/TechRpts/2009/EECS-2009-28.html
4. Zhang, Y., Sub, A., Jiang, G.: Understanding data center network architectures in virtualized environments: a view from multi-tier applications. Comput. Netw. **55**, 2196–2208 (2011)
5. Guo, C., et al.: D-cell: a scalable and fault-tolerant network structure for data centers. In: Proceedings of ACM SIGCOMM Conference on Data Communications (SIGCOMM), Seattle, WA, USA, 2008, pp. 75–86
6. Bari, M., Boutaba, R., Esteves, R., Granville, L., Podlesny, M., Rabbani, M., Zhang, Q., Zhani, M.: Data center network virtualization: a survey. In: Communications Surveys Tutorials, vol. no. 99, pp. 1–20. IEEE (2012)
7. Xu, K., Ou, Y.: Theoretical and numerical characterization of a 40 Gbps long-haul multi-channel transmission system with dispersion compensation. Digit. Commun. Netw. **1**(3), 222–228 (2015)
8. Yu, J., Zhang, J.: Recent progress on high-speed optical transmission. Digit. Commun. Netw. **2**(2), 65–76 (2016)
9. Halperin, D., Kandula, S., Padhye, J., Bahl, P., Wetherall, D.: Augmenting data center networks with multi-gigabit wireless links. In: Proceedings of ACM SIGCOMM Conference (SIGCOMM), Toronto, ON, Canada, pp. 38–49 (2011)
10. Al-Fares, M., Loukissas, A., Vahdat, A.: A scalable, commodity data center network architecture. In: Proceedings of ACM SIGCOMM Conference on Data Communications (SIGCOMM), Seattle, WA, USA, pp. 63–74 (2008)
11. Guo, C., Lu, G., Li, D., Wu, H., Zhang, X., Shi, Y., Tian, C., Zhang, Y., Lu, S.: Bcube: a high performance, server-centric network architecture for modular data centers. ACM SIGCOMM Comput. Commun. Rev. **39**(4), 63–74 (2009)
12. Guo, C., et al.: D-cell: a scalable and fault-tolerant network structure for data centers. In: Proceedings of ACM SIGCOMM Conference on Data Communications (SIGCOMM), Seattle, WA, USA, pp. 75–86 (2008)
13. Chen, K., et al.: OSA: an optical switching architecture for data center networks with unprecedented flexibility. In: Proceedings of 9th USENIX Conference on Networked System Design and Implementation (NSDI), San Jose, CA, USA, pp. 239–252 (2012)
14. Farrington, N., et al.: Helios: a hybrid electrical/optical switch architecture for modular data centers. In: Proceedings ACM SIGCOMM Conference, New Delhi, India, pp. 339–350

Wireless Sensor Networks (WSN)

Android App-based Cryptographic End-to-End Verifiable Election System

Rakesh Kumar

Abstract In this paper, we present the e-voting mechanism through an android application that assures the possibility of an accessible and safe way to acquire votes during an election process. We have described all the associated concepts like survey, methodology, and working mechanism that is required for the presented work. In order to maintain security, encryption and decryption are done through cryptographic techniques. The e-voting mechanism proposed herein is intended to be more accessible and, therefore, more suitable for comparison with other voting systems.

Keywords e-voting · Security · Cryptographic techniques · Android-based application

1 Introduction

In the present scenario, everything revolves around modern technology that is less time-consuming. Thus, the process of voting is considered to be very hectic, if it takes several days. For more information on electronic voting machines used in the electoral system in India, we refer the reader to [1, 2]. It is essential to make the voting process transparent, efficient, and user-friendly. Previously, voting processes have evolved from counting hand to hand into mechanisms that involve paper, punch card, and optical-scan machines. For historical development on e-voting, we refer the reader to [3, 4].

R. Kumar (✉)
Department of Computer Science and Engineering, Siksha 'O' Anusandhan (Deemed to be University), Bhubaneswar, Odisha 751030, India
e-mail: rakeshkumar@soa.ac.in

D. Mishra et al. (eds.), *Intelligent and Cloud Computing*,
Smart Innovation, Systems and Technologies 194,
https://doi.org/10.1007/978-981-15-5971-6_63

In [5], Rivest has addressed issues related to secure platform problem, and problems arising in providing the voter's receipt. Rivest has also provided suggestions on dissimilarity occurring between e-commerce and e-voting, wide-scale attack concern when voting from home, the threat of automated fraud parties, the simplicity and accessibility requirement of e-voting devices, etc. Clausen [6] presented a secure electronic voting system that does not depend on existing network connections, which works nicely in the absence of network connectivity.

In this work, we intend to study and investigate an android application-based e-voting mechanism. The suggested technique is to enhance the efficiency of a voting system by being localized, accessible, and adaptable with distinct versions of an operating system. This application will be advantageous as any individual all around the globe will be able to transmit his vote securely.

2 e-Voting System

In e-voting schemes, we can think about voting via SMS and the Internet as described below:

- ***Vote through message***. A voter can give his vote by sending an SMS via a mobile connection (see [7]). For such processes, an android application is developed and stored in the corresponding android phone. Both the Android system and the website will be linked to a similar database MySQL/SQL so that the voter can vote via one of the two available ways. If one tries to cast a duplicate vote, the system should deny and refuse to proceed further.
- ***Vote via Internet***. It employs electronic ballots, to permit voters to send their secured voted ballot directly to the election officials through the Internet. Researchers have repeatedly utilized electronic voting to create modern technologies, which consequently enhanced accessibility and public cooperation in a voting process.

The requirements of an e-voting system are categorized into the following:

- ***Security***. Electronic-enabled elections are open to security-related risks of the information technology. To explain the notion of security corresponding to online voting schemes, we first discuss the following security specifications.
  Privacy: The Cast votes have to be kept secret and unlinked to a voter.
  Eligibility: Only eligible voters can vote and no one votes twice.
  Individual voter verifiability: Every voter must have the option to verify their vote, whether it is correctly enclosed in the final or not.
  Correctness: The proposed protocol must constitute all votes of authorized voters accurately into the associated election result.
  Receipt freeness: No receipt is provided to the voter that can be used to convince a coercer that he voted in favor of a particular candidate.
  Robustness: The posed scheme should not tolerate any cheating/fraud by the participants or authorities.

- ***Usability***. We expect the system to be user-friendly.
  Efficiency: The system should be efficient concerning computation, storage, and communication.
- ***Legality***. From the legal point of view, a voting system must be as follows.
  Secret: One should not be able to link a vote to a voter.
  Free: Every participant must be able to vote without being forced or disturbed in any manner.

3 e-Voting on Android

Android application based e-voting mechanism provides the user an opportunity to cast their vote, and allows them to choose any particular option to a question (e.g., voted or not). Although the proposed system facilitates the voters to transmit their vote from anywhere, they have to initially provide their authentication to authorize themselves, which creates their user identity similar to any other application. The proposed system has to maintain the voter's data and their choice of vote. The authentication constraint is included to minimize the rate of fraud and to ensure that only authorized personnel are permitted to vote in an election.

3.1 System Mechanism

To implement the proposed system mechanism evidently on various types of smartphones, it is conventional to assume that each device is related to its owner via a validation database. The step-by-step architecture associated with the e-voting system is depicted in Fig. 1, and can be described as follows:
e-Voting Database (eVDB). It collects the essential data during a public voting process and is assumed to be the prime component of the formulated system. At the starting point, it only comprises the data table having the elements described below:

- The record of each ID and record for vote ID.
- Two separate columns are reserved inside the database, to store the time and date of each record.
- During the vote casting, separate columns are reserved to store the geographic state (GPS) of the smartphone.
- The final vote answer: Yes or No.
- The time, date, and address of the server and client while casting the vote.

Multiple e-Vote Collector Web Services. They obtain the linked parameters through the component of a web browser inside the android application. These three distinct web services work similarly; they collect the cast votes and fill the database with the provided information from step 1, implemented by (Servlet eV Server) Java Servlet.

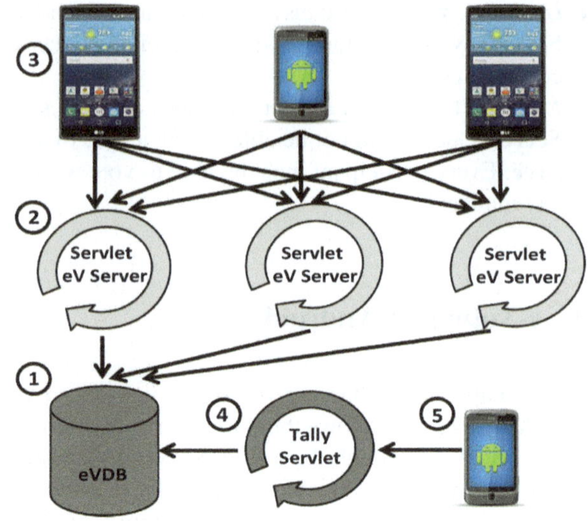

Fig. 1 Architecture of android-based e-voting application on smartphones

e-Voting App on Android Platform. It offers to answer the vital questions related to voting choices. Yes/No buttons are available to send the votes in the preferred languages like English, Hindi, etc. After casting the vote, the data is stored on the remote database, which confirms the reception of the vote.

e-Vote Tallying Web Services. These web services are developed on a Java Servlet that presents the parameters to populate the web page with the complete tally of the voting to the interested people, that is, the election authority.

e-Vote Tallying Android App. It provides the final situation and progress of the voting system, which takes the help of dynamic web pages on android application-based smartphones.

3.2 Cryptographic Protocol in the Proposed System

The cryptographic technique makes use of symmetric, asymmetric key algorithms and hash functions that ensure the privacy, fairness, receipt-freeness, and robustness properties in the system.

Asymmetric Key Algorithm. The **ElGamal** cryptosystem is used to encrypt and decrypt the votes at the time of an election process, which has two pairs of keys (public key, private key). We first describe the ElGamal cryptosystem for a subgroup G of the group of integers $\mathbb{Z}$. Let q be the order of subgroup G, where q is a large prime. It involves three steps, as described below.

- **Key generation**: Let G be a cyclic group of order q, generated by g. Take x randomly from $\{1, 2, ..., q-1\}$ and calculate $\mathbf{h = g^x \bmod q}$. Produce the public key h, G, q, g, and retain x as the *private key*, which is secret.

- **Encryption**: To encrypt a message m with a public key (G, q, g, h), take a random y from $\{1, 2, ..., q - 1\}$, then calculate $\mathbf{c_1 = g^y \bmod q}$. Find the shared secret $\mathbf{s = h^y \bmod q}$, and map the secret message m onto an element $m' G$. Again, compute $\mathbf{c_2 = m' \cdot s \bmod q}$ and the encrypted cipher text is

$$(\mathbf{c_1}, \mathbf{c_2}) = (\mathbf{g^y \bmod q}, \mathbf{m' \cdot h^y \bmod q}) = (\mathbf{g^y \bmod q}, \mathbf{m' \cdot (g^x)^y \bmod q})$$

- **Decryption**: To decrypt a cipher text (c_1, c_2) with the private key x, the decryption algorithm calculates the shared secret $\mathbf{s = c_1^x \bmod q}$, and computes $\mathbf{m' = c_2 \cdot s^{-1} \bmod q}$ which is converted back into the plain text message m. Finally, the intended message is decrypted as

$$\mathbf{c_2 \cdot s^{-1} \bmod q = m' \cdot h^y \cdot (g^{xy})^{-1} \bmod q = m' \cdot g^{xy} \cdot g^{-xy} \bmod q = m'}$$

3.3 *Working Process*

The working process (see Fig. 2) of e-voting via android is listed below:

1. Registration: Every voter is required to use his device where the e-voting application is stored. And thus, the voter has to provide his authentication like voter ID, username, password, user photo, e-mail Id, etc., that will be stored in a developed database. This stage employs **SHA-256** Hash function to encrypt the user's passwords. Consequently, with a **AES-256** key, two asymmetric key pairs are generated. One asymmetric key pair is used for blind signature and the other one to encrypt/decrypt the votes. Both the asymmetric key pairs are further encrypted with the assistance of previously generated **AES-256** keys, to store and keep them secret.
2. Sign-in: The voters are required to enter username, password, and user face recognition; those are already included during the Registration process. After validating all the data, the input password is stored in the server after being transformed using the **SHA-256** algorithm, as shown in Fig. 2.
3. Next, the associated central authority (CA) maintains the database of the voter details and also a database of voters who have already voted. The CA will check whether the voter has previously voted or not. If yes, the voter will not be allowed to vote again. If the voter has not voted earlier, the CA will verify the eligibility of the voter. If the voter authentication is valid (see Fig. 3), a One-Time Password (OTP) will be automatically generated and sent to the voter's registered e-mail address. When the voter receives this OTP, he can use it to log into the system to cast his vote.
4. After logging in, a display of the list of candidates will be available on the screen of the voter's device. Finally, the voter can choose any candidate to vote, and further confirm their vote for the candidate.
5. After casting a vote to the selected candidate, asymmetric key algorithm and hash function **SHA-256** are used to encrypt the vote. To hide the previously encrypted

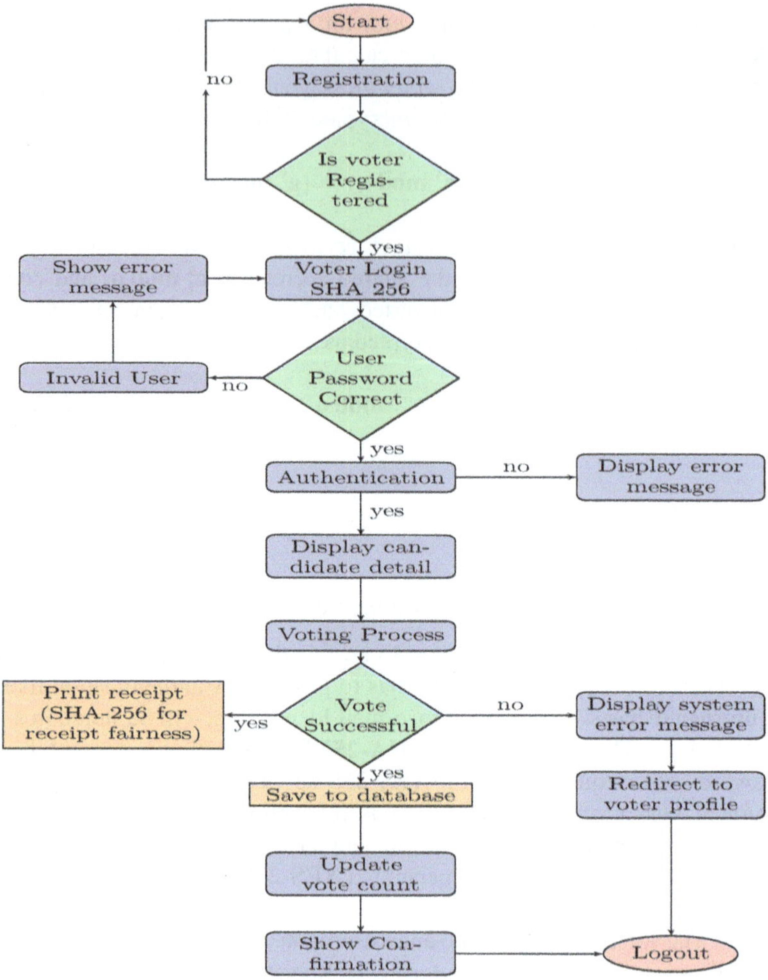

Fig. 2 Secure e-voting flowchart

vote, a blind factor is further applied, then the message is stored after being signed by the voting authority. With the assistance of **SHA-256**, a receipt is generated for the signed message together with a time stamp that authenticates the participation of an associated voter in the election.

6. After selecting a particular candidate, counting is incremented by one; a separate database is used to store the count of votes. At the last stage of the final counting process, the **ElGamal**-based blind signature scheme is used to decrypt and verify each vote. After validating all the cast votes, the final result is displayed, which has to be the same as the total number of votes.

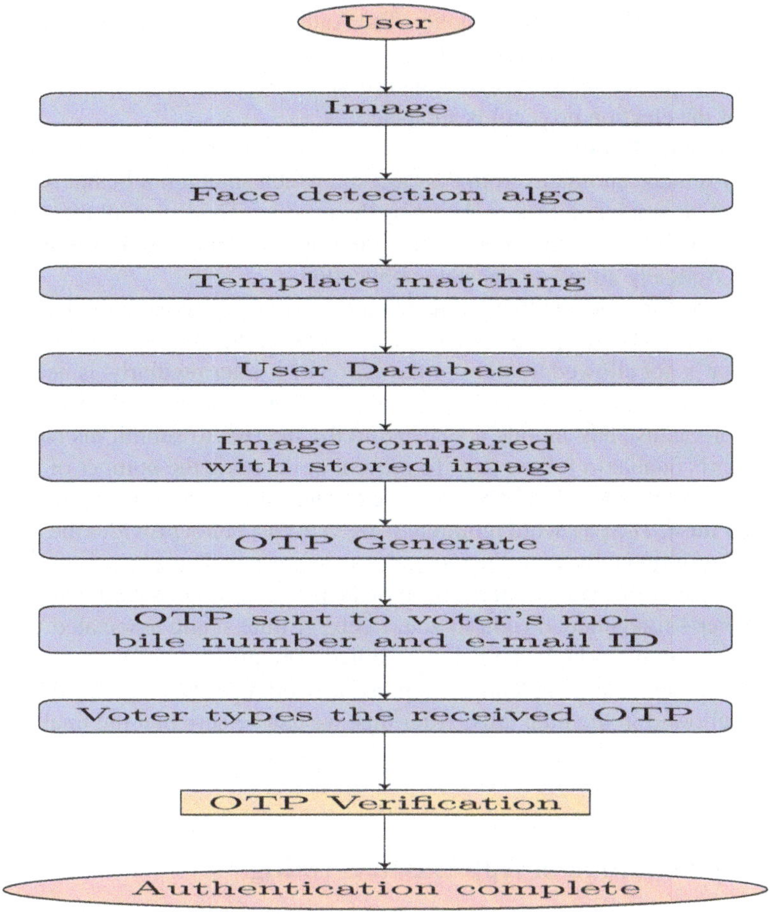

Fig. 3 Voter's authentication

4 Phase Analysis of the Proposed System

The proposed mechanism of the protocol is required to be entirely evaluated, in order to detect any disadvantages, drawbacks, and limitations. First of all, we shall analyze all the stages of the concerned coercion-resistant voting protocol.

Setup Phase. The username and the password, stored in the database, are manually handed over to the interested voters who verify their identity. When a voter tries to connect to the server, he has to provide his credentials which are matched with the one stored in the database. So, only the authorized voters can cast their votes.

Registration Phase. The correctness, verifiability, and security-related properties of a voting process depend on an accurate registration. Every voter's credential has to be uncommon, so that the correct authorization can be followed and kept secret to further prevent any type of misconduct. The Authorities verify that they are unique and thus we assume the absence of any kind of fraud or adversary in the entire registration phase. If in any case, a fraud breaks the registration phase, the complete security of the election protocol is abolished.

Candidate Slate Publication. Any adversary who manipulates the candidate slate is able to exclude candidates from the election, which should not be likely to occur, also it must be quick and easy to detect by the process. The list of all the candidates are published on the bulletin board so that one can detect any fraud, which saves the candidate from any attack on the slate publication.

Voting Phase. We consider that a voter should submit his ballots through an unidentified channel, and a fraud will not be able to find its details. The hypothesis, that any coercer is not allowed to monitor the action of a voter regularly, is needed. The assumption of an anonymous channel is accomplished by the facilitation provided that a voter can use any mobile connected to the Internet to submit his ballots, and the coercer is unable to follow this. To prevent or decrease the number of invalid or fraud votes, voters should only be able to cast one ballot per minute or so.

Tallying Phase. The password required to log in to cast ballot provides the assurance that in the resulting tally, no fallacious values are present. This phase is compulsory, as the absence of this stage will result in inclusion on unwanted data, which further seize a voter's fundamental right to vote freely. A time stamp associated with each ballot ensures that no duplicate ballot is tallied. The presence of this step sorts out ballots involving invalid credentials, at most one ballot per pre-registered credential has to be allowed in the final tally. After this, we can be sure that the final tally will be accurate.

5 Implementation of App-Based e-Voting

The **registration application** would be used by eligible citizens of India, interested to vote online on the designed App, to register for online voting. A facial record will be generated and the user's unique voter authentication records along with a personal mobile number will be registered in the user's unique record in the registration database. The registered face and its corresponding voting requisites would be verified with the real election commission database, to ensure that the voter is an authenticated and eligible citizen of India.

The **second application** would be used to generate, store, and deploy random party codes, to already registered voters, after proper biometric (facial) verification. The biometric record of the corresponding user, which should already be present in the registration database, proves that the user is an authentic one.

The **third application** will be used for voting. The servers for this application would be opened only on voting days and each user would be able to use this application, just once. This application will first perform a biometrical verification of the user, using the data already stored in the database. Then, the application would proceed to verify the user's voter identification records (e.g., Aadhar Card). Upon successful verification, an OTP would be sent to the user's registered mobile number. The application would accept and verify the OTP, see Fig. 3. Upon successful verification, the user will be allowed to enter into the voting mechanism to vote. The user then receives a "vote registered" message, and the servers are blocked for that user, using a Boolean variable.

6 Discussion on the Proposed System

In this section, we discuss the outcomes of the proposed system together with its advantages and drawbacks. The android-based system provides an interactive mechanism to develop an interface between user and database. A user can interact with MYSQL database and perform any query like retrieve, enter, view, and modify data as per the permission granted by the database administrator.

Advantages of the System

- Identification of a voter is unique: Government will provide unique voter id list, so every user can uniquely be identified by using his/her unique voter id.
- Portable system: The system we are using is portable, because we are using android-based smartphone which itself is a kind of a portable device.
- Unambiguous vote counting: Identical voters are not allowed, which assists in calculating the explicit counting of votes.
- Number of voters improves for participation; Fraud vote can't be submitted.

Disadvantages of the system

- Device dependent: This application is based on android smartphone, which results in device dependency.
- Device breakdown: There may be many reasons for device failure in android phones like hardware or system failure that can restrict the user to perform tasks through the e-voting application.

Weakness of the System. The Implementation of the system is hampered due to the following reasons:

- There are restricted resources to implement the system completely.
- There may be opposition from voters who do not believe in online voting due to security concerns.
- People from rural areas of the country may not vote through an android application because of unreliable power supply.
- People from rural areas of the country are not fully aware about an android application.

7 Conclusion

This work is intended to focus on analyzing the process of an advanced android-based electronic voting application. The proposed system is expected to encompass high usability if it gets implemented in a real-world election scenario. The mechanism described here offers a verifiable election system that is far accessible and advantageous over other systems. It will surely be helpful to the voters who want to vote online from any location, also the complete voting proceedings shall be convenient to handle. The proposed e-voting system via an android App is beneficial in minimizing any misconduct, threat of fraud votes, and recounting requirement as every step is automated by a computer. The concerned shortcomings of the presented system are also attached with the possibility of electronic failures that might occur in such electronic systems.

References

1. Hashimi, S., Komatineni, S., MacLean, D.: Pro android 2. In: Pro Android 2, pp. 10–11. Apress, New York, USA (2010)
2. Cranor, L.F, Rojas, R., Dearborn, F.: Electronic voting: encyclopedia of computers and computer history (2001)
3. Manivannan, P., Ramesh, K.: e-voting system using android smartphone. Int. Res. J. Eng. Technol. **2** (2015)
4. Mohammed, A.A., Timour, R.A.: Efficient e-voting android based system. Int. J. Adv. Res. Comput. Sci. Softw. Eng. **3** (2013)
5. Rivest, R.L.: Electronic Voting: Technical Report. Laboratory for Computer Science, Massachusetts Institute of Technology, Cambridge (2004)
6. Clausen, D., Puryear, D., Rodriguez, A.: Secure Voting Using Disconnected, Distributed Polling Devices. Department of Computer Science, Stanford University, Stanford (2000)
7. Ahmad, F., Rafay, M., Haider, K.: Mobile voting system. Int. J. Comput. Sci. Mobile Comput. (2018)
8. Chaum, D.: Secret-Ballot receipts: true voter-verifiable elections. IEEE Secur Privacy 2
9. Ryan, P.Y.A.: Pret a voter: a voter-verifiable voting system. IEEE Trans. Inf. Forensics Secur. **4** (2009)
10. Schneier, B.: Applied Cryptography: Protocols, Algorithms, and Source Code in C. Wiley (1996)
11. Stinsons, D.R.: Cryptography: Theory and Practice. CRC Press (2005)
12. Thakur, N., Ambavale, V.: e-voting using android system. Int. J. Sci. Res. **4** (2013)
13. Salih, M.N. et. al.: A design of voting system by using an Android apps for fisherman. ARPN J. Eng. Appl. Sci. (2017)

Energy Saving Delay Constraint MAC Protocol in Wireless Body Area Network

Tusharkanta Samal, Manas Ranjan Kabat, and Sushree Bibhuprada B. Priyadarshini

Abstract e-healthcare system plays a vital role in day-to-day life, particularly very useful for monitoring old and sick people. Many standards have been developed to meet above requirement. However, the standard like IEEE 802.15.4 is inefficient due to limited number of slots. The standard like IEEE 802.15.6 removes the loopholes of above standard by improving efficiency and quality of service of healthcare system. The main contribution of our proposed work is to modify the IEEE 802.15.6 standard such that more than one emergency data can be transmitted at a time and we also applied energy saving technique to save energy. All the simulation works are done in Castalia simulator. The simulation works reveal that the proposed protocol works better than existing counterparts in terms of throughput, energy consumption, and packet delivery delay.

Keywords WBAN · IEEE 802.15.6 standard · MAC · Priority queue

1 Introduction

The incredible growth of world population is a very problematic issue at present era. The vast population are facing three major challenges such as demographic increase of baby boomer, increasing number of aging population, and expensive healthcare system. Therefore, people require a healthcare system which will provide proactive

T. Samal · M. R. Kabat
Department of Computer Science and Engineering, Veer Surendra Sai University of Technology, Burla, Sambalpur 768018, Odisha, India
e-mail: samaltushar1@gmail.com

M. R. Kabat
e-mail: kabatmanas@gmail.com

S. B. B. Priyadarshini (✉)
Department of Computer Science & Information Technology, Siksha 'O' Anusandhan (Deemed to be University), Bhubaneswar, Odisha 751030, India
e-mail: bimalabibhuprada@gmail.com

D. Mishra et al. (eds.), *Intelligent and Cloud Computing*,
Smart Innovation, Systems and Technologies 194,
https://doi.org/10.1007/978-981-15-5971-6_64

wellness and prevention of diseases. These types of healthcare systems are called as e-healthcare system. The researchers have applied these magical wireless technologies in our day-to-day life to controlling, tracking, monitoring, and automation [1].

Paper [2] can be placed or implanted in the human body providing real-time data and such networks are called Wireless Body Area Network (WBAN). The IEEE 802.15.4 [3] is a standard model of medium access sub-layer that consumes less energy and provides low data rate while it transmits data. The limitation of 802.15.6 [5] model is that the nodes are awake all the time even if there is no data to transfer. Therefore, it consumes more energy. Another drawback is the highest priority data can be transmitted in Emergency Access Phase (EAP). The main contributions of the paper modify the 802.15.6 super frame structure in such a way that more than one emergency data can be transmitted. The performance of the proposed protocol is studied and compared with some existing protocol.

2 Related Works

Numerous MAC protocols have been proposed and people are doing their research to design better protocol to increase the throughput and efficiency. Rezvani and Ghorashi [6] proposed context-aware and channel-based resource allocation MAC protocol. In [7], Huq et al. proposed MEB MAC protocol. Otgonchimeg and Kwon [8] proposed HBC MAC, in which an emergency guaranteed time slot is used for transmission of emergency data. In [9], Hossain et al. adopt the EAP phase of IEEE 802.15.6 for highest priority data in EAP in WBAN, which is similar to IEEE 802.15.6 [8].

The work detailed in [10] uses beacon-enabled schedule access mode for energy analysis of IEEE 802.15.6. They also use an analytical model for finding out the device lifetime. The same authors [11] have developed integer programming to find out the lifetime of applications mentioned in ISO/IEEE 11073 using Type I scheduled access. But this analysis does not work in other applications. Motoyama [12] proposed flexible polling-based access scheme for WBAN. For energy saving, both polling and probabilistic contentions were used for channel access in [13]. But this protocol is suitable only for same data rates for all nodes. Jacob et al. [14] proposed MAC protocol in which they use the MAC sleeping schedules for contention and polling access schemes to maximize lifetime. Jacob et al. [15] proposed Green media access method in which they used IEEE 802.3az standard for energy-efficient Ethernet [16] concept to consume less energy. In this paper, our focus is on improving the reliability of emergency data frame transmission and energy saving technique in WBANs.

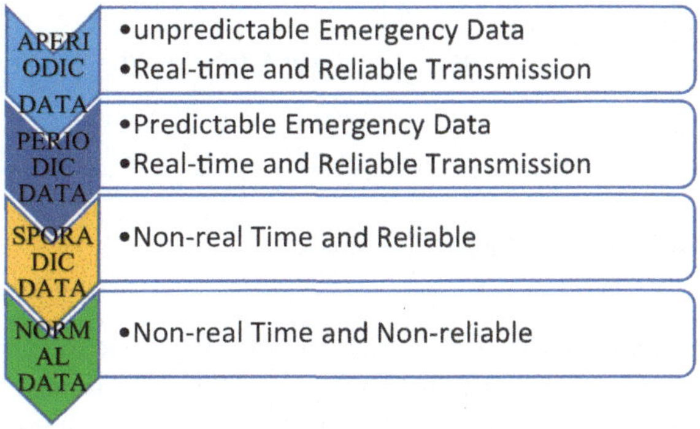

Fig. 1 Types of traffic

3 Proposed Work

The Energy-Efficient Delay Constraint MAC (PD-MAC) protocol operates on beacon-enabled mode. In this mode, the Body Coordinator's (BC) responsibility is to broadcast a beacon in the network. The beacons are sent periodically in the network.

3.1 Types of Traffic

The proposed protocol classified the patient's data into Aperiodic Data (AD), Periodic Data (PD), Sporadic Data (SD), and Normal Data (ND). The types of traffic are indicated in Fig. 1.

Aperiodic Data It is defined as the emergency data, which is critical and unpredictable.

Periodic Data It is also emergency data, which is predictable. These are generated periodically. It requires real-time and reliable transmission.

Sporadic Data It is defined as the ordinary data, occurring at irregular intervals.

Normal Data It is defined as the non-real time and non-reliable data.

3.2 Proposed Modified Superframe Structure

The proposed protocol is based on IEEE 802.15.6 model. The proposed superframe structure of ED-MAC protocol has a beacon period which indicates boundaries of superframe structure indicated in Fig. 2, one Emergency Transmission phase (ETP),

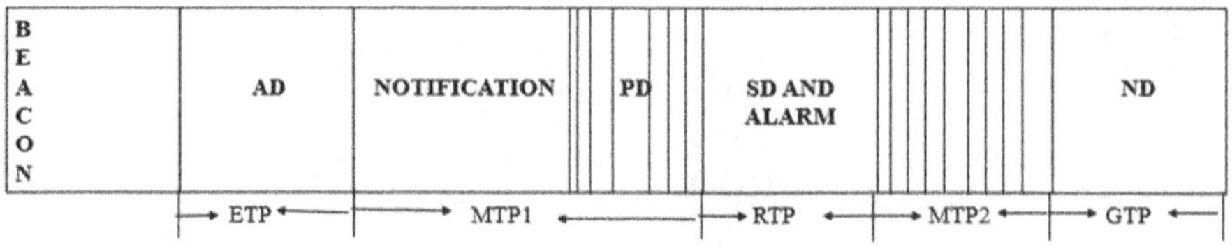

Fig. 2 ED-MAC superframe structure

two Manage Transmission Phase (MTP), one Regular Transmission Phase (RTP), and one General Transmission Phase (GTP). ETP is used for transmitting aperiodic data, i.e., for emergency life critical situation data. After transmitting Aperiodic Data (AD), notification message is transmitted during MTP1 to inform the number of Periodic Data (PD). After gaining the knowledge of number of Periodic Data (PD), they are transmitted in MTP1 in fixed time slot. Sporadic Data (SD) are generated in irregular and non-real time interval. So these types of data are transmitted in CSMA/CA with user priority 5 in RTP phase. The AD data attempt to recover from collision by sending an alarm frame to coordinator node during immediate RTP user priority 7 that alerts an emergency situation. Figure 3 illustrates the algorithm proposed and Table 1 shows the mechanism of priority queue.

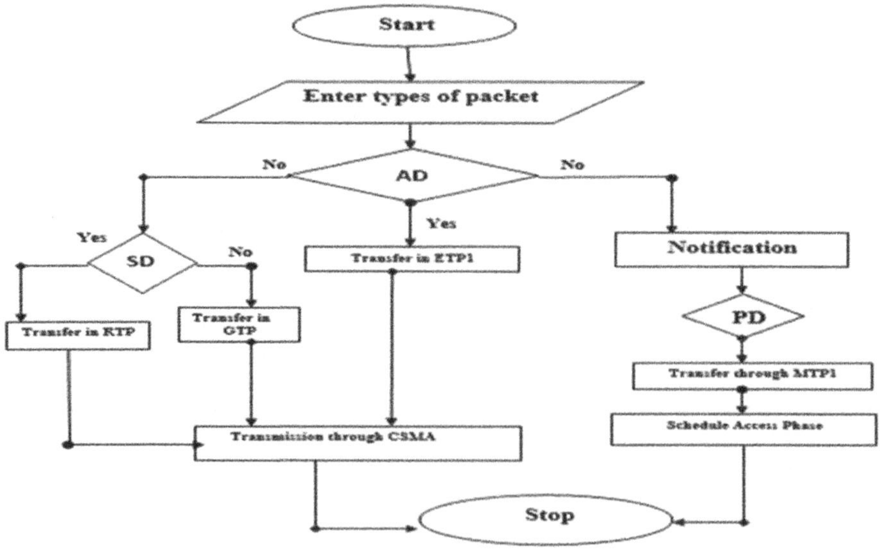

Fig. 3 Algorithm for proposed method

Table 1 Priority queue mechanism

Traffic class	User priority	Access phase	Traffic designation	CW_{Min}	CW_{Max}
AD	7	ETP, MTP	Emergency	1	1
PD	6	MTP	Emergency and periodic	2	8
SD	5	RTP	Sporadic	4	16
ND	4–0	GTP	Non-emergency	16	32

3.3 Priority Queue Mechanism

Notations:

P=Priority, AS=Alarm Signal, NF=Notification.
Input: User priority of data packet.
Output: Slot allocation for different types of packet.
Process:

Step 1: Initialize P.
Step 2: IF(P==7) THEN print AD and transmit in ETP.
Step 3: IF(P==6) THEN print PD and send NF and transmit in MTP1.
Step 4: IF(P==5) THEN print SD and transmit in RTP.
Step 5: Alarm (IF AS)

Transmit in CSMA 0.
IF ACK received for CSMA 0 THEN transmit in RTP.
ELSE transmit in CSMA 1.
IF ACK received for CSMA 1 THEN transmit in RTP.
ELSE transmit in CSMA 2.
IF ACK received for CSMA 2 THEN transmit in RTP.
ELSE discard the packet.

Step 6: FOR (P=0; P<=4; P++)

Print ND.
Transmit in GTP.

4 Performance Evaluation

For performance evaluation, we have done a comparative study among our proposed protocol ED-MAC with the existing TrysL MAC [18], BCR MAC [19], TTR MAC [20], and TP CAT [21] protocols. These above protocols have same objective as the proposed protocol. Therefore, we have selected the above protocols. The simulation was conducted in Castalia 3.2 [22], which is developed by OMNET++ Simulator. Figures 3, 4, and 5 show the results.

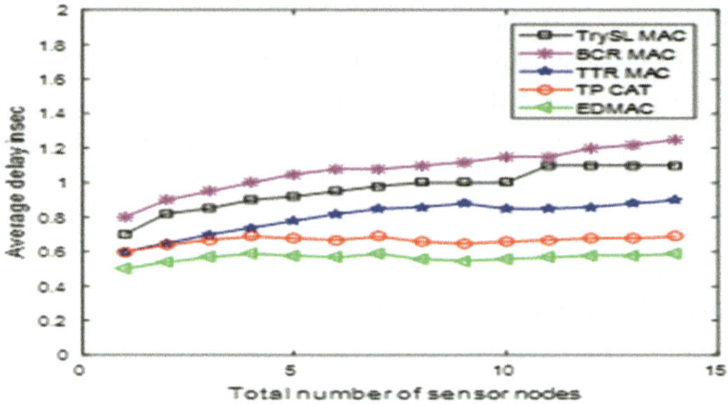

Fig. 4 Packet delivery ratio versus simulation time

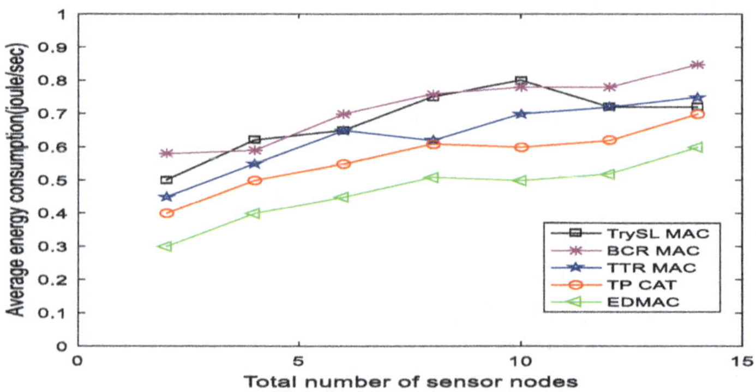

Fig. 5 Number of sensor nodes versus average energy consumption

4.1 Packet Delivery Delay

The packet delivery delay of various protocols with same objective of TrysL MAC, BCR MAC, TTR MAC, TP CAT, and ED-MAC. In case of TTR MAC, some fixed slots are reserved for transmission of emergency data. If the number of emergency data is more than number of fixed reservation slot then delay increases. The ED-MAC transmitted number of similar types of data by using respective slot.

4.2 Energy Consumption

The energy consumption of various protocols such as packet delivery delay of TrysL MAC, BCR MAC, TTR MAC, TP CAT, and ED-MAC. TrysL MAC and BCR MAC protocols consume maximum energy. The novel protocol consumes minimum energy.

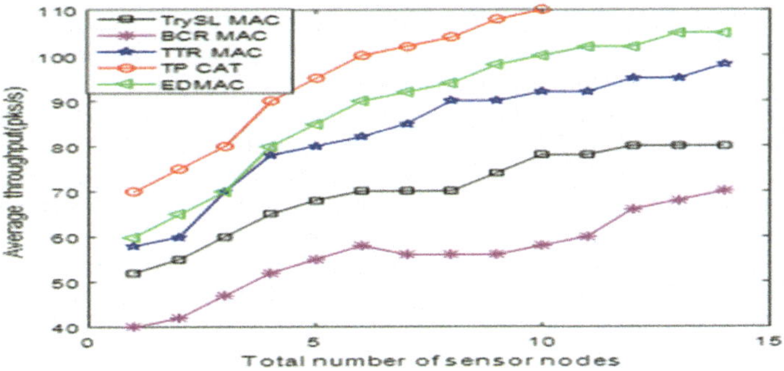

Fig. 6 Throughput versus simulation time

4.3 Throughput

CAT protocol provides specific slot for high threshold and low threshold data. TP CAT protocol has high throughput as compared to TrySL MAC, BCR MAC, and TTR MAC (Fig. 6).

5 Conclusions and Future Scope

In this paper, we have proposed an Energy saving Delay Constraint MAC (ED-MAC) protocol for carrying emergency data frame in WBAN to improve the reliability while meeting the end-to-end delay requirement. The simulation result has proved that the proposed protocol, ED-MAC protocol, works better than their counterparts..

References

1. Ullah, S., Higgins, H., Braem, B., Catre, B., Blmdia, C., Moerman, I., Salleem, S., Rahmon, Z., Kwak, K.: A compressive survey of wireless body area network. J. Med. Syst. **36**, 1065–1094 (2012)
2. Seyedi, M., Kibret, B., Lai, D., Faulkner, M.: A survey on intrabody communications for body area network applications. IEEE Trans. Biomed. Eng. **60**(8), 2067–2075 (2013)
3. Ullah, F.., Abdullah1, A., Kaiwartya, O., Kumar, S., Arshad, M.: Medium access control (MAC) for wireless body area network (WBAN): superframe structure, multiple access technique, taxonomy, and challenges. Hum. Centric Comput. Inf. Sci. 1–39 (2017)
4. Kurunathan, J.H.: Study and overview on WBAN under IEEE802.15.6, U. Porto J. Eng. **1**(1), 11–21 (2015)
5. IEEE standard for local and metropolitan area networks—Part 15.6: wireless body area networks. IEEE Standard 802.15.6 (2012)
6. Rezvani, S., Ghorashi, S.A.: Context aware and channel-based resource allocation for wireless body area networks. IET Wireless Sens. Syst. **3**(1), 16–25 (2013)

7. Huq, M.A., Dutkiewicz, E., Fang, G., Liu, R.P., Vesilo, R.: MEB MAC: improved channel access scheme for medical emergency traffic in WBAN. In: Proceedings of International Symposium on Communications and Information Technology pp. 371–376 (2012)
8. Otgonchimeg, B., Kwon, Y.: Emergency handling for MAC protocol in human body communication. EURASIP J. Wirel. Commun. Netw. **2011**(1), 786–903 (2011)
9. Hossain, M.U., Kalyan, M., Rana, M.R., Rahman, M.O.: Multidimensional traffic adaptive energy-efficient MAC protocol for wireless body area networks. In: Proceedings of 9th IEEE Int. Forum Strategic Technology, pp. 161–165 (2014)
10. Tachtatzis, C., Franco, F.D., Tracey, D.C., Timmons, N.F., Morrison, J.: An energy analysis of IEEE 802.15.6 scheduled access modes. In: Proceedings of IEEE Globecom 2010 Workshop on Mobile Computing and Emerging Communication Networks, pp. 1270–1275. IEEE, Miami (2010)
11. Tachtatzis, C., Franco, F.D., Tracey, D.C., Timmons, N.F., Morrison, J:, An energy analysis of IEEE 802.15.6 scheduled access modes for medical applications. In: Proceedings of Ad Hoc Networks: Third International ICST Conference, pp. 2019–2022. ADHOCNETS (2011)
12. Motoyama, S.: Flexible polling-based scheduling with QoS capability for wireless body sensor network. In: Proceedings of 37th Annual IEEE International Conference on Local Computer Networks-Workshops, pp. 745–752. IEEE, Florida (2012)
13. Ibarra, E., Antonopoulos, A., Kartsakli, A., Verikoukis, C.: HEH-BMAC: hybrid polling MAC protocol for WBANs operated by human energy harvesting. Telecommun. Syst. **58**(2), 111–124 (2013)
14. Jacob, A.K., Jacob, L.: Energy efficient MAC for QoS traffic in wireless body area network. Int. J. Distrib. Sens. Netw. **11**(2) (2015)
15. Jacob, A.K., Kishore, G.S., Jacob, L.: Lifetime and latency analysis of IEEE 802.15.6 WBAN with interrupted sleep mechanism. Sādhanā **42**(6), 865–878 (2017)
16. IEEE Std 802.3az-2010: https://doi.org/10.1109/IEEESTD.2010.5621025 (2010)
17. Ullah F., Abdullah, A.H., Kaiwartya, O., Cao, Y.: Traffic priority aware adaptive time slot allocation for medium access control protocol in wireless body area network. MDPI Sens. J. (2017)
18. Shin, H., Kim, Y., Lee, S.: A backoff counter reservation scheme for performance improvement in wireless body area networks. In: 12th Annual IEEE Consumer Communications Networking Conference (CCNC) (2015)
19. Yuan, X., Li, C., Yang, L., Yue, W., Zhang, B., Ullah, S.: A token-based dynamic scheduled MAC protocol for health monitoring. EURASIP J. Wireless Commun. Netw. (2016)
20. Ambigavathi, M., Sridharan, D.: Traffic priority based channel assignment technique for critical data transmission in wireless body area network. J. Med. Syst. (2018)
21. Castalia: Wireless sensor network simulator. https://castalia.fongue.nicta.com.au/index.plp/en/.accessed (2013)

QoS-Aware Unicasting Hybrid Routing Protocol for MANETs

Prabhat Kumar Sahu, Biswa Mohan Acharya, and Niranjan Panda

Abstract In recent years, the emerging technologies in the field of Networking, the wireless network, is the fastest growing and developing sector among all. And among those wireless networks, ad hoc network is attracting much attention toward Research developments and industrial developments for its most technological advances. Mobile ad hoc network (MANET) is such a type of wireless network that doesn't depend on any pre-existing infrastructure such as routers or base station. Wi-Fi technology is used for point-to-point communication without passing through an access point in MANET. Ad hoc network is basically a temporary network connection created only for some particular purpose or in any emergency case such as for military service, rescue mission team, natural disasters like floods, earthquake, etc. Congestion in the network becomes one of the major drawbacks in this type of infra structureless network. Therefore, it is quite difficult to satisfy the quality of service (QoS) for such type of a networking environment. Parameters like throughput, bandwidth, delay, jitter, etc. that influence QoS make it a more grueling task for unicasting transmission in MANETs. In this context, Zone Routing Protocol (ZRP) is treated as one of the most reliable hybrid routing protocols which can solve the failure of both proactive and reactive routing protocols to meet a certain standard. In this paper, we have considered ZRP as our primary routing protocol and have tried to discuss different QoS parameters in order to resolve congestion in the network.

Keywords MANET · QoS · Droptail · RED · ZRP

P. K. Sahu (✉)
Department of Computer Science and Engineering, Siksha O Anusandhan (Deemed to Be) University, Bhubaneswar, Odisha, India
e-mail: prabhatsahu@soa.ac.in

B. M. Acharya · N. Panda
Department of Computer Application, Siksha O Anusandhan (Deemed to Be) University, Bhubaneswar, Odisha, India
e-mail: biswaachrya@soa.ac.in

N. Panda
e-mail: niranjanpanda@soa.ac.in

D. Mishra et al. (eds.), *Intelligent and Cloud Computing*,
Smart Innovation, Systems and Technologies 194,
https://doi.org/10.1007/978-981-15-5971-6_66

1 Introduction

In recent years, the emerging technologies in the field of Networking, the wireless network, is the fastest growing and developing sector among all. And among those wireless networks, ad hoc network is attracting much attention toward Research developments and industrial developments for its most technological advances. Mobile ad hoc network (MANET) is such a type of wireless network that doesn't depend on any pre-existing infrastructure such as routers or base station. Wi-Fi technology is used for point-t-point communication without passing through an access point in MANET. "AD HOC" is a Latin word which means "for this purpose" and "Mobile" means "to move freely". As the name suggests, the mobile ad hoc network is a temporary network connection, created for some peculiar or emergency purposes and can be structured at any point of time as well as at any place without having any base station or router. Ad hoc network is basically a temporary network created only for some specific purpose or in any emergency case such as military service, rescue mission team, natural disasters like floods, earthquake, etc. The ad hoc network has a vast implementation in cab companies. Taxi cabs use this network to communicate with their headquarters as well as with their customers. The ad hoc network system is also used in sports stadiums. It helps the referees to communicate with each other and also helps the stadium viewers to share their interests and sports news on social media. In multi-hop ad hoc networks, the intermediate nodes should route the packet, i.e. they should act as a router too. With the increase in technological advancement in the area of wireless networks, the QoS becomes one of the major concerns in routing protocols. Quality of service refers to the dexterities provided by the network while transmitting data packets from one mobile node to another. Resource limitations and the complexity associated with the mobility of hosts create difficulty for getting and managing these issues in MANETs [1]. Advantages like no infrastructure, lower cost, mobility, decentralized, robust, easy to build and spontaneous infrastructures are associated with MANETs, some of the disadvantages are lower data rate, security, and energy limitation. In an ad hoc network, the routing protocol is present which instructs the nodes to decide which way to route the packets between computing devices. It is mainly used in multi-hop ad hoc networks. Besides all advantages and disadvantages, the most important/attractive property of an ad hoc networking model is its independence in a decentralized area which gives the user freedom and flexibility of access. There are still some fields/properties like security, protocols, etc. in ad hoc network in which, research is in progress, till yet not implemented properly [2].

2 Issues and Challenges for MANETs

Scalability: Scalability in ad hoc networks is the most important problematic existing issue. It is one of the most important open problems that can be broadly defined as whether the network is able to give a proper type of service to the data packets that are being transmitted between a even in the presence of the large number of nodes in the network. This problem is mainly observed in multi-hop ad hoc networks. Scalability problem mainly occurs due to the increased number of nodes or increased number of packets in the network which causes overhead, packet collision, bandwidth problem, etc. and as a result, the desired packet cannot reach to its desired destination. This issue can be solved by formulating some new protocols in an ad hoc network [3].

Security: It is a vital issue in ad hoc networks. It is usually examined to be the major "roadblock" in the commercial application. Applications like military or confidential meeting require a high degree of security passwords against enemies which sometimes is not possible in ad hoc networks. Mobility of users increases the security concern toward a wireless network. Since hackers nowadays are getting more clever, it is crucial to update the features of wireless security constantly. In multi-hop ad hoc networks since nodes use a shared radio medium to transmit the data to the desired destination in an insecure environment, they are affected by Denial of Service (DoS) attacks that are harder to track in a wireless network than in a wired network. DoS is the security attack that causes a node to use its transceiver so much that it depletes its battery. There are still several issues related to security which have not yet been solved by researchers but are important to be solved to bring the ad hoc network to a good solution.

Routing: Routing is a real challenging issue in a wireless ad hoc network. Routing in wireless ad hoc networks is nontrivial due to the highly dynamic environment. Since an ad hoc network is a combination of many wireless mobile nodes without any pre-existing infrastructure or centralized administration, while exchanging information the nodes establish routes among themselves and act as a router without any outside support.

Power supply/Energy conservation: Since an ad hoc network doesn't depend on any fixed infrastructure, its individual nodes rely on the batteries which have limited power sources. For this, the use of the available bandwidth by the nodes will be limited to consume energy efficiently. Here while sending or receiving the data, meanwhile, if the battery is depleted, then recharging is necessary.

Dynamic Topology: Due to the fixed transmission range, nodes may be in and out of range from each other which may lead to some link failure, and some new links may be established between the nodes. Due to these issues, ad hoc network faces various types of faults like transmission errors, node failure, link failure, breakages, congested nodes/links, etc.

QoS (Quality of Services): As the ad hoc network does not have a fixed topology, a lot of challenge arises which brings tedious tasks for the researchers. While transmitting the data from one device to another, a number of services are referred by QoS [1]. The network guarantees a lot of network QoS parameters like lower bandwidth, packet delay and loss, variance in packet delay, higher consumption in power, etc. The use of the ad hoc network has been extended from only army use to a lot of fields only because of its flexibility. Due to its wide use, now the users are expecting higher performance and quality of service from this network.

3 Routing Protocol

Based on different routing problems like the asymmetric link between the nodes, interference, and dynamic topology, routing protocols in MANETs are factorized into two categories as unicast and multicast routing protocols. Most of the applications of MANETs are based on unicast communication. In Unicast routing, the data packet is transmitted from one source node to one destination. Here, the sender sends one copy of data to each receiver from the source node. All the routing protocols coming under these two categories are further divided into three types such as proactive, reactive, and hybrid routing protocols [1, 4]. Proactive routing protocols are also termed as table-driven protocols where all nodes maintain a consistent view of network topology and up-to-date routing information so that they can send their information/data to their destination without facing any overhead, traffic, etc.; here, the routing path is predefined in the nodes. When a node wants to send the packet to its respective destination node, if a network topology change occurs, then the routing information gets updated throughout the network to notify the change in the routing path. Some of the Proactive Routing Protocols in MANETs are Destination Sequence Distribution Vector (DSDV), Wireless Routing Protocol (WRP), and Fisheye State Routing (FSR). Reactive routing protocols are terms as On-Demand Routing Protocols where the source node establishes a routing path to a destination only when there is a demand for it. Some Reactive MANET Routing Protocols are Ad Hoc On-Demand Distribution Vector (AODV), Dynamic Source Routing (DSR), etc. Here, it doesn't keep any up-to-date routing information which is one of the disadvantages of the Reactive Routing Protocol. Hybrid routing protocols are mainly designed to comprise the advantages of proactive and reactive routing protocols and to solve their shortcomings. Both Proactive and Reactive Routing Protocols work well in networks having a small number of nodes. As the number of nodes increases in the network, hybrid Routing Protocol performance becomes much more significant in comparison to the other two. Some Hybrid Routing Protocols are Zone Routing Protocol (ZRP), Hybrid Ad Hoc Routing Protocol (HARP), etc.

4 Unicast ZRP

ZRP can be classified as a hybrid proactive/reactive routing protocol [5]. It is mainly established to overcome the problems faced by reactive and proactive routing protocols [6]. ZRP with proactive routing design is a zone inside which some nodes are present communicating with each other. Due to this limited zone, the routing information of each node is maintained easily. Here in this case, nodes are closer to each other within the zone. In ZRP with reactive routing, nodes are present outside the zone [5]. Here, the nodes that are present inside the zone can communicate with the outer nodes with reactive routing. Since in ZRP, all nodes store their routing information in their own routing table, the communication between the nodes is performed efficiently without any network traffic or communication delay problem. ZRP is mainly used to minimize the delay in route discovery and the message traffic problem which are the main drawbacks of the reactive and proactive approaches. ZRP is of many types. Some of the most commonly used ones are tree-based and mesh-based schemes. Routing in ZRP is based on two types of routing mechanisms called Intra-Zone Routing Protocol (IARP) and Inter-Zone Routing Protocol[7] (IERP). IARP is responsible to communicate with the nodes present outside the zone. It is also known as the global reactive routing protocol. In this, first, the source node searches for the destination node. If the destination node is outside the routing zone of the source node, then the border casting process is used to discover the route for the outer zone. This border casting procedure uses Border cast Routing Protocol (BRP) to send the messages from source nodes to peripheral nodes. IERP sends the route request to the outer zone by using the topology information provided by IARP [8]. The Query-Control mechanism which can detect the stale queries has already been covered by a zone and will remove that packet from the zone before reaching to its destination or peripheral nodes [9]. Here in ZRP, often the zones overlap each other so only one route request is forwarded many times by every node. This gradually results in message traffic problems. This excess traffic problem can be avoided by ZRP by using this Query-Control mechanism. ZRP uses 3 types of Query-Control mechanisms to solve the abovementioned problems such as Query Detection, Early Termination, and Random Query Processing Delay. Advantages of ZRP not only reduce the message traffic problem occurring in the proactive protocol due to flooding of data packets but also reduce the bandwidth wastage and the overhead problem occurring in the reactive protocol. The main disadvantage of ZRP is that a large overlapping of zones may lead to network traffic problems.

5 Queue Management

From all the above scenarios, we have concluded that though Zone Routing Protocol which is a hybrid routing protocol is the better way of routing in MANETs but still it has some limitations. The most important drawback is that sending a replicate number of data packets to search for a new route which leads to network load and congestion problems. So to avoid such downside, we proposed the Multicast Zone Routing Protocol in this paper which is an extended form of Zone Routing Protocol [10]. ZRP which is a unicasting Routing Protocol helps in reducing the network load by limiting the transmission of control packets to its desired destination through multiple paths. ZRP plays a vital role in enhancing some parameters of QoS such as throughput, end-to-end delay, packet data ratio, etc. to avoid congestion and network load problems. This avoidance of 2 major problems can be done in ZRP by using a technique called Queue Management Technique. As we know, congestion is one of the most vital issues in ZRP in today's scenario. Queueing is the only effective way to manage the network load and congestion overhead. ZRP uses the Queue Management Technique to avoid such problems in the network link and thus to enhance QoS in MANETs [11]. Queue management is one of the best methods to avoid congestion. Queueing mechanism helps to determine the ordering of data packets in a queue. It helps in limiting the control packets by indicating which packet is to be transmitted and which is to be dropped. This Queue Management Technique affects the QoS parameters in terms of throughput, end-to-end delay, PDR, and packet drop. There are various types of Queue Management Techniques available such as Droptail [12], Random Early Detection (RED), Random Exponential Marking (REM), Stochastic Fair Queueing (SFQ), Adaptive Virtual Queue, etc. [13].

6 Simulation Environment and Performance Evolution

The main aim of our paper is to discuss and differentiate the QoS parameters, that is, Throughput, Packet Delivery Ratio (PDR), and End-to-End Delay of ZRP by using Queue Management: Droptail and RED. The major difference between Droptail and RED is that in RED, it takes the average queue length but in Droptail it takes directly the given queue length by considering the QoS parameters as shown in Table 1. Simulation is done through NS2 network simulator [14].

6.1 Throughput

In a wireless network, throughput can be defined as the total number of data packets transmitted from source to destination nodes. It mainly measures how much of data is being received by the receiver that is sent by the sender. As shown in Fig. 1, we can

Table 1 Performance parameters

Parameter	Values
Protocol	ZRP
Nodes considered	10, 20, 30, 40, 50
Zone radius	2
Simulation time	150 ms
Environment size	5000 × 5000
Traffic type	TCP
Antenna type	Omnidirectional
Mobility model	Random waypoint
Packet size	100 bytes

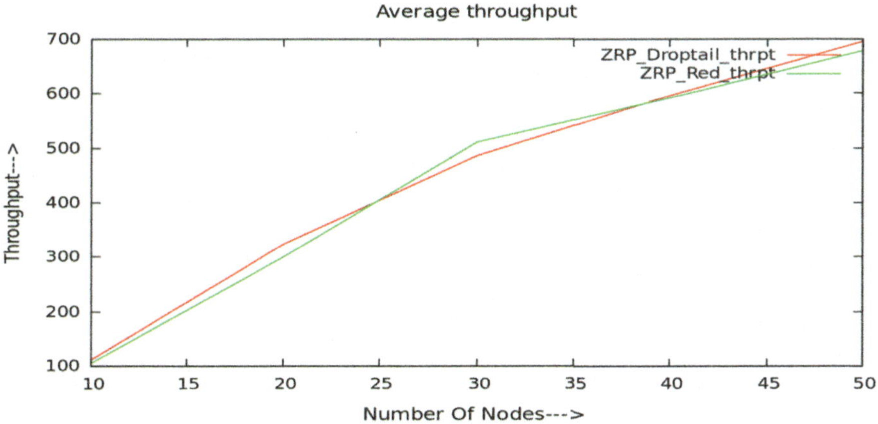

Fig. 1 Throughput versus number of nodes

clearly analyze that the throughput of the RED scheme is higher than the throughput of the Droptail scheme. As the network load becomes heavy in Droptail as the buffer exceeds beyond its maximum capacity, its throughput becomes less than that of RED, whereas in RED the network load is less as it avoids the overflow of the queue, so its throughput is high. Thus, we conclude that with the increase in network load, throughput decreases.

6.2 *End-to-End Delay*

Delay is measured in ms which is defined as the estimated time duration between the arrival and successful delivery of the packet to its desired destination. Practically, if a packet drop is less, the network congestion will be high as the packets are full in the queue and as a result, bandwidth usage will be high. We can clearly analyze from Fig. 2 that the End-to-End Delay of the Droptail scheme is much higher than the End-to-End Delay of the RED scheme. In Droptail, the packet loss is less due to which congestion problem arises. Due to this reason, the time of transmitting data packets from source to destination becomes high. Thus, we conclude that as the packet drop decreases, the End-to-End Delay increases.

6.3 *Packet Delivery Ratio (PDR)*

During data transmission, some data packets may not be delivered due to some reason like network failure and availability of corrupted in the network. In this context, the data packets are retransmitted which causes transmission delay. It is defined as the

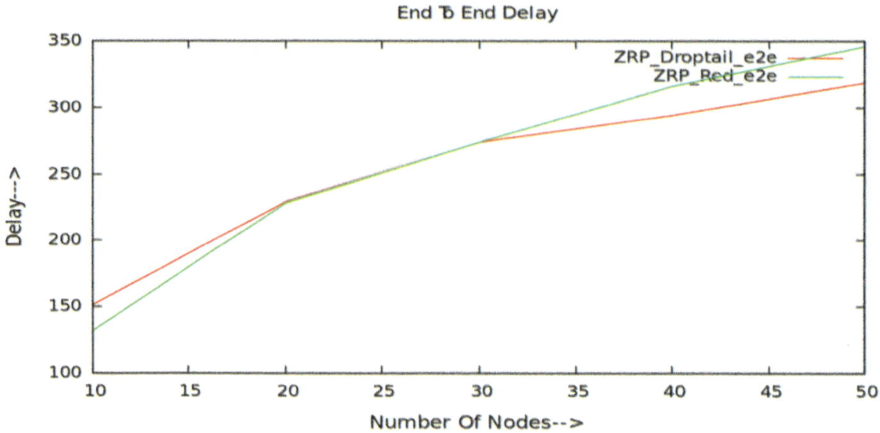

Fig. 2 Delay versus number of nodes

ratio of the data packets received by the destination node to the data packets sent by the source. We can clearly analyze from Fig. 3 that the PDR of the Droptail scheme is much higher in comparison to the RED scheme. In Droptail, the packet loss is less due to which congestion problem arises and enhances the data transmitting duration from source to destination. Thus, we conclude that as the packet drop decreases when the PDR increases.

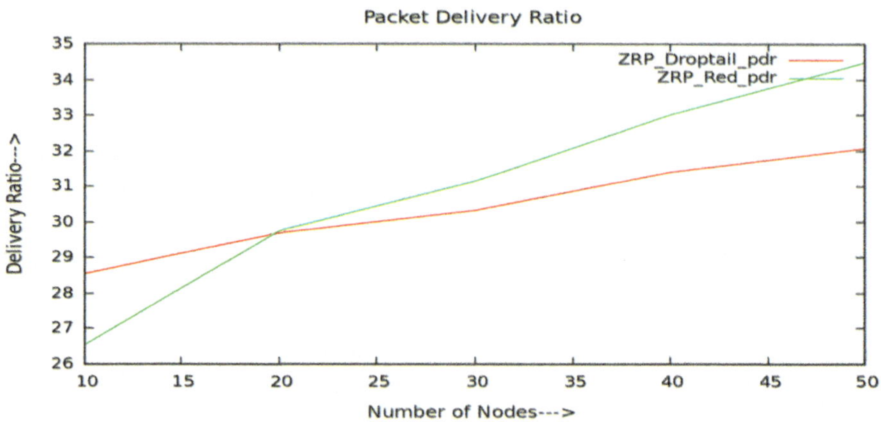

Fig. 3 PDR versus number of nodes

7 Conclusion

This paper is decorated with the performance comparison of the ZRP routing protocol using two mostly used queue management techniques such as Droptail and RED for unicasting in MANETs. We have discussed several characteristics of the mobile ad hoc networks and a hybrid routing protocol known as ZRP Routing Protocol for unicasti communication. We have also simulated the unicasting ZRP routing protocol in NS2 by defining a quintessential simulation environment with non-identical performance parameters essentially packet delivery ratio, average end-to-end delay, and average throughput. The result after simulation shows that the throughput, PDR, and delay need to be increased for a network with a huge number of nodes called a robust network. In order to carry out better performance, we introduced the concept of unicasting into the ZRP routing protocol under different queue management techniques. When the number of nodes increases, the performance of ZRP is significantly good in a unicasting environment for the Droptail queue in comparison to RED. It achieves high scalability and provides good routing performance but still has some drawbacks. So in order to overcome those drawbacks in ZRP and enhance QoS of MANETs, we analyzed the performance of some queuing techniques namely Droptail and RED.

References

1. Sahu, P.K., Pattanayak, B.K.: Quality of service based multicasting routing protocols for MANETs: a survey. Int. J. Commun. Antenna Propag. (IRECAP) **7**(N0), 5 (2017)
2. Singh, C., Gupta, V., Kaur, G.: A review paper on introduction to MANET. Int. J. Eng. Trends. Technol. (IJETT) **1**(1), 38–43 (2014)
3. Chlamtac, I., Conti, M., Liu, J.J.N.: Mobile ad hoc networking: imperatives and challenges. Ad Hoc Netw. **1**(1), 13 (2003)
4. Perkins, C., Royer, E., Das, S., Marina, K.: Performance comparison of two on-demand routing protocols for ad hoc networks. In: IEEE Personal Communications, Feb 2001
5. Ghode, S., Bhoyar, K.K.: NEMA: node energy monitoring algorithm for zone head selection in mobile ad-hoc network using residual battery power of node. In: International Conference on Wireless Communications Signal Processing and Networking (WiSPNET), pp. 1999–2004 (2016)
6. Kim, K.J., Koo, H.W.: Optimizing power-aware routing using zone routing protocol in MANET. In: IFIP International Conference on Network and Parallel Computing Workshops (NPC 2007), pp. 670–675 (2017)
7. Haas, Z.J., Pearlman, M.R., Samar, P.: Intra zone Routing Protocol (IARP). In: Internet Draft, IETF MANET Working Group, NY, July 2002
8. Xijie, Z., Chunxiuand, X., Jiaqi, X.: Hierarchical ZRP's performance vs ZRP's performance in MANET. In: IEEE International Conference on Communication Software and Networks (ICCSN), pp. 423–426 (2015)
9. Haas, Z.J., Pearlman, M.R.: The performance of query control schemes for the zone routing protocol. IEEE/ACM Trans. Netw. (TON) **9**(4), 427–438 (2001)
10. Sahu, P.K., Pattanayak, B.K.: Comparative performance analysis of ZRP routing protocol for multicasting in MANETs using Droptail and RED queue management. Far East J. Electron. Commun. **17**(2), 295–310 (2017)

11. Mohapatro, S., Sahu, P.K., Dash, P.: A hybrid active queue management technique for QoS based unicasting protocols in MANETs. Int. J. Eng. Technol. **7**(2), pp. 58–63 (2018)
12. Sahu, P.K., Acharya, B.: QoS based performance analysis of AODV and DSR routing protocols in MANET. In: 2nd International Conference on Data Science and Business Analytics, ICDSBA 201824, Article number 8588915, pp. 221–225, Dec 2018
13. Singh, D., Pattanayak, B.K., Panda, C.K.; Analysis of an improved energy balanced routing protocol for wireless sensor network. In: 2016 International Conference on Communication and Signal Processing (ICCSP), pp. 1807–1811. IEEE (2016)
14. Issariyakul, E.H.T.: Introduction to Network Simulator NS2. Springer Science and Business Media, NY (2009)

Background Modeling and Elimination Using Differential Block-Based Quantized Histogram (D-BBQH)

Satyabrata Maity, Nimisha Ghosh, Krishanu Maity, and Sayantan Saha

Abstract The proposed work uses differential block-based quantized histogram (D-$BBQH$) feature extraction technique to background modeling and elimination for foreground extraction. D-$BBQH$ uses a statistical approach to estimate the temporal consistency of pixel intensity concerning location. Besides, the block-based technique is less noise prone than the same of pixel-based processing and includes local features to estimate the background more efficiently than that of global processing. The proposed approach used a low-dimensional feature set, which requires less amount of time for background modeling or foreground extraction. A benchmark $SBMI$ dataset, which incorporates complex video data, is used to complete the experimentation of the proposed approach, and the results prove the effectiveness of the proposed approach.

Keywords Background modeling · Quantized histogram · Block-based processing · Temporal consistency

1 Introduction

The use of video information is increasing exponentially with the rapid change in technological advancement in the domain. The amount of space required to store a video is much higher than any other form of information, still, people love to watch

---

S. Maity (✉) · N. Ghosh · K. Maity · S. Saha
Department of Computer Science & Engineering, Siksha O Anusandhan (Deemed to Be) University, Bhubaneswar Odisha, India
e-mail: satyabratamaity@soa.ac.in

N. Ghosh
e-mail: nimishaghosh@soa.ac.in

K. Maity
e-mail: krishanumaity@soa.ac.in

S. Saha
e-mail: sayantansaha@soa.ac.in

D. Mishra et al. (eds.), *Intelligent and Cloud Computing*,
Smart Innovation, Systems and Technologies 194,
https://doi.org/10.1007/978-981-15-5971-6_67

the video for experiencing the fact visually. The main reason behind the volume of the video is redundant data to enrich its quality in terms of human perseverance. The huge amount of redundancy makes the process of actual information extraction manually from the video extremely difficult. This phenomenon inspires many researchers to design automatic video analysis systems.

There are two main kinds of redundancies present in video-based information: (A) Spatial redundancy: the repetition of similar information spatially by increasing the resolution of a frame to make it more visually prominent to the spectators. (B) Temporal redundancy: the replication of similar frames more to increase the smoothness in the transition of visual facts. Most of the time, redundancy is incorporated to increase the visual quality of a video. On the other hand, redundancy is negatively proportional to the information entropy. Hence, to extract the actual information, it is required to eliminate the redundancies, i.e., redundancy increases the processing overhead. A huge part of the redundancy in a video is due to the background information, which remains static throughout the scene. On the other hand, the foreground is the dynamic part with significant information entropy. Hence, it is the most required area for analysis to get through the vital information of the video. Thus, background elimination for the extraction of the foreground is a crucial step of video analysis, which is essential for various applications of video processing like human action detection, video surveillance and monitoring, medical image analysis, sports, entertainment, robotic vision, etc. The background and foreground segmentation help us in two ways, specifically (A), the removal of redundant parts results in the required area to analyze the video, (B), reduce the computational complexity for analyzing the video.

In this paper, we have proposed a background modeling technique based on the block-based probabilistic feature extraction technique, which is adaptive to change in background information. There are several environmental conditions in outdoor and indoor scenes in real-world scenarios. Some of the common challenges to the background modeling problem are bootstrapping, illumination changing, ghosting effect, dynamic background, foliage, missing or abundant objects in the scene, etc. The dataset we used to verify our results contains all these difficulties.

Human eyes can easily seance the background measuring its rigidity and are easily distinguishable from the continuous changing foreground.queryPlease check the sentences “Human eyes can easily ...” and “The changing profile stores ...” for clarity and correct if necessary. Please approve the edits made to the chapter title. The main motivation is to estimate the most statistically rigid information to model the background, and use an algorithm to eliminate it from the current frame to extract the foreground. Technically, a series of intensities are used to represent an image and the number of intensity levels is proportional to the clarity of the image/video. The human eye cannot sense all the intensity levels. Neighboring pixels with close intensity values add negligible information. On the other hand, sometimes these negligible details become the cause of false alarm. Suppose, in a video scene, where a tree with moving leaves is in the background, if we try to extract the foreground, the moving leaves will be treated as foreground. Thus pixel-level processing for background modeling is affected by minute details. Hence, a quantizing bin is used to reduce the number of levels. When a collection of close intensity levels are treated as a single bin, the bin is called a quantized bin. Let, there be L intensity levels and those are divide into Q number of bins. If $L = Q \times K$, then K is the size of each bin. This approach does not only minimize the complexity but also helps to restrict the false alarm. The quantized bin is extracted from each of the blocks in a frame. Every frame is divided into smaller-size blocks to emphasize local processing. The histogram of this quantized bin is used as the feature of each block termed as a block-based quantized bin ($BBQH$). The proposed approach tries to estimate the rigid part of a sequence of consecutive frames. Hence, it considers the unmoved features among every two consecutive frames. Thus, the feature of the proposed approach is computed by comparing each of the pairs of consecutive frames, which is called differential block-based quantized histogram (D-$BBQH$). This phenomenon reduces the effect of noise due to minute details and the block-level comparison reduces the effect of intermittent object movement but gives importance to the local values. The changing profile stores the values between every two consecutive frame pairs, a sudden change in the profile refers to the huge change in the same if this change in a longer period the proposed method re-estimate the background and thus make the procedure adaptive.

2 Related Work

A huge volume of work has been done in this domain; the interested readers are redirected to [1, 2] for completing their review in this field. The adaptive mixture model proposed by Staffur and Grimson [3] is the pioneering work in this field. The work is based on occurrence statistics of intensity values in particular pixel positions in k consecutive frames. The mixture model uses the Gaussian formula to determine the rigid and moving values of a particular location. Many other methods based on pixel-based processing have been applied with formulas like averaging [3], median [4], fuzzy-based selection, etc. Some works used frame-based global processing to estimate the rigidity like global histogram [5], optical flow-based technique [5], wavelet transform, etc. These kinds of procedures extract frame-based features to estimate the rigidity. The background is modeled by comparing those features among k consecutive frames.

Pixel-based processing is mostly affected by the noise as it considers every pixel individually. On the other hand, frame-level processing would not consider the local disparity as it focuses on the global behavior of the frames. Hence, the block-level processing is proposed that combines the advantages of both of the other strategies and reduces the shortcomings of the same. The concept is to employ the block-based feature instead of global and pixel-based processing to take care of local changes as well as to reduce the effect of noise. The key contributions, in this research work, can be summarized as follows:

- Proposal of a block-based statistical background modeling technique based on block-based temporal features.

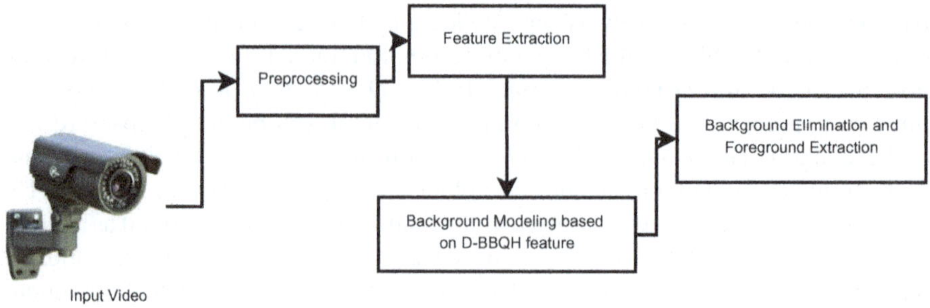

Fig. 1 Flow diagram of the proposed approach

- The newly introduced $D\text{-}BBQH$ feature vector eliminates the active parts to estimate the background. Active parts are the distinguishable parts between any consecutive frame pairs.
- The proposed method can handle several irregularities in the video data like ghosting effect, noisy conditions, bootstrapping, camouflaged conditions, etc.
- The complexity of comparing block-based features instead of pixel-based comparison is much lesser as the size of the feature vector is much lesser than the number of pixels. On the other hand counting sort of strategy helps to estimate the D-BBQH features in a single pass.

3 Proposed Work

The key operations of the proposed approach are described in Fig. 1 that include preprocessing, D-BBQH feature extraction and background modeling, background elimination and foreground extraction based on the $D\text{-}BBQH$ feature.

3.1 Preprocessing

This step is used to prepare a frame for feature extraction in a much better way as the contrast normalization removes small illumination changes and the smoothing operation reduces the intra-regional tiny details. Contrast normalization helps to extract information from low contrast image and reduces the noise due to the variation of illuminations. Contrast normalization needs two distinct values, viz., upper limit (ul) and the lower limit (ll) of the range at which the intensity values are to be normalized. The formula provided in Eq. ii is used for normalization, where hv and lv are the highest intensity and lowest intensity of the present image.

$$P_{\text{out}} = (P_{\text{in}} - \text{ll}) \left(\frac{\text{ul} - \text{ll}}{\text{hv} - \text{lv}} \right) + \text{ll} \tag{1}$$

A simple 3 × 3 fast averaging filter is used for smoothing operation for any image to reduce the intra-regional disparity; for a color image, we apply the same operation in three channels.

3.2 D-BBQH Feature Extraction and Background Modeling

3.2.1 Elimination of Inconsistent Pixels

This step eliminates the inconsistent pixels among each pair of frames. The consistency of any pixel is defined in terms of consecutive occurrence of similar intensity at the same pixel position. The similarity is checked concerning a prescribed or adaptive threshold. The differential frame includes the pixel if the absolute difference is below a threshold, otherwise, we make it '-1', which means the pixel will not be considered for feature extraction. Thus, the value -1 indicates to the pixel that it is inconsistent. We formulate the concept of temporal consistency in terms of consistency in intensity values among the consecutive frames of any pixel of the frame instead of only the occurrence probability of the same. $\Delta f(x, y, t)$ is the difference between the pixel of location (x, y) of two consecutive frames computed using Eq. 2. Equation 3 is used to compute $\Omega(x, y, t)$, the resultant value for future processing. If $\delta f(x, y, t)$ is less than a certain threshold THchg, then the value of the current frame is taken for consideration, else the value is set to be -1, as the change is greater than background consideration value. β will help to estimate the feature vector. The response computed using δ will be stored in β for further use (Fig. 2).

$$\Delta f(x, y, t) = |f(x, y, t) - f(x, y, t + 1)| \tag{2}$$

$$\Omega(x, y, t) = \begin{cases} f(x, y, t), \text{if}(\Delta(x, y, t) \geq TH_{\text{chg}} \\ -1, \text{Otherwise} \end{cases} \tag{3}$$

$$\beta_{ij}(q, 2) = p_q \times \delta^2 \tag{4}$$

3.2.2 Consistency and Occurrence Probability

The proposed approach checks the temporal consistency of any information using two parameters, specifically occurrence probability of the information (ρ_i is computed using Eq. 5), and the consistency of the information (σ_i) is computed using Eq. 6. The occurrence probability ρ_i is the probability of any information in a particular area in t consecutive frames. On the other hand , σ_i is defined in terms of the occurrence

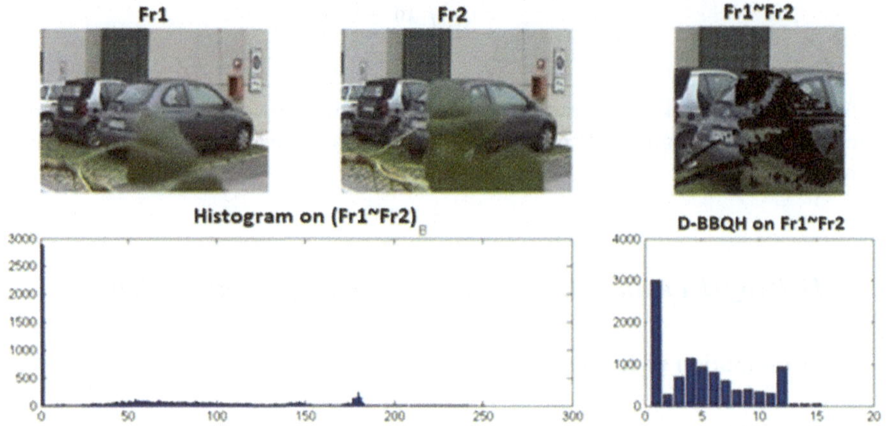

Fig. 2 Pictorial representation of consistency; clockwise from the left top corner: Frame1, Frame2, removal of inconsistent pixels to get the resultant frame (Fr1 Fr2), D-BBQH of the resultant frame, histogram of the resultant frame

in successive frames. The weighted response of temporal consistency W_i is then computed using Eq. 7.

$$\rho_i = F_i/N \tag{5}$$

$$\sigma_i = \sum(\text{Per}_{ik}), \quad \text{where } \text{Per}_{ik} = n_k(n_k + 1)/2 \tag{6}$$

$$W_i = (\rho_i \times \sigma_i) \tag{7}$$

Let a list of information in 24 consecutive frames be given in $LIST$ = [1, 1, 1, 1, 2, 3, 2, 6, 2, 1, 1, 1, 2, 4, 2, 5, 2, 3, 2, 2, 4, 3, 2, 5]. All F_i, ρ_i, σ_i and W_i are shown in the Table 1. The weighted response W_i of 1 is 4.4667, which is the highest among the list, though the occurrence probability of 2 is the highest. Hence, we can conclude that the response value of any information depends on not only its occurrence probability, but also on the consistency value. This property helps a lot to reduce several unwanted errors, which occur more frequently but do not persist for a long time.

$D\text{-}BBQH$ feature is extracted on the resultant frame of the previous operation using Eqs. 8 and 9.

$$\beta_{ij} = \left(\sum \gamma_q\right)/\beta_{ij}(q, 2), \text{ where each } \gamma_q \in q\text{th bin } of \ \beta_{ij} \tag{8}$$

$$\beta_{ij}(q, 2) = p_q + \delta^2 \tag{9}$$

Table 1 Frequency, consistency, occurrence probability and the weighted response of the corresponding sample

Sample	F_i	σ_i	ρ_i	$W_i = \sigma_i \times \rho_i$
1	7	16	0.297	4.667
2	9	10	0.375	3.75
3	3	3	0.125	0.375
4	2	2	0.083	0.1667
5	2	2	0.083	0.1667
6	1	1	0.0417	0.0417

3.2.3 Feature Analysis for Background Modeling

D-BBQH features of k consecutive frames are used to estimate the backgrounds. We compute the weight of each bin using Eq. 10, which includes consistency and occurrence probability of each bin. A threshold is used to select the *D-BBQH* features of the background frame depending on the weighted response. This feature vector of background frame β_{BAC} is used for foreground extraction.

3.3 *Background Elimination and Foreground Extraction*

It is mainly needed to extract the foreground of the current frame by eliminating the background information. Comparison is done block-wise, and if the difference between frequencies of a bin is greater than a certain threshold, the corresponding position of the foreground feature vector will store 1.

$$\beta\text{FRG}_{ij} = \begin{cases} 1\ if\ \ \beta_{ij_t}(q,2) - \beta\text{BAC}(q,3) > THi_2 \neq j_1 \\ 0\ \text{Otherwise.} \end{cases} \tag{10}$$

3.4 *Foreground Extraction*

The first foreground feature vector of a block is used to estimate the foreground area. The bin location (Bin) of every pixel is first computed, where F is the current frame and B is the number of bins. If the value of the foreground feature vector is 1 for the bin of the corresponding block, then that pixel is considered as a foreground pixel.

4 Results and Analysis

The proposed approach is verified on SBMI [14] dataset, and a comparative study with several renounced methods of related literature justifies the proficiency of the proposed approach.

4.1 Implementation Assumptions

D-BBQH features are extracted from each of the frames of a video. There is a storage, which can store *D-BBQH* features of N number of frames and it discards the *D-BBQH* features of the first frame when the D-BBQH features of $(N + 1)$th frame are needed to insert, discards the second frame when *D-BBQH* features of $(N + 2)$th frame are needed to insert and so on. Background modeling is done by merging the features of K number of D-Frames. The value of K is set to the frame rate of any video if there is no movement in the initial frames; if movement is found, it is set to twice of the same. There is no extra cost needed for feature extraction when background updating is done as the feature of the current frame is extracted for foreground extraction. There are two other thresholds namely frame differentiation threshold is set to the twice of the standard deviation of first k frames and the background bin selection threshold is set to the highest weight of the block divided by $\sqrt{2}$.

4.2 Dataset

We use Scene Background Initialization (SBMI) [6] dataset to describe the efficiency of our algorithm. The dataset includes 14 image sequences and their ground truth backgrounds with different levels of difficulties like foliage, abandoned and lost objects, late foreground movement, foreground covering certain areas in most part of the scene, etc. This dataset is adopted from the SBMI2015 workshop and this a very recent dataset as the last update was done in January 16. The challenges of the videos are given in Table 1.

4.3 Evaluation Parameters [7]

The website provides the script for evaluating results concerning six metrics, which are used in the literature for background estimation. GT denotes the ground truth image and CB is the estimated background of any background modeling approach.

Table 2 Comparative results on six evaluation parameters of the proposed approach and the same of the related research work

Approaches	Average AGE	Average pEPs	Average MS-SSIM	Average PSNR
SC-SOBS [8]	6.32	8.02	0.93	29.89
WS2006 [9]	6.50	4.69	0.934	27.605
Photomontage [10]	7.42	6.58	0.905	30.9901
CA2008 [11]	13.676	10.04	0.91	25.17
GROUSE [12]	19.166	0.359	0.915	24.37
Proposed	4.72	2.79	0.954	30.057

The six metrics are used to estimate the difference between GT and CB images and evaluate the effectiveness of the difference.

- Average Gray-level Error (AGE): It is the absolute difference between gray-level GT and CB images. This is a global estimation of differences and the smaller value is more appreciable.
- Percentage of Error Pixels (pEPs): This parameter confirms the percentage of the number of wrongly estimated background pixels by using a reasonable threshold value (20 is suggested by the link). The lower percentage of error proves the better quality of an approach.
- Peak Signal-to-Noise Ratio (PSNR): The standard formula is used to extract the PSNR of the computed background frame with a comparison to the ground truth of the same. The value provides the superiority of the information over the noise. Thus, the bigger value is more desirable.
- MultiScale Structural Similarity Index (MS-SSIM): This parameter is proposed by is used to estimate the perceived visual distortion by using structural distortion and the greater value determines the lesser distortion.

The evaluation of the results of the proposed approach with the other related research is based on the above four previously described parameters.

The proposed method used a $D\text{-}BBQH$ feature for background estimation, so the output of our method includes quantized bins instead of pixels. So when we map the features into a frame, it selects the highest bin which matches the temporal values of a location. This is needed to test our background modeling results with the same of the ground truth values. On the other hand, if we check the same concerning feature values, the evaluation procedure will not support it, though our method produces the best results in the case of AGE, MS-SSIM and PSNR values, which are used for global error estimation, multiscale structural similarity and peak signal-to noise ratio, respectively. The results in Table 2 include the average values of seven videos for several parameters of the proposed and fourteen other methods of the related research.

The proposed approach produced efficient background modeling results in spite of several difficulties of several videos enlisted in Table 2. Change in lightning conduction, background noise, jitters and presence of shadow/camouflage condition may lead to false detection. The first foreground movement is important for checking the robustness against bootstrapping. If there is no foreground movement, we do not need to store or process those frames. Finally, if we need to update the background, then two consecutive modeled background frames of any scene assist to track the abandoned or missing object as the abandoned object(s) are the new inclusion and missing object(s) are the eliminated part of the two modeled frames. The SBI dataset includes all of this kind of difficulties and our proposed approach produces effective results in spite of all the difficulties.

5 Conclusion

In this paper, we have shown a novel block-based hybrid background modeling technique in terms of frame differentiating and block-based feature extraction. The feature extraction technique is newly driven depending upon the consistency measurement of the pixel values of any block among successive frames. The results of the proposed methodology over the $SBMI$ dataset and the comparative results on the related research work show the efficiency of the same. This methodology focuses on the consistency of static information by removing the active part of any consecutive frames pair, but still the method does not provide a hundred percent accurate results, so the scope of further improvement is possible.

References

1. Goyette, N., Jodoin, P., Porikli, F., Konrad, J., Ishwar, P.: Changedetection.net: a new change detection benchmark dataset. In: 2012 IEEE Computer Society Conference on Computer Vision and Pattern Recognition Workshops, pp 1–8, June 2012
2. Stauffer, C., Grimson, W.E.L.: Adaptive background mixture models for real-time tracking. In: Proceedings. 1999 IEEE Computer Society Conference on Computer Vision and Pattern Recognition (Cat. No PR00149)
3. Ekinci, M., Gedikli, E.: Background estimation based people detection and tracking for video surveillance. In: Yaz A, Azener C (eds) Computer and information sciences—ISCIS 2003, vol 2869 (2003)
4. McFarlane, N.J.B., Schofieldi, C.P.: Machine vis. apps. In: Yaz, A., Azener, C. (eds.) Computer and Information Sciences—ISCIS 2003 (1995)
5. Sigari, M.H., Mozayani, N., Pourreza, H.R.: Concepts and application, fuzzy running average and fuzzy background subtraction (2008)
6. Maddalena, L., Petrosino, A.: Towards benchmarking scene background initialization. In: Murino, V., Puppo, E., Sona, D., Cristani, M., Sansone, C. (eds.) New trends in image analysis and processing—ICIAP Workshops (2015)
7. Maddalena, L., Bouwmans, T., Petrosino, A.: Scene background initialization: a taxonomy. Patt. Recogn. Lett. **96**, 3–11 (2017)

8. Wang, H., Suter, D.: A novel robust statistical method for background initialization and visual surveillance. In: Nayar, S.K., Shum, H.Y. (eds.) Computer Vision ACCV 2006. ACCV 2006. Lecture Notes in Computer Science, vol 3851 (2006)
9. Bouwmans, T.: Recent advanced statistical background modeling for foreground detection—a systematic survey. Recent Patents Comput. Sci. **4**, 147 (2011)
10. Maddalena, L., Petrosino, A.: The sobs algorithm: what are the limits? In: 2012 IEEE Computer Society Conference on Computer Vision and Pattern Recognition Workshops, pp 21–26, June 2012
11. Chen, C., Aggarwal, J.K.: An adaptive background model initialization algorithm with objects moving at different depths. In: 2008 15th IEEE International Conference on Image Processing, pp. 2664–2667, Oct 2008
12. Balzano, L., Wright, S.J.: On grouse and incremental svd. In: 2013 5th IEEE International Workshop on Computational Advances in Multi-Sensor Adaptive Processing (CAMSAP), pp. 1–4, Dec 2013

A Comparative Study on Unequal Clustering in Wireless Sensor Networks (WSNs)

Amrit Sahani, Abhipsa Patro, and Sushree Bibhuprada B. Priyadarshini

Abstract The sensor network is a system that is framed by the sensor nodes which are positioned either in an irregular or pre-arranged manner. Nodes accumulate information and direct it to the sink node. Sensor nodes use a high measure of vitality in communicating with different nodes in the system framework that impacts the lifetime of the system. The lifetime of the systems can be expanded by utilizing conventions that give various strategies for preserving the energy of the nodes. Clustering provides an efficient and successful method to extend the lifetime. The paper advances from the premise of formation of equal cluster, i.e. LEACH and passes onto its progression from the way of equal clustering to unequal clustering. The unequal clustering maintains a significant part in the paper and is made based on the left-out resources of energy resource of the nodes and the distance between them. Consequently, the sensor nodes which are near to the sink nodes take the formation of small clusters or we can conclude as the separation goes on diminishing the cluster size also goes on decreasing. Moreover, in this paper, certain streamlining calculations have been talked about such as LEACH, EEUC, and so on to further improve the clusters using the methodology of unequal clustering to mould it into more dynamic. Different upgradations have been done on this fundamental LEACH convention and optimal energy utilization of the system. Many of the improvements revolved in the procedure of choosing cluster head and also considering the separation between the sensor node and the sink node.

Keywords WSN · Clustering · Unequal clustering · Cluster head (CH) · Sink node · LEACH · HEED · EEUC

A. Sahani · S. B. B. Priyadarshini (✉)
Siksha 'O' Anusandhan (Deemed to be University), Bhubaneswar, Odisha 751030, India
e-mail: bimalabibhuprada@gmail.com

A. Sahani
e-mail: situnamrit@gmail.com

A. Patro
International Institute of Information Technology, Bhubaneswar, Odisha 751003, India
e-mail: patro.abhi660@gmail.com

D. Mishra et al. (eds.), *Intelligent and Cloud Computing*,
Smart Innovation, Systems and Technologies 194,
https://doi.org/10.1007/978-981-15-5971-6_68

1 Introduction

WSN is rolling out significance because of the progression in small-scale micro-electronic systems. WSN comprises of nodes that have a limited supply of energy and are useful in observing the regions that are out of human capacity. The network system framed is disseminated over an enormous zone to screen the surroundings for sound, temperature, pressure conditions and send the information over the network to different areas. Nodes interact among each other using the radio signals and are conveyed to detect, monitor and comprehend the physical world.

1.1 Network Components

The sensor network components enable the connectivity of the nodes within the system of the network framework and provide a platform for end-to-end communication platform for the application [1].

1.1.1 Gateway

It is an interface between the application and the sensing devices, and the information collected is further sent to the application through the gateway and vice versa.

1.1.2 Relay Node

These are the intermediate nodes that are in between the sender and the base station which function as a router in the network.

1.1.3 Leaf Node

These are nodes present in the network's endpoint to perform some limited tasks and build a strong interconnection among the sensor nodes by escalating some functional activities.

1.1.4 Sensor Node

The most important part of the network is that it has an interaction with the physical world and gathers data required by the client for observing and remarking the environment. There are a variety of sensors which can examine the temperature, pressure, etc. and forward these data to the end users.

Sensor nodes consist of a battery, computational controller, memory and detecting component to do its working. Each of these nodes is in connection with the sink node where the data is collected and preprocessed for further activities. Due to the tiny sensor nodes, they are easily positioned at those environments in which human interaction is not effectively conceivable.

Most of the sensors are battery worked and are being established in unfavourable environments. Substitution of batteries or sustainable source of energy is beyond the realm of imagination because of their arrangement in rough territories. There are many algorithms and conventions which are proposed for the proficient utilization of batteries and the most opted method for the structured and efficient use is clustering.

2 Clustering

A group of sensors forms a cluster where every cluster is governed by its head. The sensor from its respective clusters responds to the signals as well as forwards it to the head of the node. Cluster Head (CH) then amalgamate information which is made available to the sensor nodes in the cluster and transmits it to the base station. So in this way, every node which has to communicate to the base station first imparts the information to the cluster head, and the cluster head then sends it to the base station thus saving the energy [2].

Presently, the inquiry arises of choosing the cluster head (CH) and deciding the individual sensor nodes related to head of the cluster. There are many different techniques which have been proposed. This paper focuses to give an audit to different techniques for equal and unequal clustering with the focus of determining the strategy of cluster head. This paper concentrates on how unequal grouping of sensor nodes diminishes the utilization of energy and can be effectively used to improve the network lifetime. Moreover, unequal clustering can be upgraded by utilizing different probabilistic and heuristic approaches.

2.1 LEACH Protocol

LEACH is a protocol which uses a hierarchical levelled convention where nodes convey to head which further remits it to the sink node (base station). This convention dissipates the burden among the entire system and it offers opportunities for every node for becoming the group head. All nodes are determined by a stochastic algorithm for the formation of cluster head in the round. Nodes that are once been qualified as cluster head cannot become next time. So for N determined rounds, each node has a probability of $1/N$ for forming a head again. Sometimes the method of probability of choosing a cluster head affects the network lifetime, as the node with least energy is considered and then the cluster head dies faster as compared to other nodes. Moreover, it does not contemplate the separation of the sink node and the head node, so clusters closer to sink node emit energy out quicker in comparison to other groups reducing the life of network.

2.2 HEED Protocol

This protocol selects the cluster head considering the remaining energy of sensor nodes as well as the separation from sink node to other nodes. Residual energy decides the probability of cluster head formation. Any point when the node is in the reach of numerous clusters and then cluster head is selected depending on the minimum average reach.

Thus, this convention can utilize a multi-hop technique for the transmission of data among various cluster heads and balances out the lifetime of a network in comparison to LEACH. Some of the clusters which are near to the sink node die early as they have to handle more traffic. In this approach, it sometimes may happen that there are no sensor nodes in the cluster and single node forms a cluster head itself and has to interact directly with a base station which outcomes in energy unbalance in the network.

2.3 Weight-Based Clustering Protocol

The above-stated protocol is a weight-based grouping algorithm which demonstrates nodes in the cluster head depends directly on battery intensity and mobility. Here the selection of cluster head is delayed so that same cluster head is not selected more than once. Balancing of the load is an important feature of the algorithm.

3 Unequal Clustering

Despite the fact that the above-stated strategies of grouping assist in upgrading the system lifespan in a more prominent degree, however, the clusters shaped through such techniques were mostly of similar nature. In the survey, it is seen that the clusters which being closer to sink node use energy more, because of transmitting huge chunks of signal through them. So when the equivalent size of clusters is considered a huge amount of energy is wasted in the maintenance of intra-cluster and inter-cluster transmission. The clusters closer to the sink node die early making the network isolated, although other sensor nodes have the energy to carry out the transmission activities which leads to an improper imbalance in the network system. The need for a probabilistic and deterministic methodology came to existence along with the factor of unequal clustering mechanism. Unequal clustering categorizes into two approaches which use probabilistic and deterministic way for the selection of cluster head [3–8].

3.1 Weighted Approach

It is one of the probabilistic approaches [8–13] where the sink node is in the centre and cluster heads are organized in circles along the sink node. The nodes utilize a multi-hop technique to send data from the cluster head to the nearest neighbour which then forwards the accumulated data to the base station.

3.2 EEUC Protocol

Energy-efficient routing algorithm is a technique which uses a scattering process where the cluster heads are elected by the process of restricted conflicting competition to transmit the signal to the base node. The different sensor nodes of the clusters communicate with each other in a multi-hop process and forward it to their respective cluster head and consequently to the base station. The cluster head remits the traffic to that node which is having progressively more residual energy as a result of which the network lifetime of the system does not tend to die early.

3.3 UCRA Algorithm

Unequal clustering routing algorithm which uses a technique of remaining energy to form a cluster also evaluates the distance between the main node and the sensor nodes. The clusters which are closer to the base station remain smaller in size than those which are far away from sink node, due to which the cost of the intra-cluster mechanism is less and uses a minimal amount of energy in the communication process. Every cluster is defined with a threshold with respect to the distance from the sink node. The direct transmission between the base station only occurs when the distance is more than the given threshold value else sends signal to the cluster head having more vitality. The unequal cluster formation depends on the distance between the sink node and other sensors nodes which is given by the formula stated as follows:

$$R_i = (1 - C * \{d_{\max} - d(v_i, \text{sink})\}/d_{\max} - d_{\min}) * R_{\max} \text{ [14, 15]}$$

R_i = distance to the base station,
V_i = sensor node,
C = value ranges between 0 and 1 and

$(1 - C)R_{\max}$ to $R_{\max}$ = computational radius.

The outcomes demonstrated in Figs. 1 and 2 reveal that the UCRA algorithm is considerably more efficient than the above-mentioned techniques which use equal clustering in terms of average energy.

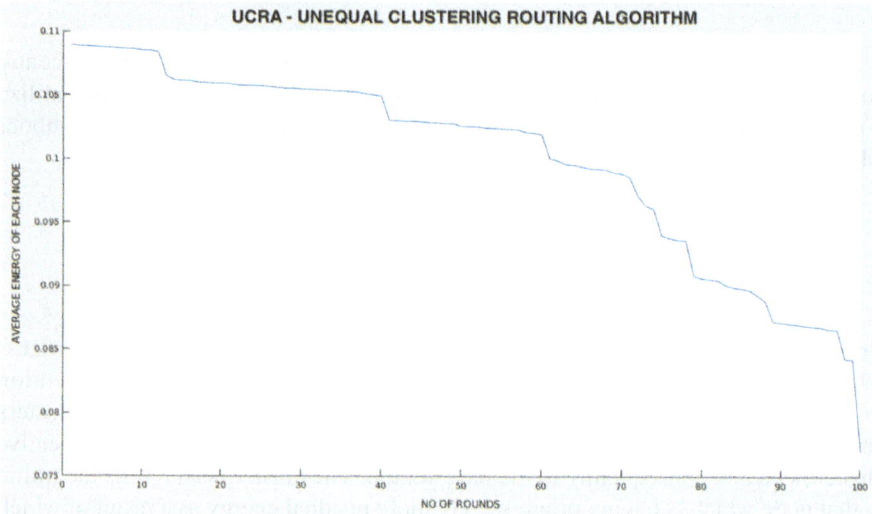

Fig. 1 Average energy of each node versus round numbers in UCRA

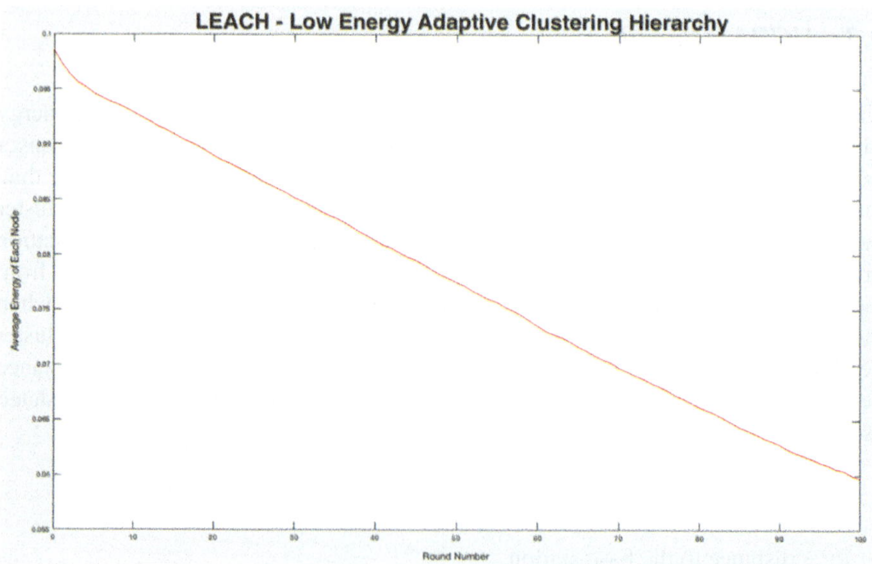

Fig. 2 Average energy of each node versus round numbers in LEACH

It is also seen that the nodes tend to be alive for more time slots in the UCRA can be depicted from (Fig. 1) algorithm rather than EEUC and LEACH can be depicted from (Fig. 2) protocol which can help when deployed in larger system network. The mechanism of unequal clustering also sustains energy in each node thus expanding the sustainability of nodes and improving the longevity of network lifespan.

4 Conclusion

Through in the current paper, we experienced significance of cluster formation and the process in which the clusters can be made more productive in terms of residual energy, sustainability and network lifetime by changing the process of selecting a cluster head which would tentatively transmit signals to and fro from the base station.

References

1. Akyildiz, I.F., Su, W., Sankarasubramaniam, Y., Cayirci, E.: A survey on sensor networks. IEEE Commun. Maga. **40**(8), 102–114 (2002)
2. Krishnamachari, B., Estrin, D., Wicker, S.: The impact of data aggregation in wireless sensor networks. In: Proceedings of IEEE International Conference on Distributed Computing Systems Workshops (ICDCSW), pp. 575–578 (2002)
3. Mhatre, V., Rosenberg, C.: Design guidelines for wireless sensor networks: communication, clustering and aggregation. Ad Hoc Netw. **2**(1), 45–63 (2004)
4. Heinzelman, W., Chandrakasan, A., Balakrishnan, H.: An application-specific protocol architecture for wireless microsensor networks. IEEE Trans. Wirel. Commun. **1**(4), 660–670
5. Younis, O., Fahmy, S.: HEED: A hybrid, energy-efficient, distributed clustering approach for ad hoc sensor networks. IEEE Trans. Mobile Comput. **3**(4), 660–669 (2004)
6. Lindsey, S., Raghavendra, C., Sivalingam, K.M.: Data gathering algorithms in sensor networks using energy metrics. IEEE Trans. Parallel Distrib. Syst. **13**(9), 924–935 (2002)
7. Choi, W., Shah, P., Das, S.K.: A framework for energy-saving data gathering using two-phase clustering in wireless sensor networks. In: Proceedings of International Conference on Mobile and Ubiquitous Systems: Networking and Services (MOBIQUITOUS), pp. 203–212 (2004)
8. iHEED: http://www.cs.purdue.edu/homes/fahmy/software/iheed/
9. Soro, S., Heinzelman, W.: Prolonging the lifetime of wireless sensor networks via unequal clustering. In: Proceedings of the 19th IEEE International Parallel and Distributed Processing Symposium (IPDPS) (2005)
10. Ye, M., Li, C.F., Chen, G.H., Wu, J.: EECS: an energy efficient clustering scheme in wireless sensor networks. In: Proceedings of IEEE international performance computing and communications conference (IPCCC), pp. 535–540 (2005)
11. Intanagonwiwat, C., Govindan, R., Estrin, D.: Directed diffusion: a scalable and robust communication paradigm for sensor networks. In: Proceedings of ACM Mobile Computing and Networking (MOBICOM), pp. 56–67 (2000)
12. Schurgers, C., Srivastava, M.B.: Energy efficient routing in wireless sensor networks. In: Proceedings of IEEE Military Communications Conference (MILCOM), vol. 1, pp. 357–361 (2001)
13. Perillo, M., Cheng, Z., Heinzelman, W.: On the problem of unbalanced load distribution in wireless sensor networks. In: Proceedings of the IEEE GLOBECOM Workshops on Wireless Ad Hoc and Sensor Networks, pp. 74–79 (2004)

14. Basagni, S.: Distributed clustering for ad hoc networks. In: Proceedings of the 4th International Symposium on Parallel Architectures, Algorithms, and Networks (I-SPAN), pp. 310–315 (1999)
15. Xia, H., Zhang, R.H., Yu, J., Pan, Z.-K.: Energy-efficient routing algorithm based on unequal clustering and connected graph in wireless sensor networks. Int. J. Wirel. Inf. Netw. **23**(2), 141–150 (2016)

Rivest Cipher 4 Cryptography and Elliptical Curve Cryptography Techniques to Secure Data in Cloud

Alakananda Tripathy, Alok Ranjan Tripathy, Smita Rath, Om Prakash Jena, and Shrabanee Swagatika

Abstract Data security is the major issue in the field of research and in the field of cloud computing; it is a major concern as data are located across the globe in the cloud. As cloud computing technology is used by government, businesses, and industries, many studies have been investigated recently to achieve authentication and the confidentiality of data while they are stored in the cloud. Cryptography plays an important role in achieving data security. The paper proposes a hybrid model that uses the symmetric RC4 algorithm to encrypt the data and the key performance encryption is secured by the ECC algorithm which is a public key algorithm. The proposed method provides high security by encrypting the session key and also needs less time for encrypting and decrypting the data than the existing model. The proposed hybrid model provides data confidentiality and also secures the session key used for encryption and decryption.

Keywords Elliptic curve cryptography · RC4 · Session key · Encryption time · Decryption time

A. Tripathy · S. Rath · S. Swagatika
Department of Computer Science and Engineering, Siksha 'O' Anusandhan (Deemed to be University), Bhubaneswar, Odisha, India
e-mail: alakanandatripathy@soa.ac.in

S. Rath
e-mail: smitarath@soa.ac.in

S. Swagatika
e-mail: shrabaneeswagatika@soa.ac.in

A. R. Tripathy (✉) · O. P. Jena
Department of Computer Science, Ravenshaw University, Cuttack, Odisha, India
e-mail: tripathyalok@gmail.com

O. P. Jena
e-mail: jena.omprakash@gmail.com

D. Mishra et al. (eds.), *Intelligent and Cloud Computing*,
Smart Innovation, Systems and Technologies 194,
https://doi.org/10.1007/978-981-15-5971-6_69

1 Introduction

Cloud computing enables the user to access the network on demand where the pool of resources like servers, storage, applications, network, and services are available with minimum efforts of the management and service provider [1]. Cloud computing enables consumers to access resources through the Internet. Distributed computing gives numerous points of interest like versatility, self-provisioning, and flexibility, yet there are some security dangers and the failure to ensure secrecy and protection of buyer information. Mobile cloud computing is utilized in maximum presently to offer computational and off-gadget stockpiling in the cloud server [2]. Cloud innovation offers numerous administrations to the clients, and this additionally brings a few strings for the client's information. Distributed storage with exceedingly brought together figuring assets makes storage faces genuine security issues [3]. The client redistributes their information in the server of the cloud that isn't completely reliable, and this prompts extreme difficulty in safeguarding information security and information proprietor's protection in the cloud [4]. To secure the information and data, encryption techniques are used. As people use cloud technology for many purposes, it is growing faster in the field of research. The use of cloud storage increases because of the maximum use of mobile phones and tablets; this leads to hampering the security of data in the context of integrity [5]. Data leakage of users occurs due to a lack of security measures in the cloud storage. Before storing the data in the cloud, the data should be secured and the problem may occur while encrypting the data. There are many security issues and challenges in the field of cloud computing. The third-party service provider stores the data in the cloud through the Internet. So, the data need to be protected to maintain confidentiality, authenticity, and integrity.

Basically, three types of cloud services are provided to store, manage, and process the data; they are as follows.

Software as a Service (SaaS):
This service is provided to the end user through a web portal like Google app, Youtube, Facebook, etc. to manage data.
Platform as a Service:
It is the service provided by the cloud to the user to create, run applications without the knowledge of the processor and memory. This service is basically provided by Aneka, Google App Engine, and Microsoft Azure.
Infrastructure as a Service (IaaS):
It is the base layer of the cloud stack which provides storage, computation, and communication with minimal cost. The companies that provide this service are Rackspace, Amazon EC2, and GoGrid.

1.1 Data Security Issues in the Cloud

The cloud is used for data storage and business for companies [6]. So the basic security issues may arise in cloud computing [7] as

1. Data Integrity: It means the data should not be tampered or altered. Cloud service uses appropriate security measures to secure the integrity of data.
2. Privacy and Confidentiality: This does not allow the data to be accessed by unauthorized users by protecting the data from loss.
3. Data Availability: The data is made available as required by the application and the end user. The data should be available in normal and in also disaster recovery operations.
4. Data Location and Relocation: As maximum data is stored from mobiles and tablets, the exact location of the data is not known to the user and customers. But for storing confidential and sensitive data, the user requests for the location or specifies the location. To fulfill the customer needs, a cloud service provider makes a contract with the user regarding space in a particular location. The major issue with this is that the user does not have any knowledge of it. To transfer the data from one location to another is again a problem. To handle these security issues, cryptography algorithms are the best which maintain the confidentiality, integrity, and authenticity of data. For protecting the data before uploading, the data is protected using the cryptographic algorithms. In this paper, we propose a hybrid algorithm that encrypts the data using RC4 and the shared key is encrypted using ECC for providing high security. Section 2 consists of related work followed by the proposed model in Sect. 3 and result analysis and conclusion in Sects. 4 and 5, respectively.

2 Related Work

This section will cover a detailed review of the related works which are referred to in formulating our proposed model. In this section, we will summarize the work on the cloud computing data security which had been done previously. The hybrid model was proposed by Abd Elminaam [8] which encrypts the data using AES and Blowfish. Both the techniques use a symmetric key algorithm and provide security, but the key used for encryption and decryption is shared for decryption. As a single key is used, the key may not be as secure. Another hybrid model was proposed by Barvekar et al. [9] which uses AES and RSA algorithms. AES is the symmetric key algorithm and RSA is a public key algorithm. The data is first encrypted using AES and then again encrypted using the RSA algorithm. This algorithm provides high security but is slow as the RSA algorithm is a public key cryptography which needs more time to generate the key. Purevjav et al. [10] proposed a hybrid cryptosystem using RSA and Blowfish which is a symmetric and asymmetric algorithm; it provides security to the data within the cloud storage regardless of the existence of attacks and leakages. It provides registration for the user in order to have the authentication during the login process to store the data. In the cloud computing environment, a secure environment was build using a method proposed by Shereek et al. [11] which uses the RSA algorithm and Fermat's theorem. RSA provides strong security but more time is needed to generate the key and this hampers the performance of RSA. Another hybrid model was proposed by Sengupta [12] which secures the data on cloud by using RSA and reduces the chance of man-in-the-middle attack use Feistel Encryption Algorithm that is DES.

An efficient, tough, and secure algorithm was proposed by Najar and Dar [13] which uses Advance Encryption Standard (AES) for encryption and RSA for key management and Secure Hash algorithm for digital signature. Timothy and Santra [14] proposed a hybrid model which uses Blowfish and RSA to solve security and privacy issues of cloud computing. The basic aim is to achieve safe transmission of data to the cloud; it uses digital signature for transmitting the data.

3 Proposed Model

Mostly, the cloud servers are used to store the confidential data; cryptographic technique is used to encrypt the information before storing it to the cloud. For better security of data with minimum time for encoding the message, we proposed a model which uses the RC4 and elliptic curve cryptography (ECC) algorithm. The elliptic curve cryptosystem is a public key algorithm, and the elliptic curve is used to generate the key to secure the session key needed by RC4 to encrypt and decrypt the data by maintaining the authenticity of data. Once the encrypted data is stored on the cloud, the cloud service provider authenticates the user requesting the data, and then decrypts the message and delivers it to the authorized user. In the proposed model, elliptic curve is used to determine the public and private keys to encode the shared key, and then the data which is taken as a string is encrypted and decrypted using the shared key.

The proposed model includes the combination of elliptic curve cryptography and RC4 algorithm which are the asymmetric and symmetric cryptographic algorithms. The elliptic curve cryptography is basically used to generate the key for encrypting the data using the RC4 algorithm. The RC4 algorithm is a symmetric key stream cipher algorithm. This methodology provides security to the key as well as to the data.

The user and the service provider involve in the communication and select a random number x from the interval $[1, n-1]$, which is the private key. The steps of the proposed algorithm are as follows:

1. Generate the private key d and e for both the sender and receiver by selecting a random number x from the interval $[1, n-1]$.

2. The public key is computed by using the private key and the base point Q given as $\mathrm{Pub_s} = d * Q$ and $\mathrm{Pub_r} = e * Q$ where $\mathrm{Pub_s}$ is the public key for the sender and $\mathrm{Pub_r}$ is the public key for the receiver.
3. The public key is communicated for decryption.
4. The shared key needed for encoding and decoding is obtained by multiplying the private key of the sender and the public key of the receiver $k = d * \mathrm{Pub_r} = e * \mathrm{Pub_S}$.
5. The key k is the shared key which is used to encrypt the data before storing it in the cloud and decrypt the encrypted data to get the original message.
6. The key k is also used to obtain the original data from the encrypted data.

In the next subsection, we explain the encryption and decryption algorithms in detail.

3.1 Encryption of Data

The data is encrypted using the RC4 algorithm before storing in the cloud. RC4 is a symmetric cipher algorithm. The data is encrypted by performing the XOR operation with the shared key. The length of the key in the RC4 algorithm is 128 bits or 256 bits long. The encryption algorithm is discussed in Algorithm 1.

Algorithm 1 Encryption Algorithm

Step 1: Let the plain text be 16 bytes long string
Step 2: A state vector S is maintained
Step 3: Steps 3 and 4 will be repeated for $i = 0$–255
Step 4: Let the state vector $S[i] = i$
Step 5: Lets take another temporary vector TV[] and initialize it to $\mathrm{TV}[i] = k[i * \mathrm{mod}(k)]$
Step 6: Set $i = 0$, $j = 0$
Step 7: Repeat steps 8 and 9 for $i = 0$–255
Step 8: Update the value of j using $j = (j + S[i] + \mathrm{TV}[i]) \bmod 256$
Step 9: Exchange the value of the state vector $S[i]$ with $S[j]$
Step 10: Encrypt the data using steps 11–14
Step 11: Update $i = (i + 1) \bmod 256$ and $j = (j + S[i]) \bmod 256$
Step 12: Exchange the values of the state vector $S[i]$ with $S[j]$
Step 13: Obtain the key $k = (S[i] + S[j]) \bmod 256$
Step 14: Compute the ciphertext $C_i = M_i \mathrm{XOR} S[k]$
Step 15: Stop.

3.2 Decryption of Data

Decryption is performed in order to obtain the data from the encrypted text. It is the reverse of the encryption process. The algorithm to decrypt the data is given in Algorithm 2.

Algorithm 2 Decryption Algorithm

Step 1: Initialize $i = 0$, $j = 0$
Step 2: Steps 3 and 4 will be repeated for $i = 0$–255
Step 3: Update the value of j using $j = (j + S[i] + \mathrm{TV}[i]) \bmod 256$
Step 4: Exchange the value of the state vector $S[i]$ with $S[j]$
Step 5: Decrypt the ciphertext using steps 6–9
Step 6: Update $i = (i + 1) \bmod 256$ and $j = (j + S[i]) \bmod 256$
Step 7: Exchange the values of the state vector $S[i]$ with $S[j]$
Step 8: Obtain the key $k = (S[i] + S[j]) \bmod 256$
Step 9: Compute the ciphertext $M_i = C_i \mathrm{XOR} S[k]$
Step 10: Stop.

4 Result

In this section, the proposed hybrid model for cloud computing is evaluated based on the encryption time, decryption time, and throughput. The encryption time of the proposed model is compared with the existing AES and blowfish symmetric key algorithm. Table 1 shows the encryption time of the different algorithms. Encryption time is the time needed to encode the data by using the key in the given time. Table 2 shows the decryption time of different algorithms with different input data in KB and time in seconds. Figures 1 and 2 shows the encryption and the decryption time of the different algorithm. Table 3 shows the throughput of the different algorithms. Higher is the throughput, better is the efficiency of the algorithm.

Table 1 Encryption Time for Different Algorithm in seconds

Input in KB	AES	Blowfish	Proposed hybrid model
125	0.01274	0.02548	0.0003503896
250	0.02567	0.05096	0.0007007836
500	0.10192	0.20384	0.014015672
750	0.20096	0.40768	0.03506919
1000	0.20384	0.411835	0.028031344

Table 2 Decryption Time for Different Algorithm in seconds

Input in KB	AES	Blowfish	Proposed hybrid model
125	0.01456	0.02912	0.0003806999
250	0.02912	0.05384	0.0009106836
500	0.05824	0.11648	0.016126678
750	0.08736	0.17032	0.04607999
1000	0.11648	0.23296	0.031231

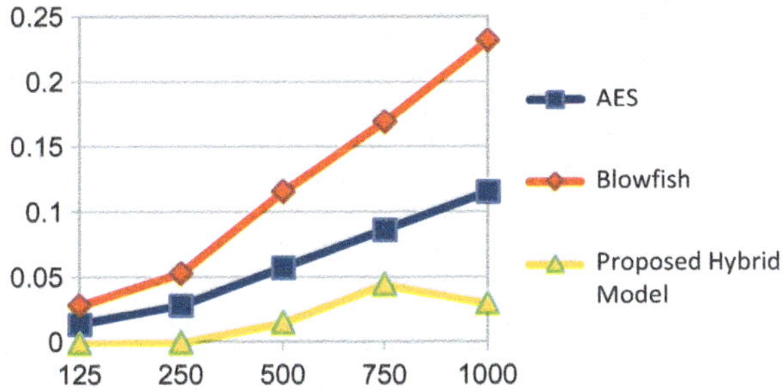

Fig. 1 Encryption time of different encryption algorithms

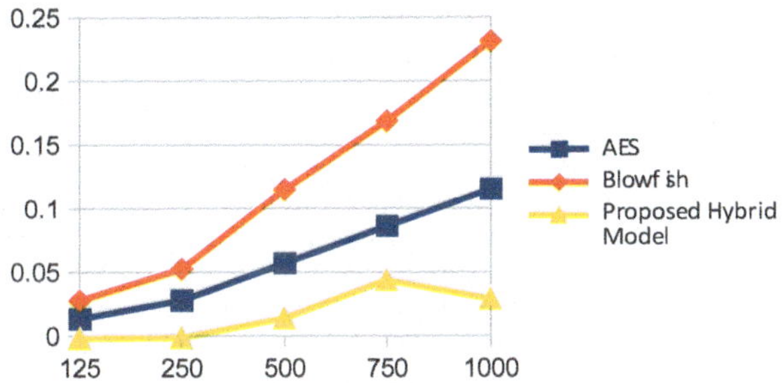

Fig. 2 Decryption time of different algorithms

Table 3 Throughput for different algorithms

Input in KB	AES	Blowfish	Proposed hybrid model
125	9811.616	4905.808	356,836.990
250	9765.625	4906.80	356,786.071
500	4906.7713	2452.904	35,674.3508
750	3733.2005	1840.03	21,386.2937
1000	4905.808	2428.363	3567.44068

5 Conclusion

Nowadays, cloud computing is one of the major fields. The cloud obtains all the requirements of the user and is dynamic, has high computing capability in different applications from business intelligence to information maintenance. In this paper, we propose a method to store data securely in the cloud by encrypting it using a hybrid model. The proposed model uses RC4 symmetric key for data encryption, and Elliptic curve cryptography is used for securing the shared key. The proposed model has higher throughput than the existing traditional symmetric key algorithm and needs less time to encrypt and decrypt alarge amount of data than the traditional model.

References

1. Rao, R.V., Selvamani, K.: Data security challenges and its solutions in cloud computing. Procedia Comput. Sci. **48**, 204–209 (2015)
2. Khan, S., Shiraz, M., Boroumand, L., Gani, A., Khan, M.K.: Towards port knocking authentication methods for mobile cloud computing. J. Netw. Comput. Appl. **97**, 66–78 (2017)
3. Fu, A., Li, Y., Yu, S., Yu, Y., Zhang, G.: DIPOR: an IDA-based dynamic proof of retrievability scheme for cloud storage systems. J. Netw. Comput. Appl. **104**, 97–106 (2018)
4. He, Q., Zhang, N., Wei, Y., Zhang, Y.: Lightweight attribute based encryption scheme for mobile cloud assisted cyber-physical systems. Comput. Netw. In Press (2018)
5. Jung, K.D., Moon, S.J., Kim, J.M.: Data access control method for multimedia content data sharing and security based on XMDR-DAI in mobile cloud storage. Multimed. Tools Appl. **76**(19), 19983–19999 (2017)
6. Kadu, C.S., Jadhav, A.A., Mandale, P.L.: Improving security of cloud environment in dynamic cloud network. Int. J. Comput. Sci. Inf. Technol. (IJCSIT) **5**(2), 1681–1684 (2014)
7. Ren, K., Wang, C., Wang, Q.: Security challenges for the public cloud. IEEE Internet Comput. **16**(1), 69–73 (2012)
8. Abd Elminaam, D.S.: Improving the security of cloud computing by building new hybrid cryptography algorithms. Int. J. Electron. Inf. Eng. **8**(1), 40–48 (2018)
9. Barvekar, R., Behere, S., Pounikar, Y., Gulhane, A.: An approach to hybrid cryptography on cloud environment. IJARIIE, **4**(2), 4626–4631 (2018)
10. Purevjav, S., Kim, K.H., Sain, M., Lee, H.: Design of hybrid cryptosystem for cloud system. In: IEEE Conference on Wireless Sensors (ICWiSe), pp. 67–70 (2015)
11. Shereek, B.M., Muda, Z., Yasin, S.: Improve cloud computing security using RSA encryption with Fermat's little theorem. IOSR J. Comput. Eng. (IOSR-JCE) **4**(2), 1–8 (2014)
12. Sengupta, N.: Designing of hybrid RSA encryption algorithm for cloud security. Int. J. Innov. Res. Comput. Commun. Eng. **3**(5), pp. 4146–4152 (2015)
13. Najar, J.M., Dar, S.B.: A new design of a hybrid encryption algorithm. Int. J. Eng. Comput. Sci. **3**(11), 9169–9171 (2014)
14. Timothy, D.P., Santra, A.K.: A hybrid cryptography algorithm for cloud computing security. In: International Conference on Microelectronic Devices, Circuits and Systems (ICMDCS), pp. 1–5. IEEE (2017)

A Survey on Various Security Issues for 5G Mobile Networks in WSN

Lucy Dash and Mitrabinda Khuntia

Abstract Wireless communication has become an integral part of mobile networks. With the increase in demand for 5G mobile wireless networks, there are many problems regarding the security of data. The wireless medium contains domain-specific information that can be used for security purpose. This paper brings a survey on security features of 5G mobile networks, which aim to bring fast authentication without compromising the components and features of data security.

Keywords 5G mobile network · Security of 5G · WSN

1 Introduction

Fifth-Generation (5G) mobile networks have high speed, high stability, flawless security, and good user experience. Researchers are going on security problems. With the advancement of technology from 1G to 4G, the major focus is on confidentiality and services like device to device communication (D2D), massive multiple inputs multiple outputs (mMIMO), etc. This paper discusses a comprehensive study on the security features of 5G mobile networks in WSN that includes data confidentiality, key management, privacy, etc.

The rapid growth of cellular network is primarily influenced by a tremendous growth in wireless user devices. By the end of 2020, 5G mobile network is expected to be launched. However, 5G requires features like secure communication, infrastructure, energy efficiency, low cost, zero latency, and high-speed connections [1]. Wireless sensor networks are typically characterized by battery-powered sensor devices. A wireless sensor network is a network consisting of autonomous devices using sensors to cooperatively monitor physical/environmental conditions such as sound, vibration,

L. Dash (✉) · M. Khuntia
Department of Computer Science and Engineering, Siksha 'O' Anusandhan (Deemed to be University), Bhubaneswar, India
e-mail: dashlucy89@gmail.com

M. Khuntia
e-mail: mitrabindakhuntia@soa.ac.in

D. Mishra et al. (eds.), *Intelligent and Cloud Computing*, Smart Innovation, Systems and Technologies 194,
https://doi.org/10.1007/978-981-15-5971-6_70

pressure, temperature, and motion/pollutants at different locations. User satisfaction and quality of experience are significantly important nowadays [1]. A collection of sensing devices that can communicate wireless are self-organizing multi-hop and ad hoc networks. WSNs have limited resources in memory, computation power, bandwidth, and energy. WSNs have a wide range of application in remote healthcare area, agriculture, etc. [1].

The rest of the paper is organized as follows: Sect. 2 presents a brief overview of the evolution of cellular networks. In Sect. 3, we discuss various challenges of 5G mobile networks. In Sect. 4, we discuss security services in 5G network, whereas different technologies used for security purpose in 5G wireless mobile networks are briefly discussed in Sect. 5. Section 6 presents the various security architectures of 5G mobile networks proposed so far, whereas Sect. 7 discusses conclusion and future work of the paper followed by the references.

2 Evolution of Cellular Network

In this section, we study the evolution and development of various generations of mobile wireless technology and the significance and advantages of one over the other. In the past few decades, mobile wireless technologies have experienced the evolution of technologies from 1G to 4G. Current research in the mobile wireless network focuses on the advance implementation of 5G technology. 5G technology is not officially implemented and is expected to be launched by the end of 2020 [2].

2.1 1G (First Generation)

First-generation mobile network refers to analog cellular network. The first commercially automated cellular network was launched in Japan by NTT (Nippon Telegraph and Telephone) in 1979. In 1984, Bell Labs (Nokia) developed modern commercial cellular technology. First-generation mobile communication technologies had limited capacity. It was used only for voice calls [2].

2.2 2G (Second Generation)

In the 1990s, the second-generation (2G) mobile phone systems came into focus. It was using the GSM (Global System for Mobile) standard. Digital transmission was the main criteria of this generation. SMS texting in the prepaid mobile system was introduced. In 2G, the digital signals required less battery power, so it helps mobile batteries to last long. These signals were environment-friendly [2].

2.3 3G (Third Generation)

The third-generation mobile network is based on high-speed data networks. The demand for the data rate was increased due to the use of 2G phones. Unable to fulfill the demand by 2G, the next-generation 3G came into focus [3]. In the mid-2000s, 3G technology was implemented, namely high-speed downlink packet access (HSDPA). There are other enhanced 3G mobile telephony communication protocols, which are coined as 3.5G, 3G+. It allows networks to have higher data transfer speeds and capacity. It provides speed from 42 up to 84 Mb/s [4].

2.4 4G (Fourth Generation)

The fourth generation of mobile networks defines the growth of mobile broadband. It has more bandwidth and services as compared to 3G. It comes with the feature of high-quality audio/video streaming over end-to-end internet protocol. It utilizes packet switching over internet, LAN or WAN networks via VoIP. Data transfer rate and accuracy are also high as compared to other generations [5] (Table 1).

Table 1 Evolution from 1G to 5G [4]

Generation	Standards	Features	Data rates	Data switching	Voice switching	QoS
1G	AMPS, TACS	Uses analog signals	N/A	Circuit	Circuit	Rudiment
2G	GSM, CDMA, EDGE, GPRS	Uses digital signals	236.8 kbps	Circuit	Circuit	Better
3G	UTMS, CDMA2000, HSPDA, EVDO	Supports multimedia content	384 kbps	Packet	Circuit	Improved
4G	LTE Advanced, IEEE 02.16 (WiMax)	Multiple high bandwidths	Up to 1 Gbps	Packet	Packet	Highly improved
5G	BDMA (Beam Division Multiple Access), FBMC (Filter Bank Multi-Carrier)	Will support ultra-dense heterogeneous network	10–50 Gbps	Packet, global roaming access	Packet	Highly improved

3 5G and Its Challenge

The next-generation (5G) mobile network is expected to meet the expectations of higher data rate with the emerging number of mobile users. Each new technology has a different requirement. 5G is designed to have low complexity and high spectral efficiency. Limitations of the conventional cellular systems gave rise to the requirement of the next generation beyond 4G, which are listed below [2]:

- Flexibility
- Scalability
- Network capacity expansion
- Performance optimization
- Economical suitability
- Handoff management
- Privacy and security of network
- Quality of Service (QoS)
- Strong infrastructure.

The growth in mobile users has led to the requirement of higher data rate, fast connectivity, and flawless network coverage. Evolution of the mobile generations from 1G to 4G has developed wireless communication. It is expected that by the end of 2020, mobile users will cross approximately 3.8 million. Revolutionary features of 5G are small cells, Massive MIMO, millimeter waves (having larger bandwidth), smart devices and device-centric connectivity, and machine to machine communication. In future, networks are expected to carry data that are 1000 times more mobile in 10 years [2].

The challenges in 5G are as follows:

3.1 Resource Allocation

In both infrastructure-based and infrastructure-less networks, it is important to provide sufficient resources to all the users, such as transmit power, available spectrum, transmission time to mitigate interference, minimize delay, and obtain adequate throughput while maximizing quality of experience [2].

3.2 Backhaul

For infrastructure-based network, operations consider backhaul as one of the most important challenges in 5G for the deployment of the small cells. Backhaul data transfer is made in three ways which include (i) wired optical fiber, (ii) wireless point-to-multipoint, and (iii) wireless point-to-point [2].

3.3 Cost

Improving the cost efficiency of deployment of the networks in 5G is important. The small cells must be developed in enormous number to take advantage of economic scale [2].

3.4 Energy Efficiency

To mitigate the inter-cell interference and save energy, energy efficiency is a major concern for the development of 5G networks. Energy is very expensive for operators and user and its production has environmental effects. Battery capacity is increasing only 1.5% per decade or 4% per year [2].

3.5 Full Duplexing

In a full-duplex system in 5G, a device can transmit and receive a signal at the same time. However, interference will also come from both the downlink and uplink [2].

4 Security Services in 5G

5G provides various security services which are discussed below.

4.1 Availability

This feature describes the availability of a service which can be used by the end-users. It means service can be available in anytime and anywhere. Jamming is a kind of deadlock which interrupts the communications between users. DoS (Denial of Service) and DDoS (Distributed Denial of Service) are kinds of attacks which prevent communication. Resource allocation can be the solution to tackle this type of attack.

4.2 Authentication

Mainly authentication is of two types which are (i) Message Authentication and (ii) Entity Authentication. Authentication schemes are basically symmetric key based on 3G cellular network. But in 5G, authentication scheme has to be more efficient as it requires low latency and high speed. For this, SDN (Software-Defined Network) is proposed. In authentication, mechanism RFID (Radio Frequency Identification) has been used. A single RFID has multiple applications.

4.3 Integrity

Duplication of data in message authentication is a major problem in the cellular network. Integrity blocks the ways for modification of message. Unauthorized attacks are responsible for message injection and data modification without the knowledge of the attacks as active attacks are very much valid. Data integrity is very much useful in real world in the field of health services and vehicle monitoring.

4.4 Confidentiality

Data confidentiality is a major aspect of 5G mobile networks. Symmetric key encryption technique is applied for preventing unauthorized attacks. Private key encryption technique cannot handle a huge number of unidentified attacks. To tackle this problem, PLS (Polynomial Local Search) can provide better options to fight with the Eavesdropping and Jamming attack.

4.5 Privacy

Privacy protection is a big concern for 5G mobile networks. The information carried in the wireless network carries private contents, position, and identity. Privacy is mostly required in smart transportation, healthcare, and various industries. Due to huge data flow in the 5G network, privacy protection is encrypted by HetNet, where a decentralized algorithm is proposed based on access point selection.

5 Different Technologies Applied for Security Purpose in 5G

This section presents the security activities from the point of view of technologies applied to 5G. The technologies applied are HetNet, Massive MIMO, D2D, SDN, and IoT which are discussed as follow:

5.1 HetNet (Heterogeneous Network)

HetNet is such a technique that provides coverage wirelessly in 5G mobile networks. This is a multi-tier system where nodes present in different tier have different features, which are coverage size, transmission power, and radio access technologies [6]. For the enhancement of the communication coverage, co-ordinated multiple transmission (CoMP) is applied. In this network, multiple base stations (BS) are selected to transmit a message. In [6], the author proposed that the base station threshold for CoMP can improve the performance security for coverage and the theoretical and simulation result conclude the same. Generally, HetNet uses a combination of systems like GSM, UMTS, and modern radio access technologies such as LTE and Wi-Fi [7].

5.2 Massive MIMO (Multiple Input Multiple Output)

Massive MIMO can enhance the security of communications, especially for macro-cell users. It uses a large number of antennas at the base station which gives high efficiency, thus helpful for a large no of users. The number of antennas in the base station affects security performance [7].

5.3 D2D (Device to Device)

In D2D communication, devices can communicate with each other without the help of connection with the base station. But, the absence of the security infrastructure makes it a less secure connection. The security problem is a major issue for radio communications and large-scale deployment. Co-operation scheme to secure communications is proposed to check the security problem [7].

5.4 SDN (Software-Defined Network)

Next-generation systems will enable various types of security issues which can be designed and analyzed by SDNs. To work for security constraint in network as well as security domain, SDN works in the decoupled control panel. It offers some key attributes which are (i) Programmability, (ii) Abstraction, and (iii) Logically centralized intelligence [7].

5.5 IoT (Internet of Things)

IoT nodes are low powered devices which require lightweight and efficient mechanism to handle the security issues. These nodes have low computation capabilities. One technology called RFID (Radio Frequency Identification) is an effective method used to deal with power consumption. It used in traffic flow automation and inventory management [7].

6 Security Architecture Schemes for 5G Mobile Networks

Due to different kinds of attacks in the wireless network, it is important to maintain security. Different type of attacks is as follows.

6.1 Jamming

It is an active type of attack which completely prevents the data communication and access of radio resources. Resource allocation technique is used for handling this type of attack [8].

6.2 DoS and DDoS

It is an active kind of attack which prevents legitimate users from communication. DoS or DDoS can attack various layers of network which completely exhaust network resources. It also attacks radio, memory, battery, and many other nodes of network. It also consumes the availability of network [8].

6.3 MITM

It is an active kind of attack. It is responsible for manipulation message between the sender and user. It replaces or modifies the control channel of the network. It attacks the availability, integrity, and confidentiality of data in the network [8].

6.4 Eavesdropping and Traffic Analysis

It is a passive kind of attack so remains unnoticed. It affects the communication so encryption technique is used to handle it. Eavesdropper has lacuna of understanding an encrypted message. Traffic analysis is used for the detection of data location. To tackle this problem, different technologies like IoT, HetNet, and Cryptography PLS are used in 5G [8].

To deal with different security attacks and issues, some security architectures are proposed which can be discussed below.

6.5 5G Wireless Network Architecture

Wireless network architecture for 5G includes one central cloud, a cloud-based heterogeneous radio access network, centralized and co-ordinated techniques for efficient and flexible resource allocation, and distributed edge cloud. Compared to the traditional cellular system, 5G is intended to provide new perspectives and changes in the wireless system. There are many technologies introduced to support performance and connectivity which are massive MIMO, D2D, SDN, etc. [7].

6.6 5G Wireless Security Architecture

The 5G wireless security architecture is a proposed model which has four different components which are Access and Mobility Management Function (AMF), Session Management Function (SMF), Unified Data Management (UDM), and Policy Control Function (PCF).

UDM manages the subscriber data and authenticity of users, while PCF provides mobility and roaming management. There are four domain security approaches which are network access security, network domain security, user domain security, and application domain security [9].

These domains provide network access and security features which give mutual authentication between a user and next generation core as well as user interface and network. It also protects against the attacks and enables secure transmission [6].

Nowadays research work is going on for authentication, data signaling, and location confidentiality.

6.7 Security Architecture for SDN/NFV Access Network

This architecture has the advantage of flexibility and independence. According to the need of user and network, security applications are designed [10]. The communication between the application and the SDN controller is based on the security layer. A disadvantage of the scheme is that network is in deadlock position when the security layer stops working. SDN, NFV (Network Function Virtualization), and edge computing enable the management of software functions as well as network service authentication [6].

7 Conclusion and Future Works

This paper gives an idea about the recent developments of 5G wireless security. The current trend is based on security services which are availability, authentication, integrity, privacy, key management, etc. In the near future, 5G will develop mobile communication network. It will be responsible for the development of an intelligent city, military security, national defense, intelligent traffic, etc. So, it is very important to enhance the network security architecture of 5G. Many research works are going on this field to implement the security on 5G wireless network.

References

1. Alkalbani, A.S., Mantoro, T.: Security comparison between dynamic & static WSN for 5G networks. In: 2017 Second International Conference on Informatics and Computing (ICIC), pp. 1–4, Nov 2017. IEEE
2. Akyildiz, I.F., Lee, W.Y., Vuran, M.C., Mohanty, S.: NeXt generation/dynamic spectrum access/cognitive radio wireless networks: a survey. Comput. Netw. **50**(13), 2127–2159 (2006)
3. Pan, F., Wen, H., Song, H., Jie, T., Wang, L.: 5G security architecture and light weight security authentication. In: 2015 IEEE/CIC International Conference on Communications in China-Workshops (CIC/ICCC), pp. 94–98, Nov 2015. IEEE
4. Panwar, N., Sharma, S., Singh, A.K.: A survey on 5G: the next generation of mobile communication. Phys. Commun. **18**, 64–84 (2016)
5. Sendra Compte, S., Lloret, J., García Pineda, M., Toledo Alarcón, J.F.: Power saving and energy optimization techniques for wireless sensor networks. J. Commun. **6**(6), 439–459 (2011)
6. Liang, X., Qiu, X.: A software defined security architecture for SDN-based 5G network. In: 2016 IEEE International Conference on Network Infrastructure and Digital Content (IC-NIDC), pp. 17–21, Sept 2016. IEEE
7. Fang, D., Qian, Y., Hu, R.: QSecurity for 5G mobile wireless networks. IEEE Access **6**, 4850–4874 (2017)

8. Bendale, S.P., Prasad, J.R.: Security threats and challenges in future mobile wireless networks. In: 2018 IEEE Global Conference on Wireless Computing and Networking (GCWCN), pp. 146–150, Nov 2018. IEEE
9. Wong, E., Grigoreva, E., Wosinska, L., Machuca, C.M.: Enhancing the survivability and power savings of 5G transport networks based on DWDM rings. IEEE/OSA J. Opt. Commun. Netw. **9**(9), D74–D85 (2017)
10. Siddiqui, M.S., Escalona, E., Trouva, E., Kourtis, M.A., Kritharidis, D., Katsaros, K., Lorenzo, M.: Policy based virtualised security architecture for SDN/NFV enabled 5G access networks. In: 2016 IEEE Conference on Network Function Virtualization and Software Defined Networks (NFV-SDN), pp. 44–49, Nov 2016. IEEE

Genetic Algorithm and Probability-Based Leach Variant Trust Management Model for WSNs

Lakshmisree Panigrahi and Debasish Jena

Abstract As trust-based data and message security is reliable and efficient in the case of resource and power constraints networks, it finds a new field of research for researchers in wireless sensor networks. Till now, a lot of researches have been carried out to set up trust management systems in the above network. Probability distribution functions have a great role in the calculation of trust and reputation. In this paper, we have introduced a trust model based on clustered routing scheme. Probability distribution functions and metaheuristic algorithms have been used in the calculation of trust values. By using MATLAB tool, our work has been compared with ECSO algorithm, which is one of the latest efficient trust scheme in WSNs and obtained much efficient results. We have experimented taking into consideration of small networks consisting of 50 and 100 sensor nodes and can claim to have the latest efficient trust model to be used for enhancing the security of small-scale WSNs.

Keywords Wireless sensor networks · Cluster-based routing · LEACH protocol · Trust models · Security

1 Introduction

Even though the original motivation of the development of WSNs is that of battlefield surveillance, nowadays these networks are used in many diverse application areas. The applications of WSNs nowadays are in various civilian and military areas like security, environmental monitoring, health monitoring, battlefields, industry, seismology, construction, agriculture, transportation, warfare, traffic control, to list the important ones. Due to the high risk of various attacks to the sensor nodes, mainly as a result of random deployment in an unattended environment, it is always advisable

L. Panigrahi (✉)
Siksha 'O' Anusandhan (Deemed to be University), Bhubaneswar, India
e-mail: lakshmisreecontact@gmail.com

D. Jena
International Institute of Information Technology, Bhubaneswar, India

D. Mishra et al. (eds.), *Intelligent and Cloud Computing*,
Smart Innovation, Systems and Technologies 194,
https://doi.org/10.1007/978-981-15-5971-6_71

to implement security mechanisms to the network. While implementing security into the tiny and remotely deployed sensor nodes, we have to rule in the fact of limited resources of the sensor nodes. These nodes are generally small in size, constrained in memory capacity, resource power, battery-operated, have the low computational capability, limited communication bandwidth, etc. Due to the above reason, it is not flexible and efficient enough to implement the traditional security mechanisms like IDS, firewalls, encryption in the case of WSNS. Trust-based models have emerged as an effective alternative to address the above issue. Since our proposed work is based on trust models and LEACH protocol, in Sect. 2.1, there is a discussion on trust models and in Sect. 2.2 on cluster-based routing in WSNs.

2 Related Work

2.1 Concept of Trust and Trust Models in WSNs

In a general sense, trust is a logical concept in between different entities of a system to ensure better communication leading to better long-term productivity. This directly emphasizes to the following phrase: "Always be within the communication circle of trustable neighbours and distract yourself from the untrustable ones." The same trust concept and TRM (Trust and Reputation Management) are used in case of WSNs. TRM manages node's trust and the component's trust in a WSN. There are many works done from decades upon the concept of TRM.

The general steps of a TRM are [1]:

1. Collect the information
2. Disseminate the information
3. Mapping of information to the trust model
4. Making decisions.

The TRM can be of two types: Centralized and Distributed. In centralized TRM, trust information is stored in a central place, while in distributed TRM, trust values are stored in the nodes of the network. The initialization of a TRM will be done in three ways with consideration of the following: (i) each node having a neutral trust rating, (ii) all nodes as untrustworthy, and (iii) each network node as trustworthy. CONFIDANT [2] uses Bayesian estimation and CORE [3] uses game theory to maintain the reputation and gives the above a statistical representation by counting selfish misbehavior of nodes. A beta and link quality indicator (LQI)-based trust model (BLTM) has been proposed in [4]. For trust calculation, we have used the trust calculation scheme of the above in our proposed model.

2.2 Cluster-Based Routing and Related Trust Models in WSNs

If considered from a power efficiency point of view, cluster-based topology and their associated routing protocols are more suitable as compared with flat topologies and their related routing techniques. For increasing energy efficiency, the different strategies followed in WSN are given below:

- Clustering and Routing
- Coverage Method
- Green Method (Fig. 1).

In cluster-based topologies, there is a tree-like arrangement of sensor nodes partitioned into different levels. Some of the popular hierarchical routing schemes are listed in Fig. 2.

One of the most popular routing schemes based on clustering is LEACH, developed in the year 2000. In spite of its many advantages, LEACH has some pitfalls which further have been rectified resulting in much efficient protocols in terms of energy efficiency, network lifetime, security, to name a few. Almost all the modified-LEACH protocols rectify pitfalls of LEACH retaining the original structure of the above protocol. So, before discussing the multipath routing protocols based on LEACH, it is a primary need to have a discussion LEACH protocol. LEACH, stands for "Low Energy Adaptive Clustering Hierarchy," developed by Heinzelman et al. [5] is a cluster-based routing protocol for WSNs. There are several rounds of operations performed in the above protocol. In each round, two phases have been introduced, one is setup phase and another is steady-state phase. In the setup phase, the all new cluster heads and their cluster members will be selected by going through a random number generation, threshold value, and broadcasting message strategy. After the setting up of all clusters in the sensor network, steady-state phase begins. In this phase, actual routing was done within every cluster using TDMA time-driven scheduling. After cluster heads accumulate all the information from their members, they send those

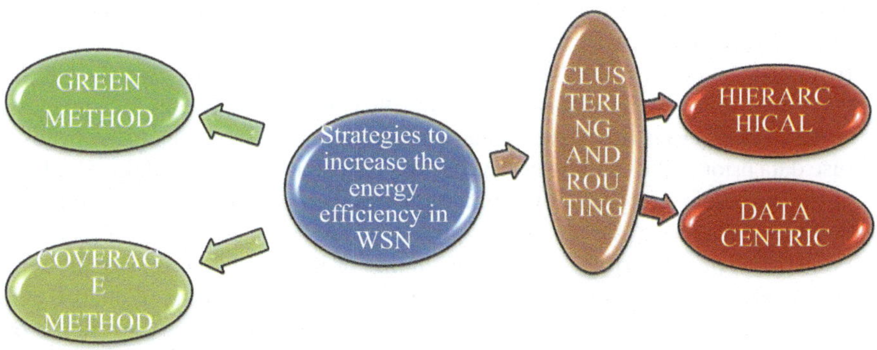

Fig. 1 Strategies for increasing energy efficiency in WSN

Fig. 2 Hierarchical routing protocols in WSNs

accumulated and aggregated data values to the sink. As previously stated, LEACH has pitfalls like fixed no of cluster heads in each round, use of a stochastic algorithm at each round in the cluster head decision-making process, uneven location of cluster heads in each round resulting in unbalanced utilization of energy, and decrease in network lifetime. So, there needed some refinements in the said protocol, without its original structure and basic operation. Several researches have been done in the above topic. It is not possible to maintain all those works at one place.

- In case of LEACH-C [6], there is a random selection of cluster heads and the selection is carried by the base station; but in case of LEACH, the sensor network nodes themselves are doing the above random selection. As compared to LEACH, there is an improvement of network lifetime in case of LEACH-C. However, there is also one shortcoming of the said protocol, i.e., LEACH-C may not be suitable for a large sensor network area.
- In another work [7] of LEACH–the variation, $T(n)$ is readjusted by considering the factors of the long-distance node and that of residual energy of the sensor nodes. After $T(n)$ is readjusted, data fusion rate is applied on the cluster heads to fuse data prior to the sending of data to the base station. Finally, for the avoidance of the excessive energy consumption by nodes, the multi-path fading model and the free-space model are adopted.
- Preparation phase performed in the beginning of first round along with the setup and steady-state phases in case of LEACH-GA [8]. In this protocol, the LEACH protocol setup and steady-state phases are used. Using a Genetic Algorithm (GA) based on cluster head optimal value probability, the above preparation phase is used for the selection of cluster heads. In the beginning, the selection of cluster

head is done by all sensor network nodes with message sending to that of the base station. At the base station, at the completion of first round, GA has been used for finding cluster heads with minimum energy consumption. Execution of preparation phase is done only once at the beginning of setup phase and steady-state phase. The above protocol minimizes sensor node's energy consumption as compared to LEACH and LEACH-C.

- According to [9], Leach-GA gives better results as compared to LEACH and LEACH-C.
- In LEACH-M [10], at the beginning of every round for the selection of cluster heads, counting of current Eres is performed by base station for every sensor node and after the above count, average energy Eavg is calculated as Eq. (1). If Eres of a cluster head is less than that of current Eavg, to dismiss old cluster heads and to vote new ones, BS announces cluster head selection news. In this protocol, the network address of every sensor node is used for the modification of cluster head threshold. DAAM of ZigBee is used to calculate the network address of the nodes.

$$\mathrm{Eavg} = \left(\sum_{i=0}^{N-1} \mathrm{Eres}(i) \right) / N \tag{1}$$

- In the case of exponential cat swarm optimization (ECSO) [11], cluster heads will be selected based on PFuzzyACO. After the selection of cluster heads, various parameters like energy, trust, distance, traffic density, delay, and link lifetime are used in multipath transmission strategy to find optimal path from source to destination during message transfer. ECSO shows high performance as compared to artificial bee colony, ACO, fractional artificial bee colony [12], Low-Energy Adaptive Clustering Hierarchy (LEACH), FuzzyACO [13], and PeSOA [14].

3 Proposed Model

Our proposed protocol is based on implementing trust computation on metaheuristic-based LEACH variant routing protocol. Our model follows the basic steps of LEACH protocol and consists of several rounds. Each round in turn, consists of two phases, setup and steady-state phases. In the setup phase, for cluster head selection, we have used LEACH-GA [8] algorithm according to the study of [9]. After the cluster heads are selected in the setup phase, we have used the data fusion rate for the transmission of messages in between the sensor nodes as well as in between the nodes of the network. In addition to the above data transmission, to forward messages, we have proposed a beta distribution methodology to compute trust in between two entities of transmission pair to achieve a secure path eliminating all malicious and doubtful nodes of the network.

Table 1 Simulation parameters

Parameter name	Value
Initial energy/node (in J)	0.5
Data aggregation energy = (/message)	5 nJ/bit
Transmitter electronics = receiver electronics	50 nJ/bit
Percentage of cluster heads	5%
Base station position (relative)	(50, 50)
Packet size	4000 bits
Number of nodes (excluding BS)	50 and 100
Proposed approach compared with	ECSO, LEACH

3.1 Simulation Results

We have simulated through MATLAB with a deployment area of 100 m × 100 m and simulation parameters as given in Table 1. Our proposed model has been shown with ECSO algorithm for showing efficiency and with LEACH for reference purpose. From [11], it has been concluded that ECSO shows good results as compared to LEACH, artificial bee colony, PSoA, and PFuzzyACO. Hence, comparing with ECSO and showing good results indicates that the proposed scheme is effective as compared to the above protocols (Figs. 3, 4, 5, 6, 7, and 8).

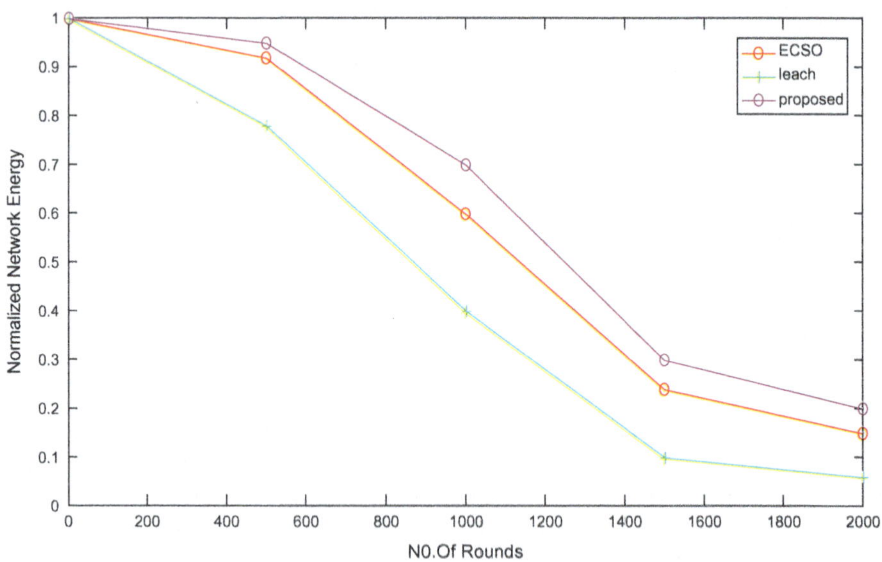

Fig. 3 Normalized network energy (50 nodes)

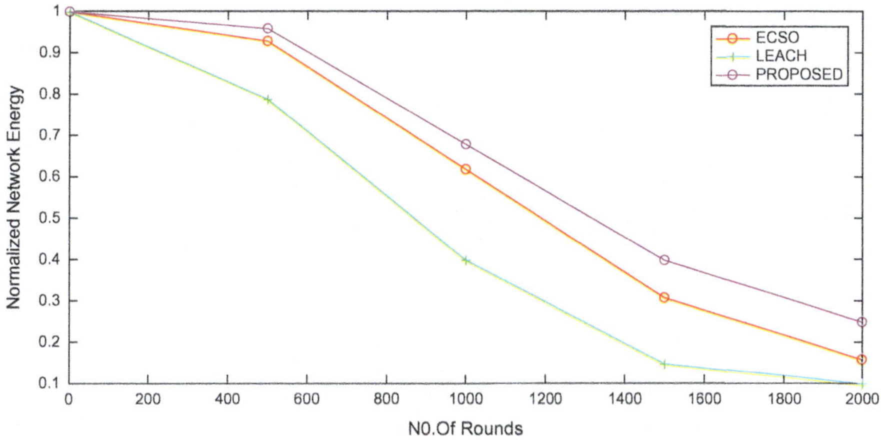

Fig. 4 Normalized network energy (100 nodes)

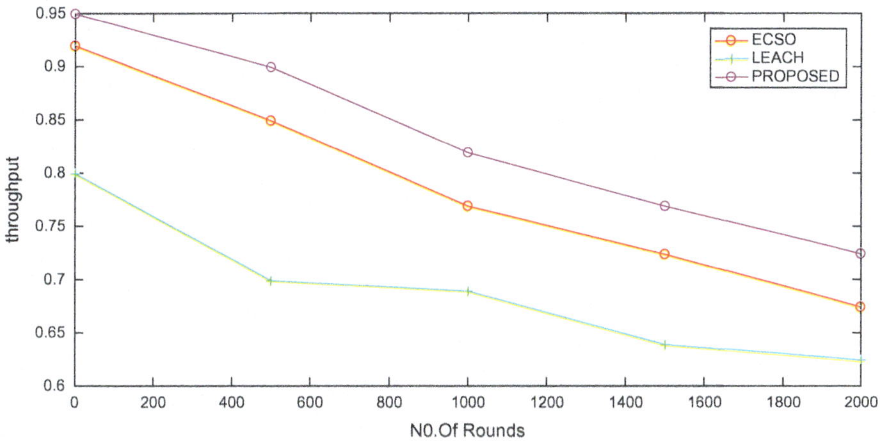

Fig. 5 Throughput (50 nodes)

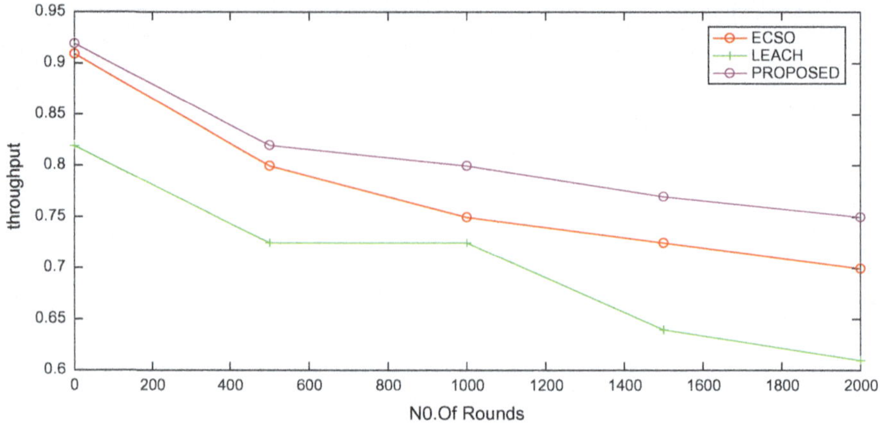

Fig. 6 Throughput (100 nodes)

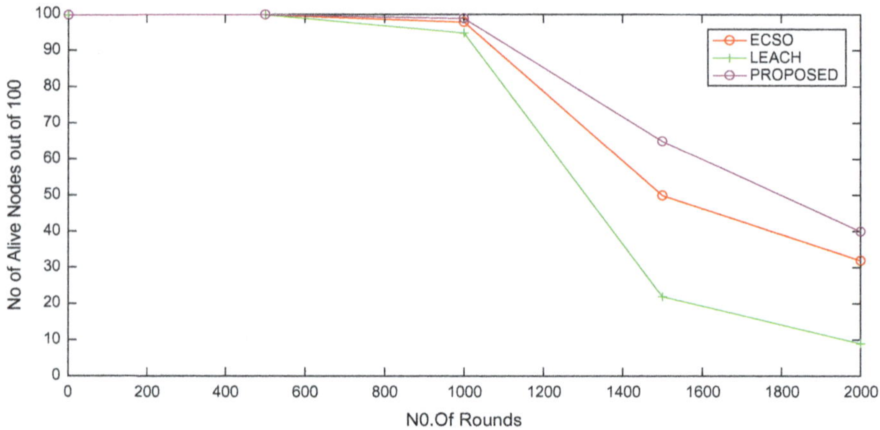

Fig. 7 No. of alive nodes (100 nodes)

4 Conclusion

In our proposed trust model, we have integrated the benefits of metaheuristic algorithm with beta distribution probability function to design an effective wireless sensor network in terms of network throughput, network lifetime, and other networking parameters. While doing simulation, we have taken a network of small size and also assumed the sensor nodes to be of homogeneous nature. The proposed work can be experimented with a bigger size and heterogeneous network in future.

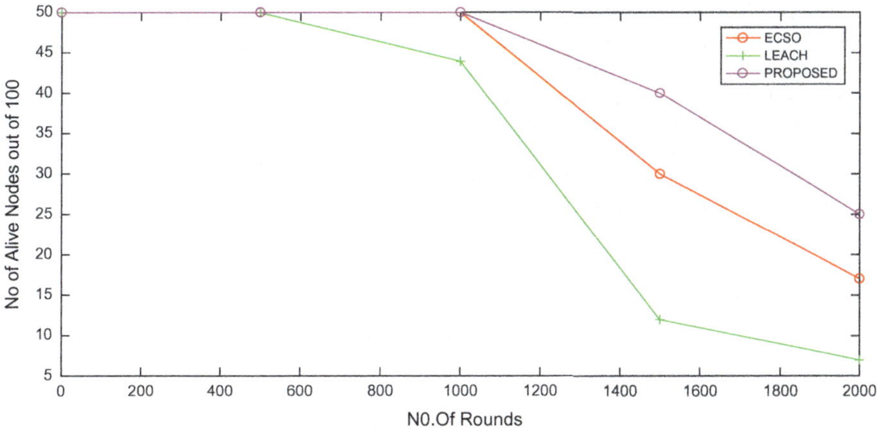

Fig. 8 No. of alive nodes (50 nodes)

References

1. Xiang, G., Jianlin, Q., Jin, W.: Research on trust model of sensor nodes in WSNs. Procedia Eng. 909–913 (2012)
2. Buchegger, S., Boudec, J.Y.L.: Performance analysis of the CONFIDANT protocol (Cooperation Of Nodes—Fairness In Dynamic Ad-hoc NeTworks). In: The 3rd ACM International Symposium Mobile Ad-hoc Networking & Computing (MobiHoc'02), Lausanne, CH (2002)
3. Michiardi, P., Molva, R.: CORE: a collaborative reputation mechanism to enforce node cooperation in mobile ad-hoc networks. In: The IFIP TC6/TC11 Sixth Joint Working Conference on Communications and Multimedia Security: Advanced Communications and Multimedia Security, Portoroz, Slovenia (2002)
4. Wu, X., Huang, J., Ling, J., Shu, L.: BLTM: beta and LQI based trust model for wireless sensor networks. IEEE Access **7** (2019). https://doi.org/10.1109/access.2019.2905550
5. Heinzelman, W.R., Chandrakasan, A., Balakrishnan, H.: Energy-efficient communication protocol for wireless microsensor networks. In: Proceedings of the Hawaii International Conference on System Sciences, Maui, Hawaii, 4–7 Jan 2000
6. Heinzelman, W., Chandrakasan, A., Balakrishnan, H.: An application-specific protocol architecture for wireless microsensor networks. IEEE Trans. Wireless Commun. **1**(4), 660–670 (2002)
7. Wu, W., Xiong, N., Wu, C.: Improved clustering algorithm based on energy consumption in wireless sensor networks. In: IET Networks. The Institution of Engineering and Technology (2017)
8. Liu, J.L., Ravishankar, C.V.: LEACH-GA: genetic algorithm-based energy efficient adaptive clustering protocol for wireless sensor networks. Int. J. Mach. Learn. Comput. **1**(1), 79–85 (2011)
9. Sivakumar, P., Radhika, M.: Performance analysis of LEACH-GA over LEACH and LEACH-C in WSN. Procedia Comput. Sci. **125**, 248–256 (2018)
10. Zhao, L., Qu, S., Yi, Y.: A modified cluster-head selection algorithm in wireless sensor networks based on LEACH. EURASIP J. Wireless Commun. Netw. **2018**, 287 (2018)
11. Kulkarni, P.K.H., Jesudason, P.M.: Multipath data transmission in WSN using exponential cat swarm and fuzzy optimisation. IET Commun. **13**(11), 1685–1695 (2019). https://doi.org/10.1049/iet-com.2018.5708

12. Youdao, Y., Wagan, R.A., Bukhari, A.H.S.: Parametric identification for fractional order model based on hybrid artificial bee colony algorithm. In: 5th International Conference on Computer Science and Network Technology (ICCSNT), Changchun, China, pp. 401–406 (2016)
13. Valdez, F., Vazquez, J.C., Gaxiola, F.: Fuzzy dynamic parameter adaptation in ACO and PSO for designing fuzzy controllers: the cases of water level and temperature control. Adv. Fuzzy Syst. 19. Article ID 1274969 (2018)
14. Hindriks, K.V., Hoogendoorn, M., Goebel, R.: Penguins search optimization algorithm (PeSOA) in applied artificial (2013)

EOQ Model with Backorder for Items with Imperfect Quality Using Cross-Selling Effects

Rashmi Rani Patro, Rojalin Patro, Srikanta Patnaik, and Binod Kumar Sahoo

Abstract This paper aims to obtain Economic Order Quantity (EOQ) model for the items of imperfect quality with backorder using cross-selling effect, where all the items are considered without their interdependency with each other. However, there exist a strong relation between different items, sale of an item depends on the sale of others. Depending on the strength of relationship between items, there may be a loss in their sales, that is, cross-selling effect. Apriori algorithm is used to generate frequent item-sets among the item-sets in an inventory data base. In case of imperfect quality items, cross-selling effects are visible clearly. Moreover, with this effect, frequent item-sets are taken individually and EOQ is computed for them along with shortage level. With the help of a numerical example, we compared results obtained with the existing models.

Keywords Inventory · Imperfect quality · Proportionate discount · Backorder · EOQ

R. R. Patro (✉) · S. Patnaik
Department of CSE, ITER, S'O'A (Deemed to be University), Bhubaneswar 751030, India
e-mail: rashmi.patro@gmail.com

S. Patnaik
e-mail: srikantapatnaik@soa.ac.in

R. Patro
Department of Mathematics, NIIS Group of Institution, Sarada Vihar, Madanpur, Khurda, Bhubaneswar 752054, India
e-mail: patro.rojalin@gmail.com

B. K. Sahoo
Department of EE, ITER, S'O'A (Deemed to be University), Bhubaneswar 751030, India
e-mail: binoditer@gmail.com

D. Mishra et al. (eds.), *Intelligent and Cloud Computing*, Smart Innovation, Systems and Technologies 194,
https://doi.org/10.1007/978-981-15-5971-6_72

1 Introduction

In business expansion and decision–making, data mining techniques play a very important role. Freshly, the technique of data mining has fascinated much of deliberation in the society and the information industry where the data continue to grow on a daily basis. The process of data mining is discovering valuable information from huge amounts of data stored in databases. Association rule mining is one of the important areas of data mining, which finds interesting association among massive amounts of data items, maintaining details regarding a business transaction. It can help for various business decision-making processes, say cross-selling and inventory control. A huge difference in economic order quantity is occurring due to the cross-selling effect in the frequent item-set.

Silver et al. [1] was the first researcher who considered the inventory model for various items organized with a precise cost system. He had not applied any valid inventory policies when the item-set correlates with each other. An inventory model was developed by Kaku [2], where he had applied data mining technique to classify inventories and also presented the frequent item-sets have totally different assets from other individual items. Initially, each item may be in different class but the frequent item-set contains multiple items in it. Moreover, there is a corelation among items in a frequent item-set. Different developers have investigated the economic order quantity broadly in realistic conditions. An inventory model for items with imperfect quality with cross-selling effects by using the association rule mining has been developed by the authors in Mittal et al. [3]. The development of the basic EOQ model has been first proposed to determine the effect of defective items by Porteus [4]. It has created an interest in the research community dealing with data mining and many EOQ models addressing the defective items have been developed. One such classical EOQ approach is to sell the defective items with a constant discount rate for the whole batch after completion of 100% screening by Salameh and Jaber [5]. During 2000–2019, different authors have considered the first model presented by [5] and build up their own ideas. Rezaei [6] was the first researcher who expanded this work by taking a shortage and backorder situation. In this paper, he initiated at the beginning of each cycle the cyclic shortage to come out of the imperfect production which effects to backorder. Wee et al. [7] proposed an EOQ model for items of imperfect quality with shortages are backordered by permitting hundred percent screening process where demand is less as compared to the screening rate. Later on, Eroglu [8] developed an EOQ model with defective items by permitting shortages with backordered and they examined the expected total profit and the effects of different levels of defectives fractions on order lots where the incoming lot contains some defectives. Chang and Ho [9] investigated an economic order quantity model with the imperfect quality items where the imperfect items are satisfying the backorder and they used renewal reward theorem to compute the expected total profit per unit time. Hsu [10] established an economic order quantity model for imperfect quality items with introducing multiple screening processes under shortage backordering. Kumar and Goswami [11] proposed a inventory model for imperfect quality items where shortages are permitted under completely backlogged. Patro [12] established an EOQ model for items with imperfect quality with backorder by introducing proportionate discount.

We presented an EOQ model for imperfect quality items by introducing cross-selling effect with shortages. We considered the situation of a transaction database with few items in a frequent item-set having relationships among them. Each lot considered for the presented model consists of some percentage of defectives. A hundred percentage screening is done for the lot, in order to obtain the total profit. Lastly Numerical example and graphical analysis is given to analyze the results and the effect of the various parameters considered for the model.

2 Notations

In this present paper, by considering the above mentioned notations a mathematical model is developed which is very close to the real world situations; because in an economic order quantity the buyer may receive an imperfect lot (Table 1).

Now, let us consider that the items of frequent item-set are delivered instantaneously with the order size of L with a unit purchasing price P_c. Consider the EOQ model with demand rate D units per unit time, fixed ordering cost O_c and the inventory holding cost H_c unit wise per unit time.β is the percentage of defectives in the received lot. A 100% screening process of the lot is conducted at a rate of w units per unit time with unit screening cost S_c. After completion of the screening process, items of poor quality are sold as a single batch at a proportionate discounted price. But the price of good quality and imperfect quality items are considered to be S_g per

Table 1 Notations

L	Size of the lot
P_c	Purchasing cost per unit
O_c	Ordering cost
H_c	Unit-wise holding cost
β	% Of defective items in L
$f(\beta)$	p.d.f. of β
S_g	Unit-wise selling price of good quality item
S_I	Unit-wise selling price of imperfect quality item
w	Rate of screening
S_c	Screening cost per unit
T	One cycle length
B_c	Shortage cost per unit
U	Expected shortage level
IM	Maximal level of inventory

unit and S_{I} per unit, respectively. Shortage is considered with a shortage cost B_{c} per unit.

3 Mathematical Model

This paper develops a differentiation between the order quantities for imperfect quality items by introducing proportionate discount in frequent item-set, considering cross-selling effects.

Let us consider the given item-set $f = \{r_1, r_2, r_3, \ldots, r_n\}$,

$$\text{Support}(r_1) = \text{Frequency}(r_1)/\text{No of total Transactions} \tag{1}$$

Confidence specifies the interrelation among items and is defined as conditional probability. conf ($r_1 \rightarrow r_2$) indicates the frequency of purchasing r_2 when r_1 is purchased.

$$\text{Conf}(r_1 \rightarrow r_2) = \text{Support}(r_1 \cup r_2)/\text{Support}(r_1) \tag{2}$$

Using apriori algorithm, we generated rules for the association of items provided, both support and confidence are more as compared to minimum support and confidence defined by the programmer. Based on this, frequent item-set can be find out.

Step 1 It scans all the transactions to count occurrences of each item to find a set of frequent 1-item-set (L_1).

Step 2 It consists of two parts: as apriori generation and support of candidate counting.

During this step, L_{k-1} is used for generation of candidate item-set C_k to find L_k, then the database is scanned and compared with candidate support count with minimum support count, i.e., satisfying the condition, $k \geq 2$.

Apriori generation function apriori_gen () is used for performing two tasks such as: join and prune. Join step performs join of L_{k-1} with L_{k-1} in order to generate potential candidates, whereas prune step uses the apriori property to discard the items that has infrequent subsets.

Cross-selling effects of items are represented with confidence among them. Impact of an item rk being stock out on another item i in frequent item-set $f = (r_1, r_2, r_3, \ldots, r_n)$, is given by probability is defined as follows:

$$\text{Prob}_{k,i} = \sum_{i=1}^{n} \text{conf}(k \rightarrow f(k, i)) \tag{3}$$

where $k = 1, 2, 3, \ldots, n$ be the items in a frequent item-set, n is the total items in a frequent item-set.

$f(k, i)$ is the subset of item i except k item in a frequent item-set.

When $i = k$, $f_{i,i} = i$ and $\mathrm{conf}(i \rightarrow i) = 1$.

The lost cost due to cross-selling effect is called as opportunity cost for an item k it is defined as

$$\mathrm{OC}_k = \sum u_i.\mathrm{prob}_{k,i} \tag{4}$$

where u_i = cost of unit item i.

Probabilistic index,a_k which is defined as:

$$a_k = \frac{\mathrm{OC}_k + H_k}{\mathrm{OC}_k} \tag{5}$$

where H_k = holding cost of item k per unit.

a_k is used to compute opportunity cost later.

Using the above notation, the following mathematical model is used for obtaining optimal lot size and the expected shortage level is developed.

Per cycle total cost is calculated by the formula written below:

$$F(L, U) = O_c + P_c L + S_c L + H_c \times \left(\frac{L(1-\beta) - U}{2} t_1 + \frac{\beta L^2}{w} \right) + \frac{U B_c}{2} t_2 \tag{6}$$

Total revenue per cycle is calculated by adding total selling price of items with good quality and imperfect quality and is given by

$$G(L, U) = S_g L(1-\beta) + S_I L\beta \tag{7}$$

Per cycle total profit $\mathrm{TP}(L, U)$ = total revenue cycle wise—total cost cycle wise

$$\mathrm{TP}(L, U) = S_g L(1-\beta) + S_I L\beta - \left\{ O_c + P_c L + S_c L + H_c \times \left(\frac{L(1-\beta) - U}{2} t_1 + \frac{\beta L^2}{w} \right) + \frac{U B_c}{2} t_2 \right\} \tag{8}$$

The formula used for the total profit ($\mathrm{TPU}(L, U)$) is given as follows:

$$\mathrm{TPU}(L, U) = D\left(S_g - S_I + \frac{H_c L}{w} \right) + D\left(S_I - \frac{H_c L}{w} - P_c - S_c - \frac{O_c}{L} \right)\left(\frac{1}{1-\beta} \right) - \left[H_c \left(\frac{L(1-\beta)}{2} - U \right) + \frac{U^2 (B_c + H_c)}{2L(1-\beta)} \right] \tag{9}$$

As, the % of defective (β) items are considered with P.D.F, TPU(L, U) can be written as follows:

$$\begin{aligned}\text{ETPU}(L, U) &= D\left(S_g - S_I + \frac{H_c L}{w}\right) \\ &+ D\left(S_I - \frac{H_c L}{w} - P_c - S_c - \frac{O_c}{L}\right)E\left(\frac{1}{1-\beta}\right) \\ &- \left[H_c\left(\frac{L(1 - E[\beta])}{2} - U\right) + \frac{U^2(B_c + H_c)}{2L(1 - E[\beta])}\right]\end{aligned} \tag{10}$$

As, it is the nonlinear problem, the optimality condition is used for obtain ETPU(L, U) given in Eq. (10), is illustrated after computing both 1st and 2nd partial derivatives of ETPU(L, U) w.r.t. L, U we got the following expression:

$$\begin{aligned}\frac{\partial \text{ETPU}(L, U)}{\partial L} &= \frac{DH_c}{w}\left(1 - E\left(\frac{1}{1-\beta}\right)\right) + \frac{DO_c}{L^2}E\left(\frac{1}{1-\beta}\right) \\ &- H_c\left(\frac{L(1 - E[\beta])}{2}\right) + \frac{U^2(B_c + H_c)}{2L^2(1 - E[\beta])}\end{aligned} \tag{11a}$$

$$\frac{\partial \text{ETPU}(L, U)}{\partial U} = H_c - \frac{U(B_c + H_c)}{L(1 - E[\beta])} \tag{11b}$$

By equating (11a) and (11b) to zero, we get

$$L^* = \sqrt{\frac{2DO_c E\left(\frac{1}{1-\beta}\right)}{H_c\left[(1 - E[\beta])\left\{\frac{B_c}{B_c + H_c}\right\} - \frac{2D}{w}\left(1 - E\left(\frac{1}{1-\beta}\right)\right)\right]}} \tag{12}$$

$$U^* = \left(\frac{H_c L(1 - E[\beta])}{B_c + H_c}\right) \tag{13}$$

$$\text{IM} = L^* - U^* \tag{14}$$

To obtain the maximal inventory level, substituting the value of Eqs. (12) and (13) in Eq. (14) we got the expression given below:

$$\text{IM} = \frac{L(B_c + H_c E[\beta])}{B_c + H_c} \tag{15}$$

We used the Hessian matrix H, for examining the second-order sufficient conditions for maximal value as follows:

$$H = \begin{bmatrix} \frac{\partial^2 \mathrm{ETPU}(L,U)}{\partial L^2} & \frac{\partial^2 \mathrm{ETPU}(L,U)}{\partial L \partial U} \\ \frac{\partial^2 \mathrm{ETPU}(L,U)}{\partial U \partial L} & \frac{\partial^2 \mathrm{ETPU}(L,U)}{\partial U^2} \end{bmatrix} \tag{16}$$

$$\frac{\partial^2 \mathrm{ETPU}(L,U)}{\partial L^2} = -\left[\frac{2DO_c E\left(\frac{1}{1-\beta}\right)}{L^3} + \frac{U^2(B_c + H_c)}{L^3(1 - E[\beta])}\right] \tag{17}$$

$$\frac{\partial^2 \mathrm{ETPU}(L,U)}{\partial L \partial U} = \frac{U(B_c + H_c)}{L^2(1 - E[\beta])} \tag{18}$$

$$\frac{\partial^2 \mathrm{ETPU}(L,U)}{\partial U \partial L} = \frac{U(B_c + H_c)}{L^2(1 - E[\beta])} \tag{19}$$

$$\frac{\partial^2 \mathrm{ETPU}(L,U)}{\partial U^2} = -\frac{(B_c + H_c)}{L(1 - E[\beta])} \tag{20}$$

The first principal minor of H is

$$|H_{11}| = -\left[\frac{2DO_c E\left(\frac{1}{1-\beta}\right)}{L^3} + \frac{U^2(B_c + H_c)}{L^3(1 - E[\beta])}\right] < 0 \tag{21}$$

The second principal minor of H is

$$|H_{22}| = \left[\frac{2DO_c E\left(\frac{1}{1-\beta}\right)}{L^3} + \frac{U^2(B_c + H_c)}{L^3(1 - E[\beta])}\right] > 0 \tag{22}$$

At the point (L^*, U^*), the H is −ve definite. This indicates, the existence of unique values of L^* (12) and U^* (13) that maximize (10).

4 Numerical Analysis

We have considered the numerical example given below for analyzing the inventory situation presented above.

We adopted the values of the parameters given below according to Mittal et al. [3] min_sup = 50%,min_conf = 60%.

$D = 50{,}000$ units/year, $O_c = 100$/cycle, $H_c = \$10$/unit/year, $S_c = 1$/unit, $P_c = 30$/unit, $B_c = \$20$/unit, $S_g = \$60$/unit, $S_I = \$20$/unit, $w = 1$unit/min (131,400 units per year), $E[\beta] = 0.02$ and $E\left[1/(1-\beta)\right] = 1.02055$.

Let us consider the inventory item-set be $I = \{r_1, r_2, r_3, r_4, r_5, r_6\}$

Table 2 An inventory transaction database

TID	Items
1500	r_1 r_3 r_4
2500	r_2 r_3 r_5
3500	r_1 r_2 r_3 r_5
4500	r_2 r_5
5500	r_4 r_6

Table 3 The inventory policy in a frequent item-set $\{r_2, r_3, r_5\}$

Item	Min_Sup (%)	Support	Demand	Unit cost
r_2	50	75	50,000	$30.00
r_3	50	75	40,000	$20.50
r_5	50	75	45,000	$45.52

An inventory transaction database is given below. By support-confidence framework, association rules are generated among the inventory transactions. Using the apriori algorithm, from the given database the frequent item-sets are obtained, which is given as (Tables 2 and 3):

$$\{r_1\}, \{r_2\}, \{r_3\}, \{r_4\}, \{r_5\}, \{r_1\ r_3\}, \{r_2\ r_3\}, \{r_2\ r_5\}, \{r_3\ r_5\}, \{r_2\ r_3\ r_5\}$$

The item-sets $\{r_2\ \ r_3\ \ r_5\}$ are the most frequent items selected from different classes of items with support and confidence greater than min_sup and min_conf, where min_sup = 50%, min_conf = 60%. According to the apriori algorithm considering these conditions, the confidence of the other frequent item-sets are estimated and given as follows (Table 4):

Table 4 Rules with confidence

Items	Confidence (%)
$r_2 \rightarrow r_3$	66.7
$r_2 \rightarrow r_5$	100
$r_2 \rightarrow r_3 \cup r_5$	66.7
$r_3 \rightarrow r_2$	66.7
$r_3 \rightarrow r_5$	66.7
$r_3 \rightarrow r_2 \cup r_5$	66.7
$r_5 \rightarrow r_2$	100
$r_5 \rightarrow r_3$	66.7
$r_5 \rightarrow r_2 \cup r_3$	66.7

Now the opportunity cost of items in frequent item-set $\{r_2\ \ r_3\ \ r_5\}$ can be calculated by formulae.

$$\begin{aligned} \text{OC}_{r_2} &= C_{r_2}.\text{conf}(r_2 \rightarrow r_2) + C_{r_3}.\{\text{conf}(r_2 \rightarrow r_3) + \text{conf}(r_2 \rightarrow r_3 \cup r_5)\} \\ &\quad + C_{r_5}.\{\text{conf}(r_2 \rightarrow r_5) + \text{conf}(r_2 \rightarrow r_3 \cup r_5)\} \\ &= 30 \times 1 + 20.5 \times \{0.667 + 0.667\} + 45.52 \times \{1 + 0.667\} \\ &= 133.22884 \end{aligned}$$

Similarly, $\text{OC}_{r_3} = 121.24368$ and $\text{OC}_{r_5} = 122.877$.

Then modify the order policy for imperfect items in a frequent item-set

Case-1 $I_{nd_{r_2}} = \frac{\text{OC}_{r_2}+H_r}{\text{OC}_{r_2}} = 1.075058824.$

Case-2 $I_{nd_{r_3}} = \frac{\text{OC}_{r_3}+H_r}{\text{OC}_{r_3}} = 1.082478526.$

Case-3 $I_{nd_{r5}} = \frac{OC_{r5}+H_r}{OC_{r5}} = 1.081382195.$

Case-1 ***For*** item r_2.

The optimal value of L^* and $\text{ETPU}(L, U)$ is given by $L^* = 1235.13, \text{ETPU}(L, U) = 1,434,470$ and $U^* = 403.476$

Therefore, $\text{EOQ} = L^*\sqrt{I_{nd}} = 1280.645077$

Case-2 ***For*** item r_3.

The optimal value of L^* and the $\text{ETPU}(L, U)$ is given by $L^* = 1107.33, \text{ETPU}(L, U) = 1,535,010$ and $U^* = 361.728$.

Therefore, $\text{EOQ} = L^*\sqrt{I_{nd}} = 1152.090806$

Case-3 ***For*** item r_5

The optimal value of L^* and the $\text{ETPU}(L, U)$ is given by $L^* = 1173.12, \text{ETPU}(z) = 578,078$ and $U^* = 383.219$ and.

Therefore, $\text{EOQ} = L^*\sqrt{I_{nd}} = 1219.921952$ The comparision values of ETPU, EOQ, L*, U* are given in Table 5 after calculation.

Table 5 Comparision table

Items	L^*	$\text{ETPU}(L, U)$	EOQ	U^*
r_2	1235.13	1,434,470	1280.645077	403.476
r_3	1107.33	1,535,010	1152.090806	361.728
r_5	1173.12	578,078	1219.921952	383.219

5 Conclusion

In this paper, we have discussed an economic order quantity inventory model for imperfect items with shortages using cross-selling effect. It is noticed that by the time the number of defectives decreases, the rate of discount increases. A hundred percentage screening process is conducted on the lot received and the defective items are sold with a constant discount rate as a single batch. It is also observed that when we considered the frequent item-set for the items of imperfect quality, the quantity of the order size increases when the average % of opportunity cost increases and vice versa. We also observed that because of cross-selling effect in a frequent item-set, the new opportunity cost may produce a little impact on the lot size. Again, we found that the result of total profit and the lot size is more or less similar to that of the total profits calculated in the work of Rezaei [6]. It is hoped that this paper can provide insights to develop the EOQ model for two different types of imperfect quality items using cross-selling effect problems.

References

1. Silver, E.A., Pyke, D.F., Peterson, R.: Inventory Management and Production Planning and Scheduling, pp. 234–235. Wiley (1998)
2. Kaku, I.: A data mining framework for classification of inventories. In: Proceedings of the 5th International Conference on Management, pp. 450–455 (2004)
3. Mittal, M., Singh, J., Aggarwal, A., Kumari., K., Yadav, M.: Ordering policy for imperfect quality item sets using cross selling effects. Int. J. Model. Optim. **4**(1), 25–30 (2014)
4. Porteus, E.L.: Optimal lot sizing, process quality improvement and setup cost reduction. Oper. Res. **34**, 137–144 (1986)
5. Salameh, M.K., Jaber, M.Y.: Economic production quantity model for items with imperfect quality. Int. J. Prod. Econ. **64**, 59–64 (2000)
6. Rezaei, J.: Economic order quantity model with backorder for imperfect quality items. In: Proceeding of IEEE International Engineering Management Conference, pp. 466–470 (2005)
7. Wee, H.W., Yu, J., Chen, M.C.: Optimal inventory model for items with imperfect quality and shortage backordering. Omega **35**(1), 7–11.13 (2007)
8. Eroglu, A., Ozdemir, A.: An economic order quantity model with defective items and shortages. Int. J. Prod. Econ. **106**, 544–49 (2007)
9. Chang, H.C., Ho, C.H.: Exact closed-form solutions for optimal inventory model for items with imperfect quality and shortage backordering. Omega **38**(3–4), 233–237 (2010)
10. Hsu, J.T., Hsu, L.F.: A note on: 'Optimal inventory model for items with imperfect quality and shortage backordering'. Int. J. Ind. Eng. Comput. **3**(5), 939–948 (2012)
11. Kumar, R.S., Goswami, A.: Fuzzy stochastic EOQ inventory model for items with imperfect quality and shortages are backlogged. AMO Adv. Model. Optim. **15**(2), 261–279 (2013)
12. Patro, R., Acharya, S., Nayak, M.M., Acharya, M.: A fuzzy inventory model with imperfect items and backorder with allowable proportionate discount. Model. Measur. Control D **39**(1), 39–46 (2018)

A Deep Learning Approach for the Classification of Pneumonia X-Ray Image

Olive Chaudhuri and Barnali Sahu

Abstract Chest diseases are terribly serious health issues within the lifetime of individuals. These include chronic impeding pulmonic disease, pneumonia, asthma, TB and alternative respiratory organ diseases. Microorganism-related respiratory illness is that the commonest sort in adults. Respiratory illness causes inflammation within the air sacs in patients' lungs. Lungs full of fluid or pus make it troublesome to breathe. Chest X-ray is the best way to check for respiratory illness diagnosing. However, reading X-ray pictures will be difficult and needs domain experience and skill. It might be nice if we will simply raise a laptop to browse the pictures and generate the results. We have incontestable practicability of classifying the chest pathologies in chest X-rays mistreatment typical and deep learning approaches. The objective of this study is to predict whether the person has a respiratory illness from his X-ray report of lungs. For this, the convolution neural networking classification model using Keras API was trained with a dataset collected from Kaggle with optimum accuracy. After fine-tuning the model for two classes, the testing accuracy attained by the classifier is 94.51% with a loss of 0.06.

Keywords Respiratory disease · Chest X-ray · Classification · Convolutional neural network · Deep learning

1 Introduction

Pneumonia is a type of severe respiratory infection that infects lungs. When a healthy person breathes in, the alveoli (small air sacs of lungs) are filled with air. When a pneumonic patient breathes in, the alveoli, already filled with pus and liquid, limit oxygen intake and make normal breathing difficult. Pneumonia causes the most

O. Chaudhuri · B. Sahu (✉)
Department of Computer Science and Engineering, Siksha 'O' Anusandhan (Deemed to be University), Bhubaneswar, Odisha, India
e-mail: Sahu.barnali08@gmail.com

O. Chaudhuri
e-mail: chaudhuriolive@gmail.com

D. Mishra et al. (eds.), *Intelligent and Cloud Computing*,
Smart Innovation, Systems and Technologies 194,
https://doi.org/10.1007/978-981-15-5971-6_73

infection-related deaths in children worldwide. It killed 808,694 children under the age of 5 in 2017, accounting for 15% of all deaths of children under 5 years old. Early detection of pneumonia is essential in its treatment. The traditional and best-suited diagnosis is through X-Ray of chest. We require highly trained pathologists and radiologists who have domain knowledge of analysing X-Ray. Classifying chest X-ray abnormalities is a highly wearing task and it takes time. Due to the shortage of such radiologists, in cases of epidemic and large-scale treatments, the diagnosis of chest X-ray is delayed. This causes further delay of treatment. One solution to this problem is to use technology for analysing medical images. Over the past decades, computer-aided systems have been developed to classify the X-ray images and help doctors. However, this could not provide a significant level of decisions on what type of disease. Many researchers have worked on the diagnosis of X-ray images using artificial intelligence models and it proved to have large success. Deep learning models have been applied to it to increase accuracy [1–3]. This success motivated us to apply deep learning models to medical imaging for diseases classification tasks. Using the multiple available chest X-ray images as a training set, we have designed a convolutional neural network (CNN) model that takes input an image of chest X-ray and outputs as two labels, normal or pneumonia. This use of technology allows the radiologist to save time in analysing X-ray images, speeds up analysis during a shortage of radiologists and also reduces human error.

In this paper, a convolution neural network model is used after studying the related research work to improve test accuracy and minimize the loss. For this purpose, the traditional neural network is modified with backpropagation model.

The introductory section is followed by the following sections: Sect. 2 describes the dataset used, the proposed model of CNN for X-ray image classification and the empirical set up. Section 3 discusses the experimental results. Section 4 draws a conclusion from the results.

2 Materials and Methods

2.1 Dataset Used

The data used in this paper are obtained from Kaggle dataset of chest X-ray images [4]. The dataset contains 5856 frontal view chest X-ray images of different patients. Sample image inputs of size 64×64 to the model with the following labels are given in Fig. 1.

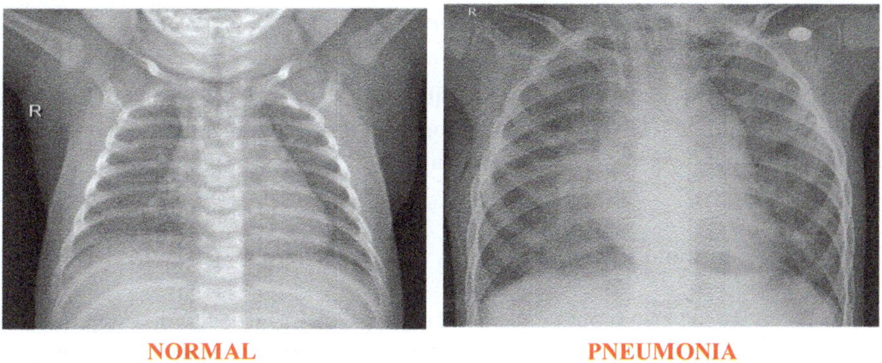

Fig. 1 Sample X-ray images of normal and pneumonia patient

2.2 *Proposed Model*

In the proposed model, the input image is taken in greyscale as a 64 × 64-pixel image into the convolution layer. In this layer, 32 feature detectors have been used with each feature detector of size 3 × 3. ReLU activation function has been used in this layer to avoid negative pixel values. In max pooling operation, a window of pool size of 2 × 2 has been chosen which outputs a pooled feature map of size 31 × 31 × 32 (Fig. 2).

All the values of the pooled feature map have been converted into a 1D vector in flattening which creates a vector length of 30,752. The input for the Artificial Neural Networks (ANN) is the flattened layer. The next layers comprise of the three hidden layers of the ANN. The first and second hidden layers consist of 120 and 60 neurons, respectively. The last layer consists of only one neuron that is called the output layer. Adadelta activation function has been used to increase accuracy and reduce the loss through binary cross-entropy and it gives us the label output of two classes as 0 and 1.

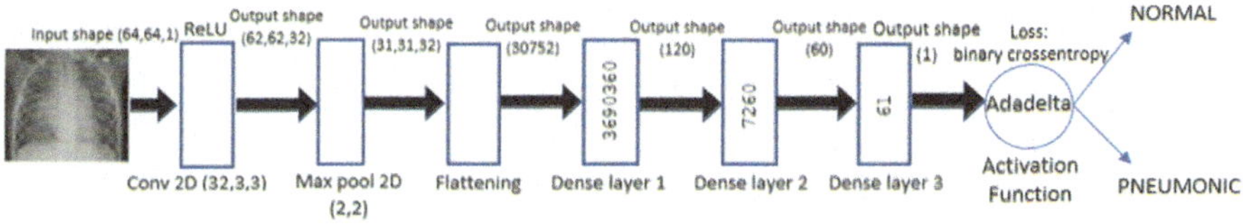

Fig. 2 Schematic representation of the proposed model for classification

2.3 *Convolutional Neural Network*

A brief description of the CNN used for this paper is given below. The vital steps involved in building a CNN are convolution, pooling, flattening and full connection and creates input layer for future ANN evaluation.

2.3.1 Convolutional Layer

In this layer, we have an input image of size, say, $r \times c$. This is the first step of evaluation of an image. Feature detectors or filters are the primary features of an image. For example, if we consider human face, then the features/filters would definitely be eyes, nose, lips and so on. In this layer, the input image is convolved with feature detectors of size $a \times a$. It is done to extract important features from the image. There are a bunch of N filters that are convolved with the image during the training process and the feature detector are updated. The output of this convolutional layer is feature maps of size $i \times i$. Figure 3 represents the feature map generation in first iteration.

$$C_j = B_i + \sum_{i=1}^{N} K_{i,j} \tag{1}$$

where B_i is the bias matrix for ith feature detector, $K_{i,j}$ is the convolutional filter of size $a \times a$ that connects to ith feature map, C_j is the jth feature map layer.

In the above example, the input image has a size of 7×7 and feature detector has a size of 3×3. For the first iteration, a window size of 3×3 equal to the size of the feature detector is chosen from the input image. Summation of multiplication of each cell of the input image window and feature detector is performed. The result

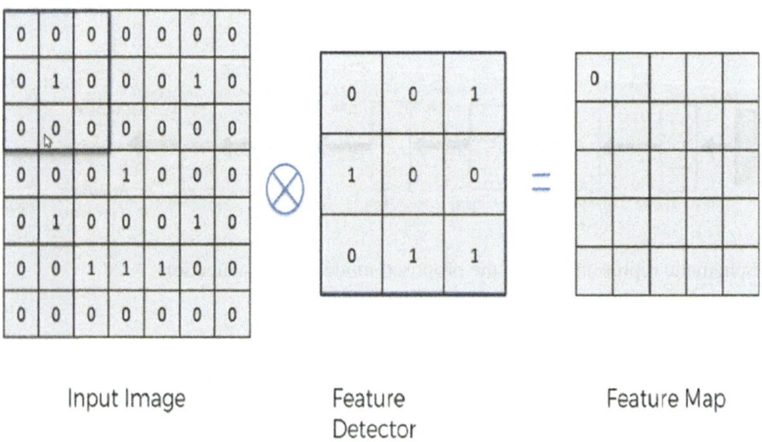

Fig. 3 Feature map generation in convolutional layer (first iteration)

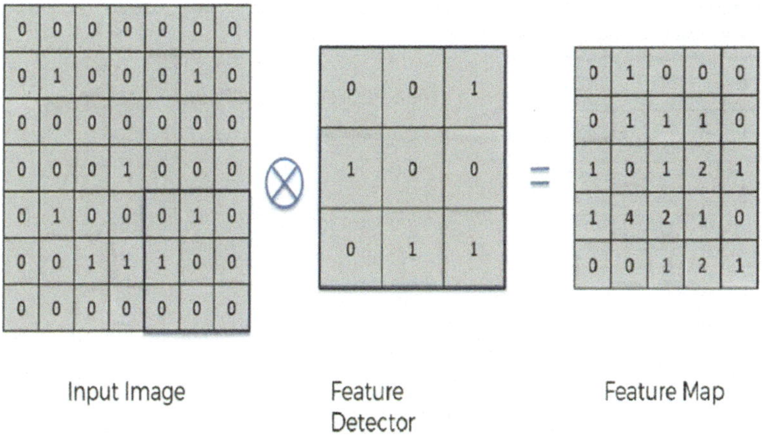

Fig. 4 Feature map generation in convolutional layer (after completion)

is then updated to the first cell of the feature map. Now, the window is shifted to the adjacent 3 × 3 area. The process is continued until we get a feature map for a feature detector. After the completion of this feature detector, we get the feature map as depicted in Fig. 4.

Similarly, *N* feature detectors/filters produce *N* feature maps as output. At the end of the convolution layer, an activation function can be applied for the nonlinear transformation of the outputs of the convolutional layer [5]. That means that we can transform the values of feature maps into the specified range using activation functions like sigmoid, ReLU, tanh, etc.

2.3.2 Pooling

Our input images may not always be in form suitable for us to extract information. It may be distorted or in different orientations. This changes the positions of the features. This problem is called spatial invariance. So, we employ a pooling operation to avoid this issue. Pooling is the solution to overcome spatial invariance and reduce the dimensions of the feature maps [6]. Many pooling operations can be performed like max pooling, min pooling, sum pooling, etc. But, max pooling is used in this paper where the maximum value of each block is the corresponding pixel value of the output image. The pooling operation is performed on a mask size of $b \times b$.

The pooled feature map generation in the first iteration of pooling layer is described in Fig. 5.

In Fig. 5, a mask size of 2 × 2 is chosen on the feature map. The maximum value within that mask is selected for output to the pooled featured map that has reduced dimensions. This process is similarly repeated to all other feature maps to get a list of pooled feature maps. After pooling operation, the pooled feature map outputs as presented in Fig. 6.

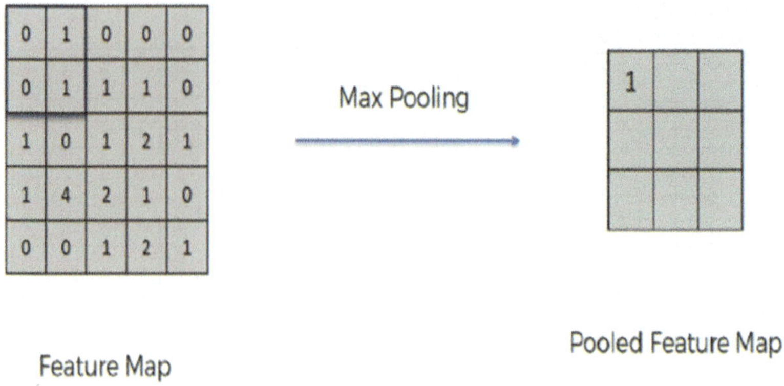

Fig. 5 Pooled feature map generation in pooling layer (first iteration)

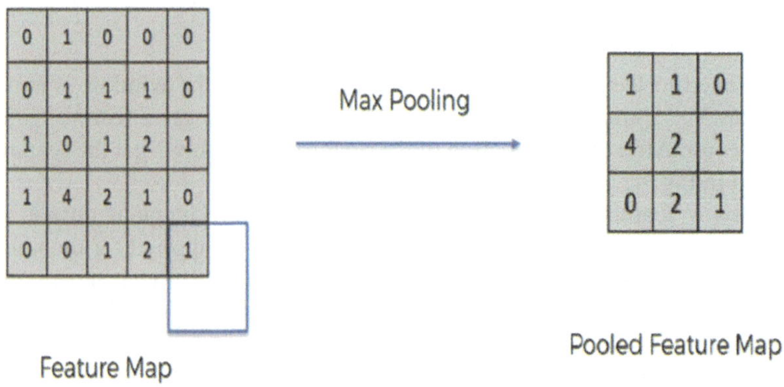

Fig. 6 Pooled feature map generation in pooling layer (after completion)

2.3.3 Flattening

Our pooled feature map is a multidimensional array. So, what it is done in the flattening layer is that the multidimensional array is simply converted to a single-dimensional array. Each cell of the single-dimensional array will contain extracted values in a vector form. The flattened vector generation from pooled feature map is presented in Fig. 7.

2.3.4 Full Connection

After flattening, the extracted vector values are the input to the ANN model. This makes the input layer for the AI model. Similar to traditional neural network models,

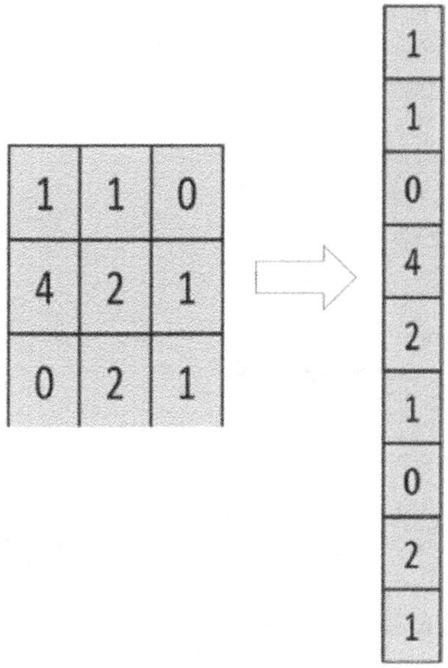

Fig. 7 Flattened vector generation from pooled feature map in the flattening layer

multiple hidden layers are added to the model with one output layer. All the connections between layers are called synapses. Each synapse is associated with a certain weight value.

$$\hat{y}_i = \sum_{i=1}^{M} w_i x_i + b, \tag{2}$$

where w_i is the weight of the ith synapse, x_i is the input value of ith input node, b is the bias, $\hat{y}_i$ is the predicted output. The resultant $\hat{y}_i$ is not in the desired form. Normalization is needed. So, the activation function is used to bring the $\hat{y}_i$ to the desired range. For example, if $\hat{y}_i$ is a negative value, it is needed to be transformed before it can be checked for accuracy. Various activation functions include sigmoid, tanh, ReLU, softmax, etc. Also, to minimize the loss/cost function, optimizers are used during compilation of the model. Optimizers may be of many types like gradient descent (includes batch, stochastic and mini-batch), Adam, Adadelta, etc. After prediction, if the predicted class does not match with the actual class, then the flow of execution needs to be backtracked and weights of synapses are changed. This is called backward propagation in neural networks. After weight adjustment, the input is trained again until the predicted classes match with the actual class. When all the input training images are trained, it is known that the predicted classes of all input images matched with the actual one. So, the model is frozen, which fixes the weight of each synapse finally that cannot be changed. In this way, a trained model is achieved (Fig. 8).

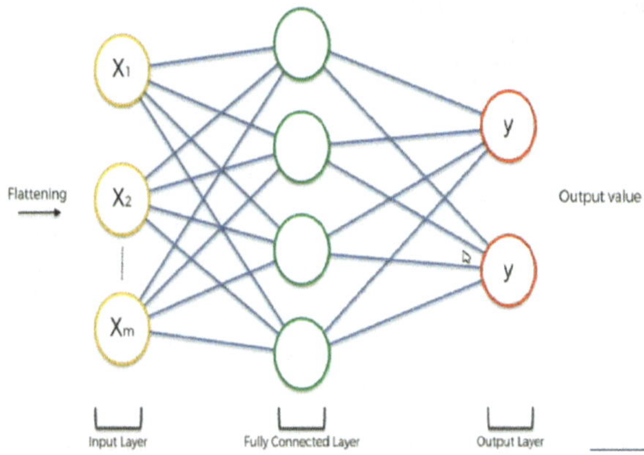

Fig. 8 Architecture of full connection with one hidden layer

3 Experimental Results and Discussion

In CNN, epochs are the number of times the training dataset is trained. An *N* epoch indicates that the training dataset has been trained *N* times. Increasing the epoch size increases, the accuracy as the data is been trained a greater number of times. But, there is a trade-off between the number of epochs and the accuracy. At a certain saturation limit, the accuracy keeps increasing with increasing epochs. Later on, the accuracy decreases. This problem is called overfitting. To reduce overfitting of training data, regularization is done mostly in the form of dropouts, i.e., few redundant nodes in the hidden layers are discarded. The other way out is to reduce the architecture complexity by reducing the number of epochs.

The overall performance of the CNN model is tested using 624 images. Table 1 shows the recognition rates obtained for the CNN using 64 $\times$ 64 pixels as input image size. In Table 1, the training and validation accuracy for different epochs are being listed.

It can be seen from Table 1 that the trained CNN model have different training and validation accuracies with varying epoch size.

Table 2 shows the performance of the proposed CNN model with different types of optimizers to update weights and minimize the loss function.

Out of the five employed optimizers, two of them, Adam and Adadelta perform well in terms of loss and accuracy.

Table 1 Training and testing accuracy for different epochs

Epochs	10	20	30	40	50	60	70
Train accuracy (%)	95.06	96.39	96.83	97.16	97.44	**97.62**	96.41
Testing accuracy (%)	89.30	90.42	92.67	93.34	93.87	**94.51**	91.17

Bold values represent the highest training and testing accuracy found in Epoch 60

Table 2 Training and validation accuracies for different optimizers

	Adam	SGD	RMSprop	Adagrad	Adadelta
Train accuracy (%)	**97.62**	95.33	96.71	95.86	**98.59**
Validation loss	**0.41**	0.44	0.51	0.41	**0.5**

Bold values represent the best training and validation accuracy for Adam optimizer

Table 3 Training and validation accuracy comparison with different state-of-the-art methods

Methods	Training accuracy (%)	Testing accuracy (%)
Our model	**97.62**	94.51
Model 1 [1]	**94.8**	93.0
Model 2 [7]	**78.73**	76.8
Model 3 [2]	**96.6**	95.6

Bold values signifies the training accuracy of our model as well as other state of the art models, which shows better result incomparision to test accuracy

From Table 3, it is apparent that the proposed model for classification is showing better results in comparison to the state of the art methods available in the literature.

4 Conclusion

In this paper, the convolution neural network is designed to classify X-ray images for diagnosis of chest diseases. Different training strategies have been employed to the dataset with different parameters and different numbers of iterations. The best yielding optimizers, activation functions and epoch sizes have been used to train the model. No manual labelling of images in datasets has been required. Moreover, CNN was trained and tested on separated datasets. With pneumonia being the major portion of patient mortality, its early detection is crucial in its treatment. X-ray is the most common medical imaging technique. So, the use of advanced medical imaging algorithms can help reduce mortality related to pneumonia. More medical imaging datasets are needed for research purposes.

References

1. Stephen, O., Sain, M., Maduh, U.J., Jeong, D.-U.: An efficient deep learning approach to pneumonia classification in healthcare. J. Healthc. Eng. **2019**, 1–7 (2019)
2. Avendi, M.R., Kheradvar, A., Jafarkhani, H.: A combined deep-learning and deformable-model approach to fully automatic segmentation of the left ventricle in cardiac MRI. Med. Image Anal. **30**, 108–119 (2016)

3. Joshi, S., Saxena, S., Khanna, N.: First steps toward CNN based source classification of document images shared over messaging app
4. https://www.kaggle.com/paultimothymooney/chest-xray-pneumonia
5. Scherer, D., Muller, A., Behnke, S.: Evaluation of pooling operations in convolutional architectures for object recognition. In: 20th International Conference on Artificial Neural Networks (ICANN), Thessaloniki, Greece, Sept 2010
6. Albawi, S., Mohammed, T.A., Al-Zawi, S.: Understanding of a convolutional neural network. In: 2017 International Conference on Engineering and Technology (ICET), Antalya, Turkey. IEEE (2017)
7. Saul, C.J., Urey, D.Y., Taktakoglu, C.D.: Early diagnosis of pneumonia with deep learning. arXiv preprint arXiv:1904.00937 (2019)

Pragmatic Study of CNN Model and Different Parameters Impact on It for the Classification of Diabetic Retinopathy

Manaswini Jena, Smita Prava Mishra, and Debahuti Mishra

Abstract Diabetic retinopathy, a cause of permanent blindness in diabetic patients is required to be detected in right time to avoid the fatal consequences. Deep learning through convolution neural network is used in this work for the automatic detection of diabetic retinopathy. The convolution neural network model basically depends on its layers of architecture and the parameters like pooling layer, activation function and learning rate which has a huge impact on the model's behaviour. Here, the study is about the response of model according to those parametric values and their types to achieve a better performative model towards the detection of diabetic retinopathy.

Keywords Fundus images · Diabetic retinopathy · Convolution neural network · Pooling layer · Learning rate · Activation function

1 Introduction

Diabetes is a very common disease in almost every country and a widely spread one whose prolonged presence in the body causes even more diseases. Diabetic retinopathy, a reason for permanent blindness is the outcome of it. So, early detection is the only solution and can save the condition of eye to some extent. Regular check-up is always advised for the diabetic and non-diabetic people to stay away from these diseases. For the automatic classification of diabetic retinopathy, many techniques are used out of which deep learning is most commonly used one. Convolution

M. Jena (✉) · D. Mishra
Department of Computer Science and Engineering, Siksha 'O' Anusandhana (Deemed to Be University), Bhubaneswar, Odisha 751030, India
e-mail: manaswinijena@gmail.com

D. Mishra
e-mail: debahutimishra@soa.ac.in

S. P. Mishra
Department of Computer Science and Information Technology, Siksha 'O' Anusandhana (Deemed to Be University), Bhubaneswar, Odisha 751030, India
e-mail: smitamishra@soa.ac.in

D. Mishra et al. (eds.), *Intelligent and Cloud Computing*, Smart Innovation, Systems and Technologies 194,
https://doi.org/10.1007/978-981-15-5971-6_74

Neural Network (CNN), a deep learning network model is specifically architected for handling image data. CNNs and its different variations are designed and have proved their efficiencies in various platforms of image processing [1–5]. Particularly, for the classification of diabetic retinopathy CNN performed well in many experiments [6–10]. The fundamental units of this multilayer perceptron network are local perceptive field, shared weight, bias and pooling [11]. In fully connected neural networks, the input is represented as a vertical line of neurons, but in CNN the input is taken as a square matrix corresponding to the pixel values of the input image, i.e. the original image is directly taken. Each connection from input layer neurons to hidden neurons bears a weight and each node in the hidden layer has a bias. The same weight and bias is used by total hidden neuron matrix, hence called as shared weight and bias. The convolutional layers generate feature maps by the local connectivity and weight sharing. The numbers of parameters are significantly reduced by the weight sharing rule which increases the efficiency and prevents over-fitting. Different layers like pooling layer, dropout layer are used to reduce the dimensionality and finally passing through a fully connected layer and softmax layer it gives the output at last. The parameters and hyper-parameters play an important role for any neural network model [12, 13]. They act like knobs during the model development. These are the pre-determined values which are set manually before starting of the training. The value of learning rate, the type of activation function, the type of pooling layer, etc. affects the model's quality and behaviour according to their types and values.

The working principle to develop a model starts from choosing these parameters, followed by executing the model (training and testing), then evaluating the output and retune the parameters if required. During these steps, we have seen the ups and down of output values, its accuracies, which may happen due to the change in any parameters or change in any values of parameters [14]. The dependencies between the parameters are not fixed for the models and it also depends on the type of data used. In this paper, the behaviour of the model according to the change in parameters and hyper-parameters are elaborated with experimental facts.

This section gives a brief introduction about this paper. Section 2 represents the theoretical background followed by description of the used dataset and the experimental setup along with model details in Sect. 3 and 4, respectively. The analytical study of the result is elaborated in Sect. 5, then finally concluding the paper in Sect. 6.

2 Background Study

CNN is a branch of deep neural networks or deep learning, having a huge applications in every field. Some basic parameters and architectural features are discussed here which affects the behaviour of the CNN model. Pooling layers, activation functions, learning rates are some preliminary parts of CNN which helps to develop a better model by tuning it as per requirement.

Different types of pooling layers are available. The basic functions of pooling layers are (a) Extracting invariant features, (b) Reducing the computational cost by reducing dimensionality and parameter, and (c) controlling over-fitting. It helps for extracting the maximum feature value along with the dominant feature value. The dimension of convolved feature maps reduces consecutively in each pooling step simultaneously by reducing the numbers of parameters, ultimately minimizing the computational time and space. Pooling helps to reduce this over-fitting by selecting a subset of features that helps to avoid the false patterns which may arise due to high number of features. The two most common types of pooling used are *Max Pooling* and *Average Pooling*. In Max pooling, the maximum value from the portion of the image covered by the *Kernel* is generated. Similarly, average pooling returns the average value of all the values from the portion covered by the kernel in the image. Root mean square pooling and min pooling are also used for pooling in CNN.

Similarly, different activation functions are used; out of which non-linear functions are more preferably. Sigmoid function is one of the most common activation functions for neural networks. It is also used in CNN along with hyperbolic tangent function, rectified linear unit function and leaky rectified linear unit function.

A scaling hyper-parameter by which the weight values are updated in each iteration in a learning model is done through the learning rate. It is the value needed to change the weights during training process. The value of it is a small positive value in the range of 0.0–1.0. It may vary differently for different models. During the training through monitoring, the learning rate is adjusted to improve the performance through adaptive learning. This study visualizes the change in accuracy and behaviour of the model with the change in learning rate values.

3 Dataset Description

A new dataset named REHLC dataset is taken for this experiment. The dataset is collected from Ruby Eye Hospital & Lasik Centre (An ISO 9001:2008 Super Speciality Eye Hospital) having Regd. No. 1617/2013 and is located at Berhampur of Ganjam district, Odisha. A fundus camera with optical resolution of 8-14 microns and 40 degrees field of view is used to capture the images. Fundus images of both male and female diabetic patients aged between 30 and 65 are collected without any intervention in the integrity of a person. The data images are separated as diseased or non-diseased image of diabetic retinopathy. It contains 102 diabetic fundus images out of which 51are diseased and 51 are non-diseased.

4 Model and Experimental Setup

The CNN model used here is having four convolution layers, four pooling layers, five activation functional layers, two fully connected layers and a softmax layer at last. The stride for all convolution layer is taken 1×1 and for all pooling layer it is taken 2×2. The images are directly taken as input for the model. For the experiment, 80% data is taken for training and 20% data is taken for testing, i.e. here out of 102 image data, 80 images are taken for training and rest 22 images are used for testing. Each image is of size 2048×1536 and is normalized to 256×256. The experimentation is performed under Linux environment by using Python through Keras and Tensorflow packages.

Relationship between the parameters is unknown and depends on various conditions. The type of data, type of model, values of parameters and type of layers—everything changes with according to experimental state. In this model, three types of pooling layers are considered, i.e. max pooling, min pooling and maxmin pooling. For each pooling, four types of activation functions are used, i.e. sigmoid function, hyperbolic tangent function, ReLU function and Leaky ReLU function. For each again four learning rate values are taken as 0.1, 0.01, 0.001 and 0.0001. Each combinations are tested in different number of epochs.

5 Result Analysis

As described above in the experimental setup each combination is tested with 25 epochs of intervals starting from 25 to 100. Using each one of the pooling types in the model, all parameters are tested. Each separate condition or separate combination of parameters gives a different confusion matrix and from confusion matrix, the accuracy is calculated. These values are stored for analysis which is clearly visualized in the following three-dimensional graphical Figs. 1, 2 and 3. One tabular representation sample is given in Table 1, which shows the result of maxmin pooling layer and LReLU as activation function with different epoch and learning rates. This pooling layer is also tested with other three activation functions as mentioned above. Similarly each pooling layer and each activation function is tested with each different epochs and learning rates. The changes in the learning rate values, the changes in the activation functions and the changes in epoch number clearly reflected their changes in output in every combination. In the x-axis number of iterations or epochs is taken, while the y-axis represents the learning rate values and finally, the accuracy is taken in z-axis. Four colours are used in the bar to represent four activation functions as mentioned in the graphs, i.e. navy blue represents ReLU activation function, sky blue represents Leaky ReLU, green represents tanh activation function and yellow is used for sigmoid activation function. Figures 1, 2 and 3 represent the three-dimensional

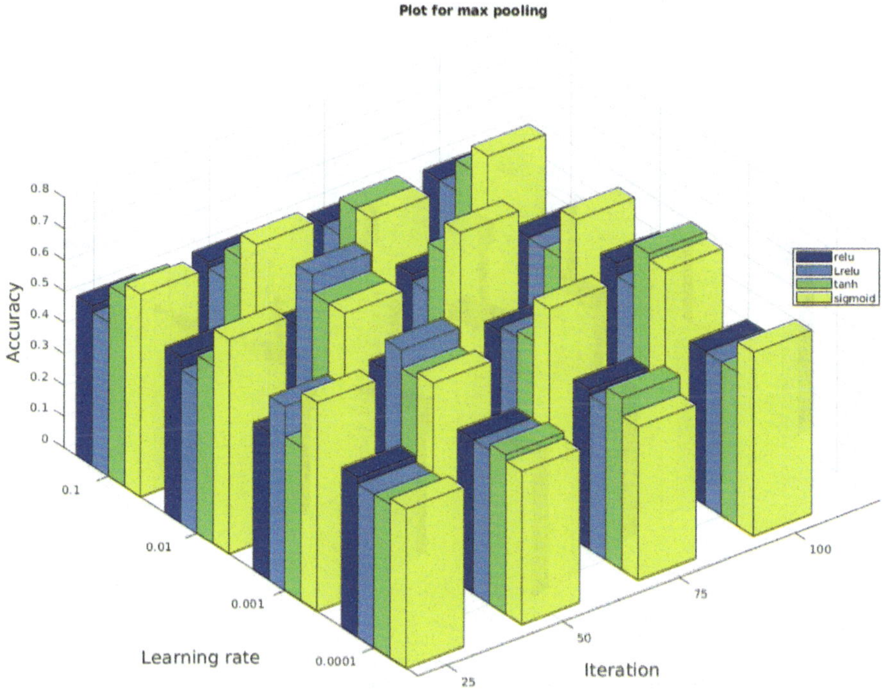

Fig. 1 Graphical representation of the result using max pooling and other parameters in the model

bar graph representation of accuracy values in the CNN model using max pooling, min pooling and maxmin pooling, respectively, along with other parameters.

The ups and downs in the colour bars show how the accuracy value increases or decreases with the change in learning rate and activation functions along with the change in number of iterations. Dependency between these parameters (*learning rate, activation functions*) and any classifier is not defined anywhere. It can only be observed by performing various experiments. This is clear from the figures that not a single dependency between parameters or models or among the type of image data are fixed. Each and every combinational condition affects the behaviour of the model in its own way.

6 Conclusion

This paper represents a pragmatic analysis about the behaviour of a CNN model, which changes with the change in values of learning rate, change in the type of activation functions and change in iterations. With modification in some values the classification accuracy increases, but not always. Some changes in parametric values or sometimes the increase in epochs may also decrease the performance of the model.

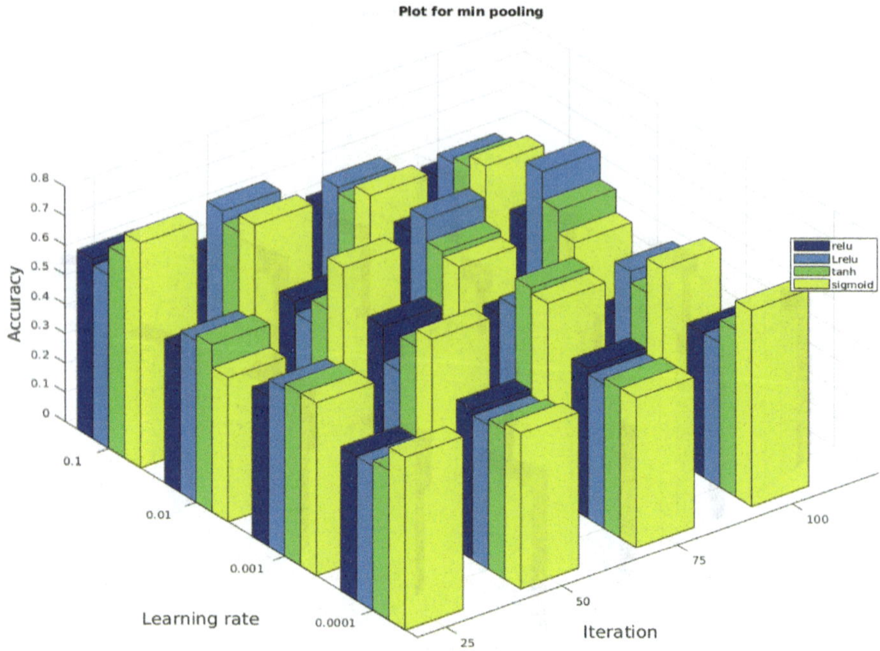

Fig. 2 Graphical representation of the result using min pooling and other parameters in the model

Here, the model's behaviour is studied by using the fundus image in CNN for diabetic retinopathy classification. So for this particular image type, the model's capacity to be a good classifier can be achieved through selecting particular combinations and using them at higher iterations. In this experiment, the maxmin pooling gives better accuracy result at 0.0001 learning rate and tanh activation function compared to other combinations.

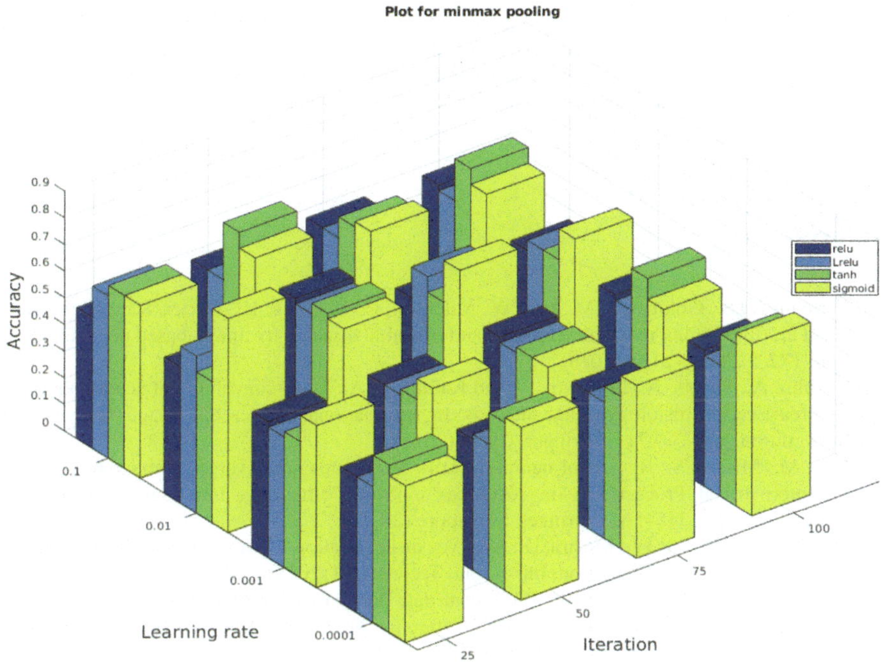

Fig. 3 Graphical representation of the result using maxmin pooling and other parameters in the model

Table 1 Result table for maxmin pooling layer with Leaky ReLU activation function

Pooling type	Activation function	Epoch	Learning rate	Accuracy
Maxmin	Lrelu	25	0.1	0.490196
Maxmin	Lrelu	25	0.01	0.607843
Maxmin	Lrelu	25	0.001	0.647059
Maxmin	Lrelu	25	0.0001	0.647059
Maxmin	Lrelu	50	0.1	0.509804
Maxmin	Lrelu	50	0.01	0.509804
Maxmin	Lrelu	50	0.001	0.72549
Maxmin	Lrelu	50	0.0001	0.666667
Maxmin	Lrelu	75	0.1	0.509804
Maxmin	Lrelu	75	0.01	0.509804
Maxmin	Lrelu	75	0.001	0.588235
Maxmin	Lrelu	75	0.0001	0.607843
Maxmin	Lrelu	100	0.1	0.509804
Maxmin	Lrelu	100	0.01	0.490196
Maxmin	Lrelu	100	0.001	0.647059
Maxmin	Lrelu	100	0.0001	0.588235

Acknowledgements This research was partially supported by Ruby Eye Hospital & Lasik Centre. We thank Dr. B.N.R. Subudhi (Retd. Professor and HOD of Dept. of Opthalmology, M.K.C.G. Medical College, Berhampur), Dr. Praveen Subudhi and Technician Bhabani Sankar Rath for their precious time to provide the fundus images data for the experiment.

References

1. Kermany, D.S., Goldbaum, M., Cai, W., Valentim, C.C., Liang, H., Baxter, S.L., McKeown, A., et al.: Identifying medical diagnoses and treatable diseases by image-based deep learning. Cell, **172**(5), 1122–1131 (2018)
2. Rakhlin, A., Shvets, A., Iglovikov, V. and Kalinin, A.A.: Deep convolutional neural networks for breast cancer histology image analysis. In: International Conference Image Analysis and Recognition, pp. 737–744. Springer, Cham (2018)
3. Jena, M., Mishra, S.: Review of neural network techniques in the verge of image processing. In: International Proceedings on Advances in Soft Computing, Intelligent Systems and Applications, pp. 345–361. Springer, Singapore (2018)
4. Jena, M., Mishra, S.P. and Mishra, D.: A survey on applications of machine learning techniques for medical image segmentation. Int. J. Eng. Technol. **10** (2018)
5. Jena, M., Dash, R., Misra, B.B.: Biological data analysis using hybrid functional link artificial neural network. In: International Conference on Swarm, Evolutionary, and Memetic Computing, pp. 88–97. Springer, Cham (2014)
6. Brown, J.M., Campbell, J.P., Beers, A., Chang, K., Ostmo, S., Chan, R.P., Dy, J., et al.: Automated diagnosis of plus disease in retinopathy of prematurity using deep convolutional neural networks. JAMA Ophthalmol. **136**(7), 803–810 (2018)
7. Jena, M., Mishra, S.P., Mishra, D.: Detection of diabetic retinopathy images using a fully convolutional neural network. In: 2018 2nd International Conference on Data Science and Business Analytics (ICDSBA), pp. 523–527. IEEE (2018)
8. Xu, Kele, Feng, Dawei, Mi, Haibo: Deep convolutional neural network-based early automated detection of diabetic retinopathy using fundus image. Molecules **22**(12), 2054 (2017)
9. Dutta, S., Manideep, B.C., Basha, S.M., Caytiles, R.D., Iyengar, N.C.: Classification of diabetic retinopathy images by using deep learning models. Int. J. Grid Distrib. Comput. **11**(1), 89–106 (2018)
10. Li, X., Pang, T., Xiong, B., Liu, W., Liang, P. and Wang, T.: Convolutional neural networks based transfer learning for diabetic retinopathy fundus image classification. In: 2017 10th International Congress on Image and Signal Processing, BioMedical Engineering and Informatics (CISP-BMEI), pp. 1–11. IEEE (2017)
11. Ashqar, B.A., Abu-Naser, S.S.: Image-Based tomato leaves diseases detection using deep learning. (2018)
12. Van Rijn, J.N., Hutter, F.: Hyperparameter importance across datasets. In: Proceedings of the 24th ACM SIGKDD International Conference on Knowledge Discovery and Data Mining, pp. 2367–2376. ACM (2018)
13. Probst, P., Bischl, B., Boulesteix, A.L.: Tunability: importance of hyperparameters of machine learning algorithms. arXiv preprint arXiv:1802.09596 (2018)
14. Albelwi, Saleh, Mahmood, Ausif: A framework for designing the architectures of deep convolutional neural networks. Entropy **19**(6), 242 (2017)

Facial Expression Recognition System (FERS): A Survey

Sonali Mishra, Ratnalata Gupta, and Sambit Kumar Mishra

Abstract Human facial expressions and emotions are considered as the fastest way of the communication medium for expressing thoughts. The ability to identify the emotional states of people surrounding us is an essential component of natural communication. Facial expression and emotion detector can be used to know whether a person is sad, happy, angry, and so on. We can better understand the thoughts and ideas of a person. This paper briefly explores the idea of recognizing the computerized facial expression detection system. First, we have discussed an overview of the Facial Expression Recognition System (FERS). Also, we have presented a glimpse of current technologies that are used for the detection of FERS. A comparative analysis of existing methodologies is also presented in this paper. It provides the basic information and general understanding of up to date state-of-the-art studies, also experienced researchers can look productive directions for future work.

Keywords Facial expressions · CNN · Deep CNN · FERS

1 Introduction

Facial expression is one of the non-verbal communications for humans to convey their message. It plays a significant role in our daily life communication. Human Facial expression and emotions play an important way for human being to socialize. They are the major channels of conveying social information among individuals. Facial expression recognition is a complex and challenging problem. Therefore, automatic

S. Mishra (✉) · S. K. Mishra
Siksha 'O' Anusandhan (Deemed to be University), Bhubaneswar, Odisha 751030, India
e-mail: smishra.nitrkl@gmail.com

S. K. Mishra
e-mail: skmishra.nitrkl@gmail.com

R. Gupta
Oriental University, Indore, India
e-mail: ratnalata@gmail.com

D. Mishra et al. (eds.), *Intelligent and Cloud Computing*,
Smart Innovation, Systems and Technologies 194,
https://doi.org/10.1007/978-981-15-5971-6_75

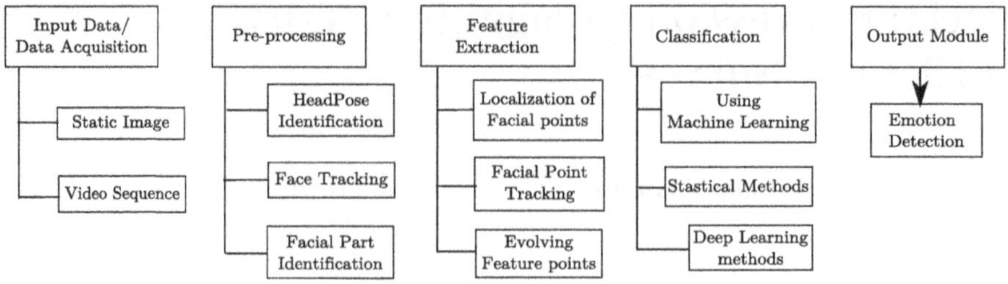

Fig. 1 Basic steps required for facial expression detection system

facial expression recognition System (FERS) is important nowadays and use many applications such as Monitoring, E-learning, Medicine, Entertainment, and Law [1].

It is challenging to implement FERS in some areas like human–computer interaction as in healthcare [2], education [3], industry [4], entertainment, law, and crime. Numerous traditional FER system requires manual feature selection methodologies like Histogram of Oriented Gradients (HoG), LBP, SIFT techniques, etc.) and then feed to a custom-designed classifier to classify expressions. However, such methods fail to produce accurate results because the attributes are extracted manually and also become difficult when the dataset is large. In general, FERS operates in three stages: preprocessing, feature extraction, and classification. Each stage of FERS has its own importance as well as complexity in the final decision step. The investigation of facial expression is the fundamental research area of psychology, and a large volume of research works has been published in recent times. However, the performance still requires potential improvement in order to meet the real-time scenario. Figure 1 describes the basic steps for the facial expression recognition system.

The rest of paper is summarized in the following way. Section 2 briefly describes the past researches carried out on Facial Expression Recognition System (FERS). Section 3 introduces the details about available datasets used for FERS. A comparison of results of the past methods is given in Sect. 4. Finally, Sect. 5 concludes the paper.

2 Related Work

In the past few years, different specialists have pulled into an automated computerized system [5] and added a different range of works until the present day in the area of FERS. Many researchers worked to overcome the problem of recognizing facial expressions. Also, research work has been conducted in the direction of creating an automated facial expression recognition system for the real-time environment. Xianzhang et al. [6] have suggested a trainable Deep Neural Network (DNN)-based frame that combines spatial and temporal information from the video. The proposed technique demonstrated an accuracy of 65.72% 42.98% on RML and eNTERFACE05 dataset, respectively. The drawback of the suggested framework is that it has a huge parameter lists, and it takes higher computation time.

Minaee and Abdolrashidi [7] have proposed an end-to-end deep learning approach based on convolutional network. Experimental results obtained on FERG, FER-2013, JAFFE, and CK+ datasets denote an overall accuracy of 92.8% , 70.0%, 92.8%, and 98.0%, respectively.

Mokhayeri and Granger [8] have proposed a new technique to attain a high level of performance in video-based Face Recognition (FR). This reconstructs a probe image utilizing an auxiliary variational dictionary and an augmented gallery dictionary. Albrici et al. [9] presented a framework that exhibits automated facial expression recognition in videos. The pre-trained CNN model in a short video clip is dedicated to analyzing the video frames. Then a LSTM network is designed to process the trajectories of the facial landmarks. The accuracy recorded for CK+ dataset was 97.4%, for MUG Dataset 95.5%, and self-created dataset was found to be 80.2%.

Zhang et al. [10] have designed a hybrid deep learning model for FER in video sequences to learn the features affectively. Their proposed technique demonstrated an accuracy of 55.85%, 73.73%, and 71.43% on BAUM-1s, RML,and MMI, dataset, respectively. The drawback of this model is that it has to be developed as a real-time FER. Liu et al. [11] have proposed an emotion recognition techniques which combine Audio-Video and facial features in both non-temporal and temporal mode. RAF and FER2013 are the image datasets used to pre-train CNN models. Their proposed technique demonstrated an accuracy of 61.87%.

Xu et al. [12] have proposed an unsupervised auto-coder and Dempster-Shafer theory fusion method to combine different prediction results based on features. The entire dataset is divided into training set, validation set, and test set. The average accuracy of the model was 59.68

Zeng et al. [13] have developed a combination of geometric and appearance characteristics of the face and are introduced to identify facial expressions due to possessing accurate and comprehensive information of emotions. For CohnKanade(CK+) database, the accuracy recorded was 95.79%. Sun et al. [14] have suggested a method to search three kinds of active regions, i.e., left eye regions, right eye regions, and mouth regions. CNN is used to extract information and classify expressions. Experiments are carried out on the CK+ database, JAFFE database, and NVIE database. The accuracy recorded was 95.36%, 96.57%, and 89.38%, respectively.

3 Dataset Used

There are sufficient data available on a public platform that incorporates many categories of possible situations for developing a Facial Expression Recognition System (FERS). This section aims at describing the publicly accessible datasets used for designing FERS. Tables 1 and 2 present a summary of these datasets, consisting of the primary reference, number of cases, number of image or video samples along with the links to access those datasets.

Table 1 An overview of Image-based FERS datasets

Dataset	Samples	Subject	Expression distribution	Access
JAFFE [15]	213 images	10	6 basic and 1 natural expression	http://www.kasrl.org/jaffe.html
FER-2013 [16]	35,887 images	N/A	6 basic and 1 natural expression	https://www.kaggle.com/c/challenges-in-representation-learning-facial-expression-recognition-challenge
FERA [7]	188	10	5 expression	https://ibug.doc.ic.ac.uk/resources/FERA15/
FERG [7]	55,767	6	6 basic and 1 natural expression	https://computervisiononline.com/dataset/1105138868
RaFD [17]	1608	67	6 basic and 1 natural expression	http://www.socsci.ru.nl:8180/RaFD2/RaFD
AffectNet [18]	450,000 images	NA	6 basic and 1 natural expression	http://mohammadmahoor.com/databases-codes/

Table 2 An overview of video-based FERS datasets

Dataset	Samples	Subject	Expression distribution	Access
CK+[19]	593 image sequence	123	8 expressions	http://www.consortium.ri.cmu.edu/ckagree/
MMI [9]	740 images and 2900 videos	25	6 basic and 1 natural expression	https://mmifacedb.eu/
AFEW 7.0 [20]	1809 videos	NA	6 basic and 1 natural	https://sites.google.com/site/emotiwchallenge/
Oulu-CASIA [21]	2880 image sequences	80	6 basic expressions	http://www.cse.oulu.fi/CMV/Downloads/Oulu-CASIA
BAUM-1s [10]	1222 video clips	31	6 basic expressions and 4 mental state	https://archive.ics.uci.edu/ml/datasets/BAUM-1
eNTERFACE05 [6]	1290 videos	43	6 expressions	

Table 3 Comparative analysis of existing methods

	Classifiers	Dataset	Accuracy rate (%)
Pan et al. [6]	HMM,GMM,SVM	RML	65.72
		eNTERFACE05	42.98
Minaee et al. [7]	CNN	JAFFE	92.8
		CK+	98.0
Mokhayeri et al. [8]	SRC	Chokepoint dataset	
		pAUC(20%)	0.8820.018
		AUPR	0.7450.019
		COX-S2V database	
		pAUC(20%)	0.8950.020
		AUPR	0.7660.017
Porcu et al. [22]	SVM	Webcam recordings	84:6
Albrici et al. [9]	Auto encoder and SOM	MUG	95.5
		CK+	97.4
Zhang et al. [10]	Linear SVM	BAUM-1s	55.85
		RML+	73.73
Sun et al. [14]	CNN	CK+	95.36
		JAFFE	96.57
		NVIE	89.38

4 Discussion and Comparison of Results

This section briefly discusses the recently developed methods in the field of FERS. The comparison of different techniques in various datasets is presented in Table 3. Experimental results showed in Table 3 indicate that the performance of Convolutional Neural Network (CNN) obtained higher accuracy on CK+ dataset. We have also highlighted the recently proposed methods. Deep convolutional neural networks have been used for feature learning in face recognition. Figure 2 represents the accuracy comparison of different methods on CK+ Dataset.

5 Conclusion

We have presented an overview of past developments in Human facial expression and emotion detection. Several challenges can be taken care of in the field of an automatic emotion detection system. Some problems are such as (1) pose and frequent head movements, (2) Presence of structural components, (3) Occlusion, (4) Image orientation, (5) Imaging conditions, (6) Subtle facial deformation, and (7) Ambiguity and uncertainty in face motion measurement. This concept must lead to a

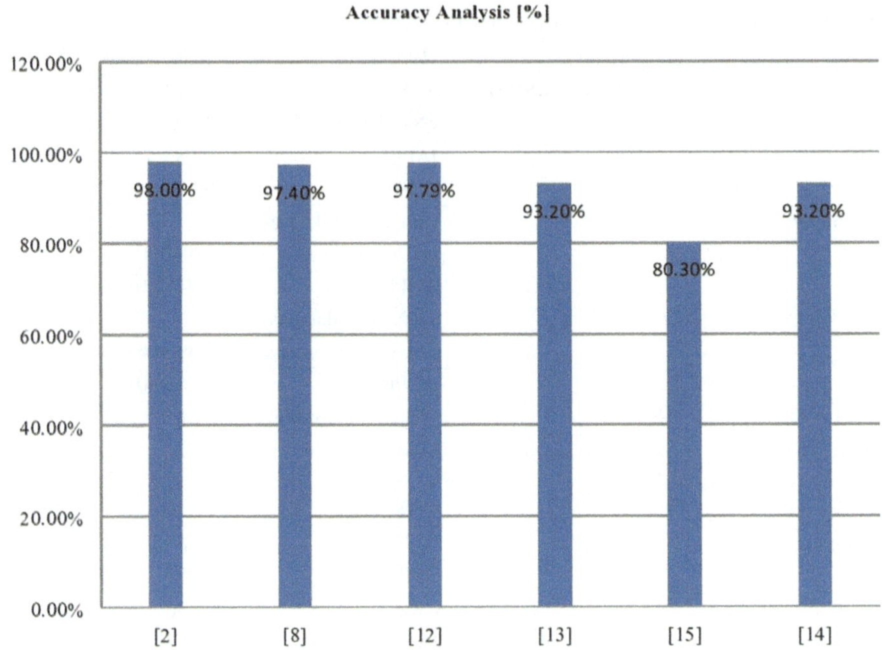

Fig. 2 Performance comparison of various methods on CK+ dataset

unique algorithm for dealing with the above problems. We have proposed model for recognition of FER and emotion recognition in video segments via a Horn Schunk method for spatial and temporal features for fine-tuned CNNs, respectively. Further, newly developed methodologies can be used to design a more reliable technique for FERS in a better way.

References

1. Dong, J., Zheng, H., Lian, L.: Dynamic facial expression recognition based on convolutional neural networks with dense connections. In: 24th IEEE International Conference on Pattern Recognition (ICPR), pp. 3433–3438 (2018)
2. Girard, J.M., Cohn, J.F., Mahoor, M.H., Mavadati, S.M., Hammal, Z., Rosenwald, D.P.: Nonverbal social withdrawal in depression: evidence from manual and automatic analyses. Image Vision Comput. **32**(10), pp. 641–647 (2014)
3. McDaniel, B., D'Mello, S., King, B., Chipman, P., Tapp, K., Graesser, A.: Facial features for affective state detection in learning environments. In: Proceedings of the Annual Meeting of the Cognitive Science Society, vol. 29 (2007)
4. Assari, M.A., Rahmati, M.: Driver drowsiness detection using face expression recognition. In: IEEE International Conference on Signal and Image Processing Applications (ICSIPA), pp. 337–341 (2011)

5. Mishra, S., Majhi, B., Sa, P.K.: A survey on automated diagnosis on the detection of Leukemia: a hematological disorder. In: 3rd IEEE International Conference on Recent Advances in Information Technology (RAIT), pp. 460–466 (2016)
6. Pan, X., Ying, G., Chen, G., Li, H., Li, W.: A deep spatial and temporal aggregation framework for video-based facial expression recognition. IEEE Access **7**, 48807–48815 (2019)
7. Minaee, S., Abdolrashidi, A.: Deep-emotion: facial expression recognition using attentional convolutional network. arXiv preprint arXiv:1902.01019 (2019)
8. Mokhayeri, F., Granger, E.: Robust video face recognition from a single still using a synthetic plus variational model. In: 14th IEEE International Conference on Automatic Face and Gesture Recognition (FG 2019), pp. 1–8) (2019)
9. Albrici, T., Fasounaki, M., Salimi, S.B., Vray, G., Bozorgtabar, B., Ekenel, H.K., Thiran, J.P.: G2-VER: Geometry guided model ensemble for video-based facial expression recognition. In: 14th IEEE International Conference on Automatic Face and Gesture Recognition (FG 2019), pp. 1–6 (2019)
10. Zhang, S., Pan, X., Cui, Y., Zhao, X., Liu, L.: Learning affective video features for facial expression recognition via hybrid deep learning. IEEE Access **7**, 32297–32304 (2019)
11. Liu, C., Tang, T., Lv, K., Wang, M.: Multi-feature based emotion recognition for video clips. In: Proceedings of the 2018 on International Conference on Multimodal Interaction, pp. 630–634. ACM (2018)
12. Xu, Q., Sun, B., He, J., Rong, B., Yu, L., Rao, P.: Multimodal facial expression recognition based on dempster-Shafer Theory fusion strategy. In: 2018 First Asian Conference on Affective Computing and Intelligent Interaction (ACII Asia) (pp. 1-5). IEEE (2018)
13. Zeng, N., Zhang, H., Song, B., Liu, W., Li, Y., Dobaie, A.M.: Facial expression recognition via learning deep sparse autoencoders. Neurocomputing **273**, 643–649 (2018)
14. Sun, A., Li, Y., Huang, Y.M., Li, Q., Lu, G.: Facial expression recognition using optimized active regions. Hum. Centric Comput. Inf. Sci. **8**(1), 33 (2018)
15. Lyons, M., Akamatsu, S., Kamachi, M., Gyoba, J.: Coding facial expressions with gabor wavelets. In: Proceedings 3rd IEEE International Conference on Automatic Face and Gesture Recognition, pp. 200–205 (1998)
16. Goodfellow, I.J., Erhan, D., Carrier, P.L., Courville, A., Mirza, M., Hamner, B., Cukierski, W., Tang, Y., Thaler, D., Lee, D.H., Zhou, Y.: Challenges in representation learning: a report on three machine learning contests. In: International Conference on Neural Information Processing, pp. 117–124. Springer, Berlin, Heidelberg (2013)
17. Langner, O., Dotsch, R., Bijlstra, G., Wigboldus, D.H., Hawk, S.T., Van Knippenberg, A.D.: Presentation and validation of the Radboud faces database. Cogn. Emot. **24**(8), 1377–1388 (2010)
18. Mollahosseini, A., Hasani, B., Mahoor, M.H.: Affectnet: a database for facial expression, valence, and arousal computing in the wild. IEEE Trans. Affect. Comput. **10**(1), 18–31 (2017)
19. Lucey, P., Cohn, J.F., Kanade, T., Saragih, J., Ambadar, Z., Matthews, I.: The extended cohn-kanade dataset (ck+): a complete dataset for action unit and emotion-specified expression. In: IEEE Computer Society Conference on Computer Vision and Pattern Recognition-Workshops, pp. 94–101 (2010)
20. Dhall, A., Ramana Murthy, O.V., Goecke, R., Joshi, J., Gedeon, T.: Video and image based emotion recognition challenges in the wild: Emotiw 2015. In: Proceedings of the ACM on International Conference on Multimodal Interaction, pp. 423–426 (2015)
21. Zhao, G., Huang, X., Taini, M., Li, S., PietikInen, M.Z.: Facial expression recognition from near-infrared videos. Image Vision Comput. **29**(9), 607–619 (2011)
22. Porcu, S., Uhrig, S., Voigt-Antons, J. N., Mller, S., & Atzori, L.: Emotional impact of video quality: self-assessment and facial expression recognition. In: IEEE 11th International Conference on Quality of Multimedia Experience (QoMEX), pp. 1–6 (2019)

A Clustering Mechanism for Energy Efficiency in the Bottleneck Zone of Wireless Sensor Networks

Dibya Ranjan Das Adhikary, Swagatika Tripathy, Dheeresh K. Mallick, and Chandrashekhar Azad

Abstract The bottleneck zone largely affects the overall performance of wireless sensor networks (WSNs). Research has shown that this issue can be handled to a large extent by effective load distribution strategies. This works aims at determining an efficient clustering scheme to achieve a better load distribution. A clustering scheme can effectively destitute load if a load of a particular area can be determined. So, the proposed approach creates an efficient clustering structure by selecting the energy-efficient node as the cluster head and then determines a cluster range appropriate by taking note of loads of its neighborhood. The simulation results showed that the proposed approach significantly improves the throughput and lifetime of the network.

Keywords Wireless sensor networks · Clustering · Fuzzy logic · Bottleneck

1 Introduction

Sensor nodes are powered by battery as the only source of energy and in most applications of WSN, the network left unattended due to the hostile environmental condition of the surroundings [1]. In the perplex of such situations, it is almost impossible to recharge or replace the batteries of the sensor nodes, whereas the surveillance nature of many sensor network applications requires a prolonged network lifetime [2]. So, for such applications, the network must thrive for the longest possible time with the

D. R. D. Adhikary (✉)
Siksha 'O' Anusandhan (Deemed to Be University), Bhubaneswar, Odisha 751030, India
e-mail: dibyadasadhikary@soa.ac.in

S. Tripathy
GIET University, Gunupur, Odisha, India

D. K. Mallick
Birla Institute of Technology, Mesra, Ranchi, Jharkhand, India

C. Azad
National Institute of Technology, Jamshedpur, Jharkhand, India

D. Mishra et al. (eds.), *Intelligent and Cloud Computing*,
Smart Innovation, Systems and Technologies 194,
https://doi.org/10.1007/978-981-15-5971-6_76

initial amount of battery power. Therefore optimization of energy consumption is the most critical research issue concerning WSNs.

Sensors consume a large portion of its energy in data communication [3]. Energy consumption in data transmission has a direct correlation with distance [4]. Thus, to decrease energy consumption, the communication distance is divided into multiple hops, and routing is performed in a multihop manner [5]. In order to provide infrastructure for energy-efficient multihop routing, data aggregation, and duty cycling, a popular mechanism known as clustering is used in WSNs [6, 7]. However, the clustering algorithms that were initially proposed, divided the network into relatively identical sized clusters thus creates an unstable energy consumption structure. Unequal clustering in such a situation is instrumental in balancing the load [8]. It creates clusters with load solely proportional to the distance from the base station. However, in non-uniformly deployed WSNs or WSNs deployed in an unfavorable terrain there exist some portions of the network which are heavily used for relaying traffics. These portions of the network are called the bottleneck zone. Due to heavy relay traffic in the bottleneck zone, the traditional unequal clustering with a multihop routing approach tends to deplete the energy of the nodes quite faster, thus require an efficient solution. This motivates to investigate an efficient unequal clustering with multihop routing approach to extend the network lifetime of WSNs with bottleneck zone.

The cause of the bottleneck zone problem in WSNs can be linked to increased congestion. Direct diffusion [9] is the first protocol to address this issue. In this protocol, the base station generates a gradient path by taking inputs from the sensor nodes and the base station selects a suitable path to minimize the congestions. Similarly, RCRT [10] is another centralized approach that minimizes congestion by maintaining a steady data rate. I2MR [11] is another such approach that resolves congestion by backpressure and multipath routing. TADR [12] is another centralized approach that minimizes the load of the congested node by diverting the traffic through idle or underloaded nodes. Hull et al. [13] proposed a scheme that uses the data rate limit, hop-to-hop flow control, and prioritization to address the congestion problem at the node level. CODA [14] is another such approach that uses back pressure to notify congestion and it uses a closed-loop multi-source regulation approach to minimize congestion. DPCC [15] is another approach that uses a rate control mechanism and dynamic back-off interval to mitigate the congestion. Similarly, ECODA [16] is another mechanism that uses a flexible priority scheme to resolve congestion. Meera et al. [17] proposed a dual queue-based cross-layer approach to mitigate the congestion at the node level. A fast congestion control [18] mechanism is proposed for mobile WSNs that make an efficient next-hop selection based on minimum queuing delay and a gravitational search-based optimal route selection to minimize the effect of congestion. Similarly, a multi-objective-based PSOGSA [19] was proposed that used rate optimization and priority-based transmission to control the congestion.

In all these approaches [9–19], the research is primarily focused on congestion arises due to traffic only. However, in a bottleneck zone, the congestion is arising due to the position of the nodes. To address this concern, an unequal clustering with a multihop routing approach is proposed.

The rest of the manuscript is organized as follows: In Sect. 2 the proposed approach has been thoroughly discussed followed by the performance analysis in Sect. 3. Finally, the finding has been concluded in Sect. 4.

2 Proposed Approach

The operations of the proposed approach are conducted on a round basis. In every round first, the clusters are formed than the sensed information is transmitted in a multihop manner. So, the overall operations of the proposed approach can be roughly divided into three phases: the neighborhood discovery phase, cluster formation phase, and data transmission phase. The neighborhood discovery phase is required only once at the start of the first round of operation. This phase begins with a Hello Message transmitted from the base station in the entire network. On receiving that message every node then calculates their distance from the base station. After that, every node transmits a reply message in its cluster range (R_{max}). On receiving a reply message node add the sender as a neighbor and record the neighborhood information.

2.1 Cluster Formation Phase

After getting the neighborhood information, the operations enter the next phase. The main goal in this phase is to select the most appropriate nodes for the role of cluster head (CH) and decide an efficient cluster range for each CH to tackle the bottleneck zone problem. The main aim of the proposed approach is to create more clusters with small clustering range in the bottleneck zone so that the relay load can be distributed. This phase is termed as cluster formation phase, where CHs are selected and the clustering range of the selected CHs is decided. The suitability of the node for the role of CH is proportional to the nodes remaining energy. So in this approach, every node waits for a variable amount of time named WaitTime before broadcast the Candidate CH Message. The calculation of WaitTime has been accomplished using quotation (1):

$$\text{WaitTime} = \max\left(\frac{1}{\text{Remaining Energy}}, T\max\right) \tag{1}$$

where T_{max} is the threshold set to put a limit to the waiting time. When the WaitTime is over, the candidate CH broadcast a Candidate CH Message consists of node id and its remaining energy, in its Actual Cluster Range (ACR). The calculation of ACR is outlined in the subsequent section. If a node that already sent a Candidate CH Message received one with higher remaining energy, it will quit the election by sending a Quit Election Message. The remaining candidate CH then send a CH

Message consists of node id and its distance to the base station, in the maximum clustering range ($R_{\max}$). The remaining node than joining the nearest CH by sending a Membership Message. After the formation of the clusters, each CH prepares a TDMA schedule for collecting data from the member nodes and broadcast it. This process guarantees that the nodes with the highest remaining energy became the CHs for the current round. This process of cluster formation has been illustrated in Algorithm 1.

Algorithm 1: The Cluster Formation Process

Input: Remaining Energy
Output: S_i as a CH or a Member Node

$S_i = \{i : i \in Sensor\ node\ set\ (S)\}$
$Nbr(S_j) = \{S_j : S_j is\ within\ the\ Transmission\ Range\ of\ S_i\}$
$R_i = \{Cluster\ range\ of\ node\ S_i\}$
$R_{max} = \{Maximum\ Cluster\ range\}$
Ere=Remaining Energy
Dbs=Distance to Base Station

1. S_i *calculate its WaitTime using equation(1).*
2. *Set CAN_CH_flag=True.*
3. *wait (WaitTime) before broadcasting a Candidate CH Message.*
4. *Set* R_i *= Actual Cluster Range (ACR) as calculated using equation (2).*
5. *if(WaitTime ==0)then*
6. *Broadcast Candidate CH Message (node_id, Ere) in* R_i;
7. *Listen channel to receive Candidate CH Message from* $Nbr(S_j)$;
8. *if (Ere* (S_i) *<Ere* (S_j)*)then*
    Broadcast Quit Election Message(node_id);
    CAN_CH_flag=False;
9. *end if*
10. *end if*
11. *if(CAN_CH_flag==True)then*
12. *Broadcast CH Message(node_id, Dbs) in* R_{max};
13. *end if*
14. *if(CAN_CH_flag==False)then*
15. *Listen channel to receive CH Message;*
16. *if(count(CH Message)==0)then*
    Broadcast CH Message(node_id, Dbs) in R_{max};
17. *else*
    Send Member Message(node_id, CH_id) to min distant CH;
18. *end if*
19. *end if*

The cluster ranges of CHs are very crucial in optimizing the energy consumption in a bottleneck zone. The proposed approach aims at creating more clusters in the bottleneck zone so that the relay load can be distributed. Bottleneck zone is the portion of the WSNs that relay more packets then the rest of the network. It is quite likely that the average remaining energy of the bottleneck zone will be quite less

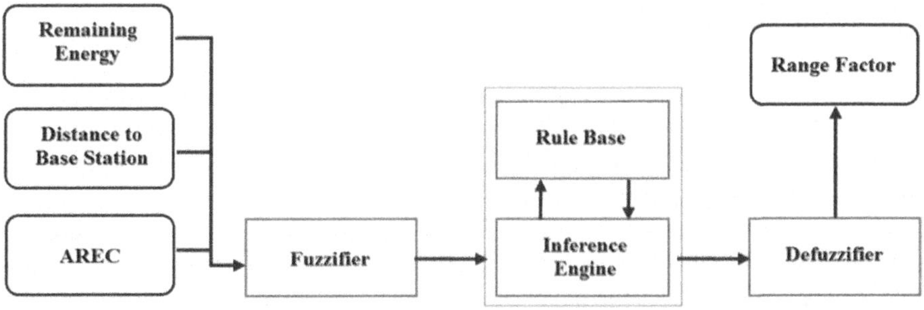

Fig. 1 Block diagram of the FIS used

than the rest of the network. So, when the cluster formation process is over, every member node will transit their remaining energy to their CHs along with the data so that the CHs can calculate the average remaining energy of its cluster and pass this information to the member nodes.

So, to select an efficient cluster range, in this proposed approach three attributes, namely, remaining energy, distance to the base station, and Average Remaining Energy of Cluster (AREC) is used. Further to handle any ambiguity and uncertainty in cluster range selection a Mamdani method [20] based fuzzy inference systems (FIS) has been used. The detailed outline of the FIS is provided in Fig. 1.

The input variables of the FIS are remaining energy, distance to the base station, and AREC. The only output of the FIS is range factor. The input variables are defined as follows:

- Remaining Energy (E_{re})—Energy level remaining of the node.
- Distance to Base Station (D_{bs})—Nodes distance (Euclidean distance) to Base Station.
- Average Remaining Energy of Cluster (AREC)—Average remaining energy as determined by the previous CH.

After getting all the three inputs, the fuzzifier module first converts the crisp inputs values to fuzzy values. Then the fuzzy inference unit applies the fuzzy “if-then” rules, to map these fuzzy input values to fuzzy output values. Finally, the defuzzification module converts these fuzzy output values into a crisp value called range factor as the output.

The range factor is the variable that decides the actual cluster range of the candidate CHs. The Actual Cluster Range (ACR) is calculated using Eq. (2) as follows:

$$\text{Affective Cluster Range (ACR)} = \text{Range Factor} \times R_{\text{max}} \tag{2}$$

2.2 Data Transfer Phase

The data transmission phase begins when the clustering process is over. At this phase, first the member nodes transmit their sensed data to their CHs and then the CH compresses the sensed data and passes this information in a multihop manner till it reaches the base station. The selection of relay is a major factor in optimizing energy consumption at the CH level. So, in this proposed approach a best-effort delivery based on relay CH's remaining energy and distance to the base station is used. Every CH calculates the next hop and forwards the gathered information. The selection of next hop is decided by Eqs. (3) and (4) given as follows:

$$\text{Next\_Hop} = \begin{cases} \text{Base Station}, & \text{if}\, \text{CH}_\text{i} \cdot D_\text{bs} \leq R\text{max} \\ \text{CH}_\text{Relay}, & \text{otherwise} \end{cases} \tag{3}$$

$$\text{CH}_\text{Relay} = \{\text{NCH}_\text{i} | \text{NCH}_\text{i} \cdot E_\text{re} \text{is maximum and} \text{NCH}_\text{i} \cdot D_\text{bs} < \text{CH}_\text{i} \cdot D_\text{bs}\} \tag{4}$$

where NCH_i is the neighboring CH, $\text{NCH}_\text{i} \cdot E_\text{re}$ is the remaining energy of NCH_i and $\text{NCH}_\text{i} \cdot D_\text{bs}$ is the distance of NCH_i from the base station. The distance constraint is used to prevent any loop to occur in the network.

3 Performance Analysis

The simulation setup and the results of the simulations are discussed in this section. The simulation of the proposed approach was conducted using MATLAB over a windows system. The simulation was conducted with 100 sets of homogeneous nodes randomly deployed over a 200 m × 200 m area. The details of the parameters used for this simulation are listed in Table 1. The energy model used for this simulation is the same as stated in Heinzelman et al. [21] and other network parameters along with

Table 1 Simulation parameters

Parameter Name	Value	Description
Area	200 × 200 m	Area of deployment
N	100	Number of nodes
Arrangement	Homogeneous	Arrangement of nodes
Distribution	Random	Distribution of nodes
Initial Energy	3 J	Initial amount of energy (Joule)
$R_{\max}$	60 m	Maximum Cluster radius
Base station location	(210,90)	Location of the base station
Transmission rate	10 Mb/s	Data transfer rate

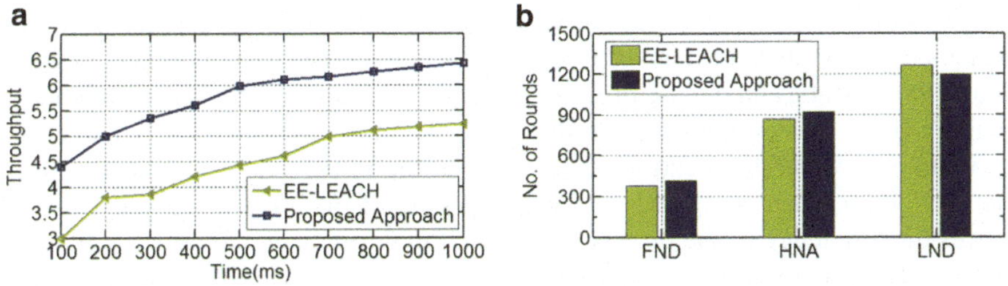

Fig. 2 **a** Analysis of throughput **b** analysis of network lifetime

the assumption about the simulation environment are the same as stated in Adhikary et al. [22]. To show the effectiveness of the proposed approach, the outcome of the simulation results is compared with EE-LEACH [23] and the efficiency of the proposed approach is evaluated based on throughput and network lifetime.

At first, the approaches are evaluated on the basis of the throughput. Throughput can be defined as the total number of successful transmission of unit data in a unit time. The throughput analysis graph is plotted between time (millisecond) and throughput. The line graph in Fig. 2a is plotted based on the throughputs of the proposed approach and existing EE-LEACH at a sample instance of 100 ms. It can be analyzed from Fig. 2a that the performance of the proposed approach not only increased in every time instance but it also way better than the existing EE-LEACH. For instance; at sample time 600 ms, the throughput of the proposed approach is 6.1321 whereas the throughput of EE-LEACH is 4.6207.

The next evaluation was placed on the basis of the network's lifetime. Network lifetime can be defined as the duration of time between the start of the simulation until the death of the last active node. The network lifetime bar graph is plotted based on the three-lifetime metrics; first node dies (FND), half of node alive (HNA), and last node dies (LND). Figure 2b plotted based on these lifetime metrics of the proposed approach and existing EE-LEACH. It can be observed in Fig. 2b is that the performance of the proposed approach is way better than the existing EE-LEACH in FND and HNA metrics. The performance of EE-LEACH is better in the LND metric but, after the death of half of the nodes, the network is not of much use.

So, it is vividly clear from the above two results that the performance of the proposed algorithm is way better in almost all the compared metrics then the existing EE-LEACH.

4 Conclusion

Energy drainage in a bottleneck zone is a major concern that largely affects the lifetime of the WSNs. However, the problem can be handled using an optimal load distribution. In this regard, an unequal distributed clustering approach has been proposed. The proposed approach makes an efficient selection of CHs using nodes remaining energy. Moreover, the FIS-based cluster range selection gives an optimal size to the clusters based on its load. Thus, the proposed method creates an optimal configuration of the clusters, and the use of fuzzy logic resolves any uncertainty in parametric quantity. This approach ensures a route that is the most energy-efficient and at the same time prevents the creation of any loop in the route. The results drawn from the simulation when compared with existing state-of-the-art protocol EE-LEACH proves the efficiency of the proposed approach in throughput and network lifetime metrics. The proposed approach assumes a static network. So the future extension of this research will focus on mobile nodes and/or base stations.

References

1. Santos, M.A., Margi, C.B.: Design and implementation of data survival in unattended wireless sensor networks. In: 30th IEEE Performance Computing and Communications Conference (IPCCC), pp. 1–6. IEEE, USA (2011)
2. Stankovic, J.A.: Research challenges for wireless sensor networks. ACM SIGBED Rev. **1**(2), 9–12 (2004)
3. Zheng, J., Jamalipour, A.: Wireless Sensor Networks: A Networking Perspective. Wiley, London (2009)
4. Zhang, Y., Li, W.: Modeling and energy consumption evaluation of a stochastic wireless sensor network. EURASIP J. Wirel. Commun. Netw. **2012**(1), 282 (2012)
5. Schurgers, C., Srivastava, M.B.: Energy efficient routing in wireless sensor networks. Military Communications Conference. MILCOM 2001, 1, pp. 357–361. IEEE, USA (2001)
6. Wen, Y.F., Lin, F.Y.S.: Energy-efficient data aggregation routing and duty-cycle scheduling in cluster-based sensor networks. In: 4th IEEE Consumer Communications and Networking Conference, pp. 95–99, IEEE, USA (2007)
7. Abbasi, A.A., Younis, M.: A survey on clustering algorithms for wireless sensor networks. Comput. Commun. **30**(14), 2826–2841 (2007)
8. Soro, S., Heinzelman, W.B.: Prolonging the lifetime of wireless sensor networks via unequal clustering. In: Proceedings of 19th IEEE International Parallel and Distributed Processing Symposium, pp. 8–16, IEEE, USA (2005)
9. Intanagonwiwat, C., Govindan, R., Estrin, D.: Directed diffusion: A scalable and robust communication paradigm for sensor networks. In: Proceedings of the 6th Annual International Conference on Mobile Computing and Networking, pp. 56–67, ACM, USA (2000)
10. Paek, J., Govindan, R.: RCRT: rate-controlled reliable transport for wireless sensor networks. In: Proceedings of the 5th International Conference on Embedded Networked Sensor Systems, pp. 305–319, ACM, Australia (2007)
11. Teo, J.Y., Ha, Y., Tham, C.K.: Interference-minimized multipath routing with congestion control in wireless sensor network for high-rate streaming. IEEE Trans. Mob. Comput. **7**(9), 1124–1137 (2008)
12. Ren, F., He, T., Das, S.K., Lin, C.: Traffic-aware dynamic routing to alleviate congestion in wireless sensor networks. IEEE Trans. Parallel Distrib. Syst. **22**(9), 1585–1599 (2011)
13. Hull, B., Jamieson, K., Balakrishnan, H.: Mitigating congestion in wireless sensor networks. In: Proceedings of the 2nd International Conference on Embedded Networked Sensor Systems, pp. 134–147, ACM, USA (2004)
14. Wan, C.Y., Eisenman, S.B., Campbell, A.T.: CODA: congestion detection and avoidance in sensor networks. In: Proceedings of the 1st International Conference on Embedded Networked Sensor Systems, pp. 266–279, ACM, USA (2003)
15. Zawodniok, M., Jagannathan, S.: Predictive congestion control protocol for wireless sensor networks. IEEE Trans. Wirel. Commun. **6**(11), 3955–3963 (2007)

16. Tao, L.Q., Yu, F.Q.: ECODA: enhanced congestion detection and avoidance for multiple class of traffic in sensor networks. IEEE Trans. Consum. Electron. **56**(3), 1387–1394 (2010)
17. Meera, S.B., Jayakumari, R.B., Senthilkumar, V.J.: Congestion control in wireless sensor networks using prioritized interface queue. In: International Conference on Recent Trends in Computational Methods, Communication and Controls, pp. 32–37, India (2012)
18. Raman, C.J., James, V.: FCC: fast congestion control scheme for wireless sensor networks using hybrid optimal routing algorithm. Cluster Comput. 1–11 (2018)
19. Singh, K., Singh, K., Aziz, A.: Congestion control in wireless sensor networks by hybrid multi-objective optimization algorithm. Comput. Netw. **138**, 90–107 (2018)
20. Mamdani, E.H., Assilian, S.: An experiment in linguistic synthesis with a fuzzy logic controller. Int. J. Man-Mach. Stud. **7**(1), 1–13 (1975)
21. Heinzelman, W.B., Chandrakasan, A.P., Balakrishnan, H.: An application-specific protocol architecture for wireless microsensor networks. IEEE Trans. Wirel. Commun. **1**(4), 660–670 (2002)
22. Adhikary, D.R.D., Mallick, D.K.: A congestion aware, energy efficient, on demand fuzzy logic based clustering protocol for multi-hop wireless sensor networks. Wirel. Pers. Commun. **97**(1), 1445–1474 (2017)
23. Arumugam, G.S., Ponnuchamy.: EE-LEACH: development of energy-efficient LEACH Protocol for data gathering in WSN. J. Wirel. Commun. Netw. **2015**, 76 (2015). https://doi.org/10.1186/s13638-015-0306-5

Survey on Hyperparameter Optimization Using Nature-Inspired Algorithm of Deep Convolution Neural Network

Rasmiranjan Mohakud and Rajashree Dash

Abstract Hyperparameters play an important role in Convolutional Neural Network (CNN). It is a tedious job for researchers to select proper hyperparameters manually for better performance on CNN. Lots of techniques are available to optimize the hyperparameter of CNN; however, some of the techniques give good result and others not. Nature-inspired algorithms as Genetic Algorithm (GA) and Particle Swarm Optimization (PSO) have put a profound effect on hyperparameter optimization in several domains. This paper presents a detailed review of the hyperparameter optimization of CNN using nature-inspired algorithms. It also includes a brief description of parameters of different CNN models used for image classification, segmentation, and styling. The review provides the researcher with the proper idea about the CNN model, its hyperparameter, and techniques used for its optimization.

Keywords Hyperparameters · CNN · GA · PSO

1 Introduction

Object recognition and image classification are the two emerging fields in current scenario that helps to recognize objects and classify the different objects from an image. Though it is easy for human beings to manually perform such task with small number of images, handling huge images for such process is really burdensome. In that angle, the machine can do the same using several intelligent techniques within fraction of time by recognizing the less complex features. Convolution Neural Network (CNN) is one of such popular techniques that uses the concept of layer-wise automatic feature extraction for object reorganization and image classification. In the

R. Mohakud (✉) · R. Dash
Institute of Technical Education and Research, Siksha 'O' Anusandhan (Deemed to be University), Bhubaneswar, Odisha 751030, India
e-mail: rashmiranjanmahakud@soa.ac.in

R. Dash
e-mail: rajashreedas@soa.ac.in

D. Mishra et al. (eds.), *Intelligent and Cloud Computing*,
Smart Innovation, Systems and Technologies 194,
https://doi.org/10.1007/978-981-15-5971-6_77

layer-wise automatic feature extraction, low-level layers extract the simple features and high-level layers extract the complex features. CNN model has different layers like input, convolution, maxpooling, activation, dropout, and dense layer. Each layer has different hyperparameters such as kernel size, the number of kernels, stride in the kernel involved in convolution, kernel size associated with maxpooling, dropout rate used in dropout, size of the dance, and so on. Other parameters like learning rate, number of convolution, number of maxpooling, and number of dance layers are the parameter of the CNN model. The performance of the CNN model totally depends on these hyperparameters.

SegNet [1], VGG [2], GoogLeNet [3], and ResNet [4] are some of the popular CNN models which have brought evolution on object detection and image classification. These models possessed complex architecture having lots of hyperparameters, but to get amazing results, these hyperparameters must be fine-tuned. Tuning hyperparameters is a very tedious task for the researcher; on the other hand, the layers of CNN is increasing day by day to cope up with huge and complex data. Reviewing some well-known CNNs, it is found that AlexNet has around 27 hyperparameters, whereas GoogLeNet, VGG-16, and ResNet-52 consist of a total 78, 57, 150 hyperparameters, respectively. It is annoying to optimize these hyperparameters manually at a reasonable cost. Different approaches are used by different researchers to optimize them. Some of them gave result up to acceptation and some are not, whereas all hyperparameter optimization using nature-inspired algorithm has commendable result.

Among a number of nature-inspired algorithms, Genetic Algorithm(GA) and Swarm Optimization (SO) are two widely applied algorithms for fine-tuning of hyperparameters of CNN. The major asset of the nature-inspired algorithm is a set of populations that can be encoded to hyperparameters of CNN, which is distributed over the search space and used to identify the changes in population for higher accuracy. The initial populations are created according to the number of hyperparameters, and then an objective function is set to find the fitness of the initial population; then, the best population is selected to modify the week population to achieve a remarkable optimum solution and then these solution sets are used for CNN model. The study of these methods can give a clear idea of their pros and cons, so that they can be suitably applied with different CNN models for solving real-world problems.

2 Literature Review

Residual network commonly known as ResNet is a 128-layer deep network, which is designed by reformulating the layers by residual function learning from input of the layer. It is used for image classification in [4]. GoogLeNet is a 22-layer architecture, which has utilized the resources of CNN efficiently inspired by Hebbian principle and multiscale processing [3]. The number of layers of CNN plays an important role. Increasing the number of layers can increase the accuracy of the network, which is effectively present by SIMON et al. popularly known as VGG network in

[2]. Semantic segmentation is one of the emerging fields, which is used for object detection and classification. SegNet is one of the effective CNN models, which is purposed by VIJAYA et al. and has encoder and decoder architecture for semantic segmentation of images [1]. WANG et al. optimized the hyperparameters of existing CNN models such as SegNet, VGG, GoogLeNet, and ResNet using a Canonical PSO (cPSO). They selected the parameters of CNN, by modifying the acceleration, fitness, and selection strategy of traditional PSO in cPSO [5]. Deep Convolution Neural Network (DCNN) is the model that is popular due to its automatic feature selection and end-to-end training. DCNN has a huge number of layers involving many hyperparameters that are hard to manually tune. MA et al. purposed an autotune method using GA for DCNN, which has divided the architecture into different blocks. They proposed an approach to select the parameters of an individual block and then encode it to integer code for initializing the population and to find the fitness using fitness function and then prepare next-generation population. Using these optimized parameters, they decide the DCNN architecture in [6]. PSO has many advantages for optimizing CNN architecture, with few disadvantages such as fix length encoding string, manual update of the dense layer parameters, position of the parameters in the encoded string, and so on. These disadvantages of PSO are handled efficiently by a Quantum-Behaved PSO (BQPSO) in [7]. The proposed technique has combined GA and PSO to avoid the remedies of PSO. Parameter optimization using GA is a popular approach that uses the basic evolutionary operations such as selection, crossover, and mutation. GA has the disadvantage that sometimes it may converge toward local optima over the iteration. As a solution to this problem, a Multilevel Evolutionary Optimization (MLEO) algorithm having two-level optimization such as within-group and between-group dynamics is proposed in [8]. Another Multiscale Multilevel Evolutionary Optimization (MSMELO) algorithm is represented in [9] by combining GPEI [10] and MELO to optimize parameters of CNN. MicroRNA plays an important role in the development, differentiation, and regulation of the cell cycle. Thousands of microRNA types are present, which are nonlinearly related to cancer. So to classify them, Al et al. purposed a CNN model having optimized parameters using GA in [11].

3 Convolution Neural Network

Object detection and image classification are two emerging fields for decades; popular techniques for it are neural network and CNN. Object detection using neural networks has the disadvantage of high cost and more time complexity due to its fully connected hidden layer. As a solution to this problem, CNN is popularly used for image classification nowadays. CNN consists of different layers as convolution, maxpool, relu, dropout, dense, and softmax. Semantic overview of a CNN is shown in Fig. 1.

Fig. 1 Schematic view of CNN model

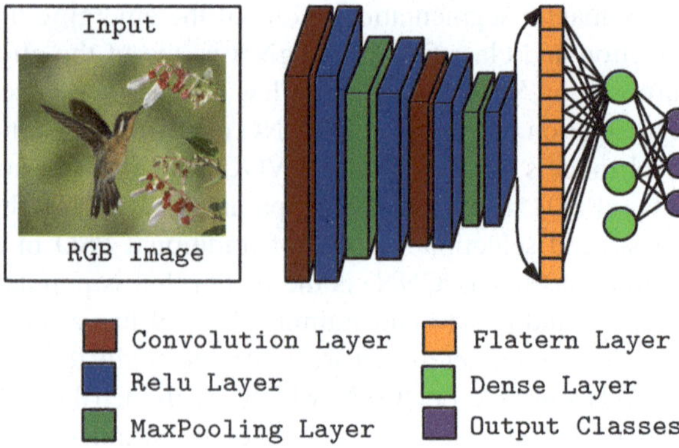

CNN model is controlled by its hyperparameter, such as the convolution layer has the number of kernels, kernel size, activation function, maxpool has kernel size, dropout has dropout rate, a dense layer has numbers of elements, numbers of layers, learning rate, and so on. To increase the accuracy of the CNN model, some researchers suggested increasing the number of layers in CNN. In case of CNN applied for automatic feature detection, simple features are detected by initial layers, followed by complex features abstracted in the later layers. For abstracting complex features, the depth of the model plays an important role. Feature selection in each layer depends on the number of kernel and kernel size. To get better features, researchers use the concept of using fewer kernels in the initial layer and large numbers of features in the latter layer. Learning rate plays an important role at the time of training the CNN model using any training algorithm like gradient descent, Adam optimizer, and gradient descent with momentum. If learning rate is too small, the gradient converges to the optimal solution slowly. If it is too large, gradient accelerates near solution; therefore, selection of proper learning rate is an important factor. Regularization of CNN is done by the dropout rate. Proper dropout rate generalizes the model. GoogLeNet having a total of 22 layers with a 5×5 kernel size of pooling layer, rectified linear unit used in convolution layer, a fully connected layer of 1024, and dropout rate of 70% is used in [3]. VGG net uses different depth networks ranging from 11 (8 convolutions and 3 fully connected) to 19 (16 convolutions and 3 fully connected) layers. In that network, the width of the convolution layer, i.e., number of kernels is varied from 64 to 512, kernel size is 3×3, dropout rate is 50%, learning rate is initialized between 10 and 2, and momentum of gradient descent is set to 0.9 [2]. SegNet is a 13-layer encoder and 13-layer decoder architecture, which has used kernel number from 64 to 512, kernel size 3×3, learning rate 0.1, and momentum 0.9 [1].

4 Nature-Inspired Algorithm for Hyperparameter Optimization

A whole lot of researches has been done to optimize the hyperparameters of neural networks. Hyperparameter annotation varies from researchers to researchers that lead to different optimization goals. Some consider hyperparameter in a confined sense, i.e., they consider hyperparameter of the existing CNN. Others consider hyperparameter in an extensive sense, i.e., they also include the number of layers, learning rate, dropout rate, and so on. Hyperparameter optimization using nature-inspired algorithm needs encoding hyperparameter, initializing population, defining an objective function to calculate the fitness of the solution, searching the best solution, and updating the worst solution according to the best solution.

WANG et al. [5] applied a cPSO to optimize the hyperparameters of the existing CNN model. cPSO-CNN has created a swarm of particles and, by encoding CNN parameter to PSO, initialized the swarm randomly using a uniform distribution function. Then, the author has selected the global and personal best of the swarm. After that, particle's positions of swarm are recalculated with both personal and environmental information. Then, the solution set is rearranged in a descendant order of fitness by updating the global and personal hyperparameter according to its fitness. Finally, the best solution is obtained after some iteration. Firstly, they have used the fitness functions which calculate the fitness value using less iteration of the CNN model. For the updated solution, they have used efficient search techniques, and finally, they have used Compound Normal Confidence (CNC) distribution for enhancing the position of the worst particle. Fitness function of cPSO has used a lightweight model of CNN which has fewer numbers of iteration than that of the full CNN model to find the fitness of the PSO solution, i.e., the hyperparameter set. They have argued that if the performance of a CNN is stable after some iteration, then there is no need for remaining iterations. Compared to the total epoch that is used in the complete training dataset, which is usually large, an iteration of CNN training only uses a batch of the same dataset with minimum computation resources. In CNN hyperparameters ranges are not even. One acceleration coefficient for all hyperparameters leads to slow in case of a large range and out of range value for small range hyperparameters. This is the reason why cPSO has used a large acceleration coefficient for a large range and small acceleration coefficient for small range hyperparameter. Updating position of the worst particle in PSO according to the global best particle position forces the solution to converge toward a local optimum with limited search space. To avoid these disadvantages, cPSO has used GA, which not only exploits the population according to the best solution; it gradually strengthens the solution over the successive iteration using crossover, mutation, and selection. A Genetic Deep Convolution Neural Network (genetic DCNN) purposed in [6] has used GA to optimize the hyperparameters. After encoding the DCNN parameter to GA, the initial population is created, and then crossover, mutation, and selection process are repeated to strengthen the fitness of the individuals.

DCNN is divided into convolution and fully connected block. Convolution block consists of convolution, pooling, activation, dropout, and batch normalization layer. Convolution layer has two parameters, i.e., the numbers of filter n and size of filter s. Batch normalization, pooling, and dropout are represented as b, p, and d, respectively. So a convolution block can be encoded as $[n, s, b, p, d]$. Fully connected layer is encoded as $[b, a, d]$ using the notations b for batch normalization, a for activation function, and d for dropout rate. After random initialization, the population is within the predefined range, fitness of the population is calculated, and the best population is selected for next generation using the roulette wheel selection scheme. For crossover a pair of DCNN randomly, select a cross-point in each DCNN and swap them to generate new DCNN. For mutation, they randomly change DCNN parameter to maintain the diversity for the next generations.

In [9], authors have proposed a multiscale and multilevel evolutionary genetic algorithm which is derived from the Multilevel Evolutionary Optimization Algorithm (MLEO) [8] and then combined it with the Gaussian Process-based Bayesian Optimization (GPEI) [10] to optimize the hyperparameter of a CNN. For a CNN with three convolutions and two fully connected layers, initial population is created by taking random values within a particular range of hyperparameter, and two best solutions are selected using the roulette wheel selection process. For a crossover, a chromosome is randomly divided into two parts; then, they are exchanged with one another. After crossover, the fitness of the new chromosome is calculated and replaced with the worst chromosome to make a new population. Mutation operation is carried out by randomly selecting one component of one individual and muted by some new value. Hyperparameter of CNN is divided into two groups. Then, both groups are submitted to fitness function and optimized to give optimum hyperparameters.

Li et al. purposed an evolution technique to evolve CNN architecture using Quantum-Behaved Particle Swarm Optimization (BQPSO) in [7]. CNN architecture is encoded to the particle by fixed-length binary string for its elements. Convolution layer is encoded to a binary string that consists of an ID followed by its parameter, like that all other elements such as maxpooling and dense are encoded, and finally, all binary strings are concatenated to represent a particle for a CNN. Initialization of the population will be conducted by the random generated binary string of fixed length and fitness calculated by running the CNN model with the selected parameter for K epoch, and then the solution is evolved by using PSO.

The CNN is popularly used in medical image processing nowadays. Due to the complexity of the medical image, it is tedious and time-consuming job to classify or segment medical images by the neural network. For these disadvantages of neural network, CNN comes into picture which solves complex image processing task efficiently. Lots of researchers do work on this field such as using a convolution neural network for microRNA in cancer classification [11], optimized the hyperparameter of CNN using the genetic algorithm, and used these optimized parameters for classification of microRNA in cancer. As a first step for decoding the CNN parameters to the genetic algorithm for a CNN of three layers, w_k , wd_k , and h_k , respectively, output channels, convolution window, and maxpooling window of $k_t h$ layer, the initial population is created by the random selection of selected hyperparameters within the predefined range, the fitness of the population is calculated by average accuracy of CNN using these hyperparameters for a certain epoch. Optimized hyperparameter is used in CNN for better accuracy.

4.1 Observation and Comparison of Different Models

Model	Dataset	Error (%)	Findings
cPSO [5] and CNN	CIFAR-10	8.67	1. Accuracy of lightweight CNN model used as fitness function 2. Different acceleration coefficients are used for large and small value parameters 3. Used GA for updating position
GA and DCNN [6]	CIFAR-10	10.48	1. Fitness function used DCNN model accuracy 2. Roulette wheel selection used for crossover 3. GA for creating next generation
MSMLEO _GPEI [9] and GA	CIFAR-10	8.6	1. Fitness function used lightweight CNN model 2. MSMLEO is used for optimized first set and GPEI used for the second set of parameters 3. Used GA for updating position
BQPSO _GPEI [7] and GA	MNIST with basic digits	0.99	1. Parameter is encoded using binary string 2. Fitness function used CNN model for predefined epoch 4. Used GA for selection and updating population for next generation
MicroRNA _GPEI [11] and GA	Cancer Genome Atlas	0.14	1. Fitness function used average accuracy of CNN model for K epoch 2. Used GA for selection and updating population for next generation

5 Conclusion

For a survey on hyperparameter optimization using the nature-inspired algorithms for convolution neural network, we have first addressed the important aspects of hyperparameter optimization such as the number of parameters, encoding technique, population initialization, fitness function, new population creation for next generation, and so on. Based on these factors, a report is presented for further study. This study may help in giving valuable insights regarding the application of currently proposed efficient nature-inspired algorithms to handle the different unravel aspects of hyperparameter tuning of CNN in other application domains.

References

1. Badrinarayanan, V., Handa, A., Cipolla, R.: Segnet: a deep convolutional encoder-decoder architecture for robust semantic pixel-wise labelling. arXiv preprint arXiv:1505.07293 (2015)
2. Simonyan, K., Zisserman, A.: Very deep convolutional networks for large-scale image recognition. arXiv preprint arXiv:1409.1556 (2014)
3. Szegedy, C., Liu, W., Jia, Y., Sermanet, P., Reed, S., Anguelov, D., Erhan, D., Vanhoucke, V., Rabinovich, A.: Going deeper with convolutions. In: Proceedings of the IEEE Conference on Computer Vision and Pattern Recognition, pp. 1–9 (2015)
4. He, K., Zhang, X., Ren, S., Sun, J.: Deep residual learning for image recognition. In: Proceedings of the IEEE Conference on Computer Vision and Pattern Recognition, pp. 770–778 (2016)
5. Wang, Y., Zhang, H., Zhang, G.: cPSO-CNN: an efficient PSO-based algorithm for fine-tuning hyper-parameters of convolutional neural networks. Swarm Evol. Comput. (2019)
6. Ma, B., Li, X., Xia, Y., Zhang, Y.: Autonomous deep learning: a genetic DCNN designer for image classification. Neurocomputing (2019)
7. Li, Y., Xiao, J., Chen, Y., Jiao, L.: Evolving deep convolutional neural networks by quantum behaved particle swarm optimization with binary encoding for image classification. Neurocomputing **362**, 156–165 (2019)
8. Akbari, Reza, Ziarati, Koorush: A multilevel evolutionary algorithm for optimizing numerical functions. Int. J. Indus. Eng. Comput. **2**(2), 419–430 (2011)
9. Cui, H., Bai, J.: A new hyperparameters optimization method for convolutional neural networks. Pattern Recogn. Lett. (2019)
10. Snoek, J., Larochelle, H., Adams, R.P.: Practical bayesian optimization of machine learning algorithms. In: Advances in Neural Information Processing Systems, pp. 2951–2959 (2012)
11. Lopez-Rincon, A., Tonda, A., Elati, M., Schwander, O., Piwowarski, B., Gallinari, P.: Evolutionary optimization of convolutional neural networks for cancer mirna biomarkers classification. Appl. Soft Comput. **65**, 91–100 (2018)

An Intelligent Type-2 Fuzzy-Based Uncertainty Reduction Technique in Mobile Ad Hoc Network

Santosh Kumar Das, Priyanshu Singh, Nihal Verma, Shom Prasad Das, and Soumen Nayak

Abstract Mobile ad hoc network (MANET) is a dynamic modern facility-based network which has no infrastructure. It is a collection of several dynamic nodes having low energy capacity of batteries. Battery energy is the main crucial parameter of the MANET. This parameter efficiently relates to other parameters of the network. Due to dynamic topology, parameters of the network vary time to time. So, network path also vary and finally uncertainty raise in the network. In this paper, a type-2 fuzzy-based uncertainty reduction process is proposed for the MANET. Type-2 fuzzy set is the extension of the fuzzy set. It is more expressive to capture imprecise information of the network and helps to reduce uncertainty in the network.

Keywords Mobile ad hoc network · Fuzzy set · Type-2 fuzzy set · Network lifetime · Routing

---

S. K. Das (✉) · P. Singh · N. Verma · S. P. Das
School of Computer Science and Engineering, National Institute of Science and Technology (Autonomous), Institute Park, Pallur Hills, Berhampur 761008, India
e-mail: sunsantosh2007@rediffmail.com

P. Singh
e-mail: priyanshrajput01@gmail.com

N. Verma
e-mail: nihalverma.10@gmail.com

S. P. Das
e-mail: shom@nist.edu

S. Nayak
Department of Computer Science and Engineering, Institute of Technical Education and Research, Siksha 'O' Anusandhan (Deemed to be University), Bhubaneswar, Odisha 751030, India
e-mail: soumennayak@soa.ac.in

D. Mishra et al. (eds.), *Intelligent and Cloud Computing*,
Smart Innovation, Systems and Technologies 194,
https://doi.org/10.1007/978-981-15-5971-6_78

1 Introduction

Mobile ad hoc network (MANET) is a rapidly growing wireless network [1, 2]. It has no infrastructure, so it is known as infrastructure-less network. It is a collection of multiple nodes that are situated in a workplace to achieve any purpose of the customer. Sometimes it is connected with a wired network to achieve the purpose. This network has several variations, such as vehicular network known as vehicular ad hoc network (VANET), wireless ad hoc network (WANET), and flying ad hoc network (FANET). This paper basically focuses on MANET, but the proposed method will also be applied simultaneously in VANET, WANET, and FANET. The network MANET has several parameters, such as energy, mobility, hop-count, distance, delay, and overhead [3, 4]. But among all parameters, energy is the main crucial parameter of the network, because everybody knows that without energy the network becomes dead and all metrics along with network lifetime is degraded.

This paper proposes an intelligent technique for reducing uncertainty of the MANET with the help of fuzzy logic. In this paper, type-2 fuzzy set is used as an intelligent technique for handling imprecise parameters of the networks. Type-2 fuzzy set is more expressive than simple fuzzy set or fuzzy set or traditional fuzzy set.

The rest of the paper is divided as follows: Section 2 highlights some existing work related to some routing techniques. Section 3 shows detailed analysis of parts of the proposed method. Finally, Sect. 4 concludes the paper.

2 Related Works

In the last few years, some works have been designed related to MANET. Some of them are as follows. Fu et al. [5] designed a multiple path system for managing the environment of the sensor network. In this paper, messages are controlled and managed in harsh situations with multiple path system. Hence, it increases the performance of sensor networks. Das and Tripathi [6] proposed a nonlinear-based routing technique for hybridization network. This is a combination of static and dynamic nodes. In this network, the determination of static and dynamic nodes is decided based on the requirement of the users. In this work, the author uses geometric programming that is nonlinear programming. This nonlinear programming helps to manage different conflicting parameters of the network and derive the optimal routing of the network. Sattari et al. [7] used an ACO-inspired algorithm for routing in VANET. In this work, several traffics are controlled and managed to find feasible paths of the network. It is also used for reliable transmission. Finally, a cellular ant-based algorithm is designed to find an optimal solution. Alaa E. Abdallah [8] proposed a smart partial flooding routing algorithm for ad hoc network. In this paper, two three-dimensional geographical routing algorithms are used for maintaining two things, like overhead and flooding in the network. Das and Tripathi [9] proposed an intelligent routing protocol for transparent heterogeneous ad hoc network. Transparent ad hoc network indicates a network that consists of two phases, that is, user and administrator. These two networks are connected by a transparent system. In this proposal, the author uses a game theory technique to optimize the network parameters. In this system, the utilities of the game are optimized with the help of linear mathematical model and enhance the network lifetime. Wilmer et al. [10] proposed a novel dynamic traffic assignment algorithm coined as reverse online algorithm for analyzing the dynamic traffic of the network. In this work, ACO uses the information collected from other neighboring vehicles, perceives the traffic, and changes route if necessary. The algorithm employs clusters to report traffic information and wipes out after the report is broadcasted. Hayes and Ali [11] proposed a routing protocol named as RASeR for mobile sensor networks. In this paper, topology changing issue is solved with the help of global time-division multiple accesses using fixed nodes. In [12], the authors designed an energy-based routing system for WANET where each node is mobile. In this protocol, the author uses an extended version of the fuzzy logic known as intuitionistic fuzzy logic. The proposed method uses some network parameters for optimizing and reducing its uncertainty. It also uses two decision-making techniques such as entropy and preference ranking organization method for enrichment of evaluations. These two methods are ranking-based technique that helps in finding intelligent route of the network. Boukhatem et al. [13] proposed an intersection-based vehicular routing protocol known as IRQV based on ACO. It dynamically chooses the best interaction for data transfer among the adjacent interactions with the use of greedy algorithm. This paper concentrates on three different processes: (i) maintaining QoS at network survey for local global end-to-end connection, (ii) selection of optimal path, and (iii) selection for terminal intersection. Singh et al. [14] designed a distributed routing algorithm for WANET. The proposed work was based on two existing routing protocols such as DSDV and AODV where performance of these protocols is analyzed based on pause time of the nodes mobility. In [15], the authors designed an artificial intelligent system for routing in WANET. In this work, the ad hoc network that is used is hybrid, which is a fusion of static and dynamic nodes to resolve the issue of the user. Here, the authors used some of the artificial techniques such as geometric programming, fuzzy logic, and optimset. Also, in the proposed method, some of the conflicting objectives function is used to optimize the network metrics. Finally, all the metrics help to enhance the network lifetime of the hybrid ad hoc network. Chatterjee and Das [16] proposed a reactive routing protocol called E-Ant-DSR that is designed by merging the DSR protocol with ACO. The paper mainly focuses on increasing the efficiency of routing with decrease in congestion and maintaining the links. In [17], the authors designed a fuzzy-based routing technique for ad hoc network. This routing technique is based on several constraints of the network where the nature of the each constraint is conflicting to one another. So, all the parameters are combined with the help of fuzzy inference system. The proposed fuzzification and defuzzification method helps to estimate the network parameters and enhance the network lifetime of the network. Eiza et al. [18] proposed a routing algorithm for VANET named as S-AMCQ. In this algorithm, ACO is used for detecting feasible paths and determining the traffic in VANET with multiple constrains for QoS. This algorithm uses the advanced graph models for detecting route control messages. Hence, it provides better security. Ani et al. [19] proposed an algorithm that prevents congestion during data transmission named as QoRA. It takes care of congestion using two components: (i) QoRA which runs parallel in each node to detect the optimal path with respect to the QoS requirements, and (ii) use of SNMP and MIB for obtaining relevant information. In [20], the authors designed an opportunistic-based routing protocol for WANET. The proposed work reduces the congestion of the network by reducing the energy of the nodes. It also helps to maintain backlog traffic of the network and helps to increase the bandwidth of the network. Finally, it helps to reduce average end-to-end delay and routing overhead, and enhances the throughput of the network.

Although the above-mentioned works are good with respect to MANET, WANET and VANET for different working protocol, but the proposed work is based on uncertainty reduction in MANET by incorporating several parameters of the network in terms of type-2 fuzzy set elements.

3 Proposed Work

In this section, the proposed method is illustrated briefly. We know that MANET is a collection of several types of imprecise information such as energy, distance, mobility, hop-count, delay, overhead, and bandwidth. The nature of these parameters is either in increasing order or decreasing order, like (i) parameters such as energy and bandwidth behave in increasing order, whereas (ii) parameters such as distance, mobility, hop-count, and overhead behave in decreasing order. In this paper, type-2 fuzzy set is used to optimize the network lifetime of the MANET.

Let $X = x_1 \times x_2 \times x_3 \times \ldots \times x_n$, where X is a product set of different universal set. In this set, $A_1, A_2, A_3 \ldots A_n$ are different fuzzy set belonging to type-1. We know that the logical "and" operator works as set theoretic intersection or min operator, and the logical "or" operator works as set theoretic union or max operator. In fuzzy set, membership function is crisp but between 0 and 1. This membership function may not be fixed, sometimes 0.5 or 0.8, and so on. So, that may be again fuzzy. Hence, ordinary fuzzy set is known as type-1 fuzzy set, and the type-2 fuzzy set is a fuzzy set which has the membership function as type-1 fuzzy set between 0 and 1. Therefore, type-5 fuzzy set is a fuzzy set which has the membership function as type m−1 fuzzy set. Figure 1 shows mapping between type-1 and type-2 fuzzy set, where X is a set of elements xi that helps to produce a new extended fuzzy set as type-2 fuzzy by using Eq. (1) that represents different elements of energy parameter as type-1 fuzzy set.

$$Y = f(x_1, x_2, x_3, \ldots, x_n) \tag{1}$$

Then the extension principle defines a fuzzy set B in Y as given by Eq. (2).

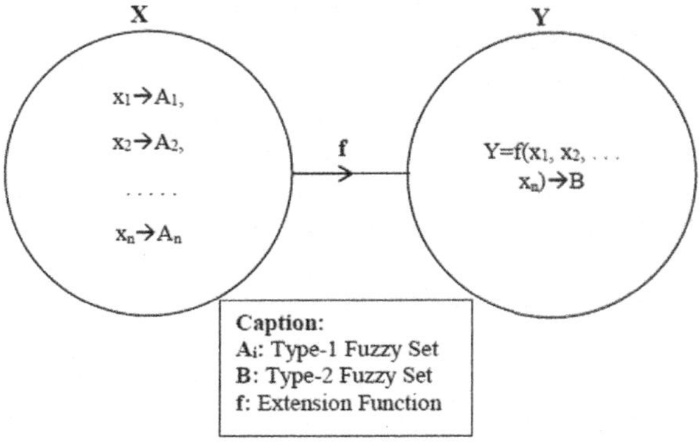

Fig. 1 Mapping between type-1 and type-2 fuzzy set

$$B = \{(y, \mu_B(y)\ |y = f(x_1, x_2, x_3, \ldots, x_n)\} \tag{2}$$

where $x_1, x_2, x_3, \ldots, x^n \in x$ and μ_B (y) as given by Eq. (3).

$$\mu_B(y) = \begin{cases} \sup\min\{\mu_{A_1}(x_1), \mu_{A_2}(x_2) \ldots\ldots \mu_{A_n}(x_n)\}, & \text{if}(x_1, x_2 \ldots x_n) \in f^{-1}(y) \text{ where } f^{-1}(y) \neq \phi \\ 0, & \text{Otherwise} \end{cases} \tag{3}$$

Let, for example, $n = 1$; here the value of n indicates one type-1 fuzzy set. So, the extension principle is given by Eq. (4).

$$B = \{(y, \mu_B(y)|y = f(x)\} \tag{4}$$

where $x \in X$ and μ_B (y) is as given in Eq. (5).

$$\mu_B(y) = \begin{cases} \sup(\mu_A(x)), & \text{if } f^{-1}(y) \neq \phi \text{ and } x \in f^{-1}(y) \\ 0, & \text{Otherwise} \end{cases}. \tag{5}$$

Let $A = \{(-1, 0.5), (0, 0.8), (1, 1), (2, 0.4)\}$ and finalize equation shown in Eq. (6)

$$f(x) = x^2 \tag{6}$$

Now, the elements of B are 0, 1, and 4 but the membership value of degree of truth is evaluated by membership functions. The new equation is as given by Eq. (7).

$$B = \{(0, \mu_B(0)), (1, \mu_B(1)), (4, \mu_B(4))\} \tag{7}$$

In Eq. (7), use only sup due to single fuzzy set, that is, A, and derive membership values as given by Eqs. (8)–(10).

$$\mu_B(0) = \underset{0 \in f^{-1(0)}}{\text{Sup}} \{0.8\} = 0.8 \tag{8}$$

$$\mu_B(1) = \operatorname*{Sup}_{(1,-1)\in f^{-1}(1)}\{0.5, 1\} = 1 \tag{9}$$

$$\mu_{\mathrm{B}}(4) = \operatorname*{Sup}_{2\in f^{-1}(4)}\{0.4\} = 0.4 \tag{10}$$

Now, optimize energy set B as given in Eq. (11).

$$B = \{(0, 0.8), (1, 1), (4, 0.4)\} \tag{11}$$

4 Conclusion and Future Work

In this paper, an intelligent technique is proposed for reducing uncertainty of the MANET. The proposed technique used type-2 fuzzy set. The fuzzy set is a multi-valued-based soft-computing technique used to capture imprecise information and reduce the uncertainty of the network. Type-2 is an extension of type-1 or generic fuzzy set. It is more expressive than type-1 fuzzy set. In this paper, it helps to derive the degree of truth between two ranges, so that flexibility of the network is increased and uncertainty of the network is reduced. Future work is to complete this work with respect to multiple parameters and design an intelligent equation for the same purpose.

References

1. Saini, T.K., Sharma, S.C.: Prominent unicast routing protocols for Mobile Ad hoc Networks: criterion, classification, and key attributes. Ad Hoc Netw. **89**, 58–77 (2019)
2. Zhong, X., Chen, F., Guan, Q., Fei, J., Yu, H. (2019). On the distribution of nodal distances in random wireless Ad hoc network with mobile node. Ad Hoc Netw
3. Meddeb, R., Jemili, F., Triki, B., Korbaa, O.: Anomaly-based behavioral detection in mobile Ad-Hoc networks. Procedia Comput. Sci. **159**, 77–86 (2019)
4. Sarkar, D., Choudhury, S., Majumder, A.: Enhanced-Ant-AODV for optimal route selection in mobile Ad-Hoc network. J. King Saud Univ. Comput. Inf. Sci. (2018)
5. Fu, X., Fortino, G., Pace, P., Aloi, G., Li, W.: Environment-fusion multipath rout-ing protocol for wireless sensor networks. Inf. Fusion **53**, 4–19 (2020)
6. Das, S.K., Tripathi, S.: Energy efficient routing formation algorithm for hybrid Ad-Hoc network: a geometric programming approach. Peer-to-Peer Netw. Appli. **12**(1), 102–128 (2019)
7. Sattari, M.R.J., Malakooti, H., Jalooli, A., Noor, R.M.: A dynamic vehicular traffic control using ant colony and traffic light optimization. In: Advances in Systems Science, pp. 57–66. Springer, Cham (2014)
8. Abdallah, A.E.: Smart partial flooding routing algorithms for 3D ad hoc networks. Proc. Comput. Sci. **94**, 264–271 (2016)
9. Das, S.K., Tripathi, S.: Adaptive and intelligent energy efficient routing for transparent heterogeneous ad-hoc network by fusion of game theory and linear programming. Appl. Intell. **48**(7), 1825–1845 (2018)

10. Arellano, W., Mahgoub, I.: Reverse online algorithm for the dynamic traffic assignment inspired by ant colony optimization for VANETs. Int. J. Swarm Intel. Evol. Comput. **3**(111), 2 (2014)
11. Hayes, T., Ali, F.H.: Robust Ad-Hoc sensor routing (RASeR) protocol for mobile wireless sensor networks. Ad Hoc Netw. **50**, 128–144 (2016)
12. Das, S.K., Tripathi, S.: Intelligent energy-aware efficient routing for MANET. Wirel. Netw. **24**(4), 1139–1159 (2018)
13. Li, G., Boukhatem, L., Martin, S.: An intersection-based QoS routing in vehicular ad hoc networks. Mob. Netw. Appl. **20**(2), 268–284 (2015)
14. Singh, J., Bhardwaj, M., Sharma, A.: Experimental analysis of distributed routing algorithms in ad hoc mobile networks. Proc. Comput. Sci. **57**, 1411–1416 (2015)
15. Das, S.K., Tripathi, S.: Energy efficient routing formation technique for hybrid ad hoc network using fusion of artificial intelligence techniques. Int. J. Commun. Syst. **30**(16), e3340, pp. 1–16 (2017)
16. Chatterjee, S., Das, S.: Ant colony optimization based enhanced dynamic source routing algorithm for mobile Ad-Hoc network. Inf. Sci. **295**, 67–90 (2015)
17. Yadav, A.K., Das, S.K., Tripathi, S.: EFMMRP: design of efficient fuzzy based multi-constraint multicast routing protocol for wireless Ad-Hoc network. Comput. Netw. **118**, 15–23 (2017)
18. Eiza, M.H., Owens, T., Ni, Q.: Secure and robust multi-constrained QoS aware routing algorithm for VANETs. IEEE Trans. Dependable Secure Comput. **13**(1), 32–45 (2015)
19. Al-Ani, A.D., Seitz, J.: QoS-aware routing in multirate Ad hoc networks based on ant colony optimization. Net. Protoc. Algorithms **7**(4), 1–25 (2015)
20. Qin, Y., Li, L., Zhong, X., Yang, Y., Ye, Y.: Opportunistic routing with admission control in wireless ad hoc networks. Comput. Commun. **55**, 32–40 (2015)

Study of Cost-Sensitive Learning Methods on Imbalanced Datasets

Neelam Rout, Debahuti Mishra, and Manas Kumar Mallick

Abstract Imbalanced data having skewed class distribution is very common in many real-world applications. The imbalance class problem affects the performance of the system. In this paper, cost-sensitive learning approaches are used with three different types of datasets which are yeast 4, yeast 5, and yeast 6 to deal with the class-imbalanced problem. The performance metrics accuracy, TPR, TNR, and *G*-Mean are taken as evaluation measures of the learning classifiers. Boosting ensemble technique is an attractive method to tackle the class imbalance problem where the rare instances are the class of interest. Cost-sensitive boosting with improved C5.0/See5 as base classifier (CSAdaC2IC5.0) is used to solve the imbalance problem. Finally, the proposed method has given the best values for the training datasets using the performance metrics that are accuracy, TPR, and *G*-Mean.

Keywords Classification · Imbalanced data · Cost-sensitive learning methods · Boosting method · C4.5 · C5.0/see5 · Performance metrics

1 Introduction

Classification is an important task of data mining and knowledge discovery in databases [1]. The imbalanced class has occurred when one class is represented by a large number of samples, while the other one is represented by only a few. Rare events (the class of interest of imbalanced data) refer to events that occur much less frequently than commonly occurring events. Software defects, cancer gene

N. Rout (✉) · D. Mishra · M. K. Mallick
Department of Computer Science and Engineering, Siksha 'O' Anusandhan (Deemed to be University), Bhubaneswar, Odisha 751030, India
e-mail: neelamrout@soauniversity.ac.in

D. Mishra
e-mail: debahutimishra@soauniversity.ac.in

M. K. Mallick
e-mail: director.iter@soauniversity.ac.in

D. Mishra et al. (eds.), *Intelligent and Cloud Computing*,
Smart Innovation, Systems and Technologies 194,
https://doi.org/10.1007/978-981-15-5971-6_79

expressions, sentimental analysis, natural disasters, modern manufacturing plants, and fraudulent credit card transactions are the areas where the imbalanced problem has shown. Existing methods have proposed based on data sampling, algorithm modification, and cost-sensitive learning for class imbalance problems. Cost-sensitive learning is a type of learning that takes the misclassification costs into consideration and the goal is to minimize the total cost [1, 2]. Ensemble methods (known as multiple classifiers or committee) are one of the most promising research areas [3]. In this work, three types of imbalanced datasets with four different types of cost-sensitive learning algorithms are used, then the evaluation process of these classifiers are done using performance metrics. Then, a method based on cost-sensitive boosting with improved C5.0/See5 as base classifier (CSAdaC2IC5.0) is used.

The rest of the paper is set up as follows: Sect. 2, Sect. 3, and Sect. 4 represent the related work, the proposed model, and the experimental results and analyses. Finally, conclusions are concluded in Sect. 5.

2 Literature Survey

The authors [4] are used as a constructive procedure to enhance SVM's standard loss function to optimize the classifier for class costs. Risk minimization and probability elicitation concepts are used to show the Bayes consistency. Then the primal and the dual objective functions are analyzed in a risk minimization framework. Finally, the authors are extended the classifier to provide superior generalization performance. The focus of the paper [5] is to learn about meta-techniques. The authors [6] are proposed an ECS-DBN method where it uses adaptive differential evolution to optimize the misclassification costs. In paper [7], the authors are proposed an objective function called accrual function which quantifies the difference in performance and size of two ensembles to determine the relative cost-effectiveness. In [8], the authors have designed three ensemble classification methods that collect information from sensors to diagnose abnormalities in the breast. In the paper [9], 38 datasets have used to show the performance of cost-sensitive classifiers to prove a balanced class distribution is more favorable than an imbalanced class distribution. In this paper [10], the authors have focused on designing modifications to support vector machines (SVMs) and GSVM-RU is efficient as well as speeds up among all the applied methods.

3 The Proposed Model for Cost-Sensitive Learning Methods

Figure 1 represents the proposed model for cost-sensitive learning methods. Here, three different types of datasets, i.e., yeast 4, yeast 5, and yeast 6 are taken as input

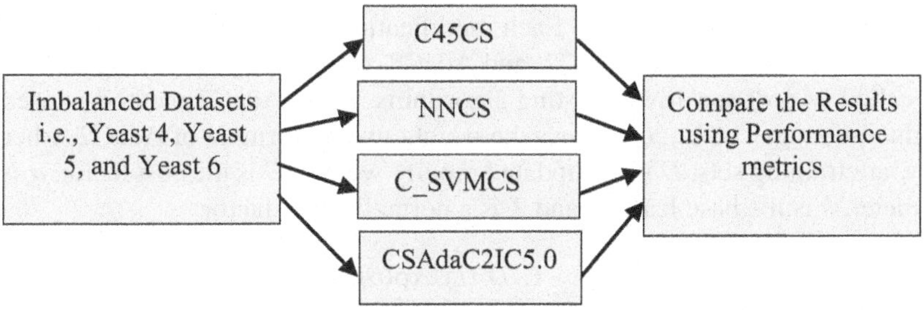

Fig. 1 The proposed model for cost-sensitive learning

to the cost-sensitive learning methods and finally, compared the results using performance metrics. There are four cost-sensitive learning methods are used which are given below.

C4.5 Cost-Sensitive (C45CS) [11, 12]: This method uses a Cost-Sensitive tree.

Multilayer Perceptron with Backpropagation Training Cost-Sensitive (NNCS) [13, 14]: This method uses the effect of sampling and threshold-moving in training cost-sensitive neural networks.

C-SVM Cost-Sensitive (C_SVMCS) [10, 15]: This technique is used to appropriately tackle the problem of class imbalance due to misclassification of cost and to speed up the model to classify unknown samples.

3.1 Proposed Method

Cost-sensitive boosting with improved C5.0/See5 as base classifier (CSAdaC2IC5.0): Cost-sensitive learning method is used for the varying costs for misclassification types. During the model building, it uses cost matrix and this learning algorithm fall in three categories [5]. In this method, boosting ensemble learning is chosen. Ensemble method is used to construct more than one classifier from the original data and then combine their prediction when classifying unknown instances. The advantage of it is to improve the generalization capacity through bias error, the intrinsic error, and the variance. Bagging, AdaBoost, and Random forest are all successful ensemble learning methods [16]. The advantages of boosting algorithms are (a) applicable to most classification system, (b) automatically excludes the extra learning cost to get best class distribution, (c) resampling through weighting sample results, and (d) reduce the bias error. It gives more weights to minority or rare class to improve the classification performance. So, here the concept of AdaBoost algorithm [5] is used. The weighting strategy of cost-sensitive boosting algorithm is used to increase weights of misclassified instances and decrease weights of correctly classified instances until the weight distribution between correctly classified samples and misclassified samples are even in each iteration. There are three ways to update the weights of AdaBoost that are outside the exponent, inside the exponent, both inside and outside the exponent. Each modification of AdaBoost is treated as new boosting algorithm AdaC1, AdaC2, and AdaC3, respectively. These algorithms are also called as cost-sensitive boosting algorithms. Here, AdaC2 method is used for weight updating. The Eq. (1) shows the weight update formula of AdaC2, where x_i and y_i are training sets, D is the updated modify weight, C is the cost factor, α is the parameter, h is the base learner, and Z is a normalization factor.

$$D^{t+1}(i) = \frac{C_i D^t(i)\exp(\alpha_t h_t(x_i) y_i)}{Z_t} \tag{1}$$

C4.5 [15] is a popular algorithm developed by Ross Quinlan which is an extension of the ID3 algorithm. It has rank one position in the top 10 algorithms in data mining, so researchers [1, 3, 12] are used as the base classifier. But in this paper, an improved version of C4.5, i.e., C5.0/See5 algorithm [17, 18] is used as the base classifier for boosting method. It has several advantages in terms of speed, memory usage, support for boosting method to improve accuracy, small decision tree, usage of winnowing method, and weighting strategy.

4 Experimental Results and Analyses

The experiment is done through Keel [19] and MATLAB [20]. There are three different types of datasets used [21], i.e., yeast 4, yeast 5, and yeast 6 having the imbalance ratio (IR) 28.1, 32.73, and 41.4. Table 1 shows the confusion matrix for binary classification [22]. Here, TP (true positive) represents the total number of minority class instances which are correctly classified, TN (true negative) represents the total number of majority class instances which are correctly classified, FP (false positive) represents the total number of minority class instances which are misclassified, and FN (false negative) represents the total number of majority class instances which are misclassified. To measure the performance of the classifiers, here four different kinds of equations are used which are accuracy (Eq. 2), TPR (Eq. 3), TNR (Eq. 4), and G-Mean (Eq. 5) [23, 24]. Though the class of interest is minority class so it is taken as positive. Here, fivefold cross-validation [22] method is used where four folds are used as training datasets and one is for test purpose.

$$\text{Accuracy} = \frac{\text{TP} + \text{TN}}{\text{TP} + \text{TN} + \text{FN} + \text{FP}} \tag{2}$$

Table 1 Confusion matrix

	Predicted majority	Predicted minority
Actual majority	True Negative (TN)	False Positive (FP)
Actual minority	False Negative (FN)	True Positive (TP)

$$\text{TPR(True positive ratio)} = \frac{\text{TP}}{\text{TP} + \text{FN}} \tag{3}$$

$$\text{TNR(True negative ratio)} = \frac{\text{TN}}{\text{TN} + \text{FP}} \tag{4}$$

$$G - \text{Mean (Geometric mean)} = \text{TPR} \times \text{TNR} \tag{5}$$

For most classification problems accuracy is a typical evaluation criterion and it represents the proportion of instances which are correctly classified to the total number of instances. In this paper, TPR, TNR, and *G*-Mean are used for performance evaluation. True Positive Ratio (TPR) calculates the proportion of minority class samples which are correctly classified to the total number of minority class samples. True Negative Ratio (TNR) calculates the proportion of majority class samples which are correctly classified to the total number of majority class samples. Then, *G*-mean measures the balance between true positive ratio and true negative ratio.

For the training datasets, CSAdaC2IC5.0 method has given the best value, i.e., 96.39% using accuracy and for the testing datasets, C45CS method has given the best value, i.e., 95.71% among all (Table 2).

For the training datasets, CSAdaC2IC5.0 method has given the best value, i.e., 96.55% using TPR and for the testing datasets; C45CS method has given the best value, i.e., 96.26% among all (Table 3).

Table 2 Results of training and testing datasets using accuracy

Datasets	Training datasets				Testing datasets			
	C45CS	NNCS	C_SVMCS	CSAdaC2IC50	C45CS	NNCS	C_SVMCS	CSAdaC2I C5.0
	Accuracy (%)							
Yeast 4	94.63	50.91	86.87	93.65	91.71	51.35	86.39	90.03
Yeast 5	98.62	55.24	93.18	99.53	97.71	53.71	93.33	98.52
Yeast 6	95.22	41.85	89.64	96.01	97.71	41.78	89.35	93.60
Averages	96.15	49.33	89.89	96.39	95.71	48.94	89.69	94.05

Table 3 Results of training and testing datasets using TPR

Datasets	Training datasets				Testing datasets			
	C45CS	NNCS	C_SVMCS	CSAdaC2I C5.0	C45CS	NNCS	C_SVMCS	CSAdaC2I C5.0
	TPR (%)							
Yeast 4	94.43	49.28	86.87	93.42	93.16	50.03	86.74	91.35
Yeast 5	98.58	54.11	92.97	99.84	97.99	52.57	93.13	99.17
Yeast 6	5.93	41.05	89.72	96.39	97.65	41.06	89.44	94.69
Averages	96.31	48.14	89.85	96.55	96.26	47.88	89.77	95.07

For the training datasets, C_SVMS method has given the best value, i.e., 90.57% using TNR and for the testing datasets, C_SVMS method has given the best value, i.e., 87.39% among all (Table 4).

For the training datasets, CSAdaC2IC5.0 method has given the best value, i.e., 92.95% using *G*-Mean and for the testing datasets, C_SVMS method has given the best value, i.e., 88.50% among all (Table 5).

Figures 2 and 3 are used to show the average results of training and testing datasets through performance metrics.

Table 4 Results of training and testing datasets using TNR

Datasets	Training datasets				Testing datasets			
	C45CS	NNCS	C_SVMCS	CSAdaC2I C5.0	C45CS	NNCS	C_SVMCS	CSAdaC2I C5.0
	TNR (%)							
Yeast 4	100	87.75	85.29	100	50.98	88.24	76.47	52.94
Yeast 5	100	92.05	100	89.20	88.64	90.91	100	77.27
Yeast 6	65.71	75.00	86.43	80.00	100	71.43	85.71	48.57
Averages	88.57	84.93	<u>90.57</u>	89.73	79.87	83.52	<u>87.39</u>	59.59

Table 5 Results of training and testing datasets using *G*-Mean

Datasets	Training datasets				Testing datasets			
	C45CS	NNCS	C_SVMCS	CSAdaC2I C5.0	C45CS	NNCS	C_SVMCS	CSAdaC2I C5.0
	G-Mean (%)							
Yeast 4	97.18	65.76	86.11	96.66	68.92	66.44	81.44	69.54
Yeast 5	99.29	70.58	96.42	94.37	93.19	69.13	96.50	87.54
Yeast 6	79.40	55.48	88.06	87.82	98.82	54.16	87.56	67.82
Averages	91.95	63.94	90.19	<u>92.95</u>	86.97	63.24	<u>88.50</u>	74.96

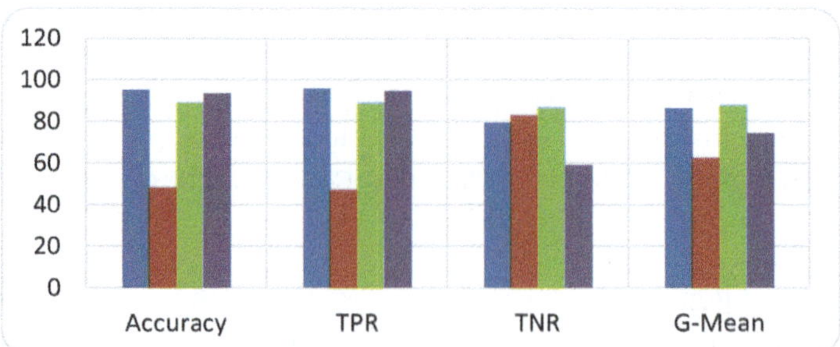

Fig. 2 Represents the average results of training datasets through performance metrics

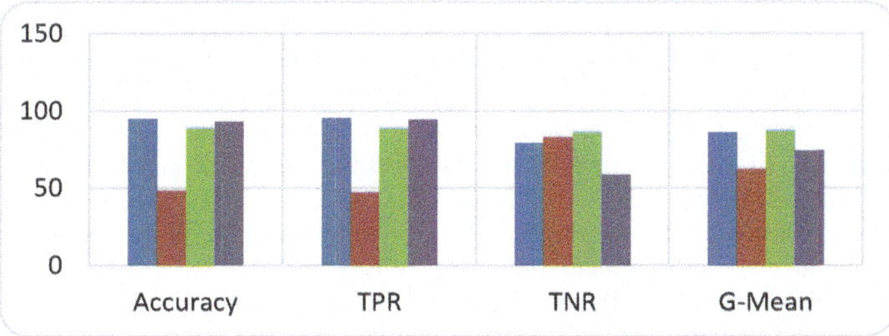

Fig. 3 Represents the average results of testing datasets through performance metrics

5 Conclusions

In this paper, different types of cost-sensitive learning algorithms for improving the classification of imbalanced data have examined. The proposed method cost-sensitive boosting with improved C5.0/See5 as base classifier (CSAdaC2IC5.0) has given the best values for the training datasets using the performance metrics that are accuracy, TPR, and *G*-Mean. In the future, the cost-sensitive methods could also be applied to highly imbalanced ratio data as well as vigorous data.

References

1. Haixiang, G., Yijing, L., Shang, J., Mingyun, G., Yuanyue, H., Bing, G.: Learning from class-imbalanced data: review of methods and applications. Expert Syst. Appl. **1**(73), 220–239 (2017)
2. Wong, M.L., Seng, K., Wong, P.K.: Cost-sensitive ensemble of stacked denoising autoencoders for class imbalance problems in business domain. Expert Syst. Appl. **141**, 112918 (2020)
3. Krawczyk, Bartosz, Minku, Leandro L., Gama, João, Stefanowski, Jerzy, Woźniak, Michał: Ensemble learning for data stream analysis: a survey. Inf. Fusion **37**, 132–156 (2017)
4. Hamed, M-S., Vasconcelos, N., Iranmehr, A.: Cost-sensitive support vector machines. arXiv preprint arXiv:1212.0975 (2012)
5. Sun, Y., Kamel, M.S., Wong, A.K., Wang, Y.: Cost-sensitive boosting for classification of imbalanced data. Pattern Recogn. **40**(12), 3358–3378 (2007)
6. Zhang, C., Tan, K.C., Li, H., Hong, G.S.: A cost-sensitive deep belief network for imbalanced classification. IEEE Trans. Neural Netw. Learn. Syst. **30**(1), 109–122 (2018)
7. Bhardwaj, M., Bhatnagar, V., Sharma, K.: Cost-effectiveness of classfication ensembles. Pattern Recogn. **57**, 84–96 (2016)
8. Li, Y., Porter, E., Santorelli, A., Popović, M., Coates, M.: Microwave breast cancer detection via cost-sensitive ensemble classifiers: Phantom and patient investigation. Bio Med. Sign. Process. Control **31**, 366–376 (2017)
9. Liu, X.Y., Zhou, Z.H.: The influence of class imbalance on cost-sensitive learning: an empirical study. In: 6th International Conference on Data Mining (ICDM'06). IEEE, Hong Kong, China, pp. 970–974 (2016)
10. Tang, Y., Zhang, Y.Q., Chawla, N.V., Krasser, S.: SVMs modeling for highly imbalanced classification. IEEE Trans. Syst. Man Cybern. Part B (Cybern.) **39**(1), 281–288 (2008)
11. Ting, K.M.: An instance-weighting method to induce cost-sensitive trees. IEEE Trans. Knowl. Data Eng. **3**, 659–665 (2002)

12. Quinlan, J.R.: C4. 5: programs for machine learning. Morgan Kaufmann, San Francisco (1993)
13. Zhou, Z.-H., Liu, X.-Y.: Training cost-sensitive neural networks with methods addressing the class imbalance problem. IEEE Trans. Knowl. Data Eng. **18**(1), 63–77 (2005)
14. Rojas, R.: Fuzzy logic. In: Neural Networks, pp. 287–308. Springer, Berlin, Heidelberg (1996)
15. Vanhoeyveld, J., Martens, D.: Imbalanced classification in sparse and large behaviour datasets. Data Min. Knowl. Discov. **32**(1), 25–82 (2018)
16. Friedman, J., Hastie, T., Tibshirani, R.: Special invited paper. Additive logistic regression: a statistical view of boosting. Annal. Stat. 337–374 (2000)
17. C4.5 Homepage.: https://en.wikipedia.org/wiki/C4.5_algorithm. Last Accessed 10 Sep 2019
18. Siknun, G.P. and Sitanggang, I.S.: Web-based classification application for forest fire data using the shiny framework and the C5. 0 algorithm. Proce. Environ. Sci. **33**, 332–339 (2016)
19. Alcalá-Fdez, J., Fernández, A., Luengo, J., Derrac, J., García, S., Sánchez, L., Herrera, F.: Keel data-mining software tool: data set repository, integration of algorithms and experimental analysis framework. J. Multiple-Valued Logic Soft Comput. **17** (2011)
20. Van Der Heijden, F., Duin, R.P., De Ridder, D., Tax, D.M.: Classification, Parameter Estimation and State Estimation: An Engineering Approach Using MATLAB. Wiley, London, 2005
21. KEEL Homepage.: http://www.keel.es/. Last Accessed 21 Sep 2019
22. Han, Jiawei, Pei, Jian, Kamber, Micheline: Data mining: concepts and techniques. Elsevier, Amsterdam (2011)
23. Gu, Q., Li, Z., Cai, Z.: Evaluation measures of the classification performance of imbalanced data sets. International Symposium on Intelligence computation and Applications, pp. 461–471. Springer, Berlin, Heidelberg (2009)
24. Wong, M.L., Seng, K., Wong, P.K.: Cost-sensitive ensemble of stacked denoising autoencoders for class imbalance problems in business domain. Exp. Syst. Appl. **141**, 112918 (2020)

Game Theory Based Optimal Decision-Making System

Santosh Kumar Das, AbhilekhNath Das, Harsh Kumar Sinha, Shom Prasad Das, and Soumen Nayak

Abstract In the modern era, the application of decision-making situation is increasing rapidly due to variations of customer requirement and satisfaction. There are several decision solutions also available in the market. These solutions may be crisp or fuzzy in nature. But most of the solutions are based on crisp nature. Day by day conflicting situations are rising frequently. So, there is a need to design an optimal decision-making solution for both crisp and fuzzy situation. In this paper, an intelligent decision-making situation is proposed that incorporates both cooperative and noncooperative situation in the environments of crisp and fuzzy. Finally, the proposed method was simulated and compared through Python language in terms of payoff metrics for outperforming the results.

Keywords Decision-making system · Game theory · Crisp logic · Fuzzy logic · Payoff-matrix

S. K. Das (✉) · A. Das · H. K. Sinha · S. P. Das
School of Computer Science and Engineering, National Institute of Science and Technology (Autonomous), Institute Park, Pallur Hills, Berhampur 761008, India
e-mail: sunsantosh2007@rediffmail.com

A. Das
e-mail: abhilekhnathdas111@gmail.com

H. K. Sinha
e-mail: harsh826001@gmail.com

S. P. Das
e-mail: shom@nist.edu

S. Nayak
Department of Computer Science and Engineering, Institute of Technical Education and Research, Siksha 'O' Anusandhan (Deemed to be University), Bhubaneswar, Odisha 751030, India
e-mail: soumennayak@soa.ac.in

D. Mishra et al. (eds.), *Intelligent and Cloud Computing*,
Smart Innovation, Systems and Technologies 194,
https://doi.org/10.1007/978-981-15-5971-6_80

1 Introduction

In the past few decades, the applications of decision-making system has increased rapidly due to its growing advantages [1–4]. There are several advantages and disadvantages of a proper decision-making system. However in comparison to disadvantage, advantages are more that is reflected in the companies. Most of the companies formalize their different strategies in terms of profit and opponent company formalizes their strategies in term of loss. Both have a specific goal, the first company wants to become a winner and second company wants to become a looser. And the combination of both provides an optimal result that is known as the value of the game. Decision-making system is rapidly used in several areas that are uncountable due to advancement of the technology. Some of them as a wireless network, there are several variations such as Wireless Ad hoc Network (WANET), Vehicular Ad hoc Network (VANET), Mobile Ad hoc Network (MANET), Flying Ad hoc Network (FANET), machine learning and data prediction, image processing, information retrieval, thread detection, cybercrime analysis and solution, etc.

In this paper, WANET is considered as a domain that is used for a decision-making system. The proposed method is used to compare both games as cooperative and noncooperative in the environments of crisp and fuzzy logics.

The rest of the paper is divided as follows. Section 2 highlights some existing work related to some routing techniques. Section 3 shows detailed analysis parts of the proposed method. Finally, in Sect. 4 concludes the paper.

2 Related Works

In the past few years, some works have been designed related to wireless networks [5, 6]. Each work consists of some methodology that helps to optimize the network metrics efficiently. Some of them as follows. Sun et al. [7] proposed a technique for optimization of the nodes deployment for WSN using ACO. In this technique, the authors apply two techniques like culture algorithm with the ant colony algorithm. The proposed technique is used to solve coverage and connectivity issues and enhance the network lifetime of the WSN. Das and Tripathi [8] proposed a nonlinear-based routing technique for a hybridization network. This is a combination of static and dynamic nodes. In this network, static and dynamic nodes determination is decided based on the requirement of the users. In this work, the author uses geometric programming that is nonlinear programming. This nonlinear programming helps to manage different conflicting parameters of the network and derive the optimal routing of the network. Sharma and Grover [9] proposed a model in WSN for energy-efficient routing protocol. The key technique of this model is modified ACO, which is an extension of earlier ACO. The proposed technique handles the utilization of the node batteries of the WSN. This efficient utilization helps to control nonuniform signal transformation during operation. Das and Tripathi [10] proposed an intelligent routing protocol for transparent heterogeneous ad hoc network. Transparent ad hoc network indicates a network that consists of two phases, i.e., user and administrator. These two networks are connected by a transparency system. In this proposal, the author uses a game theory technique to optimize the network parameters. In this system, the utilities of the game are optimized with the help of linear mathematical model and enhance the network lifetime. Kaur and Mahajan [11] proposed a metaheuristic-based efficient routing protocol for WSN. The proposed method is the fusion of two metaheuristic techniques as ACO and PSO. This efficient routing protocol is based on the energy capacity of the nodes. The proposed method helps in data aggregation of the network by reducing the residual energy capacity of the nodes. Liao and Wu [12] proposed a method for sensor nodes deployment in WSN using ACO. The proposed technique helps to increase the efficient monitoring of the network by expanding coverage areas. This method helps to increase the search space of the network and enhance the network lifetime of network. In [13], the authors designed an energy-based routing system for WANET where each node is mobile. In this protocol, the author uses an extended version of the fuzzy logic, i.e., known as intuitionistic fuzzy logic. The proposed method uses some network parameters for optimizing and reducing its uncertainty. It also uses two decision-making techniques such as entropy and preference ranking organization method for enrichment of evaluations. These two methods are ranking-based technique which helps the intelligent route of the network. Gholami et al. [14] proposed an ANN based-localization method for WSN. The proposed method first analyzes the ambient conditions of the network and investigate blockage parts of the network. This process helps to estimate different types of uncertainties of the network efficiently. Finally, it validates in a practical environment and this validation shows that outperform network metrics. In [15], the authors designed an artificial intelligent system for routing in WANET. In this work, the ad hoc network, i.e., used hybrid which is a fusion of static and dynamic nodes to resolve the issue of the user. Here, the author uses some of the artificial techniques such as geometric programming, fuzzy logic, optimset. The author uses in this proposal some of the conflicting objectives function to optimize the network metrics. Finally, all the metrics help to enhance the network lifetime of the hybrid ad hoc network. Alarifi and Tolba [16] proposed an adaptive neural network based optimization technique for WSN. This is based on the reinforcement-based learning technique. So, it uses adaptive Q-learning method. It helps to design clustering in the network that helps to reduce the energy consumption of the nodes and data collection from sensor nodes to the BS. Eldhose and Jisha [17] proposed a node aggregation method for WSN. This is based on clustering method where each active cluster is decided by ANN. This neural network method helps to process the data packet periodically and send to the BS efficiently. In [18], the authors designed a fuzzy-based routing technique for the ad hoc network. This routing technique is based on several constraints of the network where the nature of each constraint is conflicting one to another. So, all the parameters are combined with the help of a fuzzy inference system. The proposed fuzzification and defuzzification method help to estimate the network parameters and enhance the network lifetime of the network. Chang et al. [19] proposed a frailty detection system for WSN using ANN. It is totally a home-based system. The main purpose of this proposal to automatically manage the personal information of a home. It consists of several information such as ePad, eChair, eScale, etc. Finally, it helps to predict frailty and test overall system performance. In [20], the authors designed a multicast routing system for WANET. In this proposal, the selected parameters are energy and distance. Both the parameters are very crucial in terms of network lifetime. These two parameters are fuzzified in terms of generating linguistic variables. These linguistic variables help to design the fuzzy rule base system of the network. It helps to design the fuzzy output parameter, which is based on two input parameters, i.e., energy and distance. The name of the output parameter is reward. It helps to remove ambiguity of the route selection and estimates uncertainties of the network. In [21], the authors designed a fuzzy-based multicast routing system WANET using four parameters hop-count, packet, energy, and route lifetime. In these parameters, energy, hop-count, and packet are input parameters and route lifetime is the output parameter. Among all input parameters, energy is the crucial parameter. So, using this energy parameter, fuzzy decision-maker enhances the network lifetime of the ad hoc network. Finally, this method is validated numerically as well as simulation to outperform the results.

3 Proposed Work

The proposed model is based on optimal decision-making technique based on game theory. In game theory, there are two types of situations such as cooperative and noncooperative, both are further divided into two types as crisp and fuzzy models. Details analysis parts given as following subsections.

3.1 Crisp Cooperative Situation

In this technique, the players are completely involved in the decision-making process or don't take part at all. They make a decision with certainty. Let there are two companies as Company 1 ($C1$) and Company 2 ($C2$). Table 1 shows a matrix that indicates how much they get benefitted in two strategies such as Strategy 1 (Cooperation) and Strategy 2 (Noncooperation). Finally, a comparison is done for both companies whether they should go for cooperation or noncooperation.

Table 1 Crisp cooperative modeling

		Company 2 ($C2$)	
Company 1 ($C1$)		Cooperation	Noncooperation
	Cooperation	(5, 5)	(6, 2)
	Noncooperation	(2, 6)	(2, 2)

If $C1$ cooperates then the payoff of $C1$ for Strategy 1 as δ1 shown in Eq. 1. It indicates a situation for cooperation. If $C1$ not cooperates then the payoff of $C1$ for Strategy 2 as δ2 shown in Eq. 2. It indicates a situation for noncooperation. If $C2$ cooperates then the payoff of $C2$ for Strategy 1 as δ3 shown in Eq. 3. It indicates a situation for cooperation. If $C2$ not cooperates then the payoff of company $C2$ for Strategy 2 shown in Eq. 4.

$$\delta 1 = C1(0, 0) * 0.5 + C1(0, 1) * 0.5 = 5 * 0.5 + 6 * 0.5 = 5.5 \quad (1)$$

$$\delta 2 = C1(1, 0) * 0.5 + C1(1, 1) * 0.5 = 2 * 0.5 + 2 * 0.5 = 2 \quad (2)$$

$$\delta 3 = C2(0, 0) * 0.5 + C2(1, 0) * 0.5 = 5 * 0.5 + 6 * 0.5 = 5.5 \quad (3)$$

$$\delta 4 = C2(0, 1) * 0.5 + C2(1, 1) * 0.5 = 2 * 0.5 + 2 * 0.5 = 2 \quad (4)$$

So, it has been observed maximum payoff is in Eqs. (1) and (3). Hence, considering both cooperation and noncooperation for both companies, it is clearly visible that cooperating with each other will be the choice for both of them.

3.2 *Fuzzy Cooperative Situation*

In fuzzy cooperative game theory, the players involved do not participate completely and the decision of cooperation is dependent upon the level of participation where the level of participation lies between the ranges 0–1. It is also known as the degree of involvement as like in fuzzy set theory degree of truth. Table 2 shows the details of this modelling. In this modelling, fuzzy degree of involvement is denoted as $FC1$ for $C1$ and $FC2$ for $C2$. Here, the payoff matrix demonstrating the benefits of both countries are the same as the previous matrix used for crisp cooperative game theory. A new matrix is shown below demonstrating the Degree of Involvement (DOI) in cooperation or noncooperation by both countries.

If $C1$ cooperates then the payoff value shown as Eq. (5) as δ5. If $C1$ does not cooperate then the payoff value is as shown in Eq. (6) which is δ6. If $C2$ cooperates then the payoff value is as shown in Eq. (7) which is δ7. If $C2$ does not cooperate

Table 2 Fuzzy cooperative modeling

		DOI company 2($FC2$)	
DOI company 1 ($FC1$)		Cooperation	Noncooperation
	Cooperation	(0.8, 0.8)	(0.6, 0.3)
	Noncooperation	(0.3, 0.6)	(0.2, 0.2)

then the payoff value is as shown in Eq. (8) which is δ8.

$$\delta 5 = C1(0,0) * FC1(0,0) + C1(0,1) * FC1(0,1) = 5 * 0.8 + 6 * 0.6 = 7.6 \quad (5)$$

$$\delta 6 = C1(1,0) * FC1(1,0) + C1(1,1) * FC1(1,1) = 2 * 0.3 + 2 * 0.2 = 1.0 \quad (6)$$

$$\delta 7 = C2(0,0) * FC1(0,0) + C2(1,0) * FC1(1,0) = 5 * 0.8 + 6 * 0.6 = 7.6 \quad (7)$$

$$\delta 8 = C2(0,1) * FC1(0,1) + C2(1,1) * FC1(1,1) = 2 * 0.3 + 2 * 0.2 = 1.0 \quad (8)$$

3.3 Crisp Noncooperative Situation

In crisp noncooperative game theory, the players are completely involved in decision-making process or do not take part at all. They make decision with certainty. Let there are two companies $C1$ and $C2$. Table 3 shows a matrix that demonstrate the details of this situation with respect to two strategies.

If $C1$ cooperates then the payoff given in Eq. (9) as δ9. If C1 does not cooperate then the payoff value is as given in Eq. (10) as δ10. If $C2$ cooperates then the payoff value is as given in Eq. (11) as δ11. If $C2$ does not cooperate then the payoff value is as given in Eq. (12) as δ12.

$$\delta 9 = C1(0,0) * 0.5 + C1(0,1) * 0.5 = 3 * 0.5 + 2 * 0.5 = 2.5 \quad (9)$$

$$\delta 10 = C1(1,0) * 0.5 + C1(1,1) * 0.5 = 6 * 0.5 + 8 * 0.5 = 7 \quad (10)$$

$$\delta 11 = C2(0,0) * 0.5 + C2(1,0) * 0.5 = 3 * 0.5 + 2 * 0.5 = 2.5 \quad (11)$$

$$\delta 12 = C2(0,1) * 0.5 + C2(1,1) * 0.5 = 6 * 0.5 + 8 * 0.5 = 7 \quad (12)$$

Table 3 Crisp noncooperative modeling

		Company 2 ($C2$)	
Company 1 ($C1$)		Cooperation	Noncooperation
	Cooperation	(3, 3)	(2, 6)
	Noncooperation	(6, 2)	(8, 8)

Table 4 Fuzzy noncooperative modeling

		DOI Company 2 ($C2$)	
DOI Company 1 ($C1$)		Cooperation	Noncooperation
	Cooperation	(0.3, 0.3)	(0.2, 0.6)
	Noncooperation	(0.6, 0.2)	(0.9, 0.9)

So, it has been observed that maximum payoff is in Eqs. (9) and (11). Hence, payoff values by considering both cooperation and noncooperation for both the companies, it is clearly visible that non-cooperating with each other will be the choice for both of them.

3.4 *Fuzzy Noncooperative Situation*

In fuzzy noncooperative game theory, the players involved do not participate completely and the decision of noncooperation is dependent upon the level of participation. Level of participation lies between 0 and 1. Here, the payoff matrix demonstrates the benefits of both countries are the same as the previous matrix used for crisp cooperative game theory. A new matrix is shown in Table 4 that demonstrates the DOI in cooperation or noncooperation by both countries.

If $C1$ cooperates then the payoff value as shown in Eq. (13) as $\delta 13$. If $C1$ does not cooperate then the payoff value is as shown in Eq. (14) as $\delta 14$. If $C2$ cooperates then the payoff value as shown in Eq. (15) as $\delta 15$. If $C2$ does not cooperate then the payoff value is as shown in Eq. (16) as $\delta 16$.

$$\delta 13 = C1(0,0) * FC1(0,0) + C1(0,1) * FC1(0,1) = 3 * 0.3 + 2 * 0.2 = 1.3 \tag{13}$$

$$\delta 14 = C1(1,0) * FC1(1,0) + C1(1,1) * FC1(1,1) = 6 * 0.6 + 8 * 0.9 = 10.8 \tag{14}$$

$$\delta 15 = C2(0,0) * FC1(0,0) + C2(1,0) * FC1(1,0) = 3 * 0.3 + 2 * 0.2 = 1.3 \tag{15}$$

$$\delta 16 = C2(0,1) * FC1(0,1) + C2(1,1) * FC1(1,1) = 6 * 0.6 + 8 * 0.9 = 10.8 \tag{16}$$

So, it has been observed maximum payoff is in Eqs. (13) and (15). Hence, payoff values considering both cooperation and noncooperation for both companies, it is clearly visible that non-cooperating with each other will be the choice for both of them.

4 Conclusion

The proposed method helps to illustrate the decision-making situations efficiently in terms of two basic and important environments like cooperative, and noncooperative. Both situations are important in real life to predict and analyze the strategy of solution. It helps to derive the different constraints of decision-making problem. It also helps to deal with players' strategies and methodologies of the game in different scenarios and produce their payoff values with respect to the payoff matrix. This matrix helps to produce optimal as well as a feasible solution to the decision that applies to the real-life scenario to solve the different issues.

References

1. Sañudo, R., Moura, J.L., Ibeas, A.: Decision making system for stopping high speed trains during emergency situations. Procedia Soc. Behav. Sci. **162**, 429–438 (2014)
2. Cavicchiolo, M.E., Cavallin, F., Staffler, A., Pizzol, D., Matediana, E., Wingi, O.M., Trevisanuto, D.: Decision making and situational awareness in neonatal resuscitation in low resource settings. Resuscitation **134**, 41–48 (2019)
3. Kotenko, I., Saenko, I., Ageev, S.: Hierarchical fuzzy situational networks for online decision-making: application to telecommunication systems. Knowl. Based Syst. (2019)
4. Dutta, S., Sadhu, P.K., Reddy, M.J.B., Mohanta, D.K.: Role of microphasor measurement unit for decision making based on enhanced situational awareness of a modern distribution system. In: Decision Making Applications in Modern Power Systems, pp. 181–199. Academic Press (2020)
5. Das, S.K., Samanta, S., Dey, N., Kumar, R.: Design frameworks for wireless networks (Doctoral dissertation, Department of Computer Science, City University of Hong Kong), pp. 1–439 (2019)
6. De, D., Mukherjee, A., Das, S.K., Dey, N.: Nature-inspired computing for wireless sensor networks. Springer Tracts in Nature-Inspired Computing (2019)
7. Sun, X., Zhang, Y., Ren, X., Chen, K.: Optimization deployment of wireless sensor networks based on culture–ant colony algorithm. Appl. Math. Comput. **250**, 58–70 (2015)
8. Das, S.K., Tripathi, S.: Energy efficient routing formation algorithm for hybrid ad-hoc network: a geometric programming approach. Peer Peer Netw. Appl. **12**(1), 102–128 (2019)
9. Sharma, V., Grover, A.: A modified ant colony optimization algorithm (mACO) for energy efficient wireless sensor networks. Optik Int. J. Light Electron Opt. **127**(4), 2169–2172 (2016)
10. Das, S.K., Tripathi, S.: Adaptive and intelligent energy efficient routing for trans-parent heterogeneous ad-hoc network by fusion of game theory and linear programming. Appl. Intell. **48**(7), 1825–1845 (2018)
11. Kaur, S., Mahajan, R.: Hybrid meta-heuristic optimization based energy efficient protocol for wireless sensor networks. Egypt. Inf. J. **19**(3), 145–150 (2018)
12. Liao, W.H., Kao, Y., Wu, R.T.: Ant colony optimization based sensor deployment protocol for wireless sensor networks. Expert Syst. Appl. **38**(6), 6599–6605 (2011)
13. Das, S.K., Tripathi, S.: Intelligent energy-aware efficient routing for MANET. Wireless Netw. **24**(4), 1139–1159 (2018)
14. Gholami, M., Cai, N., Brennan, R.W.: An artificial neural network approach to the problem of wireless sensors network localization. Robot. Comput. Integr. Manuf. **29**(1), 96–109 (2013)
15. Das, S.K., Tripathi, S.: Energy efficient routing formation technique for hybrid ad hoc network using fusion of artificial intelligence techniques. Int. J. Commun. Syst. **30**(16), e3340, pp. 1–16 (2017)

16. Alarifi, A., Tolba, A.: Optimizing the network energy of cloud assisted internet of things by using the adaptive neural learning approach in wireless sensor networks. Comput. Ind. **106**, 133–141 (2019)
17. Eldhose, E.K., Jisha, G.: Active cluster node aggregation scheme in wireless sensor network using neural network. Procedia Technol. **24**, 1603–1608 (2016)
18. Yadav, A.K., Das, S.K., Tripathi, S.: EFMMRP: design of efficient fuzzy based multi-constraint multicast routing protocol for wireless ad-hoc network. Comput. Netw. **118**, 15–23 (2017)
19. Chang, Y.C., Lin, C.C., Lin, P.H., Chen, C.C., Lee, R.G., Huang, J.S., Tsai, T.H.: eFurniture for home-based frailty detection using artificial neural networks and wireless sensors. Med. Eng. Phys. **35**(2), 263–268 (2013)
20. Das, S.K., Yadav, A.K., Tripathi, S.: IE2M: design of intellectual energy efficient multicast routing protocol for ad-hoc network. Peer Peer Netw. Appl. **10**(3), 670–687 (2017)
21. Das, S.K., Tripathi, S., Burnwal, A.P.: Fuzzy based energy efficient multicast routing for ad-hoc network. In: Proceedings of the 2015 Third International Conference on Computer, Communication, Control and Information Technology (C3IT), pp. 1–5, IEEE (February, 2015)

A Static Approach for Access Control with an Application-Derived Intrusion System

Shaunak Chattopadhyay, Sonali Mishra, and Sambit Kumar Mishra

Abstract In the era of cyberspace, enforcing an Intrusion Detection System (IDS) and Firewall on a system is a common practice among network administrators or engineers. But, with the due time, just implementing IDS and firewall isn't just enough to secure our systems, especially with the present trend of spreading new malware attacks. Its quite easy to victimize a machine, even with IDS and firewalls enforced on the networks by easily uploading shells in the form of pdf, jpg, txt, etc. Due to which machine can easily be victimized without much effort, for this, we probe to apply a new approach to overcome this anomaly. Understandably, with the increasing demand for IoT devices in the market, safeguarding these devices are also a big challenge. Motivated by this problem, we try to perform inspections to maintain stability and functionality by adding code that allows the application to keep track of operating constraints of the application during an attack. Hence, in the background of this, we discuss intrusion detection systems, firewalls, and applicability. Further, we tend to identify open challenges in this direction.

Keywords Intrusion detection system (IDS) · Firewall · Access control · Hacker

1 Introduction

With the increasing popularity and constant growth in the field of networking technologies, security stands off as a big concern at this stage. It tends to be believed that two ecosystems are to be existent in future approaches. Where one would be dealing

S. Chattopadhyay (✉) · S. Mishra · S. K. Mishra
Siksha 'O' Anusandhan (Deemed to be University), Bhubaneswar, Odisha, India
e-mail: shaunak.chattopadhyay99@gmail.com

S. Mishra
e-mail: sonalimishra@soa.ac.in

S. K. Mishra
e-mail: skmishra.nitrkl@gmail.com

D. Mishra et al. (eds.), *Intelligent and Cloud Computing*,
Smart Innovation, Systems and Technologies 194,
https://doi.org/10.1007/978-981-15-5971-6_81

with the biology of living beings, and the other would lead to deal with technologies revolving around the human. Obviously, the question arises that living beings can protect themselves from being harmed, but how come these types of equipment protect themselves from being tampered. While the harm caused can be both physically and internally read the networking system. Currently, Intrusion Detection System (IDS) and Firewall are broadly used in (Educational and Corporate) organizations and deployed at strategical points to monitor the traffic and disrupt any malpractices [1, 2].

In this paper, we focus on how to protect a network internally. In a network, deployment of IDS and Firewalls are not enough to protect as insider attacks are a big challenge for security experts [1]. Especially where data management is an issue, for example, in an organization where huge legacy codes are being maintained, and then modifying them is a significant burden. This is very costly and time-consuming in terms of getting the proper outcome. Our approach aims to provide a vivid layer of extra protection for functional vulnerabilities and defective applications of IDS and Firewalls [2]. Henceforth, to overcome these shortcomings, we intend to contribute a more secure and safe environment in the field of Cybersecurity.

In this study, we propose a new model to implement security in the machines, where indeed threats are widely inside the network. This study tends to improvise the outlook of the security researcher to focus on the threats present inside the network ecosystem and initiating a new problem to look after in the current scenario. In our proposed methodology, we tend to provide a broad aspect of mitigating these certain attack vectors with a secure and straightforward approach to overcoming the problems. This ascertains a pervasive issue day in and day out for a vulnerability analyst to overcome.

The remaining of this paper is organized as follows. Section 2 introduces the background of past works on improvements in IDS and Firewall along with the classifications of IDS and Firewall. In Sect. 3, we discuss how our model can be applied to solve the challenges in network security. We then discuss some open issues in Sect. 4 and conclude our work with some future directions in Sect. 5.

2 Related Works and Prerequisite

An application-derived Intrusion System is more of an extra layer of security padding enforced to a system, and security researchers are often coming up with new ideas as such [3–6]. Still, neither of these raised or concern the fact of inside network attacks, where the attacker is present and actively launching the attacks in the network. While in this paper, we focus on how to provide an extra layer of security, not just only to add an extra layer, but also to inculcate a new problem into the research domain. Prominently, in a survey, it was found that 21% of the attacks are launched inside the network [7], which also draws our attention to frauds and espionage [8], if in this scenario, gaining access into a system is not an issue for the attacker. Then, indeed, we need to focus our attention on resolving the problem to stop attackers from maintaining access in this scenario. The basic prerequisite required for this work is as follows including the classification of Firewall and IDS.

2.1 *Classifications of Firewalls*

A firewall is a device or software that is established in an internal network of an organization, which connects to the outside of the network. It enforces a set of rules to filter unauthorized packets to flow into the network to prevent unwanted internet access by a user present inside the private network. The firewall has the ability to provide in-depth control over what type of packets are entering into the network likely, inbound, and outbound from the system. And determine whether it should be passed or blocked. The firewall is needed to (1) validate access in the network, (2) record the events, and (3) control the traffic in the network.

There are three types of firewall: (1) packet filtering, (2) stateful inspection, and (3) proxy server enabled firewall. A packet filtering firewall analyzes each of the packets inbound and outbound of the network to determine if any security policies/rules are being violated or not. A stateful inspection firewall is a dynamic technology that reviews all packets establishing a connection by verifying firewall protocol in sequential order. If the packet goes through any protocol without verification, the packet is denied. If the connection is accepted, the session is established in the firewall's stateful inspection table [9]. A proxy firewall is an application level of the firewall working as a server for incoming packets for the client within the internal network. While sending a request to the server, it acts as a client on an external network. It acts as an intermediate between the internal and external network connection [10–12].

2.2 *Classifications of Intrusion Detection Systems*

An intrusion detection system (IDS) is a device or software application that monitors a network or system for malicious activity or policy violations [13]. Broadly, IDS can be divided into Passive IDS and Active IDS. The passive IDS is a system that is deployed into the network to analyze and monitor the network packets flowing. It creates a log of suspicious activities for the network administrator. The active IDS is a system that deployed into the network to analyze and monitor the network packets flowing. It also does create a log of suspicious activities as well as takes some preventive action to sabotage the attacker from compromising the network. Intrusion detection systems are deployed in the network to log any malicious activity which should not be happened to take place in the network. The data collected in either reported or collected centrally using a Security Information and Event Management System (SIEM) [14]. The following three types of IDSs we need to know are the following:

1. Host-based intrusion detection system;
2. Network-based intrusion detection system;
3. Behavioral-based intrusion detection system.

2.2.1 Host-Based Intrusion Detection System (HIDS)

A HIDS is responsible for monitoring the inbound and outbound packet flowing to analyze any suspective activity. It also takes a snapshot of present system files and maps it with the previous snapshot. If the critical system files have tampered, an alert is sent to the administrator to investigate.

2.2.2 Network-Based Intrusion Detection System (NIDS)

A NIDS is responsible for analyzing network packet flowing from and to all devices in the network. NIDS are usually deployed at a different strategic position. NID Systems are capable of inspecting signatures for similar packets to link and reject the harmful detected packets which have a malicious record in the NIDS.

2.2.3 Behavioral-Based Intrusion Detection System (BIDS)

A BIDS is responsible for analyzing the network packet flowing with previously monitored patterns of anomaly identified in any previous attack vectors that happened previously in the network.

3 Proposed Method

Researcher proposed a few mechanisms to protect the networks from various strategical points [8]. An attacker may have a wide range of attacks to victimize the machine. These attacks can be broadly categorized into different possible attack vectors on a network, they are as follows.

- IP spoofing;
- Time to live headers;
- Packet flooding;
- Steganography.

3.1 IP Spoofing

With the above threats, an attacker is likely to permute the set of attacks to gain access to the network. While all of these methods are not new to security researchers, but still, these attacks are often the cause of data breaches. IP spoofing is a technique to reconnaissance a network, different types of attack such as Decoy Scan (a scan available in Nmap) is used by the attacker to gain information about the network. For this, they utilized dead IPs, which provide scope to scan the network by obfuscating their own IP.

3.2 Time to Live Headers (Fragmentation)

After scanning, the attacker object remains either to gain access to the network or to collapse the network. They often use the technique of packet fragmentation to overbear the network. While fragmentation of the packet isn't the concern in this type of attack, but the crafted packets that are sent over the networks are crucial more often. There are two types of attack vector as follows.

1. **UDP/ICMP fragmentation attacks**—These attack vectors involve sending circulation of malicious packets of more than the Maximum Transmission Unit (MTU) in a network. Hence, leading to unable to reassemble, resulting in the crashing of the server.
2. **TCP fragmentation attacks**—This attack vector is one of the most renowned attacks, which was predominantly common in older windows distribution. This kind of TCP/IP stack is targeted in a similar way as above leading to packet overlapping, which overflows the buffer, and leads to a Buffer Overflow. Usually, network administrators often tend to keep Block Fragmented IP packets off by default because no one would ever bother to drop packets of a client sitting far from the server by tightening security, leaving a major flaw in the network.

3.3 Packet Flooding

Although packet flooding has a similar approach, on a large scale sending numerous numbers of fraudulent/ICMP packets to cause huge traffic to interfere with the usual flow in the network will cause the death of the server. These types of attacks are often mentioned as the ping of death.

3.4 Steganography

The severity of the problem increases when attack vectors have a disguise, i.e., steganography, to victimize a network. It is double-edged which can be used for enforcing security, as well; on the other hand, it is used to exploit by attaching malware with media files to hide the payload. Steganography has been an enormous threat to security experts [15]. Since these vulnerabilities remain as a threat to security modules, here we would like to propose a derived approach to impose an IDS like a layer of security encapsulated in an application for our Machines, namely, an Application-derived Intrusion System. Its main purpose would be to create an alarm for any intrusion in the system and do something to defend itself from the attack vectors. Although most antiviruses are responsible for mitigating such types of scenarios. But in an organization where a bunch of legacy codes running from generation to generation where inbuilt functional vulnerabilities are a concern rather than outside network threat. In contrast, the inside attacks are a big problem, and also upgrading these modules is near to impossible, leading the whole organization system to collapse. But also security is a concern, to be aware of the issues involved in writing unexploitable code because they would require a lot of modification in their codes and in such scenarios this model would perfectly fit. We suggest something like the below code snippet.

1	#For BufferOverflow
2	If(request.length >Threshold value in X) { //X is referred as network
3	IP = getIp()
4	Details = getDetails(IP)
5	Mac = ExtractMac(Details)
6	Block_Src(Mac) }
7	else
8	{// request accepted}

The above pseudocode works as follows. At (1), if a request is found more significant than the accepted threshold value in the network. Then, at (2), we will find its IP from the request, and with IP in Address Resolution Protocol (ARP) table, we will fetch the details. Then, at (5), the mac id will be generated with the details we get from step (4). And then at (6), we will block for further not responding to any requests.

1	#For User privilege Escalation
2	If(request == 'login') {
3	//Check state machine
4	If(user_loggedin == true) {

(continued)

(continued)

1	#For User privilege Escalation
5	Block_requests(for some time period) //Race condition

6	Alert() //Attack in progress
	}

7	}

In the above pseudocode, at step (2), if a user tries to log in to the machine remotely, then the system checks the validity of the user at step (4). If the machine is already logged in, then its an attack at step (5), and requests are to be blocked for some predefined time. And at step (6), an alert message would be sent.

Also, the monitoring system is a must as it would be monitoring:

- Number of attempts
- Types of attempt

If the security logs show a massive amount of brute-forcing of possible data in the server (DDoS), then verification for each user is required to prevent any crashes of the server.

1	If (request from x) //x is a machine {
2	MACofX = getMAC(x);
3	Found = false;
4	While (All the entries in Intruder's Table){
5	If (MACofX == PreviousEntry) {
6	BlockRequest (MAC Return); } }
7	While (All the entries in Legal Table){
8	If (MACofX == PreviousEntry){
9	Found = true;
10	count = getcount (MACofX); // Count > than no. of times expected
11	addinIntruder'sTable (MACofX);}
12	else {count++;}
	}
13	If (Found = false)
14	addinLegalTables (MacofX);

With this approach, we will be auditing each machine in the network and create a separate log for each machine to use it in the database for further mitigations in similar attack vectors [16]. At step (1) of the above code, we accept the request and at step (2), get the mac id of the x. At step (3), we set the variable found to false to check at step (4) for all table entries in intruders list if mac of z matches any entry. If so, then at step (5), the system denies any such requests. In step (7), we again check for all matches of x in the legal table, and if found in step (9), then the found value is set to be true. In step 10, the count variable stores the number of requests from x, and check the threshold value. Then, at step (11), we add it to the Intruders Table. If the intruder doesn't exist, then the counter value will be increased. If x has no previous violation, then x is added in legal tables in step (14).

4 Discussion

The above code snippets are useful for any developers with legacy code handling or dealing with any IoT devices where the working pattern is similar to any low-level applications in a system. These codes still remain a secondary layer of protection for an application working in the system connected to a network; this shall only ensure secure padding but also introduce a new barrier for the attacker to evade past the security of the system instead of existing backdoors/vulnerabilities. The problem with this model is if new types of techniques are evolved to execute a similar attack vector, it would require constant updating in the codes. Also, using this model won't be enough for a system to be completely secure as this code focuses on specific attack vectors. A model can secure even IoT devices by embedding these code snippets alongside the dedicated working algorithms, as theses code are considerably light in nature and often useful to mitigate any mentioned attack.

5 Conclusion and Future Works

In this work, a new problem domain is introduced, which is the primary motive. Including this, we have provided a preliminary solution for the mitigation. Therefore, we would like to add further solutions for more such attack vectors, which are as such common in the security domain, but still remains a challenge to mitigate them. From the above explanation, we like to conclude that the proposed approach is indeed a simple, secure, and easy to implement the model. We shall also be making a full-fledged application in the future to summarize all the snippets in a module, which would make it more aggregated and to get a clearer and wiser algorithm out of it.

References

1. Meng, W., Tischhauser, E.W., Wang, Q., Wang, Y., Han, J.: When intrusion detection meets blockchain technology: a review. IEEE Access **6**, 10179–10188 (2018)
2. Axelrod, C.W.: The need for functional security testing. STSC CrossTalk. J. Defense Softw. Eng. 17–21 (2011)

3. Osken, S., Yildirim, E.N., Karatas, G., Cuhaci, L.: Intrusion detection systems with deep learning: a systematic mapping study. In: IEEE Scientific Meeting on Electrical-Electronics & Biomedical Engineering and Computer Science (EBBT), pp. 1–4, Apr 2019
4. Ugochukwu, C.J., Bennett, E.O.: An intrusion detection system using machine learning algorithm. Int. J. Comput. Sci. Math. Theory **4**(1), 39–47 (2018)
5. Yuan, H., Zheng, L., Qiu, S., Peng, X., Liang, Y., Hu, Y., Deng, G.: Design and implementation of enterprise network security system based on firewall. In: The International Conference on Cyber Security Intelligence and Analytics, pp. 1070–1078. Springer, Cham, Feb, 2019
6. Vetriselvi, V., Shruti, P.S., Abraham, S.: Two-level intrusion detection system in sdn using machine learning. In: International Conference on Communications and Cyber Physical Engineering, pp. 449–461. Springer, Singapore, Jan 2018
7. Gaigole, M.S., Kalyankar, M.A.: The study of network security with its penetrating attacks and possible security mechanisms. Int. J. Comput. Sci. Mobile Comput. **4**(5), 729 (2015)
8. McCormac, A., Parsons, K., Butavicius, M.: Preventing and profiling malicious insider attacks (No. DSTO-TR-2697). Defence Science and Technology Organization Edinburgh (Australia) Command Control Communications and Intelligence Division (2012)
9. Culbert, P.: U.S. Patent No. 7,512,781. Washington, DC: U.S. Patent and Trademark Office (2009)
10. Zalenski, R.: Firewall technologies. IEEE Potentials **21**(1), 24–29 (2002)
11. Mishra, S.K., Sahoo, B., Jena, S.K.: A secure VM consolidation in cloud using learning automata. Recent Findings in Intelligent Computing Techniques, pp. 617–623. Springer, Singapore (2019)
12. Puthal, D., Ranjan, R., Nanda, A., Nanda, P., Jayaraman, P.P., Zomaya, A.Y.: Secure authentication and load balancing of distributed edge datacenters. J. Parallel Distrib. Comput. **124**, 60–69 (2019)
13. Waidyarathna, D.W.Y.O., Nayantha, W.V.A.C., Wijesinghe, W.M.T.C., Abeywardena, K.Y.: Intrusion detection system with correlation engine and vulnerability assessment. Int. J. Adv. Comput. Sci. Appl. **9**(9), 365–370 (2018)
14. Athira, A.B., Pathari, V.: Standardisation and classification of alerts generated by intrusion detection systems. Int. J. Cybern. Inf. (IJCI) **5**(2), 21–29 (2016)
15. Mazurczyk, W., Caviglione, L.: Steganography in modern smartphones and mitigation techniques. IEEE Commun. Surveys Tutor. **17**(1), 334–357 (2014)
16. Prakash, A., Satish, M., Bhargav, T.S.S., Bhalaji, N.: Detection and mitigation of denial of service attacks using stratified architecture. Procedia Comput. Sci. **87**, 275–280 (2016)

A Survey on Transfer Learning

Santisudha Panigrahi, Anuja Nanda, and Tripti Swarnkar

Abstract To facilitate learning in a target domain, transfer learning borrows knowledge from a source domain. What and how to transfer are two main issues that need to be addressed in transferring learning. Different transfer learning algorithms result in different knowledge transferred between them for a couple of domains. To find the optimal transfer learning algorithm that maximizes learning efficiency in the target domain, scientists need to investigate all current computationally intractable transfer learning algorithms exhaustively. A sub-optimal algorithm is selected as a trade-off, which in an ad hoc way requires considerable expertise. In instructional psychology, meanwhile, it is commonly recognized that people enhance the transfer of teaching abilities to decide what to transfer. This paper discusses what is transfer learning, the different transfer learning techniques, future scope, and applications of it.

Keywords Transfer learning · Machine learning · Domain adaptation · Deep learning · Computer vision

1 Introduction

Inspired by the ability of humans to transfer knowledge through tasks, transfer learning aims to leverage knowledge from a source domain to enhance the learning efficiency or reduce the number of labeled examples in a target domain.

S. Panigrahi (✉)
Department of Computer Science and Engineering, Siksha 'O' Anusandhan (Deemed to be University), Bhubaneswar, Odisha 751030, India
e-mail: santisudha.nanda@gmail.com

A. Nanda
Department of Electrical and Electronics Engineering, Siksha 'O' Anusandhan (Deemed to be University), Bhubaneswar, Odisha 751030, India

T. Swarnkar
Department of Computer Application, SOA Deemed to be University, Bhubaneswar, Odisha 751030, India

D. Mishra et al. (eds.), *Intelligent and Cloud Computing*,
Smart Innovation, Systems and Technologies 194,
https://doi.org/10.1007/978-981-15-5971-6_83

Transfer learning has demonstrated its broad applicability in image classification [1], sentiment classification [2], dialog systems [3], and urban computing [4].

In many areas of information engineering, including classification, regression and clustering, data mining, and machine learning techniques substantial results have already been made (e.g., [5, 6]). Most machine learning techniques, however, only function well if the training and testing data are derived from the same feature space and distribution. In numerous applications of the real world, retrieving the required training data and restoring the models is costly or impossible. Reducing the necessity and strength to re-collect the training information would be nice. In such instances, it would be beneficial to transfer knowledge across task domains.

There are numerous examples of knowledge engineering where transfer learning can be really helpful. For web document classification [7–9], the main goal is to classify a given web document into several predefined groups. In [10], the labeled example is university web pages linked to class information attained through prior manual labeling efforts. There may be a scarcity of labeled training data for a classification task on a freshly created website where data features or data distributions are dissimilar. We cannot, therefore, apply directly to the new website the web page classifiers learned on the university website. In such instances, it would be beneficial if we could transfer classification information to the new domain.

Another example is the issue of sentiment classification, where the job is to classify the reviews on a product automatically. It is important to collect several product reviews for this classification task to annotate them. Then a classifier must be equipped with their matching labels based on the review. But, the review results may be very different for different product types. To achieve a very good performance in classification, a significant amount of data must be collected to train the review classification system for an individual sample. Though data labeling is an expensive process, it is required to select a classification model which is trained on certain products to aid know how to classify for some other products to minimize the effort to annotate feedback for different products. Transfer training can save a considerable amount of effort for labeling in such cases [11]. A great deal of work has been done on transfer learning for reinforcement learning (e.g., [12, 13]).

The rest of this paper is structured as follows. We first give a common overview and describe the definition of transfer learning. Then the history of transfer learning and categorization of it into three different types are discussed. Finally, application and future scope in various field and conclusion are provided.

2 Overview

Traditional data mining and machine learning algorithms use statistical models which are trained on prior collected labeled or unlabeled training data to predict future data [14–16]. Zhu and Xindong [17] have explored how the noisy group tag problems can be treated. Yang et al. found cost-sensitive learning [18] when further analysis for potential samples could be carried out. So, most of the researchers believed that

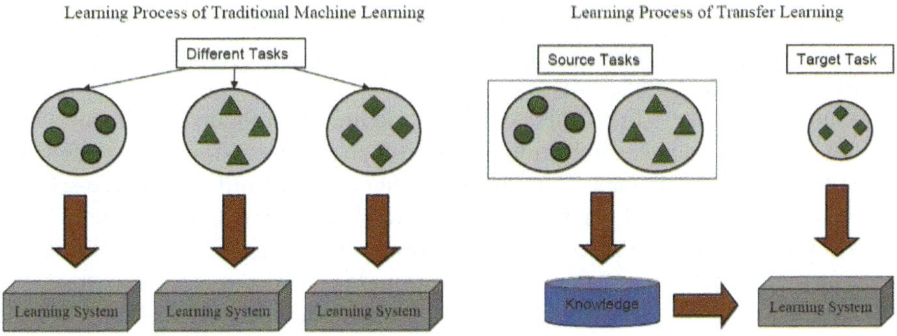

Fig. 1 The learning processes between traditional machine learning and transfer learning

the labeled and unlabeled data distributions are the same. Transfer learning supports various domains, tasks, and distributions used in training and testing. We see several examples in the real world utilizing transfer learning. For example, learning to play the electric organs will help to play the keyboard.

The research of transfer learning is driven by the concept that humans can use previously learned knowledge intelligently to seek quicker and enhanced approaches to new problems. The multi-task learning framework [19] is a closely related learning strategy for transferring learning, which aims to learn multiple tasks concurrently even if they are distinct. A traditional multi-task learning approach is to discover the specific features that every distinct task can take advantage from. Multi-task learning, instead of concurrently learning all the source and target tasks, concerns most about the target task.

Figure 1 illustrates the difference between the traditional machine learning and transfer learning techniques. For traditional machine learning technique, the tasks try to learn separately from scratch, whereas transfer learning technique tries to transfer the information from some prior tasks to a new task if the target has less high-quality target data.

2.1 Notions and Definitions

Transfer learning is a research topic in machine learning which emphasizes on storing the acquired information when resolving one issue and putting it to a different but related issue.

In transfer learning, the pre-trained weights and learned features of a previously trained model are taken to predict the new classes. The benefits of transfer learning are:

1. An extremely large training dataset is not required.
2. Less computing power is needed and only the weights of the last few layers have to be trained.

So, transfer learning is the capability to utilize existing information from the source task in the target task. The following three critical questions need to be answered during the transfer learning process.

What to transfer—In the first step, it is required to find answers about which portion of the knowledge can be transferred from the source to the target to enhance the target task's performance. Thus, the researchers try to identify the source-specific portion of knowledge and what is common among the source and the target.

When to transfer—There may be situations where information transfer for its sake can make matters worse than improving things (also known as negative transfer). So, all must try to use transfer learning to enhance the performance/results of the target task and not reduce it.

How to transfer—This consists of changes to the existing algorithms and different techniques of actually transferring the knowledge through domains/tasks.

3 Process of Transfer Learning

While training a deep neural network, the main aim is to find the optimal values on each filter matrix so that when the image is propagated through the network, the output activations can be used to determine the category to which the image belongs accurately. Gradient descent is used to find the values of filter matrix.

The filters are learning to identify colors and certain horizontal and vertical lines for the first few layers of ConvNet.

Gradually, the next few layers learn to identify trivial types using lines and colors learned in former layers.

Then the layers that follow learn to recognize textures, so pieces of items of pets, vehicles, etc., are recognized

Finally, the filters of the last layer get triggered by complete items like pets, vehicles, etc. By using a pre-trained network to do transfer learning, basically, a few dense layers are added at the end of the pre-trained network and learn which arrangement of these trained features help to recognize the artifacts in the new dataset.

So simply it is required to train a few dense layers. In addition, a combination of these features is used that have already been learned to identify new objects. All this helps to make the training process very quick and needs much less training data than a scratch-based ConvNet training.

For computer vision tasks, some pre-trained models are

- VGG16
- VGG19
- InceptionV3
- Xception
- Resnet-50.

4 Transfer Learning Categorization

Depending on the description of transfer learning, they are divided into three types: *"inductive transfer learning," "transductive transfer learning" and "unsupervised transfer learning."*

1. The source and the target tasks are different in inductive transfer learning. To persuade an unbiased predictive model to be used in the target domain, some labeled data are required. Depending on the labeled and unlabeled data, it can be further categorized into two cases.
   1.1. In this case, many labeled data are obtainable for the source domain. So inductive transfer learning is comparable to multi-task learning and achieves higher performance in the target task.
   1.2 In this case, for the source domain, no labeled data are available. So it is a self-taught learning suggested by Raina et al. [20]. The knowledge of the source domain may not be used directly for the target domain.
2. In the transductive transfer learning, the target and source tasks are same but the domains are different. Few labeled data are available in the target domain, while many labeled data are available in the source domain. It is categorized into two types:
   2.1. Different feature spaces for both the domains.
   2.2. Same feature space between the domains. Here, the learning is associated with domain adaptation for transferring knowledge in text classification [21] and sample selection bias [22] where assumptions are related.
3. In unsupervised transfer learning, the target job is different but related to source job.

No labeled data for both domains are available here. However, it focuses on solving unsupervised learning tasks in the target domain, such as clustering, dimensionality reduction, and density estimation [23, 24]. Table 1 represents the different settings of transfer learning with the related areas the task they perform.

Based on "what to transfer," the three categories of transfer learning can be described in four cases.

The first case is "instance-based transfer learning" [25–28], which means that definite portions of the information in the source domain may be used again by re-weighting for learning in the target domain.

A second case is "feature-representation transfer approach" [29–32]. The instinctive idea behind this case is to learn a "good" feature representation for the target domain.

A third case is "parameter-transfer approach" [33–35], which assumes that certain parameters or former distributions of the hyper-parameters of the models are shared by the source and the target tasks.

Table 1 Different settings of transfer learning

Transfer learning settings	Related areas	Source domain labels	Target domain labels	Tasks
Inductive transfer learning	Multi-task learning	Available	Available	Regression, Classification
	Self-taught learning	Un available	Available	Regression, Classification
Transductive transfer learning	Domain adoption, Sample selection bias	Available	Available	Regression, Classification
Unsupervised transfer learning		Un available	Un available	Clustering, Dimensionality reduction

Last case is the "relational knowledge-transfer" [36], that manages transfer learning for relational domains. The information to be transferred is, therefore, the data relationship. This background has currently been dominated by statistical quantitative learning techniques [37, 38]. Figure 2 shows the hierarchical structure of transfer learning categories.

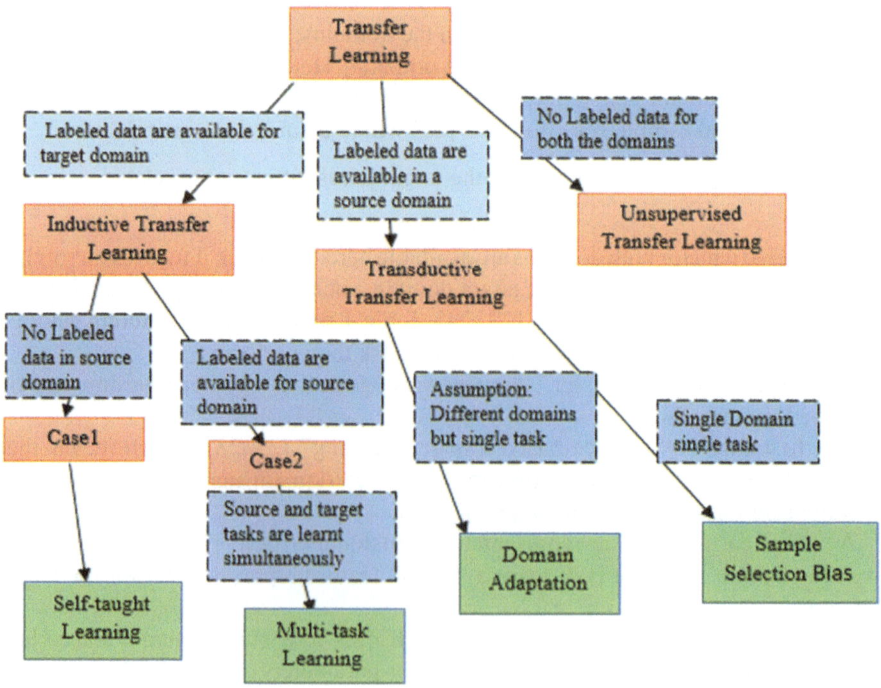

Fig. 2 Overview of different settings of transfer learning

5 Application and Future Scope

Deep learning is certainly one of the unique algorithms used to quite effectively reap the benefits of transferring knowledge. The following are a few examples:

- **Transfer Learning for NLP**: Textual data pose all sorts of challenges when it comes to ML and deep learning. Typically, these are transformed or vectorized using various techniques. Word embedding is designed using various training datasets, with Word2vec and Fast Text (a tool for text classification). These are used by moving the information from the source tasks in various tasks, such as sentiment analysis and document classification. In addition to this, new models such as the Universal Sentence Encoder and BERT (an NLP pre-training approach) certainly present a variety of future possibilities.
- **Transfer Learning for Audio/Speech**: Deep learning has been used effectively for tasks based on audio information, similar to domains such as NLP and Computer Vision. For example, Automatic Speech Recognition (ASR) models developed for English are successfully used, for other languages, such as German, to enhance speech recognition efficiency. Automated speaker recognition is also another instance that has greatly helped transfer training.
- **Transfer Learning for Computer Vision**: Deep learning was used successfully with various CNN architectures for different computer vision tasks, such as object recognition and classification. Yosinski and his co-authors (https:/arxiv.org/abs/1411.1792) present their results on how the lower layers act as traditional extractors of computer vision features, such as edge detectors, while the final layers function against specific tasks. Transfer learning can also solve more complex computer vision problems like image captioning.

6 Conclusion

Some current trends in transfer learning are analyzed in this review. The three types of transfer learning, namely "inductive transfer learning," "transductive transfer learning," and "unsupervised transfer learning" are discussed. The former two environments were based on most of the previous works. Unsupervised transfer learning in the future can attract more and more attention. In addition, every transfer learning methods can be divided into four settings focusing on the learning "what to transfer." These include the learning of "instance-transfer," "feature-representation-transfer," "parameter transfer," and "relational-knowledge-transfer."

Transfer learning methods will be commonly used in the future to address other difficult problems, such as video recognition, a study of social networks, and logical inference.

References

1. Long, M., Yue, C., Jianmin, W., Michael, I.J.: Learning transferable features with deep adaptation networks. arXiv:1502.02791 (2015)
2. Blitzer, J., Ryan, M., Fernando, P.: Domain adaptation with structural correspondence learning. In: Proceedings of the 2006 Conference on Empirical Methods in Natural Language Processing, pp. 120–128. Association for Computational Linguistics (2006)
3. Mo, K., Shuangyin, L., Yu, Z., Jiajun, L., Qiang, Y.: Personalizing a dialogue system with transfer learning. arXiv:1610.02891 (2016)
4. Wei, Y., Zheng, Y., Yang, Q.: Transfer knowledge between cities. In: KDD, pp. 1905–1914, 9(2016)
5. Wu, X., Vipin, K.J., Ross, Q., Joydeep, G., Qiang, Y., Hiroshi, M., Geoffrey, J.M. et al.: Top 10 algorithms in data mining. Knowl. Inf. Syst. **14**(1), 1–37 (2008)
6. Yang, Q., Xindong, W.: 10 challenging problems in data mining research. Int. J. Inf. Technol. Decis. Mak. **5**(04), 597–604 (2006)
7. Fung, Gabriel, P.C., Jeffrey, X.Y., Hongjun, L., Philip, S.Y.: Text classification without negative examples revisit. IEEE Trans. Knowl. Data Eng. **18**(1), 6–20 (2005)
8. Al-Mubaid, H., Syed, A.U.: A new text categorization technique using distributional clustering and learning logic. IEEE Trans. Knowl. Data Eng. **9**, 1156–1165 (2006)
9. Sarinnapakorn, K., Miroslav, K.: Combining subclassifiers in text categorization: a dst-based solution and a case study. IEEE Trans. Knowl. Data Eng. **19**(12), 1638–1651 (2007)
10. Dai, W., Qiang, Y., Gui-Rong, X., Yong, Y.: Boosting for transfer learning. In: Proceedings of the 24th International Conference on Machine Learning, pp. 193–200. ACM (2007)
11. Blitzer, J., Mark, D., Fernando, P.: Biographies, bollywood, boom-boxes and blenders: Domain adaptation for sentiment classification. In: Proceedings of the 45th Annual Meeting of the Association of Computational Linguistics, pp. 440–447 (2007)
12. Ramon, J., Kurt, D., Tom, C.: Transfer learning in reinforcement learning problems through partial policy recycling. In: European Conference on Machine Learning, pp. 699–707. Springer, Berlin, Heidelberg (2007)
13. Taylor M.E., Peter, S.: Cross-domain transfer for reinforcement learning. In: Proceedings of the 24th International Conference on Machine Learning, pp. 879–886. ACM (2007)
14. Yin, X., Jiawei, H., Jiong, Y., Yu Philip, S.: Efficient classification across multiple database relations: A crossmine approach. IEEE Trans. Knowl. Data Eng. **6**, 770–783 (2006)
15. Kuncheva, L.I., Juan, J.R.: Classifier ensembles with a random linear oracle. IEEE Trans. Knowl. Data Eng. **19**(4), 500–508 (2007)
16. Baralis, E., Silvia, C., Paolo, G.: A lazy approach to associative classification. IEEE Trans. Knowl. Data Eng. **20**(2), 156–171 (2007)
17. Zhu, X., Xindong, W.: Class noise handling for effective cost-sensitive learning by cost-guided iterative classification filtering. IEEE Trans. Knowl. Data Eng. **10**, 1435–1440 (2006)
18. Yang, Q., Charles, L., Xiaoyong, C., Rong, P.: Test-cost sensitive classification on data with missing values. IEEE Trans. Knowl. Data Eng. **5**, 626–638 (2006)
19. Caruana, R.: Multitask learning. Mach. Learn. **28**(1), 41–75 (1997)
20. Raina, R., Alexis, B., Honglak, L., Benjamin, P., Ng, A.Y.: Self-taught learning: transfer learning from unlabeled data. In: Proceedings of the 24th International Conference on Machine Learning, pp. 759–766. ACM (2007)
21. Daume III, H., Daniel, M.: Domain adaptation for statistical classifiers. J. Artif. Intell. Res. **26**, 101–126 (2006)
22. Zadrozny, B.: Learning and evaluating classifiers under sample selection bias. In: Proceedings of the Twenty-First International Conference on Machine Learning, p. 114. ACM (2004)
23. Dai, W., Qiang, Y., Gui-Rong, X., Yu, Y.: Self-taught clustering. In: Proceedings of the 25th International Conference on Machine Learning, pp. 200–207. ACM (2008)
24. Wang, Z., Yangqiu, S., Changshui, Z.: Transferred dimensionality reduction. In: Joint European Conference on Machine Learning and Knowledge Discovery in Databases, pp. 550–565. Springer, Berlin, Heidelberg (2008)

25. Dai, W., Xue, G., Yang, Q., Yu, Y.: Transferring Naive Bayes classifiers for text classification. In: Proceedings of the 22rd AAAI Conference on Artificial Intelligence, Vancouver, British Columbia, Canada, July 2007, pp. 540–545
26. Quionero-Candela, J., Sugiyama, M., Schwaighofer, A. Lawrence, D.N: Dataset Shift in Machine Learning. The MIT Press (2009)
27. Jiang, J., Zhai, C.: Instance weighting for domain adaptation in NLP. In: Proceedings of the 45th Annual Meeting of the Association of Computational Linguistics, June 2007, pp. 264–271. Association for Computational Linguistics, Prague, Czech Republic
28. Liao, X., Xue, Y., Carin, L.: Logistic regression with an auxiliary data source. In: Proceedings of the 21st International Conference on Machine Learning, Bonn, Germany, August 2005, pp. 505–512
29. Dai, W., Xue, G., Yang, Q. Yu, Y.: Co-clustering based classification for out-of-domain documents. In: Proceedings of the 13th ACM SIGKDD International Conference on Knowledge Discovery and Data Mining, San Jose, California, USA, August 2007
30. Ando, K.R., Zhang, T.: A high-performance semi-supervised learning method for text chunking. In: Proceedings of the 43rd Annual Meeting on Association for Computational Linguistics. Association for Computational Linguistics, Morristown, NJ, USA, pp. 1–9 (2005)
31. Blitzer, J., McDonald, R., Pereira, P.: Domain adaptation with structural correspondence learning. Proceedings of the Conference on Empirical Methods in Natural Language, Sydney, Australia, July 2006, pp. 120–128
32. Daumé III, H.: Frustratingly easy domain adaptation. Proceedings of the 45th Annual Meeting of the Association of Computational Linguistics, Prague, Czech Republic, June 2007, pp. 256–263
33. Wang, C., Mahadevan, S.: Manifold alignment using Procrustes analysis. In: Proceedings of the 25th International Conference on Machine Learning, July 2008, pp. 1120–1127. ACM, Helsinki, Finland
34. Bonilla, E.M., Chai, K., Williams, C.: Multi-task gaussian process prediction. In: Proceedings of the 20th Annual Conference on Neural Information Processing Systems, pp. 153–160. MIT Press, Cambridge, MA (2008)
35. Gao, J., Fan, W., Jiang, J., Han, J.: Knowledge transfer via multiple model local structure mapping. In: Proceedings of the 14th ACM SIGKDD International Conference on Knowledge Discovery and Data Mining, August 2008, pp. 283–291. ACM, Las Vegas, Nevada
36. Mihalkova, L., Huynh, T., Mooney, J.R.: Mapping and revising markov logic networks for transfer learning, In: Proceedings of the 22nd AAAI Conference on Artificial Intelligence, Vancouver, British Columbia, Canada, July 2007, pp. 608–614
37. Mihalkova, L., Mooney, J.R.: Transfer learning by mapping with minimal target data. In: Proceedings of the AAAI-2008 Workshop on Transfer Learning for Complex Tasks, Chicago, Illinois, USA, July 2008
38. Davis, J., Domingos, P.: Deep transfer via second-order markov logic. In: Proceedings of the AAAI-2008 Workshop on Transfer Learning for Complex Tasks, Chicago, Illinois, USA, July 2008

A Survey on Energy Awareness Mechanisms in ACO-Based Routing Protocols for MANETs

Niranjan Panda, Prabhat Ku Sahu, Manoranjan Parhi, and Binod Ku Pattanayak

Abstract Mobile Ad Hoc Network (MANET) is slowly emerging to be one of the most innovative and challenging areas of wireless networking due to its increasing demand in various areas of application including wireless sensor network (WSN), military application, disaster relief, collaborative and distributed computing, and health and business. Still, the energy-constrained nature of the wireless mobile nodes imposes many restrictions during the design and implementation of MANET routing protocols. This paper focuses on the energy awareness issue during routing protocol design in MANET. Several energy-aware routing protocols have been proposed in past and, out of them, we have considered energy-aware Ant Colony Optimization (ACO) routing protocols for MANET for a comprehensive review. From this paper, researchers can get a brief idea about the functionality of energy-aware routing protocols, mainly using ACO algorithm and use that to propose new mechanisms to handle the energy awareness issue in a MANET.

Keywords MANET · ACO · Optimal route discovery · Energy-aware routing

---

N. Panda (✉) · P. K. Sahu · M. Parhi · B. K. Pattanayak
Department of Computer Science and Engineering, Siksha 'O' Anusandhan (Deemed to be University), Bhubaneswar, Odisha 751030, India
e-mail: niranjanpanda@soa.ac.in

P. K. Sahu
e-mail: prabhatsahu@soa.ac.in

M. Parhi
e-mail: manoranjanparhi@soa.ac.in

B. K. Pattanayak
e-mail: binodpattanayak@soa.ac.in

D. Mishra et al. (eds.), *Intelligent and Cloud Computing*, Smart Innovation, Systems and Technologies 194,
https://doi.org/10.1007/978-981-15-5971-6_85

1 Introduction

A significant development has been observed in MANET's wireless communication domain in recent years. In its real-world application, routing in a MANET plays a vital role. In contrast, MANET routing is synonymous with a collection of issues that must be resolved prior to its implementation. The concerns include meeting the Quality of Service (QoS) guidelines, energy awareness, and security problems. The primary intention of the routing protocols is to find an optimum and cost-effective path, linking the sender and the receiver nodes, which may additionally deplete the node's energy. A further transmission of information across the same path will violently impact the lifespan of the network forcing many nodes to drop out of battery power. Selecting multiple paths joining a couple of nodes helps to spread the information transmission informally across the chosen paths through various nodes and spreads the network load reducing node energy constraints and minimizing the lifetime of the network. It also improves network efficiency, minimizes overheads incurred by data transmission and the expense of restoring alternative paths after connection breakdown, and eliminates communication delay when sending data along multiple paths. Multipath routing protocols, therefore, show a better performance as compared to single-path routing protocols when transmitting data from source to destination. However, battery-powered nature mobile node and the reduced rate of progress in battery technology make MANET more challenging to be used in broad scenarios. Therefore, energy awareness in terms of energy spent by communicating nodes and energy deviation across the network nodes to be minimized for extending the network lifetime.

The flexible and dynamic feature of ACO algorithms makes it ideal during routing to take care of the complex issues. This is a heuristic approach as the result produced affects all the ants present within the network. ACO offers better solutions to choose the shortest path in comparison to the Genetic Algorithm (GA) and Simulated Annealing (SA) [1]. Within the process, the probabilistic movement made by the ants helps them to discover new paths and reevaluate the previously traversed paths. The artificial ants move toward the optimal direction based on the value of the pheromone amassing, and their dispersion helps the machine to forget the old information and avoid the rapid convergence to the suboptimal solution. ACO's actions inspired many researchers to use ACO metaheuristics to model protocols.

The rest of the paper is structured as follows. Section 2 outlines the ACO metaheuristics. Sect. 3 provides a review with a tabular analysis of different energy-aware routing protocols based on ACO. The paper is concluded with future work in the Sect. 4.

2 Ant Colony Optimization

Swarm intelligence strategies [2] are triggered by social bugs and creatures interacting and are utilized to solve different real-life issues. Numerous methods are suggested that make ACO [3] one of the most popular and widely used techniques of optimization in view of the biotic ants' behavior.

Artificial ants cooperate in ACO metaheuristics to find optimal solutions to challenging problems of combinatorial optimization. In order to use the ACO technique to solve a real-life optimization problem, the problem is depicted initially as a construction graph, the confinements of the problem should be portrayed as the ants may simply continue forward on the arcs interfacing the vertices. Each arc related to the pheromone trails (a steady memory of the ant foraging procedure) and a heuristic value (an earlier awareness of the problem happening or run-time information provided by various sources). The ants follow a resolution on the likelihood of using the network as a component of the heuristic information as well as the neighborhood pheromone trails. It can also be cognizant of the private memory of the ant and the requirements of the problem.

Normally, the simple ACO algorithm can be represented as the trading of three essential methods, building ants solutions, updating pheromones, and daemon behavior. Numerous variations of ACO calculations are proposed in past, out of which Ant System (AS) [4], MAX−MIN Ant System (MMAS) [5] and Ant Colony System (ACS) [6] are the widely used ones.

3 Energy-Aware ACO Routing Protocols

ACO-based routing protocols can be classified into several categories based on constraints such as optimal routing, quality of service (QoS), energy awareness, geographical position, secure communication, etc. In [7], we have analyzed the effect of attacks on basic routing protocols, and a survey on various secure routing protocols was carried out in [8]. However, here in our investigation, we center just on the energy-aware routing protocols. Some energy-aware routing protocols for MANET based on ACO routing algorithm has been detailed within this section. Also, a comparison of the energy-aware routing protocols is presented in a tabular form in Table 1 within this section based on various parameters.

3.1 Energy-Aware Ant-Based Routing

Misra et al. suggested an ant swarm-inspired Energy-Aware Ant-based Routing (EAAR) [9] protocol for wireless ad hoc networks based on AntHocNet routing protocol along with an energy efficiency concept to increase the longevity of a node

Table 1 Comparison of various Energy-Aware ACO routing protocols

Year	2010	2013	2015	2016	2016	2017
Authors	Misra et al.	Woungang et al.	Vijayalakshmi et al.	Zhou et al.	Prabaharan and Ponnusamy	Rangaraj et al.
Protocols compared	AODV, MMBCR, AntHocNet	EEAODR	AntHocNet, LAR, MMP, R-ACO1	AOMDV, EAAR	Normal ACO	AODV, ACO-ALEEP
Simulation tools used	GloMoSim	GloMoSim	NS-2	NS-2	Self-designed simulator	Not specified
Techniques used	AntHocNet routing and MRB	Monitors energy across the paths	MMP algorithm	Average energy and minimum energy measurement of the route	SA and randomness to select an optimal path, trust value to provide security	Duty cycle algorithm to minimize the energy drain rate and FDRPSO algorithm with energy fitness function
Trans method	Broadcast	Broadcast	Unicast	Broadcast	Unicast	Unicast
Ant Agents used	FANT, BANT	RREQ, RREP	FANT, BANT	FANT, BANT	FANT	FANT
Routing method	Reactive	Reactive	Proactive	Proactive	Reactive	Reactive
Protocol	EAAR	ACO-EEAODR	AntHoc MMP	ACECR	Hybrid ACO	ACO-FDRPSO
Performance metrics used	Packet drop, Dead node, Energy consumption	Network life, Energy consumption,	PDR, Control overhead, Delay	Delay, Routing load, Energy consumption	Path finding time, Level of randomness and reuse, Distance covered	Throughput, Packet drop, Residual energy, PDR

concerning battery life in a multipath routing scenario. This algorithm uses three tables: a neighbor table that contains the list of nodes in the neighborhood, a seen table that contains the details of packets received, and a routing table that contains routing pheromone values. Route discovery is done using the ant agent FANTs and BANTs as in AntHocNet. Initially, the sender broadcasts FANTs toward the destination and when a reactive FANT reaches the destination, it waits for other FANTs to come for a time period equal to a constant time correspond to end-to-end delay of the earliest FANT received. After that, FANTs are converted into BANTs and transmitted toward the source retracing the paths. During transmission, the routing pheromone table entries are updated in terms of Maximum–minimum Remaining Battery energy (MRB) and the hop count of the link. When the BANTs reach the source waits for time period equals to a constant time of end-to-end delay of the first BANT received for other BANTs to come, afterward they are killed and multiple routes are established. Further, data transmissions are carried out using pheromone values available in the routing table in a probabilistic manner. Route maintenance is done using the pheromone reinforcement and evaporation. Link failures are handled by initiating a new route discovery process by the source.

The authors simulated their proposed algorithm using GloMoSim tool and PARSEC [10] compiler with various parameters. Compared results are shown with reference to various protocols like AODV, AntHocNet, and Min–Max Battery Cost Routing (MMBCR) [11]. This protocol performs well and gives minimal energy consumption for a network, but due to its longer time taken for the best route selection, it shows less efficiency in networks where nodes are extremely motile in nature.

3.2 Ant Swarm-Inspired Energy-Efficient Ad Hoc On-Demand Routing

Woungang et al. proposed an Ant Swarm-inspired Energy-Efficient Ad hoc On-Demand Routing (ACO-EEAODR) [12] protocol for MANET, based on the approach of ACO learning algorithm. This protocol aims to use the network in such a way that each node must possess the energy to sustain within the network. This protocol selects an optimal path based on the parameters such as remaining battery power and the number of hops. At each node, pheromone values are updated according to the remaining battery power. As energy-efficient routing protocol, selection of optimal path is not distributed equally among the two parameters and emphasizes more on choosing a path with higher remaining battery power rather than a path with a minimum number of hops in a ratio seven each to three. Next hop availability metric is defined as a function of local heuristic and pheromone determined stochastically rather than following a probabilistic approach as the next hop is usually known. Since the protocol uses an on-demand approach for data transmission, the route discovery begins only when a node requests to start a communication or data transfer and it ends when all the data get successfully transmitted or after getting a link failure due to unavailability of the route from source to destination. Ant agents in the form of route requests (RREQs) and route replies (RREPs) are used as control packets for route establishment and maintenance during communication.

The authors used GloMoSim as a simulation tool for examination and evaluation of the behavior of the routing protocol. Energy consumed during path selection and network lifetime is used as the performance metrics for evaluation of the protocol with respect to Energy-Efficient Ad hoc On-Demand Routing (EEAODR) protocol [13]. Results show that the protocol provides an optimum result regarding an increase in lifetime of the entire network. As the emphasis is given only toward energy efficiency and other features like path robustness and energy consumption may be affected.

3.3 Robust Energy-Efficient ACO Routing Algorithm for Multi-hop Ad Hoc Networks

A robust energy-efficient ACO routing algorithm for multi-hop Ad hoc networks in MANETs has been proposed by Vijaylakshmi et al. named as AntHocMMP [14] that relies on ant agents for establishing optimal paths from source to destination using Max–Min Path (MMP) algorithm [15] as a basis. This protocol works on a two-phase approach, the first phase is termed as the selection phase, whereas the second phase as an energy conservation phase. In the selection phase, a number of paths are established from source to destination using MMP approach. MMP algorithm computes the path using two metrics energy path cost and energy level. The energy level is categorized as a saturation energy level when it ends up zero and the threshold energy level is a minimal measure of energy that ought to be kept up to keep the node alive. Energy path cost among source and destination node is resolved utilizing two kinds of cost functions; transmission cost is the aggregate amount of the energy spent during the travel through different paths and reception cost, is the energy spent in relation to the Euclidean distance. In the energy conservation phase, FANTs are sent on all possible available paths using an on-demand approach. BANT retraces the path traveled by the FANT till the source. The pheromone values are updated by the FANTs as well as BANTs for every node on the path during their travel and on reaching the source, an optimal path is selected. This phase aims to raise the lifespan of a network by keeping the energy of each node beyond the threshold level. This protocol in the first phase selects all possible relative paths without considering the pheromone values across the nodes and uses the ACO algorithm further in the second phase to find an optimal path out of the selected ones. An adaptive retransmission approach is followed for the detection of link failure and selection of a new path during data transmission. During link failure, the announcement messages are redirected to all participating nodes and alternate links are found out using MMP algorithm.

AntHocMMP algorithm is simulated in NS 2.34 environment and its function is evaluated using metrics like packet delivery ratio, average end-to-end delay, hop-count, and control overhead with variable speed of node, communication range, and the number of nodes. The authors concluded that AntHocMMP shows an outstanding performance in comparison with Robustness-ACO, LAR, AntHocNet, and MMP routing protocol.

3.4 Ant Colony-Based Energy-Control Routing

Ant Colony-based Energy-Control Routing (ACECR) [16] protocol is proposed by Zhou et al. for MANETs that computes the residual battery energy of nodes for path selection taking into account average and minimum energy requirements of the route. During route discovery, FANT travels from source to destination and keeps the rundown of nodes visited. BANT returns up to the source node following the path discovered by the FANT and during its travel, update the routing pheromone table at every intermediary node. The pheromone values are updated considering the average energy of the path, minimum energy of nodes on the path, and hop count across the path traveled by the BANT. When the BANT reaches the source, route discovery process completes and multiple paths get established. Further, data transmissions are done with a probability computed using the pheromone table entries. If no entries are available within the routing pheromone table, then a random selection among the neighbors is done.

This protocol is tested by the authors using three different types of mobility models such as a random waypoint, random walk, and reference point. The outcomes are compared with other routing protocols like Ad hoc On-demand Multipath Distance Vector (AOMDV) [17], EAAR, etc., and found to be showing a better performance in all scenarios as far as average end-to-end delay, routing load ratio, and energy consumption because of the reduction in route rediscovery, reduction in control packets for route maintenance, and selection of a path with higher residual energy on general view.

3.5 Hybrid ACO Routing

A secure and energy-efficient MANET routing incorporating trust values using Hybrid ACO [18] has been suggested by Prabaharan and Ponnusamy that aims to provide energy efficiency in terms of effective load distribution along with security for routing in MANET. This protocol introduces the concepts of Simulated Annealing (SA) and randomness to select an optimal path from one node to another. A trust value is assigned initially to each node within the network as pheromone value that provides security during communication. A FANT selects the next hop comparing the trust values of 25 nodes in two-hop distance (five of its neighbors to be at level one or one-hop distance and each neighbor with its five neighbors to be at level two or two-hop distance) using the SA algorithm. Out of the selected nodes, the best path is chosen (path consisting of two nodes, one from level one and another from level two) based upon the trust value of nodes, calculated using the weighted sum method. Once the node gets selected, the FANT moves to that node updating the trust values or pheromone values across the links. Repeating the process, the destination node is selected. A route with the shortest path from source to destination can be selected or an optimal route can be found using the approach. This protocol distributes the load across the network nodes enhancing the network stability.

The authors carried out the simulation and the result compared with the standard ACO algorithm executing Ant Density strategy. Analysis of results implies that the proposed routing protocol shows a better efficiency than across the other protocols in terms of load distribution and security management within the network. But, this approach takes more time to determine the routing path and the routing path found is of more distance in comparison with the paths found using the regular ACO algorithm.

3.6 ACO-FDRPSO

Rangaraj et al. proposed a power-aware multipath routing algorithm for MANET combining the features of Hybrid ACO [18] and Fitness Distance Ratio-based Particle Swarm Optimization (FDRPSO) [19] algorithms named as ACO-FDRPSO [20]. This algorithm optimizes the performance of MANET by selecting the optimal path in terms of energy efficiency. ACO extends the node's lifetime using the duty cycle algorithm, whereas FDRPSO minimizes the energy consumption across the network nodes. Duty cycle algorithm has been used to minimize the energy drain rate by increasing the Time-To-Live (TTL) of the nodes. Route discovery is accomplished by the ants selecting the nodes with a higher amount of residual energy as per the duty cycle algorithm. Ants are sent toward the destination from source and again traverse back to the source by visiting every one of the intermediary nodes. During their travel, ants collect the path information, update the pheromone value at each node they visit thereby, and completing a traversal update the information at the source establishing multiple paths connecting source and destination. After the path establishment, data transmission is initiated by choosing an optimal path with higher pheromone concentration. During transmission, each node goes into a sleep state when no transmission is there to perform and turns its state to active when any transmission gets initiated. The path information is forwarded to the FDRPSO algorithm as an initial population and successive energy-efficient paths are selected using an energy fitness function by the algorithm. Path maintenance is carried out by updating the residual energy across each node using ACO after every k number of transmissions.

Authors simulated their proposed routing algorithm ACO-FDRPSO and analyzed its performance using the execution measurements like throughput, packet delivery ratio, packet drop, and residual energy under a variable number of nodes and node velocity. The results obtained are compared with the existing AODV and Autonomous Localization-based Eligible Energetic Path with ACO (ACO-ALEEP) [21] routing protocols, and it was concluded by the authors that an improvement has been achieved by the proposed protocol over the compared protocols.

4 Conclusion and Future Work

Energy-conscious ACO routing protocols for MANET were discussed in detail in this paper and evaluated based on their working concepts, prospects, and consequences. In our discussion, we concentrated only on energy conscious as a major concern during the development of the routing protocol, while other considerations such as flexibility, reliability, and QoS can also be considered. Comparison and evaluation of the routing protocols tested have been published, showing that routing protocols based on ACO perform better than traditional routing protocols. Ultimately, it can be concluded that the current MANET scenario requires a multi-path routing algorithm hybrid, reliable, efficient, flexible, safe, and energy conscious to address the disadvantages of existing routing protocols. The content of this paper offers the basic idea of energy-conscious routing strategies based on ACO and can be useful for researchers in the development of new energy-conscious routing protocols.

References

1. Alhamdy, S.S., Noudehi, A.N., Majdara, M.: Solving traveling salesman problem (TSP) using ants colony (ACO) algorithm and comparing with tabu search, simulated annealing and genetic algorithm. J. Appl. Sci. Res **8**(1), 434–440 (2012)
2. Dorigo, M., Birattari, M., Blum, C., Clerc, M., Stützle, T., Winfield, A. (eds.). Ant Colony Optimization and Swarm Intelligence: 6th International Conference, ANTS 2008, Brussels, Belgium, 22–24 Sept 2008, Proceedings vol. 5217. Springer (2008)
3. Dorigo, M., Blum, C.: Ant colony optimization theory: a survey. Theor. Comput. Sci. **344**(2–3), 243–278 (2005)
4. Dorigo, M., Maniezzo, V., Colorni, A.: Ant system: optimization by a colony of cooperating agents. IEEE Trans. Syst. Man Cybern. B Cybern. **26**(1), 29–41 (1996)
5. Stützle, T., Hoos, H.H.: MAX–MIN ant system. Future Gener. Comput. Syst. **16**(8), 889–914 (1996)
6. Dorigo, M., Gambardella, L.M.: Ant colony system: a cooperative learning approach to the traveling salesman problem. IEEE Trans. Evol. Comput. **1**(1), 53–66 (1997)
7. Panda, N., Pattanayak, B.K.: Analysis of blackhole attack in AODV and DSR. Int. J. Electr. Comput. Eng. **8**(5), 3093–3102 (2018)
8. Panda, N., Pattanayak, B.K.: ACO-based secure routing protocols in MANETs. In: Patnaik, S., Ip, A., Tavana, M., Jain, V. (eds.) New Paradigm in Decision Science and Management. Advances in Intelligent Systems and Computing, vol. 1005, pp. 195–206. Springer, Singapore (2020)
9. Misra, S., Dhurandher, S.K., Obaidat, M.S., Gupta, P., Verma, K., Narula, P.: An ant swarm-inspired energy-aware routing protocol for wireless ad-hoc networks. J. Syst. Softw. **83**(11), 2188–2199 (2010)

10. Bosilca, G., Bouteiller, A., Danalis, A., Faverge, M., Hérault, T., Dongarra, J.J.: Parsec: exploiting heterogeneity to enhance scalability. Comput. Sci. Eng. **15**(6), 36–45 (2013)
11. Singh, S., Woo, M., Raghavendra, C.S.: Power-aware routing in mobile ad hoc networks. In: MobiCom, Oct 1998, vol. 98, pp. 181–190 (1998)
12. Woungang, I., Obaidat, M.S., Dhurandher, S.K., Ferworn, A., Shah, W.: An ant-swarm inspired energy-efficient ad hoc on-demand routing protocol for mobile ad hoc networks. In: IEEE International Conference on Communications (ICC) June 2013, pp. 3645–3649. IEEE (2013)
13. Dhurandher, S.K., Misra, S., Obaidat, M.S., Bansal, V., Singh, P.R., Punia, V.: EEAODR: an energy-efficient ad hoc on-demand routing protocol for mobile ad hoc networks. Int. J. Commun Syst. **22**(7), 789–817 (2009)
14. Vijayalakshmi, P., Francis, S.A.J., Dinakaran, J.A.: A robust energy efficient ant colony optimization routing algorithm for multi-hop ad hoc networks in MANETs. Wirel. Netw. **22**(6), 2081–2100 (2016)
15. Ponnuswamy, V., Anand John Francis, S., Dinakaran, J.A.: Max-min-path energy-efficient routing algorithm—A novel approach to enhance network lifetime of MANETs. In: Chatterjee, M., Cao, J., Kothapalli, K., Rajsbaum, S. (eds.) Distributed Computing and Networking. ICDCN 2014. Lecture Notes in Computer Science, vol. 8314, pp. 512–518. Springer, Berlin, Heidelberg (2014)
16. Zhou, J., Tan, H., Deng, Y., et al.: Ant colony-based energy control routing protocol for mobile ad hoc networks under different node mobility models. J Wirel. Commun. Netw. **2016**(1), 105 (2016)
17. Yuan, Y., Chen, H., Jia, M.: An optimized ad-hoc on-demand multipath distance vector (AOMDV) routing protocol. In: Asia-Pacific Conference on Communications, Oct 2005, pp. 569–573. IEEE (2005)
18. Prabaharan, S.B., Ponnusamy, R.: Secure and energy efficient MANET routing incorporating trust values using hybrid ACO. In: International Conference on Computer Communication and Informatics (ICCCI), Jan 2016. pp. 1–8. IEEE (2016)
19. Peram, T., Veeramachaneni, K., Mohan, C.K.: Fitness-distance-ratio based particle swarm optimization. In: Proceedings of the 2003 IEEE Swarm Intelligence Symposium. SIS'03, pp. 174–181. IEEE (2003)
20. Rangaraj, J., Anitha, M.: Implementing energy optimization by a novel hybrid technique for performance improvement in mobile ad hoc network. Int. J. Appl. Eng. Res. **12**(22), 12029–12035 (2017)
21. Vallikannu, R., George, A., Srivatsa, S.K.: Autonomous localization based energy saving mechanism in indoor MANETs using ACO. J. Discrete Algor. **33**, 19–30 (2015)

A Comprehensive Study of MeanShift and CamShift Algorithms for Real-Time Face Tracking

Smriti Gupta, Kundan Kumar, Sabita Pal, and Kuntal Ghosh

Abstract Human face detection is an essential area of computer vision that harness the digital vision for detecting, recognizing, and tracking human faces. Face detection and tracking is widely being used in different applications like security monitoring, home video surveillance, etc. This paper demonstrates an effective human face tracking system for a real-time video stream. In particular, we investigate the two widely used algorithms, MeanShift and CamShift algorithms, for face tracking. However, face detection is performed only in the first video frame using basic Haar cascade features and AdaBoost classifiers. The advantages and drawbacks of MeanShift and CamShift algorithms are discussed in detail with the help of real-time video frames. Results show that the CamShift algorithm is more relevant in head and face tracking, which makes the algorithm highly robust and practically adaptive.

Keywords Human face detection · Haar-like features · AdaBoost classifier · MeanShift algorithm · CamShift algorithm

1 Introduction

With the advent of technology, the digital vision has reached the masses. With the easy access to digital vision, the need arises to harness this immense quantity of data for our benefit. Human face detection is one such area which harnesses digital vision

S. Gupta · K. Kumar (✉) · S. Pal · K. Ghosh
Department of Electronics and Communication Engineering ITER, Siksha 'O' Anusandhan (Deemed to Be University), Bhubaneswar, Odisha 751030, India
e-mail: erkundanec@gmail.com

S. Gupta
e-mail: smriti3009@gmail.com

S. Pal
e-mail: sabitapal@soa.ac.in

K. Ghosh
e-mail: kuntalghosh@soa.ac.in

D. Mishra et al. (eds.), *Intelligent and Cloud Computing*,
Smart Innovation, Systems and Technologies 194,
https://doi.org/10.1007/978-981-15-5971-6_86

for detecting, recognizing, and tracking of a human face. It is widely being used in various situations like security monitoring, home video surveillance, military, transportation, etc. [1]. For instance, face detection and tracking can be applied to different public places like airports, shopping complexes, etc. for counting the number of visitors and tracking their movement. The human face is a unique and non-replicable feature which can be used to identify individuals [2]. However, in the case of real-time applications, in any case, there are extreme variations like cluttered background, different illuminations, facial expressions, variable moving speed, etc. In these scenarios, utilization of human face detection itself a very challenging task [3, 4].

One of the most straightforward approaches to conquering these challenges is by removing the background and keep concentration on the facial regions [5]. Multiple techniques can be used for face detection such as detecting faces using color or motion, identifying faces in a limited area of pixels, Haar cascade classifiers, etc. An appearance-based method is known to provide the best result with a higher detection rate [6]. The cascaded algorithm proposed by Viola and Jones [7] based on the AdaBoost face detection method detects the face with high precision. Modified boosting algorithms such as Real AdaBoost [8], Gentle AdaBoost [9], and Float Boost [10] have improved the detection performance significantly. In literature, the mean shift (MeanShift) algorithm has heavily been used for face tracking [11]. It is a non-parametric and iterative procedure that relocates and exploits the density function's local maximum value. The maximum value for a search window is computed after initializing the part of the distribution. The search window is then relocated to the maximum value and is iterated till convergence [12]. The standard MeanShift algorithm deals with only static distribution. It has a fixed window size, which results in improper detection when an object is either too close or too far from the camera. Therefore, to solve these problems, the Continuously Adaptive Mean Shift (CamShift) algorithm comes into existence [13]. The CamShift exploits an adaptive search window. The size of the search window for the next video frame is a function of zero moment of the window in the current frame that increases the accuracy of target tracking as the tracking box is adjusted adaptively, unlike MeanShift [6]. Comaniciu *et.al.* [11] have proposed a new approach toward target localization which realizes MeanShift as kernel-based object tracking.

This paper describes a face detection and tracking system for real-time applications specifically for the indoor environment. Here, we introduce a face tracking framework, as shown in Fig. 1 using MeanShift and CamShift algorithms. For a fully automated system, Haar feature-based cascade classifier is used to detect a face in the first frame after that detected frame acts as a search window to track the face and head in consecutive video frames. Further, MeanShift and CamShift process the consecutive frames for face tracking that use the back-projection image generated from the probability distribution of the feature map of the search window.

The rest of the paper is segmented as follows. The face detection and tracking framework is described in Sect. 2. Object detection and tracking approaches are presented in Sects. 2.1 and 2.2, respectively. Section 3 discusses the experimental results, and finally, Sect. 4 concludes the paper.

2 Face Detection and Tracking Framework

In this section, a general framework for face detection and tracking (shown in Fig. 1) is explained in detail. In general, any object tracking algorithm is performed in two significant steps. In the first step, the object is detected in the first frame; thereafter, the detected object is tracked using model updating in consecutive frames. In this work, for face detection, a robust face detection approach based on the Haar cascade classifier is adopted, which is extremely fast for real-time applications [7].

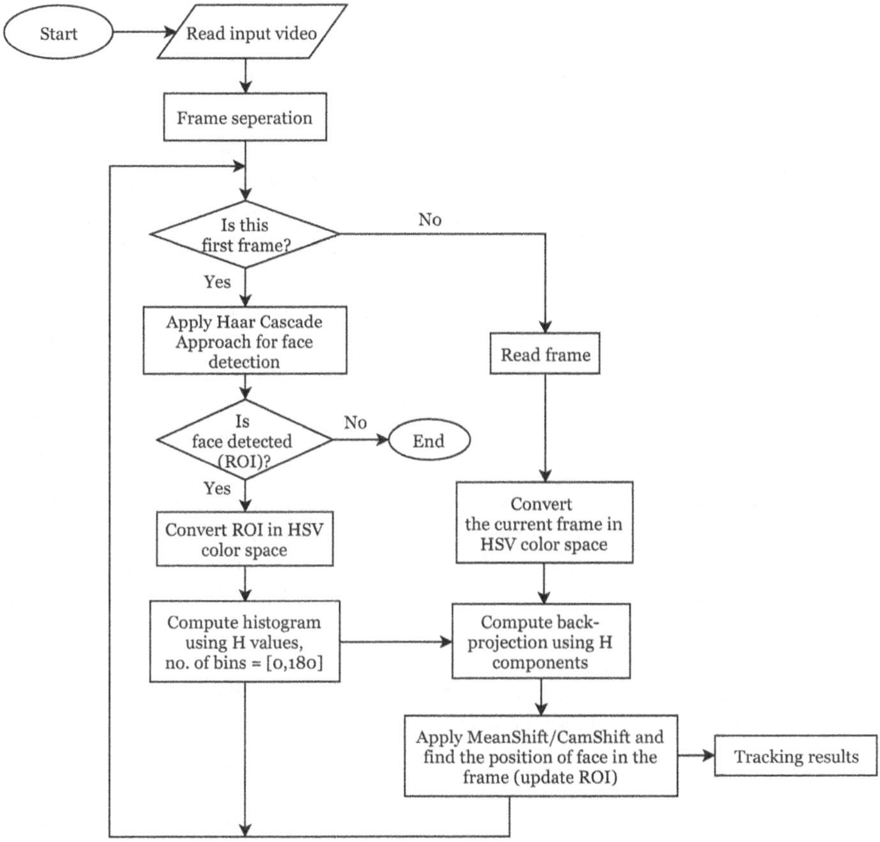

Fig. 1 A general framework for face detection and tracking

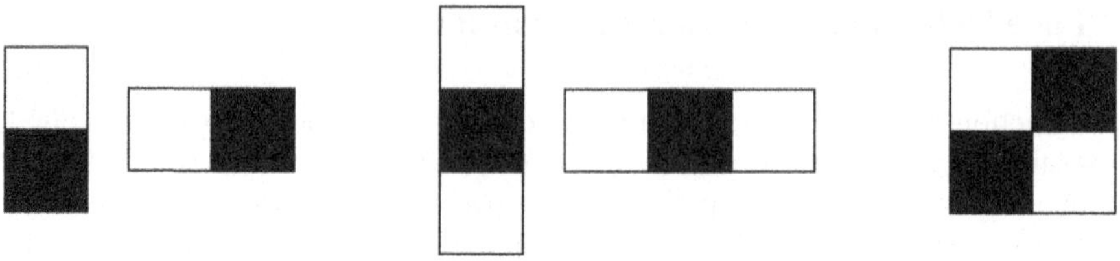

Fig. 2 Common Haar features

2.1 Face Detection Using Haar Cascade Approach

Haar feature-based cascade classifiers is an algorithm for object disclosure which is capable of detecting a face as well. It is based on Haar-like features that use the change in contrast values between a rectangular group of pixels for determining light and dark areas [14]. Usually, Haar-like rectangle features consider a change in the intensity (near edges) at different scales, which helps to extract robust features from face images. Viola–Jones algorithm for face detection is performed in four significant steps: (i) Haar feature selection, (ii) creating an integral image, (iii) AdaBoost training, and (iv) cascading classifiers. In [7], Viola–Jones have considered a standard pixel window of size 24×24 in which almost 160,000 rectangle Haar-like features are possible. After further processing, out of 160,000 Haar features, around 3150 rectangle features are selected to discriminate between the face and non-face windows. To compute these features, a very efficient integral image is used to make the algorithm computationally faster. Figure 2 shows the Haar-like features, which is considered for feature extraction from the 24×24 windows. Further, the cascade classifier is designed using the AdaBoost approach, which produces a strong classifier by combining multiple week classifiers.

After complete training of the cascade classifier, the face is detected in the first video frame. The detected face in the first frame is considered as search window (region of interest (ROI)), which needs to be tracked in consecutive frames.

2.2 Face Tracking

Face tracking is one of the most researched areas of digital vision. Accurate tracking of the face has become a challenging task due to the changing appearance of the target, which is due to their non-rigid configuration and 3D movement. Other obstructions such as occlusion, the abrupt change in environmental conditions like illumination, or interference with background increase the difficulties furthermore. A face tracking algorithm is robust when it is adaptive to changing appearance and scene conditions.

For face tracking, MeanShift and its improved version CamShift algorithms are widely used in the domain. Both the algorithms are explained in more detail in the next two sections.

2.2.1 MeanShift Algorithm

The MeanShift algorithm is a non-parametric, iterative procedure that tries to track the distribution of a region. It tries to find the mode of the feature distribution in a consecutive video frame that is why it is also called a mode-seeking algorithm. As discussed earlier, MeanShift-based face tracking algorithm considers detected facial region as the initial search window. The algorithm search in the neighborhood in the next frame has a similar probability distribution as the search window. Further, it maximizes some similarity function, and the process is repeated for individual consecutive video frames until the termination criteria meet.

The series of steps required for performing face tracking using MeanShift is as follows [13]:

1. Select an initial mean location p_k of the search window, which acts as target distribution to be tracked.
2. Compute new mean location within the search window in the next frame by using Eq. (1).

$$\bar{p}_k(W) = \frac{1}{|W|}\sum_{j \in W} p_j, \quad (1)$$

   where W is the detected search window, p_k is the center data point of the initial search window, and $\bar{p}_k(W)$ is the computed mean position within the search window in the next frame. The position of mean climbs the gradient $f(p)$ to find the mode of the distribution using Eq. (2):

$$\bar{p}_k(W) - p_k \approx \frac{f'(p_k)}{f(p_k)}, \quad (2)$$

   where $f(p_k)$ is the continuous probability density function over the real-valued vector p_k.
3. Move the center of the search window to the updated mean location $\bar{p}_k$ calculated by Eq. (1).
4. Repeat steps 2 and 4 till convergence.

In the domain, convergence of MeanShift is still a debatable topic; however, the algorithm stated here converges to $f'(p_k) \cong 0$, i.e., MeanShift algorithm converges at $f'(p_k) \cong 0$. MeanShift algorithm tracks a fixed size search window in consecutive video frames. Consequently, it fails when the object is too close or too far away from the camera. In these conditions, the original target may be lost and fails to track exact object size [3]. The algorithm may also fail to track an object moving too fast.

2.2.2 CamShift Algorithm

The CamShift is a modification of MeanShift that allows window size estimation. The primary difference between the two is that CamShift is based on dynamic distribution, which recomputes the probability distribution for each frame, while MeanShift is based on static distribution [15]. The CamShift adaptively adjusts the track window size; accordingly, the distribution pattern of target updates during tracking. It calculates zero moment and first-order moments of the search window to obtain centroid position coordinates [16].

The zero moment of the search window M_{00} is measured using Eq. (3):

$$M_{00} = \sum_x \sum_y I(x, y). \tag{3}$$

Similarly, first-order moments M_{10} and M_{01} are obtained by

$$M_{10} = \sum_x \sum_y xI(x, y) \tag{4}$$

$$M_{01} = \sum_x \sum_y yI(x, y). \tag{5}$$

Centroid within the search window is represented as (x_c, y_c)

$$x_c = \frac{M_{10}}{M_{00}}; \; y_c = \frac{M_{01}}{M_{00}}. \tag{6}$$

Here, $I(x, y)$ is the pixel value at the video frame location (x, y). The steps involved in the CamShift algorithm are as follows [13]:

1. Select an initial mean location of the MeanShift search window.
2. Iterate the MeanShift algorithm to compute and store the zeroth moment (distribution area) and centroid location using Eqs. (3) and (6), respectively.
3. The function of zero moment acts as the size of the search window for the next video frame.
4. Repeat steps 2 and 3 till convergence.

3 Results and Discussions

We have implemented and tested the algorithms on real-time video captured using a webcam. The captured video has 124 video frames having resolution 540 × 360 pixels. For the evaluation purpose, we have created a ground truth video by drawing bounding box manually over the face in each frame. We have recorded the rectangle

(a) (b)

Fig. 3 **a** Face detection using Haar feature-based cascade classifier in the first video frame **b** back-projection of the first frame using histogram of ROI

coordinates to compute the Intersection over Union (IoU) for each frame. The IoU is defined as the ratio of area of overlap and area of union between ground truth bounding box and predicted bounding box.

Initially, video frames are separated in a set of consecutive frames. After that, Haar feature-based cascade classifier is applied to the first video frame for face detection as a Region of Interest (ROI), as shown in Fig. 3a. The detected ROI serves as an initial search window for further processing using MeanShift- and CamShift-based tracking algorithms. Before applying MeanShift and CamShift, ROI is converted from RGB (Red, Green, and Blue) color space to HSV (Hue, Saturation, and Value) color space. Further, the Hue component is utilized to compute the histogram of the search window. The ROI image and corresponding histogram are shown in Fig. 4a, b, respectively. The bin size of the histogram is considered as 180. From Fig. 4, we can observe that most of the facial feature lies in the range of [0,25]. The computed histogram is further used to obtain the back-projection image of the first frame, which is shown in Fig. 3a. Here, each pixel of the image is related to the corresponding bin value of the histogram. Histogram bin values are rescaled to the new range [0,255]. The pixel belonging to the ROI histogram is mapped as visible intensities, and the rest of the pixels are assigned to zero, which makes background dark and highlights the only region of interest in the back-projected image. After that, the next frame is read, and back-projection is applied to that too. Finally, MeanShift and CamShift algorithms are applied to track the search window in the consecutive frames resulting in the face as tracked in successive frames. This process is repeated until the last frame of the video. The tracked results for a few frames at certain intervals are shown in Figs. 5 and 6 using MeanShift and CamShift algorithms, respectively. Each frame shows the ground truth box shown by a green rectangle (thin line) and a bounding box predicted by the algorithm shown by a blue rectangle (thick line).

To evaluate the tracking algorithms, Intersection over Union (IoU) is computed for individual frames. We set a threshold at 0.5 over computed IoU to obtain the accuracy of algorithms for captured video. If the IoU exceeds the threshold, then

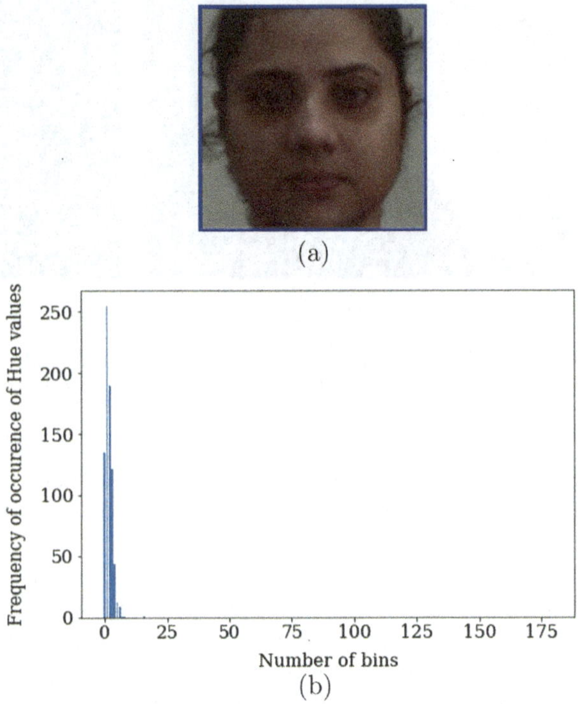

Fig. 4 **a** Region of interest/search window, **b** corresponding Hue channel histogram

face is predicted correctly in the current frame; otherwise, the frame is counted as false prediction. However, the accuracy is computed using Eq. (7):

$$\text{accuracy} = \frac{\text{no. of correctly predicted frames}}{\text{total number of frames}} \times 100. \tag{7}$$

After evaluating each video frame, we obtained the accuracy of MeanShift and CamShift algorithms as 79.83 % and 91.12 %, respectively. Accuracy results show that the CamShift algorithm has better capability to track the face in consecutive frames compared to MeanShift algorithm. The CamShift has an advantage of adaptive window size due to which it can detect the faces in different scenarios, such as face too close or too far from the camera or object moving too fast where MeanShift fails. But in some cases, when background interferes with the face, the tracking cannot be done correctly, and ambiguous results were generated.

Fig. 5 Face detection and tracking in video frames utilizing MeanShift algorithm

4 Conclusion

In this paper, Haar feature-based cascade classifier is used for face detection, which yields the result with high accuracy. MeanShift and CamShift algorithms have been successfully applied for face tracking in real-time video. Through implementation, we conclude that the CamShift algorithm is more robust and gives better results compared to the MeanShift algorithm due to its adaptive window property. Furthermore, CamShift is capable of tracking the face close or far from the camera. However, the performance of these algorithms is unreliable when another object interferes with the tracked object with a similar color distribution. In this case, it follows the wrong object and loses the information of an actual tracked object. By considering the features independent of color or multidimensional histograms, tracking can be further improved, that is, our future work.

Fig. 6 Face detection and tracking in video frames utilizing CamShift algorithm

References

1. Cong, D., Shi, P., Zhou, D.: An improved camshift algorithm based on RGB histogram equalization. In: 2014 7th International Congress on Image and Signal Processing. pp. 426–430. IEEE (2014)
2. Cuimei, L., Zhiliang, Q., Nan, J., Jianhua, W.: Human face detection algorithm via haar cascade classifier combined with three additional classifiers. In: 2017 13th IEEE International Conference on Electronic Measurement and Instruments (ICEMI) (2017)
3. Hongling, W., Bo, Y., Guodong, T., Aidong, M.: Object tracking by applying mean-shift algorithm into particle filtering. In: 2009 2nd IEEE International Conference on Broadband Network and Multimedia Technology. pp. 550–554 (2009)
4. Chen, S., Ma, X., Zhang, S.: Adaboost face detection based on haar-like intensity features and multi-threshold features. In: 2011 International Conference on Multimedia and Signal Processing. vol. 1, pp. 251–255 (2011)
5. Viraktamath, S.V., Katti, M., Khatawkar, A., Kulkarni, P.: Face detection and tracking using openCV. SIJ Trans. Comput. Netw. Commun. Eng. **4**(3), 1–6 (2016)
6. Cong, D., Shi, P., Zhou, D.: An improved camshift algorithm based on RGB histogram equalization. In: 2014 7th International Congress on Image and Signal Processing. pp. 426–430 (2014)

7. Viola, P., Jones, M.: Robust real-time face detection. In: Proceedings Eighth IEEE International Conference on Computer Vision. ICCV 2001. vol. 57, pp. 137–154 (2001)
8. Schapire, R.E., Singer, Y.: Improved boosting algorithms using confidence-rated predictions. In: Proceedings of the 11th Annual Conference on Computational Learning Theory, vol. 37, pp. 80–91 (1998)
9. Freund, Y., Schapire, R.E.: Experiments with a new boosting algorithm. In: ICML'96 Proceedings of the Thirteenth International Conference on International Conference on Machine Learning. pp. 148–156 (1996)
10. Li, S., Zhang, Z.: Floatboost learning and statistical face detection. IEEE Trans. Pattern Anal. Mach. Intell. **26**(9), 1112–1123 (2004)
11. Comaniciu, D., Ramesh, V., Meer, P.: Kernel-based object tracking. IEEE Trans. Pattern Anal. Mach. Intell. **25**(5), 564–575 (2003)
12. Sun, L., Wang, B., Ikenaga, T.: Real-time non-rigid object tracking using camshift with weighted back projection. In: 2010 International Conference on Computational Science and Its Applications. pp. 86–91 (2010)
13. Bradski, G.R.: Computer vision face tracking for use in a perceptual user interface. In: WACV (1998)
14. Wilson, P.I., Fernandez, J.: Facial feature detection using haar classifiers. J. Comput. Sci. Coll. **21**(4), 127–133 (2006)
15. Allen, J.G., Xu, R.Y.D., Jin, J.S.: Object tracking using camshift algorithm and multiple quantized feature spaces. VIP '05 Proceedings of the Pan-Sydney Area Workshop on Visual information Processing pp. 3–7 (2004)
16. Kim, M., Kumar, S., Pavlovic, V., Rowley, H.: Face tracking and recognition with visual constraints in real-world videos. In: 2008 IEEE Conference on Computer Vision and Pattern Recognition. pp. 1–8 (2008)

Blockchain: Applications and Challenges in the Industry

Amrit Sahani, Jaswant Arya, Abhipsa Patro, and Sushree Bibhuprada B. Priyadarshini

Abstract Blockchain technology has brought potential security features for the industrial communities. The capability of blockchain can be seen in solving the industrial challenges as in sharing the transaction records with security, interoperability between the devices, transparency in the supply chain and logistics industry, etc. Blockchain provides the distributed, decentralized, and secured transaction ledger. The omitting intermediaries' property of blockchain reduces the effective cost, while the distributed and decentralized property has shown the application possibility in various cases. This paper has described the potential applications of blockchain and their drawbacks in the various industrial sectors. However, blockchain applications have several challenges which are to be solved.

Keywords Blockchain · Cryptography · Supply chain

1 Introduction

Blockchain technology has established, optimized, and secured industrial transaction processes. It has the use cases in the financial sector, healthcare sector, supply chain, and logistics sector. To establish trust among the participants is always problematic because it requires the trust and centralized system does not guarantee the

A. Sahani · S. B. B. Priyadarshini (✉)
Siksha 'O' Anusandhan (Deemed to be University), Bhubaneswar, Odisha 751030, India
e-mail: bibhupradabimala@gmail.com

A. Sahani
e-mail: situnamrit@gmail.com

J. Arya
National Institute of Technology, Shillong, Meghalaya, India
e-mail: jaswantarya442@gmail.com

A. Patro
International Institute of Information Technology, Bhubaneswar, India
e-mail: patro.abhi660@gmail.com

D. Mishra et al. (eds.), *Intelligent and Cloud Computing*,
Smart Innovation, Systems and Technologies 194,
https://doi.org/10.1007/978-981-15-5971-6_87

trust. The sharable record or ledger with security is guaranteed by blockchain for various organizations. The record cannot be altered by using blockchain technology. Blockchain-based system provides the traceability and transparency in the transactions. The smart contract version of blockchain allows the parties to share the ledger with consistency without any centralized third party [1]. The agreement is registered in the smart contract having several conditions. The conditions are executed automatically when the conditions are satisfied. There may be various use cases including supply chain, lab testing in the medical for a patient. Blockchain technology integrates the benefits of decentralization, distribution of the records with the consistency. Blockchain platform invites the parties for various collaborations to achieve different tasks without needing any central authority. Blockchain feature of securing the entities invite the business to adopt it. It keeps the record of entries in the chronological order using a timestamp. The use of blockchain can be seen in the industry for making a survey on the applications, challenges and opportunities, financial sector, logistics, health care, etc. Many papers are describing the blockchain applications in energy, robotics, and industries, etc. This paper has optimized all the applications and challenges into one paper. It also highlights the possible applications of blockchain.

Several papers are discussing the consensus algorithms and their scalability issues briefly. In this paper, we have discussed the major used algorithms and their challenges. This paper has not described the architecture briefly but the basic concept has been included. The industry-related applications are described in deep. The fundamental concept of blockchain and its functionality also has been discussed. Blockchain has shown its potential in several domains.

1.1 Blockchain Technical Fundamentals

Blockchain consists of the blocks connected in the chronological order. Each block has its parent block hash to verify the connection. If the parent hash does not match with the previous block hash, then the blockchain is supposed to change by an attacker. Therefore, the consistency is measured with the hash values.

1.2 Block Structure

A block is similar to a data structure or object which contains the information regarding the block header and contains the transactions. This block has block version, timestamp, nBIts, Nonce, parent block hash with other several details like Merkle tree root hash, etc. A block contains the transactions according to its predefined block size.

1.3 Digital Signature

Asymmetric cryptography is used to validate the identity of the record or transaction. The blockchain system user has one private key and one public key to show the identity as its owner. The transactions are broadcasted in the whole network with the sign of ownership. First time, the transaction is signed and the next time, it is verified. Elliptic curve digital signature algorithms are one of the algorithms used in a digital signature.

1.4 Consensus Mechanism

To verify the consistency of the transactions, there are a few algorithms that are used. These algorithms are based on the agreed mechanism among the users. The ledger of the transaction record is share among all the users. This ledger is in the form of blocks or chain of blocks. To reach on the consensus among a large number of the trustless participants, proof of work, proof of stake, proof of concept, PBFT, etc. are used. Proof of work requires the solution of a mathematical puzzle to get the authenticity of creating a block. Proof of stake requires proving the maximum stake investment to create the block. Practical byzantine fault tolerance or PBFT is based on the solution of byzantine faults. It has multiple rounds to reach on the consensus. Three-consensus process has three steps. Successful verification of each step leads to enter in the next step. Delegated proof of stake or DPOS has the new feature of making a smaller set of the miner. It allows delegates to verify and create the new blocks. A smaller set of delegates makes the block verification faster. Ripple consensus works on the sub-networks. The server nodes participants in the consensus and the client transfers the funds. The Unique Node List is used while verifying and adding the transaction in the blockchain. There are many other algorithms used for consensus.

The following Error: Reference source not found gives a summary of the consensus mechanism used in blockchain (Table 1).

Table 1 Comparison of the various consensus mechanism used in blockchain

Consensus algorithm	Example	User identity
Proof of work	Bitcoin	Public
Proof of stake	Peercoin	Public
PBFT	Hyperledger fabric	Private
DPOS	Bitshares	Public
Ripple	Ripple	Public

2 Applications of Blockchain in Various Industries/Applications

Blockchain technology has various application sectors including financial sector, health care, land registration and records, big data, IoT, supply chain, and logistics and manufacturing industry, etc.

2.1 Finance Industry

Bitcoin was the first and most successful crypto-currency which forced financial world to understand the blockchain. Financial sector uses the third parties to confirm the reliability of the users, identifying the double spending, registering the users, verify the transaction records, and contacting to their end-users as an agent or intermediaries. The blockchain technology brings the feature of registration, validation, and finding the duplicity or double spending of the transactions. Double spending allows the user to spend the money more than what he has in the account. This problem has been solved in blockchain as all the financial transactions are verified before the actual adding into the blocks. Blockchain has an immutable ledger of all the financial records. It means that the records can be appended to the ledger only. There are a few major applications of blockchain as digital currencies and stock trading.

2.2 Digital Currencies

The cryptocurrencies are the digital currencies that use the cryptographic hash algorithms and blockchain technology. The blockchain is used to verify the ownership of the currencies and validate the transactions. The first and most successful cryptocurrency Bitcoin was released in January 2019 [2]. Stock Trading. The intermediaries increase the cost of the end product. The centralized and trusted authority maintains the records of all the transactions. The blockchain technology features such as decentralization, transparency, security with the cryptographic algorithm, and auditability have benefited the stock market in reducing the additional cost. ZERO [2] and Chain [2] are the two successful examples of blockchain in the stock market.

Financial Settlements: Blockchain supports the industry in validation and verification of the financial settlements. With the combination of other applications, blockchain is very useful. RBC was one of the organizations which implemented the blockchain technology to secure and fast their financial settlement process [3]. Blockchain supports the insurance market by providing various automated policies to the insurance companies. It helps in submitting and processing the claims using the smart contract [4]. The processing of insurance claims can be made faster and cheaper in using smart contracts. The terms and conditions can be saved in the conditions of the smart contract program. When these conditions are satisfied, the program executes the further process accordingly. The blockchain features reduce the cost and overhead of the adminstration [5].

2.3 Supply Chain and Logic

The management of delivering raw material to reaching the final product to the customers involves many processes. The multiple organizations are involved to make the complete process using various channels. The various channels and the multiple organizations increase the complexities of the supply chain logistics industry. The activities and sub-activities of the organizations are to be synchronized during the whole process. The planning, coordinating, monitoring, validating, and verifying the various activities can be optimized using blockchain technology. The shared and distributed mechanism of blockchain reduces the complexity of whole supply chain. Hijro [6] and Provenance [7] are the blockchain-based supply chain and logistics applications.

3 Blockchain Challenges

Blockchain technology has proven its potential in various sector but still, it suffers from the following challenges.

3.1 Scalability

Blockchain technology requires multiple parties to validate and verify the transaction records added into the blocks. This complexity to mine the block increases as the number of transactions increases. The size of the block is to be fixed during blockchain development. And, the time to create a block also depends upon the number of user and transactions [8].

3.2 Security and Privacy

The security and privacy are to keep using the cryptography and continuous verification of the nodes. But, there are a few cases which show the vulnerability in the blockchain application Bitcoin [9, 10] has the attacking the Bitcoin network.

3.3 Blockchain Challenges Regarding Regulations

The blockchain technology integrates all transaction process, registration, validation, and verification of the user and their transaction. But, the regulation is still required to solve the disputes.

4 Conclusion

In this paper, we have described a few possible directions of blockchain applications in the various sectors. The utilization of blockchain technology has been demonstrated for the financial sector, logistics sector, etc. We have written the challenges of blockchain technology in the current scenario.

References

1. Zheng, Z., et al.: Blockchain challenges and opportunities: a survey. Int. J. Web Grid Service. 352–375 (2018)
2. del Castillo, M.: Chain is now working on six 'Citi-Sized' blockchain networks. https://www.coindesk.com/chain-now-working-six-citi-sized-blockchain-networks. Accessed June 2017
3. Suberg, W.: Hyperledger Blockchain 'Shadows' Canadian Bank's International Payments. Cointelegraph Website. https://cointelegraph.com/news/ameri-caneconomic-association-holds-presentations-on-crypto-blockchain (2017). Accessed Jan 2019
4. Cohn, A., West, T., Parker, C.: Smart after all: blockchain, smart contracts, parametric insurance, and smart energy grids. Georgetown Law Technol. Rev. **1**(2), 273–304 (2017)
5. Mainelli, M., von Gunten, C.: Chain of a lifetime: how blockchain technology might transform personal insurance. Long Finance, London, UK (2014). http://archive.long-finance.net/images/Chain_Of_A_Lifetime_December2014.pdf
6. Hijro: https://hijro.com/. Accessed Nov 2018
7. Steiner, A., Baker, J.: Blockchain: the solution for transparency in product supply chains. In: White Paper. Project Provenance Ltd., London, UK (2015)
8. Karame, G.: On the security and scalability of bitcoin's blockchain. In: Proceedings of ACM SIGSAC Conference on Computers and Communications Security (CCS), pp. 1861–1862. New York, NY, USA (2016)
9. Dale, B.: This college freshman is out to 51% attack your cryptocurrency. Coindesk. https://www.coindesk.com/this-college-freshman-is-out-to-51-attackyour-cryptocurrency. Accessed 14 Jan 2019
10. Porru, S., Pinna, A., Marchesi, M., Tonelli, R.: Blockchain-oriented software engineering: Challenges and new direction. In: Proceedings of 39th International Conference on Software Engineering Companion, pp. 169–171 (2017)

Application of Big Data Problem-Solving Framework in Healthcare Sector—Recent Advancement

Sushreeta Tripathy and Tripti Swarnkar

Abstract In the field of engineering and medical science domains, organized, semi-organized, and unstructured data are rapidly growing in recent years. The healthcare sector has been tackled by requirements of big data being generated by different origin, which are outstanding for delivering high volumes of heterogeneous information. Big data analysis on the healthcare domain relays on the utilization of convenient tools and architecture. In the recent past a lot of analyses have been conducted on application-specific healthcare foundation, which results in data-precise competence for managing sources of information ranging from digital health documents to pictures of different body parts. In this manuscript, we have focused on different critical avenues that exist in the electronic health repository from the point of view of different contributors. Authors studied on the different massive data groundwork underlying sources of data, problem-solving capability, and different application areas.

Keywords Big data · Healthcare · Electronic health documents · Analytics

1 Introduction

Data are produced by a scope of dissimilar applications, appliances, and landscape analysis for the reasons such as climate forecasting, disaster control, crime identification, medical science, railway reservation system, risk handling in defense, and so on. In a recent scenario, massive data have been combined with basic technologies

S. Tripathy (✉)
Department of Computer Science and Information Technology, Siksha 'O' Anusandhan (Deemed to Be University), Bhubaneswar, Odisha 751030, India
e-mail: sushreetatripathy@soa.ac.in

T. Swarnkar
Department of Computer Application, Siksha 'O' Anusandhan (Deemed to Be University), Bhubaneswar, Odisha 751030, India

D. Mishra et al. (eds.), *Intelligent and Cloud Computing*,
Smart Innovation, Systems and Technologies 194,
https://doi.org/10.1007/978-981-15-5971-6_88

and different enterprises, including Google, Facebook, and IBM, which fetch meaningful information from the massive amount of data collection [1–3]. An era of open information in the health sector is now underway.

The manuscript is organized as follows: (i) Sect. 2 contains brief introductions about big data; (ii) Sect. 3 is about the importance of big data processing in the health domain; (iii) Sect. 4 briefly describes the influence of big data in the health sector; (iv) finally, the conclusion is drawn in Sect. 5.

2 Definition of Big Data

The growth rate of data and information has doubled due to the advancement of technologies and the internet. Huge amount of data are easily available due to increase in the binned practice of data assembly. Every moment, progressive data are being generated, which need to be stored and analyzed in order to extract more valuable information. Massive data introduce large, complex data sets that are beyond the capabilities of a normal database to store, manage, and process. The characteristics of six V's in huge data need a new type of data analytics methods [4].

The massive amount of data in the health sector include huge numbers of data from different healthcare substructures by gathering, handling, surveying, imaging, and producing information for making a conclusive decision. The vast data in medical management are connected with six features, that is, volume (hugeness of data), variety (available of data from different sources), velocity (frequency of data generating), veracity (correctness and accuracy of information), variability (handling data with its fluctuations), and value (inherit meaningful information from massive amount of data) [5, 6]. The six V's of massive data are displayed in Fig. 1.

The vast data sources of healthcare sector are categorized as: (i) organized data, (ii) semi-organized data, and (iii) unstructured data. Organized data obey defined data and format. Its best examples are different types of diseases, its symptoms, information about laboratory test, patient's admission records, prescribed medicines, and billing details. Semi-organized data is formed with the nominal form along with the self-classifying character. Massive data are produced from appliances such as photoelectric cells for the fruitful recording of patient's behavior. Unregulated data have no basic form, which include handwritten prescriptions, clinical reports, patient discharge reports, and so on. The analysis of medical data to accomplish beneficial inputs for different collaborators (medicals, patients, clinicians, and drugstore) is a burdensome job due to massive amount of information derived from different sources [7].

Fig. 1 Big data associated with its six V's

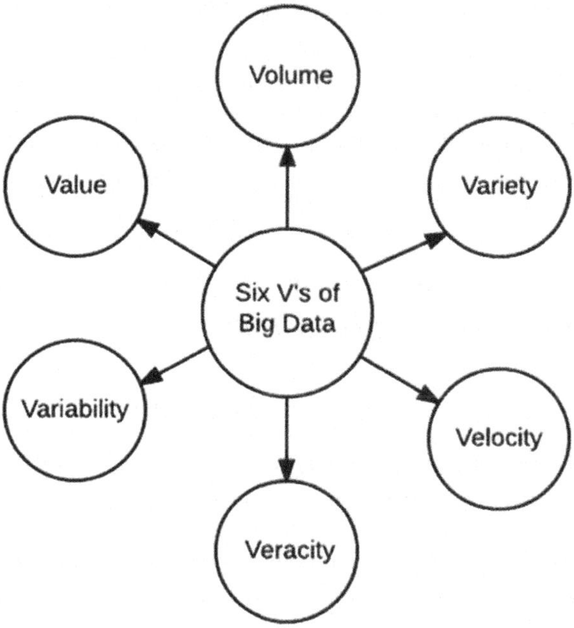

3 Analysis Stages of Huge Data in Health Sector

Analysis of huge data has the potential to bring innovation in the methods, practice, and equipment of healthcare providers, to enhance cognizance from their clinical databank and make a proclaimed conclusion. Healthcare analytics of huge data has been divided into five stages: data acquisition, storage, management, analytics, and visualization. The analysis stages of huge data in the health sector have been shown in Fig. 2.

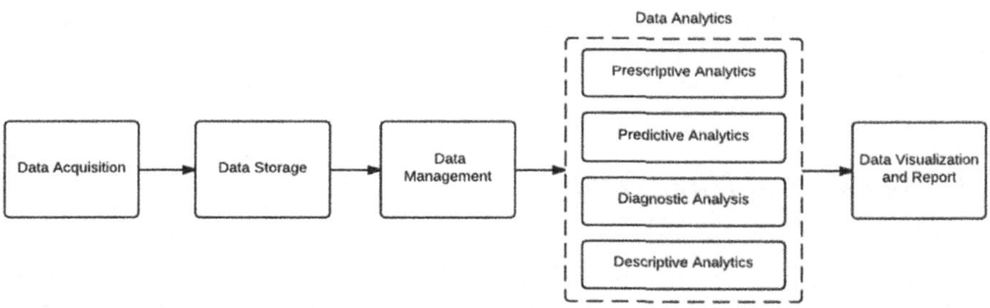

Fig. 2 Process of huge healthcare data

3.1 Acquisition of Huge Data

The huge data in the health sector have been classified as organized, semi-organized, or unregulated, and it can also be acquired from main sources (e.g., medical decision management, digital reports of patients health, medical X-ray image, social media data, and web data) and other sources (laboratories, insurance companies, and pharmacies) [8].

3.2 Storage of Huge Data

The data repository acts as an important phase in huge data analysis. As the dimension of information in the health sector is expanding, we require a well-organized and massive warehouse program. The vast cloud is the best auspicious tool which gives elasticity and proficiency to enter into data. It also boosts the ability for scalable analytics solutions, awareness, and driving value. Cloud assessing is a strong and auspicious tool to accumulate a vast amount of data and achieve in huge scale, and evaluate which is complex in nature [9].

3.3 Management of Huge Data

Managing huge data in the health sector includes categorizing, noise removing, retrieval, mining knowledge, and lawful control of data. Also it contains the approach of verifying whether any inconsistencies in information or there is some suspicion data present in the huge databank. Now deletion of such data is required [10]. It provides benefits in the assessment risk of patients, an individualized discharge plan. Apache Ambari and Catalog are the most important data management tools provided by different companies. Data retrieval is a process of extracting meaningful data from a vast healthcare repository. Analysis of huge data in the health sector contains data fetching and recovery of meaningful information.

3.4 Analytics of Huge Data

The technique used for converting sensitive data into information is called analytics. Analytics of healthcare information has been categorized as diagnostic, expressive, prognostic, and authoritative analysis [11]. Hence, the current study in industry and by academicians shows that distributers can achieve 15–20% magnification of ROI by placing analytics of massive data.

4 Influence of Huge Data in Healthcare

The trading industry always gives emphasis to discover the behavioral patterns of customers to develop innovative business services. In the health domain, the application of massive data serves analytical methods and machine learning approaches [8] for the provision of justifiable answers such as the application of personalized medical care and treatment plans [12]. The feature of massive health data is also precise and divided into three attributes, namely databank, security, and variety, instead of volume, velocity, and variety.

The databank stores patients' healthcare records updated in collaborators premises such as medicals. The attributes of security focus on the additional supervision required in preserving health sector information. Attributes of different types specify the existence of health data in many forms, such as organized, semi-organized, and unstructured. The influence of massive data in the health sector results in recognizing new sources of data like social media, telemedicine, and devices which are wearable. Additionally, the surveying of legacy sources contains past reports of patients, clinical test data, and index of medicine usefulness. It generates meaningful information from sources for health sector researchers toward attaining novel healthcare outcomes when a combination of these data sources and analysis is mixed with each other [13–15].

Figure 3 presents the classical patient care ecosystem along with its significant collaborators and different sources of data (organized/semi-organized/unregulated) [16]. The healthcare outcomes could be provided in a profitable way along with much upgraded standard, although collaborators are exerting collaboratively and pass their information successfully.

4.1 Healthcare Collaborators and Sources of Huge Data

It describes in what way different domains of massive data supplies to the needs of well-known collaborators (patients, clinicians, health workers, drug, and health sector insurers) in attaining beneficial outcomes of medical care. It also provides suitable analytical approach to fetch valuable health maintenance interesting samples from the stipulated sources of data.

Patients. Every time patient expectations are high for availing cost-effective healthcare services with individualized suggestions. Additionally, the physician's diagnosis inside the clinic, the need for hours is to obtain healthcare information across electronic platforms mainly social sites and clinical health records. The patients are able to couple with identical people in obtaining details like symptom of various diseases, reaction, medicine details, hospitalization, and clinical test with better confidentiality provided by massive data sources [17]. The platform can be used to store important health symptoms like heart beat, temperature, and blood pressure of the body, and flow of this information into an integrated databank for

Fig. 3 The big data perspective patient healthcare ecosystem

generating periodical health alerts. Patients can use telemedicine facilities for their medical care, and it needs directly through massive data sources available in the social media platforms, clinical forums, and so on, without visiting hospitals.

Medical Professionals. A large quantity of data are being produced from different levels of disease analysis and diagnostic procedures of patients. It assists the patient care suppliers to offer better and improvised treatment plans to the patients. The vast sources of massive data are produced during the execution of treatment plans by the healthcare system. It includes classes of different diseases, laboratory outcomes, clinical hand-written report, X-ray imaging, and sensor appliances that focus on patient's behavior under divergent scenarios. When such massive data sources are considered for constructing a repository of clinical disease, it provides the benefit of public health observation and provides quicker reply by successful analytics and solution of various diseases.

Medical Workers. A healthcare provider basically depends on the outcomes of massive data sources for managing patient's disease and its treatment plans. Analysis of prescriptive and predictive outcomes improved along with the proficiency in data analytics for evaluating the strength of associations among index of patient content and utilized benefits. In addtion, the distribution of resources and optimization procedures may be employed on the basis of accessible massive data toward achieving the personnel needs for separate departments of medical. The tactical operators of the medicals can also employ the awareness data of different areas to determine the grouping of different sections to optimize the use of medical equipment which is highly expensive. Evolution of descriptive framework based on post-treatment information produced by mobile, e-mail, documents would also make easy on the improvement in healthcare systems.

Pharmaceutical and Researchers in the Clinic. The impact of massive data can bring reform in healthcare area of pharmaceutical and clinical research. Omics and clinical usage of huge data [18] assist to construct a predictive framework for realizing the medicine and biological activities. Analysis of fitness data from different sources of huge data assists drug agencies to survey the result of designed medicines with brief trials.

Healthcare Supporters. Massive data open new analytical avenues for the advantage of the health sector insurers. Advocating suitable health schemes for clients on the basis of different attributes such as different age group, history of the family, and category of service gives advantage for the insurer as well as for the customer. Analysis of the formless data from insurance claiming history through predictive approaches permits the insurance company to assume a sequence of authentic claims and curious outliers for reducing the price of misuse. Huge data along with cloud computing facilitate the insurance company to launch the latest and novel business frameworks, such as manipulation-based insurance by observing the information about client action recorded in real time.

5 Conclusion

A large amount of data analysis diminishes the gap between organized and unstructured data sources. The principles of massive data are heavily relied on the idea of depth of information and more insights, and thus we can gain meaningful information and make predictions for future events. However, in a short span of time, we have witnessed a spectrum of analytics presently in use that has shown significant impacts on the decision-making and performance of the healthcare sector. The growth of medical data from different areas has imposed computational experts to design innovative schemes to analyze and interpret such a vast amount of data within a given time period. Therefore, it is crucial for technologists and professionals to understand evolving conditions. This reading addressed the main analysis of data governance in health sector. Depth analysis of the data control components may be a research field in the future.

References

1. Gandomi, A., Haider, M.: Beyond the hype Big data concepts, methods, and analytics. Int. J. Inf. Manage. **35**(2), 137–144 (2015)
2. Driscoll, A., Daugelaite, J., Sleator, R.D.: Big data, Hadoop and cloud computing in genomics. J. Biomed. Inf. **46**(5), 774–781 (2013)
3. Chen, C.P., Zhang, C.Y.: Data-intensive applications, challenges, techniques and technologies: a survey on big Data. Inf. Sci. **275**, 314–347 (2014)
4. Bello-Orgaz, G., Jung, J.J., Camacho, D.: Social big data: recent achievements and new challenges. Inf. Fusion **28**, 45–59 (2016)
5. Leventhal, R.: Trend: big data. big data analytics: from volume to value. Healthc. Inf. Bus. Maga. Inf. Commun. Syst. **30**(2), 12–14 (2013)
6. Jin, X., Wah, B.W., Cheng, X., Wang, Y.: Significance and challenges of big data research. Big Data Res. **2**(2), 59–64 (2015)
7. Ola, O., Sedig, K.: The challenge of big data in public health: an opportunity for visual analytics. Online J. Pub. Health Inf. **5**(3), 223 (2014)
8. Al-Jarrah, O.Y., Yoo, P.D., Muhaidat, S., Karagiannidis, G.K., Taha, K.: Efficient machine learning for big data: a review. Big Data Res. **2**(3), 87–93 (2015)
9. Hashem, I.A.T., Yaqoob, I., Anuar, N.B., Mokhtar, S., Gani, A., Khan, S.U.: The rise of "big data" on cloud computing: review and open research issues. Inf. Syst. **47**, 98–115 (2015)
10. Archenaa, J., Anita, E.M.: A survey of big data analytics in healthcare and government. Proce. Comput. Sci. **50**, 408–413 (2015)
11. Kaisler, S., Armour, F., Espinosa, J.A., Money, W.: Big data: issues and challenges moving forward. In: 2013 46th Hawaii International Conference on System Sciences, pp. 995–1004. IEEE (2013)
12. Jee, K., Kim, G.H.: Potentiality of big data in the medical sector: focus on how to reshape the healthcare system. Healthc. Inf. Res. **19**(2), 79–85 (2013)
13. Zhou, L., Pan, S., Wang, J., Vasilakos, A.V.: Machine learning on big data: opportunities and challenges. Neurocomputing **237**, 350–361 (2017)
14. Tripathy, S.: Performance analysis of several machine learning approaches used in the diagnosis of mammograms. Int. J. Innovative Technol. Exploring Eng. **8**(10), 228–232 (2019)
15. Tripathy, S., Tripti, S.: Imaging and machine learning techniques used for early identification of cancer in breast mammogram. Int. J. Recent Technol. Eng. **8**(3), 7376–7383 (2019)
16. Palanisamy, V., Thirunavukarasu, R.: Implications of big data analytics in developing healthcare frameworks–a review. J. King Saud Univ. Comput. Inf. Sci. (2017)
17. Mancini, M.: Exploiting big data for improving healthcare services. J. e- Learn. Knowl. Soc. **10**(2) (2014)
18. Wu, P.Y., Cheng, C.W., Kaddi, C.D., Venugopalan, J., Hoffman, R., Wang, M.D.: Omic and electronic health record big data analytics for precision medicine. IEEE Trans. Biomed. Eng. **64**(2), 263–273 (2016)

A GALA Based Hybrid Gene Selection Model for Identification of Relevant Genes for Cancer Microarray Data

Barnali Sahu and Alok Kumar Jagadev

Abstract Relevant gene selection in microarray data analysis is a challenging task due to the diverse complexity of different cancers. In this study, a novel evolutionary method based on genetic algorithm combined with learning automata is proposed for the identification of relevant genes for cancer classification. In the first stage of gene selection *k*-means clustering algorithm is applied to cluster the genes of pre-processed microarray data, then on each cluster gene ranking techniques were applied to find the top-scored feature subsets of each cluster. A feature pool was constructed by aggregating all top-scored feature subsets as a result of reducing microarray data dimensionality. In stage two further dimension reduction occurs by applying the GALA algorithm with modified operators. For classification SVM classifier was applied to measure the classification accuracy. The proposed method was benchmarked upon two popular cancer microarray datasets.

Keywords Microarray data · Gene selection · Classification · Genetic algorithm · Learning automata

1 Introduction

Microarray data plays a significant role in today's era in clinical diagnosis. It allows researchers to simultaneously measure the expression levels of thousands of genes in a microarray experiment [1]. Microarray data are usually represented as a two-dimensional matrix, representing the number of samples and genes in the matrix row and column. The dataset consists mainly of thousands of genes and a small number of samples, contributing to the "curse of dimensionality problem". To overcome the mentioned issue feature selection techniques are applied. The acquired genes

B. Sahu (✉)
Department of Computer Science, Siksha 'O' Anusandhan (Deemed to Be University), Bhubaneswar, Odisha 751030, India
e-mail: Sahu.barnali08@gmail.com

A. K. Jagadev
School of Computer Science, KIIT (Deemed to Be University), Bhubaneswar, Odisha, India

D. Mishra et al. (eds.), *Intelligent and Cloud Computing*, Smart Innovation, Systems and Technologies 194,
https://doi.org/10.1007/978-981-15-5971-6_89

can be used in many ways, including medical diagnosis and prognosis. One of the main application of microarray data is classification of tissue samples into healthy or cancerous tissues. There are several gene selection methods are available in the literature among them filter method [2] and wrapper method [3] are termed as classical feature selection techniques. The goodness of both filter and wrapper feature section techniques have been combined to develop a hybrid feature selection technique [4], which can improve the classification accuracy with less computational time. In this approach, the process of gene selection uses a fitness function to calculate the ability to discriminate against a gene and to distinguish the different labels. In the current work a two stage gene selection approach followed by a metaheuristic wrapper technique has been used, which is a combination of Genetic algorithm and Learning automata [5] to find the optimal feature subset.

The introduction section is followed by Sect. 2, containing the information about the data set used and the proposed GALA based hybrid model has been elucidated. Section 3, describes the experimental results and finally Sect. 4, deals with the concluding remarks and findings of the study.

2 Materials and Methods

Microarray data is generally characterized as an NXM matrix, where *N* signifies a number of samples and *M* represents the number of genes. In the current study two microarray data sets were used to estimate the proposed hybrid feature selection model for gene selection and cancer classification. The first data set is prostate cancer data [6], which is the most intricate data set among others. It contains 136 samples and 12,600 genes. The second data set is diffuse large B-cell Lymphoma(DLBCL) [7] obtained from Stanford University. Using this a number of analysis has been performed which reflects in the experimental results. In the following section, the proposed model is described in detail.

2.1 Proposed Model

The proposed hybrid model of feature selection consists of two-stage feature selection strategy. In the first stage *k*-means clustering algorithm for attribute clustering is employed followed by the application of gene ranking techniques on each gene cluster to get the top-scored genes from each cluster. Later the top-scored genes have been aggregated to formulate the feature pool. The original dimension of the data set was reduced to the number of features in the feature pool. In the next stage the dimension of the data set is reduced by the application of a novel evolutionary algorithm. The data set with the optimal feature subset is divided into training and testing data. SVM classifier has been used to classify the test data and the classification accuracy has been measured (Fig. 1).

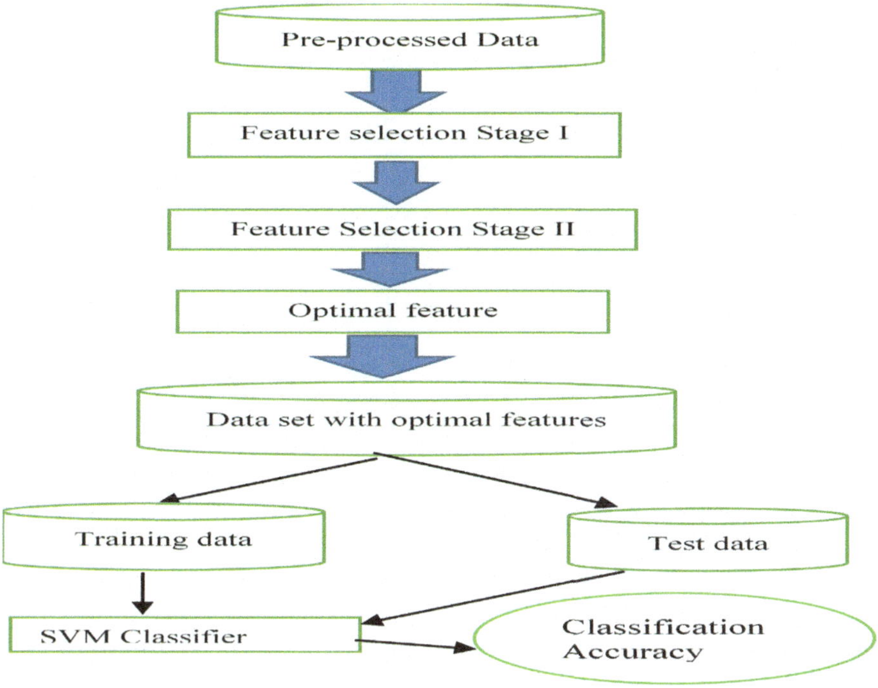

Fig. 1 Schematic representation of the proposed hybrid feature selection and classification model

2.2 *First Stage Gene Selection*

Gene selection in microarray data classification is treated as a preprocessing method in several data mining applications to minimize the search space for reducing the time complexity while increasing the chance to find high-quality solutions. In this stage, most hybrid methods exploit a filter method to score and pick the genes that statistically reveal more information. The use of such filtering techniques helps to exclude irrelevant and redundant characteristics, thereby improving the issue of high dimensionality of microarray data. In the current study three supervised scoring methods have been used to reduce the current high dimension of the microarray data. The feature scoring techniques used are Significant Analysis of Microarray(SAM) [8], Signal to Noise Ratio(SNR) [9] and t-statistics [10]. The process of selecting the first phase gene set is discussed in the following algorithm [11] (Fig. 2).

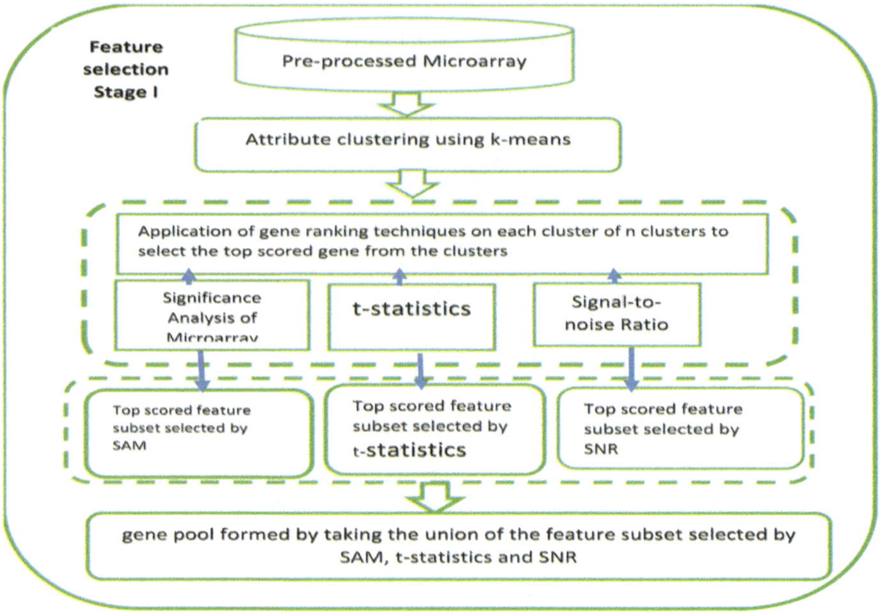

Fig. 2 Schematic representation of first stage gene selection

Algorithm 1

Input: Preprocessed Microarray data (D)
Output: gene pool
Step1: Apply the k-means algorithm for attribute clustering of D.
Step 3: Initialize C, F (C= number of clusters, *F*=number of filters)
Step4: for each f , $1 \le f \le |F|$, do
Step 5: for each c, $1 \le c \le |C|$, do
Step 6: apply a filter f to cluster c, and find the topmost scored gene a, A_{fc} =a
Step 7:end for
Step 8:end for
Step 9: for each filter *f*, $1 \le f \le |F|$ do
Step 10: for each cluster c, $1 \le c \le |C|$ do
Step 11: Generate attribute matrix $E = (A_{fc})$, where A_{fc} , topmost scored attribute generated while filter f is applied on cluster *c*.
Step 12: end for
Step 13: end for
Step 14: Represent each row of E by $E_f = \{A_{f1}, A_{f2}, \ldots\ldots, A_{fn}\}, 1 \le f \le |F|$
Step 15: find $E_{new} = \bigcup_{i=1}^{n} Af_i$
Step 16: Let H= E_f^{new} , where H is the resultant gene pool

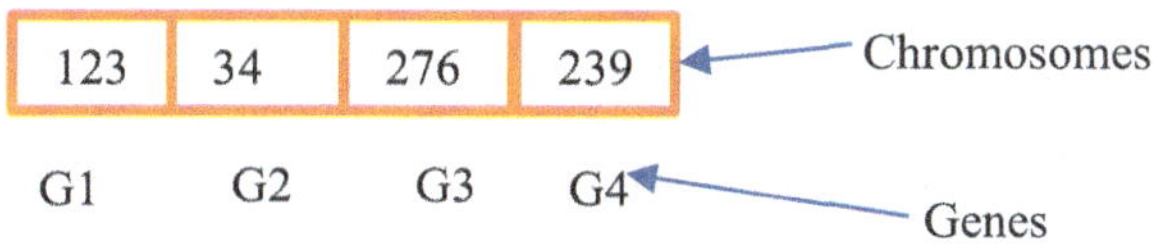

Fig. 3 Instance of an arbitrary chromosome

2.3 *Second Stage Gene Selection Using a GALA Algorithm*

In this examination a metaheuristic search algorithm for example GA and learning automata has been coordinated to deliver an ideal quality subset for two malignant growth datasets. A Learning automata(LA) is an abstract system with an adaptive decision making unit that learns how to select an action through repeated interactions with its environment. The environment then determines the behaviour selected and generates a proper response using a valid signal. In LA, an action is selected, penalized and rewarded based on predefined rules for each learning iteration. Let V is the environment and consists of α, β and c, where $\alpha = \{\alpha_1, \alpha_2, \ldots, \alpha_r\}$ is the set of inputs,$\beta = \{\beta_1, \beta_2, \ldots, \beta_r\}$ is the set of outputs and $c = \{c_1, c_2, \ldots, c_r\}$ is the group of penalty rates.

Contrarily, a genetic algorithm tries to obtain an optimal solution dependent on fitness function. In GALA, the reward and penalty operators are combined with GA. Thus, an activity in LA is equal to a GA chromosome, and an automation action is proportionate to each GA gene. At first GA relegates a score to the chromosomes through fitness function. The certified chromosome will be conveyed if a score for quality region can be relegated in a chromosome. LA, where every automation is identical to a chromosome and each activity is equal to a gene, is utilized in the present methodology. Gene places are identified by rewarding and penalizing operators; thereby increasing the speed of convergence.

Initial Population The primary population contains chromosomes generated at random. Each chromosome comprises of various genes from the input data set. Figure 3 is a sample representation of a chromosome.

There are four genes in the above chromosome. Every number within a gene corresponds to the index of the selected gene from the data set.

The Fitness Function All the data sets used in the study have been divided into five equal parts for calculating the fitness value. In the process, one set of data is treated as a test set and rest is considered as a training set and the process is repeated five times. A support vector machine is used for fitness calculation. The resultant fitness is the average classification accuracies from all iteration.

Crossover Operator Crossover operators are responsible for generating new chromosomes by changing some parts of the current chromosomes using the operator. The crossover operator takes multiple parent solutions and generates the same number of children solutions. A single point crossover is used in the current study.

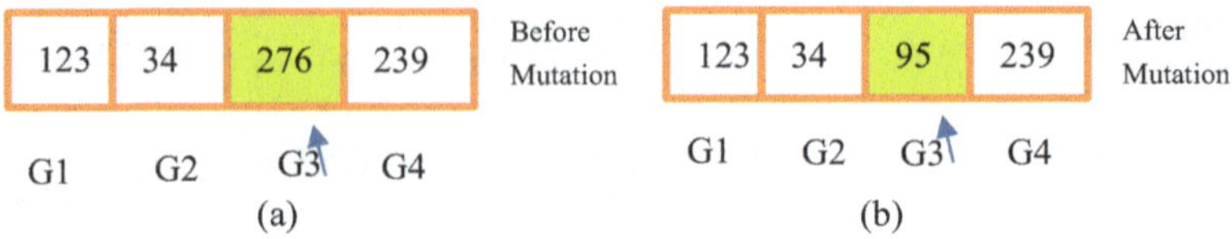

Fig. 4 Sample of a mutation operator. **a** Post-mutation chromosome **b** chromosome state after mutation

Mutation Operator The mutation operator is extended over two generations to maintain genetic verity, since the solution can vary fully from the prior result. Consequently, GA can be customized to achieve superior results. In the current study, the type of mutation operator used is ordered dependent mutation (Fig. 4).

Penalty and Reward Operators Memory size has been specified for every action of the automaton. At first, the size of the memory was set to zero. The scale of an action memory indicates the significance. It means that the most important action gets more memory. The calculations of reward and penalty are done as bellow.

1. Take an automaton.
2. Calculate the fitness value for the automaton chosen, considering it to be x.
3. Choose an arbitrary action from the new automata and set a random value for it.
4. Evaluate the fitness of modified automaton, and consider it to be y.
5. if $x \geq y$, the automata is rewarded and penalize it if $x < y$.
6. Reward: do memory size + 1 and do not amendment the intervention costs if the memory size > 0. Do not amendment the size of the memory or switch the action value to a random value if the memory size = 0 (Figs. 5 and 6).

Selection Operator and Elitism Elitism operator passes the amount of chromosomes to the consequent generation. In the present study 10% of individuals with the maximum fitness values are elect to be enraptured to the following generation, avoiding the crossover and mutation operators and choosing the remaining individuals via a roulette wheel selection operator.

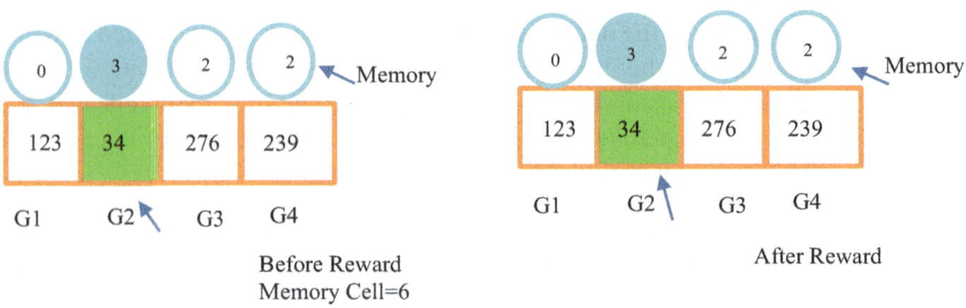

Fig. 5 Reward operator on G2

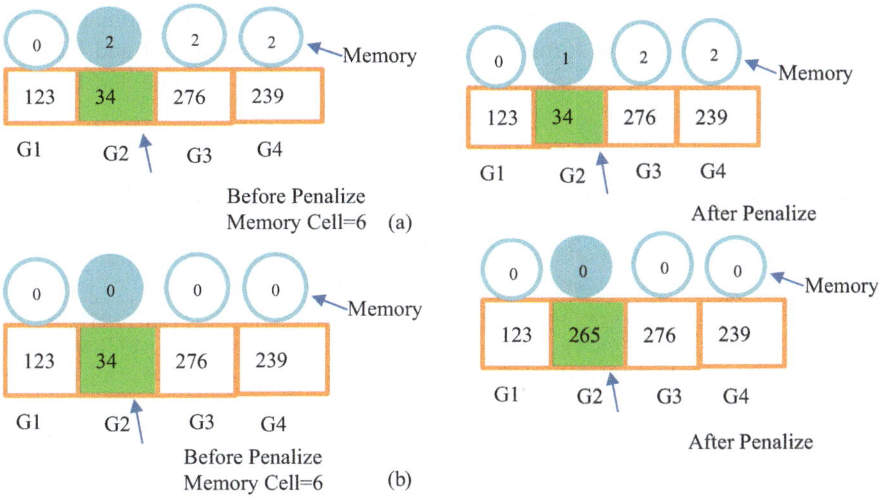

Fig. 6 Gene G2's penalty operator, **a** reduced size of the memory gene G2 by one. **b** changed value of Gene G2 to a random available value

Algorithm 2 (GALA)
Input: Data set
Output: a subset of genes
Step 1: Initialize population
Step2: Calculate fitness function
For every chromosome do;
2.1. split initial dataset to 5 parts
2.2. for every dataset create model using train data and calculate accuracy using SVM classifier
2.3. Fitness=test accuracy1+test accuracy2+test accuracy 3+test accuracy 4+test accuracy 5
Step 3: Make selection using the selection operator
Step 4: Execute cross over
Step 5: Execute mutation
Step 6: Execute reward/penalize
Step 7: If reproduction occurs
Step7.1. find the fitness of new chromosome else go to step 8
Step8: Return subset of genes

3 Experimental Results

After the first stage of feature selection the gene pool constructed is depicted in Table 2. The 20 top-scored features selected by each individual ranking methods have been listed in Table 1. The first column of Table 2 consists of the feature pool of Prostate cancer data and the second column consists of a feature pool of the

DLBCL dataset. Table 3, first and second column consists of the optimal feature subset selected by the GALA method for the Prostate cancer dataset and DLBCL dataset respectively. The accuracy of the SVM classifier for Prostate and DLBCL data set is listed in Table 4.

Table 1 Top scored features selected by SNR, SAM and t-test from 20 clusters for Prostate Cancer Dataset and DLBCL

Method	Column IDs of selected features of prostate cancer data	Column IDs of selected features of DLBCL data
KM + SNR	[1011, 1690, 379, 1026, 1282, 1215, 1362, 760, 889, 606, 1138, 1213, 152, 342, 1893, 918, 2107, 1737, 44, 432]	[14, 905, 2176, 1371, 2261, 2359, 383, 1345, 1719, 1028, 1506, 573, 1092, 270, 973, 564, 2076, 1995, 715, 550]
KM + SAM	[1011, 1690, 379, 1026, 1282, 1215, 1362, 337, 895, 606, 1138, 1213, 152, 342, 1893, 918, 2107, 1737, 44, 432]	[14, 905, 2176, 1371, 2261, 2359, 383, 1191, 1016, 2062, 2550, 573, 1092, 904, 973, 564, 2076, 1995, 1942, 550]
KM + t-test	[1363, 188, 379, 1026, 1292, 1215, 1362, 1858, 1071, 606, 1138, 1213, 1553, 498, 1893, 918, 2107, 1737, 1616, 432]	[998, 905, 2176, 1371, 2261, 2359, 383, 1191, 1845, 1028, 2550, 573, 1092, 662, 973, 564, 2076, 1995, 715, 550]

Table 2 Feature pool constructed by combining the features selected by SNR, SAM and t-test for prostate cancer data and DLBCL dataset

Column Ids of feature pool for prostate cancer dataset	Column Ids of feature pool for DLBCL dataset
1011 1363 1690 188 379 1026 1282 1292 1215 1362 760 337 1858 889 895 1071 606 1138 1213 152 1553 342 498 1893 918 2107 1737 44 1616 432	14 998 905 2176 1371 2261 2359 383 1345 1191 1719 1845 1016 1028 2062 1506 2550 573 1092 270 662 904 973 564 2076 1995 715 1942 550

Table 3 Optimal feature selected by GALA using feature pool of prostate dataset and DLBCL data set

Column IDs of prostate data	Column IDs of DLBCL data
8 15 27 30	1 3 6 8 10 11 12 17 18 19 29

Table 4 Comparison of classification accuracy (for number of genes) of proposed model with other existing hybrid models

Models	DLBCL	Prostate cancer
Proposed model	97.4026(11)	98.0392(4)
PLSDR [12]	93.0(20)	91.7(20)
DRFO-CFS [13]	93.33(11)	85.299(113)
LDA-GA [14]	99.1(3)	99.5(3)

4 Conclusion

In this paper, a hybrid gene section approach using GALA is introduced. It applies the goodness of both filter and wrapper technique to identify the most discriminating genes from the cancer microarray data for classification. The SVM classifier was applied as the fitness function as well as the classification model in our approach. The findings obtained through the proposed approach shows remarkable results for Prostate as well as DLBCL datasets.

References

1. Lee, C.-P., Leu, Y.: Appl. Soft Comput. **11**, 208–213 (2011)
2. Cancer, E., Xue, B., Zhang, M.: Differential evolution for filter feature selection based on information theory and feature ranking. Knowl. Based Syst. **140**, 103–119 (2018)
3. Wang, A., An, N., Yang, J., Chen, G., Li, L., Alterovitz, G.: Wrapper-based gene selection with markov blanket. Comput. Biol. Med. **81**, 11–23 (2017)
4. Aziz, R., Verma, C.K., Srivastava, N.: A novel approach for dimension reduction of microarray. Comput. Biol. Chem. **71**, 161–169 (2017)
5. Motieghader, H., Najafi, A., Sadeghi, B., Masoudi-Nejad, A.: A hybrid gene selection algorithm for microarray cancer classification using the genetic algorithm and learning automata. Inf. Med. Unlocked **9**, 246–254 (2017)
6. Singh, D., et al.: Gene expression correlates of clinical prostate cancer behavior. Cancer Cell **1**(2), 203–2099 (2002)
7. Alizadeh, A.A., et al.: Distinct types of diffuse large B-cell lymphoma identified by gene expression profiling. Nature **403**(6769), 503–511 (2000). Available at: http://www.nature.com/doifinder/10.1038/35000501
8. Tushar, V., Tibshirani, R., Chu, G.: Significance analysis of microarrays applied to the ionizing radiation response. Proc. Natl. Acad. Sci. U.S.A. **98**(9), 5116–21 (2001)
9. Sahu, B., Mishra, D.: A novel approach for selecting informative genes from gene expression data using signal-to-noise-Ratio and t-statistics. Proc. Int. Conf. Comput. Commun. Technol. **2**(4), 1414–1419 (2011)
10. Sahu, B., Dhehuri, S., Jagadev, A.K.: Identification of top-ranked features using consensus affinity of state-of-the-art methods. Soft Comput. Theor. Appl. **583**, 293–305 (2018)
11. Sahu, B., Dehuri, S., Jagadev, A.K.: Feature selection model based on clustering and ranking in Pipeline for microarray data. Inf. Med. Unlocked **9**, 107–122 (2017)
12. Li, G.-Z., Zeng, X.-Q., Yang, J.Y., Yang, M.Q.: Partial least squares based dimension reduction with gene selection for tumour classification. In: Proceedings of the 7th IEEE International Conference on Bioinformatics and Bioengineering, IEEE, pp. 1439–1444 (2007)
13. Bolón-Canedo, V., Sánchez-Maroño, N., Alonso-Betanzos, A.: Distributed feature selection: an application to microarray data classification. Appl. Soft Comput. **30**, 136–150 (2015)
14. Bonilla-Huerta, E., Duval, B., Hernández, J.C.H., Hao, J.K., Morales-Caporal, R.: Hybrid filter-wrapper with a specialized random multi-parent crossover operator for gene selection and classification problems. In: Bio-Inspired Computing and Applications, pp. 453–461. Springer (2012)

Conclusion

References

A Brief Survey on Placement of Access Points in Ultra-Dense Heterogeneous 5G Networks

Satya R. Das, Koushik Sinha, Nandini Mukherjee, and Bhabani P. Sinha

Abstract Ultra-dense heterogeneous networks involving LTE, Wi-Fi, and D2D communications appear to be the dominant technologies for many more years to come for achieving the distinct capabilities that 5G or higher generation networks are expected to provide. LTE and Wi-Fi devices follow different protocols in grabbing the communication channel. Because of this heterogeneity and also the proximity of various types of devices due to ultra densification, there will be increased channel interferences among the communicating devices which often degrade the overall performance of the network. Judicious placement of the access points (APs) or base stations (BSs) to cover a given geographical region in such an ultra-dense network taking the transmitting powers and interfering distances into account constitutes an important research problem of current interest. We present here a brief survey on the various techniques for the placement of APs or BSs under such a scenario of ultra-dense heterogeneous networks.

Keywords Cellular networks · 5G communication · Access point · Base station · Placement problem · Massive MIMO

S. R. Das (✉) · B. P. Sinha
Department of Computer Science and Engineering, Siksha 'O' Anusandhan Deemed To Be University, Bhubaneswar, Odisha, India
e-mail: satyadas@soa.ac.in

K. Sinha
Department of Computer Science, Southern Illinois University, Carbondale, USA

N. Mukherjee
Department of Computer Science and Engineering, Jadavpur University, Kolkata, India

D. Mishra et al. (eds.), *Intelligent and Cloud Computing*, Smart Innovation, Systems and Technologies 194,
https://doi.org/10.1007/978-981-15-5971-6_90

1 Introduction

The future of networking promises unlimited sharing of data, accessible from everywhere and at all times for everyone and everything. As per a CISCO report, more than 500 billion connected devices, especially in the context of the Internet of Things (IoT), are expected to become operational by the end of 2030. This translates into more than six connected devices per person, encompassing both human and machine communications. Many such dense and massive networks of resource-constrained wireless devices will interconnect using ad hoc links, such as those found in wireless sensor networks (WSNs). Such ultra-dense and massive ad hoc networks will increasingly be entrusted with sensing and managing highly complex ecosystems to render various complex data-driven services to humans as well as automated systems in a hyper-connected world. 5G, the next generation all-IP cellular networks, is envisioned to be the answer to efficiently handling such a tsunami of data coming from diverse sources. In many ways, 5G networks will be unique in comparison to its predecessor cellular networks. 5G, the cornerstone of future generation cellular networks has ambitious plans of achieving very low communication latencies, manifold improvements in network throughputs, and battery life of networked devices. The vision of 5G is to serve as an umbrella network for a connected society in which sensors, vehicles, drones, wearable devices, etc., will connect with one another and interact with human end-users to provide innovative and complex services such as smart homes, smart cities, smart vehicles, industrial process control, tele-surgery, environment monitoring, and security, among other things.

To achieve such a dramatic improvement in user experience, 5G will use multiple path-breaking technologies, hitherto unseen in cellular networks. The introduction of ultra-dense network of small cells and D2D communication holds the promise of finally realizing the holy grail of creating a unified networking paradigm of wireless ad hoc networks and cellular networks to usher in a smart digital society. Resource-constrained devices typically organize themselves into stand-alone ad hoc networks due to cost, transmission power, and computational capability constraints. 5G's device-to-device (D2D) communication models will make it possible for devices in a dense wireless network to self-organize into a massive ad hoc mesh networks. Using D2D communication models for ad hoc network-based connectivity between devices will improve spectrum usage through improved interference avoidance, reduce transmission power requirements (and hence extend battery life of devices), and provide network coverage to areas without cellular connectivity. Simultaneously, service area coverage by ultra-dense deployment of small cells in 5G will enable the energy-constrained devices to directly communicate with the small cell base stations (BSs) using short links comparable to those in ad hoc networks.

Unfortunately, with the increase in network densification, the inter-cell interference is also assumed to be aggravated significantly due to proximity of transmitting pairs using the same frequency, thereby offsetting the gains from D2D-based communication and shorter links to BSs. Compared to a density of about 4–5 BSs/km^2 in 3G and about 8–10 BSs/km^2 in 4G cellular networks (e.g., Long-Term Evolution-

Advanced or LTE-A), the density of 5G BSs is highly expected to come up to 40–50 BSs/km^2 due to the technology choices of massive MIMO, and utilization of the unused, high-frequency millimeter-wave spectrum band, ranging from 3 to 300 GHz (due to the propagation characteristics of millimeter-wave, transmission range of millimeter-wave communications has to be restricted to 100 m for most parts).

Therefore, unless the placement of the BSs is judiciously done, such ultra-densification of the cellular network is expected to severely affect end-to-end packet latencies and battery life of devices participating in a D2D-based ultra-dense ad hoc network. Unfortunately, the discussion on ultra-dense cellular networks is still in the infant stage. So far, ultra-dense cellular networks have been studied only under simple scenarios and primarily some basic features of 5G networks have been evaluated. In view of the potential benefits that such an ultra-dense integrated ad hoc-cellular network can provide, we would like to present the technological requirements in the next section.

2 Background Ideas

5G wireless communication is the next phase (fifth generation) of mobile telecommunication beyond the current 4G LTE (Long-Term Evolution) standards, targeted to be operational around 2020. One of the main driving forces in moving to 5G standard is the exponential growth in demand for fast as well as bandwidth-hungry wireless communication experienced in many real-life applications, e.g., video, online-gaming, and so on. Also, the recent trend of Internet of Things (IoT), where a large number of smart devices are connected to each other through wired and wireless links and communicate over the Internet, also demands a higher generation of communication standards than the existing 4G. The desired performance goals, which a 5G communication network is expected to achieve, are given as follows:

1. spectral efficiency significantly enhanced compared to 4G;
2. improved coverage;
3. much higher volume of traffic than that in 4G;
4. higher data rates of 1 Gbps simultaneously to many users at one place, e.g., on the same floor of an office;
5. several hundreds of thousands of simultaneous connections for wireless sensors;
6. significantly reduced latency;
7. about a tenfold increase in battery life.

To achieve these objectives, 5G needs to provide extremely high broadband speed, ultra-low end-to-end latency, ultra-reliable web Connectivity, and ultra-low energy for communication which depend on the success of the following technologies: (i) millimeter-wave communication, (ii) massive multi-input multi-output (MIMO) techniques, (iii) heterogeneous networking and ultra densification, (iv) Device-to-Device (D2D) communication, and (v) a paradigm shift from the conventional data communication techniques for ultra- low-latency and reliable communication.

Millimeter-wave communication can be used for direct short-range communications with relatively low multi-user interference (MUI), which can support simultaneous communications over the same radio spectrum. Such millimeter-wave communication can provide a very high data rate (multi-Gbps) in 30–300 GHz band and uses directional antennas to cope with large propagation loss. Relays may be used in case line-of-sight (LOS) paths are obstructed.

To multiply capacity without multiplying the spectrum, one of the recent techniques is the use of the Multiple-Input Multi-Output (MIMO) technique which uses a very large array of antennas at each base station to simultaneously serve a large number of users enabling transmitting and receiving over the same radio channel. In a MIMO system, the complexity associated with the wireless nodes quickly increases with the increase in the number of antennas and data streams.

Different types of wireless networks (e.g., WLAN, WiMAX, Wi-Fi, and E-UTRAN) usually operate under different access technologies and protocols. A combination of such wireless networks along with LTEs may be used to maintain the services to a large number of users over a wide region, in order to improve the user experience, reducing bottlenecks in Radio Access Network (RAN), constituting what is called a heterogeneous network (HetNet). Such a heterogeneous network architecture helps us in efficient routing, resource allocation, and load balancing while catering services to a large number of users with varying quality of experience (QoE) requirements and optimizing the network performance. The introduction of ultra-dense network of small cells and D2D communication helps in realizing the challenge of creating a unified networking paradigm of wireless ad hoc networks and cellular networks to usher in a smart digital society. The density of BSs in 3G and 4G (LTE-A) cellular network is about 4–5 BSs/km^2 and 8–10 BSs/km^2, respectively. For seamless coverage, the BS density of 5G is highly expected to come up to 40–50 BS/km^2. Therefore, the future 5G cellular network is an ultra-dense cellular network. D2D is considered as a radio technology that enables devices to communicate directly without routing the data paths through a network infrastructure in order to improve (i) capacity gain, (ii) user data rate gain, (iii) reduced end-to-end latency, (iv) low energy consumption, and (v) better load balancing.

In a cellular network, each cell is under the control of a base station which communicates simultaneously with all mobile terminals within the cell, and also passes the traffic to the Mobile Switching Center (MSC). The base station is connected to the mobile terminals through a radio (wireless) link. Different types of base stations are set up according to the coverage needed, which are as follows:

(i) macro cells, covering a large area which are usually situated in rural areas,
(ii) micro cells, usually formed in suburban and urban areas controlled by low-power base stations,
(iii) pico cells, usually formed inside the buildings controlled by smaller base stations providing more localized coverage in areas with many users where the network quality is otherwise poor.

To cope with the demand from the ever-increasing huge number of cellular users, one possible approach is to reduce the cell size even from that of the pico cells, resulting in the concept of femto cells which are usually controlled by very low-power base stations or access points (AP) that may be typically used inside a building. As it follows, the transmission power of the APs controlling a femto cell will also be very small due to their small coverage areas. Thus, along with catering services to a large number of cellular users, this concept of using femto cells would also lead to reducing transmitter energy of the APs as well as the mobile terminals under their control. However, reduction of cell size may be associated with the undesirable effect of increased channel interferences among the constituent devices/mobile terminals unless appropriate measures are taken to mitigate such interferences. Two types of channel interference are generally taken into consideration: (i) co-channel interference, which occurs when the same channel is being simultaneously used by more than one mobile user closely located in the cellular structure, and (ii) adjacent channel interference, which happens due to simultaneously assigning some closely spaced channels to more than one mobile user. It may be noted that the degree of such interference caused by a mobile terminal MT_i at another mobile terminal MT_j depends on several factors like transmission power of MT_i, distance between MT_i and MT_j, and antenna gain of MT_j. Assigning frequency channels to the cells with minimum bandwidth possible as per their demands and avoiding the channel interference constraints is known as the channel assignment problem (CAP). In generalized formulation, the CAP is similar to a very well-known NP-complete problem that is a graph-coloring problem.

3 Placement of Access Points

The placement problem deals with a judicious placement of access points (APs) or base stations (BSs) in a 5G ultra-dense heterogeneous network of small cells so that a given deployment area will be covered without leaving any holes or uncovered regions by all such APs (BSs), using minimum transmission power and bandwidth consumed by all such access points. In essence, this problem can be treated as a multi-objective multi-constrained optimization problem.

3.1 Existing Works on Placement Problem

Authors in [1] have presented an exhaustive review of various design updates of the radio access network (RAN) such as heterogeneous RAN, and intelligent antennas, and technological requirements for 5G cellular network such as millimetric-wave physical, massive MIMO technologies, model for new channel estimation, and beam-forming antenna. In [2], a distributed network architecture is described for 5G ultra-dense cellular networks based on massive MIMO and millimeter-wave communication technologies. In addition, investigations are done using back-haul network capacity and energy efficiency to determine the limits of densification required in

5G. A fair discussion is found in [21] on capacity growth, and it has been observed that advanced network architectures like heterogeneous networks can be used with the convergence of communication and information to increase the system capacity. Primitive aspects of ultra-dense networks like positives of network-wise coordination, finding the requirement for infrastructure density, potential implementation in 5G cellular network, etc., are enlighted in [22]. Major challenging issues of 5G networks such as network traffic management, workload and cost related to network, deployment of antennas in cells, issues related to high data rate and capacity, and reuse of radio frequency spectrum are discussed in [23]. Two major technologies such as IP-based vertical handover and massive MIMO-OFDMA are discussed in the proposed architecture in [24] for ultra-dense heterogeneous 5G networks. A technique for computing the proportion of two types of stations for optimally placing them has been presented in [25]. Their method has been tested for random scenarios as well as for a real example taken from Hurricane Katrina. The test results substantiate the strength of LAPs so far as the high bandwidth utilization and ability to cover wider areas along with portability and adaptability to height are concerned. The number of base stations needed to cover the desired region is smaller with these LAPs than that required with other existing techniques. Authors in [26] have proposed a placement technique which considers minimizing the transmission power of all APs as well as the bandwidth requirement through reuse of channels used by various APs. The strategic regular placement of APs in their proposed technique minimizes the wastage power of the APs and eases up the channel reuse mechanism for avoiding co-channel and adjacent channel interferences. It has been shown that while fully covering a given rectangular region, the wastage of transmission power in the inside bulk region with intersecting and touching circles by their placement strategy is 81.4% and 20.77%, respectively. Authors in [27] have considered placement in an indoor environment. First, they have proposed a simple technique to convert a given rectangular room to be covered to a virtual one based on the assumption of omnidirectional signal propagation from any access point and nonuniform attenuation inside the room and through its bounding walls. Then they have considered coverage of this virtual region by circles where a circle represents the transmission zone of an AP in two-dimension. It has been shown that while covering a square region, the wastage of transmission energy will be smaller with intersecting circles (20.77%) than that with the touching circles (81.4%) in covering the inner bulk region. However, the wastage increases to higher values in both cases to cover the boundary region. When the size of the region becomes very high compared to the radius of the circles used for covering, the wastage, however, asymptotically tends to 20.77% while using the intersecting circles in covering the region. An algorithm has also been presented based on their idea which adjusts the radius of the circle (transmission zone of each AP) relative to the dimension of the square region to be covered, to minimize the wastage in transmission energy of the APs. In [28], a grid-based approximation algorithm has been proposed for placing the access points with a minimum number of access points so that the received signal-to-noise ratio (SNR) at each location is sufficient to achieve the desired reliability in communication (in terms of bit error rate (BER)). The time complexity of their placement algorithm is $O(n_{opt} \log n_{opt})$, where

n_{opt} is the optimum number of access points needed to cover the region according to their technique. Authors in [29] have proposed a placement technique which first decomposes the given placement problem into several subproblems then and tackles each of these subproblems to solve the original problem. Their method is very suitable for large-scale HetNets with optimal deployment of the base stations. They have observed that some HetNet cells may have strong correlations due to inter-cell interference. Based on this observation, they have proposed a machine learning (ML) -based correlation grouping approach so that the cells are grouped together based on their mutual interferences. Both the analytical results and simulation studies based on this technique indicate that their proposed solution outperforms the existing algorithms based on random grouping as well as the evolutionary algorithms that identify interrelated parameters by monitoring the changes in the objective function based on system performance. The effects of access point density and bandwidth partitioning on the performance of dense wireless networks have been investigated in [30], while considering interference among them due to their proximity. They have shown that the user rate outage probability depends on the number of subchannels used (due to bandwidth partitioning) and the particular way how these subchannels are being utilized by the multiple access system. They have found a lower bound on the number of such subchannels (due to bandwidth partitioning) for large values of access point density. Also, they have found an upper bound on the minimum access point density required for an asymptotically small user rate outage probability.

4 Future Prospects

The impact of ultra-dense deployment of BSs has not been explored sufficiently in the literature in the context of efficient multi- hop data routing strategies with Quality of Service (QoS) guarantees in an integrated ad hoc-cellular network that intelligently uses the wireless network induced by the small cell BSs or APs. Similarly, another area of significant interest would be to study the benefits that such an ultra-dense hybrid network can provide for multi-hop energy-efficient communication in a network of battery-powered devices through the exploitation of correlation between data arising from spatial proximity of the devices (e.g., in WSNs). Due to ultra densification in 5G cellular networks with small cell sizes, it is really a challenge for the researchers to place the APs, particularly in a restricted indoor environment, where there are several constraints due to various obstacles where signals are obstructed or reflected/scattered from their surfaces.

5 Conclusion

We have discussed the technological requirements for achieving the desired goals of 5G and beyond in the form of millimeter-wave communication, MIMO, and ultra-dense heterogeneous networks. For optimal placement of access points in a 5G cellular network, we have presented a brief survey of the state-of-the-art strategies related to base station placement based on novel paradigms, such as geometric placements, machine learning-related techniques, and grid-based approximation algorithms.

References

1. Agiwal, M., Roy, A., Saxena, N.: Next generation 5G wireless networks: a comprehensive survey. IEEE Commun. Surv. Tutorials **18**(3), 1617–1655 (2016)
2. Ge, X., Tu, S., Mao, G., Wang, C.-X., Tao, H.: 5G ultra-dense cellular networks. IEEE Wirel. Commun. 72–79 (2016)
3. Liu, J., Sheng, M., Liu, L., Li, J.: How Dense is Ultra-Dense for Wireless Networks: From Far- to Near-Field Communications. arXiv:1606.04749 (2016)
4. Al-Dulaimi, A., Al-Rubaye, S., Cosmas, J., Anpalagan, A.: Planning of ultra-dense wireless network. IEEE Netw. **31**(2), 90–96 (2017)
5. Sulyman, A.I., Alwarafy, A., MacCartney, G.R., Rappaport, T.S., Alsanie, A.: Directional radio propagation path loss models for millimeter-wave wireless networks in the 28, 60, and 73-GHz bands. IEEE Trans. Wirel. Commun. **15**(10), 6939–6947 (2016). Oct
6. Meguerdichian, S., Koushanfar, F., Potkonjak, M. and Srivastava, M.B.: Coverage Problems in Wireless Ad Hoc Sensor Networks. In: Proceedings of IEEE INFOCOM (2001)
7. Bai, X., Kumar, S., Xuan, D., Yun, Z., Lai, T.H.: Deploying wireless sensors to achieve both coverage and connectivity.In: Proceedings of ACM International Symposium Mobile Ad Hoc Networking and Computing (MobiHoc) (2006)
8. Ghosh, S.C., Sinha, B.P., Das, N.: Channel assignment using genetic algorithm based on geometric symmetry. IEEE Trans. Vehicul. Technol. **52**(4), 860–875 (2003). July
9. Ghosh, S.C., Sinha, B.P., Das, N.: A new approach to efficient channel assignment for hexagonal cellular networks. Int. J. Found. Comput. Sci. **14**(3), 439–463 (2003). June
10. Zhang, H., Hou, J.: Maintaining sensing coverage and connectivity in large sensor networks. J. Ad Hoc Sens. Wirel. Netw. **1**, 89–124 (2005)
11. Luo, C.J., Tang, B., Zhou, M.T., Cao, Z.: Analysis of the wireless sensor networks efficient coverage. In: Proceedings of International Conference on Apperceiving Computing and Intelligence Analysis (ICACIA), pp. 194197 (2010)
12. Poe, W.Y., Schmitt, J.B.: Node deployment in large wireless sensor networks: coverage, energy consumption, and worst-case delay. In: Conference on Asian internet engineering. ACM, New York, p. 7784 (2009)
13. Sheu, J.P., Yu, C.H., Tu, S.C.: A distributed protocol for query execution in sensor networks. In: Proceedings of IEEE Wireless Communications and Networking Conference, vol. 3, pp. 1824–1829 (2005)
14. Ke, W., Liu, B., Tsai, M.: The critical-square-grid coverage problem in wireless sensor networks is NP-complete. Int. J. Comput. Geom. Appl. **55**(9), 2209–2220 (2011)
15. Saha, D., Das, N.: Distributed area coverage by connected set cover partitioning in wireless sensor networks. In: Proceedings of First International Workshop on Sustainable Monitoring Through Cyber- Physical Systems (SuMo-CPS), ICDCN (2013)
16. Sakai, K., Sun, M., Ku, W., Lai, T.H., Vasilakos, A.V.: A framework for the optimal k-coverage deployment patterns of wireless sensors. IEEE Sens. J. **15**, 7273–7283 (2015)
17. Das, G.K., Fraser, R., Lpez-Ortiz, A., Nickerson, B.G.: On the discrete unit disk cover problem. Int. J. Comput. Geom. Appl. **22**, 407–420 (2012)
18. Saha, D., Das, N.: Self-organized area coverage in wireless sensor networks by limited node mobility. Innovations Syst. Softw. Eng. NASA J. **12**, 227–238 (2016)
19. Gamst, A.: Some lower bound for a class of frequency assignment problems. IEEE Trans. Vehi. Tech. **35**, 8–14 (1986)

20. Sung, C.W., Wong, W.S.: Sequential packing algorithm for channel assignment under co-channel and adjacent channel interference constraint. IEEE Trans. Vehicul. Technol. **46**, 676–686 (1997)
21. Li, Q.C., Niu, H., Papathanassiou, A.T., Wu, G.: 5G network capacity: key elements and technologies. IEEE Vehicul. Technol. Maga. (2014)
22. Verma, G., Mandal, M., Rohela, S.: Ultra-Dense Networks in 5G A Survey Paper
23. Cengiz, K., Aydemir, M.: Next-generation infrastructure and technology issues in 5G systems. J. Commun. Softw. Syst. **14**(1) (2018)
24. Tahir, S.: A novel architecture for 5G ultra dense heterogeneous cellular network. In: Int. J. Adv. Comput. Sci. and Appl. **9**(11) (2018)
25. Košmerl, J., Vilhar, A.: Base stations placement optimization in wireless networks for emergency communications. In: ICC'14-W11, IEEE Conference on Advances in Public Safety and Emergency Communication
26. Audhya, G.K., Sinha, K., Majumder, P., Das, S.R. and Sinha, B.P.: Placement of access points in an ultra-dense 5G network with optimum power and bandwidth. In: IEEE Wireless Communications and Networking Conference 15-18 Apr 2018, Barcelona, Spain
27. Audhya, G.K., Sinha, K., Majumder, P., Das, S.: Placement of access points with minimal wastage of transmission power in an indoor environment. In: IEEE Intenational Conference on Advanced Networking and Telecommunication Systems. Bhubaneswar, India (2017)
28. Rengarajan, B., de Veciana, G.: Optimizing placement of wireless access points
29. Lee, S., Lee, S., Kim, K., Kim, Y.H.: Base station placement algorithm for large-scale LTE heterogeneous networks. PLOSONEI (2015). https://doi.org/10.1371/journal.pone.0139190
30. Stefanatos, S., Alexiou, A.: Access point density and bandwidth partitioning in ultra dense wireless networks. IEEE Trans. Commun. **62**(9) (2014)

Qualitative Analysis of Recursive Search Algorithm: A Case Study

Sangram Panigrahi, Kesari Verma, and Priyanka Tripathi

Abstract This paper presents a qualitative analysis of the Recursive Search Algorithm (RSA) technique. The RSA is a land cover change detection technique, which determines that a particular land cover is changed or not. It is a data mining approach and established on the segmentation technique. Furthermore, the RSA is a land cover change detection technique and identifies the land cover change by finding the sudden gains and drop-offs in a vegetation index time series data. The proficiency of the RSA technique is exhibited by using qualitative analysis. To exhibit the qualitative analysis, the RSA method is applied to the Naya Raipur Area data set, which is an enhanced vegetation index time series data set. It has been noticed that the significant land cover changes have been happening in the Naya Raipur Area.

Keywords Data mining · Land cover change detection · RSA · EVI time series

1 Introduction

NASA's earth observation satellites are acquiring huge amounts (more than 10 terabits data daily) of temporal, spatial, and spatiotemporal data by using both passive and active sensors. These enormous amounts of satellite data available in the public domain, which is prompting the researcher to explore some robust, efficient, and cost-effective solution for the earth science domain problem. Furthermore, now the

---

S. Panigrahi (✉)
Department of CSIT, ITER, Siksha 'O' Anusandhan (Deemed to be University), Bhubaneswar, India
e-mail: sangrampanigrahi@soa.ac.in

K. Verma · P. Tripathi
Department of Computer Applications, National Institute of Technology Raipur, Raipur, India
e-mail: keshriverma@gmail.com

P. Tripathi
e-mail: ptripathi.mca@nitrr.ac.in

D. Mishra et al. (eds.), *Intelligent and Cloud Computing*,
Smart Innovation, Systems and Technologies 194,
https://doi.org/10.1007/978-981-15-5971-6_91

researchers acknowledge the challenges in analyzing large-scale earth science data and translate some earth science domain problems into computer science problems and provide computational solutions.

In time series data, if any, significant changes that happen over time is called an event. For example, a change in the form of vegetation coverage over time could be an indication of changes in the land cover (for example, a modification in land cover type of barren land to forest land), while instability in the temporal pattern of precipitation may indicate a climate issue (e.g., flood or drought situation). Mining the time series render crucial evidence about the cause of a pattern change in time series data. Hence, change detection in time series using data mining technique is an active area of research in the earth science domain at the same time as the computer science domain.

In the present age, the land use and land cover change (LULC) detection is an imperative area of research, because the LULC change tremendously affects the local and/or regional environment [1], also held responsible for several unusual environmental effects [2–5]. The quantifiable cognition of LULC change is facilitated to induce an effective plan for the protection and appropriate usage of natural resources. Therefore, it is prompting scientists to invent several land cover change detection methods.

In recent times, the data mining based methods have seemed as an eventual data-driven techniques [6] to determine land cover change. The data mining based methods [7–14] identify the land cover change by finding the sudden gains and drop-offs or gradual changes in a vegetation index time series data.

The RSA [15] is an uncomplicated, effective, and segmentation based land cover change detection technique. To ascertain a specific location is changed or not, the RSA technique takes the input as a vegetation index time series and gives output as a change score. Subsequently, the algorithm compares the change score with the stipulated threshold value. If the change score is larger than the threshold, then the location is recognized as a changed location otherwise the location is declared as unchanged. The proficiency of the RSA technique had been exhibited by using the quantitative technique [15]. To exhibit the quantitative analysis, the assessment results of the RSA technique were scientifically evaluated and compared with the results of existing data mining based techniques [15, 16], i.e., Modified Lunetta, Cumulative sum MEAN (CUSUM MEAN), Yearly Delta (YD), and Recursive Merging (RM) technique. The RSA algorithm witnesses the deviation in time series data and as well as deciding the change time, based on the temporal resolution (i.e., every 16 days or a month) available in the time series and label it with the change type (whether sudden gains or drop-offs type).

In this paper, the proficiency of the RSA technique is exhibited by using qualitative analysis. To exhibit the qualitative analysis, the RSA method is applied to the Naya Raipur Area data set, which is an enhanced vegetation index time series data set. It has been noticed that the significant land cover changes have been happening in the Naya Raipur Area.

The rest of the paper is organized as follows: in Sect. 2 the details about the data set and the pre-processing techniques are discussed. Section 3 a concise explanation of the RSA algorithm is presented. Section 4 presents the proficiency of the RSA technique by using qualitative analysis. Finally, Sect. 5 summarizes the present study and draw some conclusions.

2 Data Set

The remote sensing data act as a central component to deal with the natural resources by managing land use and land cover changes on a worldwide scale. The Moderate Resolution Imaging Spectrometer (MODIS) vegetation index (VI) data products (i.e., Normalized Difference vegetation index (NDVI) and Enhanced vegetation index (EVI) data) is used to monitor the vegetation variations and land cover change detection [17]. The vegetation index is the greenness measure of vegetation at a specific location and it is a very important ecosystem parameter, which is utilized for phenologic monitoring; vegetation classification, and land cover, i.e., drought detection, deforestation, land degradation, land cover change detection and monitoring of the detected land cover change [18]. The MODIS Land Discipline Group renders NDVI and EVI data with different spatial resolutions (i.e., 250 m, 500 m, 1 km, and 5.6 km (0.05^0)) and temporal resolutions (i.e., every 16 days and a month) from the year 2000 to till date. These data sets also incorporate quality assurance (QA) data, which clarifies that under which state of affairs a pixel (i.e., location) is acquired and processed [19, 20]. Furthermore, these data sets are available in the public for downloading free of cost, by the Land Processes Distributed Active Archive Center (LP DAAC) [21].

The Naya Raipur Area data set consist of real EVI time series data and uses for evaluating. The following section explains the brief descriptions about the Naya Raipur Area data set and the preprocessing techniques undertaken for the elimination of poor-quality data.

2.1 Naya Raipur Area Data Set

The Chhattisgarh state is established in November 2000. The state government decided to construct a well-planned new capital city (i.e., Naya Raipur) in Chhattisgarh. The Naya Raipur has widened over 8000 hectares roughly. In Naya Raipur, about 50% of the entire obtain land is being developed for establishing a forest on previously unforested land, roads, parks, greenbelts, water resources for the community, and so on; roughly 23% of the total land would be reserved for educational organizations, offices, and theater, and so on; and almost 30% of the whole land to be used for housing purpose [22]. Many infrastructures have already been built, i.e., parks, recreation centers, housing subdivisions, institutional buildings, stadiums, etc., as a consequence the crucial land cover changes have been happening in the Naya Raipur area. Thus, the Naya Raipur Area of Chhattisgarh, India, has been preferred to study,

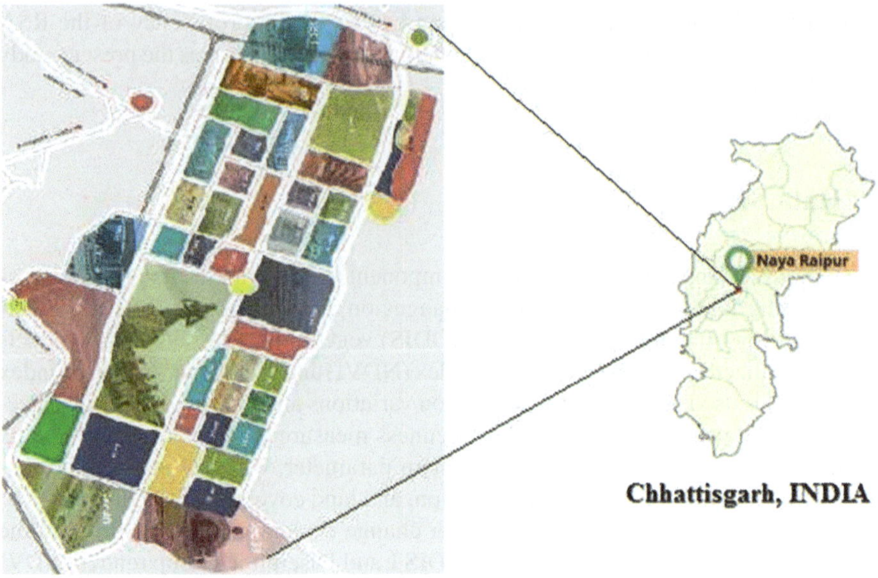

Fig. 1 Naya Raipur Area of Chhattisgarh, India

which is shown in Fig. 1 [23]. The EVI time series data (MOD13Q1) of Naya Raipur Area have been downloaded from LP DAAC [21], from periods February 2000 to December 2018. Principally, the EVI data set of Naya Raipur Area has comprised 1287 locations (i.e., time series data). To eradicate the poor-quality measurements, the pre-processing techniques have been employed on Naya Raipur Area data set.

Compute the monthly EVI value by taking the average of the available biweekly data in a month. Subsequently, prepare the time series data set. First, discard all the time series from the data set, which have low QA value. After that, remove all the time series from the data set which consist of EVI value beyond the ranges from 0 to 0.9 (on the suggestion of earth science domain experts). Finally, exclude all the time series from the data set which are comprising missing value [10]. After the preprocessing, the resultant Naya Raipur Area data set comprising 973 locations and each location has EVI time series with length 188, which is representing about 15 years and 8 months, monthly EVI value from February 2000 to September 2015.

3 Algorithms for Land Cover Change

The several data mining techniques have urged to detect the land cover change by finding out the fluctuations in vegetation index time series data [16]. These techniques take advantage of the inherent feature present in vegetation index time series data, i.e., seasonality. The data mining based land cover change detection algorithms receive vegetation index time series data set as input and deliver the change score list as output. Each change score indicates a specific location. To decide a specific location is changed or not, the algorithm compares the change score with the stipulated threshold value; If the change score is larger than the threshold, then the location is recognized as a changed location otherwise the location is declared as unchanged. The subsequent section furnishes the concise summarization of Recursive Search Algorithms, which is a data mining based land cover change detection method.

3.1 Recursive Search Algorithm

The RSA [15] is an uncomplicated, effective, and segmentation-based land cover change detection technique. The time series means a specific location. The fundamental idea of the RSA technique is first breaking the entire time series into various segments and identify the two successive segments in such a manner that the values held by the individual segment are almost alike while the values contained in two different segments are being considerably dissimilar. For which the RSA algorithm first partitions the entire time series into annual segments and evaluate the difference among every two consecutive annual segments by means of city-block distance. The computation of distance is engaging the seasonality present in the time series data. The two successive annual segments difference specifies the degree of change. The RSA method makes use of two-way searching strategy to determine the most prominent successive annual segment difference and its corresponding index. The brief description of the RSA method is presented as follows:

Algorithm:

Step 1: RSA technique first divides the whole time series into annual segments.

Step 2: It uses two-way searching technique.

- *Step 2.1*: Computes first consecutive annual segmentation difference value and last consecutive annual segmentation difference value and then compare. The consecutive annual segmentation difference is computed by using city-block distance.
  - *Step 2.1.1*: Store the largest difference value and its corresponding index value in S and j, respectively.
- *Step 2.2*: Compute the second consecutive annual segmentation difference value and second last consecutive annual segmentation difference value and compares, then select the largest difference value among them.
- *Step 2.3*: The largest difference value further compares with S value.
  - *Step 2.3.1*: If the largest difference value is greater than the S value, then replaces the S and j values by largest difference value and its corresponding index value.
  - *Step 2.3.2*: Otherwise, S and j values are unchanged.

Step 3: This comparison process is called recursively and finds the two consecutive annual segments with high difference value.

Step 3.1: From the resultant successive annual segments, the first annual segment is named as unchanged segment (i.e., j^{th} annual segment) and the second annual segment is identified changed segment (i.e., $(j+1)^{th}$ annual segment), which are depicted in Fig. 2.

Step 4: RSA method allots a change score (CS) to a time series.
$CS = |S_{diff}| = |S_{org} - S_{avg}|$
Here, S_{org} is the annual sum of $(j+1)^{th}$ annual segment and S_{avg} is the annual sum average of 1st to j^{th} annual segments. In the time series data, the 1st to j^{th} segments are seemed alike pattern and which are distinct from the pattern of $(j+1)^{th}$ segment.

Step 5: Based on CS, the RSA algorithm determines the particular location is changed or not. Furthermore, the RSA technique specifies the type of change.

Step 5.1: If the allotted CS is greater than the stipulated threshold value, then that location is recognized as a changed location.

Step 5.2: The type of change of a specific location is found out from the sign of the S_{diff} value, i.e., if S_{diff} value is with a $+ve$ sign, then the location is tagged as positive change; If S_{diff} value is with a $-ve$ sign, then the location is tagged as negative change; otherwise (i.e., $S_{diff} = 0$), the location is tagged as no change.

Step 6: The RSA method decides the change point or change time (shown in Fig. 2) by engaging a confidence percentage method.

Step 6.1: To fix the change point, the RSA method, first calculate the absolute value of the month wise differences between j^{th} annual segment and $(j+1)^{th}$ annual segment, after that calculate the absolute value of the month-wise difference between $(j+1)^{th}$ annual segment and $(j+2)^{th}$ annual segment.

Step 6.2: Subsequently the resulting values are stored in Y.

Step 6.3: The RSA method believes that the change point lies in the before change segment (i.e., j^{th} annual segment), thus it finds the confidence of the first 12 elements of Y.

Step 6.3.1: The confidence of an element (i.e., Y_i) is decided by studying the number of times an element (i.e., Y_i) greater than to its succeeding 12 elements (i.e., Y_{i+1} to Y_{i+12}). The change point is a point with the confidence percentage greater than the stipulated threshold (Th) value and highest differential value (such as largest Y_i value).

The RSA technique focuses on the sudden change and only finding out the most significant alteration in a time series data. Furthermore, this technique label the change time series by means of the change type, i.e., abrupt increase or decrease. It too determines the change time about the real change time, with precision adequate to the temporal resolutions (usually, a 16 days or a month) of input time series. This technique is scalable to handle big data set.

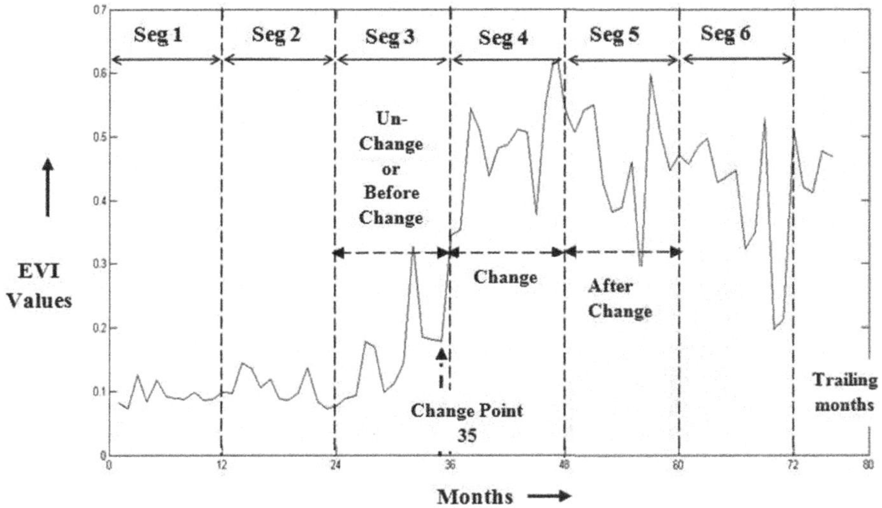

Fig. 2 The RSA algorithm determines the change segment and change point [15]

4 Result and Discussion

The qualitative evaluation technique is employed to examine and comprehend the performance of the RSA algorithm. For qualitative analysis, the Naya Raipur Area data set has been employed. The particulars of the Naya Raipur Area data set were explained in Sect. 2. The results of qualitative analysis are investigated subsequently.

4.1 Qualitative Analysis and Discussion

This section exhibits the qualitative analysis outcomes of the RSA technique. The Chhattisgarh state is established in November 2000. The state government decided to construct a well planned new capital city (i.e., Naya Raipur) in Chhattisgarh [23]. Many infrastructures have already been built, i.e., parks, recreation centers, housing subdivisions, institutional buildings, stadiums, etc., as a consequence the crucial land cover changes have been happening in the Naya Raipur area. Thus, the Naya Raipur Area of Chhattisgarh, India, has been preferred to study.

To exhibit the qualitative analysis, the RSA method is applied to the Naya Raipur Area data set, which is an enhanced vegetation index time series data set. As an outcome, it has returned a list of change scores. A change score value from the list is referring to a particular location. Then, the algorithm compares the change score with the stipulated threshold value; if the particular location's change score value is higher than the stipulated threshold value, then the location is acknowledged as changed or else that location is declared as unchanged. To decide the specific threshold value,

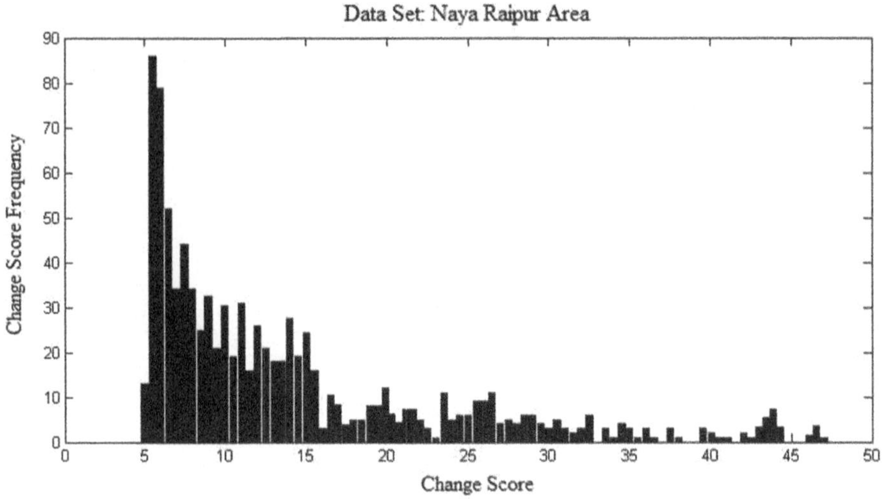

Fig. 3 Histogram of change scores for the Naya Raipur Area data set

first, the histogram of change scores is plotted by means of the change scores list produced by the algorithm. After that, decide the threshold value by choosing the change score from the histogram, when the number of occurrences of change score has been coming down suddenly [15, 24]. The histogram of change score of the Naya Raipur Area data set is presented in Fig. 3 and the determined threshold value is 16. The researcher observed that the larger part of the locations in the data set does not specify any changes. Later the result of exceptionally chosen locations (i.e., the locations with extremely high change score) is manually scrutinized which had a change score bigger than a threshold of 24; the researcher found that various locations are significant for exciting land cover changes, while others locations correspond to the alteration of the crop patterns in farmlands. The few numbers of exciting land cover changes that took place in the Naya Raipur Area data set are demonstrated in Fig. 4 and Fig. 5 (Note that the apiece figure is evidence for different time series, which are represented the collection of spatially neighbor locations with description.)

Construction of a new sector: Figure 4 depicts the time series and map for a location where sector 19 was built. The sector 19 area of Naya Raipur is consisting of many office buildings. In Fig. 4a, prior to 2009, the time series for that location indicates the vegetation with the seasonal rise and drop-offs. Furthermore, in the first quarter of the year 2009, the sudden decreases in vegetation take place, after which the vegetations are comparatively low and invariable, which is also exhibited in Fig. 4a. The same land site which was used as an agricultural land in the past, which is shown in Fig. 4b. Figure 4c shows the initial phase of building construction at sector 19. After completion of most of the construction work the sector 19 is looking like as shown in Fig. 4d. It is an illustration of land cover change, which is due to urbanization of the farming land.

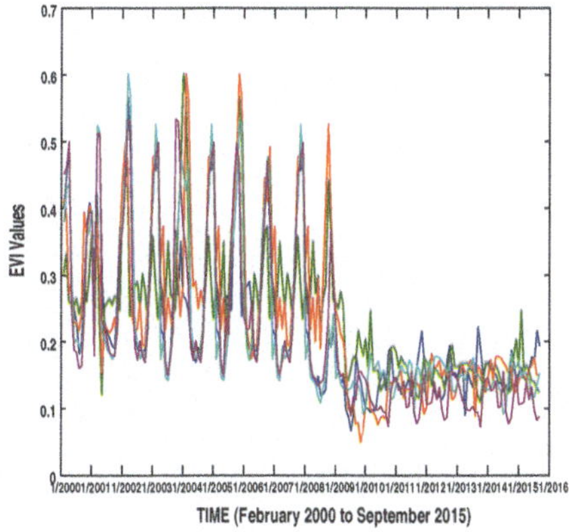

(a) Time series corresponds to Sector 19 of Naya Raipur.

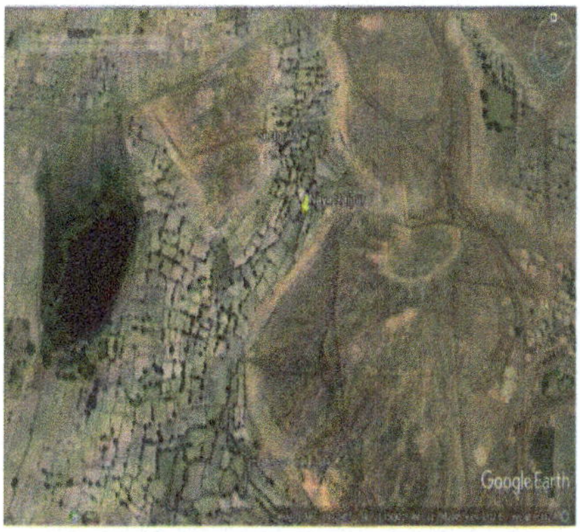

(b) Before constructions the site was agricultural land, image on March 2007.

(c) Initial phase of construction at Sector 19, image on April 2009.

(d) After construction at Sector 19, image on November 2013.

Fig. 4 Change detection due to construction at sector 19 in the Naya Raipur City. *Image Source* Google Earth

Construction of a cricket stadium: Figure 5 depicts the time series and map for a location, where a new cricket stadium was constructed in the Naya Raipur. The construction of the cricket stadium began in the year 2001 and it was completed in 2008. This location's time series indicates the vegetation with seasonal growth and reduce prior to 2008 and then in 2008 unexpected jump took place in vegetation; after which the vegetation is reasonably homogeneous and stable, which is depicted in Fig. 5a. The vegetation index value is too low, which is corresponding to a barren land. The initial and intermediate stage of construction of the cricket stadium are shown in Fig. 5b and Fig. 5c, respectively. After construction of cricket stadium, which is shown in Fig. 5d; the area is planted with grasses, which is the reason

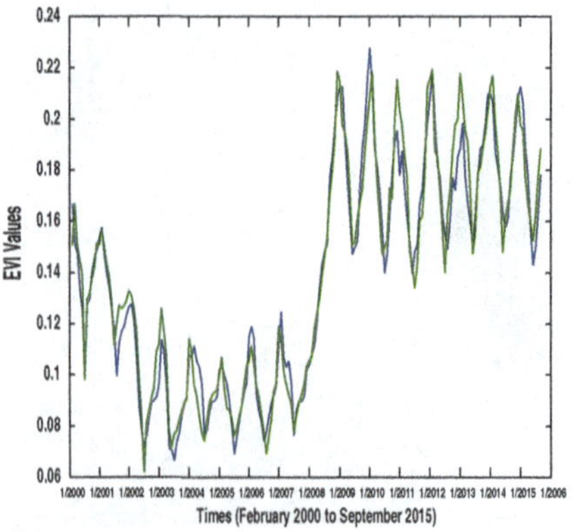

(a) Time series corresponds to the cricket stadium in Naya Raipur, CG, India.

(b) Initial phase of construction of cricket stadium, image on November 2003.

(c) Intermediate phase of construction of cricket stadium, image on march 2007.

(d) After final constructions of cricket stadium, image on March 2009.

Fig. 5 Change detection due to construction cricket stadium in the Naya Raipur City. *Image Source* Google Earth

for the enhancement in vegetation index and is evidenced at the time series of that location. It is an instance of land cover change where the barren land is transformed into a cricket stadium due to the development of urban area.

5 Conclusion

In this paper, the RSA algorithm has been discussed briefly. Furthermore, the proficiency of the RSA technique is exhibited by using qualitative analysis. To exhibit the qualitative analysis, the RSA method is applied to the Naya Raipur Area data set. The study found that significant land cover changes have been happening in the

Naya Raipur Area. The RSA algorithm detects the land cover change by finding the sudden changes (i.e., unexpected gains or drop-offs type) and single most substantial changes in a vegetation index time series data. Besides many of the merits, the RSA algorithm has several shortcomings too. First of the weakness is that the RSA algorithm is incapable to detect change, if an event happened on the first and last annual segment of time series. The second weakness is that this technique always determines the change point at earlier change segment, but possibly the change point may take place in the near the beginning time of the change segment. The third weakness is that the RSA algorithm does not exercise the spatial information present in the EVI data set, however Earth Science data sets show evidence of significant spatio-temporal auto-correlation.

References

1. Briassoulis, H.: Land-Use, Land-cover Changes and Global Aggregate Impacts, Land Use, Land Cover and Soil Sciences. Encyclopedia of Life Support Systems (EOLSS-UNESCO) Publications, Oxford (2004)
2. Dickinson, R.E, Kennedy, P.: Impacts on regional climate of Amazon deforestation. Geophys. Res. Lett. **19**(19), 1947–1950 (1992)
3. Henderson-Sellers, A., Dickinson, R.E., Durbidge, T.B., Kennedy, P.J., McGuffie, K., Pitman, A.J.: Tropical deforestation: modeling local to regional scale climate change. J. Geophy. Res. Atmos. **98**(D4), 7289–7315 (1993)
4. Panigrahi, S., Verma, K., Tripathi, P., Sharma, R.: Knowledge discovery from earth science data. In: 2014 4th International Conference, Communication Systems and Network Technologies (CSNT), IEEE, pp. 398–403 (2014)
5. Sharma, R., Verma, K., Panigrahi, S., Nirupama, N.: A multi-tier hazard: Part II meteorological analysis. Nat. Hazards **76**(1), 271–281 (2015)
6. Chen, M.S.: Han J and Yu P S Data mining: an overview from a database perspective. IEEE Trans. Knowl. Data Eng. **8**(6), 866–883 (1996)
7. Boriah, S., Kumar, V., Steinbach, M., Potter, C., Klooster, S.: Land cover change detection: a case study. In: Proceedings of the 14th ACM SIGKDD International Conference on Knowledge Discovery and Data Mining, KDD08: pp. 857–865 (2008)
8. Boriah, S., Kumar, V., Steinbach, M.S., Tan, P.N., Potter, C., Klooster, S.A.: Detecting ecosystem disturbances and land cover change using data mining. Next Generation of Data Mining. vol. 2, pp. 29–46, CRC Press, CH (2008)
9. Boriah, S.: Time series change detection: algorithms for land cover change. Ph.D. thesis, Department of CSE, University of Minnesota, pp. 1–160 (2010)
10. Boriah, S., Mithal, V., Garg, A., Kumar, V., Steinbach, M., Potter, C., Klooster, S.: A comparative study of algorithms for land cover change. In: The proceeding of the 2010 Conference on Intelligent Data Understanding, CIDU2010, pp. 175–187 (2010)
11. Chamber, Y., Garg, A., Mithal, V., Brugere, I., Lau, M., Krishna, V., Boriah, S., Steinbach, M., Kumar, V., Potter, C., Klooster, S.: A novel time series based approach to detect gradual vegitation changes in forests. In: Proceeding of the 2011 Conference on Intelligent Data Understanding, CIDU2011, pp. 248–262 (2011)
12. Chen, X.C., Karpatne, A., Chamber, Y., Mithal, V., Lau, M., Steinhaeuser, K., Boriah, S., Steinbach, M., Kumar, V., Potter, C.S. and Klooster, S.A., Abraham, T., Stanley, J.D., Castilla-Rubio, J.C.: A new data mining framework for forest fire mapping. In: IEEE Conference on Intelligent Data Understanding, CIDU2012, pp. 104–111 (2012)

13. Garg, A., Manikonda, L., Kumar, S., Krishna, V., Boriah, S., Steinbach, M.S., Toshnival, D., Kumar, V., Potter, C., Klooster, S.A.: A model-free time series segmentation approach for land cover change detection. In: Proceedings of the 2011 NASA Conference on Intelligent Data Understanding, CIDU2011, pp. 144–158 (2011)
14. Panigrahi, S., Verma, K., Tripathi, P.: Land cover change detection using focused time delay neural network. Soft Comput. **23**(17), 7699–7713 (2019)
15. Panigrahi, S., Verma, K., Tripathi, P.: An efficient approach to detect sudden changes in vegetation index time series for land change detection. IETE Tech. Rev. **33**(5), 539–556 (2016)
16. Panigrahi, S., Verma, K., Tripathi, P.: Data mining algorithms for land cover change detection: a review. Sadhana J. **42**(12), 2081–2097 (2017)
17. Huete, A.R., Didan, K., Miura, T., Rodriguez, E.P., Gao, X., Ferreira, L.G.: Overview of the radiometric and biophysical performance of the MODIS vegetation indices. Remote Sens. Environ. **83**(1), 195–213 (2002)
18. Huete, A.R., Justice, C., Leeuwen, W.V.: MODIS vegetation index (MOD13) algorithm theoretical basis document, Ver. 3. Vegetation Index and Phenology Lab, Department of Environmental Sciences, University of Virginia (1999). Online: https://pdfs.semanticscholar.org/2204/b55a9ad69e8b69d19e88ad1f0e1f81a5d72b.pdf. Last Accessed on August 2019
19. MODIS overview. (Online). https://lpdaac.usgs.gov/products/modis_products_table/modis_overview. Last Accessed on August 2019
20. MODIS user guide. (Online) www.gscloud.cn/userfiles/file/MOD13_UserGuide.pdf. Last Accessed on August 2019
21. Land processes distributed active archive center (LP DAAC). Online https://lpd-aac.usgs.gov/dataset_discovery/modis/modis_products_table. Last Accessed on August 2019
22. Naya Raipur, India. (Online) https://en.wikipedia.org/wiki/Naya_Raipur. Last Accessed on August 2019
23. Naya Raipur map, India. (Online) http://www.nayaraipur.gov.in/user/map.aspx. Last Accessed on August 2019
24. Panigrahi, S., Verma, K., Tripathi, P.: Optimal threshold value determination for land change detection. Int. Arab J. Inf. Technol. **16**(2), 1–10 (2019)

Author Index

D. Mishra et al. (eds.), *Intelligent and Cloud Computing*,
Smart Innovation, Systems and Technologies 194,
https://doi.org/10.1007/978-981-15-5971-6

The manufacturer's authorised representative in the EU is Springer Nature Customer Service Centre GmbH, Europaplatz 3, 69115 Heidelberg, Germany. If you have any concerns regarding our products, please contact ProductSafety@springernature.com

Printed and bound by CPI Group (UK) Ltd, Croydon, CR0 4YY
30/09/2026
02225292-0020